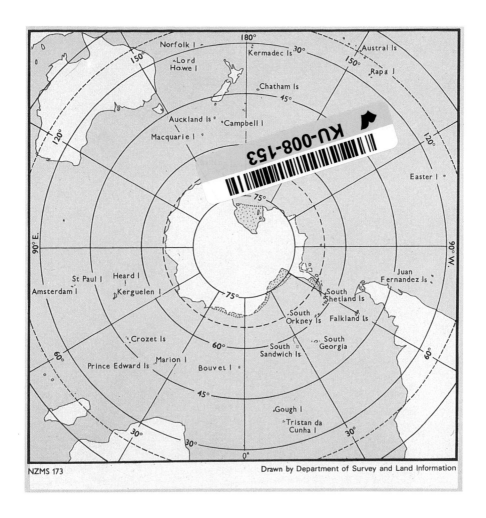

NZMS 173

Drawn by Department of Survey and Land Information

THE
SOUTHERN ISLANDS

FLORA
OF
NEW ZEALAND

THE GRASSES

Manaaki Whenua Press acknowledges a generous grant from the
Miss E.L. Hellaby Indigenous Grasslands Research Trust
towards the cost of publishing this Flora.

The Trust was established in 1959 by Eleanor Lillywhite Hellaby, daughter of William Hellaby a pioneer of New Zealand's meat industry, to benefit the pastoral industry through fellowships for the study of a seemingly neglected field of research — indigenous grasslands.

In addition to fellowships the Trust has paid field expenses, assisted with the publication of floras, contributed to conference travel, and purchased areas of tussock grasslands for reservation and conservation.

Research policy is determined by a Board of Governors; assets of the Trust are managed by the Guardian Trust.

Applications for grants may be made to the Board of Governors, Hellaby Trust,
C/o P.O. Box 295 Dunedin.

FLORA
OF
NEW ZEALAND

VOLUME V

GRAMINEAE

by

E. Edgar
and
H. E. Connor

With contributions by W. R. Sykes and M. I. Dawson

Manaaki
Whenua
PRESS

2000
Manaaki Whenua Press, Lincoln, New Zealand

CATALOGUING IN PUBLICATION

EDGAR, E. (Elizabeth)
>Flora of New Zealand. Volume V. Gramineae
/ by E. Edgar and H.E. Connor, with contributions by
W.R. Sykes and M.I. Dawson. - Lincoln,
Canterbury, N.Z. : Manaaki Whenua Press, 1999.

>ISBN 0-478-09331-4

>I. Connor, H.E. (Henry Eamonn). II. Title.

>UDC 582.542.1(931)

Cover design by P.A. Brooke.
Set in 10.5 pt Baskerville and printed on 80gsm offset paper.

Illustrations by S.B. Malcolm, K.R. West, and P.A. Brooke,
with photographs by P.N. Johnson.

Typeset and printed by The Caxton Press, Christchurch, New Zealand.

Publication of this volume was supported by the Foundation for Research,
Science and Technology.

Published by Manaaki Whenua Press, Landcare Research,
P.O. Box 40, Lincoln 8152, New Zealand.

To commemorate

JOHN BUCHANAN F.L.S.
1819–1898

author and illustrator
of
the first book on New Zealand Grasses

 N three bookes therefore, as in three gardens, all our Plants are bestowed ; sorted as neere as might be in kindred and neighbourhood.

The first booke hath Grasses, Rushes, Corne, Reeds... And thus hauing giuen thee a generall view of this Garden, now with our friendly Labors wee will accompanie thee and leade thee through a Grasse-plot, little or nothing of many Herbarists heretofore touched; and begin with the most common or best knowne Grasse, which is called in the Latine *Gramen Pratense* ; and then by little and little conduct thee through most pleasant gardens and other delightfull places, where any Herbe or Plant may be found..."

John Gerard (The Herball, 1597)

CONTENTS

CONTENTS

ACKNOWLEDGMENTS

We are grateful to colleagues from many institutions in several countries for specialist information.

United Kingdom: We are glad to have this opportunity to express our appreciation of the kind welcome and generous assistance extended to each of us by the late Dr C. E. Hubbard on visits to the Royal Botanic Gardens, Kew, and of his subsequent helpful correspondence. We also thank Dr W. D. Clayton and Dr S. A. Renvoize of Kew for invaluable discussion and are most grateful to Dr Renvoize for help with difficult specimens. Dr P. M. Smith, University of Edinburgh, kindly examined New Zealand specimens of *Bromus* and Mr E. J. Clement F.L.S. provided information on *Sporobolus.*

Europe: We have long corresponded with Dr J. F. Veldkamp, Rijksherbarium, Leiden, on problems of taxonomy and nomenclature in grass genera common to New Zealand and Malesia; we thank him for much useful information. We also thank Professor Dr F. Albers, Westfälische Wilhelms-Universität, Münster, for permission to cite his chromosome counts of some New Zealand species of *Deschampsia.*

America: We are grateful to the late Dr T. R. Soderstrom and to Dr R. J. Soreng, Smithsonian Institution, for far-ranging discussion of relationships within Gramineae, and also to Dr M. E. Barkworth, Utah, in particular for her insight into Stipeae and Hordeeae. We also thank Dr S. G. Aitken, Ottawa for helpful discussion of *Festuca* and other genera. Dr O. Mattei and Dr M. Muñoz-Schick, Chile, examined New Zealand specimens of *Bromus* and gave us information on *Bromus* sect. *Ceratochloa.*

South Africa: Dr N. P. Barker, Rhodes University, and Professor H. P. Linder, Bolus Herbarium, assisted in discussions about austral Arundinoideae, and Dr G. A. Verboom, University of Cape Town, about tribe Ehrharteae.

Australia: Dr S. W. L. Jacobs, Royal Botanic Gardens, Sydney, has for many years accorded us unfailing help and encouragement in our endeavour to clarify the taxonomy and nomenclature of closely related New Zealand and Australian grasses; we thank him for his friendship and for his careful refereeing of several revisions, sometimes at very short notice. For generous assistance and information we are indebted to Dr J. H. Ross, Melbourne, Dr B. K. Symon, Brisbane, and Drs T. D. Macfarlane and L. Watson, Western Australia.

New Zealand: We thank the many colleagues who have assisted us over the years in diverse ways. Our special debt to the late Mr A. P. Druce is clearly apparent in the very many pages of this Flora in which his name appears, as collector of type specimens, or of specimens requiring special mention because they add a new dimension to the diversity within a taxon. His wide-ranging collections of grass specimens at CHR provide a wealth of comparative material and his keen field observations led to the recognition of many new taxa.

To the late Dr M. B. Forde we are indebted for collaboration in the revision of *Agrostis* and *Bromus*, for reviewing the draft text of *Lolium*, as well as for much information on the distribution and spread of chloridoid and panicoid grasses within New Zealand. Her untimely death deprived us of a trusted and joyous source of information and inspiration.

We are especially grateful to Mr A. J. Healy for his expert knowledge of naturalised grasses gained over many years, and for sharing his recent observations of these plants; to Dr E. J. Godley, Director, Botany Division, DSIR, 1958–1980, for constant encouragement and for information on historical specimens collected from the Subantarctic Islands; to Dr W. Harris for his concern to promote the completion of the New Zealand Flora series; to Mr A. E. Esler and Mr W. R. Sykes, for detailed comments on a draft text of subfamilies Chloridoideae and Panicoideae; and to Dr E. H. C. McKenzie, for identification of fungal infections which may give rise to puzzling malformations. We also thank Mr Sykes for contributing the text for subfamily Bambusoideae and Mr M. I. Dawson for preparing the list of chromosome numbers.

For specimens, information and encouragement we are indebted to many colleagues; for information on grasses in North Island we especially thank Dr B. D. Clarkson, Mr E. K. Cameron, Mr P. J. de Lange, and Mr C. C. Ogle; for South Island we thank Dr G. W. Bourdôt, Mr S. Courtney, Mr G. T. Jane, Dr P. N. Johnson, Dr W. G. Lee, Professor A. F. Mark, Dr B. P. J. Molloy, Dr A. V. Stewart, Dr P. Wardle, Dr P. A. Williams, and Mr H. D. Wilson.

Herbaria: We have, during many trips abroad, enjoyed the hospitality of herbaria in Europe, the United Kingdom, North and South America, Asia, Israel, and Australia and express our thanks to the Directors and staff of these institutions which we visited to study type and critical specimens and to inspect grasses unknown to us: B, BM, BOG, BRI, CANB, CORD, FI, GOET, K, L, LE, LINN, MEL, MO, NSW, P, SGO, SI, UPS, US, W.

For the loan of type specimens and other collections we thank the Directors and Curators of the following herbaria in New Zealand and overseas: AK, AKU, BM, C, HO, K, L, LY, MPN, NSW, NZFRI, OTA, P, W, WAIK, WELT, Z. We greatly appreciate the assistance of Miss B. H. Macmillan, then at CHR, for facilitating all these loans and caring so meticulously for the specimens; and we also thank Dr M. J. Parsons, Herbarium Keeper, CHR, and his assistant, Miss K. A. Ford, for their kind help and support.

For technical assistance we thank especially Mrs B. A. Matthews who first helped us with this project and compiled an invaluable inventory of the protologues of all grass taxa described from New Zealand; we acknowledge the subsequent assistance of Mrs M. A. O'Brien, Miss J. Francis, Ms J. E. Shand, all Botany Division, and lastly of Ms E. S. Gibb, Landcare Research, who also helped us with editing the text and then went on to prepare it for printing.

We are deeply grateful to the late Mrs M. E. Blackmore, Librarian, Botany Division, DSIR, and latterly to Mrs M. Bowden, Librarian, Landcare Research and her staff, for bibliographic assistance. Dr I. Breitwieser, Mr M. I. Dawson, Miss K. A. Ford, Ms E. S. Gibb, Mr D. Glenny, Mr P. B. Heenan and Dr S. J. Wagstaff, all Landcare Research, helped to collate references for the Annals of Taxonomic Research.

Acknowledgments for illustrations to Ms S. B. Malcolm, Dr P. N. Johnson, Mr K. R. West, and Ms P. A. Brooke are made under that section of the Flora, and we thank Ms Brooke for the cover design.

We thank Miss A. White and former typists at the Canterbury Agriculture and Science Centre, Lincoln, and Mrs W. Weller and her staff at Landcare Research, Lincoln, for typing draft text.

The co-operation of Mr G. I. Comfort, Manaaki Whenua Press, Landcare Research, Lincoln, and of The Caxton Press, Christchurch, is gratefully acknowledged.

Henry Connor wishes particularly to acknowledge the excellent technical support of A. H. MacRae and the late A. W. Purdie during the experimental studies on some of the grasses described in this Flora. He is appreciative of the University of Canterbury offering an intellectual home since 1982 to allow him to work on the Flora, firstly at the Centre for Resource Management, then directed by the late Dr J. A. Hayward, and latterly in the Department of Geography, where he is an Honorary Fellow.

In this final volume of *Flora of New Zealand: Tracheophyta* Elizabeth Edgar wishes to express her profound gratitude for the opportunity that was granted to her as an inexpert, untried, junior at Botany Division to collaborate with Dr L. B. Moore, Mr A. J. Healy and Dr H. E. Connor in the writing of three volumes of the *Flora of New Zealand*. To learn from and work with these outstanding dedicated botanists has been a tremendous privilege.

PREFACE

1. HISTORICAL

"Is there a fatherless book, an orphan volume in this world? A book that is not the descendant of other books? . . . Is there creation without tradition?" In *How I wrote one of my Books* Carlos Fuentes (1988) posed these questions. This grass Flora is the direct descendant of John Buchanan's *Indigenous Grasses of New Zealand*, the first New Zealand book devoted solely to grasses, originally published in three parts — Plates 1–20 in 1878, Plates 21–40 June 1879, and Plates 41–58, 17(2), 26(2), and 36(2) in June 1880, and printed in Imperial Quarto (14½ inches × 10 inches) because the plates were to be nature-printed. The half-bound edition weighs 2.66 kg. As James Hector remarked in his Preface to the 1880 Royal Octavo (9½ inches × 6 inches) complete edition, the large size of the original was inconvenient, but the grasses would now be illustrated at half their natural size. The arrangement of the text differs between the 1880 Imperial Quarto (price 3 guineas) and the Royal Octavo (price 7s.6d.) editions, particularly in the placing of the nomenclaturally important Addenda et Corrigenda. The differing pagination of the indexes to Popular Names and to Genera and Species is not critical, but the omission of Buchanan's "Remarks on the Distribution of Grasses in New Zealand" of pages 8–10 of the Introduction denied general access to one aspect of its original purpose.

Thus Buchanan's illustrated *Manual*; but Buchanan was guided by J. D. Hooker's *Handbook of the New Zealand Flora* (1864, 1867) with its 26 genera of grasses, because Buchanan dealt with the same 26 genera although he increased them to 28 by plates for *Stipa petriei* and *Deyeuxia scabra* as Addenda. Buchanan is a Hooker descendant; his lineage is secure. Hooker's *Handbook* was "Published under the Authority of the Government of New Zealand." Buchanan's was "Published by Command"; there was discussion about it, in the House of Representatives (Parliamentary Debates 20: 217–219, 340 [1876]) notably by Sir George Grey K.C.B., and its preparation was resolved in Parliament on 29 June 1876.

Our Flora may also be interpreted as the linear successor of H. H. Allan's *Introduction to the Grasses of New Zealand* (1936), price 4s. Allan's purpose was "to provide a means of recognising the different species" and to encourage ". . . others to join the present small but enthusiastic band of collectors." That band provided the basis for this Flora. A more correct interpretation might be that *Flora Volume V* is the final step in the sequence from Allan's *Introduction*. The compass is the same. Allan included 76 genera and both native and naturalised species; he provided keys often as workable now as then. In 160 pages there is ample introduction, as one of us remembers well, and there are 102 figures. Figures 1–4 are explanatory of grass features, 5–102 are species habit or detail; some include several taxa, e.g., in fig.12 there are six taxa, in fig. 13 there are 10 taxa. What has never been discussed is the number of plates drawn by V. D. Zotov whom Allan acknowledges as providing ". . . the detail drawings of indigenous grasses." By our count Zotov drew 27 plates incorporating 130 taxa. There were 72 habit drawings and 387 details of spikelets, florets, and flowers; often the detail is too small, and the captions all lack any explanation. Those drawings of native grasses were never used again even though Allan assured us that ". . . they were prepared for another purpose."

H. H. Allan in his *Handbook of the Naturalised Flora of New Zealand* (1940) listed the grasses in two parts, 87 naturalised species, and a further 60 species in the Appendix, some of which were ". . . non-persistent escapes from cultivation." Five were thought to have fairly good claim to naturalisation: *Avena strigosa, Entolasia marginata, Glyceria striata, Nassella trichotoma, Oryzopsis miliacea*; all are now.

Flora of New Zealand Vol. V Gramineae will meet the modern requirements of a flora — descriptions, distributions, keys, uncertainties, modern classification, notes, reference to allied studies, and to the relevant revisions. It will inform New Zealand about the grasses resident here, regardless of their status as native or naturalised entities. It will tell the world outside New Zealand of the kinds of grasses here, especially those with which they are unacquainted. It will tell other parts of the world about the variation, behaviour, ecology, and distribution of their grasses in the New Zealand environment.

We are unable to meet some of the requirements of a flora that Funk (1993) elaborated, such as that of solely monophyletic taxa, and a phylogenetic arrangement of them; we do not know either, but opinions are widely available in up-to-date summaries by Hsiao et al. (1999) and by Soreng and Davis (1998). We do know that we tired of referees' comment on our revisions "Where is the cladogram?" We do endeavour, however, to provide comparable and general descriptions, and synopses where the groups are large. Endemism is clearly designated; but not always can we indicate the closest relative of any taxon because that is not the stage of evolution of our flora writing. No-one should have difficulty with the detection of the circumpolar, the subantarctic, or the Pacific elements, and thus the subregions within the New Zealand Botanical Region.

Nowhere in this Flora will there be a philosophical discussion of the Gramineae — the absence of epiphytism, parasitism, succulence, or heterostyly for example. The evolution of the endemic grass florula is not attempted here, and although well aware of the Gondwanaland influences we find discussion of them unsuited to a flora. Nor will we debate the outcomes of studies deriving from molecular biology; those conclusions are readily available and recent as in Soreng and Davis (1998) and Hsiao et al. (1999). We choose only to note two novel grass universals, the small *trnT* inversion in cpDNA (Doyle et al. 1992), and the crimp mark on the leaf-blades indelibly present through the action of the ligule (Espie, Connor and McCracken 1992), and note also the suggestion of the pathway to one morphological character that defines the subfamily Panicoideae (Le Roux and Kellogg 1999).

The role of experiment as an adjunct to taxonomy came to the fore after World War II. At Botany Division, D.S.I.R., experimental projects included growing plants from different localities and habitats in a common garden, and the production of interspecific and intraspecific hybrids in several grass genera. One example of large scale experiment was in *Cortaderia*, a danthonioid genus of five endemic species, the toetoes, and two naturalised species, the pampas grasses of South American origin.

Gynodioecism where the populations consist of plants bearing hermaphrodite flowers and plants with female (male sterile) flowers is present in all endemic species of *Cortaderia*, except in *C. turbaria* from the Chatham Is. Extensive experiments involving just more than 13,000 plants in cultivation mostly of South Island *C. richardii*, but including the other endemics, attempted to interpret the genetics of male sterility. Because each generation of plants from controlled pollination first flowered after four years these species were not as ideal subjects for experiment as the naturalised pampas grass, *C. selloana*, which comes to flower in its first growing season. Families of *C. selloana*, totalling 17,000 plants, were raised, and analysis of the frequencies of male steriles showed that sterility was under recessive genic control, the double recessive at any of up to three complementary loci causing male sterility. No adequate solution accounted for male sterility in endemic species other than to note that control was genic, probably recessive, and common to all four species.

Trials with grasses in cultivation may not always be simple or rapid. From seeds *Festuca novae-zelandiae* and *F. matthewsii* come to flower in their second growth year, and *Chionochloa* spp. raised from seeds come into flower in their third or fourth year. Some endemic species of *Poa* flower in their first year, but not all, and endemic species of *Elymus* also come to flower in their first year from seeds. Large scale experiments with native grasses have not always been easily or quickly managed.

2. TAXONOMIC TREATMENT

In this volume of the *Flora of New Zealand* we have followed the taxonomic system that W. D. Clayton and S. A. Renvoize set out in their *Genera Graminum: Grasses of the World* (1986). Amendments to their scheme and some nomenclatural variation especially of tribal names will be evident, but our departures are so insignificant as to cause little disquiet and every agrostologist will easily find the related group. At the generic level the defence of our choices, should defence ever be needed, is usually in one of our revisions that have preceded this volume.

Volume V in the *Flora of New Zealand* series departs from Vols I–IV, as they from each other, by presenting endemic, indigenous, and naturalised taxa equally and without any discriminating sign. Volume V follows the style of Vols I and II by presenting the full synonymy for endemic species, and for indigenous species as it affects New Zealand; for naturalised taxa it offers the best binomial or polynomial, just as in Volumes III and IV.

The only generic name of uncertain application is *Kampmannia* Steud., *Syn. Pl. Glum. 1*: 34 (1854) with the single species *K. zeylandica* Steud. The type, said to be of New Zealand origin, has been lost and the protologue does not allow of any positive interpretation.

Two issues of concern to taxonomists, though not necessarily so to all users, are nomenclature and typification.

Typification

The typification of all New Zealand endemic and indigenous taxa is indicated throughout; holotypes are signified and for lectotypes the designators and references are given in full. Typification of a further 11 taxa is listed on p. xxxiv; we do it in this Flora to avoid a separate, and probably later, publication. Some indigenous taxa are based on New Zealand plants; in those, should the necessity for lectotypification arise, it is seen as a local responsibility; by contrast, for indigenous taxa based, for example, on Australian collections, we refrain from typification on the grounds that the responsibility is best assumed by botanists with local knowledge. Thus for dioecious *Spinifex sericeus*, shared with Australia, the R. Brown sheet 6147, East Coast, Broad Sound, BM! comprises both sex forms of which the female is a solitary inflorescence, the lectotypification by Craig (1984) is incomplete.

The rare neotype, or replacement of a lectotype known to have been destroyed, is clearly defined. It is our belief that this emphasis on and clear identification of holotypes and lectotypes will benefit future New Zealand agrostographers.

Nomenclature

For endemic or indigenous taxa the appropriate synonymy is given. Endemic taxa are no problem, but for indigenous species, those shared perhaps with Australia, the holotype or lectotype may be of Australian provenance, and the synonymy may not correctly apply to New Zealand plants, e.g., in *Hierochloe redolens* the binomial *Disarrenum antarcticum* Labill., *Nov. Holl. Pl. Spec. 2*: 83 t. 232 (1807), with its subsequent combinations, is based on an Australian provenance. *Melica magellanica* Desr. *in* Lamarck *Encycl. Méth. Bot. 4*: 72 (1797) with its subsequent combinations, is based on Commerson's specimen from the Straits of Magellan. Both names are treated by all authorities as synonymous with *H. redolens*. We have not seen the types for either, and having limited ourselves to material originating in New Zealand, have not included these names in our synonymy.

We have endeavoured to verify, unify, correct, and bring up-to-date all nomenclatural references; in doing so there will be some differences — dates in particular — from those commonly found in earlier works. *Poa, Agrostis,* and *Rytidosperma* are examples of our approach to nomenclature; the status of the plant in New Zealand determines our treatment. Synonymy of naturalised species is never given, nor is typification indicated except by way of offering older or historical names.

Hawksworth (1992) asked how much time taxonomists spent on nomenclatural matters, time that could well have been better devoted to taxonomic studies. Our answer for our studies is simple — no longer than necessary. Homotypic synonyms are relatively easily determined and, apart from customary verification of date and place, often pose no major problem other than to ensure that the epithet in question is not a later homonym and that transfers were not precluded by earlier names. Heterotypic synonymy involves taxonomic judgement, and it is the taxonomy not the nomenclature that is demanding. We have devoted our time to taxonomy and attended to the nomenclature as its consequence. Never for a moment would we deny that some nomenclatural issues were not a source of concern — worry, more often. In *Lachnagrostis* where many combinations were illegitimate but the varietal epithets were valid, confusion may seem to rule even though it never rises above initial uncertainty.

Autonyms, names generated automatically as the consequence of subdivisions of a taxon — species in this Flora — apply in several genera; we clearly indicate their use.

The names of New Zealand grasses reflect nothing of indigenous origin at generic level, and at specific level "toetoe" and "unarede" are the only uses of Māori words. *Chionochloa flavicans* forma *temata* invokes a Māori place name — even if its author was moved to its use by the name of a local wine estate. Rakiura, the Māori name for Stewart Island is reflected in *Poa aucklandica* subsp. *rakiura*. No political overtones, no conservation ploy, no historic person, event, or place is called into commemoration even though there were many opportunities. Our grass nomenclature is singularly free of overtones.

3. STRUCTURE OF THE GRASS FLORA

An analysis of the structure of the grass flora is in Table 1; there arranged by tribes, the number of species in each, and their classification into endemic, indigenous, naturalised, or transient (zeta entries, ζ) is shown. Of these four classes only the last may need definition — these are the species found once, or perhaps twice, and not collected again. Their share is 10% of the entries in this Flora, and of course increases the incidence of the alien elements here.

The endemic florula is 157 species, *c.* 38% of the total, and has its highest concentration of taxa in three tribes where the frequencies are almost equal — Poeae 49 spp.; Agrostideae 45 spp.; Danthonieae 43 spp. Addition of the indigenous element is only conspicuous in Agrostideae, the remaining taxa are spread very thinly among the tribes.

On biogeographic grounds the expectation for indigenous Bambuseae must be low, and in oligotypic tribes the chances for an indigene or endeme are greatly reduced. The signal lack is of chloridoid and panicoid endemics; the significant entry is Agrostideae in all classes — endemics, indigenous, and naturalised taxa. World-wide that tribe has an estimated 1000 spp. mostly of temperate areas.

The naturalised grass florula exceeds the total of the endemic and indigenous components, and the evidence of the number of species which had escaped from cultivation or lived a short time after diaspores were first dispersed is indicative of a further probable increase in alien grasses.

It may be of no great significance but only in the tribes Poeae, Agrostideae and Danthonieae does the number of endemic and indigenous species exceed the alien element (Table 1).

Endemic Grasses

Genera There are five endemic genera of New Zealand grasses, two of them are monotypic and the others are oligotypic: *Anemanthele* 1 sp.

Table 1. Structure and composition of the New Zealand grass flora

Tribe	No. of genera	Endemic	Indigenous	Naturalised	Transient	n
1. Bambuseae	9			16		16
2. Ehrharteae	3	5	2	4		11
3. Oryzeae	2			2	1	3
4. Nardeae	1			1		1
5. Stipeae	5	2	1	14		17
6. Poeae	12	49	4	32	5	90
7. Hainardieae	2			3		3
8. Meliceae	2			5	1	6
9. Agrostideae	28	45	15	35	8	103
10. Bromeae	1			15	5	20
11. Brachypodieae	1			2	1	3
12. Hordeeae	10	10	1	17	2	30
13. Arundineae	2			2		2
14. Danthonieae	5	43	3	12		58
15. Aristideae	1			1	1	2
16. Chlorideae	5	2		6	1	9
17. Leptureae	1		1			1
18. Eragrostideae	5			8	8	16
19. Paniceae	15		3	42	9	54
20. Isachneae	1		1			1
21. Andropogoneae	10	1		9	4	14
	121	157	31	226	46	460

(Stipeae), *Pyrrhanthera* 1 sp. (Danthonieae), *Simplicia* 2 spp. (Agrostideae), *Stenostachys* 3 spp. (Hordeeae), *Zotovia* 3 spp. (Ehrharteae).

Linder and Verboom (1996) reduced *Pyrrhanthera* to synonymy in *Rytidosperma*, and Willemse (1982) concluded that *Zotovia* spp. (as *Petriella*) should be included in *Ehrharta*. The generic status of *Anemanthele, Simplicia,* and *Stenostachys* has not been assailed. None is diploid, and all have discernible relatives except *Simplicia*.

Species The frequency of endemic species, arranged by tribes, is in Table 1, but that simple analysis hides the detail. High concentrations of endemic species (n ⩾ 10) are: *Agrostis* 10 spp. (Agrostideae), *Chionochloa* 22 spp. (Danthonieae, *c.* 90% of the genus), *Lachnagrostis* 10 spp. (Agrostideae, *c.* 50% of the genus), *Poa* 37 spp. (Poeae), *Rytidosperma* 15 spp. (Danthonieae). This is a small assemblage; the high frequencies are in cosmopolitan genera such as *Poa* and *Agrostis*, and in austral danthonioids.

At lower levels of endemism, 4–9 species per genus, there are: *Cortaderia* 5 spp. (Danthonieae), *Deschampsia* 4 spp. (Agrostideae), *Deyeuxia* 4 spp. (Agrostideae), *Elymus* 7 spp. (Hordeeae), *Festuca* 9 spp. (Poeae), *Hierochloe* 6 spp. (Agrostideae), *Trisetum* 9 spp. (Agrostideae). Apart from the preponderance of agrostidoids, 23 of 45 taxa, this assemblage is undistinguished; all are cosmopolitan genera except austral *Cortaderia* and *Deyeuxia*.

Single endemic species are few: *Achnatherum* (Stipeae), *Australopyrum* (Hordeeae), *Dichelachne* (Agrostideae), *Imperata* (Andropogoneae). Of these *Australopyrum* is the only known endemic pooid diploid in New Zealand (2*n* = 14), and *Achnatherum petriei* is one species in a genus of *c.* 500 spp.

Calcicoles Molloy (1994) identified the endemic grasses that are specialists of limestone or detrital marble habitats. This list now includes: *Australopyrum calcis, Chionochloa flavescens* subsp. *lupeola, C. flavicans* forma *temata, C. spiralis, Dichelachne lautumia, Elymus sacandros, Festuca deflexa, F. luciarum, F. multinodis, Poa acicularifolia, P. spania, P. sudicola, P. xenica, Simplicia laxa*. By Molloy's criteria *Poa schistacea* and *Rytidosperma tenue* of southern South Island schists could also be included here. Four of these are in the Threatened Plants list below.

Three species of ultramafic soils may be mentioned here as there is no other particular place to list them; they are *Chionochloa defracta*, *Festuca ultramafica* and *Trisetum serpentinum*.

Conservation of Endemic Species

Among the many plants at conservation risk in New Zealand there are some grasses (de Lange et al. 1999). No grass is presumed now extinct but the class "Threatened" which includes 108 taxa whose survival is a matter of concern and priority contains several grass species. We use the classification of de Lange et al. (1999) for the details below.

Critically endangered (3 grasses among 24 taxa): *Amphibromus fluitans* (Agrostideae), *Australopyrum calcis* subsp. *calcis* (Hordeeae), *Poa spania* (Poeae).

Endangered (4 grasses among 31 taxa): *Chionochloa flavicans* forma *temata* (Danthonieae), *Cortaderia turbaria* (Danthonieae), *Puccinellia raroflorens* (Poeae), *Simplicia laxa* (Agrostideae).

Vulnerable (2 grasses among 52 taxa): *Australopyrum calcis* subsp. *optatum* (Hordeeae), *Deschampsia cespitosa* (Agrostideae).

Should the group "Declining" be added to the preceding, the only entry is *Austrofestuca littoralis* (Poeae).

Naturalised Grasses

The naturalised element exceeds the total of endemic and indigenous taxa (Table 1). Tribes with significantly high representation are Paniceae 42 spp., Agrostideae 35 spp., Poeae 31 spp.; in total they comprise 48% of the naturalised taxa. At a second level of high frequency are Hordeeae 18 spp., Bambuseae 16 spp., and Stipeae 14 spp. — tribes in which endemism is very low or nil except for Hordeeae. The most striking element here is the number of panicoids — seven species each of *Panicum, Pennisetum,* and *Setaria* and six in

Digitaria and *Paspalum*; most of these have North Island distribution but many extend well into South Island. Indigenous panicoids, by contrast, number three including the coastal psammophilous *Spinifex sericeus.*

Most naturalised species in Stipeae are of Australian provenance especially as *Austrostipa* spp., but South American *Nassella* has contributed three species. *Poa* with 10 naturalised species has the largest number of naturalised Poeae, all European except two from Australia. Other taxa in Poeae are principally European. Agrostidoid genera with naturalised species total 14, but only in *Agrostis* 4 spp., *Avena* 5 spp., and *Phalaris* 5 spp. are the numbers relatively high. *Bromus* (Bromeae) represented by 15 naturalised species and five transients, is the genus with most naturalised taxa; most are Eurasian (7 spp.) and South American (5 spp.).

Weediness is not our chief concern, but some genera were destined to be weeds of arable land, e.g., *Elytrigia repens,* or weeds of pastoral land, e.g., species of *Austrostipa*, and *Nassella, N. trichotoma* in particular. *Zizania* and *Spartina* are notorious in aquatic systems. Species of *Critesion* (Hordeeae) are prone to damage livestock, meat and pelts, through their penetrating awns.

About 40% of the naturalised species are Eurasian taxa; some of them probably arrived here via Australia, our nearest neighbour, but just how many cannot be determined. From Australia itself 35 species are naturalised here, half of them as species of *Austrostipa* and *Rytidosperma.* The other southern continents, Africa and South America, make equal but smaller contributions though the genera are rarely species rich here except *Bromus* as already remarked, and five African species of *Pennisetum.*

Two naturalised species of *Cortaderia* (Danthonieae) are especially undesirable in plantation forests notably so in North Island and Nelson; one of them *C. selloana,* pampas grass, had earlier enjoyed a reputation for both shelter and fodder for cattle. In the popular days of pampas grass culture one selection, "MacLeans", was of a tall, vigorous, large white-plumed, female line; seeds were never set because no pollen bearing plants were planted with them. Later commercial releases of pampas grass comprised both female and pollen-producing plants like the natural populations at some North Island sites; seed set was inevitable and the consequences are evident.

Surveillance Grasses

Some grasses are considered potentially so ecologically dangerous that they have been designated Surveillance Grasses. Under the Biosecurity Act 1993 and its amendments, the sale, propagation, and distribution of particular grasses is prohibited. Species such as *Phragmites australis*, common reed, *Zizania latifolia*, Manchurian wild rice, *Pennisetum macrourum*, African feather grass, and the two naturalised species of *Cortaderia*, the pampas grasses *C. selloana* and *C. jubata*, are among 10 or more subject to Plant Pest Management Strategies under that Act. Currently the Surveillance List contains reference to species of *Stipa*; in this Flora there are no species included in *Stipa sens. strict.* and species formerly included there are now referred to *Achnatherum, Austrostipa*, and *Nassella*.

Johnson grass, *Sorghum halepense* (Andropogoneae), is a weed of national concern, classified under the Biosecurity Act as a Notifiable Organism, one that must be reported to the Ministry of Agriculture and Forests. This species has the highest ranking for biological success and weediness among the grasses discussed in Esler et al. (1993).

4. CHROMOSOME NUMBERS

In the list below (p. 1–9) prepared by M. I. Dawson, Landcare Research, Lincoln, there are chromosome counts for 88 species among 21 genera; most of these (*c.* 50%) are in polytypic *Poa* and *Chionochloa*. Other counts are in genera where intensive work had been done, e.g., *Elymus, Festuca, Cortaderia*. By our estimate the chromosome numbers are known for *c.* 50% of endemic and indigenous species. The area of greatest deficiency lies in Agrostideae where no chromosome numbers are known for *Deyeuxia* or *Lachnagrostis*. Few numbers are recorded for New Zealand *Rytidosperma* (Danthonieae), by comparison with the detail and the wide range of levels of infraspecific polyploidy described in Australia (Abele 1959; Brock and Brown 1961).

Two significant features emerge from the data in Dawson's list — endemic *Australopyrum calcis* (Hordeeae) is diploid ($2n = 14$) as are its Australian congeners, and dioecious *Spinifex sericeus* (Paniceae), shared with eastern Australia is also diploid ($2n = 18$). All other taxa are polyploids. In subantarctic *Poa litorosa* $2n = c. 263–265$ (Hair and Beuzenberg 1961) or $2n = c. 266$ (Hair 1968); in *Festuca contracta* $2n = c. 170$ (Moore 1960); no other chromosome number of New Zealand grasses approaches these levels of ploidy; in both *Poa chathamica* and *P. cockayneana* $2n = 112$ ($16x$) and this level is not reached elsewhere in New Zealand grasses either. The *Poa* count of $2n = c. 265$ is the highest chromosome number recorded in any grass in the world.

Where there is little cytological information, e.g., *Dichelachne*, a count in a second species is confirmatory, and in *Microlaena*, cytologically unknown until recently, the four counts are coincident ($2n = 48$). For the endemic genus, *Zotovia* (Ehrharteae), the counts, even though one is approximate, are concordant with *Microlaena* and South African species of *Ehrharta*. The count in endemic *Anemanthele lessoniana* is inexact, but approximates to an expectation in Stipeae where $x = 11$. Chromosome numbers $2n = 28$ are of no particular assistance in assessing the true relationships of *Simplicia*, a ditypic endemic genus.

When Hair (1966) discussed the overall frequency of polyploidy in New Zealand grasses, he used much of his data which were to be published later; he could not have known about diploid *Australopyrum calcis*, nor that *Chionochloa* would not be a dibasic genus with $x = 6$ and $x = 7$, but exclusively based on $x = 7$. For the rest his assessment holds true — 98% of the grasses are polyploid — but there is no way of knowing which among them are palaeopolyploids.

5. GRASSES AND ANIMAL HEALTH

Pastures in New Zealand are of diverse structure and their products are a substantial part of New Zealand's GDP. Some of the grasses used in these pastures may cause seasonal and acute disorders in livestock; these were discussed in Connor (1977). But since that time advances in animal husbandry have replaced many of the data there. Ryegrass staggers, a neuromuscular disease, is now known to be caused by toxins produced by the fungal endophyte *Neotyphodium lolii* (Fletcher and Harvey 1981). The main toxin is the tremorgen lolitrem B (Gallagher et al. 1984). Ryegrass, *Lolium perenne,* is thus the host of the disease-producing organism. A full set of papers on ryegrass staggers and on the associated fescue toxicosis is in *Neotyphodium/Grass Interactions* (1997 Bacon and Hill, Eds). Fescue toxicosis, one syndrome known locally as fescue foot, has been associated in New Zealand with grazing of *Festuca arundinacea* since the 19th century. The alkaloid ergovaline, synthesised by the endophyte *N. coenophialum,* is a vasorestrictor and the chief associate of the syndrome in animals; ergovaline is also synthesised by ryegrass.

Neotyphodium was isolated from indigenous *Echinopogon ovatus* and other Australian species of *Echinopogon*; it is a fungus which is serologically related to *Neotyphodium lolii* and other species of *Epichloe* and *Neotyphodium* which cause staggers in livestock. This is the first time an endophytic species of *Neotyphodium* has been identified in grasses native to New Zealand and Australia (Miles et al. 1998).

Although the name "phalaris staggers" is thought to be an inaccurate terminology for the suite of disorders of livestock caused by alkaloids in *Phalaris aquatica*, phalaris staggering is one syndrome which also results from grazing *P. arundinacea*, reed canary grass.

Facial eczema, an important photosensitivity disease of livestock, is caused by toxins produced by the fungus *Pithomyces chartarum*, but photosensitisation has been caused here by the grasses Japanese millet, *Echinochloa esculenta* and *Panicum miliaceum*, broomcorn millet.

The first grass shown to produce hydrocyanic acid was *Cortaderia selloana*, pampas grass. All species of *Cortaderia* in New Zealand, endemic and naturalised, are cyanophoric (Tjon Sie Fat 1979), but none has any deleterious effect on livestock. The glycoside there is triglochinin. In *Glyceria maxima*, reed sweet grass, hydrocyanic acid concentrations were high enough to poison cattle, just as in Sudax, a commercial hybrid involving *Sorghum sudanense*, Sudan grass.

In *The Poisonous Plants in New Zealand* (Connor 1977) oat, *Avena sativa*, is included because it is an accumulator of nitrates even to very high levels. These nitrates are reduced to toxic nitrites in the rumen of sheep and cattle, leading consequently to methaemoglobinaemia. Oats, like ryegrass and tall fescue, are vehicles for the accumulation and transport of toxins but not their syntheses.

6. FLORA OF AUSTRALIA, POACEAE

Flora of Australia Vol. 43 and Vol. 44, Poaceae, are well advanced in the editing process and expected to start appearing in 2000 (Orchard 1999, and pers. comm.). The coincidence of publication of grass floras on both sides of the Tasman Sea should evoke some excitement. Their grass flora, much larger than ours, may contain some conflicts in treatment but there will be many correspondences, and these may allow comparisons to be made that were formerly impossible, thus pleasing Funk (1993).

We take this opportunity to congratulate Australian agrostographers on the completion of a task three times larger than ours.

7. DE AUCTORIBUS

This, the last volume of the *Flora of New Zealand* series, unlike Buchanan's *Indigenous Grasses* is unheralded by Parliamentary discussion or entitled "By command", but should bear the annotation "Published in the Fulfilment of Promises." Those promises were never originally made by either author, but became accepted by them when external circumstances altered. Both of us have worked on this Flora in our retirement from official positions at the then Botany Division, Department of Scientific and Industrial Research, Connor since late 1982 and Edgar since late 1988. We have, between us, published fifteen revisions involving 18 genera (p. xxxii), all of which were the necessary precursors to this modern Flora, and none of them gives the appearance of ease of completion. V. D. Zotov had earlier treated *Simplicia, Zoysia,* and *Hierochloe*; these were re-examined and resulted in some adjustments reported here. *Deschampsia* needed attention, too, especially the question of typification of seven taxa.

Is there a millennial messianic message in this *liber peroptatus*? It must be that the next grass to become naturalised in New Zealand is not yet established. The latest on that list is *Achnatherum caudatum* whose arrival from South America has been awaited by one of us (H.E.C.) for more than 40 years; we record it here on p. 65.

Although the authorship of this Flora is not that planned in the 1950s, Elizabeth Edgar, the senior author has spent her whole professional life writing floras, or preparing revisions of monocotyledonous Glumiflorae in preparation for floras both in New Zealand and Australia. Her associate author, Henry Connor, has no such experience at flora writing but is more practised at grasses. Despite this broad experience, there is still what Hilaire Belloc in his pilgrimage book *The Path to Rome* (1902) called the "Difficulty of Ending". We eschew gratuitous exhortations on the benefits of cladistic analyses or the singular results of molecular biology; unsolicited advice of this sort is all too common, and is invaluable. Belloc's preface — "Praise of this Book" — is silent on the point but in "Introductory Rambling" he gives a clear method for an ending. One, he wrote, ". . . is to rummage about one's manuscripts till one has found a bit of Fine Writing (no matter upon what subject), . . . to introduce a row of asterisks, and then paste on to the paper below these the piece of Fine Writing one has found."

* * * * * * * * * * * * * * * * * * *

What is the *meaning* of the differences that separate the Gramineae so delicately, yet so definitely, from any other order, and that so prevail that a grass remains a grass, however freely the type may vary? To attribute these differences to genic constitution is an 'explanation' of a merely descriptive kind; it enables us, indeed, to assign a place to them in the mental framework which we impose upon reality, but in so doing we have shelved, not solved, the problem. The mystery abides. (Agnes Arber *The Gramineae*, 1934).

Elizabeth Edgar
Henry Eamonn Connor

Christchurch
SS Thomas More and John Fisher
June 1999

REFERENCES

Abele, K. 1959: Cytological studies in the genus *Danthonia. Trans. Roy. Soc. S. Aust. 82*: 163–173.

Allan, H. H. 1936: An introduction to the grasses of New Zealand. *N.Z. DSIR Bull. 49.*

Allan, H. H. 1940: A handbook of the naturalized flora of New Zealand. *N.Z. DSIR Bull. 83.*

Bacon, C. W. and N. S. Hill (Eds). 1997: *Neotyphodium/Grass Interactions.* New York, Plenum Press.

Belloc, H. 1902: *The Path to Rome.* London, Allen and Unwin.

Brock, R. D. and J. A. M. Brown. 1961: Cytotaxonomy of Australian *Danthonia. Aust. J. Bot. 9*: 69–91.

Buchanan, J. 1878: *Indigenous Grasses of New Zealand.* t. 1–20. Wellington, James Hughes.

Buchanan, J. 1879: *idem*, t. 21–40.

Buchanan, J. 1880: *idem*, t. 41–58, 17(2), 26(2), 36(2).

Buchanan, J. 1880: *idem*, consolidated.

Clayton, W. D. and S. A. Renvoize. 1986: *Genera Graminum; Grasses of the World.* London, H.M.S.O.

Connor, H. E. 1977: *The Poisonous Plants in New Zealand.* Wellington, Government Printer.

Craig, G. F. 1984: Reinstatement of *Spinifex sericeus* R.Br. and hybrid status of *S. alterniflorus* Nees (Poaceae). *Nuytsia 5*: 67–74.

de Lange, P. J. et al. 1999: Threatened and uncommon plants of New Zealand. *N.Z. J. Bot. 37*: 603–628.

Doyle, J. D. et al. 1992: Chloroplast DNA inversions and the origin of the grass family (Poaceae). *Proc. Nat. Acad. Sci. 89*: 7722–7726.

Esler, A. E., Liefting, L. W. and P. D. Champion. 1993: *Biological Success and Weediness of the Noxious Plants of New Zealand.* Ministry of Agriculture and Fisheries.

Espie, P. R., Connor, H. E. and I. J. McCracken. 1992: Leaf-blade crimping in grasses: a new measure of growth. *Experientia 48*: 91–94.

Fletcher, L. R. and I. C. Harvey. 1981: An association of a *Lolium* endophyte with ryegrass staggers. *N.Z. Vet. J. 29*: 185–186.

Fuentes, C. 1988: *How I wrote one of my Books.* London, Andre Deutsch.

Funk, V. A. 1993: Uses and misuses of floras. *Taxon 42*: 761–772.

Gallagher, R. T. et al. 1984: Tremogenic neurotoxins from perennial ryegrass causing ryegrass staggers disorder of livestock: structure elucidation of lolitrem B. *J. Chem. Soc. Chem. Comm. 1984*: 614–616.

Hair, J. B. 1966: Biosystematics of the New Zealand flora, 1945–1964. *N.Z. J. Bot. 4*: 559–595.

Hair, J. B. 1968: Contributions to a chromosome atlas of the New Zealand flora — 12. *Poa* (Gramineae). *N.Z. J. Bot. 6*: 267–276.

Hair, J. B. and E. J. Beuzenberg. 1961: High polyploidy in New Zealand *Poa*. *Nature 189*: 160.

Hawksworth, D. L. 1992: The need for a more effective biological nomenclature for the 21st century. *Bot. J. Linn. Soc. 109*: 543–567.

Hooker, J. D. 1864: *Handbook of the New Zealand Flora*. Part 1. London, Reeves.

Hooker, J. D. 1867: *idem*. Part II.

Hsiao, C. et al. 1999: A molecular phylogeny of the grass family (Poaceae) based on the sequence of nuclear ribosomal DNA (ITS). *Aust. Syst. Bot. 11*: 667–688.

Le Roux, L. G. and E. A. Kellogg. 1999: Floral developments and the formation of unisexual spikelets in the Andropogoneae (Poaceae). *Amer. J. Bot. 86*: 354–366.

Linder, H. P. and G. A. Verboom. 1996. Generic limits in the *Rytidosperma* (Danthonieae, Poaceae) complex. *Telopea 6*: 597–626.

Miles, C. O. et al. 1998: Endophytic fungi in indigenous Australasian grasses associated with toxicity to livestock. *App. Envir. Microbiol. 64*: 601–606.

Molloy, B. P. J. 1994: Observations on the ecology and conservation of *Australopyrum calcis* (Triticeae: Gramineae) in New Zealand. *N.Z. J. Bot. 32*: 37–51.

Moore, D. M. 1960: Chromosome numbers of flowering plants from Macquarie Island. *Bot. Not. 113*: 185–191.

Orchard, A. E. 1999: ABRS Report. *Aust. Syst. Bot. Soc. Newsletter 98*: 4–5.

Soreng, R. J. and J. I. Davis. 1998: Phytogenetics and character evaluation in the grass family (Poaceae): simultaneous analyses of morphological and chloroplast DNA restriction site character sets. *Bot. Rev. 64*: 1–88.

Tjon Sie Fat, L. 1979: Cyanogenesis in some grasses. IV. The genus *Cortaderia. Proc. Kon. Nederl. Akad. Wetensch. C. 82*: 165–170.

Willemse, L. P. M. 1982: A discussion of the Ehrharteae (Gramineae) with special reference to the Malesian taxa formerly included in *Microlaena. Blumea 28*: 181–194.

REVISIONS

The full list of revisions carried out by us and the late V. D. Zotov in preparation for Vol. V Gramineae, classified by tribes is:

Ehrharteae

1998: E. Edgar and H. E. Connor. *Zotovia* and *Microlaena* : New Zealand ehrhartoid Gramineae. *N.Z. J. Bot. 36*: 565–586.

Stipeae

1989: S. W. L. Jacobs, J. Everett, H. E. Connor and E. Edgar. Stipoid grasses in New Zealand. *N.Z. J. Bot. 27*: 569–582.

Poeae

1986: E. Edgar. *Poa* in New Zealand. *N.Z. J. Bot. 24*: 425–503.

1996: E. Edgar. *Puccinellia* Parl. (Gramineae: Poeae) in New Zealand. *N.Z. J. Bot. 34*: 17–32.

1998: H. E. Connor. *Festuca* (Poeae: Gramineae) in New Zealand. I. Indigenous taxa. *N.Z. J. Bot. 36*: 329–367.

Agrostideae

1971: V. D. Zotov. *Simplicia* T. Kirk (Gramineae). *N.Z. J. Bot. 9*: 539–544.

1973: V. D. Zotov. *Hierochloe* R. Br. (Gramineae) in New Zealand. *N.Z. J. Bot. 11*: 561–580.

1982: E. Edgar and H. E. Connor. *Dichelachne* (Gramineae) in New Zealand. *N.Z. J. Bot. 20*: 303–309.

1991: E. Edgar and M. B. Forde. *Agrostis* L. in New Zealand. *N.Z. J. Bot. 29*: 139–161.

1995: E. Edgar. New Zealand species of *Deyeuxia* P.Beauv. and *Lachnagrostis* Trin. (Gramineae: Aveneae). *N.Z. J. Bot. 33*: 1–33.

1998: E. Edgar. *Trisetum* Pers. (Gramineae: Aveneae) in New Zealand. *N.Z. J. Bot. 36*: 539–564.

1999: E. Edgar and E. S. Gibb. *Koeleria* Pers. (Gramineae: Aveneae) in New Zealand. *N.Z. J. Bot. 37*: 51–61.

Hordeeae

1982:   Á. Löve and H. E. Connor. Relationships and taxonomy
        of New Zealand wheatgrasses. *N.Z. J. Bot. 20*: 169–
        186.

1993:   H. E. Connor, B. P. J. Molloy and M. I. Dawson. *Australopyrum*
        (Triticeae: Gramineae) in New Zealand. *N.Z. J. Bot.
        31*: 1–10.

1994:   H. E. Connor. Indigenous New Zealand Triticeae:
        Gramineae. *N.Z. J. Bot. 32*: 125–154.

Danthonieae

1963:   V. D. Zotov. Synopsis of the grass subfamily Arundinoideae
        in New Zealand. *N.Z. J. Bot. 1*: 78–136.

1979:   H. E. Connor and E. Edgar. *Rytidosperma* Steudel
        (*Notodanthonia* Zotov) in New Zealand. *N.Z. J. Bot. 17*:
        311–337.

1991:   H. E. Connor. *Chionochloa* Zotov (Gramineae) in New
        Zealand. *N.Z. J. Bot. 29*: 219–282.

Chlorideae

1971:   V. D. Zotov. *Zoysia* Willd. (Gramineae) in New Zealand.
        N.Z. J. Bot. 9: 639–644.

Mixtae

1999:   E. Edgar and H. E. Connor. Species novae graminum
        Novae-Zelandiae I. *N.Z. J. Bot. 37*: 63–70.

TYPIFICATION IN THIS FLORA

Within this Flora eleven taxa have been lectotypified, or assigned new types after original specimens were lost or destroyed; all are in genera we have not separately revised except for one in a synonym in *Poa*.

Poeae:

> sub *Austrofestuca littoralis*: the lectotype for the synonym *Schedonorus littoralis* var. *β minor* Hook.f., p. 86.

> sub *Poa novae-zelandiae* : the lectotype for the synonym *P. foliosa* var. γ Buchanan, p. 174.

Agrostideae:

> *Amphibromus fluitans* Kirk, p. 298.

> sub *Deschampsia cespitosa*: the lectotype for the synonym *D. cespitosa* var. *macrantha* Hack., p. 308.

> *D. chapmanii* Petrie, p. 309.

> sub *D. chapmanii*: the lectotype for the synonym *D. novae-zelandiae* Petrie, p. 309.

> *D. gracillima* Kirk, p. 312.

> *D. pusilla* Petrie, p. 313.

> *D. tenella* Petrie, p. 313.

> sub *D. tenella* : the lectotype for the synonym *D. tenella* var. *procera* Petrie, p. 313.

Chlorideae:

> *Zoysia pauciflora* Mez : the lectotype designated by A. Chase and C. D. Miles was destroyed at B. Its duplicate at WELT is the new lectotype, p. 517.

CHECKLISTS OF NATURALISED GRASSES

The purpose of the Checklists, eight of which were published, was to record for use in advance of the Flora (i) taxa naturalised here; (ii) region of origin; (iii) distribution in New Zealand, and (iv) the date of the first record of appearance here, and (v) provide keys to native and naturalised genera, to naturalised species, and to native species where they occur with naturalised species in the same genus. The only genera for which there are no published data are *Achnatherum* (Stipeae) and *Festuca* and *Schedonorus* (Poeae); all are attended to below.

Bambusoideae

> Sykes, W. R. 1996: Checklist of bamboos (Poaceae) naturalised in New Zealand. *N.Z. J. Bot. 34*: 153–156.

Pooideae

> Edgar, E., O'Brien, M. A. and H. E. Connor. 1991: Checklist of pooid grasses naturalised in New Zealand. 1. Tribes Nardeae, Stipeae, Hainardieae, Meliceae, and Aveneae. *N.Z. J. Bot. 29*: 101–116.

> Connor, H. E. and E. Edgar. 1994: Checklist of pooid grasses naturalised in New Zealand. 2. Tribe Triticeae. *N.Z. J. Bot. 32*: 409–413.

> Forde, M. B. and E. Edgar. 1995: Checklist of pooid grasses naturalised in New Zealand. 3. Tribes Bromeae and Brachypodieae. *N.Z. J. Bot. 33*: 35–42.

> Edgar, E. and E. S. Gibb. 1996: Checklist of pooid grasses naturalised in New Zealand. 4. Tribe Poeae. *N.Z. J. Bot. 34*: 147–152.

Panicoideae

> Edgar, E. and J. E. Shand. 1987: Checklist of panicoid grasses naturalised in New Zealand; with a key to native and naturalised species. *N.Z. J. Bot. 25*: 343–353.

> Edgar, E. 1998: Supplement to checklist of panicoid grasses naturalised in New Zealand. *N.Z. J. Bot. 36*: 163.

Mixtae

Edgar, E., Connor, H. E. and J. E. Shand. 1991: Checklist of oryzoid, arundinoid and chloridoid grasses naturalised in New Zealand. *N.Z. J. Bot. 29*: 117–129.

FIRST RECORDS

Distribution of the six taxa below is in the body of the Flora.

Achnatherum caudatum (Trin.) S.W.L.Jacobs et J.Everett

First Record: New record, CHR 518502, CHR 518503, *H. E. Connor & J. Clapham*, North Canterbury (Amberley), Christchurch City.
Region of Origin: S. America.

Festuca filiformis Pourr.

First record: Allan, H. H. *N.Z. DSIR Bull. 83*: 311 (1940), as *Festuca capillata* Lam.
Region of Origin: Europe.

Festuca ovina subsp. *hirtula* (W.G.Travis) M.J.Wilk.

First record: Armstrong, J. F. *T.N.Z.I. 4*: 289 (1872), as *Festuca ovina* L.
Region of Origin: Europe.

Festuca rubra L.

First record: Armstrong J. F. *T.N.Z.I. 4*: 289 (1872).
Region of Origin: Europe.

Schedonorus phoenix (Scop.) Holub

First record: Kirk, T. *T.N.Z.I. 3*: 161 (1871), as *Festuca elatior* L.; known also as *Festuca arundinacea* Schreb.
Region of Origin: Europe.

Schedonorus pratensis (Huds.) P.Beauv.

First record: Kirk, T. *T.N.Z.I. 3*: 161 (1871), as *Festuca pratensis* L., known also as *Festuca elatior* L.
Region of Origin: Europe.

REPRODUCTIVE BIOLOGY

During the last 45 years particular attention was paid to the reproductive systems active in endemic and indigenous New Zealand grasses, in the belief that such studies would reflect in some measure the morphological variation encountered in the field and in the herbarium. A series "Breeding Systems in New Zealand Grasses" was published in 12 parts, but for many species data on compatibility reactions were included in other studies. In all, the breeding systems of about 60 species among 20 genera are known, and among them self-compatible hermaphroditism is at a very high frequency. Dioecism is in five species of *Poa* and in *Spinifex sericeus*. Andromonoecism characterises all species of *Hierochloe* in New Zealand, gynomonoecism occurs in three species of *Poa*, and gynodioecism is present in all species of *Cortaderia* except *C. turbaria* from Chatham Is. Self-compatibility is easily recognised in chasmogamous-cleistogamous *Dichelachne* and *Microlaena*; two species of this latter also produce cleistogenes. Cleistogamy is certain for two species of *Festuca*, and two of *Deyeuxia*, but only tentatively suggested for some *Poa* spp.

No endemic species is known to be apomictic, but *Poa*, where apomictic phenomena abound (Anton and Connor 1995), has not been subject to any critical embryological examination. For the other polytypic genera, *Chionochloa* — a genus with periodic flowering and mast seeding — and *Rytidosperma*, self-compatibility is known in about seven species of the former, but none of the latter has been subject to experiment.

Papers listed below deal with the reproductive biology of native species, but much is known about the breeding systems of naturalised grasses, e.g., the genetics of incompatibility in *Lolium* and *Phalaris*; autonomous precocious apomixis in *Cortaderia jubata*; self-compatibility in *Vulpia* spp. and *Briza* spp.; monoecism in *Zizania latifolia*; cleistogamy in *Sieglingia* and in *Nassella* spp. especially where *N. neesiana* also forms cleistogenes in the axils of leaves at the base of a tiller. Details for these and other taxa are in Connor (1979, 1981, 1987).

In a review of the reproductive biology of grasses Connor (1979) continued to express the concern that C. E. Hubbard of Kew had raised twenty-five years earlier: the omission of the description of the flowers in many treatments and in many new species definitions. Surprisingly, this failure persists, often associated with carpological deficiencies; yet the flowers and their resultant seeds eventually validate taxonomic predictions. We make no apology for the floral descriptions here which may not be needed by some users but which are vital to floral biologists and evolutionists, and after all only complete the species description. In this Flora we endeavour, especially for endemic and indigenous taxa, to include data on lodicules, stamens, and gynoecia, their dimorphism and heteromorphism where these occur, and their distribution between flowers when they are separate. We endeavour to clarify the reproductive biology of taxa where uncertainty existed, and indicate the presence of cleistogamy, protogyny, chasmogamy, asexuality, and sterility where these occur.

Anton, A. M. and H. E. Connor. 1995: Floral biology and reproduction in *Poa. Aust. J. Bot. 43*: 577–599.

Connor, H. E. and A. B. Cook. 1952: Note on the breeding system of *Dichelachne crinita* (Linn. f.) Hook. f. in New Zealand. *N.Z. J. Sci. Tech. A34*: 369–371.

Connor, H. E. and A. B. Cook. 1955: The breeding system of New Zealand fescue-tussock *Festuca novae-zelandiae* (Hack.) Cockayne. *N.Z. J. Sci. Tech. A37*: 103–105.

Connor, H. E. 1957: Breeding systems in some New Zealand grasses. *N.Z. J. Sci. Tech. A38*: 742–751.

Connor, H. E. and E. D. Penny. 1960: Breeding systems in New Zealand grasses II. Gynodioecy in *Arundo richardii. N.Z. J. Agric. Res. 3*: 725–727.

Connor, H. E. 1960: Breeding systems in New Zealand grasses. III. Festuceae, Aveneae, Agrostideae. *N.Z. J. Agric. Res. 3*: 728–733.

Connor, H. E. 1963: Breeding systems in New Zealand grasses. IV. Gynodioecism in *Cortaderia. N.Z. J. Bot. 1*: 258–264.

Connor, H. E. 1965: Breeding systems in New Zealand grasses. V. Naturalised species of *Cortaderia. N.Z. J. Bot. 3*: 17–23.

Connor, H. E. 1965: Breeding systems in New Zealand grasses. VI. Control of gynodioecism in *Cortaderia richardii. N.Z. J. Bot. 3*: 233–242.

Connor, H. E. 1965: Breeding systems in New Zealand grasses. VII. Periodic flowering of snow tussock *Chionochloa rigida*. *N.Z. J. Bot.* 4: 392–397.

Connor, H. E. 1973: Breeding systems in *Cortaderia* (Gramineae). *Evolution* 27: 663–678.

Connor, H. E. and B. A. Matthews. 1977: Breeding systems in New Zealand grasses. VIII. Cleistogamy in *Microlaena*. *N.Z. J. Bot.* 15: 531–534.

Connor, H. E. 1979: Breeding systems in the grasses: a survey. *N.Z. J. Bot.* 17: 547–574.

Connor, H. E. 1981: Evolution of reproductive systems in the Gramineae. *Ann. Miss. Bot. Gard.* 68: 48–74.

Connor, H. E. 1984: Breeding systems in New Zealand grasses. XI. Sex ratios in dioecious *Spinifex sericeus*. *N.Z. J. Bot.* 22: 569–574.

Connor, H. E. 1987: Reproductive biology in the grasses. *In* Soderstrom, T. R. et al. (Eds) *Grass Systematics and Evolution*. Washington, D. C., Smithsonian Institution Press, pp. 117–132.

Connor, H. E. 1988: Breeding systems in New Zealand grasses. X. Species at risk for conservation. *N.Z. J. Bot.* 26: 163–167.

Connor, H. and D. Charlesworth. 1989: Genetics of male sterility in gynodioecious *Cortaderia* (Gramineae). *Heredity* 63: 373–382.

Connor, H. E. 1990: Breeding systems in New Zealand grasses. XI. Gynodioecism in *Chionochloa bromoides*. *N.Z. J. Bot.* 28: 59–65.

Connor, H. E. 1998: Breeding systems in New Zealand grasses. XII. Cleistogamy in *Festuca*. *N.Z. J. Bot.* 36: 471–476.

McKone, M. J., Thom, A. L. and D. Kelly. 1997: Self-compatibility in *Chionochloa pallens* and *C. macra* (Poaceae) confirmed by hand pollination of excised styles. *N.Z. J. Bot.* 35: 259–262.

ANNALS OF TAXONOMIC RESEARCH ON INDIGENOUS NEW ZEALAND GYMNOSPERMAE AND ANGIOSPERMAE 1987–1996

(Together with some earlier titles not listed
in Volumes 1, 2, 3 and 4)

1803– Lambert, A. B. *A description of the genus Pinus*, illustrated with figures,
1824   directions relative to the cultivation, and remarkes on the uses of the
several species... London, J. White. 2 vols. Vol. 1, authored by Lambert,
Vol. 2, authored by D. Don. *Dacrydium cupressinum* Lamb. described.

1803– Lambert, A. B. *A description of the genus Pinus*, with directions relative to
1824   the cultivation, and remarkes on the uses of the several species: also
descriptions of many other new species of the family of Coniferae.
Illustrated with figures. London, Weddell. 2 vols. Referred to as the
"third edition". *Podocarpus totara* G.Benn. ex D.Don in Lamb.
described.

1915   Domin, K. Beiträge zur Flora und Pflanzengeographie Australiens.
Biblioth. Bot. 85: 1–551. Includes spp. common to N.Z. and Australia.

1928   Howarth, W. O. The genus *Festuca* in New Zealand. *J. Linn. Soc. Lond. 48*:
57–77.

1954   de Wet, J. M. J. Stomatal size as a cytological criterion in *Danthonia*.
Cytologia 19. 176–181. Includes (as *Danthonia* spp.) 7 N.Z. taxa now in
Chionochloa or *Rytidosperma*: *R. nigricans* (as *D. semiannularis*),
R. australe, C. bromoides, C. conspicua subsp. *cunninghamii* (as
D. cunninghamii), *C. crassiuscula, C. oreophila, C. ovata*.

1956   Connor, H. E. Interspecific hybrids in New Zealand *Agropyron*. *Evolution
10*: 415–420.

1960   Connor, H. E. Breeding systems in New Zealand grasses. III. Festuceae,
Aveneae and Agrostideae. *N.Z. J. Agric. Res. 3*: 728–733.

1961   Connor, H. E. A tall-tussock grassland community in New Zealand. *N.Z.
J. Sci. 4*: 825–835. Includes nomenclatural note on use of some names
in *Danthonia*.

1963   Tateoka, T. Notes on some grasses. XIII. Relationship between Oryzeae
and Ehrharteae, with special reference to leaf anatomy and histology.
Bot. Gaz. 124: 264–270. Includes N.Z. spp. of *Microlaena*.

1965   Mark, A. F. Flowering, seeding, and seedling establishment of narrow-
leaved snow-tussock, *Chionochloa rigida*. *N.Z. J. Bot. 3*: 180–193.

       Mark, A. F. Effects of management practices on narrow-leaved snow
tussock *Chionochloa rigida*. *N.Z. J. Bot. 3*: 300–319. Discussed mast years
in *Chionochloa*.

1966   Connor, H. E. Breeding systems in New Zealand grasses. VII. Periodic
flowering of snow tussock, *Chionochloa rigida*. *N.Z. J. Bot. 4*: 392–397.

1968   Mark, A. F. Factors controlling irregular flowering in four alpine species
       of *Chionochloa. Proc. N.Z. Ecol. Soc. 15*: 55–60.
1969   Fineran, B. A. The flora of the Snares Islands, New Zealand. *T.R.S.N.Z.
       (Bot.) 3*: 237–270.
       Mark, A. F. Ecology of snow tussocks in the mountains of New Zealand.
       Vegetatio 18: 289–306. Discusses reproduction and mast years in
       Chionochloa.
1970   Connor, H. E., Bailey, R. W. and K. F. O'Connor. Chemical composition of
       New Zealand tall-tussocks (*Chionochloa*). *N.Z. J. Agric. Res. 13*: 534–554.
       Scott, D. Relative growth rates under controlled temperatures of some
       New Zealand indigenous and introduced grasses. *N.Z. J. Bot. 8*: 76–81.
       Recognised a high altitude ecotype of *Festuca novae-zelandiae.*
1972   Bailey, R. W. and H. E. Connor. Structural polysaccharides in leaf blades
       and sheaths in the arundinoid grass *Chionochloa. N.Z. J. Bot. 10*: 533–544.
       Connor, H. E. and R. W. Bailey. Leaf strength in four species of *Chionochloa*
       (Arundineae). *N.Z. J. Bot. 10*: 515–532.
1975   White, E. G. An investigation and survey of insect damage affecting
       Chionochloa seed production in some alpine tussock grasslands. *N.Z.
       J. Agric. Res. 18*: 163–178.
1977   Buurman, J. Contribution to the pollen morphology of the Bignoniaceae
       with special reference to the tricolpate type. *Pollen Spores 19*: 447–519.
       Includes *Tecomanthe speciosa.*
       Godley, E. J. and D. H. Smith. Kowhais and their flowering. *Ann. J. Roy.
       N.Z. Inst. Hort. 5*: 24–31.
       Williams, P. A., Mugambi, S. and K. F. O'Connor. Shoot properties of nine
       species of *Chionochloa* Zotov (Gramineae). *N.Z. J. Bot. 15*: 761–765.
1978   Payton, I. J. and D. J. Brasch. Growth and non-structural carbohydrate
       reserves in *Chionochloa rigida* and *C. macra*, and their short-term
       response to fire. *N.Z. J. Bot. 16*: 435–460.
       Townrow, J. E. S. The genus *Stipa* L. in Tasmania. Part 3 — Revised
       taxonomy. *Pap. Proc. Roy. Soc. Tasmania 112*: 227–287. Lectotypified
       Stipa stipoides, indigenous to N.Z. and Australia; now *Austrostipa
       stipoides.*
       Williams, P. A. et al. Macro-elements within shoots of tall-tussocks
       (*Chionochloa*), and soil properties on Mt Kaiparoro, Wairarapa, New
       Zealand. *N.Z. J. Bot. 16*: 255–260.
       Williams, P. A. et al. Macro-element composition of tall-tussocks
       (*Chionochloa*) in the South Island, New Zealand and their relationship
       with soil chemical properties. *N.Z. J. Bot. 16*: 479–498.
1979   Lieu, S. M. Organogenesis in *Triglochin striata. Canad. J. Bot. 57*: 1418–1438.
       Payton, I. J. and A. F. Mark. Long-term effects of burning on growth,
       flowering, and carbohydrate reserves in narrow-leaved snow tussock
       (*Chionochloa rigida*). *N.Z. J. Bot. 17*: 43–54.

1980 Primack, R. B. Variation in the phenology of natural populations of montane shrubs in New Zealand. *J. Ecol. 68*: 849–862.

1982 Fosberg, F. R. A preliminary conspectus of the genus *Leptostigma* (Rubiaceae). *Acta Phytotax. Geobot. 33*: 73–83. Makes new combination *Leptostigma setulosum* (Hook.f.) Fosberg for *Nertera setulosa* Hook.f.

Fosberg, F. R. and M. H. Sachet. Micronesian Poaceae: critical and distributional notes. *Micronesica 18*: 45–102. Includes *Oplismenus*.

Veldkamp, J. F. *Agrostis* (Gramineae) in Malesia and Taiwan. *Blumea 28*: 199–228. Comments on N.Z. varieties in *Deyeuxia forsteri* and *D. filiformis*.

1984 Pauptit, R. A. et al. The structure of 19aH-lupeol methyl ether from *Chionochloa bromoides*. *Aust. J. Chem. 37*: 1341–1347.

Schoen, D. J. and D. G. Lloyd. The selection of cleistogamy and heteromorphic diaspores. *Biol. J. Linn. Soc. 23*: 303–322.

1985 Knowles, B. and C. Ecroyd. Species of *Cortaderia* (pampas grasses and toetoe) in New Zealand. *Forest Res. Inst. Bull. 105*: 1–24.

1986 Clayton, W. D. and S. A. Renvoize. *Genera Graminum; Grasses of the World*. London, H.M.S.O.

Payton, I. J. et al. Nutrient concentrations in narrow-leaved snow tussock (*Chionochloa rigida*) after spring burning. *N.Z. J. Bot. 24*: 529–537.

Vickery, J. W., Jacobs, S. W. L. and J. Everett. Taxonomic studies in *Stipa* (Poaceae) in Australia. *Telopea 3*: 1–132.

1987 Barkworth, M. E. and J. Everett. Evolution in the Stipeae: identification and relationships of its monophyletic taxa. *In* Soderstrom, T. R. et al. (Eds) *Grass Systematics and Evolution*. Washington, D. C., Smithsonian Institution Press, pp. 251–264.

Breuer, B. et al. Fatty acids of some Cornaceae, Hydrangeaceae, Aquifoliaceae, Hamamelidaceae and Styracaceae. *Phytochem. 26*: 1441–1445. Includes *Corokia macrocarpa* and *Griselinia littoralis*.

Collet, C. and M. Westerman. Interspecies comparison of the highly-repeated DNA of Australasian *Luzula* (Juncaceae). *Genetica 74*: 95–103.

Conert, H. J. Current concepts in the systematics of the Arundinoideae. *In* Soderstrom, T. R. et al. (Eds) *Grass Systematics and Evolution*. Washington, D. C., Smithsonian Institution Press, pp. 239–250.

Connor, H. E. Reproductive biology in the grasses. *In* Soderstrom, T. R. et al. (Eds) *Grass Systematics and Evolution*. Washington, D. C., Smithsonian Institution Press, pp. 107–132.

Connor, H. E. and E. Edgar. Name changes in the indigenous New Zealand flora, 1960–1986 and Nomina Nova IV, 1983–1986. *N.Z. J. Bot. 25*: 115–170.

Cook, C. D. K. and M. S. Nicholls. A monographic study of the genus *Sparganium* (Sparganiaceae). Part 2. Subgenus *Sparganium*. *Bot. Helvet. 97*: 1–35. Includes *S. subglobosum*.

1987 Cristofolini, G. Serological relationships among Sophoreae, Thermopsideae and Genisteae (Fabaceae). *Bot. J. Linn. Soc. 94*: 421–432.

Dawson, M. I. Contributions to a chromosome atlas of the New Zealand flora. 29, Myrtaceae. *N.Z. J. Bot. 25*: 367–369.

Druce, A. P., Williams, P. A. and J. C. Heine. Vegetation and flora of Tertiary calcareous rocks in the mountains of Western Nelson, New Zealand. *N.Z. J. Bot. 25*: 41–78.

Edgar, E. and H. E. Connor. Varietal grass names and combinations published by Hackel in Cheeseman's Manual of the New Zealand Flora, 1906. *N.Z. J. Bot. 25*: 455–457.

Eichler, Hj. Nomenclatural and bibliographical survey of *Hydrocotyle* L. (Apiaceae), 1. *Feddes Repert. 98*: 1–51. 2. *Feddes Repert. 98*: 145–196. 3. *Feddes Repert. 98*: 273–351.

Endress, P. K. The Chloranthaceae: reproductive structures and phylogenetic position. *Bot. Jahrb. Syst. 109*: 153–226. Includes *Ascarina lucida.*

Foo, L. Y. Phenylpropanoid derivatives of catechin, apicatechin and phylloflavan from *Phyllocladus trichomanoides. Phytochem. 26*: 2825–2830.

Friis, I., Immelman, K. and C. M. Wilmot-Dear. New taxa and combinations in Old World Urticaceae. *Nord. J. Bot. 7*: 125–126. Makes new combination *Australina pusilla* subsp. *muelleri* (Wedd.) Friis & Wilmot-Dear based on *A. muelleri* from Victoria.

Garnock-Jones, P. J. What is *Leonohebe? Hebe Newsletter 4*: 19–22.

George, A. S. (Ed.) *Flora of Australia. 45, Hydatellaceae to Liliaceae.* Canberra, AGPS. A very few spp. are common to N.Z. and Australia.

Grey-Wilson, C. Plant portraits: 82, *Clematis marmoraria. Kew Mag. 4*: 116–119.

Haase, P. Ecological studies on *Hoheria glabrata* (Malvaceae) at Arthur's Pass, South Island, New Zealand. *N.Z. J. Bot. 25*: 401–409. Contains much phenological information.

Heads, M. New names in New Zealand Scrophulariaceae. *Bot. Soc. Otago Newsletter 5*: 4–11.

Jones, D. L. and M. A. Clements. Reinstatement of the genus *Cyrtostylis* R. Br. and its relationship with *Acianthus* R. Br. (Orchidaceae). *Orchadian 2*: 156–160.

Larsen, P. O. et al. Relationship between subfamilies, tribes, and genera in Iridaceae inferred from chemical characters. *Biochem. Syst. Ecol. 15*: 575–579. Includes *Libertia grandiflora.*

Lloyd, D. G. and C. J. Webb. The reinstatement of *Leptinella* at generic rank, and the status of the 'Cotuleae' (Asteraceae, Anthemideae). *N.Z. J. Bot. 25*: 99–105.

Lorimer, S. D. and R. T. Weavers. Foliage sesquiterpenes and diterpenes of *Podocarpus spicatus. Phytochem. 26*: 3207–3215.

1987 Moore, L. B. *Pomaderris* revisited. *Tane 31*: 139–143.

Ng, K. M. et al. The biochemical systematics of *Tetradium, Euodia* and *Melicope* and their significance in the Rutaceae. *Biochem. Syst. Ecol. 15*: 587–593. Includes *M. simplex* and *M. ternata.*

Nicora, E. G. and Z. E. Rúgolo de Agrasar. *Los Generos de Gramineas de America Austral.* Buenos Aires, Editorial Hemisferio Sur S. A.

Nyananyo, B. L. and V. H. Heywood. A new combination in *Lyallia* (Portulacaceae). *Taxon 36*: 640–641. *Lyallia* and *Hectorella* are placed in the one genus *Lyallia.* Makes the new combination *Lyallia caespitosa* (Hook.f.) Nyananyo et Heywood based on *Hectorella caespitosa* Hook.f.

Ogle, C. A rarely seen native grass, *Amphibromus fluitans. Wellington Bot. Soc. Bull. 43*: 29–32.

Orchard, A. E. A revision of the *Coprosma pumila* (Rubiaceae) complex in Australia, New Zealand and the subantarctic islands. *Brunonia 9*: 119–138.

Patel, R. N. Wood anatomy of the dicotyledons indigenous to New Zealand. 16, Lauraceae. *N.Z. J. Bot. 25*: 477–488.

Philipp, M. Reproductive biology of *Geranium sessiliflorum.* 2, The genetics and morph frequencies of the leaf colour polymorphism. *N.Z. J. Bot. 25*: 309–313.

Philipson, W. R. *Corynocarpus* J. R. & G. Forst. — an isolated genus. *Bot. J. Linn. Soc. 95*: 9–18.

Philipson, W. R. The treatment of isolated genera. *Bot. J. Linn. Soc. 95*: 19–25. Several examples from the N.Z. flora.

Philipson, W. R. A classification of the Monimiaceae. *Nord. J. Bot. 7*: 25–29.

Quinn, C. J. The Phyllocladaceae Keng — a critique. *Taxon 36*: 559–565.

Rama, T. V., Rao, P. N. and R. S. Rao. Somatic chromosome morphology in the genus *Iphigenia. Cytologia 52*: 434–444. $2n = 22$ in Indian spp. in contrast to $2n = 20$ for *Iphigenia novae-zelandiae.*

Rodriguez, R. and C. Marticorena. Las especies del genero *Luzuriaga* R. et P. *Gayana, Bot. 44*: 3–15.

Sampson, F. B. Stamen venation in the Winteraceae. *Blumea 32*: 79–89.

Selvaraj, R. Karyomorphological studies in south Indian Rubiaceae. *Cytologia 52*: 343–356. Includes *Coprosma baueri* and *C. lucida.*

Sykes, W. R. The parapara, *Pisonia brunoniana* (Nyctaginaceae). *N.Z. J. Bot. 25*: 459–466.

Wardle, P. *Dracophyllum* (Epacridaceae) in the Chatham and subantarctic islands of New Zealand. *N.Z. J. Bot. 25*: 107–114.

Webb, C. J. and E. J. Beuzenberg. Contributions to a chromosome atlas of the New Zealand flora. Corrections and additions to number 21, Umbelliferae (*Hydrocotyle*). *N.Z. J. Bot. 25*: 371–372.

1987 Webb, C. J. and J. Littleton. Flower longevity and protandry in two species of *Gentiana* (Gentianaceae). *Ann. Missouri Bot. Gard. 74*: 51–57. Studied *G. saxosa* and *G. serotina.*

Webby, R. F., Markham, K. R. and B. P. J. Molloy. The characterization of New Zealand *Podocarpus* hybrids using flavonoid markers. *N.Z. J. Bot. 25*: 355–366.

Webster, R. D. *The Australian Paniceae (Poaceae)*. Berlin and Stuttgart, J. Cramer. Includes species also indigenous to New Zealand.

1988 Adsersen, A., Adsersen, H. and L. Brimer. Cyanogenic constituents in plants from the Galápagos Islands. *Biochem. Syst. Ecol. 16*: 65–77. Includes *Ipomoea pes-caprae.*

Anderson, Ø. M. Proportions of individual anthocyanins in the genus *Metrosideros. Biochem. Syst. Ecol. 16*: 535–539.

Arroyo, S. C. and B. E. Leuenberger. Leaf morphology and taxonomic history of *Luzuriaga (Philesiaceae)*. *Willdenowia 17*: 159–172.

Bevalot, F. et al. Coumarins from three *Phebalium* species. *Biochem. Syst. Ecol. 16*: 631–633. Mentions *P. nudum.*

Burns-Balagh, P. and P. Bernhardt. Floral evolution and phylogeny in the tribe Thelymitreae (Orchidaceae: Neottioideae). *Pl. Syst. Evol. 159*: 19–47.

Chalk, D. *Hebes and Parahebes.* Christchurch, Caxton Press.

Clarkson, B. D. A natural intergeneric hybrid, *Celmisia gracilenta* × *Olearia arborescens* (Compositae) from Mt Tarawera, New Zealand. *N.Z. J. Bot. 26*: 325–331.

Conner, L. N. and A. J. Conner. Seed biology of *Chordospartium stevensonii. N.Z. J. Bot. 26*: 473–475.

Connor, H. E. Breeding systems in New Zealand grasses. 10, Species at risk for conservation. *N.Z. J. Bot. 26*: 163–167.

Craig, J. L. and A. M. Stewart. Reproductive biology of *Phormium tenax*: a honeyeater-pollinated species. *N.Z. J. Bot. 26*: 453–463.

Doyle, M. F. and R. Scogin. A comparative phytochemical profile of the Gunneraceae. *N.Z. J. Bot. 26*: 493–496.

Druce, A. P. and W. R. Sykes. A new species of *Crassula* L. in New Zealand. *N.Z. J. Bot. 26*: 477–478.

Dubcovsky, J. and A. Martinez. Phenetic relationships in the *Festuca* spp. from Patagonia. *Canad. J. Bot. 66*: 468–478. Includes *F. contracta.*

Garnock-Jones, P. J. and P. N. Johnson. *Iti lacustris* (Brassicaceae), a new genus and species from southern New Zealand. *N.Z. J. Bot. 25*: 603–610.

Garnock-Jones, P. J. and B. Jonsell. *Rorippa divaricata* (Brassicaceae), a new combination. *N.Z. J. Bot. 26*: 479–480.

Hind, D. J. N. and C. Jeffrey. *Brachycome* Cass. corr. Cass. and *Lagenophora* Cass. corr. Cass. are correct. *Kew Bull. 43*: 329–331.

1988   McEwen, W. M. Cone and seed phenology in several New Zealand conifer tree species. *Tuatara 30*: 66–76.

McKone, M. J. and C. J. Webb. A difference in pollen size between the male and hermaphrodite flowers of two species of Apiaceae. *Aust. J. Bot. 36*: 331–337. *Lignocarpa diversifolia* of northern South Id, N.Z. and *Gingidia harveyana* of south-eastern Australia were studied.

Nelson, E. C. Holotype of *Olearia semidentata* Decne. *N.Z. J. Bot. 26*: 464–466.

Norton, D. A. Recent changes in the names of New Zealand tree and shrub species. *Roy. N.Z. Inst. Hort. Ann. J. 15*: 33–35.

Nyananyo, B. L. Leaf anatomical studies in the Portulacaceae (Centrospermae) with regards to photosynthetic pathways. *Folia Geobot. Phytotax. 23*: 99–101. Includes *Hectorella* and *Lyallia*.

Nyananyo, B. L. The systematic significance of seed morphology and anatomy in the Portulacaceae (Centrospermae). *Folia Geobot. Phytotax. 23*: 275–279. Includes *Lyallia*.

Patel, R. N. Wood anatomy of the dicotyledons indigenous to New Zealand. 17, Tiliaceae. *N.Z. J. Bot. 26*: 337–343.

Philipson, W. R. Seedling and shoot morphology of the New Zealand species of *Nothofagus* (Fagaceae). *N.Z. J. Bot. 26*: 401–407.

Philipson, W. R. and M. N. Philipson. A classification of the genus *Nothofagus* (Fabaceae). *Bot. J. Linn. Soc. 98*: 27–36.

Rendle, H. and B. G. Murray. Breeding systems and pollen tube behaviour in compatible and incompatible crosses in New Zealand species of *Ranunculus* L. *N.Z. J. Bot. 26*: 467–471.

Rogers, G. Parentless *Pittosporum turneri*. *Bull. Wellington Bot. Soc. 44*: 26–36.

Rudall, P. J. and H. P. Linder. Megagametophyte and nucellus in Restionaceae and Flagellariaceae. *Amer. J. Bot. 75*: 1777–1786. Includes *Empodisma minus* and *Sporodanthus traversii*.

Saleh, N. A. M. et al. The chemosystematics of local members of the subtribe Gnaphaliinae (Compositae). *Biochem. Syst. Ecol. 16*: 615–617. Includes *Pseudognaphalium luteoalbum*.

Savill, M. G., Bickerstaffe, R. and H. E. Connor. Interspecific variation in epicuticular waxes of *Chionochloa*. *Phytochem. 27*: 3499–3507.

Seberg, O. Taxonomy, phylogeny, and biogeography of the genus *Oreobolus* R.Br. (Cyperaceae), with comments on the biogeography of the South Pacific continents. *Bot. J. Linn. Soc. 96*: 119–195.

Seberg, O. Leaf anatomy of *Oreobolus* R.Br. and *Schoenoides* Seberg (Cyperaceae). *Bot. Jahrb. Syst. 110*: 187–214.

Short, P. S. Two new species of *Brachyscome* Cass. (Compositae: Astereae), with a note on the orthography of the generic name. *Muelleria 6*: 389–398.

1988   Smith, A. C. *Flora Vitiensis Nova: A New Flora of Fiji (Spermatophytes only). 4.* Lawai, Hawaii, National Tropical Botanical Garden.

Sorsa, P. Pollen morphology of *Potamogeton* and *Groenlandia* (Potamogetonaceae) and its taxonomic significance. *Ann. Bot. Fennici* 25: 179–199. Includes *P. cheesemanii, P. ochreatus* and *P. pectinatus.*

Sykes, M. T. and J. B. Wilson. An experimental investigation into the response of some New Zealand sand dune species to salt spray. *Ann. Bot.* 62: 159–166.

Sykes, W. R. Kermadec ngaio (*Myoporum,* Myoporaceae). *N.Z. J. Bot.* 25: 595–601.

Sykes, W. R. A new species of *Senecio* from New Zealand. *N.Z. J. Bot.* 25: 611–613.

Sykes, W. R. Notes on New Zealand *Plantago* species. *N.Z. J. Bot.* 26: 321–323. Describes *P. obconica* W.Sykes and makes new combinations *P. triandra* subsp. *masonae* and *P. spathulata* subsp. *picta.*

Vink, W. Taxonomy in *Winteraceae. Taxon* 37: 691–698.

Webb, C. J. Notes on the *Senecio lautus* complex in New Zealand. *N.Z. J. Bot.* 26: 481–484. Describes new sp. *S. marotiri* C.Webb and makes new combination *S. carnosulus* (Kirk) C.Webb.

Webb, C. J. *Gnaphalium laterale,* a new species for New Zealand. *N.Z. J. Bot.* 26: 485–487.

Webb, C. J., Sykes, W. R. and P. J. Garnock-Jones. *Flora of New Zealand. 4.* Christchurch, Botany Division, DSIR.

Weberling, F. The architecture of inflorescences in the Myrtales. *Ann. Missouri Bot. Gard.* 75: 226–310. Includes N.Z. spp. of *Epilobium* and *Metrosideros.*

Wiegleb, G. Notes on pondweeds — outlines for a monographical treatment of the genus *Potamogeton* L. *Feddes Repert.* 99: 249–266.

Wilson, K. L. *Polygonum sensu latu* (Polygonaceae) in Australia. *Telopea* 3: 177–182. Makes new combination in *Persicaria* for *Polygonum decipiens.*

1989   Adolphi, K., Seybold, S. and L. A. S. Johnson. (952) Proposal to conserve 8878 *Brachycome* Cass. (*Asteraceae*). *Taxon* 38: 511–513.

Anderberg, A. A. Phylogeny and reclassification of the tribe Inuleae (Asteraceae). *Canad. J. Bot.* 67: 2277–2296.

Bannister, P. Nitrogen concentration and mimicry in some New Zealand mistletoes. *Oecologica* 79: 128–132.

Browning, J. Studies in Cyperaceae in southern Africa. 14, A reappraisal of *Scirpus nodosus* and *S. dioecus. S. Afr. J. Bot.* 55: 422–432.

Connor, H. E. and D. Charlesworth. Genetics of male-sterility in gynodioecious *Cortaderia* (Gramineae). *Heredity* 63: 373–382.

Craig, J. L. A differential response to self pollination: seed size in *Phormium. N.Z. J. Bot.* 27: 583–586.

Dahlgren, G. An updated angiosperm classification. *Bot. J. Linn. Soc. 100:* 197–203.

1989   Dawson, M. I. Contributions to a chromosome atlas of the New Zealand flora. 30, Miscellaneous species. *N.Z. J. Bot. 27*: 163–165.

Delph, L. F. and C. M. Liverly. The evolution of floral color change: pollinator attraction versus physiological constraints in *Fuchsia excorticata. Evolution 43*: 1252–1262.

Dickison, W. C. Stem and leaf anatomy of the Alseuosmiaceae. *Aliso 12*: 567–578.

Druce, A. P. *Coprosma waima* (Rubiaceae) — a new species from northern New Zealand. *N.Z. J. Bot. 27*: 119–128.

Endress, P. K. and L. D. Hufford. The diversity of stamen structures and dehiscence patterns among Magnoliidae. *Bot. J. Linn. Soc. 100*: 45–85. Includes *Pseudowintera*.

Fountain, D. W., Holdsworth, J. M. and H. A. Outred. The dispersal unit of *Dacrycarpus dacrydioides* (A.Rich.) de Laubenfels (Podocarpaceae) and the significance of the fleshy receptacle. *Bot. J. Linn. Soc. 99*: 197–207.

Gadek, P. A., Bruhl, J. J. and C. J. Quinn. Exine structure in the 'Cotuleae' (Anthemideae, Asteraceae). *Grana 28*: 163–178. Includes N.Z. spp.

George, A. S. (Ed.) *Flora of Australia. 3, Hamamelidales to Casuarinales.* Canberra, AGPS. Includes *Parietaria debilis* and *Australina pusilla.*

Godley, E. J. The flora of Antipodes Island. *N.Z. J. Bot. 27*: 531–563.

Jacobs, S. W. L. et al. Stipoid grasses in New Zealand. *N.Z. J. Bot. 27*: 569–582.

Johnson, P. N. and P. A. Brooke. *Wetland Plants in New Zealand.* Wellington, DSIR Publishing.

Jones, D. L. and M. A. Clements. Reinterpretation of the genus *Genoplesium* R. Br. (Orchidaceae: Prasophyllinae). *Lindleyana 4*: 139–145. Makes new combinations *G. nudum* and *G. pumilum.*

Lee, W. G. and M. Fenner. Mineral nutrient allocation in seeds and shoots of twelve *Chionochloa* species in relation to soil fertility. *J. Ecol. 77*: 704–716.

Macmillan, B. H. *Acaena juvenca* and *Acaena emittens* (Rosaceae) — two new species from New Zealand. *N.Z. J. Bot. 27*: 109–117.

Markham, K. R. et al. Support from flavonoid glycoside distribution for the division of *Dacrydium* sensu lato. *N.Z. J. Bot. 27*: 1–11.

Martinez, S. El genero *Azorella* (Apiaceae—Hydrocotyloideae) en la Argentina. *Darwiniana 29*: 139–178. Includes *Azorella selago.*

Mathew, P. M. and B. Vijayavalli. Cytology of species of *Cordyline* and *Dracaena* from south India. *Cytologia 54*: 573–579. Includes chromosome count of *Cordyline indivisa.*

Murray, B. G., Braggins, J. E. and P. D. Newman. Intraspecific polyploidy in *Hebe diosmifolia* (Cunn.) Cockayne et Allan (Scrophulariaceae). *N.Z. J. Bot. 27*: 587–589.

1989   Ogle, C. *Sebaea ovata* (Gentianaceae) and its habitat near Wanganui. *Bull. Wellington Bot. Soc. 45*: 92–99.

Orchard, A. E. *Azorella* Lamarck (Apiaceae) on Heard and Macquarie Islands, with description of a new species, *A. macquariensis. Muelleria 7*: 15–20.

Patel, R. N. Wood anatomy of the dicotyledons indigenous to New Zealand. 18, Elaeocarpaceae. *N.Z. J. Bot. 27*: 325–335.

Philipson, W. R. Semihypogeal germination in two New Zealand genera: *Dysoxylum* and *Syzygium. N.Z. J. Bot. 27*: 311–312.

Rendle, H. and B. G. Murray. Chromosome relationships and breeding barriers in New Zealand species of *Ranunculus. N.Z. J. Bot. 27*: 437–448.

Short, P. S., Wilson, K. E. and J. Nailon. Notes on the fruit anatomy of Australian members of the Inuleae (Compositae). *Muelleria 7*: 57–79. Makes some comparisons with N.Z. spp.

Spence, J. R. and W. R. Sykes. Are *Plantago novae-zelandiae* L. Moore and *P. lanigera* Hook. f. (Plantaginaceae) different? *N.Z. J. Bot. 27*: 499–502.

Stanley, T. D. and E. M. Ross. *Flora of south-eastern Queensland.* Vol. 3. Brisbane, Queensland Department of Primary Industries. Includes species also indigenous to New Zealand.

Taylor, P. The genus *Utricularia* — a taxonomic monograph. *Kew Bull. Addit. Ser. 14*: 1–724.

Thompson, J. A revision of the genus *Leptospermum* (Myrtaceae). *Telopea 3*: 301–448.

Tomlinson, P. B., Takaso, T. and J. A. Rattenbury. Cone and ovule ontogeny in *Phyllocladus* (Podocarpaceae). *Bot. J. Linn. Soc. 99*: 209–221.

Tomlinson, P. B., Takaso, T. and J. A. Rattenbury. Developmental shoot morphology in *Phyllocladus* (Podocarpaceae). *Bot. J. Linn. Soc. 99*: 223–248.

Vincent, P. L. D. and F. M. Getliffe Norris. An SEM study of the external pollen morphology in *Senecio* and some related genera in the subtribe Senecioninae (Asteraceae: Senecioneae). *S. Afr. J. Bot. 55*: 304–309.

Webb, C. J. *Senecio esleri* (Asteraceae), a new fireweed. *N.Z. J. Bot. 27*: 565–567.

Wells, P. M. and R. S. Hill. Leaf morphology of the imbricate-leaved Podocarpaceae. *Aust. Syst. Bot. 2*: 369–386.

Wilson, C. M. and D. R. Given. *Threatened Plants of New Zealand.* Wellington, DSIR Publishing.

Wilson, K. L. Proposal to conserve 468 *Scirpus* L. (Cyperaceae) with *S. sylvaticus* L. as type. *Taxon 38*: 316–320.

Ziman, S. N. and C. S. Keener. A geographical analysis of the family Ranunculaceae. *Ann. Missouri Bot. Gard. 76*: 1012–1049.

1990   Allen, R. B. and K. H. Platt. Annual seedfall variation in *Nothofagus solandri* (Fagaceae), Canterbury, New Zealand. *Oikos 57*: 199–206.

Ball, P. W. Some aspects of the phytogeography of *Carex. Canad. J. Bot. 68*: 1462–1472. Includes *Carex pyrenaica.*

Bernard, F. A., Bernard, J. M. and P. Denny. Flower structure, anatomy and life history of *Wolffia australiana* (Benth.) den Hartog & van der Plas. *Bull. Torrey Bot. Club 117*: 18–26.

Bruhl, J. J. and C. J. Quinn. Cypsela anatomy in the 'Cotuleae'. (Asteraceae—Anthemideae). *Bot. J. Linn. Soc. 102*: 37–59. Includes some N.Z. spp. of *Leptinella.*

Carolin, R. Nomenclatural notes, new taxa and the systematic arrangement in the genus *Scaevola* (Goodeniaceae) including synonyms. *Telopea 3*: 477–515.

Connor, H. E. Breeding systems in New Zealand grasses. 11, Gynodioecism in *Chionochloa bromoides. N.Z. J. Bot. 28*: 59–65.

Crisci, J. V. and P. E. Berry. A phylogenetic reevaluation of the old world species of *Fuchsia* (Onagraceae). *Ann. Missouri Bot. Gard. 77*: 517–522.

Delph, L. F. Sex-differential resource allocation patterns in the subdioecious shrub *Hebe subalpina. Ecology 71*: 1342–1351.

Delph, L. F. The evolution of gender dimorphism in New Zealand *Hebe* (Scrophulariaceae) species. *Evol. Trends Pl. 4*: 85–97.

Delph, L. F. Sex-ratio variation in the gynodioecious shrub *Hebe strictissima* (Scrophulariaceae). *Evolution 44*: 134–142.

Feuer, S. Pollen aperture evolution among the subfamilies Persoonioideae, Sphalmioideae and Carnarvonioideae (Proteaceae). *Amer. J. Bot. 77*: 783–794. Includes *Toronia toru.*

Garnock-Jones, P. J. Typification of *Ranunculus* names in New Zealand. *N.Z. J. Bot. 28*: 115–123.

Garnock-Jones, P. J., Briggs, B. G. and F. Ehrendorfer. (989) Proposal to conserve 7599a *Parahebe* W.Oliver against *Derwentia* Raf. (Scrophulariaceae). *Taxon 39*: 536–537.

George, A. S. (Ed.) *Flora of Australia. 18, Podostemaceae to Combretaceae.* Canberra, AGPS. Includes new combination *Kelleria laxa* (Cheeseman) M.Heads (Thymelaeaceae).

Gill, L. S. and H. G. K. Nyawuame. Phylogenetic and systematic value of stomata in Bicarpellatae (Bentham et Hooker sensu stricto). *Feddes Repert. 101*: 453–498. Includes *Ipomoea pes-caprae.*

Green, P. S. Notes relating to the floras of Norfolk and Lord Howe Islands, III. *Kew Bull. 45*: 235–255. Regards N.Z. *Planchonella novozelandica* as taxonomically identical with *Pouteria costata* (Endl.) Baehni of Norfolk Id.

Harden, G. J. (Ed.) *Flora of New South Wales. 1.* Kensington, NSW Univ. Press.

1990 Hayman, A. R. and R. T. Weavers. Terpenes of foliage oils from *Halocarpus bidwillii*. *Phytochem. 29*: 3157–3162.

Heads, M. J. A revision of the genera *Kelleria* and *Drapetes* (Thymelaeaceae). *Aust. Syst. Bot. 3*: 595–652.

Hilu, K. W. and A. Esen. Prolamins in systematics of Poaceae subfam. Arundinoideae. *Pl. Syst. Evol. 173*: 57–70. Includes *Rytidosperma penicillatum* and *R. unarede* (as *Danthonia*).

Jacobs, S. W. L. Notes on Australian grasses (Poaceae). *Telopea 3*: 601–603. Discusses *Austrofestuca*.

Kravtsova, T. I. The carpological characteristics of members of the genus *Parietaria* (Urticaceae): the secondary covers and evolutionary trends in fruit development. *Bot. Zhurn. 75*: 1497–1508. Mentions *Parietaria debilis*.

Lee, W. G., Hodgkinson, I. J. and P. N. Johnson. A test for ultraviolet reflectance from fleshy fruits of New Zealand plant species. *N.Z. J. Bot. 28*: 21–24.

Les, D. H. and D. J. Sheridan. Biochemical heterophylly and flavonoid evolution in North American *Potamogeton* (Potamogetonaceae). *Amer. J. Bot. 77*: 453–465. Includes *Potamogeton pectinatus*.

Maze, K. M. and R. D. B. Whalley. Sex ratios and related characteristics in *Spinifex sericeus* (Poaceae). *Aust. J. Bot. 38*: 153–160.

McKone, M. J. Characteristics of pollen production in a population of New Zealand snow-tussock grass (*Chionochloa pallens* Zotov). *New Phytol. 116*: 555–562.

Middleton, D. J. and C. C. Wilcock. A critical examination of the status of *Pernettya* Gaud. as a genus distinct from *Gaultheria* L. *Edinb. J. Bot. 47*: 291–301. New combinations in *Gaultheria* for spp. formerly placed in *Pernettya*.

Middleton, D. J. and C. C. Wilcock. Chromosome counts in the genus *Gaultheria* and related genera. *Edinb. J. Bot. 47*: 303–313. Includes *Gaultheria antipoda*.

Molloy, B. P. J. and E. D. Hatch. *Thelymitra tholiformis* (Orchidaceae) — a new species endemic to New Zealand, with notes on associated taxa. *N.Z. J. Bot. 28*: 105–114.

Nyananyo, B. L. Tribal and generic relationships in the Portulacaceae (Centrospermae). *Feddes Repert. 101*: 237–241. Discusses *Lyallia* and *Hectorella*.

Patel, R. N. Wood anatomy of the dicotyledons indigenous to New Zealand. 19, Gesneriaceae. *N.Z. J. Bot. 28*: 85–94.

Patel, R. N. Wood anatomy of the dicotyledons indigenous to New Zealand. 20, Cunoniaceae. *N.Z. J. Bot. 28*: 347–355.

Paton, A. A global taxonomic investigation of *Scutellaria* (Labiatae). *Kew. Bull. 45*: 399–450.

1990   Paton, A. The phytogeography of *Scutellaria* L. *Notes Roy. Bot. Gard. Edinb.* *46*: 345–359. Comments on *Scutellaria novae-zelandiae.*

Philipson, W. R. and B. P. J. Molloy. Seedling, shoot, and adult morphology of New Zealand conifers. The genera *Dacrycarpus, Podocarpus, Dacrydium* and *Prumnopitys. N.Z. J. Bot. 28*: 73–84.

Robson, N. K. B. Studies in the genus *Hypericum* L. (Guttiferae) 8. Sections 29. *Brathys* (part 2) and 30. *Trigynobrathys. Bull. Brit. Mus. (Nat. Hist.)* Bot. ser. *20*: 1–151. Includes *H. gramineum* and *H. japonicum.*

Sattler, R. and R. Rutishauser. Structural and dynamic descriptions of the development of *Utricularia foliosa* and *U. australis. Canad. J. Bot. 68*: 1989–2003.

Shanmukha Rao, S. R. Leaf architecture in relation to taxonomy: *Ipomoea* L. *Feddes Repert. 101*: 611–616. Includes *I. pes-caprae.*

Simon, B. K. *A Key to Australian Grasses.* Brisbane, Queensland Dept. of Primary Industries.

Sonck, C. E. A new *Taraxacum* species, *T. castellanum,* from New Zealand. *Ann. Bot. Fennici 27*: 277–279.

St George, I. and D. McCrae (Eds). *The New Zealand Orchids: Natural History and Cultivation.* Dunedin, N.Z. Native Orchid Group.

Webb, C. J. *Gentiana lilliputiana* (Gentianaceae), a new, minute, annual from New Zealand. *N.Z. J. Bot. 28*: 1–3.

Webb, C. J. New Zealand species of *Hydrocotyle* (Apiaceae) naturalised in Britain and Ireland. *Watsonia 18*: 93–95.

Webb, C. J., Johnson, P. N. and W. R. Sykes. *Flowering Plants of New Zealand.* Christchurch, DSIR Botany.

Wheeler, D. J. B., Jacobs, S. W. L. and B. E. Norton. *Grasses of New South Wales.* Ed. 2. Armidale, University of New England.

Wiegleb, G. The importance of stem anatomical characters for the systematics of the genus *Potamogeton* L. *Flora 184*: 197–208.

1991   Anderberg, A. A. Taxonomy and phylogeny of the tribe *Gnaphalieae* (Asteraceae). *Op. Bot. 104*: 1–195. Recognises *Euchiton* as distinct from *Gnaphalium* and makes new combinations for N.Z. spp.; makes new combinations in *Ozothamnus* for N.Z. spp. of *Helichrysum.*

Barker, W. R. A taxonomic revision of *Mazus* Lour. (Scrophulariaceae) in Australia. *In* Banks, M. R. et al. (Eds) *Aspects of Tasmanian Botany — a Tribute to Winifred Curtis.* Hobart, Roy. Soc. Tasmania. pp. 85–94. Recognises 2 spp. for N.Z.: *M. novaezeelandiae* W.R.Barker and *M. radicans* (Hook.f.) Cheeseman.

Bittrich, V. and M. do Carma E. Amaral. Proanthocyanidins in the testa of Centrospermous seeds. *Biochem. Syst. Ecol. 19*: 319–321. Includes *Tetragonia tetragonioides.*

1991   Bremer, B. and R. K. Jansen. Comparative restriction site mapping of chloroplast DNA implies new phylogenetic relationships within Rubiaceae. *Amer. J. Bot. 78*: 198–213. Includes *Coprosma pumila.*

Brown, M. J. A synopsis of the genus *Plantago* L. in Tasmania. *In* Banks, M. R. et al. (Eds) *Aspects of Tasmanian botany — a tribute to Winifred Curtis.* Hobart, Roy. Soc. Tasmania. pp. 65–74. Includes *Plantago triantha.*

Bruhl, J. J. Comparative development of some taxonomically critical floral/inflorescence features in Cyperaceae. *Aust. J. Bot. 39*: 119–127. *Eleocharis acuta, Schoenoplectus validus* and *Lepidosperma laterale* were among spp. examined.

Bruhl, J. J. and C. J. Quinn. Floral morphology and a reassessment of affinities in the 'Cotuleae' (Asteraceae). *Aust. Syst. Bot. 4*: 637–654.

Chapman, A. D. *Australian Plant Name Index.* Canberra, AGPS.

Clements, M. A., Jones, D. L. and P. B. J. Molloy. Recently named Australian orchid taxa: 2. *Thelymitra. Lindleyana 6*: 59–60.

Connor, H. E. *Chionochloa* Zotov (Gramineae) in New Zealand. *N.Z. J. Bot. 29*: 219–282.

Cooper, R. C. and R. C. Cambie. *New Zealand's Economic Native Plants.* Auckland, Oxford University Press.

Delph, L. F. and D. G. Lloyd. Environmental and genetic control of gender in the dimorphic shrub *Hebe subalpina. Evolution 45*: 1957–1964.

Duke, N. C. A systematic revision of the mangrove genus *Avicennia* (Avicenniaceae) in Australasia. *Aust. Syst. Bot. 4*: 299–324. N.Z. plants are treated as *Avicennia marina* var. *australasica* (Walp.) Moldenke.

Edgar, E. and M. B. Forde. *Agrostis* L. in New Zealand. *N.Z. J. Bot. 29*: 139–161.

Fineran, B. A. Root hemi-parasitism in the Santalales. *Bot. Jahrb. Syst. 113*: 277–308.

Garnock-Jones, P. J. Seed morphology and anatomy of the New Zealand genera *Cheesemania, Ischnocarpus, Iti, Notothlaspi,* and *Pachycladon* (Brassicaceae). *N.Z. J. Bot. 29*: 71–82.

Garnock-Jones, P. J. Gender dimorphism in *Cheesemania wallii* (Brassicaceae). *N.Z. J. Bot. 29*: 87–90.

Gibson, N. The anatomy and morphology of four Tasmanian cushion species. *In* Banks, M. R. et al. (Eds) *Aspects of Tasmanian botany — a tribute to Winifred Curtis.* Hobart, Roy. Soc. Tasmania. pp. 231–238. Includes *Donatia novae-zelandiae* and *Phyllachne colensoi.*

Heenan, P. *Olearia* ×*matthewsii* 'Highland Mist': a new interspecific and cultivar name. *Hort. N.Z. 2*: 13–14.

Hill, R. S. and J. Read. A revised infrageneric classification of *Nothofagus* (Fagaceae). *Bot. J. Linn. Soc. 105*: 37–72.

1991   Huynh, K.-L. The flower structure in the genus *Freycinetia*, Pandanaceae (part 1) — Potential bisexuality in the genus *Freycinetia*. *Bot. Jahrb. Syst. 112*: 295–328.

Jansen, R. K., Michaels, H. J. and J. D. Palmer. Phylogeny and character evolution in the Asteraceae based on chloroplast DNA restriction site mapping. *Syst. Bot. 16*: 98–115. Includes *Gnaphalium luteoalbum*.

Kapil, R. N. and A. K. Bhatnagar. Embryological evidence in angiosperm classification and phylogeny. *Bot. Jahrb. Syst. 113*: 309–338. Mentions *Corokia* and *Donatia*.

Lord, J. M. Pollination and seed dispersal in *Freycinetia baueriana*, a dioecious liane that has lost its bat pollinator. *N.Z. J. Bot. 29*: 83–86.

Macmillan, B. H. *Acaena rorida* and *Acaena tesca* (Rosaceae) — two new species from New Zealand. *N.Z. J. Bot. 29*: 131–138.

Macmillan, B. H. *Acaena pallida* (Kirk) Allan (Rosaceae) in Tasmania and New South Wales, Australia. *In* Banks, M. R. et al. (Eds) *Aspects of Tasmanian botany — a tribute to Winifred Curtis*. Hobart, Roy. Soc. Tasmania. pp. 53–55.

Middleton, D. J. Infrageneric classification of the genus *Gaultheria* L. (Ericaceae). *Bot. J. Linn. Soc. 106*: 229–258.

Middleton, D. J. Ecology, reproductive biology and hybridization in *Gaultheria* L. *Edin. J. Bot. 48*: 81–89.

Middleton, D. J. Taxonomic studies in the *Gaultheria* group of genera of the Tribe Andromedeae (Ericaceae). *Edin. J. Bot. 48*: 283–306.

Murray, D. R. and A. W. D. Larkum. Seed proteins of the seagrass *Zostera capricorni*. *Aquatic Bot. 40*: 101–108.

Oginuma, K. et al. Karyomorphology of *Coriaria* (Coriariaceae): taxonomic implications. *Bot. Mag. Tokyo 103*: 297–308. Includes N.Z. spp.

Patel, R. N. Wood anatomy of the dicotyledons indigenous to New Zealand. 21, Loranthaceae. *N.Z. J. Bot. 29*: 429–449.

Pennington, T. D. *The Genera of Sapotaceae*. Kew, Royal Botanic Gardens. *Planchonella* is regarded as a synonym of *Pouteria*, and the name *Pouteria costata* is reinstated for the N.Z. sp.

Smith, A. C. *Flora Vitiensis Nova: A New Flora of Fiji (Spermatophytes only)*. 5. Lawai, Hawaii, National Tropical Botanical Garden.

Stringer, S. and J. G. Conran. Stamen and seed cuticle morphology in some *Arthropodium* and *Dichopogon* species (Anthericaceae). *Aust. J. Bot. 39*: 129–135. Includes *Arthropodium candidum* and *A. cirratum*.

Sytsma, K. J., Smith, J. F. and P. E. Berry. The use of chloroplast DNA to assess biogeography and evolution of morphology, breeding systems, and flavonoids in *Fuchsia* sect. *Skinnera* (Onagraceae). *Syst. Bot. 16*: 257–269.

Todzia, C. A. and R. C. Keating. Leaf architecture of the Chloranthaceae. *Ann. Missouri Bot. Gard. 78*: 476–496. Includes *Ascarina lucida*.

1991   Tomlinson, P. B., Braggins, J. E. and J. A. Rattenbury. Pollination drop in relation to cone morphology in Podocarpaceae: a novel reproductive mechanism. *Amer. J. Bot. 78*: 1289–1303.

Umaderi, I. and M. Daniel. Chemosystematics of the Sapindaceae. *Feddes Repert. 102*: 607–612. Includes *Dodonaea viscosa.*

van der Ham, R. W. J. M. The apertural system in Nephelieae pollen (Sapindaceae): form, function and evolution. *Acta Bot. Neerl. 40*: 162–163. *Alectryon* pollen mentioned.

Wardle, P. *Vegetation of New Zealand.* Cambridge, Cambridge University Press.

Webb, C. J. and M. J. A. Simpson. Seed morphology in relation to taxonomy in New Zealand species of *Weinmannia, Ackama,* and the related South American *Caldcluvia paniculata* (Cunoniaceae). *N.Z. J. Bot. 29*: 451–453.

1992   Barker, W. R. New Australasian species of *Peplidium* and *Glossostigma* (Scrophulariaceae). *J. Adelaide Bot. Gard. 15*: 71–74. Describes *Glossostigma cleistanthum* for central and S.E. Australia and N.Z.

Belcher, R. O. The genus *Senecio* (*Compositae*) on Lord Howe and Norfolk Islands. *Kew Bull. 47*: 765–773. All spp. are endemic; a number of names previously ascribed to taxa from these islands are excluded including *S. lautus.*

Belcher, R. O. Rediscovery of *Senecio australis* Willd. (Asteraceae) after nearly two centuries. *Taxon 41*: 235–252. *Senecio australis* Willd. was based on a Forster specimen said to be from N.Z. Specimens from Norfolk Id are shown to be identical with the holotype of *S. australis* and distinct fron *S. lautus.*

Bergin, D. O. and M. O. Kimberley. Provenance variation in *Podocarpus totara. N.Z. J. Ecol. 16*: 5–13.

Bergman, B., Johanson, C. and E. Söderback. The *Nostoc - Gunnera* symbiosis. *New Phytol. 122*: 379–400. Includes summary of the genus *Gunnera.*

Briggs, B. G. and F. Ehrendorfer. A revision of the Australian species of *Parahebe* and *Derwentia* (Scrophulariaceae). *Telopea 5*: 241–287. Discusses N.Z. *Parahebe.*

Carlquist, S. Pit membrane remnants in perforation plates of primitive dicotyledons and their significance. *Amer. J. Bot. 79*: 660–672. Includes several N.Z. spp.

Chinnock, R. J. New taxa and combinations in the Myoporaceae. *J. Adelaide Bot. Gard. 15*: 75–79. *Eremophila debilis* (Andr.) Chinnock a new name for *Myoporum debile.*

Chuang, T. I and R. Ornduff. Seed morphology and systematics of Menyanthaceae. *Amer. J. Bot. 79*: 1396–1406. Includes *Liparophyllum gunnii.*

1992   Connor, H. E. The botany of change in tussock grasslands in the
        MacKenzie Country, South Canterbury, New Zealand; the 1992
        McCaskill Memorial Lecture. *Review 49*: 1–31.

        Delph, L. F. and D. G. Lloyd. Environmental and genetic control of
        gender in the dimorphic shrub *Hebe subalpina. Evolution 45*: 1957–
        1964.

        Dubcovsky, J. and A. J. Martinez. Cytotaxonomy of the *Festuca* spp. from
        Patagonia. *Canad. J. Bot. 70*: 1134–1140. Includes *Festuca contracta.*

        Enright, N. J. Factors affecting reproductive behaviour in the New
        Zealand nikau palm, *Rhopalostylis sapida* Wendl. et Drude. *N.Z. J. Bot.
        30*: 69–80.

        Espie, P. R., Connor, H. E. and I. J. McCracken. Leaf-blade crimping in
        grasses: a new measure of growth. *Experientia 48*: 91–94.

        Gandolfo, M. A. and E. J. Romero. Leaf morphology and a key to species
        of *Nothofagus* Bl. *Bull. Torrey Bot. Club 119*: 152–166.

        Glossner, F. Ultraviolet patterns in the traps and flowers of some carnivorous
        plants. *Bot. Jahrb. Syst. 113*: 577–587. Includes *Drosera binata.*

        Haase, P. Isozyme variability and biogeography of *Nothofagus truncata*
        (Fagaceae). *N.Z. J. Bot. 30*: 315–328.

        Haase, P. Isozyme variation and genetic relationships in *Phyllocladus
        trichomanoides* and *P. alpinus* (Podocarpaceae). *N.Z. J. Bot. 30*: 359–363.

        Harden, G. J. (Ed.) *Flora of New South Wales 3.* Kensington, NSW Univ. Press.

        Harris, W., Porter, N. G. and M. I. Dawson. Observations on biosystematic
        relationships of *Kunzea sinclairii* and on an intergeneric hybrid *Kunzea
        sinclairii* × *Leptospermum scoparium. N.Z. J. Bot. 30*: 213–230.

        Heads, M. J. Taxonomic notes on the *Hebe* (Scrophulariaceae) complex
        in the New Zealand mountains. *Candollea 47*: 583–595.

        Huynh, K.-L. The flower structure in the genus *Freycinetia*, Pandanaceae
        (part 2) — Early differentiation of the sex organs, especially of the
        staminodes; and further notes on the anthers. *Bot. Jahrb. Syst. 114*:
        417–441. Mentions *Freycinetia banksii.*

        Huynh, K.-L. and F. B. Sampson. Flower structure in *Freycinetia banksii*
        (Pandanaceae) of New Zealand. *Bot. Helv. 102*: 175–191.

        Imaichi, R. and K. Okamoto. Comparative androecium morphogenesis
        of *Sicyos angulatus* and *Sechium edule* (Cucurbitaceae). *Bot. Mag. Tokyo
        105*: 539–548.

        Jaaska, V. Isoenzyme variation in the grass genus *Elymus* (Poaceae).
        Hereditas 117: 11–22.

        Kamelina, O. P. On the embryology of the genus *Ixerba* in relation to its
        systematic position. *Bot. Zhurn. 77 (12)*: 112–117. (In Russian).

        Kelly, D. et al. Mast seeding of *Chionochloa* (Poaceae) and pre-dispersal
        seed predation by a specialist fly (Diplotoxa, Diptera: Chloropidae).
        N.Z. J. Bot. 30: 125–133.

1992 Linder, H. P. The gynoecia of Australian Restionaceae: morphology, anatomy and systematic implications. *Aust. Syst. Bot. 5*: 227–245. Includes *Empodisma minus.*

Lloyd, D. G. and M. S. Wells. Reproductive biology of a primitive angiosperm, *Pseudowintera colorata* (Winteraceae) and the evolution of pollination systems in the *Anthophyta. Plant Syst. Evol. 181*: 77–95.

Middleton, D. J. A chemotaxonomic survey of flavonoids and simple phenols in the leaves of *Gaultheria* L. and related genera (Ericaceae). *Bot. J. Linn. Soc. 110*: 313–324.

Murray, B. G., Cameron, E. K. and L. S. Standring. Chromosome numbers, karyotypes, and nuclear DNA variation in *Pratia* Gaudin (Lobeliaceae). *N.Z. J. Bot. 30*: 181–187.

Nicolson, D. H. Seventy-two proposals for the conservation of types of selected Linnaean generic names, the report of Subcommittee 3C on the lectotypification of Linnaean generic names. *Taxon 41*: 552–583. Includes discussion on lectotypification of *Agrostis, Anemone, Apium, Atriplex, Carex, Convolvulus, Elymus, Galium, Gnaphalium, Passiflora, Rubus, Scutellaria, Sophora.*

Nyananyo, B. L. Pollen morphology in the Portulacaceae (Centrospermae). *Folia Geobot. Phytotax. 27*: 387–400. Includes *Lyallia.*

Patel, R. N. Wood anatomy of the dicotyledons indigenous to New Zealand. 22, Proteaceae. *N.Z. J. Bot. 30*: 415–428.

Pedral, J. (1050) Proposal to amend 2261 *Suaeda*, nom. cons. (Chenopodiaceae). *Taxon 41*: 337–338.

Philbrick, C. T. and L. M. Bernadello. Taxonomic and geographic distribution of internal geitonogamy in New World *Callitriche* (Callitrichaceae). *Amer. J. Bot. 79*: 887–890. Notes that Old World, including N.Z., spp. lacked internal geitonogamy.

Philipp, M. Reproductive biology of *Geranium sessiliflorum.* 3. Population ecology of two populations and three leaf colour morphs. *N.Z. J. Bot. 30*: 151–161.

Rahn, K. Trichomes within the Plantaginaceae. *Nord. J. Bot. 12*: 3–12. Includes *Plantago raoulii, P. spathulata, P. triandra, P. triantha* and *P. uniflora.*

Robertson, A. W. The relationship between floral display size, pollen carryover and geitonogamy in *Myosotis colensoi* (Kirk) Macbride (Boraginaceae). *Biol. J. Linn. Soc. 46*: 333–349.

Rowan, D. D. and G. B. Russell. 3β-methoxyhop-22(29)-ene from *Chionochloa cheesemanii. Phytochem. 31*: 702–703.

Rudall, P. and S. Morley. Embryo sac and early post-fertilization development in *Thismia (Burmanniaceae). Kew Bull. 47*: 625–632. *T. rodwayi* is mentioned.

Salmon, J. T. *A field guide to the alpine plants of New Zealand.* 3rd ed. Auckland, Godwit Press.

1992 Seawright, A. A. and B. L. Smith. The indigenous poisonous plants of Australia and New Zealand. In James, L. F. et al. Poisonous plants. Proc. Third Int. Symp. Ames. pp. 55–62.

Shanmukha Rao, S. R. and M. Leela. Chemotaxonomy of some *Ipomoea* L. (Convolvulaceae). *Feddes Repert. 103*: 351–355. Includes *I. pes-caprae*.

Smith, P. J. A revision of the genus *Wahlenbergia* (Campanulaceae) in Australia. *Telopea 5*: 91–175. N.Z. plants are mentioned.

Sykes, W. R. Two new names in *Macropiper* Miq. (Piperaceae) from New Zealand. *N.Z. J. Bot. 30*: 231–236.

Tobe, H., Suzuki, M. and T. Fukuhara. Pericarp anatomy and evolution in *Coriaria* (Coriariaceae). *Bot. Mag. Tokyo 105*: 289–302.

Veldkamp, J. F. Miscellaneous notes on Southeast Asian Gramineae VII. *Blumea 37*: 227–237. Published a superfluous name *Lachnagrostis avenacea* (J.F.Gmel.) Veldkamp for *L. filiformis*.

Vincent, P. L. D. and F. M. Getliffe. Elucidative studies on the generic concept of *Senecio* (Asteraceae). *Bot. J. Linn. Soc. 108*: 55–81.

Walsh, N. A new combination in *Pomaderris* (Rhamnaceae) in New Zealand. *N.Z. J. Bot. 30*: 117–118.

Watson, L. and M. J. Dallwitz. *The Grass Genera of the World.* Cambridge, University Press.

Webb, C. J. Sex ratios from seed in six families of *Scandia geniculata* (Apiaceae). *N.Z. J. Bot. 30*: 401–404.

West, M. M., Lott, J. N. A. and D. R. Murray. Studies of storage reserves in seeds of the marine angiosperm *Zostera capricorni. Aquatic Bot. 43*: 75–85.

Wilkinson, H. P. Leaf anatomy of the Pittosporaceae R. Br. *Bot. J. Linn. Soc. 110*: 1–59.

Williams, P. Ecology of the endangered herb *Scutellaria novae-zelandiae. N.Z. J. Ecol. 16*: 127–135.

Wilson, H. D. and P. J. Garnock-Jones. Two new species names in *Olearia* (Asteraceae) from New Zealand. *N.Z. J. Bot. 30*: 365–368.

Yoda, K. and M. Suzuki. Comparative wood anatomy of *Coriaria. Bot. Mag. Tokyo 105*: 235–245. Investigated N.Z. spp.

Yukawa, T. et al. Existence of two stomatal shapes in the genus *Dendrobium* (Orchidaceae) and its systematic significance. *Amer. J. Bot. 79*: 946–952. Includes *D. cunninghamii*.

1993 Amor-Prats, D. and J. B. Harborne. New sources of ergoline alkaloids within the genus *Ipomoea. Biochem. Syst. Ecol. 21*: 455–462. Includes *I. pes-caprae*.

Barrett, S. C. H., Eckert, C. G. and B. C. Husband. Evolutionary processes in aquatic plant populations. *Aquatic Bot. 44*: 105–145. Includes measurements of electrophoretically detectable variation at isozyme loci in *Lemna minor, Potamogeton pectinatus, Zostera capricorni, Z. novozelandica*.

1993   Breitwieser, I. Comparative leaf anatomy of New Zealand and Tasmanian
       Inuleae (Compositae). *Bot. J. Linn. Soc. 111*: 183–209.

       Breitwieser, I. and J. M. Ward. Systematics of New Zealand Inuleae
       (Compositae - Asteraceae). 3, Numerical phenetic analysis of leaf
       anatomy and flavonoids. *N.Z. J. Bot. 31*: 43–58.

       Bremer, K. New subtribes of the Lactuceae (Asteraceae). *Novon 3*: 328–
       330. Lactuceae subtribe Sonchinae K.Bremer includes *Embergeria*, and
       Kirkianella.

       Brummitt, R. K. Report of the Committee for Spermatophyta: 38. *Taxon
       42*: 687–697.
       (726) Rejection of *Epilobium junceum* — recommended.
       (931) Conserve *Scirpus* L. with a new type, *S. lacustris* L. —
       recommended.
       (952) Conserved spelling "Brachycome" — not recommended;
       spelling *Brachyscome* is preferred.

       Brummitt, R. K. Report of the Committee for Spermatophyta: 39. *Taxon
       42*: 873–879.
       (989) Conserve 7759a *Parahebe* W.R.B.Oliv. against *Derwentia* —
       recommended.

       Cambie, R. C. et al. Triterpenes from *Elingamita johnsonii* G.T.Baylis
       (Myrsinaceae). *N.Z. J. Bot. 31*: 425–426.

       Cameron, E. K. et al. Threatened and local plant lists (1993 revision).
       N.Z. Bot. Soc. Newsletter 32: 14–28.

       Clifford, H. T. Dispersal of fleshy diaspores in the seed floras of the
       South Island (New Zealand) and Tasmania. *Aust. Syst. Bot. 6*: 449–
       455.

       Connor, H. E., Molloy, B. P. J. and M. I. Dawson. *Australopyrum* (Triticeae:
       Gramineae) in New Zealand. *N.Z. J. Bot. 31*: 1–10.

       Dawson, M. I. Contributions to a chromosome atlas of the New Zealand
       flora. 31, *Clematis* (Ranunculaceae). *N.Z. J. Bot. 31*: 91–96.

       Dawson, M. I. et al. Contributions to a chromosome atlas of the New
       Zealand flora. 32, *Raoulia* (Inuleae—Compositae (Asteraceae)). *N.Z.
       J. Bot. 31*: 97–106.

       Dillon, M. O. and M. Muñoz-Schick. A revision of the dioecious genus
       Griselinia (Griseliniaceae) including a new species from the coastal
       Atacama Desert of northern Chile. *Brittonia 45*: 261–274.

       Forster, P. I. A taxonomic revision of the genus *Peperomia* Ruiz & Pav.
       (Piperaceae) in mainland Australia. *Austrobaileya 4*: 93–104. Includes
       Peperomia tetraphylla.

       Garnock-Jones, P. J. Phylogeny of the *Hebe* complex (Scrophulariaceae:
       Veronicae). *Aust. Syst. Bot. 6*: 457–479.

       George, A. S., Orchard, A. E. and H. J. Hewson (Eds) *Flora of Australia. 50,
       Oceanic Islands 2*. Canberra, AGPS. Includes flora of Macquarie Id.

1993   Green, P. S. Notes relating to the floras of Norfolk and Lord Howe Islands, IV. *Kew Bull. 48*: 307–325. Includes *Macropiper excelsum* subsp. *psittacorum*.

Gould, K. S. Leaf heteroblasty in *Pseudopanax crassifolius*: functional significance of leaf morphology and anatomy. *Ann. Bot. 71*: 61–70.

Haase, P., Breitwieser, I. and J. M. Ward. Genetic relationships of *Helichrysum dimorphum* (Inuleae—Compositae (Asteraceae)) with *H. filicaule, H. depressum* and *Raoulia glabra* as resolved by isozyme analysis. *N.Z. J. Bot. 31*: 59–64.

Harden, G. J. (Ed.) *Flora of New South Wales. 4.* Kensington, NSW Univ. Press.

Hatch, D. *Corybas longipetalus* (Hatch) Hatch *comb. et stat. nov. N.Z. Native Orchid Group J. 47*: 6.

Heads, M. J. Biogeography and biodiversity in *Hebe*, a South Pacific genus of Scrophulariaceae. *Candollea 48*: 19–60.

Hill, R. S. and G. J. Jordan. The evolutionary history of *Nothofagus* (Nothofagaceae). *Aust. Syst. Bot. 6*: 111–126.

Ho, T.-N. and S.-W. Liu. New combinations, names and taxonomic notes on *Gentianella* (Gentianaceae) from South America and New Zealand. *Bull. Nat. Hist. Mus. Lond. (Bot.) 23*: 61–65. Makes new combinations for twenty N.Z. spp. formerly placed in *Gentiana*.

Holzapfel, S. and H. W. Lack. New species of *Picris* (Asteraceae, Lactuceae) from Australia. *Willdenowia 23*: 181–191. Describes 5 new spp. of which *Picris burbidgei* S.Holzapfel also occurs in N.Z.

Huynh, K.-L. Some new distinctive features between *Freycinetia banksii* Cunn. (Pandanaceae) of New Zealand and *F. baueriana* Endl. of Norfolk Island. *Candollea 48*: 501–510.

Irwin, J. B. Notes on seven forms of *Corybas rivularis. N.Z. Native Orchid Group J. 47*: 7–9.

Jacobs, S. W. L., McClay, K. L. and B. K. Simon. Review of *Dichelachne* (Gramineae) in Australia. *Telopea 5*: 325–328. All N.Z. spp. also occur in Australia.

Kellogg, E. A. and L. Watson. Phylogenetic studies of a large data set. 1. Bambusoideae, Andropogonodae, and Pooideae (Gramineae). *Bot. Rev. 59*: 273–343. Explicitly excludes *Australopyrum*.

Keogh, J. A. and P. Bannister. Transoceanic dispersal in the amphiantarctic genus *Discaria*: an evaluation. *N.Z. J. Bot. 31*: 427–430.

Kores, P. J., Molvray, M. and S. P. Darwin. Morphometric variation in three species of *Cyrtostylis* (Orchidaceae). *Syst. Bot. 18*: 274–282. Includes N.Z. *C. oblonga* and *C. reniformis*.

Lakshminara Yana, K. and P. Srinivasa Rao. Embryological investigations in some species of *Ipomoea. Taiwania 38*: 109–116. Includes *I. pes-caprae*.

1993   Les, D. H. and Philbrick, C. T. Studies of hybridization and chromosome number variation in aquatic angiosperms: evolutionary implications. *Aquatic Bot. 44*: 181–228. Includes *Callitriche petriei, Cotula coronopifolia, Lemna minor, Ruppia polycarpa, Potamogeton pectinatus, Wolffia australiana.*

Lord, J. M. Does clonal fragmentation contribute to recruitment in *Festuca novae-zelandiae? N.Z. J. Bot. 31*: 133–138.

Lu, B.-R. Meiotic studies of *Elymus nutans* and *E. jaquemontii* (Poaceae, Triticeae) and their hybrids with *Pseudoroegneria spicata* and seventeen *Elymus* species. *Pl. Syst. Evol. 186*: 193–212.

Lu, B.-R. and R. von Bothmer. Meiotic analysis of *Elymus caucasicus, E. longearistatus,* and their interspecific hybrids with twenty-three *Elymus* species (Triticeae, Poaceae). *Pl. Syst. Evol. 185*: 33–53.

Martin, P. G. and J. M. Dowd. Using sequences of *rbc*L to study phylogeny and biogeography of *Nothofagus* species. *Aust. Syst. Bot. 6*: 441–447.

Martinez, S. Relaciones feneticas entre las especies del genero *Azorella* (Apiaceae, Hydrocotyloideae). *Darwiniana 32*: 159–170. Includes *Azorella selago.*

Martinez, S. Sinopsis del genero *Azorella* (Apiaceae, Hydrocotyloideae). *Darwiniana 32*: 171–184.

Martinsson, K. The pollen of Swedish *Callitriche* (Callitrichaceae) — trends towards submergence. *Grana 32*: 198–209. Includes *Callitriche stagnalis.*

Metcalf, L. *The cultivation of New Zealand plants.* Auckland, Godwit Press.

Middleton, D. J. A systematic survey of leaf and stem anatomical characters in the genus *Gaultheria* and related genera (Ericaceae). *Bot. J. Linn. Soc. 113*: 199–215.

Moar, N. T. *Pollen Grains of New Zealand Dicotyledonous Plants.* Lincoln, Manaaki Whenua Press.

Molloy, B. *Corybas longipetalus* (Hatch) Hatch *nom. illeg. N.Z. Native Orchid Group J. 48*: 6–7.

Posluszny, U. and Charlton, W. A. Evolution of the helobial flower. *Aquatic Bot. 44*: 303–324. Discusses floral structure of *Potamogeton, Triglochin* including *T. striata,* and *Ruppia.*

Reid, A. R. and B. A. Bohm. External and vacuolar flavonoids of *Brachyglottis cassinioides. Biochem. Syst. Ecol. 21*: 746.

Simon, B. K. *A Key to Australian Grasses,* ed. 2. Brisbane, Queensland Dept. of Primary Industries.

Snogerup, S. A revision of *Juncus* subgen. *Juncus* (Juncaceae). *Willdenowia 23*: 23–73. Makes new combination *Juncus kraussii* var. *australiensis* (Buchanan) Snogerup for the Australasian plants formerly known as *J. maritimus* var. *australiensis.*

St George, I. The Pacifc genus *Earina. Orchidian 11*: 56–65. Summarised history of genus with brief descriptions of the six species.

1993   Strong, M. T. New combinations in *Schoenoplectus* (Cyperaceae). *Novon*
        3: 202–203. Does not accept the separation of *Bolboschoenus* and
        Schoenoplectus. The combination *Schoenoplectus fluviatilis* (Torrey)
        M.T.Strong is based on *Scirpus maritimus* var. *fluviatilis* Torrey.

        Suh, Y. et al. Molecular evolution and phylogenetic implications of
        internal transcribed spacer sequences of ribosomal DNA in
        Winteraceae. *Amer. J. Bot. 80*: 1042–1055. Includes *Pseudowintera
        axillaris* and *P. colorata*.

        Swenson, U. The identity of *Abrotanella christensenii* Petrie (Asteraceae).
        Compositae Newsletter 23: 3–6. *A. christensenii* is regarded as synonymous
        with Australian *Solenogyne gunnii* which has been recorded for N.Z. as
        a naturalised sp.

        Sydes, M. A. and D. M. Calder. Comparative reproduction biology of two
        sun-orchids: the vulnerable *Thelymitra circumsepta* and the widespread
        T. ixioides (Orchidaceae). *Aust. J. Bot. 41*: 577–589.

        Sykes, W. R. Reinstatement of *Pseudopanax kermadecensis* (W.R.B. Oliv.)
        Philipson (Araliaceae). *N.Z. J. Bot. 31*: 19–20.

        Sykes, W. R. and P. J. de Lange. *Plectranthus parviflorus* Willd. (Lamiaceae)
        in New Zealand. *N.Z. J. Bot. 31*: 11–14.

        Talavera, S., García-Murillo, P. and J. Herrera. Chromosome numbers
        and a new model for karyotype evolution in *Ruppia* L. (Ruppiaceae).
        Aquatic Bot. 45: 1–13. Includes *Ruppia megacarpa* and *R. polycarpa*.

        Thompson, J. A revision of the genus *Swainsona* (Fabaceae). *Telopea 5*:
        427–582.

        Torres, A. M. Revisión del género *Stipa* (Poaceae) en la provincia de Buenos
        Aires. *Monografía 12*. Prov. Buenos Aires, Comisión Invest. Cien.

        Vanvinckenroye, P. et al. A comparative floral developmental study in
        Pisonia, Bougainvillea and *Mirabilis* (Nyctaginaceae) with special
        emphasis on the gynoecium and floral nectaries. *Bull. Jard. Bot. Nat.
        Belg. 62*: 69–96. *Pisonia brunoniana* was studied.

        Ward, J. M. Systematics of New Zealand Inuleae (Compositae —
        Asteraceae). 1, A numerical phenetic study of the species of *Raoulia*.
        N.Z. J. Bot. 31: 21–28.

        Ward, J. M. Systematics of New Zealand Inuleae (Compositae —
        Asteraceae). 2, A numerical phenetic study of *Raoulia* in relation to
        allied genera. *N.Z. J. Bot. 31*: 29–42.

        Webb, C. J. and D. Kelly. The reproductive biology of the New Zealand
        flora. *Trends Ecol. Evol. 8*: 442–447.

        Webb, C. J. and P. E. Pearson. The evolution of approach herkogamy
        from protandry in New Zealand *Gentiana* (Gentianaceae). *Pl. Syst.
        Evol. 186*: 187–191.

        Wilson, H. D. and T. Galloway. *Small-leaved Shrubs of New Zealand*.
        Christchurch, Manuka Press.

1993   Xiang, Q.-Y. et al. Phylogenetic relationships of *Cornus* L. sensu lato and putative relatives inferred from *rbc*L sequence data. *Ann. Missouri Bot. Gard. 80*: 723–734. *Corokia* and *Griselinia* previously placed in Cornaceae by some authors are only distantly related to *Cornus*.

1994   Al-Shammary, K. I. A. and R. J. Gornall. Trichome anatomy of the Saxifragaceae s.l. from the southern hemisphere. *Bot. J. Linn. Soc. 114*: 99–131.

Battjes, J., Chambers, K. L. and K. Bachmann. Evolution of microsporangium numbers in *Microseris* (Asteraceae: Lactuceae). *Amer. J. Bot. 81*: 641–647. Includes *Microseris scapigera*.

Brummitt, R. K. Report of the Committee for Spermatophyta: 40. *Taxon 43*: 113–126.
(1032) Authorship of Donatiaceae recommended to be attributed to Chandler 1911 instead of Dostál 1957.
Baumea Gaudich. (Cyperaceae) and *Baumia* Engl. & Gilg. (Scrophulariaceae) deemed not confused but the Committee thought that *Baumea* Gaudich. was not in current use.

Brummitt, R. K. Report of the Committee for Spermatophyta: 40. *Taxon 43*: 271–277.
(1050) validating author for the genus *Suaeda* should be J.F. Gmelin.

Chambers, H. L. and K. E. Hummer. Chromosome counts in the *Mentha* collection at the USDA–ARS National Clonal Germplasm Repository. *Taxon 43*: 423–432. Includes *Mentha cunninghamii*.

Clearwater, M. J. and K. S. Gould. Comparative leaf development of juvenile and adult *Pseudopanax crassifolius. Canad. J. Bot. 72*: 658–670.

Conacher, C. A. et al. Morphology, flowering and seed production of *Zostera capricorni* Aschers. in subtropical Australia. *Aquatic Bot. 49*: 33–46.

Conn, B. J. and P. G. Richards. A new species of *Oxalis* section Corniculatae (Oxalidaceae) from Australasia. *Aust. Syst. Bot. 7*: 171–181. *Oxalis thompsoniae* Conn et Richards is described for Papua New Guinea, eastern Australia, Lord Howe Id, and N.Z.

Connor, H. E. Indigenous New Zealand Triticeae: Gramineae. *N.Z. J. Bot. 32*: 125–154.

Culham, A. and R. J. Gornall. The taxonomic significance of naphthoquinones in the Droseraceae. *Biochem. Syst. Ecol. 22*: 507–515. Includes most N.Z. spp.

Curtis, W. M. and D. I. Morris. *The Students' Flora of Tasmania. 4B, Angiospermae: Alismataceae to Burmanniaceae.* Hobart, St David's Park Publishing.

Endress, P. K. Shapes, sizes and evolutionary trends in stamens of Magnoliidae. *Bot. Jahrb. Syst. 115*: 429–460. *Macropiper excelsum* included.

1994   Everett, J. New combinations in the genus *Avicennia* (Avicenniaceae). *Telopea* 5: 627–629. Makes the new combination *Avicennia marina* subsp. *australasica* (Walp.) J.Everett and discusses the typification.

Fuller, G. Observations on the pollination of *Corybas* "A". *N.Z. Native Orchid Group J.* 53: 18–22. Described later (1996) as *Corybas iridescens* Irwin et Molloy.

Garnock-Jones, P. J. *Heliohebe* (Scrophulariaceae — Veroniceae), a new genus segregated from *Hebe*. *N.Z. J. Bot.* 31: 323–339.

Garnock-Jones, P. J. and B. D. Clarkson. *Hebe adamsii* and *H. murrellii* (Scrophulariaceae) reinstated. *N.Z. J. Bot.* 32: 11–15.

Haase, P. Genetic relationships and inferred evolutionary divergence in the New Zealand taxa of *Nothofagus* — results from isozyme analysis. *Aust. Syst. Bot.* 7: 47–55.

Heads, M. A biogeographic review of *Parahebe* (Scrophulariaceae). *Bot. J. Linn. Soc. 115*: 65–89. Makes new combinations in *Parahebe*.

Heads, M. J. Morphology, architecture and taxonomy in the *Hebe* complex (Scrophulariaceae). *Bull. Mus. Nat. Hist. Nat.* Sect. B, *Adansonia 16*: 163–191.

Heads, M. J. Biodiversity and biogeogaphy in New Zealand *Ourisia* (Scrophulariaceae). *Candollea 49*: 23–36.

Heads, M. J. Biogeography and biodiversity in New Zealand *Pimelea* (Thymelaeaceae). *Candollea 49*: 37–53.

Heads, M. J. Biogeographic studies in New Zealand Scrophulariaceae: tribes Rhinantheae, Calceolarieae and Gratioleae. *Candollea 49*: 55–80.

Heads, M. J. Biogeography and evolution in the *Hebe* complex (Scrophulariaceae): *Leonohebe* and *Chionohebe*. *Candollea 49*: 81–119.

Heenan, P. B. The status of names in *Hebe* published by Professor Arnold Wall in 1929. *N.Z. J. Bot.* 32: 521–522.

Hennion, F., Fiasson, J. L. and K. Gluchoff-Fiasson. Morphological and phytochemical relationships between *Ranunculus* species from Iles Kerguelen. *Biochem. Syst. Ecol.* 22: 533–542. Includes *R. biternatus*.

Herrick, J. F. and E. K. Cameron. Annotated checklist of type specimens of New Zealand plants in the Auckland Institute and Museum Herbarium (AK): part 5. Dicotyledons. *Rec. Auckland Inst. Mus.* 31: 89–173.

Holzapfel, S. A revision of the genus *Picris* (Asteraceae, Lactuceae) s.l. in Australia. *Willdenowia 24*: 97–218. Includes *Picris burbidgei* S.Holzapfel which also occurs in N.Z.

Irwin, J. B. *Corybas rivularis* — one species or several? *Wellington Bot. Soc. Bull. 46*: 48–53.

Jones, D. L. New species of Orchidaceae from south-eastern Australia. *Muelleria 8*: 177–192. *Pterostylis tasmanica* described as new for southern Victoria, Tasmania and N.Z.

1994 Kelly, D. The evolutionary ecology of mast seeding. *Trends Ecol. Evol. 9*: 465–470.

Kosenko. V. N. Pollen morphology of the families Phormiaceae, Blandfordiaceae and Doryanthaceae. *Bot. Zhurn. 79(7)*: 1–12. (In Russian).

Kravtsova, T. I. and D. V. Geltman. Pericarp and seed coat anatomy and ultrasculpture in representatives of *Urtica* (Urticaceae). *Bot. Zhurn. 79(2)*: 27–44. (In Russian). Includes *Urtica ferox.*

Large, M. F. and D. J. Mabberley. The pollen of *Dysoxylum* (Meliaceae): the demise of *Pseudocarapa. Bot. J. Linn. Soc. 116*: 1–12.

Lavin, M., Doyle, J. J. and J. D. Palmer. Evolutionary significance of the loss of the chloroplast-DNA inverted repeat in the Leguminosae subfamily Papilionidae. *Evolution 44*: 390–402. Includes *Carmichaelia.*

Lord, J. M. Variation in *Festuca novae-zelandiae* (Hack.) Cockayne germination behaviour with altitude of seed source. *N.Z. J. Bot. 32*: 227–235.

Molloy, B. P. J. Reinstatement of *Corybas orbiculatus* (Colenso) L.B. Moore. *N.Z. Native Orchid Group J. 51*: 12–14.

Molloy, B. P. J. and G. N. Bawden. Taxonomic and conservation status of *Geum divergens* Cheeseman (Rosaceae). *N.Z. J. Bot. 32*: 119–124.

Molloy, B. P. J. and A. P. Druce. A new species name in *Melicytus* (Violaceae) from New Zealand. *N.Z. J. Bot. 32*: 113–118.

Molloy, B. P. J. and I. M. St George. A new species of *Drymoanthus* (Orchidaceae) from New Zealand, and typification of *D. adversus. N.Z. J. Bot. 32*: 415–421.

Molloy, B. P. J. and C. J. Webb. Taxonomy and typification of New Zealand *Geum* (Rosaceae). *N.Z. J. Bot. 32*: 423–428.

Orchard, A. E. and A. J. G. Wilson. *Flora of Australia. 49, Oceanic Islands, 1.* Canberra, AGPS. Many Norfolk Id spp. are shared with N.Z.

Patel, R. N. Wood anatomy of the dicotyledons indigenous to New Zealand. 23, Myrtaceae — subfam. Leptospermoideae (part 1). *N.Z. J. Bot. 32*: 95–112.

Poole, A. L. and N. M. Adams. *Trees and shrubs of New Zealand.* Revised ed. Lincoln, Manaaki Whenua Press.

Reid, A. R. and B. A. Bohm. Vacuolar and exudate flavonoids of New Zealand *Cassinia* (Asteraceae: Gnaphalieae). *Biochem. Syst. Ecol. 22*: 501–505.

Soják, J. Notes on *Potentilla* (Rosaceae) X–XII. —X. The section *Dumosae.* XI. The *P. microphylla* and *P. stenophylla* groups (sect. *Pentaphylloides*). XII. Key to the taxa of *P.* sect. *Pentaphylloides* (*Anserina*). *Bot. Jahrb. Syst. 116*: 11–81. Treats the N.Z. taxon as *P. anserina* subsp. *anserinoides* (Raoul) Soják.

Stock, P. A. and W. B. Silvester. Phloem transport of recently-fixed nitrogen in the *Gunnera - Nostoc* symbiosis. *New Phytol. 126*: 254–266. Studies involved *Gunnera monoica.*

1994   Taylor, C. M. Revision of *Tetragonia* (Aizoaceae) in South America. *Syst. Bot. 19*: 575–589. Includes *T. tetragonioides.*

Tomlinson, P. B. Functional morphology of saccate pollen in conifers with special reference to Podocarpaceae. *Int. J. Pl. Sci. 155*: 699–715.

Tucker, S. C. Floral ontogeny in Sophoreae (Leguminosae: Papilionoideae): II. *Sophora* sensu lato (Sophora group). *Amer. J. Bot. 81*: 368–380. Includes *Sophora microphylla.*

Verboom, G. A., Linder, H. P. and N. P. Barker. Haustorial synergids: an important character in the systematics of danthonioid grasses (Arundinoideae: Poaceae)? *Amer. J. Bot. 81*: 1601–1610.

Walsh, N. G. and T. G. Entwisle (Eds) *Flora of Victoria. 2, Ferns and Allied Plants, Conifers and Monocotyledons.* Melbourne, Inkata Press.

Webb, C. J. Pollination, self-incompatibility and fruit production in *Corokia cotoneaster* (Escalloniaceae). *N.Z. J. Bot. 32*: 385–392.

Webster, G. L. Synopsis of the genera and suprageneric taxa of Euphorbiaceae. *Ann. Missouri Bot. Gard. 81*: 33–144. Accepts *Poranthera* and *Oreoporanthera* as distinct genera within subfamily Phyllanthoideae, Tribe Antidesmeae, subtribe Porantherinae. Attempted to validate Köhler's publication of subtribe Porantherinae in 1965 but the correct citation would be Euphorbiaceae subtribe Porantherinae (Müll.Arg.) G.L.Webster *Ann. Missouri Bot. Gard. 81*: 53 (1994). Retains the spelling *Omalanthus.*

Weiller, C. M., Crowdon, R. K. and J. M. Powell. Morphology and taxonomic significance of leaf epicuticular waxes in the Epacridaceae. *Aust. Syst. Bot. 7*: 125–152.

Westphalen, G. and J. G. Conran. Chromosome numbers in the *Arthropodium-Dichopogon* complex (Asparagales: Anthericaceae). *Taxon 43*: 377–381. Includes chromosome counts of *Arthropodium candidum* and *A. cirratum.*

Williams, S. E., Albert, V. A. and M. W. Chase. Relationships of Droseraceae: a cladistic analysis of *rbc*L sequence and morphological data. *Amer. J. Bot. 81*: 1027–1037. Includes *Drosera binata* and *D. spathulata.*

Wilson, K. L. New taxa and combinations in the family Cyperaceae in eastern Australia. *Telopea 5*: 589–625. Discusses the type specimens of *Lepidosperma filiforme* and *L. laterale* which also occur in N.Z.

1995   Adams, L. G. *Chionogentias* (Gentianaceae) a new generic name for the Australasian 'snow-gentians' and a revision of the Australian species. *Aust. Syst. Bot. 8*: 935–1011. Makes new combinations for some N.Z. spp. formerly placed in *Gentiana.*

Anton, A. M. and H. E. Connor. Floral biology and reproduction in *Poa* (Poeae: Gramineae). *Aust. J. Bot. 43*: 577–599.

Barker, N. P., Linder, H. P. and E. H. Harley. Polyphyly of the Arundinoideae (Poaceae): evidence from *rbc*L sequence data. *Syst. Bot. 20*: 423–435.

1995   Bloor, S. J. A survey of extracts of New Zealand indigenous plants for selected biological activities. *N.Z. J. Bot. 33*: 523–540.

Browning, J. and K. D. Gordon-Gray. Studies in Cyperaceae in southern Africa 27: a contribution to knowledge of spikelet morphology in *Epischoenus* and the relationship of this genus to *Schoenus*. *S. Afr. J. Bot. 61*: 147–152. Discusses spikelet structure of *Schoenus; S. apogon* and *S. brevifolius* were among the spp. studied.

Browning, J., Gordon-Gray, K. D. and S. Galen Smith. Studies in Cyperaceae in southern Africa. 25: *Schoenoplectus tabernaemontani*. *S. Afr. J. Bot. 61*: 39–42. Unites *Schoenoplectus validus* with *S. tabernaemontani.*

Bruhl, J. J. Sedge genera of the world: relationships and a new classification of the Cyperaceae. *Aust. Syst. Bot. 8*: 125–305.

Butz Huryn, V. M. Use of native New Zealand plants by honey bees (*Apis mellifera* L.): a review. *N.Z. J. Bot. 33*: 497–512.

Cameron, E. K. et al. New Zealand Botanical Society threatened and local plant lists (1995 revision). *N.Z. Bot. Soc. Newsletter 39*: 15–28.

Clark, L. G., Zhang, W. and J. F. Wendel. A phylogeny of the grass family (Poaceae) based on *ndhF* sequence data. *Syst. Bot. 20*: 436–460.

Conn, B. J. Taxonomic revision of *Logania* section *Logania* (Loganiaceae). *Aust. Syst. Bot. 8*: 585–665. Discusses N.Z. *Logania depressa.*

Dawson, M. I. Contributions to a chromosome atlas of the New Zealand flora. 33, Miscellaneous species. *N.Z. J. Bot. 33*: 477–487.

Dijkgraaf, A. C., Lewis, G. D. and N. D. Mitchell. Chromosome number of the New Zealand puriri, *Vitex lucens* Kirk. *N.Z. J. Bot. 33*: 425–426.

Edgar, E. New Zealand species of *Deyeuxia* P.Beauv. and *Lachnagrostis* Trin. (Gramineae: Aveneae). *N.Z. J. Bot. 33*: 1–33.

Garay, L. A. and E. A. Christenson. *Danhatchia* a new genus for *Yoania australis*. *Orchadian 11*: 469–471.

Gardner, R. O. Some seed in *Melicytus* (Violaceae). *J. Auck. Bot. Soc. 50*: 35–37.

Garnock-Jones, P. J. and D. A. Norton. *Lepidium naufragorum* (Brassicaceae), a new species from Westland, and notes on other New Zealand coastal species of *Lepidium*. *N.Z. J. Bot. 33*: 43–51.

Godley, E. J. and P. E. Berry. The biology and systematics of *Fuchsia* in the South Pacific. *Ann. Missouri Bot. Gard. 82*: 473–516. *Fuchsia procumbens* is placed in a new section *Procumbentes.*

Heenan, P. B. Typification of names in *Carmichaelia, Chordospartium, Corallospartium*, and *Notospartium* (Fabaceae - Galegeae) from New Zealand. *N.Z. J. Bot. 33*: 439–454.

Heenan, P. B. A taxonomic revision of *Carmichaelia* (Fabaceae-Galegeae) in New Zealand. Part 1. *N.Z. J. Bot. 33*: 455–475.

Heenan, P. B. Typification of *Clianthus puniceus* (Fabaceae-Galegeae). *N.Z. J. Bot. 33*: 561–562.

1995 Hofstra, D. E., Adam, K. D. and J. S. Clayton. Isozyme variation in New Zealand populations of *Myriophyllum* and *Potamogeton* species. *Aquatic Bot.* *52*: 121–131.

Jacobs, S. W. L., Everett, J. and M. E. Barkworth. Clarification of morphological terms used in the *Stipeae* (*Gramineae*), and a reassessment of *Nassella* in Australia. *Taxon 44*: 33–41.

Jane, G. T. and W. R. Sykes. *Wilsonia backhousei* (Convolvulaceae) in New Zealand. *N.Z. Bot. Soc. Newsletter 42*: 4–6. Plants of Australian *W. backhousei* in Tasman Bay are regarded as indigenous to N.Z.

Kores, P. J. A systematic study of the genus *Acianthus* (Orchidaceae: Diurideae). *Allertonia 7*: 87–220.

Länger, R., Pein, I. and B. Kopp. Glandular hairs in the genus *Drosera* (Droseraceae). *Pl. Syst. Evol. 194*: 163–172. Includes all N.Z. spp.

Linder, H. P. and M. D. Crisp. *Nothofagus* and Pacific biogeography. *Cladistics 11*: 5–32.

Macmillan, B. H. *Nertera villosa* B.H.Macmill. et R.Mason (Rubiaceae), a new species from New Zealand. *N.Z. J. Bot. 33*: 435–438.

Mark, A. F. and N. M. Adams. *The New Zealand alpine plants*. 2nd ed. Auckland, Godwit.

Molloy, B. P. J. Two new species of *Leucogenes* (Inuleae: Asteraceae) from New Zealand, and typification of *L. grandiceps*. *N.Z. J. Bot. 33*: 53–63.

Molloy, B. P. J. *Manoao* (Podocarpaceae), a new monotypic conifer genus endemic to New Zealand. *N.Z. J. Bot. 33*: 183–201.

Murray, B. G. and P. J. de Lange. Chromosome numbers in the rare endemic *Pennantia baylisiana* (W.R.B.Oliv.) G.T.S.Baylis (Icacinaceae) and related species. *N.Z. J. Bot. 33*: 563–564.

Nelson, E. C. *Lophomyrtus aotearoana* E.C.Nelson: a new Latin name for the New Zealand native myrtle, ramarama. *N.Z. J. Bot. 33*: 557–559. The name was illegitimate; see Garnock-Jones & Craven (1996) and Nic Lughadha (1996).

Ohyama, M. et al. Occurrence of prenylated flavonoids and oligostilbenes and its significance for chemotaxonomy of genus *Sophora* (Leguminosae). *Biochem. Syst. Ecol. 23*: 669–677. Includes *S. prostrata* and *S. tetraptera*.

Patel, R. N. Wood anatomy of the dicotyledons indigenous to New Zealand. 24, Fabaceae — subfam. Faboideae (part 1). *N.Z. J. Bot. 33*: 121–130.

Patel, R. N. Wood anatomy of the dicotyledons indigenous to New Zealand. 25, Myrtaceae — subfam. Myrtoideae (part 1). *N.Z. J. Bot. 33*: 541–555.

Petterson, J. A., Williams, E. G. and M. I. Dawson. Contributions to a chromosome atlas of the New Zealand flora. 34, *Wahlenbergia* (Campanulaceae). *N.Z. J. Bot. 33*: 489–496.

1995   Plunkett, G. M. et al. Phylogenetic relationships between Juncaceae and Cyperaceae: insights from *rbcL* sequence data. *Amer. J. Bot. 82*: 520–525. Includes *Rostkovia magellanica.*

Reid, A. and B. A. Bohm. Flavonoids of *Raoulia* Hook.f. ex Raoul (Asteraceae: Gnaphalieae). *Biochem. Syst. Ecol. 23*: 209–210.

Smith, S. G. New combinations in North American *Schoenoplectus, Bulboschoenus, Isolepis* and *Trichophorum* (Cyperaceae). *Novon 5*: 97–102. *Schoenoplectus validus* is conspecific with *S. tabernaemontani,* the older name.

Swenson, U. Systematics of *Abrotanella,* an amphi-Pacific genus of Asteraceae (Senecioneae). *Plant. Syst. Evol. 197*: 149–193. Describes *A. fertilis* Swenson from N.Z.

Thompson, P. N. and R. J. Gornall. Breeding systems in *Coriaria* (Coriariaceae). *Bot. J. Linn. Soc. 117*: 293–304.

1996   Baeza P., C. M. Los géneros *Danthonia* DC. y *Rytidosperma* Steud. (Poaceae) en América — Una revisión. *Sendtnera 3*: 11–93. Discusses characters distinguishing *Rytidosperma* from *Danthonia.*

Beever, R. E. and S. L. Parkes. Self-incompatibility in *Cordyline australis* (Asteliaceae). *N.Z. J. Bot. 34*: 135–137.

Browning, J. and K. D. Gordon-Gray. A gynophore in *Desmoschoenus* (Cyperaceae). *N.Z. J. Bot. 34*: 131–134.

Cambie, R. C., Pan, Y. J. and B. F. Bowden. Flavonoids of the barks of *Melicope simplex* and *Melicope ternata. Biochem. Syst. Ecol. 24*: 461–462.

Chase, M. W., Rudall, P. J. and J. G. Conran. New circumscriptions and a new family of asparagoid lilies: genera formerly included in Anthericaceae. *Kew Bull. 51*: 667–680. *Arthropodium* and *Cordyline* assigned to the newly circumscribed family Lomandraceae.

Chinnock, R. J. To the limits of *Disphyma* (Aizoaceae: Ruschioideae) and beyond. *Aloe 33*: 59–61. Capsule of *D. australe* compared with other spp.

Clarkson, B. The distribution of *Androstoma* (*Cyathodes*) *empetrifolia* in the North Island. *N.Z. Bot. Soc. Newsletter 46*: 17–18.

Clarkson, B. D. and P. J. Garnock-Jones. *Hebe tairawhiti* (Scrophulariaceae): a new shrub species from New Zealand. *N.Z. J. Bot. 34*: 51–56.

Clarkson, B. D. and P. J. Garnock-Jones. A new hebe species: *Hebe tairawhiti* B.D.Clarkson et Garn.-Jones. *N.Z. Bot. Soc. Newsletter 44*: 13–14.

Connor, H. E. Breeding systems in Indomalesian *Spinifex* (Paniceae: Gramineae). *Blumea 41*: 445–454. Mentions *S. sericeus* in discussion.

Crayn, D. M. et al. Delimitation of Epacridaceae: preliminary molecular evidence. *Ann. Bot. 77*: 317–321.

Dawson, J. W. and R. Lucas. *New Zealand coast and mountain plants: their communities and lifestyles.* Wellington, Victoria University Press.

1996   de Lange, P. J. *Hebe bishopiana* (Scrophulariaceae) — an endemic species of the Waitakere Ranges, west Auckland, New Zealand. *N.Z. J. Bot. 34*: 187–194.

de Lange, P. J., Norton, D. A. and B. P. J. Molloy. A revised checklist of New Zealand mistletoe (Loranthaceae) hosts. *N.Z. Bot. Soc. Newsletter 44*: 15–24.

Delph, L. F. and D. G. Lloyd. Inbreeding depression in the gynodioecious shrub *Hebe subalpina* (Scrophulariaceae). *N.Z. J. Bot. 34*: 241–247.

Edgar, E. *Puccinellia* Parl. (Gramineae: Poeae) in New Zealand. *N.Z. J. Bot. 34*: 17–32.

Esser, H. J. (1248) Proposal to conserve the name *Homalanthus* (Euphorbiaceae) with a conserved spelling. *Taxon 45*: 555–556.

Gardner, R. O. Fruit and seed of *Beilschmiedia* (Lauraceae) in New Zealand. *Blumea 41*: 245–250.

Garnock-Jones, P. J. Addendum. *N.Z. J. Bot. 34*: 427. Provides valid publication of *Heliohebe lavaudiana* (Raoul) Garn.-Jones. The publication of this combination in 1993 was invalid because the full citation of the basionym was inadvertently omitted.

Garnock-Jones, P. J. and L. A. Craven. The correct scientific name in *Lophomyrtus* (Myrtaceae) for ramarama is *L. bullata* Burret. *N.Z. J. Bot. 34*: 279–280.

Garnock-Jones, P. J. and R. Elder. Nomenclatural validation of *Coprosma pseudocuneata* (Rubiaceae). *N.Z. J. Bot. 34*: 139–140.

Garnock-Jones, P. J., Timmerman, G. M. and S. J. Wagstaff. Unknown New Zealand angiosperm assigned to Cunoniaceae using sequence of the chloroplast *rbc*L gene. *Pl. Syst. Evol. 202*: 211–218.

Gasson, P. Wood anatomy of the Elaeocarpaceae. *In* Donaldson, L. et al. (Eds) *Recent advances in wood anatomy*. Rotorua, New Zealand Forest Research Institute. pp. 47–71.

Harris, W. Genecological aspects of flowering patterns of populations of *Kunzea ericoides* and *K. sinclairii* (Myrtaceae). *N.Z. J. Bot. 34*: 333–354.

Heads, M. J. Biogeography, taxonomy and evolution in the Pacific genus *Coprosma* (Rubiaceae). *Candollea 51*: 381–405.

Heenan, P. B. *Uncinia obtusifolia* (Cyperaceae), a new species of hooked sedge in New Zealand. *N.Z. J. Bot. 34*: 11–15.

Heenan, P. B. *Epilobium petraeum* (Onagraceae), a new species of alpine willow-herb from New Zealand. *N.Z. J. Bot. 34*: 41–45.

Heenan, P. B. A taxonomic revision of *Carmichaelia* (Fabaceae-Galegeae) in New Zealand. Part 2. *N.Z. J. Bot. 34*: 157–177.

Heenan, P. B. Identification and distribution of the Marlborough pink brooms, *Notospartium carmichaeliae* and *N. glabrescens* (Fabaceae – Galegeae), in New Zealand. *N.Z. J. Bot. 34*: 299–307.

1996   Heenan, P. B., Webb, C. J. and P. N. Johnson. *Mazus arenarius* (Scrophulariaceae), a new, small-flowered, and rare species segregated from *M. radicans. N.Z. J. Bot. 34*: 33–40.

Irwin, B. Further notes on *Corybas rivularis. Bull. Wellington Bot. Soc. 47*: 55–58.

Jacobs, S. W. L. and J. Everett. *Austrostipa*, a new genus, and new names for Australian species formerly included in *Stipa* (Gramineae). *Telopea 6*: 579–595. Makes new combinations *Achnatherum petriei* (Buchanan) S.W.L.Jacobs et J.Everett for *Stipa petriei* and *Austrostipa stipoides* (Hook.f.) S.W.L.Jacobs et J.Everett for *Stipa stipoides.*

Jones, D. L. Reinstatement of *Caladenia alpina* R.S.Rogers (Orchidaceae) as distinct from *Caladenia lyallii* Hook.f. and the description of *Caladenia cracens*, a related new species from southern Tasmania. *Muelleria 9*: 41–50. *C. lyallii* is regarded as endemic to N.Z.

Jones, D. L. Resolution of the *Prasophyllum alpinum* R.Br. (Orchidaceae) complex in mainland south-eastern Australia, Tasmania and New Zealand. *Muelleria 9*: 51–62. *P. colensoi* is endemic to N.Z.

Kellogg, E. A., Appels, R. and R. J. Mason Gomer. When grass genes tell different stories: the diploid genera of Triticeae (Gramineae). *Syst. Bot. 21*: 321–347. Includes *Australopyrum.*

King, M. J., Vincent, J. F. V. and W. Harris. Curling and folding of leaves of monocotyledons: a strategy for structural stiffness. *N.Z. J. Bot. 34*: 411–416.

Kron, K. A. Phylogenetic relationships of Empetraceae, Epacridaceae, Ericaceae, Monotropaceae, and Pyrolaceae: evidence from nuclear ribosomal 18s sequence data. *Ann. Bot. 77*: 293–303.

Linder, H. P. and G. A. Verboom. Generic limits in the *Rytidosperma* (Danthonieae, Poaceae) complex. *Telopea 6*: 597–627.

Molloy, B. P. J. A new species name in *Phyllocladus* (Phyllocladaceae) from New Zealand. *N.Z. J. Bot. 34*: 287–297.

Molloy, B. P. J. and B. D. Clarkson. A new, rare species of *Melicytus* (Violaceae) from New Zealand. *N.Z. J. Bot. 34*: 431–440.

Molloy, B. P. J. and J. B. Irwin. Two new species of *Corybas* (Orchidaceae) from New Zealand, and taxonomic notes on *C. rivularis* and *C. orbiculatus. N.Z. J. Bot. 34*: 1–10. Describes *Corybas iridescens* and *C. papa.*

Morris, M. W., Stern, W. L. and W. S. Judd. Vegetative anatomy and systematics of subtribe Dendrobiinae (Orchidaceae). *Bot. J. Linn. Soc. 120*: 89–144.

Morvan, J. La valeur morphologique du cone femelle chez *Dacrydium cupressinum* Sol. Podocarpacees. *Phytomorphology 46*: 305–316.

Nic Lughadha, E. *Lophomyrtus bullata* is the correct name for the New Zealand ramarama. *Kew Bull. 51*: 815–817.

1996 Nicolson, D. H. (1233) Proposal to conserve the name *Lagenophora* (Compositae) with a conserved spelling. *Taxon 45*: 341–342.

Nowicke, J. W. Pollen morphology, exine structure and the relationships of Basellaceae and Didiereaceae to Portulacaceae. *Syst. Bot. 21*: 187–208. Supports the inclusion of Hectorellaceae in suborder Portulacineae.

Orchard, A. E. and A. Wilson. *Flora of Australia. 28, Gentianales.* Melbourne, CSIRO Australia. Includes *Sebaea ovata* and compares 2 endemic Tasmanian spp. of *Chionogentias* with N.Z. *C. corymbifera* and *C. cerina.*

Oskolski, A. A. A survey of the wood anatomy of the Araliaceae. *In* Donaldson, L. et al. (Eds) *Recent advances in wood anatomy.* Rotorua, New Zealand Forest Research Institute. pp. 99–119.

Peña, R. C. and B. K. Cassels. Phylogenetic relationships among Chilean *Sophora* species. *Biochem. Syst. Ecol. 24*: 725–733. Includes *S. microphylla*, *S. prostrata* and *S. tetraptera* and mentions *S. microphylla* var. *fulvida* and *S. chathamica.*

Powell, J. M. et al. A re-assessment of relationships within Epacridaceae. *Ann. Bot. 77*: 305–315.

Rahn, K. A phylogenetic study of the Plantaginaceae. *Bot. J. Linn. Soc. 120*: 145–198.

Rance, C. and B. Rance. *Gunnera hamiltonii* new information on phenology and distribution. *N.Z. Bot. Soc. Newsletter 44*: 8–9.

Sampson, F. B. Pollen morphology and ultrastructure of *Laurelia*, *Laureliopsis* and *Dryadodaphne* (Atherospermataceae [Monimiaceae]). *Grana 35*: 257–265.

Sarker, S. D. et al. The genus *Ourisia* (Scrophulariaceae): a potential source of phytoecdysteroids. *Biochem. Syst. Ecol. 24*: 803–804.

St George, I., Irwin, B. and D. Hatch. *Field guide to the New Zealand Orchids.* Wellington, New Zealand Native Orchid Group.

Stace, H. M. and S. H. James. Another perspective on cytoevolution in Lobelioideae (Campanulaceae). *Amer. J. Bot. 83*: 1356–1364. Includes *Isotoma fluviatilis.* New Zealand *Pratia* and *Hypsela* also mentioned in discussion.

Steinke, E., Williams, P. G. and A. E. Ashford. The structure and fungal associates of mycorrhiza in *Leucopogon parviflorus* (Andr.) Lindl. *Ann. Bot. 77*: 413–419. Includes root hair structure of *L. parviflorus* and comparison with hyphal forms in *Pernettya macrostigma.*

Svitashev, S. et al. A study of 28 *Elymus* species using repetitive DNA sequences. *Genome 39*: 1093–1101.

Swenson, U. *Abrotanella rostrata* (Asteraceae, Senecioneae) — a new species for New Zealand. *N.Z. J. Bot. 34*: 47–50.

Sykes, W. R. and C. J. West. New records and other information on the vascular flora of the Kermadec Islands. *N.Z. J. Bot. 34*: 447–462.

1996   Thiele, H. L., Clifford, H. T. and R. W. Rogers. Diversity in the grass pistil and its taxonomic significance. *Aust. Syst. Bot. 9*: 903–912. Includes *Microlaena stipoides* and *Dichelachne micrantha.*

Webb, C. J. Seed production in *Gunnera hamiltonii. Conservation Advisory Sci. Notes*: 133.

Webb, C. J. A rose by any other name: two problems of scent in the naming and typification of New Zealand plants. *N.Z. J. Bot. 34*: 281–283. *Coprosma, Scandia* and *Anisotome.*

Webb, C. J. The breeding system of *Pennantia baylisiana* (Icacinaceae). *N.Z. J. Bot. 34*: 421–422.

Weiller, C. M. Reassessment of *Cyathodes* (Epacridaceae). *Aust. Syst. Bot. 9*: 491–507.

Weiller, C. M. Reinstatement of the genus *Androstoma* Hook.f. (Epacridaceae). *N.Z. J. Bot. 34*: 179–185. The genus is reinstated for plants latterly known as *Cyathodes empetrifolia.*

Wilson, H. D. *Wild Plants of Mount Cook National Park.* 2nd ed. Christchurch, Manuka Press.

SUBJECT INDEX TO ANNALS OF TAXONOMIC RESEARCH

Works cited in the Annals of Taxonomic Research in Volume V of the *Flora* are listed here (by year and author) according to the families with which they deal. The families are arranged alphabetically within the major plant groups Gymnospermae, Dicotyledones and Monocotyledones. Letters in brackets after the authors' names classify the papers as follows:

A Anatomy; C Chemotaxonomy; F Floristics; G Genetics; K Karyology (Cytology); M Morphology; MB Molecular Biology; P Pollen morphology; R Reproductive biology; T Taxonomy.

GYMNOSPERMAE

PHYLLOCLADACEAE: 1987 Foo (C); Quinn (T). 1989 Tomlinson et al. (2M). 1996 Molloy (T).

PODOCARPACEAE: 1803-1824 Lambert (2T). 1987 Lorimer & W. (C); Webby et al. (C, T). 1989 Fountain et al. (M, R); Markham et al. (C, T); Wells & H. (M). 1990 Hayman & W. (C); Philipson & M. (M). 1991 Tomlinson et al. (R, M). 1992 Bergin & K. (F); Haase (G). 1994 Tomlinson (M). 1995 Molloy (T). 1996 Morvan (M).

DICOTYLEDONES

AIZOACEAE: 1991 Bittrich & C. (C). 1994 Taylor (T). 1996 Chinnock (M, T).

ALSEUOSMIACEAE: 1989 Dickison (A).

APIACEAE: 1987 Eichler (T). 1988 McKone & W. (K, R). 1989 Martinez (T); Orchard (T). 1990 Webb (T). 1992 Webb (R). 1993 Martinez (R, T: T). 1996 Webb (T).

ARALIACEAE: 1993 Gould (A, M); Sykes (T). 1994 Clearwater & G. (M). 1996 Oskolski (A).

ASTERACEAE: 1987 Lloyd & W. (T). 1988 Clarkson (T); Hind & J. (T); Nelson (T); Saleh et al. (C); Short (T); Sykes (T); Webb (2T). 1989 Adolphi et al. (T); Anderberg (T); Gadek et al. (P, A); Short et al. (A); Vincent & G. (P, M); Webb (T). 1990 Bruhl & Q. (A); Sonck (T). 1991 Anderberg (T); Bruhl & Q. (M, T); Heenan (T); Jansen et al. (MB, T). 1992 Belcher (2T); Vincent & G. (T); Wilson & G. (T). 1993 Breitwieser (A, T); Breitwieser & W. (A, C, T); Bremer (T); Brummitt (T); Dawson et al. (K); Haase et al. (T, G); Holzapfel & L. (T); Reid & B. (C); Swenson (T); Ward (2T). 1994 Battjes et al. (G, A); Holzapfel (T); Reid & B. (C). 1995 Molloy (T); Reid & B. (C); Swenson (T). 1996 Nicolson (T); Swenson (T).

MONOCOTYLEDONES

(T). Watson & D. (T, M). 1993 Connor et al. (T); Jacobs et al. (T); Kellogg & W. (T, M); Lord (G, R); Lu (G, K); Lu & B. (G, K); Simon (T); Torres (T). 1994 Connor (T); Lord (R); Verboom et al. (A, M, T). 1995 Anton & C. (R); Barker et al. (MB, T); Clark et al. (MB, T); Edgar (T); Jacobs et al. (M, T). 1996 Baeza (T); Connor (R); Edgar (T); Jacobs & E. (T); Kellogg et al. (G, T); Linder & V. (T); Svitashev et al. (MB); Thiele et al. (M, T).

POTAMOGETONACEAE: 1988 Sorsa (P, M, T); Wiegleb (T). 1990 Lees & S. (C, G); Wiegleb (A, T). 1993 Barrett et al. (G); Posluszny & C. (M). 1995 Hofstra et al. (G).

RESTIONACEAE: 1988 Rudall & L. (A, T). 1992 Linder (M, A, T).

RUPPIACEAE: 1993 Talavera et al. (K).

SPARGANIACEAE: 1987 Cook & N. (T).

THISMIACEAE: 1992 Rudall & M. (M, A).

ZOSTERACEAE: 1991 Murray & L. (C). 1992 West et al. (C). 1993 Barrett et al. (G). 1994 Conacher et al. (M, R).

TAXA MIXTA

Including Several Families (Dicotyledons, Monocotyledons, Gymnosperms)

ANATOMY: 1992 Carlquist.

CHEMOTAXONOMY: 1989 Bannister. 1991 Cooper & C. 1992 Seawright & S. 1995 Bloor.

CYTOLOGY: 1989 Dawson. 1993 Les & P. 1995 Dawson.

FLORISTIC WORKS: 1969 Fineran. 1987 Druce et al.; George. 1988 Smith; Sykes & W.; Webb et al. 1989 Godley; Johnson & B.; Stanley & R.; Wilson & G. 1990 Harden; Webb et al. 1991 Smith; Wardle. 1992 Connor; Harden; Salmon. 1993 Cameron et al.; Clifford; George et al.; Harden; Wilson & G. 1994 Curtis & M.; Orchard & W.; Poole & A.; Walsh & E. 1995 Cameron et al.; Mark & A. 1996 Dawson & L.; Sykes & W.; Wilson.

MORPHOLOGY: 1987 Philipson. 1989 Bannister; Endress & H. 1990 Lee et al. 1993 Wilson & G.

PALYNOLOGY: 1993 Moar.

REPRODUCTIVE BIOLOGY: 1980 Primack. 1984 Schoen & L. 1988 McEwen. 1993 Clifford; Metcalf; Webb & K. 1994 Kelly. 1995 Butz Huryn.

TAXONOMY: 1915 Domin. 1987 Connor & E.; Druce et al.; George; Nicora & R.; Philipson. 1988 Norton; Smith; Webb et al. 1989 Dahlgren; Stanley & R. 1990 Harden. 1991 Chapman; Smith. 1992 Harden; Nicolson. 1993 George et al.; Harden. 1994 Curtis & M.; Herrick & C.; Orchard & W.; Walsh & E. 1996 Sykes & W.

ILLUSTRATIONS

In this Flora there are three kinds of illustrations; first are 24 line drawings of flowers and their surrounding organs, 21 reflecting the tribes of grasses in New Zealand, and a further three to illustrate subtribes of significance to New Zealand agrostology. All were drawn by Sabrina B. Malcolm except Figure 1 drawn by Pat A. Brooke. Secondly there are 16 paintings of florets of species of *Rytidosperma*, the originals by Keith R. West were intended to accompany our 1979 paper on *Rytidosperma* (*N.Z.J. Bot. 17*: 311–337) but were judged at the time to be expensive to print and to add limited scientific value to the text. They have lain awaiting a suitable opportunity and vehicle; twenty years later part of our original plan is attained and West's paintings are published.

An essential part of any flora is the presentation of accurate habitat and ecological details for each of the taxa described. In the third form of illustration in this Flora, the colour photographs (Plates 1–10) of Peter N. Johnson illustrate how both native and naturalised grasses may dominate landscapes. An array of habitat and portrait photographs is indicative of the many environments which grasses occupy, and of their diversity in growth habit. Ecologically important genera such as *Chionochloa* and *Poa* are featured by a selection of their species, and the bases of tillers in species of *Chionochloa* show diagnostic characters just as West's paintings (Plates 11 and 12) do for *Rytidosperma* florets.

Our gratitude to our illustrators can only match their quality.

FIGURES

PLATES

END PAPERS

CHROMOSOME NUMBERS OF INDIGENOUS NEW ZEALAND GRAMINEAE

M. I. Dawson, Landcare Research, Lincoln, New Zealand

This list provides referenced chromosome counts in grasses indigenous to the New Zealand Botanical Region. Macquarie Island is part of that region, so the chromosome numbers of Moore (1960) are included. Counts made from plants occurring naturally outside the New Zealand Botanical Region are not listed, even if those taxa are considered also to occur in New Zealand, thus for example, the many counts of *Deschampsia cespitosa* and *Trisetum spicatum* from other countries have been excluded.

Taxa are listed alphabetically by genus, species, and subspecies; natural and artificial hybrids are excluded. Chromosome numbers cited as "n" represent the gametic count determined at meiosis, those cited as "$2n$" represent the somatic count determined at mitosis. Counts are cited here as presented in the original publications, in preference to secondary sources such as chromosome indexes (e.g., Darlington & Wylie 1955; *Index to Plant Chromosome Numbers* series) or review papers (e.g., Hair 1966).

Comments on the validity of certain counts or identifications are in the reference column. Many early publications do not cite herbarium voucher specimens, so the identification of the material examined cannot always be confirmed. Herbarium numbers are cited for several of my unpublished counts, and elsewhere to clarify ambiguities.

I am indebted to Drs H. E. Connor and E. Edgar for checking this list and for providing the taxonomic and nomenclatural treatments. However, any errors or omissions remain mine.

	n	$2n$	
Agrostis			
dyeri Petrie	21	–	Beuzenberg & Hair 1983:14
magellanica Lam.	42	–	Beuzenberg & Hair 1983:14
magellanica Lam.	–	72	Moore 1960:187
muelleriana Vickery	21	–	Beuzenberg & Hair 1983:14; as *A. subulata*. See Edgar & Forde 1991:143

Agrostis	n	2n	
muscosa Kirk	21	–	Beuzenberg & Hair 1983:14
personata Edgar	21	–	Beuzenberg & Hair 1983:14; as *A. dyeri* & *A. parviflora.* See Edgar & Forde 1991:143
Anemanthele			
lessoniana (Steud.) Veldkamp	–	40–44	M. I. Dawson, unpubl. (CHR 514915)
Australopyrum			
calcis Connor et Molloy			
subsp. *calcis*	–	14	Connor et al. 1993:2,8
calcis subsp. *optatum* Connor et Molloy	–	14	Connor et al. 1993:2,8
Austrofestuca			
littoralis (Labill.) E.B.Alexeev	–	28	Hair 1968:271; as *Poa triodioides*
littoralis (Labill.) E.B.Alexeev	–	28	Beuzenberg & Hair 1983:14; as *Festuca littoralis*
Chionochloa			
acicularis Zotov	–	c.42	Dawson 1989:164
antarctica (Hook.f.) Zotov	–	42	Beuzenberg & Hair 1983:14
antarctica (Hook.f.) Zotov	–	42	Dawson 1989:164
australis (Buchanan) Zotov	–	42	Dawson 1989:164
australis (Buchanan) Zotov	–	36	Calder 1937:7; as *Danthonia australis.* See Zotov 1963:89 & Connor 1991:224
beddiei Zotov	–	42	Dawson 1989:164
bromoides (Hook.f.) Zotov	–	42	Dawson 1989:164
conspicua (G.Forst.) Zotov			
subsp. *conspicua*	21	42	Calder 1937:7; as *D. cunninghamii.* See Zotov 1963:89
crassiuscula (Kirk) Zotov	–	36	Calder 1937:7; as *D. crassiuscula.* See Zotov 1963:89 & Connor 1991:224
flavescens Zotov	21	–	Calder 1937:7; as *D. raoulii* var. *flavescens.* See Zotov 1963:89
flavicans Zotov	–	42	Dawson 1989:164
macra Zotov	–	42	Dawson 1989:164
oreophila (Petrie) Zotov	–	42	Dawson 1989:165
oreophila (Petrie) Zotov	–	36	Calder 1937:7; as *D. oreophila.* See Zotov 1963:89 & Connor 1991:224
ovata (Buchanan) Zotov	–	c.42	Dawson 1989:165
pallens subsp. *cadens* Connor	–	42	Dawson 1989:165; as *C. pallens* (CHR 439665)
pallens subsp. *pilosa* Connor	–	42	Dawson 1989:165; as *C. pallens* (CHR 437929)

	n	2n	
Chionochloa			
rigida (Raoul) Zotov	–	42	Calder 1937:7; as *D. raoulii.* See Zotov 1963:89
rubra Zotov	–	42	Calder 1937:7; as *D. raoulii* var. *rubra.* See Zotov 1963:89
spiralis Zotov	–	42	Dawson 1989:165
teretifolia (Petrie) Zotov	–	42	Calder 1937:7; as *D. ovata.* See Zotov 1963:89
Cortaderia			
fulvida (Buchanan) Zotov	–	90	Hair & Beuzenberg 1966:256
richardii (Endl.) Zotov	–	90	Hair & Beuzenberg 1966:256
splendens Connor	–	90	E. J. Beuzenberg in Connor 1971: 521, 523
toetoe Zotov	–	90	Hair & Beuzenberg 1966:256
Deschampsia			
cespitosa (L.) P.Beauv.	–	26	Beuzenberg & Hair 1983:14
cespitosa (L.) P.Beauv.	–	26	F. Albers, *hic comm.*
chapmanii Petrie	–	26	F. Albers, *hic comm.*
chapmanii Petrie	–	28	Moore 1960:187
tenella Petrie	–	26	F. Albers, *hic comm.*
Dichelachne			
crinita (L.f.) Hook.f.	–	70	Beuzenberg & Hair 1983:14
lautumia Edgar et Connor	–	70(+2f)	M. I. Dawson in Edgar & Connor 1999:68 (CHR 517725)
Echinopogon			
ovatus (G.Forst.) P.Beauv.	21	–	Beuzenberg & Hair 1983:14
Elymus			
apricus Á.Löve et Connor	–	42	Hair et al. 1967:187; as *Agropyron scabrum* (Group Otago). See Connor 1994:130–131
apricus Á.Löve et Connor	–	42	Löve & Connor 1982:170,182,183
enysii (Kirk) Á.Löve et Connor	–	28	J. B. Hair in Connor 1954:316; as *A. enysii.* See Connor 1994:131–132
enysii (Kirk) Á.Löve et Connor	–	28	Hair 1966:584; as *A. enysii.* See Connor 1994:131–132
enysii (Kirk) Á.Löve et Connor	14	28	Hair et al. 1967:186; as *A. enysii.* See Connor 1994:131–132
enysii (Kirk) Á.Löve et Connor	–	28	Löve & Connor 1982:170,183
enysii (Kirk) Á.Löve et Connor	–	4x[28]	Svitashev et al. 1996:1094,1099
enysii (Kirk) Á.Löve et Connor	–	28	M. I. Dawson, unpubl.
falcis Connor	–	42	Hair et al. 1967:186; as *A. scabrum* (Group Tekapo II). See Connor 1994:132–134

Elymus	n	2n	
falcis Connor	–	42	Löve & Connor 1982:170; as *A. scabrum* (Group Tekapo II). See Connor 1994:132–134
multiflorus (Hook.f.) Á.Löve et Connor	–	42	J. B. Hair in Connor 1954:317; as *A. kirkii.* See Connor 1994:134–137
multiflorus (Hook.f.) Á.Löve et Connor	–	42	J. B. Hair in Darlington & Wylie 1955: 453; as *A. kirkii.* See Connor 1994:134–137
multiflorus (Hook.f.) Á.Löve et Connor	–	42	Hair 1966:584; as *A. kirkii.* See Connor 1994:134–137
multiflorus (Hook.f.) Á.Löve et Connor	21	42	Hair et al. 1967:186; as *A. kirkii* & *A. kirkii* var. *longisetum.* See Connor 1994:134–137
multiflorus (Hook.f.) Á.Löve et Connor	–	42	Löve & Connor 1982:170,183
solandri (Steud.) Connor	–	42	Hair 1953:215; as *A. scabrum.* See Connor 1994:140–142
solandri (Steud.) Connor	–	42	J. B. Hair in Darlington & Wylie 1955: 453; as *A. scabrum* (sexual). See Connor 1994:140–142
solandri (Steud.) Connor	–	42	Hair 1966:584; as *A. scabrum.* See Connor 1994:140–142
solandri (Steud.) Connor	21	42	Hair et al. 1967:186–187; as *A. scabrum.* See Connor 1994:140–142
solandri (Steud.) Connor	–	42	Löve & Connor 1982:170; as *A. scabrum.* See Connor 1994:140–142
solandri (Steud.) Connor	–	42	Lu 1993:196; as *E. scaber.* See Connor 1994:140–142
solandri (Steud.) Connor	–	42	Lu & Bothmer 1993:37; as *E. scaber.* See Connor 1994:140–142
tenuis (Buchanan) Á.Löve et Connor	–	56	J. B. Hair in Connor 1954:318; as *A. tenue.* See Connor 1994:142–143
tenuis (Buchanan) Á.Löve et Connor	–	56	J. B. Hair in Darlington & Wylie 1955: 453; as *A. scabrum* var. *tenue.* See Connor 1994:142–143

	n	2n	
Elymus			
tenuis (Buchanan) Á.Löve et Connor	–	56	Hair 1966:584; as *A. tenue.* See Connor 1994:142–143
tenuis (Buchanan) Á.Löve et Connor	28	56	Hair et al. 1967:187; as *A. tenue.* See Connor 1994:142–143
tenuis (Buchanan) Á.Löve et Connor	–	56	Löve & Connor 1982:170,183
Festuca			
contracta Kirk	–	c.170	Moore 1960:187; as *F. erecta*
coxii (Petrie) Hack.	–	8*x* [56]	J. B. Hair in Connor 1968:295
coxii (Petrie) Hack.	–	56	Beuzenberg & Hair 1983:14
matthewsii subsp. *latifundii* Connor	–	42	J. B. Hair in Connor 1968:295; as *F. matthewsii*
matthewsii subsp. *latifundii* Connor	–	42, 42+B	Beuzenberg & Hair 1983:14; as *F. matthewsii*
multinodis Petrie et Hack.	–	8*x* [56]	J. B. Hair in Connor 1968:295
multinodis Petrie et Hack.	–	56	Beuzenberg & Hair 1983:14
novae-zelandiae (Hack.) Cockayne	–	42	J. B. Hair in Connor 1968:295
novae-zelandiae (Hack.) Cockayne	–	42	Beuzenberg & Hair 1983:14
novae-zelandiae (Hack.) Cockayne	–	42	M. I. Dawson, unpubl.
Microlaena			
avenacea (Raoul) Hook.f.	–	48	M. I. Dawson in Edgar & Connor 1998:566,577 (CHR 495047, CHR 495048)
carsei Cheeseman	–	48	M. I. Dawson in Edgar & Connor 1998:566,578
polynoda (Hook.f.) Hook.f.	–	48	E. J. Beuzenberg in Connor & Edgar 1986:427
stipoides (Labill.) R.Br.	–	48	M. I. Dawson in Edgar & Connor 1998:565,583 (CHR 514916, CHR 514917)
stipoides (Labill.) R.Br.	–	42	J. B. Hair in Tothill & Love 1964:21. Count undocumented; see Edgar & Connor 1998:566

	n	*2n*	
Oplismenus			
hirtellus subsp. *imbecillis* (R.Br.)			
U.Scholz	–	54	M. I. Dawson, unpubl. (CHR 495040)
Poa			
acicularifolia Buchanan			
subsp. *acicularifolia*	–	28	Hair 1968:269; as *P. acicularifolia.* See Edgar 1986:444
anceps G.Forst. subsp. *anceps*	–	28	Hair 1968:269; as *P. anceps*
anceps subsp. *polyphylla* (Hack.)			
Edgar	–	28	Hair 1968:271; as *P. polyphylla.* See Edgar 1986:452
astonii Petrie	–	28	Hair 1968:269
aucklandica Petrie			
subsp. *aucklandica*	–	28	Hair 1968:269; as *P. aucklandica* (CHR 102389, CHR 102390). See Edgar 1986:467
aucklandica subsp. *campbellensis*			
(Petrie) Edgar	–	28	Hair 1968:269; as *P. aucklandica* (CHR 102391). See Edgar 1986:467
breviglumis Hook.f.	–	28	Hair 1968:269; as *P. breviglumis* var. *breviglumis* & *P. breviglumis* var. *brockiei.* See Edgar 1986:474
buchananii Zotov	–	28	Hair 1968:271; as *P. sclerophylla.* See Edgar 1986:480
chathamica Petrie	–	112	Hair 1968:269
cita Edgar	–	84	Hair 1968:270; as *P. laevis.* See Edgar 1986:447
cockayneana Petrie	–	112	Hair 1968:269
colensoi Hook.f.	–	28	Hair 1968:269
dipsacea Petrie	–	28	Hair 1968:269; as *P. cheesemanii* & *P. dipsacea.* See Edgar 1986:460
foliosa (Hook.f.) Hook.f.	–	28, 29	Moore 1960:187
foliosa (Hook.f.) Hook.f.	–	28	D. M. Moore in Hair 1968:270; as *P. hamiltonii*
foliosa (Hook.f.) Hook.f.	–	28	Hair 1968:269
hesperia Edgar	–	28	Hair 1968:269; as *P. colensoi* (CHR 101785). See Edgar 1986:442
imbecilla Spreng.	–	28	Hair 1968:271; as *P.* sp. (CHR 102415). See Edgar 1986:471
kirkii Buchanan	–	28	Hair 1968:270
lindsayi Hook.f.	–	28	Hair 1968:270
litorosa Cheeseman	–	263–265	Hair & Beuzenberg 1961:160

	n	2n	
Poa			
litorosa Cheeseman	–	c.266	Hair 1968:270
maniototo Petrie	–	28	Hair 1968:270
matthewsii Petrie	–	28	Hair 1968:270; as *P. matthewsii* var. *matthewsii, P. matthewsii* var. *minor, P. matthewsii* var. *tenuis.* See Edgar 1986: 470–471
novae-zelandiae Hack.	–	28	Hair 1968:270,271; as *P. novae-zelandiae* var. *humilior, P. novae-zelandiae* var. *laxiuscula,* & *P. novae-zelandiae* var. *wallii.* See Edgar 1986:436
pusilla Berggr.	–	28	Hair 1968:271; as *P. pusilla* & *P. seticulmis.* See Edgar 1986:453
pygmaea Buchanan	–	28	Hair 1968:271
ramosissima Hook.f.	–	28	Hair 1968:271
subvestita (Hack.) Edgar	–	28	Hair 1968:270; as *P. novae-zelandiae* var. *subvestita.* See Edgar 1986:437
tennantiana Petrie	–	56	Hair 1968:271
Puccinellia			
macquariensis (Cheeseman) Allan et Jansen	–	28	Moore 1960:187
Rytidosperma			
buchananii (Hook.f.) Connor et Edgar	36	c.72	Calder 1937:7; as *Danthonia buchanani*
gracile (Hook.f.) Connor et Edgar	12	24	Calder 1937:7; as *D. gracilis*
nigricans (Petrie) Connor et Edgar	12	24	Calder 1937:7; as *D. nigricans*
setifolium (Hook.f.) Connor et Edgar	12	24	Calder 1937:7; as *D. setifolia*
Simplicia			
buchananii (Zotov) Zotov	–	28	J. B. Hair & B. E. Groves in Zotov 1971:544
buchananii (Zotov) Zotov	–	28	M. I. Dawson, unpubl.
laxa Kirk	–	28	J. B. Hair & B. E. Groves in Zotov 1971:541
laxa Kirk	–	28	Beuzenberg & Hair 1983:14
laxa Kirk	–	28	M. I. Dawson, unpubl. (CHR 483932)
Spinifex			
sericeus R.Br.	–	18	Hair & Beuzenberg 1966:256; as *S. hirsutus*

Stenostachys	n	2n	
gracilis (Hook.f.) Connor	14	28	Hair et al. 1967:187; as *Cockaynea gracilis*. See Connor 1994:146
gracilis (Hook.f.) Connor	–	28	Löve & Connor 1982:170,184; as *Elymus narduroides*. See Connor 1994:146
laevis (Petrie) Connor	–	28	Hair et al. 1967:187; as *C. laevis*. See Connor 1994:146–148
laevis (Petrie) Connor	–	28	Löve & Connor 1982:170,184; as *E. laevis*. See Connor 1994:146–148
Trisetum			
lepidum Edgar et A.P.Druce	–	28	Beuzenberg & Hair 1983:14; as *T. antarcticum*. See Edgar 1998:553
Zotovia			
colensoi (Hook.f.) Edgar et Connor	–	48	M. I. Dawson in Edgar & Connor 1998:566,571 (CHR 483931)
thomsonii (Petrie) Edgar et Connor	–	c.48	M. I. Dawson in Edgar & Connor 1998:566,573

REFERENCES

Beuzenberg, E. J. and Hair, J. B. 1983: Contributions to a chromosome atlas of the New Zealand flora — 25. Miscellaneous species. *N.Z. J. Bot. 21*: 13–20.

Calder, J. W. 1937: A cytological study of some New Zealand species and varieties of *Danthonia. J. Linn. Soc. (Bot.) 51*: 1–9.

Connor, H. E. 1954: Studies in New Zealand *Agropyron*. Parts I and II. *N.Z. J. Sci. Tech. 35B*: 315–343.

Connor, H. E. 1968: Interspecific hybrids in hexaploid New Zealand *Festuca*. *N.Z. J. Bot. 6*: 295–308.

Connor, H. E. 1971: *Cortaderia splendens* Connor sp. nov. (Gramineae). *N.Z. J. Bot. 9*: 519–525.

Connor, H. E. 1991: *Chionochloa* Zotov (Gramineae) in New Zealand. *N.Z. J. Bot. 29*: 219–283.

Connor, H. E. 1994: Indigenous New Zealand Triticeae: Gramineae. *N.Z. J. Bot. 32*: 125–154.

Connor, H. E. and Edgar, E. 1986: Australasian alpine grasses: diversification and specialization. (Chapter 25). *In* Barlow, B. A. (Ed.) Flora and fauna of alpine Australasia — ages and origins. Melbourne, CSIRO.

Connor, H. E., Molloy, B. P. J. and Dawson, M. I. 1993: *Australopyrum* (Triticeae: Gramineae) in New Zealand. *N.Z. J. Bot. 31*: 1–10.

Darlington, C. D. and Wylie, A. P. 1955: Chromosome atlas of flowering plants. London, George Allen & Unwin Ltd.

Dawson, M. I. 1989: Contributions to a chromosome atlas of the New Zealand flora — 30. Miscellaneous species. *N.Z. J. Bot. 27*: 163–165.

Edgar, E. 1986: *Poa* L. in New Zealand. *N.Z. J. Bot. 24*: 425–503.

Edgar, E. 1998: *Trisetum* Pers. (Gramineae: Aveneae) in New Zealand. *N.Z. J. Bot. 36*: 539–564.

Edgar, E. and Connor, H. E. 1998: *Zotovia* and *Microlaena*: New Zealand Ehrhartoid Gramineae. *N.Z. J. Bot. 36*: 565–586.

Edgar, E. and Connor, H. E. 1999: Species novae graminum Novae-Zelandiae I. *N.Z. J. Bot. 37*: 63–70.

Edgar, E. and Forde, M. B. 1991: *Agrostis* in New Zealand. *N.Z. J. Bot. 29*: 139–161.

Hair, J. B. 1953: The origin of new chromosomes in *Agropyron*. *Heredity 6 (Suppl.)*: 215–233.

Hair, J. B. 1966: Biosystematics of the New Zealand flora, 1945–1964. *N.Z. J. Bot. 4*: 559–595.

Hair, J. B. 1968: Contributions to a chromosome atlas of the New Zealand flora — 12. *Poa* (Gramineae). *N.Z. J. Bot. 6*: 267–276.

Hair, J. B. and Beuzenberg, E. J. 1961: High polyploidy in New Zealand *Poa*. *Nature 189*: 160.

Hair, J. B. and Beuzenberg, E. J. 1966: Contributions to a chromosome atlas of the New Zealand flora — 7. Miscellaneous Families. *N.Z. J. Bot. 4*: 255–266.

Hair, J. B., Beuzenberg, E. J. and Pearson, B. 1967: Contributions to a chromosome atlas of the New Zealand flora — 9. Miscellaneous families. *N.Z. J. Bot. 5*: 185–196.

Löve, Á. and Connor, H. E. 1982: Relationships and taxonomy of New Zealand wheatgrasses. *N.Z. J. Bot. 20*: 169–186.

Lu, B.-R. 1993: Meiotic studies of *Elymus nutans* and *E. jacquemontii* (Poaceae, Triticeae) and their hybrids with *Pseudoroegneria spicata* and seventeen *Elymus* species. *Pl. Syst. Evol. 186*: 193–212.

Lu, B.-R. and Bothmer, R. von 1993: Meiotic analysis of *Elymus caucasicus, E. longearistatus*, and their interspecific hybrids with twenty-three *Elymus* species (Triticeae, Poaceae). *Pl. Syst. Evol. 185*: 35–53.

Moore, D. M. 1960: Chromosome numbers of flowering plants from Macquarie Island. *Bot. Not. 113*: 185–191.

Svitashev, S.; Salomon, B.; Bryngelsson, T. and Bothmer, R. von 1996: A study of 28 *Elymus* species using repetitive DNA sequences. *Genome 39*: 1093–1101.

Tothill, J. C. and Love, R. M. 1964: Supernumerary chromosomes and variation in *Ehrharta calycina* Smith. *Phyton 21*: 21–28.

Zotov, V. D. 1963: Synopsis of the grass subfamily Arundinoideae in New Zealand. *N.Z. J. Bot. 1*: 78–136.

Zotov, V. D. 1971: *Simplicia* T. Kirk (Gramineae). *N.Z. J. Bot. 9*: 539–544.

GRAMINEAE Juss., 1789

nom. altern. **Poaceae Barnhart, 1895**

SYNOPSIS

I SUBFAMILY BAMBUSOIDEAE

Spikelets 1–many-flowered, usually laterally compressed. Glumes 2–4; lemma 5–10-nerved, awnless or rarely shortly awned; lodicules 3, usually large; stamens 3–6, rarely more; hilum linear. Inflorescence a panicle, fascicle, raceme or spike. Perennials, woody or herbaceous. Leaves with membranous ligule; leaf-blade tessellate or not. **1. BAMBUSEAE** (*Bambusa, Chimonobambusa, Himalayacalamus, Phyllostachys, Pleioblastus, Pseudosasa, Sasa, Sasaella, Semiarundinaria*)

II SUBFAMILY EHRHARTOIDEAE

Spikelets 1–3-flowered, 2 lower ∅ upper ♀ or ♂ and ♀; laterally compressed; disarticulating above reduced or 0 glumes; lemma 5–10-nerved, awned or awnless; palea hyaline 0–7-nerved; lodicules 2 or 0; stamens 6–1; stigmas plumose branches to base or not. Inflorescence a panicle or raceme. Leaves with cross-veins or not.

A. Inflorescence a panicle or sometimes a unilateral raceme. Spikelets 3-flowered with 2 lower florets ∅ and reduced to lemmas, upper ♀. Glumes 2, persistent, < or > florets. Palea hyaline, nerves 0, 1, 2 or rarely 3–5. Lodicules 2 or 0. Stamens 6 or 4, 3, 2, or 1. **2. EHRHARTEAE** (*Ehrharta, Microlaena, Zotovia*)

B. Inflorescence a panicle. Spikelets 1-flowered, ♀, or 3-flowered with 2 lower florets ∅ and reduced to lemmas, upper ♀, or plant monoecious and spikelets unisexual. Glumes 0, or discernible as obscure lips at pedicel apex. Palea nerves 3–7, keel central. Lodicules 2. Stamens 6 sometimes fewer. **3. ORYZEAE** (*Leersia, Zizania*)

III SUBFAMILY POOIDEAE

Spikelets 1–many-flowered, laterally compressed. Lemma several-nerved, awn apical or dorsal, or 0. Lodicules 2–3, membranous or hyaline. Stamens 3, rarely fewer. Stigmas usually 2. Caryopsis usually ellipsoid to fusiform; embryo ⅓–⅙ length of caryopsis; hilum linear to round. Herbs, usually with linear leaf-blades. Ligule membranous. Inflorescence usually a panicle or raceme.

A. Ovary apex glabrous, or hairy without an appendage

1. Inflorescence a single unilateral raceme. Spikelets 1-flowered. Glumes much reduced; lower rim-like, upper almost 0. Lemma 3-nerved, with weak middorsal and 2 strong lateral keels. Lodicules 0. **4. NARDEAE** (*Nardus*)

2. Inflorescence an open or contracted panicle. Spikelets 1-flowered. Glumes usually > floret. Lemma 5–9-nerved, rounded, membranous to crustaceous, terete to lenticular, awned. Lodicules 2 or 3. **5. STIPEAE** (*Achnatherum, Anemanthele, Austrostipa, Nassella, Piptatherum*)

3. Inflorescence a panicle, sometimes a spike or unilateral raceme. Spikelets 2–many-flowered. Glumes usually ≤ lowest lemma. Lemma (3)–5–7–(13)-nerved, awned or awnless. Lodicules 2. **6. POEAE** (*Austrofestuca, Briza, Catapodium, Cynosurus, Dactylis, Festuca, Lolium, Poa, Puccinellia, Schedonorus, Vulpia*)

4. Inflorescence a single bilateral raceme. Spikelets 1–2-flowered. Glumes appressed to rachis, usually exceeding and covering floret(s). Lemma 3–5-nerved, entire, awnless. Lodicules 2. **7. HAINARDIEAE** (*Hainardia, Parapholis*)

5. Inflorescence a panicle, or raceme-like. Spikelets several–many-flowered. Glumes often papery, rarely > adjacent lemmas. Lemma 5–9–(13)-nerved, rounded. Lodicules 2, usually connate. **8. MELICEAE** (*Glyceria*)

6. Inflorescence a panicle. Spikelets 1–several-flowered. Glumes usually > adjacent lemmas, often = spikelet. Lemma (3)–5–11-nerved, often with geniculate, dorsal awn. Lodicules 2, free. **9. AGROSTIDEAE**

(a) Spikelets 1-flowered, ♀. **AGROSTIDINAE** (*Agrostis, Alopecurus, Ammophila, Deyeuxia, Dichelachne, Echinopogon, Gastridium, Lachnagrostis, Lagurus, Pentapogon, Phleum, Polypogon, Simplicia*)

(b) Spikelets 2–several-flowered

    (i) Spikelets with 2–several ♀ florets, or occasionally with 1 ♀ and 1 ♂ floret. **AVENINAE** (*Aira, Amphibromus, Arrhenatherum, Avena, Deschampsia, Holcus, Koeleria, Trisetum*)

    (ii) Spikelets 3-flowered, the 2 lower florets ♂ or ∅, the uppermost ♀. **PHALARIDINAE** (*Anthoxanthum, Hierochloe, Phalaris*)

B. Ovary capped by fleshy hairy appendage

    1. Inflorescence a panicle. Spikelets several–many-flowered. Glumes < lowest lemma. Lemma 5–15-nerved, usually awned, apex bifid. Lodicules 2, glabrous. **10. BROMEAE** (*Bromus*)

    2. Inflorescence spike-like. Spikelets 5–20-flowered. Glumes < lowest lemma. Lemma 3–9-nerved, sometimes awned, apex entire. Lodicules 2, ciliate. **11. BRACHYPODIEAE** (*Brachypodium*)

    3. Inflorescence a single bilateral raceme, with spikelets alternate in 2 opposite rows, single or in groups of 2–3. Spikelets 1–many-flowered. Glumes shorter or narrower than lowest lemma leaving much of it exposed. Lemma 5–11-nerved, awnless or awned. Lodicules 2, ciliate. **12. HORDEEAE** (*Australopyrum, Elymus, Elytrigia, Hordeum, Leymus, Secale, Stenostachys, Thinopyrum, Triticum*)

IV SUBFAMILY ARUNDINOIDEAE

Spikelets usually 2–many-flowered, sometimes 1-flowered, laterally compressed, with fragile rachilla; glumes 2, well-developed; lemma (1)–3–9-nerved; palea 2-nerved; lodicules 2, cuneate, fleshy, truncate; stamens 3, or 2; stigmas 2. Inflorescence usually a panicle. Ligule a line of hairs.

A. Spikelets 2–several-flowered; florets all ♀ or the lowest spikelet ♂ or ∅; lemma 3–7-nerved, ± acute or prolonged to a straight awn from tip. Glumes 3–5-nerved. Plants large, reed-like, leaves cauline. Inflorescence a large plumose panicle.   **13. ARUNDINEAE** (*Arundo, Phragmites*)

B. Spikelets 2–several-flowered; florets ♀ rarely unisexual; lemma entire or bilobed, with or without a straight or geniculate terminal awn, or awned from sinus. Plants tufted or tussock-forming. Inflorescence usually a small or large plumose panicle. **14. DANTHONIEAE**

   1. Florets usually ♀; lemma nerves 7–9, anastomosing, bidentate or bilobed at apex, often with a central geniculate awn from sinus. Glumes 3–13-nerved. Plants tufted, or forming small to large tussocks, leaves mainly basal. Inflorescence variable from a single spikelet to a small or large plumose panicle. **DANTHONIINAE** (*Chionochloa, Pyrrhanthera, Rytidosperma, Sieglingia*)

   2. Florets unisexual, or ♀; lemma 3-nerved, bidentate and awned from sinus or drawn out to an awn from tip, awn straight. Glumes 1-nerved. Plants forming large tussocks, leaves mostly basal. Inflorescence a large plumose panicle. **CORTADERIINAE** (*Cortaderia*)

V SUBFAMILY ARISTIDOIDEAE

Spikelet 1-flowered, rachilla prolongation absent; glumes 1–many-nerved; lemma 1–3-nerved, concealing palea, 3-awned apically with or without a column; inflorescence paniculate. **15. ARISTIDEAE** (*Aristida*)

VI SUBFAMILY CHLORIDOIDEAE

Spikelets 1-flowered, or 2–many-flowered, laterally compressed; rachilla fragile; florets ♀, or ♂ and ♀ on dioecious plants. Glumes 2, or rarely the lower minute or suppressed. Lemma 1–3-nerved. Palea 2-nerved. Lodicules 2, cuneate, fleshy, usually truncate. Stamens (1)–3. Stigmas 2. Caryopsis ovoid, fusiform or globose; pericarp adhering to seed or becoming loose; embryo large, often ⅓–¾ length of caryopsis; hilum punctiform or occasionally elliptic. Annual or perennial herbs. Ligule very variable, membranous, or membranous with a ciliate fringe, or a line of hairs. Leaf-blade linear or convolute. Inflorescence a panicle, or of 1 to several racemes.

A. Spikelets with 1 ♀ flower; inflorescence racemose, sometimes reduced to a single spikelet

  1. Inflorescence of tough unilateral racemes, sometimes reduced to a single spikelet. **16. CHLORIDEAE** (*Chloris, Cynodon, Spartina, Zoysia*)

  2. Inflorescence a single fragile bilateral raceme. **17. LEPTUREAE** (*Lepturus*)

B. Spikelets with many ♀ flowers and inflorescence paniculate or racemose, or spikelets 1-flowered and inflorescence paniculate. **18. ERAGROSTIDEAE** (*Eleusine, Eragrostis, Sporobolus*)

VII SUBFAMILY PANICOIDEAE

Spikelets solitary, or paired, or in triplets, usually dorsally compressed, all alike, or differing in sex, size, shape and structure. Fertile spikelets usually 2-flowered, the lower floret ♂, ∅, or ♀, the upper floret ♀ or ♀; rachilla not prolonged; rarely plants dioecious. Glumes or upper lemma indurated. Lodicules 2, cuneate, fleshy, truncate. Stamens usually 3, rarely fewer. Stigmas usually 2. Caryopsis with large embryo. Ligule usually a membranous rim with dense ciliate fringe, but sometimes entirely membranous (*Axonopus, Digitaria, Paspalum*) and occasionally suppressed (*Echinochloa*).

A. Spikelets solitary, or in pairs or triplets on pedicels of equal length; glumes usually herbaceous or membranous, lower usually smaller and often suppressed, upper usually similar in size and texture to lemma of lower floret; lemma of upper floret varying in texture from papery to very tough and rigid

  1. Spikelets falling entire at maturity. **19. PANICEAE** (*Axonopus, Cenchrus, Digitaria, Echinochloa, Entolasia, Oplismenus, Panicum, Paspalum, Pennisetum, Sacciolepis, Setaria, Spinifex, Stenotaphrum, Urochloa*)

  2. Spikelets disarticulating above glumes at maturity. **20. ISACHNEAE** (*Isachne*)

B. Spikelets in pairs or triplets with one sessile or shortly pedicelled, the other(s) longer pedicelled; glumes ± rigid, = spikelet and firmer than lemmas; lemma of upper floret hyaline or membranous, often awned. **21. ANDROPOGONEAE** (*Andropogon, Bothriochloa, Imperata, Miscanthus, Sorghum, Themeda*)

KEY TO TRIBES

(For genera present in New Zealand and represented by native or naturalised species)

1 Bamboo with woody culms .... **1. BAMBUSEAE** (Fig. 1)
Herbaceous grasses, or cane-like, or reeds *2*

2 Spikelets 2-flowered, dorsally compressed; lower floret ♂ or Ø, upper floret ♀ (in dioecious *Spinifex* florets all ♂, or Ø and ♀) *3*
Spikelets 1–many-flowered; or 2-flowered, laterally compressed, both florets ♀ or the upper floret Ø or ♂ *5*

3 Spikelets in pairs, or in triplets, one sessile or shortly pedicelled and the other(s) longer pedicelled; glumes ± rigid, firmer than lemmas ....
.... **21. ANDROPOGONEAE** (Fig. 24)
Spikelets solitary, or if in pairs or triplets then on pedicels of equal length; glumes herbaceous or membranous, similar to lemma of lower floret, less rigid than ± indurated lemma of upper floret *4*

4 Spikelets falling entire at maturity .... **19. PANICEAE** (Fig. 22)
Spikelets disarticulating above persistent glumes .. **20. ISACHNEAE** (Fig. 23)

5 Spikelets with 1 ♀ floret (in monoecious *Zizania* florets ♂ and ♀) with or without additional ♂ or Ø florets *6*
Spikelets with 2 or more ♀ florets (in dioecious *Poa* florets all ♂ or all ♀) .. *18*

6 Glumes 0 or discernible as obscure lips, or lower glume rim-like *7*
Glumes both or 1 well-developed *8*

7 Inflorescence a panicle; lemma 5–10-nerved .... **3. ORYZEAE** (Fig. 3)
Inflorescence a raceme; lemma 3-nerved .... **4. NARDEAE** (Fig. 4)

8 Inflorescence a panicle, lax or spike-like *9*
Inflorescence of 1–many racemes (reduced to a single spikelet in some *Zoysia*) *14*

9 Spikelets with 2 ♂ or Ø florets below ♀ floret *10*
Spikelets with 1 ♀ floret only, or with at most 1 additional ♂ floret *11*

10 Lemmas of 2 lower ∅ florets well-developed, coriaceous; awn terminal or 0 .... **2. EHRHARTEAE** (Fig. 2)
Lemmas of 2 lower florets scale-like (*Phalaris*), or lemmas well-developed, membranous to coriaceous, and awn dorsal or subapical..........................
.... **9. AGROSTIDEAE** subtribe **PHALARIDINAE** (Fig. 11)

11 Ligule membranous *12*
Ligule a line of hairs.. *13*

12 Lemma terete to gibbous, firmly membranous to crustaceous, usually enfolding and concealing palea; awn terminal .... **5. STIPEAE** (Fig. 5)
Lemma not indurated or clasping palea; awn usually dorsal or 0 ....
.... **9. AGROSTIDEAE**
12a ♀ floret solitary .... **AGROSTIDINAE** (Fig. 9)
♀ floret accompanied by 1 ♂ floret .... **AVENINAE**

13 Lemma 3-awned, and concealing palea .... **15. ARISTIDEAE** (Fig. 18)
Lemma awnless, palea not concealed .... **18. ERAGROSTIDEAE**

14 Spikelets ± sunken in cavities of rachis, ± concealed by upper glume *15*
Spikelets not sunken in rachis, nor concealed by glumes *16*

15 Ligule membranous, minutely ciliate; spikelets edgewise to rachis, glume 1; lateral nerves of lemma almost to tip .... **17. LEPTUREAE** (Fig. 20)
Ligule membranous, glabrous; spikelets broadside to rachis and glumes 2, *or* lateral nerves of lemma < ½ its length .... **7. HAINARDIEAE** (Fig. 7)

16 Spikelets with 2 ∅ florets below ♀ floret............................ **2. EHRHARTEAE**
Spikelets without or rarely with 1 ∅ floret below ♀ floret *17*

17 Lemma 1–3–(5)-nerved; racemes 1–many .... **16. CHLORIDEAE** (Fig. 19)
Lemma 5–9-nerved; racemes single, bilateral ..... **12. HORDEEAE** (Fig. 14)

18 ♀ spikelet accompanied by ∅ spikelets (*Cynosurus*) .... **6. POEAE**
♀ spikelet not accompanied by ∅ spikelets *19*

19 Inflorescence of several racemes .... **18. ERAGROSTIDEAE**
Inflorescence a panicle or a single, sometimes compound raceme *20*

20 Ligule membranous, sometimes with a ciliolate apex *21*
Ligule a line of hairs.. *27*

21 Inflorescence a spicate raceme .... **12. HORDEEAE**
Inflorescence a panicle *22*

22 Spikelets 2–(3)-flowered (rarely 3–6-flowered); upper glume exceeding adjacent floret, *or* awn dorsal, geniculate, recurved or straight, sometimes subapical or 0 (some *Deschampsia, Koeleria*) ....
.... **9. AGROSTIDEAE** subtribe **AVENINAE** (Fig. 10)
Spikelets 2–many-flowered; upper glume usually ≤ adjacent floret and awn terminal, subapical or 0 .... *23*

23 Ovary capped by hairy appendage; stigmas subterminal *24*
Ovary lacking a hairy appendage; stigmas usually terminal *25*

24 Inflorescence a lax panicle; lemmas usually awned, apex bifid; lodicules glabrous .... **10. BROMEAE** (Fig. 12)
Inflorescence a spike-like panicle; lemmas sometimes awned, apex entire; lodicules ciliate .... **11. BRACHYPODIEAE** (Fig. 13)

25 Lemma keeled or rounded, apex usually firm, terminally awned or awnless .... **6. POEAE** (Fig. 6)
Lemma rounded, apex hyaline or scarious, awnless *26*

26 Lemma ± flat, nerves 7, parallel, prominent, usually not extending into wide hyaline upper margin; lodicules connate .... **8. MELICEAE** (Fig. 8)
Lemma basally thickened and cordate (*Briza*) *or* nerves 5–(7), usually extending to upper margin and midnerve often minutely excurrent (*Puccinellia*); lodicules free .... **6. POEAE**

27 Glumes 0–1–(3)-nerved; lemma 1–3-nerved ....
.... **18. ERAGROSTIDEAE** (Fig. 21)
Glumes 1–13-nerved; lemma 3–9-nerved *28*

28 Awn from tip, straight; lemma 1–5- or 5–7-nerved ....
.... **13. ARUNDINEAE** (Fig. 15)
Awn from between lobes or teeth, usually geniculate; lemma 3- or 7–9-nerved .... **14. DANTHONIEAE**

28a Glumes 3–13-nerved; lemma 7–9-nerved .. **DANTHONIINAE** (Fig. 16)
Glumes 1-nerved; lemma 3-nerved .... **CORTADERIINAE** (Fig. 17)

I SUBFAMILY BAMBUSOIDEAE

Spikelets 1–many-flowered, usually laterally compressed. Glumes 2–4; lemma 5–10-nerved, awnless or rarely shortly awned; lodicules 3, usually large; stamens 3–6, rarely more; hilum linear. Inflorescence a panicle, fascicle, raceme or spike. Perennials, woody or herbaceous. Leaves with membranous ligule; leaf-blade tessellate or not.

Tribe: 1. Bambuseae

1. BAMBUSEAE *

Evergreen grasses, usually with woody culms (bamboos), sometimes herbaceous (not in N.Z.), occasionally climbing. Growth either sympodial (determinate) with short and thick rhizomes (pachymorph) which form dense clumps or monopodial (indeterminate) with short to long ± slender rhizomes (leptomorph) which form ± diffuse thickets. Culm nodes marked by 1 or 2 prominent rings, the lower representing the scar from a modified sheathing leaf (culm-sheath), internodes usually hollow. Culm-sheath with sheath proper, membranous ligule and often auricles and associated oral bristles, and a short, thickened, sometimes caducous sheath-blade. Branches 1 to many, often only from middle and upper nodes (nodes often many-branched because of rebranching at base from the single bud). Leaf-sheath with ligules and often auricles and oral bristles. Leaf-blade petiolate, often tessellate (cross veins forming a lattice). Inflorescence of panicles, racemes, or spikes often subtended by a leaf-like spathe; sometimes a condensed fascicle with branches very contracted and small bracts subtending the spikelets. Spikelets all similar, 1–many-flowered. Glumes (0)–2–(4) (florets often ∅ in countries of introduction). Palea exposed or ± enfolded by lemma. Lodicules usually 3. Stamens 3–6 (in N.Z.). Gynoecium: style 1; stigmas 1–3. Fruit usually a caryopsis; hilum linear.

Since flowering is so irregular, bamboos are usually identified by vegetative features. As far as possible the generic descriptions below are comprehensive for species naturalised in N.Z. Other genera are in cultivation here; a few of these are mentioned because they have been sometimes confused with taxa treated as naturalised.

---

* by W.R. Sykes

Flowering in bamboos is usually infrequent and irregular, often many years elapsing between flowering periods, although some species flower annually. Synchronous flowering is a feature of many species, following which all the plants may die or, as with most species in N.Z., parts of the clump die. Sometimes this happens for several years in succession. However some species have never flowered in this country since they were introduced, and some have probably never flowered anywhere in cultivation.

The treatment of Bambuseae here includes genera accepted by Stapleton, C. *Bamboos of Nepal: an illustrated guide* (1994), and by McClintock, D. *in* Walters, S. M. et al. (Eds) *European Garden Flora Vol. II* (1984).

1 Culm internodes flattened or grooved on alternate sides or obtusely quadrangular, at least in upper culm *2*
   Culm internodes terete, occasionally with a few uppermost internodes flattened on alternate sides *4*

2 Culm internodes above lowest branches obtusely quadrangular; basal nodes with short hard or spine-like adventitious rootlets; branches 3 at each node, subequal .... **Chimonobambusa**
   Culm internodes above lowest and middle (or sometimes only upper) branches flattened and/or grooved on alternate sides; basal nodes lacking spine-like adventitious rootlets; branches 2–(3) at each node, unequal (excluding rudimentary branch if present), later sometimes clustered .. *3*

3 Culm internodes flattened or grooved on alternate sides throughout or to lowest branch-bearing node; nodes with a 2-branched system which appears diffuse .... **Phyllostachys**
   Culm internodes terete, some flattened or grooved on alternate sides; nodes ultimately with up to 8 ± dense branches .... **Semiarundinaria**

4 Middle culm nodes usually with one branch; culm-sheaths persistent; rhizomes monopodial *5*
   Middle culm nodes with several to many branches; culm-sheaths deciduous, sometimes tardily so; rhizomes sympodial or monopodial *7*

5 Culm 3–5 m; lower and middle culm-sheaths > internodes; oral bristles smooth, or 0 .... **Pseudosasa**
   Culm < 2 m (in wild and cultivated N.Z. plants); lower and middle culm-sheaths < internodes; oral bristles scabrid throughout or below, rarely 0 .... *6*

6 Culms ascending; oral bristles scabrid along their whole length .... **Sasa**
   Culms strictly erect; oral bristles scabrid only near base .... **Sasaella**
7 Culms > 3 m × > 1 cm diam. (often very much higher and thicker); new
   culms appearing in winter; leaf-blade not tessellate; inflorescence a dense
   or capitate spike .... **Bambusa**
   Culms usually < 3 m × < 1 cm diam.; new culms appearing in late spring and
   summer; leaf-blade generally tessellate; inflorescence a diffuse panicle.. 8
8 Rhizomes running, sometimes shortly so, branching always monopodial; plant
   dense but not caespitose; culm-sheaths ± persistent; leaf-blade obviously
   tessellate .... **Pleioblastus**
   Rhizomes very short, branching sympodial; plant densely caespitose; culm-
   sheaths deciduous; leaf-blade not obviously tessellate .... **Himalayacalamus**

BAMBUSA Schreb., 1789 *nom. cons.*

Medium-sized to large bamboos, rarely climbing, with very short
sympodial rhizomes forming dense clumps. Culms terete. Culm-sheaths
deciduous, broad, triangular, thick; auricles usually present. Branches
many at each node, sometimes with branch thorns. Leaf-blades small
to medium-sized, usually not tessellate. Inflorescence stems often short,
with dense tufts of 1–many spikelets at nodes. Spikelets sessile, 2–many-
flowered. Glumes 1–3. Stamens 6. Stigmas usually 3.

Type species: *B. arundinacea* Retz.

c. 120 spp. in tropical Asia, Malesia and tropical America. Naturalised
spp. 2.

The largest genus of bamboos but little cultivated in N.Z.

1 Culms usually 3–5 m × 1–2 cm diam.; leaf-blade linear-lanceolate, 0.6–1.5 cm
   wide, abaxially glaucous, puberulent .... **1. multiplex**
   Culms usually 10–13 m × 4–8 cm diam.; leaf-blade oblong-lanceolate,
   1.5–2.5 cm wide, abaxially deep green and glabrous .... **2. oldhamii**

1. B. multiplex (Lour.) Schult. et Schult.f., *Syst. Veg.* ed. 16, *7*: 1350 (1830).

Dense clump-forming; habit spreading. Culms very numerous,
3–5 m × 1–2 cm diam. (except in small cultivated forms), many

uniformly deep green but always some with prominent yellow stripes or bands. Culm-sheath with caducous black hairs; ligule short, inconspicuous; auricles 0; sheath-blade triangular. Branches several at each node, often branched secondarily, forming a tuft, the central branch largest. Leaf-sheath with long white apical bristles. Ligule abbreviated. Auricles present. Leaf-blade 6–15 × 0.6–1.5 cm, linear-lanceolate, acuminate, glaucous, abaxially finely puberulent. Inflorescence culms short, arising from periphery of clump, very slender. Spikelets sessile, several-flowered, in small distant clusters; bracts ensheathing clusters 1.3–1.8 cm. Lemma = palea. Lodicules membranous.

Naturalised from south China.

N.: Hamilton East (Waikato River bank). Roadside bank and nearby gully, in scrub dominated by naturalised spp.; probably originating from cast out garden refuse.

The plant sometimes cultivated in North Id as *B. eutuldoides* McClure appears to be indistinguishable from *B. multiplex* 'Alphonse Karr'. Both, therefore, have some culms longitudinally striped yellow and green, a feature found in other spp. of bamboo, but not in any growing wild in N.Z.

2. B. oldhamii Munro, *Trans. Linn. Soc. 26*: 109 (1868).

Oldham's bamboo

Large, clump-forming. Culm usually 10–13 m (often taller in cultivation) × 4–8 cm diam., not narrowing for many metres, dark green. Culm-sheath very large, ephemeral dark appressed hairs on the larger shoots only; ligule very short; auricles very small or 0; sheath-blade to *c.* 13 × 10 cm, broadly triangular. Branches several from each node, usually ≤ 1 m long (plant appearing fastigiate), 1 branch much stouter than others. Leaf-sheath glabrous. Oral bristles prominent. Auricles small. Ligule small. Leaf-blade 8–18 × 1.5–3 cm, ± oblong-lanceolate, glabrous, abaxially green, adaxially darker and ± shining green. Inflorescence culms to *c.* 1 m, from periphery of clump. Spikelets sessile, in small clusters, 3–4 cm, laterally flattened with glumes appearing plaited, shining purple or greenish purple.

Naturalised from south China.

N.: Hamilton (Awatere Avenue and Hunter Gully near the Waikato River). In gully amongst willows and along part of adjacent river bank and roadside.

Flowering began in summer 1994/95 and continues.

Bambusa oldhamii is commonly cultivated in warmer parts of North Id, especially as an orchard windbreak because of its dense semi-fastigiate, non-invasive habit. The small wild population near the Waikato River most probably originated from cultivation on the cliff top above.

Bambusa oldhamii has been confused with *Dendrocalamus latifolius* Munro which is cultivated in N.Z., but the latter has generally larger leaf-blades, auriculate culm-sheaths, and rather small, narrow triangular or oblong-triangular sheath-blades.

CHIMONOBAMBUSA Makino, 1914

Small to medium-sized bamboos of rather diffuse habit, with rhizomes monopodial and running freely. Culm-internodes grooved or flattened, at least on one side; nodes very swollen, lower nodes often with rootlet thorns. Culm-sheaths ± persistent; sheath-blades very small. Branches 3 at each node, subequal; smaller secondary branches often present. Inflorescence diffuse. Spikelets many-flowered, sessile. Stamens 3. Stigmas 2.

Type species: *C. marmorea* (Mitford) Makino

Nearly 20 spp. in eastern Asia, Indochina, westwards to India. Naturalised sp. 1.

1. C. quadrangularis (Franceschi) Makino, *Bot. Mag. Tokyo 28*: 153 (1914).   square-stemmed bamboo

Diffuse bamboo with extensively running rhizomes. Culm 3–6 m, dark green; nodes swollen and prominent; internodes obtusely quadrangular in cross section; lower internodes less obviously so; lowest nodes with short indurated adventitious rootlets which become spiny. Culm-sheath to 12 × 8 cm, triangular, tardily deciduous, purple with dark purplish tessellations; auricles and oral bristles 0; sheath-blades very small. Branches spreading widely and ± drooping above, sometimes purplish. Leaf-sheath with long bristles at apex. Leaf-blade 9–24 × 0.7–2 cm, narrow lanceolate or linear-lanceolate, deep green on both surfaces, tessellate, ± drooping.

Naturalised from China.

N.: Wanganui District (Aird property near Fordell). Plantation of introduced, mainly deciduous trees in small valley where it has spread for some distance from the original planting.

Occasionally cultivated in N.Z. *Chimonobambusa quadrangularis* is very easily distinguished in N.Z. from any other bamboo by its quadrangular stems and the spiny lowermost culm nodes.

Also known in N.Z. as *Phyllostachys quadrangularis* (Franceschi) Rendle and *Tetragonocalamus angulatus* (Munro) Nakai.

The type sp., *C. marmorea* (Mitford) Makino, marble bamboo, is sometimes cultivated, but although it has running rhizomes it has not yet been reported as growing wild. *C. marmorea* has pencil-thick culms only *c.* 1 m high and culm-sheaths attractively marbled brown and white. Neither *C. marmorea* nor *C. quadrangularis* is known to have flowered in N.Z.

HIMALAYACALAMUS Keng f., 1983

Small to medium-sized bamboos with short sympodial rhizomes, forming a dense clump. Culm-internodes hollow, terete, rather slender, nodes not very prominent. Culm-sheaths tardily deciduous, ≈ internodes, glabrous, ± concolorous, broad and little narrowed distally; margins hairy. Leaf-sheath without auricles and oral bristles. Ligule pubescent. Leaf-blade not obviously tessellate. Inflorescence a large racemose panicle or raceme. Spikelets 1–(2)-flowered, one pedicelled, the other sessile. Glumes 2, strongly 4–8-nerved. Stamens 3. Stigmas 3.

Type species: *H. falconeri* (Munro) Keng f.

7 spp. in the Himalaya. Naturalised sp. 1.

The genus *Himalayacalamus* was described by Keng, P.C. *J. Bamboo Res. 2(1)*: 23 (1983), based on *Thamnocalamus falconeri* Munro which was previously also commonly referred to as *Arundinaria falconeri* (Munro) Benth. et Hook.f.

1. H. falconeri (Munro) Keng f., *J. Bamboo Res.* 2(1): 24 (1983).

fountain bamboo, fairy bamboo

Clump-forming, fountain-like; rhizomes not running. Culms initially erect and wand-like, later arching over, 2–2.5 m, glaucous-white when young, becoming medium to deep green. Culm-sheath uniformly green or purplish green; ligule 1–5 mm, puberulent; auricles and bristles 0; sheath-blades linear-subulate. Branches many at each node, small, densely tufted. Leaf-sheath green or purplish, glabrous excepting margins sometimes ciliate. Ligule prominent, oblong, minutely puberulent. Leaf-blade usually 4–12–(15) × 0.4–0.8–(1.3) cm, narrow lanceolate or linear-lanceolate, long acuminate. Panicles large and diffuse. Spikelets 8–12 mm, often purple or strongly flushed purple. Glumes < florets, unequal, minutely puberulent; lower ¾–⅚ length of upper. Lodicules *c.* 2 × 2 mm, ciliate. Caryopsis 6–7 mm, blackish.

Naturalised from the Himalaya.

N.: Waikato River (Hamilton), Taranaki (Urenui). River bank, amongst scrub and introduced trees, scattered plants for *c.* 100 m. Sometimes regenerates in gardens and nurseries in the vicinity of cultivated plants.

Fountain bamboo is the only bamboo in N.Z. which has spread beyond its place of cultivation by regeneration from seed. The latest flowering period in N.Z. was 1992–1998, and some plants have flowered and died. These periods occur at approximately 15–20 year intervals, after which the plants usually die.

Himalayacalamus falconeri is the most popular cultivated bamboo in N.Z. in recent times.

PHYLLOSTACHYS Siebold et Zucc., 1843 *nom. cons.*

Tall or medium-sized, often diffuse; rhizomes short or long with monopodial branching; shoots arising from lateral buds; culms well spaced. Culm nodes prominent; internodes hollow, flattened or grooved on alternate sides above each node, usually terete below lowest branches. Culm-sheath deciduous, usually < internodes, often darkly spotted. Branches usually 3 per node, central branch small, rudimentary or 0, the others markedly unequal. Leaf-sheath often with oral bristles and auricles. Leaf-blade distinctly tessellate, apex characteristically dying long before remainder of leaf-blade (at least in N.Z.). Inflorescence a panicle, composed of many spike-like clusters. Spikelets subtended by large ± foliaceous bracts, usually 5–13-flowered. Glumes (0)–2–(3). Lodicules 3. Stamens 3. Stigmas 3.

Type species: *P. bambusoides* Siebold et Zucc.

50–60 spp. in eastern Asia, westwards to the Himalaya, southwards to Indochina, all except 2 or 3 spp. indigenous or endemic to China. Naturalised spp. 3.

Culm height in N.Z. is generally very variable; plants in cool eastern areas of South Id are much shorter than many of those in warmer parts of North Id. Some species bear culm-sheath auricles which may be ephemeral and lost even before the deciduous sheath falls, and on smaller culm shoots (especially those at the ends of running rhizomes) auricles are often absent. Oral bristles around the ligule are also often deciduous.

> *1* Culm-sheath without auricles or bristles; at least some culms with congested and rather swollen lower internodes; nodes with a swollen ring beneath
> .... **1. aurea**
> Culm-sheath with auricles and bristles, at least when young; culms with ± evenly spaced internodes throughout; nodes lacking a swollen ring *2*
>
> *2* Culm-sheath pubescent; nodes with 2 prominent ridges .... **3. nigra**
>   *2a* Culm becoming black or blotched with black after the first season; culm-sheath often dark-mottled towards apex .... var. **nigra**
>   Culm remaining green; culm-sheath uniformly pale purplish ....
> .... var. **henonis**
> Culm-sheath glabrous; nodes with one prominent ridge ...... **2. bambusoides**

1. P. aurea Rivière et C.Rivière, *Bull. Soc. Acclim. Sér. 3 (5)*: 262 (1878).

walking stick bamboo

Fairly dense thicket-forming with rhizomes running extensively. Culm usually 3–6 m (often taller in cultivation), glaucous when young, green during the first year, yellow afterwards, erect, with white waxy ring beneath nodes when young; nodes on some shoots congested and sometimes zigzag towards base; internodes swollen and often asymmetric. Culm-sheath light mauve-brown or pinkish brown, mottled dark purplish, especially towards apex; auricles and oral bristles 0; sheath-blade narrow linear, flat, often with greenish pink marginal band. Branches spreading widely to give an open network. Leaf-sheath with long oral bristles. Auricles 0. Ligule minutely puberulent. Leaf-blade usually 5–13–(15) × 0.7–2–(2.7) cm, lanceolate, abaxially glaucescent, acuminate.

Naturalised from south-eastern China.

N.: scattered in several places – North Auckland, Auckland City, Hamilton area and occasionally elsewhere in the Waikato, Bay of Plenty; K.: Raoul Id.

Phyllostachys aurea has been widely planted as far south as Canterbury but is not as aggressive there as in warmer areas. Flowering not reported for N.Z.

There are other spp. of *Phyllostachys* in N.Z. with yellow stems but *P. aurea* is easily recognisable by the swollen and asymmetric internodes and zigzag nodes in the lower part of some culms in a clump. The cultivated *P. edulis* (Carrière) Houz. 'Heterocycla' also has lower internodes on some culms strongly swollen on one side and grossly asymmetric, giving rise to zigzag nodes. However, the culm-sheaths are hairy abaxially in *P. edulis* but glabrous in *P. aurea*.

2. P. bambusoides Siebold et Zucc., *Abh. Akad. Muench. 3* (2): 745 (1843).

Dense thicket-forming with rhizomes running extensively. Culm 10–12 m, green, later yellow, curving over slightly towards apex; nodes with one prominent ridge, otherwise inconspicuous, none congested. Culm-sheath glabrous, light brownish yellow or greenish brown, mottled dark brownish; auricles present, at least when young; oral bristles deciduous; sheath-blade ± linear. Branches dense, primary branches at *c.* 45°. Leaf-sheath with conspicuous auricles and long oral bristles. Leaf-blade 8–14–(20) × 1–2.7–(4.5) cm, lanceolate or ovate-lanceolate, abaxially glaucescent. Inflorescence to 30 cm, narrowly paniculate. Spikelets subtended by sheath-like bracts with leafy blades. Glumes 2.3–2.7 cm, including apical bristle, ± purple, glabrous or nearly so.

Lemma and palea 1.5–2 cm, minutely puberulent, ± tinged purple, aristate. Lodicules *c.* 2 mm, ± obliquely lanceolate. Stamens long exserted; anthers *c.* 1 cm. Style and stigmas purple.

Naturalised from south and central China.

N.: Wanganui (east bank of Wanganui River, at base of St John's Hill where it has spread into a plantation forming a large stand).

Phyllostachys bambusoides has been widely planted in North Id and most probably has spread to a minor extent elsewhere. It is not as common in cultivation as *P. aurea* and *P. nigra*.

Phyllostachys bambusoides is the tallest sp. in the genus and is much taller in Asia than in N.Z. In N.Z. it is rarely as aggressive as the other two spp. treated here. The description of spikelets was compiled from cultivated specimens including fresh material from the latest flowering phase in 1995.

3. P. nigra (Lodd.) Munro, *Trans. Linn. Soc. 26*: 38 (1867).

Open thicket-forming with rhizomes running extensively. Culm 4–7 m (often much taller in cultivation), green, remaining so or becoming black or heavily blotched black in 2nd year, erect, with white waxy ring beneath nodes when young; nodes all ± evenly spaced. Culm-sheaths uniformly purple or purple-flushed, sometimes dark blotched above; auricles on main shoots prominent, dark; oral bristles long, caducous; sheath-blade linear, undulate. Branches green or black, spreading widely (giving an open network). Leaf-sheath with very short auricles. Oral bristles present. Leaf-blade usually 5–9–(13)×0.5–2–(3) cm, lanceolate, abaxially glaucescent, adaxially moderately shining, acuminate.

Naturalised from China.

Flowering in *P. nigra* may have occurred in N.Z. in the early 1960s. An incomplete flowering specimen collected in Wanganui at that time may be *P. nigra* but is too incomplete for precise determination.

　　var. **nigra**　　　　　　　　　　　　　　　　black bamboo

Culms becoming black or heavily blotched black in 2nd year. Culm-sheath often with dark blotches towards apex. Branches black.

N.: Hamilton (Awatere Avenue), Bay of Plenty (McClaren's Falls near Tauranga, Opotiki), Wanganui (base of St John's Hill). River banks, plantations.

A popular ornamental bamboo cultivated at least as far south as Canterbury. It is taller and more aggressive in warmer areas. The variation in black mottling on some culms recalls descriptions of forma *punctata* (Bean) Makino and cultivar 'Boryana', names sometimes applied to cultivated plants. It is impractical to recognise these as separate entities in stands treated as wild because wholly or partly black culms may occur in the same stand.

var. **henonis** (Mitford) Rendle, *J. Linn. Soc. Bot. 36*: 443 (1904).

Henon bamboo

Culms remaining green. Culm-sheath purple or purple-flushed, concolorous. Branches green.

N.: Wanganui (Wanganui River bank), Bay of Plenty (Kutarere). River bank.

Phyllostachys nigra var. *henonis* is not as widely cultivated in N.Z. as var. *nigra* and often is not recognised as being a variety of *P. nigra*; var. *henonis* is more vigorous than var. *nigra*. Var. *henonis* has usually been known as *P. mitis* in N.Z.

PLEIOBLASTUS Nakai, 1925

Very small to medium-sized with rhizomes monopodial, short or elongated. Culms dense or diffuse, sometimes very slender; nodes prominent; internodes terete, ± hollow, fistula sometimes very narrow. Culm-sheaths ± persistent, < internodes, concolorous; auricles 0; oral bristles glabrous. Branches usually 1–7 per node. Leaf-sheath with oral bristles usually present (at least when young). Leaf-blade usually conspicuously tessellate, abaxially usually partly glaucous or glaucesent and partly green. Inflorescence racemose or paniculate. Spikelets 5–13-flowered. Lodicules 3, ciliate. Stamens 3. Stigmas 3.

Type species: *P. communis* (Makino) Nakai

c. 20 spp. in China and Japan. Naturalised spp. 5.

Pleioblastus was formerly included in *Arundinaria* L. but the latter has been greatly reduced and some authors consider it to be a monotypic genus confined to North America.

Species of *Pleioblastus* are sometimes difficult to identify and *P. chino* has often been confused with the larger related sp. *P. simonii* (Carrière) Nakai (≡ *Arundinaria simonii* (Carrière) Rivière et

C.Rivière). Although no sp. of *Pleioblastus* has more than very minor status in the wild, the two variegated species treated here are sometimes aggressive and troublesome within gardens.

1 Leaf-blade concolorous; culm usually > *c.* 1.5 m high; branches > 5 at middle
and upper nodes *2*
Leaf-blade variegated; culm < 1.5 m high; branches 1–2 at middle and upper
nodes *4*

2 Leaf-blade 4–8 mm wide; length 15–22 × width; ligule 2–3 mm; leaf-sheath
upper margin oblique .... **3. gramineus**
Leaf-blade usually 7–15 mm wide; length to 10 × width; ligule 1–2 mm; leaf-
sheath upper margin horizontal *3*

3 Culm *c.* 2 m; internodes terete .... **2. chino**
Culm 2.5–4 m; uppermost internodes flattened on one side, lower and middle
internodes terete .... **4. hindsii**

4 Leaf-blade striped medium green and yellow or yellowish-green above, 1.5–
3 cm wide on main culms; leaf-sheath puberulent .... **1. auricomus**
Leaf-blade striped dark green and white or cream above, mostly 0.6–1.3 cm wide
on main culms; leaf-sheath glabrous except on margins .... **5. variegatus**

1. P. auricomus (Mitford) D.C.McClint., *Bamboo Soc. Newsl. 12*: 11
(1991).

Small, often forming a dense stand but not clump-forming; rhizomes running extensively. Culm usually 0.5–1.5 m, purple or green, very slender, puberulent when young, erect; internodes terete. Culm-sheath with few oral bristles; sheath-blade ovate, puberulent, ciliolate. Branches 1 at lower or middle nodes and 1–2 at upper nodes. Leaf-sheath finely puberulent. Ligule abbreviated. Oral bristles few and caducous. Leaf-blade usually 10–18 × 1.5–3 cm, narrowly elliptic-lanceolate, finely puberulent especially abaxially, either uniformly green or, more usually, longitudinally striped with pale green, greenish yellow or yellow, some leaves mainly yellow with narrow green streaks.

Naturalised from Japan.

N.: Auckland; S.: Nelson. Footpaths and garden surrounds, roadsides outside gardens.

Although indigenous to Japan it is only known in cultivation there. Flowering has not been reported from N.Z.

Also known in N.Z. as *Arundinaria auricoma* Mitford, *Bambusa viridi-striata* Regel, and *Pleioblastus viridi-striatus* (Regel) Makino.

2. P. chino (Franch. et Sav.) Makino, *J. Jap. Bot. 3*: 23 (1926).

Dense thicket-forming; rhizomes running. Culm usually 1.5–3 m, purplish when young, becoming deep green, arching over towards apex; internodes terete. Culm-sheath green, glabrous. Branches 3–6 from middle and upper nodes; main branches branching again and bearing additional small branchlets at the base, forming a dense cluster. Leaf-sheath glabrous, oblique at apex. Ligule 3–4 mm, appressed puberulent. Oral bristles few, caducous. Leaf-blade 6–18–(30) × 0.7–1.5–(2.3) cm, narrowly lanceolate, often yellowish green above, abaxially ± glaucescent, long acuminate. Inflorescence composed of small spikes (2–4 cm) at the tips of lateral branchlets, with main rachis densely pubescent. Spikes subtended by sheaths ≪ leaf-sheaths. Spikelets few, stongly antrorse, 1–1.5 cm. Glumes strongly tessellate.

Naturalised from Japan.

N.: Wanganui (Wanganui River bank); S.: Canterbury (Otahuna near Tai Tapu, Selwyn River bed). In old shrubberies, especially in moist habitats.

Flowering occurred in N.Z. in the 1962–1964 period and available specimens are from Tai Tapu, Canterbury. The general vigour of the plants does not seem to have been much affected.

Also known in N.Z. as *Arundinaria chino* (Franch. et Sav.) Makino and *A. simonii* var. *chino* (Franch. et Sav.) Makino.

3. P. gramineus (Bean) Nakai, *J. Arn. Arb. 6*: 146 (1925).

Rather dense thicket-forming; rhizomes running ± extensively. Culms *c.* 2 – *c.* 3 m, erect, green; internodes terete. Sheath-blades very narrowly triangular. Branches many from each middle and upper node. Leaf-sheath oblique at apex. Ligule 2–3 mm, minutely puberulent, often purplish. Leaf-blade 8–18–(22) × 0.4–0.8 cm (to 1.5 cm wide on strong young culms on outside of clump), linear-lanceolate, ± uniformly green, finely but conspicuously tessellate, long acuminate with finely tapering tips.

Naturalised from Japan, Ryuku Islands.

N.: Wanganui (east side of Wanganui River); Hauraki district (Hikutaia Stream). Established on river bank in Wanganui during the 1960s but that population was subsequently destroyed. Today this bamboo is only seen in a few specialist collections.

Pleioblastus gramineus has narrower leaves than any other species of *Pleioblastus* treated here, the ratio of length to width being usually 18–22 : 1. Leaves of this sp. are said to have a twist towards the apex, a feature which cannot be determined from dried material. Flowers not observed in N.Z.

4. P. hindsii (Munro) Nakai, *J. Arn. Arb.* 6: 146 (1925).

Medium-sized forming dense thickets, rhizomes running extensively. Culm usually 4–6 m, dark green at first, later often completely yellow, erect; nodes not prominent; internodes terete except for upper ones flattened on one side. Culm-sheath greenish, brownish green or brownish pink; sheath-blade green, 3–6 cm, linear-subulate. Branches ± erect, several at each node, forming dense clusters. Leaf-sheath apex oblique. Ligule short. Oral bristles conspicuous. Leaf-blade partly or wholly drooping, 6–15–(18) × 0.6–1.5–(2.2) cm, ± lanceolate, uniformly dark green, acuminate.

Naturalised from Japan.

N.: Wanganui, banks of Wanganui River, east side in and near Kowhai Park.

Pleioblastus hindsii was originally planted in the Wanganui R. area and now forms a dense narrow band for over a kilometre on top of the bank. Not known to be wild elsewhere in N.Z. although sometimes cultivated.

Flowers not reported from N.Z.

The record of *P. hindsii* from the Selwyn River bed in Canterbury, Healy, A. J. *in* Knox, G. A. (Ed.) *Nat. Hist. Canty* 310 (1969) is in error and is most probably based on *P. chino.*

Also known in N.Z. as *Arundinaria hindsii* Munro.

5. P. variegatus (Miq.) Makino, *J. Jap. Bot. 3*: 23 (1926).

Small with slender running rhizomes, sometimes plant forming dense stands. Culm 30–60 cm, green, very slender. Culm-sheaths green; oral bristles present. Branches 1–2 at middle and lower nodes or often culms unbranched. Leaf-sheath glabrous except on margins. Ligule very small. Leaf-blade usually 7–18 × 0.6–1.3 cm, narrow lanceolate, acuminate, dark green with prominent longitudinal white or pale cream stripes or bands of varying widths, abaxially uniformly pubescent.

Naturalised.

N.: Wanganui, sandy roadside near airport; Rotorua, Lake Okareka, roadside by old quarry.

Known in Japan only in cultivation.

Pleioblastus variegatus is not as common in cultivation as *P. auricomus* but both have almost certainly escaped in localities other than those recorded here. Flowers not reported in N.Z.

Also known in N.Z. as *Arundinaria fortunei* (Van Houtte) Rivière and *A. variegatus* (Miq.) Makino.

INCERTAE SEDIS   A very small bamboo cultivated at Lincoln in 1981 from a garden escape at Patutahi, Gisborne (CHR 367710), may be *P. humilis* (Mitford) Nakai. The material consists of one flower-bearing and several non-flowering culms, all under 40 cm high. The glabrous, uniformly green, leaf-blade and glabrous leaf-sheath (excepting oral bristles), distinguish it from the spp. of *Pleioblastus* treated here.

PSEUDOSASA Nakai, 1925

Tall or medium-sized with short or running monopodial rhizomes. Culms dense, forming thickets; nodes not prominent; internodes hollow, terete. Branches 1 per node, sometimes 2–3 at upper nodes, absent from lower nodes. Culm-sheath persistent, > internodes but ≤ upper internodes, unspotted; auricles absent; oral bristles usually absent, smooth if present. Leaf-sheath without auricles and oral bristles. Leaf-blade tessellate. Inflorescence a racemose panicle. Spikelets 3–8-flowered. Glumes 2. Lodicules 3. Stamens 3–(4). Stigmas 3. Fig. 1.

Type species: *P. japonica* (Steud.) Nakai

6 spp. in eastern Asia (3 in Japan, 3 in Taiwan). Naturalised sp. 1.

Pseudosasa is closely related to *Sasa* but the culms are taller than any *Sasa*, at least in N.Z. (wild or cultivated). The upper culm nodes of *Pseudosasa* often bear 2 or 3 branches but there is only one in *Sasa*. In N.Z. *P. japonica* produced flowers for many years until the early 1990s whereas no spp. of *Sasa* seem to have flowered here since the early 1970s.

1. P. japonica (Steud.) Makino, *J. Jap. Bot.* **2**: 15 (1920).

arrow bamboo

Medium-sized and forming dense thickets; rhizomes running extensively from periphery. Culm 3–5 m, dark green; banded white just below nodes. Culm-sheath green or greenish purple, eventually very pale, internodes almost all hidden in lower and middle parts of culm; ligule 2–3 mm, broad and prominent, truncate. Branches 1 per node, sometimes 2–3 at upper nodes. Leaf-sheath glabrous, often purplish above. Ligule minutely puberulent. Leaf-blade (1)–3–5–(7) on each branch, to 25–(30) × 3.5–(4) cm, broadly linear or linear-oblong, abaxially upper ⅓ green and lower ⅔ glaucescent, adaxially shining dark green, acuminate. Inflorescence ± purple, at least on exposed side. Spikelets 4–9 cm, flattened; florets to *c.* 12, distichous. Glumes 2. Lemma 10–13 mm, including short awn, ciliate, obscurely tessellate. Fig. 1.

Naturalised from Japan, south Korea.

N.: North Auckland (Cavalli Is, Ruawai), Auckland, Waikato (especially Hamilton area), Bay of Plenty, Rotorua, Wanganui, Manawatu, Wellington area; S.: Nelson (especially Nelson City area), Buller, Westland (southwards to Hokitika), Marlborough and Canterbury (usually coastal). Roadsides, riverbanks, in or around plantations, especially near garden boundaries, in scrub and on forest margins, abandoned garden sites and waste places.

Fig. 1 *Pseudosasa japonica*
I, inflorescence, × ½; **S**, spikelet, and **St**, stamens, × 3; **G₁**, **G₂**, glumes, × 4; **L**, lemma, × 4;
P, palea dorsal view, × 4; **Lo**, lodicules, × 4; **O**, ovary, styles and stigmas, subtended by
lodicules and anther filaments, × 10

Pseudosasa japonica is by far the commonest and most widespread naturalised bamboo in N.Z. and has escaped from cultivation on the Volcanic Plateau and in most lowland areas southwards except in southern areas of the South Id. This roughly corresponds to its main regions of cultivation. *Pseudosasa japonica* is also the sp. most commonly known to flower, and flowered abundantly during the 1980s. The thickets do not usually die completely after flowering although large parts often do. Flowering continues for several years but viable seed seems to be uncommon and no regeneration from seed so far has been reported.

SASA Makino et Shibata, 1901

Short or, more rarely, medium-sized with monopodial rhizomes. Culm diffuse, slender; nodes ± prominent with conspicuous white waxy bloom just below; internodes hollow, terete, ascending. Culm-sheath persistent, < internodes, unspotted. Branches 1 at each node, about as thick as culm, absent from lower and most middle nodes. Oral bristles usually present and scabrid throughout. Leaf-blade large, distinctly tessellate. Inflorescence paniculate. Spikelets 4–10-flowered. Glumes 2. Stamens 6. Stigmas 3.

Type species: *S. albo-marginata* (Miq.) Makino et Shibata

20–50 spp. in eastern Asia, the exact number varying according to the interpretation of generic limits. Naturalised sp. 1.

Although very few spp. of *Sasa* are cultivated in N.Z., they generally are amongst the hardiest bamboos, extending northwards naturally to extreme eastern Siberia and northern Japan.

1. S. palmata (Burb.) E.G.Camus, *Les Bambusées* 25 (1913).

Short with rhizomes extensively running, sometimes forming large stands but not dense clumps. Culm 1–3 m, permanently light green and ± concolorous or very strongly mottled and blotched dark purplish brown. Culm-sheath glaucous-white when young, at maturity obscuring much of the culm. Branches only from upper nodes. Leaf-sheath without oral bristles. Ligule 2–3 mm, truncate, densely puberulent. Leaf-blade ± horizontal, (10)–14–38 × (2)–4–9 cm, elliptic, narrowly ovate-elliptic, or lanceolate-elliptic, abaxially glaucous or glaucescent, glabrous, tessellate, adaxially shining deep green. Inflorescence of small panicles, each composed of a few short racemes, 2–4 cm. Spikelets 10–14 mm, 5–10-flowered. Lemmas 10–14 mm, tessellate.

Naturalised from Japan, south Korea, Sakhalin Id.

N.: Wanganui (base of St John's Hill and Kowhai Park); S.: North Canterbury (Rangiora). Old and neglected gardens and parks, in and around shrubberies or bamboo plantations.

Another sp. *S. veitchii* (Carrière) Rehder, of similar habit and form to *S. palmata*, has white-margined leaves which eventually become pale brown after maturity, and is cultivated occasionally but has not been reported wild. *Sasa palmata* last flowered in N.Z. in the second half of the 1960s and first half of the 1970s. Flowering was then widespread.

The name *Sasa senanensis* forma *nebulosa* (Makino) Rehder is incorrectly applied to a form of *S. palmata* with dark mottled culms. In the St John's Hill population of *S. palmata*, the culms were sometimes uniformly light green and sometimes dark-mottled, this being unconnected with age.

This sp. has also been known in N.Z. as *Bambusa palmata* Burb.

SASAELLA Makino, 1929

Small with rhizomes moderately to strongly monopodially branching, often forming large stands. Culms dense or diffuse, slender, erect from the base; nodes solid; internodes terete, thick-walled. Culm-sheaths deciduous, often tardily so, concolorous. Branches 1, rarely 2–3, at nodes, concentrated in middle and lower part of culm. Oral bristles of leaf-sheath scabrid at base, smooth above. Leaf-blade prominently tessellate. Inflorescence paniculate. Spikelets 5–10. Glumes 2. Lodicules 3. Stamens 6. Stigmas 3.

Type species: *S. ramosa* (Makino) Makino

13 spp. from Japan. Naturalised sp. 1.

1. S. ramosa (Makino) Makino, *J. Jap. Bot. 6*: 15 (1929).

Small bamboo, sometimes forming a dense stand; rhizomes long and mostly forming a tangled network. Culm to 1 m, erect, usually green or pale green, sometimes purplish; nodes with one slender branch or occasionally with a second very small branch. Culm-sheaths glabrous. Leaf-sheath hairy at first and with raised longitudinal ribs, becoming ± glabrous and smooth. Leaf-blade usually 7–13 × 0.8–2.2 cm, lanceolate with aristate apex, initially uniformly green, later often becoming whitish to pale brown near margins above; abaxially ± pubescent.

Naturalised from Japan.

S.: Christchurch, New Brighton, Beach Road. Old neglected garden on coastal sand, occupying many square metres.

Little planted in N.Z., this small bamboo is notorious in Europe under the name *Arundinaria vagans* for its very invasive rhizome system, which often extends far beyond the original planted stock. Although flowering has not been reported for N.Z. it has been noted in Europe from 1981 onwards.

Also known in N.Z. as *Arundinaria ramosa* Makino and *A. vagans* Gamble.

SEMIARUNDINARIA Nakai, 1925

Medium-sized with short monopodial rhizomes. Culms dense; nodes prominent; internodes hollow, lower and middle internodes terete, upper internodes flattened and slightly grooved on alternate sides above each node. Culm-sheath tardily deciduous. Culms with branches usually 3 per node when young, later clustered and up to 8. Leaf-sheath with oral bristles. Leaf-blade distinctly tessellate. Inflorescence of fascicled spikes, subtended by sheathing bracts; each spike consisting of 1 to few spikelets with 2–(7) florets. Glumes 0–1. Lodicules 3. Stamens 3. Stigmas 3.

Type species: *S. fastuosa* (Mitford) Nakai

c. 10 spp. in eastern Asia, 5 spp. in Japan. Naturalised sp. 1.

1.   S. fastuosa (Mitford) Makino, *J. Jap. Bot. 2*: 8 (1918).

Thicket-forming with rhizomes shortly creeping. Culm usually 4–6 m × 1.3–2 cm diam. below, upper internodes flattened, glaucous-white when young, becoming olive-green and often reddish brown on exposed parts, erect, semi-fastigiate. Culm-sheath glabrous, to *c.* 20

× 8 cm on large culms, abaxially greenish, or ± entirely green on small peripheral culms, adaxially shining light purple; ligule *c.* 2 mm, ciliate; oral bristles 0; sheath-blade 4–11 cm, narrow linear. Leaf-sheath with oral bristles. Ligule truncate. Leaf-blade 8–18–(21) × 1–2–(2.3) cm, linear-lanceolate, finely long acuminate. Spikes 3–7 cm, purplish. Lemma 1.8–2.2 cm, aristate. Lodicules 4–5 mm, ciliate. Anthers long-exserted, 7–8 mm.

Naturalised from Japan.

N.: Waikato (near Rangiriri). Roadside bank, also extending into pasture. Sometimes cultivated at least as far south as Canterbury, especially for a windbreak.

Flowering occurred in N.Z. in 1964–1965 and available specimens are from Canterbury. It is not known if the plants subsequently died.

II SUBFAMILY EHRHARTOIDEAE

Spikelets 1–3-flowered, 2 lower ∅ upper ♀ or ♂ and ♀; laterally compressed; disarticulating above reduced or 0 glumes; lemma 5–10-nerved, awned or awnless; palea hyaline 0–7-nerved; lodicules 2 or 0; stamens 6–1; stigmas plumose branches to style base or not. Inflorescence a panicle or raceme. Leaves with cross-veins or not.

Tribes: 2. Ehrharteae; 3. Oryzeae

2. EHRHARTEAE

Annual or perennial. Ligule usually membranous, sometimes reduced to a rim, or a row of hairs. Collar frequently with tuberculate hairs. Leaf-blade linear, flat to convolute or setaceous, cross-veins present or 0. Inflorescence a panicle or sometimes a unilateral raceme. Spikelets all alike, laterally compressed, 3-flowered with 2 lower florets ∅ and reduced to lemmas, upper floret ♀; disarticulation above glumes but not between florets; rachilla usually not prolonged. Glumes 2, persistent, < or > florets, sometimes very small, membranous. Lemma of ∅ florets coriaceous, awned or awnless, both, or at least the upper, = lemma of ♀ floret, awn, if present, ± strict, terminal; lemma of ♀ floret firmly cartilaginous to coriaceous, 5–7-nerved, entire, awnless. Palea hyaline, with 0, 1, 2, or rarely 3–5 nerves. Lodicules 2 or 0. Stamens 6 or 4, 3, 2, or 1. Caryopsis ellipsoid; embryo small; hilum linear, epiblast absent.

4 genera of temperate Australasia and South Africa.

The affinities of the tribe Ehrharteae are incompletely resolved but they are clearly an associate of the Oryzeae [Tateoka, T. *Nucleus 3*: 81–110 (1960), *Bot. Gaz. 124*: 264–270 (1963); Clayton and Renvoize (1986 op. cit.)]. The tribe was misplaced in the Bambusoideae [Kellogg, E. A. and Campbell, C. S. *in* Soderstrom, T. R., et al. (Eds) *Grass Systematics and Evolution* pp. 310–324 (1987); Soreng, R. J. and Davis, J. I. *Bot. Rev. 64*: 1–85 (1998)].

Willemse, L. P. M. *Blumea 28*: 181–194 (1982) treated *Microlaena* as part of a broader *Ehrharta*; names were already available there for two New Zealand species. *Petriella* Zotov non Curzi was similarly fused in *Ehrharta* where two names were also available. Here we accept three genera: *Ehrharta*, with four naturalised species; *Microlaena* with four species, indigenous and/or endemic; and *Zotovia* for three endemic species two of which Zotov, V. D. *T.R.S.N.Z. 73*: 235 (1943) had included in his invalid *Petriella*.

1 Branching extravaginal; perennial; stamens 2–4 .... **Microlaena**
  Branching intravaginal; annual or perennial; stamens 2 or 6 2

2 Spikelets shining, glabrous; stamens 2; leaf-blade cross-veins prominent;
  rachilla prolongation minute .... **Zotovia**
  Spikelets variously ornamented; stamens 6; leaf-blade lacking cross-veins;
  rachilla prolongation 0 .... **Ehrharta**

EHRHARTA Thunb., 1779 *nom. cons.*

Annual or perennial. Ligule membranous, short, or a ring of hairs. Leaf-blade flat to convolute. Inflorescence usually a contracted panicle, rarely a raceme, often secund. Spikelets pedicelled; disarticulation above glumes but not between florets; rachilla not prolonged. Glumes equal or unequal, usually ≤ spikelet and becoming widely reflexed. Lemmas of ∅ florets usually unequal, glabrous or hairy, one or both often transversely rugose, awned or awnless; lower lemma ≤ glumes, upper usually > glumes, boat-shaped, frequently with basal appendages or tufts of hairs. Lemma of ♀ floret usually shorter, awned or awnless, very often with a knob-like appendage at base forming a hinge with the appendage of the upper ∅ lemma. Palea ≈ lemma of ♀ floret, narrowly 2-nerved. Lodicules cuneate, finely nerved below. Stamens 6 or 3. Stigma branches aspergilliform. Caryopsis elliptic; embryo ¼–⅕ length of caryopsis; hilum ≈ caryopsis.

Type species: *E. capensis* Thunb.

c. 25 spp. of Africa. Naturalised spp. 4.

1 Glumes < florets; upper ∅ lemma rugose or hispid *2*
  Glumes ± enclosing florets; upper ∅ lemma hairy *3*

2 Upper ∅ lemma transversely rugose, glabrous; apex hooded .... **2. erecta**
  Upper ∅ lemma hispid, shining below; apex long awned .... **3. longiflora**

3 Upper glume glabrous; spikelet 5–7 mm .... **1. calycina**
  Upper glume hairy at margins; spikelet 10–20 mm .... **4. villosa**

1. E. calycina Sm., *Pl. Icon. Ined.* t. 33 (1790).

Erect, tufted perennials, 50–80 cm, with rather soft culms. Leaf-sheath finely striate, smooth, or rarely densely short-pubescent between nerves, often purplish. Ligule 1.5–3 mm, abaxially minutely pubescent-scabrid, lacerate. Collar deep purple-tinged, hairs few to *c.* 1.5 mm. Leaf-blade 4–15–(20) cm × 2–6 mm, flat, rather stiff, abaxially finely scabrid towards acute tip, adaxially glabrous, except for scattered hairs above ligule and near the closely scabrid margins. Culm *c.* 30–60 cm, sometimes slightly geniculate at base, nodes deep purple, internodes glabrous. Inflorescence a rather narrowly branched panicle, (7)–15–25 cm; branches and pedicels almost filiform, glabrous, often purplish. Spikelets 5–7 mm, pale green to whitish, sometimes purple-tinged. Glumes ± equal and ± enclosing spikelet, 4–6 mm, 5–7-nerved, firmly membranous, elliptic-oblong, subobtuse to subacute, glabrous. Lemmas of ∅ florets dissimilar in size and shape, hinge-like appendage at base, subcoriaceous, with long, straight hairs, lateral nerves faint; lower ∅ lemma 4–5 mm, > hairs,

3–5-nerved, narrow-linear, tip subobtuse to slightly mucronate, callus hairs few; upper ∅ lemma 5–6 mm, = hairs, 5–7-nerved, elliptic-oblong, with 2 minute, hyaline, oval flaps beside basal appendages, upper keel ± thickened and short-ciliate, tipped by a ± strong mucro to 1 mm. Lemma of ♀ floret 4–5 mm, 7–(9)-nerved, subcoriaceous, obtuse, awnless, glabrous to sparsely hairy. Palea hyaline, glabrous, except at base. Lodicules *c.* 1 mm, glabrous. Stamens 6; anthers 2.5–3.5 mm. Gynoecium: ovary 1.0 mm; stigma-styles 2 mm. Caryopsis *c.* 3 × 1 mm.

Naturalised from South Africa.

N.: Auckland (Waimauku), Manawatu (Bulls, Foxton, Waitarere Beach), Wairarapa (Porangahau). Stabilised sand dunes or sandy places.

2. E. erecta Lam., *Encycl. Méth. Bot. 2*: 347 (1786).   veld grass

Loosely tufted, soft-leaved, sometimes sprawling perennials (20)–30–80–(140) cm, with rather weak culms sometimes rooting at lower nodes. Leaf-sheath finely striate, glabrous or shortly hairy. Ligule 1–4 mm, glabrous, lacerate. Collar hairs 1–2.0 mm. Leaf-blades 6–16 cm × 2–7.5–(12) mm, flat, rather thin, short, fine sometimes scattered hairs especially near margin, finely nerved but midrib obvious abaxially, margins finely scabrid or hairy, tip acute. Culms (15)–20–70 cm, usually ± procumbent at base and geniculate-ascending above, or erect, internodes glabrous. Inflorescence usually a rather narrowly branched panicle (5)–10–16 cm; branches and pedicels slender, smooth to finely pubescent-scabrid, pedicels finely hairy. Spikelets 3–4 mm, very pale green. Glumes < lemmas, unequal, membranous, ovate, subobtuse, smooth, tip minutely ciliate; lower 1.5–2 mm, 3–5-nerved, upper 2.2–2.7 mm, 5-nerved. Lemmas of ∅ florets (2.5)–3–3.5 mm, coriaceous, elliptic-oblong, obtuse, awnless, faintly 5-nerved, glabrous; lower ∅ lemma < upper, usually smooth rarely transversely rugose near tip, shining; upper ∅ lemma transversely rugose above and with a thick basal hinge-like appendage. Lemma of ♀ floret 2.5–3.2 mm, faintly 7-nerved, subcoriaceous, ovate, obtuse, awnless, glabrous. Palea 2.5 mm, hyaline, glabrous, 2-nerved. Lodicules 0.5–0.7 mm, nerved, glabrous. Stamens 6; anthers 0.7–1.2 mm. Gynoecium: ovary 0.5 mm; stigma-styles 1.5 mm. Caryopsis 2–2.7 × *c.* 1 mm.

Naturalised from South Africa.

N.: throughout Auckland Province, Taranaki (Moturoa Id, Hawera), Wellington Province; S.: Nelson (Abel Tasman National Park), Canterbury (Akaroa, Christchurch, Ashburton). Stabilised sand dunes, and waste places, or ruderal.

In the last 10–15 years *E. erecta* has spread rapidly, especially in the Palmerston North and Wellington areas, where it has become abundant.

3. E. longiflora Sm., *Pl. Icon. Ined.* t. 32 (1790).   annual veld grass

Erect, tufted annuals, 60–100 cm. Leaf-sheath obviously nerved, keeled, submembranous, with sparse short hairs near base. Ligule *c.* 2 mm, deeply lacerate. Collar with hairs to 2.0 mm. Leaf-blade to 20 cm × 10 mm, flat, glabrous or with scattered long hairs, margins scaberulous. Culm 25–50–(80) cm, geniculate at base, internodes minutely scaberulous just below panicle and nodes. Panicle to 20 cm, narrow, open, with few, short, slender branches; pedicels filiform, long hairy. Spikelets 14–20 mm, light purplish green. Glumes unequal, < spikelets, membranous, margins ciliate; lower 2.5–3.5 mm, 5-nerved, elliptic, narrowed above to acute tip, upper *c.* 4.5 mm, 9-nerved, broadly ovate, tip mucronate. Lemmas of ∅ florets similar in shape, indistinctly 5–7-nerved, coriaceous, scabrid, conspicuously awned to *c.* 9 mm and with a hinge-like appendage at base; lower < upper, with 3 basal tufts of hairs to 2 mm, upper with 2 less conspicuous tufts. Lemma of ♀ floret 6–7 mm, 7-nerved, elliptic, subcoriaceous, glabrous, awnless. Palea 5–6 mm, subhyaline, glabrous but finely ciliate-scabrid on keel below. Lodicules 1–1.5 mm, hair-tipped. Stamens 6; anthers *c.* 1.5 mm. Gynoecium: ovary 1 mm; stigma-styles 1.75 mm. Caryopsis 4–5 × 1–1.5 mm.

Naturalised from South Africa.

N.: Wanganui. Waste places and weed of gardens and their surrounds; locally common.

4. E. villosa (L.f.) Schult. et Schult.f., *Syst. Veg.* 7: 1374 (1830).

pyp grass

Strongly rhizomatous perennials to 200 cm, with very stout naked culms branched above and bearing few leaves reduced to sheaths at base and at lower nodes, leaves at upper nodes longer, with involute blades. Leaf-sheath glabrous, finely many-nerved, light brown or purplish. Ligule *c.* 1 mm, truncate with ciliate fringe. Collar thick, hairs to 2 mm. Leaf-blade to 16 cm, strongly involute, abaxially glabrous, adaxially strongly ribbed and finely pubescent-scabrid, narrowed to fine, semipungent tip. Culm to 2 m, nodes purplish to blackish, internodes glabrous. Panicle to 25 cm, narrow, erect, rather lax; rachis and short, filiform, flexuose branchlets glabrous, fascicled below; pedicels filiform, smooth, becoming sparsely scabrid above. Spikelets 10–15 mm, oblong, very pale greyish brown. Glumes subequal, ± enclosing spikelet, 10–14 mm, 5–7-nerved, sub-membranous, oblong, subacute, glabrous with margins ciliolate near tip. Lemmas of ∅ florets similar in shape, 7-nerved, narrow-oblong, entire at base, coriaceous, densely villous, mucro to 2 mm; lower ∅ lemma < upper, callus hairs to 3 mm; upper ∅ lemma bearded on each side above the contracted base. Lemma of ♀ floret ≈ ∅ lemma, 7-nerved, oblong, subcoriaceous, truncate, less villous than ∅ lemmas. Palea 9 mm, membranous, glabrous. Lodicules *c.* 2 mm, glabrous. Stamens 6; anthers 6–8 mm. Gynoecium: ovary 1.0–1.3 mm; stigma-styles to 3.5 mm. Caryopsis *c.* 6.5 × 2.5 mm.

Naturalised from South Africa.

N.: Wanganui (Turakina Beach); Hawke's Bay (Taikura near Blackhead Point). Sand dunes and sandy places.

Originally planted at Turakina Beach in the early 1960s as a sand dune binder. A note on AK 216398 *A. E. Esler* Turakina Beach, 1969, indicated that it was showing a tendency to increase.

MICROLAENA R.Br., 1810

Rhizomatous, caespitose or scrambling perennial, branching at nodes. Branching extravaginal. Auricles 0–1–2, fringed with long cushion-based hairs. Spikelets laterally compressed, solitary, pedicelled, in open laxly branched panicles or slender racemes, falling entire from above persistent, unequal, 1–3-nerved, glumes; 3-flowered, 2 lower reduced to unequal, awned, scabrid or glabrous, indurated, 3–7-nerved, ∅ lemmas, the lowermost always with a bearded callus. ♀ floret < ∅ lemma, chasmogamous or cleistogamous; lemma prickle-toothed at apex otherwise glabrous, keeled, 3–5-nerved, shortly awned, mucronate or awnless. Palea membranous, 1-nerved, ≤ lemma, often enfolded by upper ∅ lemma. Lodicules 2, membranous, entire or lobed, strongly nerved, sometimes hairy on margins, 0 in cleistogamous flowers. Stamens 1–4; anthers tailed, much reduced in cleistogamous flowers. Styles free; stigmata plumose, naked below or branches to base. Caryopsis free, linear, compressed; embryo ± ⅙ of caryopsis; hilum linear = caryopsis. Axillary, single-spikeleted cleistogenes occur in *M. polynoda* and *M. stipoides*. Fig. 2.

Type species: *M. stipoides* (Labill.) R.Br.

c. 5 spp. native to Australasia, New Guinea, Indonesia, Pacific Islands. Endemic spp. 2, indigenous spp. 2.

The N.Z. spp. were revised by Edgar, E. and Connor, H. E. *N.Z. J. Bot. 36*: 565–586 (1998).

1 Callus hairs present on lower ∅ lemma; stamens 2–4 .... *2*
  Callus hairs present on both ∅ lemmas; stamens 2 .... *3*

2 Glumes ± equal and distant from base of florets .... **4. stipoides**
  Glumes unequal and covering base of florets .... **3. polynoda**

3 Lower ∅ lemma ⅔ as long as upper; usually caespitose .... **1. avenacea**
  Lower ∅ lemma ½ as long as upper; shoots with fine, elongate, long
  internodes .... **2. carsei**

1. M. avenacea (Raoul) Hook.f., *Handbk N.Z. Fl.* 320 (1864) ≡ *Diplax avenacea* Raoul, *Ann. Sci. Nat.* Ser. 3, Bot. *2*: 116 (1844) ≡ *Microlaena avenacea* (Raoul) Hook.f. var. *avenacea* (autonym, Zotov 1943 op. cit. p. 235) ≡ *Ehrharta diplax* F.Muell., *Fragm. 7*: 90 (1870); Holotype: P! *Raoul* Akaroa in umbrosis.   bush rice grass

Caespitose wide-leaved perennial from short rhizome with long culms and large inflorescences. Branching extravaginal, occasionally with stolons rooting at nodes; cataphylls broad and shining. Leaf-sheath smooth or sometimes scabrid; keel stout, often prickled towards apex. Ligule to 0.5 mm, membranous, crenate or shallowly toothed, broadly triangular, ciliolate. Auricle 1, ±5 mm, margin hairs 2–4 mm. Leaf-blade to 50 cm × 6–12 mm wide, bright green, keel strong, sometimes scabrid

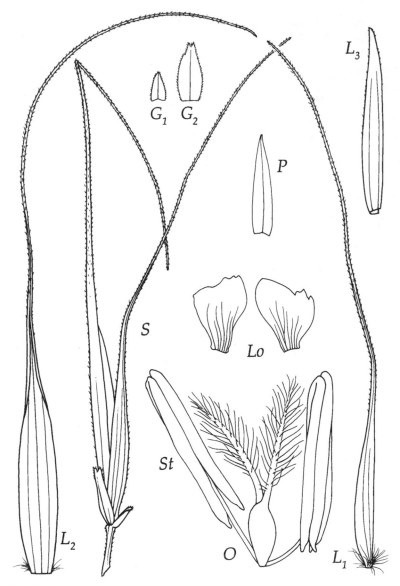

Fig. 2   *Microlaena avenacea*

S, spikelet, × 9; **G₁**, **G₂**, glumes, × 9; **L₁**, **L₂**, lemma of ∅ florets, × 9, **L₃**, lemma of ♀ floret, × 9; **P**, palea dorsal view, × 9; **Lo**, lodicules, × 24; **O**, ovary, styles and stigmas, × 24; **St**, stamens, × 24.

Plate 1. Native grasses dominating landscapes. **A** *Chionochloa antarctica* on subantarctic Auckland Island; **B** The snow tussocks *Chionochloa crassiuscula* subsp. *torta* (foreground) and *C. teretifolia* in subalpine zone, Borland Valley, Fiordland; **C** Fescue tussock, *Festuca novae-zelandiae*, in montane zone, Mackenzie Basin.

Plate 2. Naturalised grasses dominating landscapes. **A** Hay bales cut from a paddock that is largely perennial ryegrass, *Lolium perenne*, Motatapu, Otago; **B** Pasture with creeping bent grass, *Agrostis stolonifera*, bordering drain, Karamea; **C** *Spartina anglica* in estuarine bay head, Havelock, Marlborough Sounds.

below; sharp, fine, close-set, dense antrorse prickles on margins above becoming smooth below except for auricle hairs extending for *c.* 1 cm; abaxially rough occasionally scabrid, occasional prickle-teeth near ligule. Culm to 1.25 m; internodes laterally compressed and conspicuously, longitudinally grooved, smooth to finely prickle-toothed. Panicle 25–65 cm, nodes widely spaced, branches filiform, > internodes; nodes with short branches of close-set spikelets and long flexuous branches to 20 cm, naked below; rachis longitudinally grooved, finely scabrid to ± smooth, branches and pedicels finely prickle-toothed. Spikelets 18–30 mm, solitary, on short or long pedicels. Glumes unequal, 1-nerved occasionally 3-nerved, margins ciliate, entire or 2 or 3-toothed especially upper; lower to 1 mm, = callus hairs, upper 1–2.5 mm > callus hairs. Lower ∅ lemma 9–15–25 mm, 5-nerved, prickle-toothed especially on keel, adaxially hairy above, awn > ½ lemma length; callus hairs to 1 mm; upper ∅ lemma 12–23–27 mm, 3-nerved, scabrid above, awn > ½ lemma length; callus hairs 1–2 mm. ♂ lemma 5–7–12 mm, 3-nerved, awnless and ciliate or occasionally shortly (0.5–2 mm) awned from bifid apex; callus hairs few and short. Palea 3–4–5 mm, membranous, acute, upper margins and apex ciliate. Lodicules 2, 0.6–1.2 mm, > ovary, lobed occasionally flabellate, truncate and finely erose, occasionally lobes long hairy, persistent or falling with caryopsis. Stamens (1)–2, anthers (1)–2–3–(4) mm. Gynoecium: ovary 1 mm; stigma-styles 1–1.5–2.5 mm, styles naked below. Caryopsis 2–3.5–5 mm, often falling free; embryo 0.5–0.7 mm. $2n = 48$. Fig. 2. Plate 6C.

Indigenous.

N.: throughout; S.: Nelson and west of Main Divide to Fiordland, less frequent in east; St.; A. Forests; sea level to 1300 m.

Willemse, L. P. M. *Blumea 28*: 181–194 (1982) treated New Zealand, Fijian and Tahitian plants as *Ehrharta diplax* var. *diplax,* and those of New Guinea and Irian Jaya as var. *giulianettii* (Stapf) L.P.M.Willemse. The autonym *M. avenacea* var. *avenacea* was generated by Zotov, V. D. *T.R.S.N.Z. 73*: 235 (1943); the need for it does not arise here as varietal rank is not used.

2. M. carsei Cheeseman, *T.N.Z.I. 47*: 47 (1915) ≡ *M. avenacea* var. *carsei* (Cheeseman) Zotov, *T.R.S.N.Z. 73*: 235 (1943); Lectotype: AK 1308! *H. Carse* forests at Kaiaka, Mangonui County, Dec 1913 (designated by Edgar and Connor 1998 op. cit. p. 578).

Rhizomatous, perennial with fine elongate stem internodes with tufts of narrow leaves from which a long slender shortly-branched inflorescence emerges. Rhizome short; branching extravaginal; cataphylls 1–2 cm, bases swollen, shining, keeled, acute. Internodes elongate, slender, shining glabrous, stramineous to brown. Leaf-sheath glabrous, keel stout, margin membranous. Ligule 0.75–1 mm, broadly triangular. Auricles 2, margin hairs 1.5–2–2.5 mm. Leaf-blade 8–15–25 cm × 4–5–6 mm wide, bright green, keel stout; margins sharp, coarsely prickled in close-set columns, becoming smooth below except for some short, stiff hairs near ligule. Culm to 60 cm, laterally compressed, internodes longitudinally grooved, glabrous. Panicle 20–25–35 cm, narrow, slender, nodes ± close, branches ± appressed, erect, filiform, ≥ internodes; nodes with short branches of close-set spikelets and longer branches naked below; rachis longitudinally grooved, rachis, branches and pedicels with fine prickles. Spikelets 15–23 mm, numerous, solitary, on short or long pedicels. Glumes unequal, covering the base of spikelet, margins ciliate; lower 0.2–0.6 mm, 1-nerved, < callus hairs, upper 0.75–1.0–1.7 mm, 3-nerved, triangular acute or irregularly lobed or erose, > callus hairs. Lower ∅ lemma 5–10–16 mm, 3–5-nerved, prickles on keel and above smooth elsewhere, adaxially hairy above, awn ½ lemma length, sometimes ≤ ♂ lemma; callus hairs to 0.5 mm; upper ∅ lemma 14–20–25 mm, 3–5-nerved, densely prickle-toothed on keel and above, smooth elsewhere, adaxially hairy above, awn ½ lemma length; callus hairs to 1.5 mm. ♂ lemma 5–6–10 mm, 3–5-nerved, awn 0 or short (0.1–0.5 mm) between small lobes, margins ciliate above, apex deeply bifid (0.4–0.8 mm) and ciliate. Palea 3–3.5–4.5 mm, membranous, hyaline, 1-nerved, apex ciliate; often enclosed by upper ∅ lemma. Lodicules 2, 0.8–1.0–1.5 mm, > ovary, lobed, several nerves. Stamens 2; anthers 1.0–1.7–2.2 mm, yellow. Gynoecium: ovary 0.65–0.75 mm; stigma-styles 1.0–1.3–2.2 mm, naked below. Caryopsis 3–3.5–4 mm; embryo 0.5–0.6 mm. $2n = 48$.

Endemic.

N.: Northland, south of Muriwhenua peninsula to Waipoua forest and Trounson Park (35°00' – 35°40' S). Forest floor to 400 m.

3. M. polynoda (Hook.f.) Hook.f., *Handbk N.Z. Fl.* 320 (1864) ≡ *Diplax polynoda* Hook.f., *Fl. N.Z. 1*: 290 (1853) ≡ *Ehrharta multinoda* F.Muell., *Fragm.* 7: 90 (1870) *nom. illeg.*; Lectotype: K! *Colenso 1557* N. Zealand [base of Ruahine Range] (designated by Edgar and Connor 1998, op. cit. p. 580). = *Microlaena ramosissima* Colenso, *T.N.Z.I. 21*: 105 (1889); Lectotype: WELT 21966! *W. Colenso* Dannevirke H.B. (designated by Edgar and Connor 1998 op. cit. p. 580).

Scrambling, bambusiform, rhizomatous perennial with stout, bare internodes below but leafy above, emitting cataphyll-surrounded shoots at rhizome and at nodes; inflorescence short, infrequent; cleistogenes common along stems. Leaf-sheath keeled, glabrous, < culm internodes, margins membranous, apex with conspicuous tufts of hairs 1.5–3–5 mm sometimes absent. Ligule 0.2–0.3–0.5 mm, erose, ciliate. Auricles 0. Collar thick, fringed by some hairs from sheath apex. Leaf-blade 10–30 cm × 2–5 mm wide, becoming long and fine, midrib evident, glabrous; margins scabrid, prickles spaced throughout but fewer below. Culms to 1.5 m × 2–2.5 mm diam., many noded, glabrous or hairy, or densely long hairy below nodes, longitudinally grooved. Inflorescence a narrow raceme 6–19 cm, internodes short, 3–10 solitary spikelets, or panicle to 15 cm of short and long branches at nodes, > internodes, up to 20 spikelets; rachis densely hairy at nodes sometimes throughout, branches and pedicels hairy. Spikelets 14–17 mm, on short or long pedicels. Glumes unequal, 1-nerved, occasionally bifid, margins and apex ciliate; lower 0.5–1–2 mm, ≤ callus hairs, upper 2–3–5 mm, > callus hairs. Lower ∅ lemma 8–11–15 mm, 3–5-nerved, finely prickle-toothed, margin shortly (0.5–0.75 mm) hairy; adaxially finely hairy, awn < ⅕ lemma length; callus hairs 0.5–1.0 mm; upper ∅ lemma 12–14–18 mm, 5–7-nerved, keel prickle-toothed, margin ciliate, finely prickle-toothed, awn < ½ lemma length; callus hairs few. ♂ lemma 6–7.5–10 mm, 5–7-nerved, awn 0 or 0.3–0.5 mm from shortly lobed ciliate apex, finely prickle-toothed above, glabrous below, keel toothed. Palea 4.5–5.5–6.5 mm, membranous, becoming acute; keel, upper margins and apex long ciliate. Lodicules 2, 0.8–1.0–1.75 mm, irregularly lobed or entire, sometimes hair-tipped. Stamens 4; anthers 1.0–2.5–4.5 mm, yellow. Gynoecium: ovary 0.7–1.3 mm;

stigma-styles 1.0–2–3.5 mm, stigmatic branches to apex of ovary. Caryopsis 3–4.5–5.5 mm; embryo 1 mm. $2n = 48$. Chasmogamous and cleistogamous.

Endemic.

N.: Northland, offshore islands; central and eastern regions; S.: Nelson and in east from Marlborough, to Otago Peninsula but sporadic. Forest, scrub and rocky bluffs; sea level to 400 m.

Willemse (1982 op. cit.) compared *M. polynoda* with *M. stipoides* var. *breviseta* Vickery of Australia to which it bears no resemblance; *M. polynoda* is not a name "of doubtful application" as Willemse thought. This comment by Clayton and Renvoize (1986 op. cit. p. 77) is perplexing: "Cleistogenes occur in *E. stipoides* [Connor, H. E. and Matthews, B. A. *N.Z. J. Bot. 15*: 531–534 (1977)], where they have been mistaken for a distinct species".

Cleistogenes: Axillary cleistogenes in swellings at culm nodes are common — as many as seven at consecutive nodes — and preclude the possibility of axillary branches forming there. Cleistogenes are longer than aerial spikelets. Glumes, or bracts, 1 or 2, subtend the asymmetrically grooved or flattened spikelet. There are no basal hairs. Lower ∅ lemma 10–14–20 mm, awned, shining, smooth except for teeth on keel; upper ∅ lemma 10–15–20 mm awned, shining, glabrous. ♀ lemma 7–9.5–12 mm, shining, glabrous, awned. Palea 4.5–5.5–7 mm, narrow, hyaline, ciliate, sometimes adherent to caryopsis. Lodicules absent. Stamens 2–(3–4); anthers 0.4–0.7–1.3 mm. Gynoecium: ovary 1.0 mm; stigma-styles 1.5–2.5–2.8 mm. Caryopsis 2.8–5.5–7.0 mm, longer than in aerial flower. Pollen germinates in the loculi and pollen tubes grow out onto the stigmatic branches. Details are in Connor and Matthews (1977 op. cit.). Hubbard, C. E. *Icon. Pl. (5th Series) 3*: t. 3209 (1930) had recorded the cleistogenes much earlier almost as an aside to his description of *Cleistochloa subjuncea*. Specimens in cultivation of Marlborough Sounds provenance produced both aerial chasmogamous flowers and cleistogenes, as did plants from the Rakaia Gorge, Canterbury.

On some specimens only cleistogenes are present. For further discussion see Schoen, D. J. *Amer. J. Bot. 71*: 711–719 (1984) and Schoen, D. J. and Lloyd, D. G. *Biol. J. Linn. Soc. 23*: 303–322 (1984), and for a commentary see Connor, H. E. *in* Soderstrom, T. R. et al. (Eds) *Grass Systematics and Evolution* (1987).

This species has sometimes been confused with the endemic *Anemanthele lessoniana* (Stipeae) which is also bambusiform in its habit.

4. M. stipoides (Labill.) R.Br., *Prodr.* 210 (1810) ≡ *Ehrharta stipoides* Labill., *Nov. Holl. Pl. 1*: 91, t. 118 (1805) ≡ *Microlaena stipoides* (Labill.) R.Br. var. *stipoides* [autonym, Domin *Biblioth. Bot. 85*: 366 (1915)]; Holotype: FI! *Labillardière* Tasmania, Capite van - Dieman.

var. **stipoides**

Rhizomatous perennial with extravaginal branching, and much branched, multinoded stems of many cauline, often stiff hairy leaves, and often purpled spikelets on erectly branched narrow panicles. Branching extravaginal; rhizomes sometimes short; cataphylls chaffy. Leaf-sheath tight, short hairy or glabrous, overlapping node above, keel obscure. Ligule to 0.5 mm, finely erose, ciliate. Auricles 2, small, margin hairs (*c.* 3 mm) few. Collar short stiff hairy. Leaf-blade 5–25 cm × 2–5 mm wide, abaxially and adaxially abundantly stiff hairy sometimes more so abaxially, midrib prominent; margins with close-set, sharp teeth, retrorse below antrorse above, sometimes thickened, white, undulating. Culm to 85 cm, internodes many, glabrous, sometimes longitudinally grooved. Panicle 10–25 cm, narrow, slender, internodes short, branches erect to 5 cm, > internodes, rachis longitudinally grooved; branches and pedicels finely scabrid. Spikelets 25–35 mm, on slender finely scabrid pedicels; ∅ lemmas often purple or purple-suffused. Glumes ± equal, 1-nerved, entire, often purple, margins ciliate, distant from ∅ lemmas, < callus hairs; lower to 0.8 mm, upper to 1.5 mm. Callus 1–1.5–2 mm, clothed in hairs to 2 mm. Lower ∅ lemma 12–20–30 mm, 5–7-nerved, scabrid, margin short (0.25 mm) hairy, awn > ½ lemma length; upper ∅ lemma 17–26–35 mm, prickle-toothed throughout or glabrous below and prickled above, 5–7-nerved, awn > ½ lemma length; callus hairs 0. ♂ lemma 5–7–9 mm, 5–7-nerved, shining, smooth except for teeth on upper ½ of keel, shortly awned (0.1–1.0 mm) from shortly lobed ciliate apex, margins hyaline; when cleistogamous often included by upper ∅ lemma. Palea membranous, hyaline, produced into a mucro or becoming acute, apex ciliate; dimorphic, 4–6–8 mm in chasmogamous flowers, 2–3–4 mm in cleistogamous flowers. Lodicules 2, 1–2–2.7 mm in chasmogamous flowers occasionally hair-tipped, falling with caryopsis; 0 in cleistogamous flowers. Stamens 2–3–4; anthers 1.5–2.5–4.2 mm in chasmogamous flowers, (1)–2–3–4, 0.2–0.4–1.0 mm in cleistogamous flowers. Gynoecium: ovary 0.8–1.2 mm; stigma-styles 1.2–2.5–4.0 mm in chasmogamous flowers, 0.8–1.4–2.0 mm in cleistogamous flowers, stigma branches reaching ovary. Caryopsis laterally compressed, finely wrinkled, often falling free, 3.5–4.5–5.5 mm in chasmogamous flowers, 2–3.8–5.8 in cleistogamous flowers; embryo 0.6–1.0 mm. $2n = 48$.

Indigenous.

N.: throughout; S.: throughout, less frequent in Canterbury, Otago and Westland, absent in Fiordland; St.; K.: Raoul Id; Ch. Open lowland forest and manuka-kanuka scrub, ruderal and in rough pasture; sea level to 1300 m.

Also indigenous to Australia and Malesia.

Cleistogenes: Axillary single spikeleted cleistogenes occur regularly (Connor and Matthews 1977 op. cit.). Cleistogenes are smaller than aerial spikelets, asymmetrically grooved, awns 2–4 mm, lower ∅ lemma = fertile lemma, vesture prominent only on main nerve; callus and callus hairs very small. Anthers as for cleistogamous flowers; stigma-styles shorter. Caryopsis as for cleistogamous flowers.

There is every reason to believe that Australian plants are here as migrants, but it may be impossible to differentiate between them and indigenous plants; we are certain that *M. stipoides* var. *breviseta* Vickery, *Telopea 1*: 43 (1975), is not here.

REPRODUCTIVE BIOLOGY   Chasmogamous and cleistogamous flowering are recognised in *M. polynoda* and *M. stipoides* (Connor and Matthews 1977 op. cit.) where cleistogenes are present sometimes at high frequency. Cleistogamy also occurs in aerially borne inflorescences of *M. stipoides*. There are no estimates of the frequencies in natural populations of chasmogamous and cleistogamous plants.

Dimorphism in lodicule, anther and palea lengths in *M. polynoda* and *M. stipoides* are described by Edgar and Connor (1998 op. cit.) as well as the frequencies of stamen number. There is an interspecific dimorphism for the distribution of stigmatic hairs on the style; stigmatic hairs reach to the base of the style in *M. polynoda* and *M. stipoides* but in *M. avenacea* and *M. carsei* the styles are naked below. The aspergilliform stigmas found in *Ehrharta* are unknown in *Microlaena*.

ZOTOVIA Edgar et Connor, 1998

≡ *Petriella* Zotov, *T.R.S.N.Z. 73*: 235 (1943) non Curzi (1930).

Rhizomatous perennials, compact, cushion-like, with close-set leaves, sometimes forming tufts from trailing culms; branching intravaginal. Leaf-sheath cross-veinlets few. Auricles fringed with some long fine hairs, or auricles 0. Ligule a minute ciliolate rim, or ciliate. Leaf-blade flat or inrolled, cross-veinlets prominent. Inflorescence a contracted panicle or reduced to a raceme. Spikelets laterally compressed, shortly pedicelled, falling entire from above persistent subequal glumes; 3-flowered, 2 lower reduced to unequal awned ∅ lemmas; uppermost floret ♀, awnless; rachilla minutely prolonged. Glumes ≤ ½ length of florets, ovate, subcoriaceous, smooth; lower usually faintly 3-nerved, upper usually faintly 5-nerved. ∅ lemmas narrowly ovate-lanceolate, slightly narrowed above to a short, relatively thick, scabrid awn, cartilaginous, abaxially smooth or scarcely scabrid, adaxially minutely pubescent, keel smooth or scabrid, shortly hairy callus at base; lower ∅ lemma 5–7-nerved; upper ∅ lemma longer, usually 7-nerved. ♀ lemma ≥ lower ∅ lemma, ≤ upper ∅ lemma, slightly broader, ovate-lanceolate, awnless, subcoriaceous, smooth, but sometimes ± scabrid on midnerve, faintly 5–7-nerved, callus smooth. Palea membranous with 2 close-set central nerves. Lodicules 2, finely many-nerved in lower ½. Stamens 2. Styles free naked below; stigmata plumose. Caryopsis narrow-oblong; embryo *c.* ⅕ caryopsis; hilum ≈ caryopsis.

The flowers of *Z. colensoi* are correctly illustrated by Buchanan, J. *Indig. Grasses. N.Z.* t. 1 (1878) as are those of *Z. thomsonii* by the same artist in Petrie, D. *T.N.Z.I. 12*: t. 10 (1880).

Type species: *Z. colensoi* (Hook.f.) Edgar et Connor

3 spp., endemic to New Zealand; usually subalpine to alpine, in damp ground.

The treatment here is that of Edgar, E. and Connor, H. E. *N.Z. J. Bot. 36*: 568–586 (1998).

1 Plant trailing, (4)–15–25–(45) cm; inflorescence a curved or drooping contracted panicle .... **2. colensoi**
Plant forming low mats 2–4–(8) cm; inflorescence a small erect few-spikeleted raceme *2*

2 Ligule a minute rim, sparsely ciliolate; spikelet pedicels glabrous; glumes 1–2 mm, ± unequal .... **3. thomsonii**
Ligule tapered, obviously ciliate; spikelet pedicels scabrid-pubescent; glumes 2–3.5 mm, ± equal .... **1. acicularis**

1. Z. acicularis Edgar et Connor, *N.Z. J. Bot. 36*: 569 (1998); Holotype: CHR 108276! *L.J. Metcalf* Lyall Bay, Thompson Sound, Fiordland, wet shady cliffs, *c.* 2200 ft, 22 Jan 1958.

Small tufts 3–13 cm from long fine rhizomes, of many dense fine leaves with needle-like tips subtending short racemose inflorescences at first partly concealed among leaves, later borne well above on elongate stems; branching intravaginal. Leaf-sheath to 8 mm, keeled, glabrous, wider than blade, lobed apically. Auricles 0. Ligule *c.* 0.5 mm, tapered and longer at centre, noticeably ciliate. Leaf-blade 1–2.6 cm × 0.5 mm diam., tightly inrolled, light green with cream cross-veinlets, glabrous, pinched into curved, pale tip; margins with minute prickle-teeth. Culm very slender, minutely pubescent below inflorescence. Raceme *c.* 1 cm, of 3–6 spikelets; pedicels with dense, minute prickle-teeth. Spikelets 5–7.5 mm, laterally compressed. Glumes ± equal, 2–3.5 mm, *c.* ½ length of spikelet, lanceolate, subacute, keeled above, glabrous, margins with minute prickle-teeth; lower 1-nerved, upper 3-nerved. ∅ lemmas unequal, lanceolate, long-narrowed above, keel and margins minutely prickle-toothed; lower ∅ lemma 5–5.5 mm, 5-nerved, callus hairs 1–1.5 mm; upper ∅ lemma 6–6.8 mm, 3-nerved, callus hairs 1.5–2 mm. ♀ lemma 3–4 mm, narrow elliptic, glabrous, tip truncate, erose to almost lacerate, ciliate. Palea 2.5–2.75 mm, apex ciliate. Lodicules *c.* 0.75 mm, flabellate, erose. Anthers 1.1–1.5 mm. Gynoecium: ovary *c.* 0.6 mm; stigma-styles *c.* 1 mm, stigmas plumose. Caryopsis not seen.

Endemic.

S.: Fiordland: Thompson Sound (Lyall Bay), Secretary Id (All Round Point), and Lake Mike. Rock or wet shady cliffs; 670–1100 m.

The membranous tapered ciliate ligule of *Z. acicularis* contrasts markedly with the scarcely visible, rim-like sparsely ciliolate ligules of *Z. colensoi* and *Z. thomsonii*. The upper part of the flowering culm and the spikelet-pedicels are quite densely minutely pubescent or prickle-toothed in *Z. acicularis* whereas they are smooth or almost so in the other two species of *Zotovia*. The glumes in *Z. acicularis* are *c.* ½ the length of the spikelet whereas in *Z. colensoi* and *Z. thomsonii* the glumes are *c.* ⅓ the length of the spikelet.

2. Z. colensoi (Hook.f.) Edgar et Connor, *N.Z. J. Bot. 36*: 571 (1998) ≡ *Ehrharta colensoi* Hook.f., *Fl. N.Z. 1*: 288, t. 65A (1853) ≡ *Microlaena colensoi* (Hook.f.) J.C.Sm., *T.N.Z.I. 43*: 253 (1911)

≡ *Petriella colensoi* (Hook.f.) Zotov, *T.R.S.N.Z. 73*: 236 (1943) *comb. illeg.*; Holotype: K! *Colenso 1568* N. Zealand, [dry short grass, growing in tufts near top of Ruahine Range].

Perennial, (4)–15–(45) cm, forming narrow tufts with yellow-green, papery leaves from long, trailing culms, narrowly branching intravaginally and bare below except for remains of old sheaths. Leaf-sheath smooth, cross-veinlets ± visible, cream. Ligule 0.1–0.2 mm, a truncate minutely ciliolate rim. Leaf-blade 4–10 cm × 1.5 mm, ± erect, linear, flat, becoming inrolled, ribs and cross-veinlets cream, abaxially smooth, adaxially minutely scabrid on ribs, long-tapering to acicular tip, margins minutely scabrid. Panicle (1)–3–5–(7) cm, loosely spike-like, curved or drooping; branches smooth to slightly scabrid. Spikelets (6)–7–10 mm, shortly pedicelled. Glumes obtuse to acute, to truncate and shortly mucronate, margins finely ciliate above; lower 2–3.7 mm, upper 2.5–4 mm. ∅ lemmas abaxially smooth, sometimes cross-veinlets visible, ± scabrid on keel, adaxially minutely pubescent above, awn 2–3 mm, scabrid; callus hairs dense, 2–3 mm; lower ∅ lemma 5–6–(8) mm; upper ∅ lemma (5)–6–9 mm. ♂ lemma 4–5.5 mm, obtuse tip ciliolate, occasionally obliquely truncate and very shortly mucronate, keel smooth. Palea 2.5–4 mm. Rachilla prolongation to 0.7 mm, minutely hairy at apex, or reduced to a minute knob tipped by fine hairs. Lodicules 1–1.5 mm, cuneate, erose to lacerate. Anthers *c.* 2 mm. Gynoecium: ovary to 1 mm; stigma-styles to 1.5 mm. Caryopsis *c.* 2.5 × 1 mm. $2n = c.$ 48.

Endemic.

N.: Pouakai Range, Taranaki, north-eastern and southern mountains; S.: west of Main Divide extending east into Canterbury and Otago. Snow-tussock grassland, on margins of snow-patch areas and at forest margins, often hanging down damp rocks near streams or on wet shaded banks or cliffs, occasionally in herbfields; 1000–2000 m. FL Nov–Feb. FT Jan–May.

The leaf-blades are shorter and narrower than usual, 1–3 cm × 0.4–0.7 mm, in two collections from Nelson in rock crevices: CHR 180026 *M. J. A. Simpson 5091* Fifth Basin, Travers Range, 14 Feb 1967; CHR 204960 *M. J. A. Simpson 5705* Waiau Pass, Spenser Range, 12 Jan 1970. The spikelets in these plants are small, *c.* 6 mm, with all parts proportionally smaller than usual.

There are five Petrie sheets of *Z. colensoi* with rather short leaves and emerging panicles collected from the Longwood Range, Southland: WELT 76455a,b, CHR 4077, CHR 6802, CHR 292290. These create the impression of a local population of such plants, but all are identical and undoubtedly from a single plant.

3. Z. thomsonii (Petrie) Edgar et Connor, *N.Z. J. Bot. 36*: 573 (1998) ≡ *Ehrharta thomsonii* Petrie, *T.N.Z.I. 12*: 356, t. 10 (1880) ≡ *Microlaena thomsonii* (Petrie) Petrie, *in* Chilton *Subantarct. Is N.Z. 2*: 472 (1909) ≡ *Petriella thomsonii* (Petrie) Zotov, *T.R.S.N.Z. 73*: 236 (1943) *comb. illeg.*; Lectotype: WELT 76439a! *D. Petrie* Frazer Peaks, Port Pegasus, Stewart Id, 1878 (No 1018 to Hackel) (designated by Edgar and Connor 1998 op. cit. p. 573).

Rhizomatous perennials, with culms branching freely intravaginally and forming light green, dense, low mats (1.5)–2–4–(8) cm high. Leaf-sheath smooth, usually cream, cross-veinlets few. Ligule almost a line. Leaf-blade 0.5–1.8 cm × 1.4–2–(3) mm, spreading, usually flat, margins incurved but sometimes more inrolled, lanceolate, glabrous, cream cross-veinlets visible, only slightly tapering to short, acute tip, margins minutely scabrid. Culm erect, only slightly projecting beyond leaves at flowering but later elongating to *c.* 8 cm. Inflorescence an erect raceme 0.5–1–(1.5) cm, of (2)–4–7 spikelets, occasionally reduced to a single spikelet. Spikelets (3.5)–4.5–6 mm, on short, smooth pedicels appressed to rachis. Glumes obtuse, margins smooth, rarely ciliate above; lower 1–1.5 mm, upper 1.2–2 mm. ∅ lemmas abaxially smooth, sometimes cross-veinlets visible, adaxially minutely pubescent above, awn scabrid, 1–1.5 mm; callus hairs dense, *c.* 1 mm; lower ∅ lemma 3–4.5 mm; upper ∅ lemma 3.5–5.5 mm. ♀ lemma 2.3–5 mm, acute or obliquely truncate, occasionally mucronate, smooth, keel sometimes minutely scabrid near tip. Palea 2.5–3.5 mm, keel scabrid near base. Rachilla prolongation to 0.5 mm, or reduced to a minute knob. Lodicules *c.* 0.8–1 mm, ± elliptic to cuneate, erose. Anthers 1–1.7 mm. Gynoecium: ovary *c.* 0.8 mm; stigma-styles 1–1.3 mm. Caryopsis 2–2.5 × *c.* 0.7 mm. $2n = c.$ 48.

Endemic.

S.: north-west Nelson to southern Westland, very local, more common in Fiordland and Southland; St.; A (one record only). Bogs, wet, open, scrub-tussock grassland, wet rock crevices, and cushion herbfield; 350–1500 m, nearly to sea level in Stewart Id. FL Aug–Jan. FT Oct–Mar.

3. ORYZEAE

Herbaceous. Ligule membranous. Leaf-blade usually linear without cross-veins. Inflorescence a panicle with ♂ spikelets, or plant monoecious and spikelets unisexual. Spikelets 1-flowered, or 3-flowered with 2 lower florets ∅ and reduced to lemmas (not in N.Z.), usually laterally compressed; rachilla not prolonged. Glumes 0, or as obscure lips at tip of pedicel. Lemma membranous to coriaceous, 5–10-nerved, with straight terminal awn or awnless. Palea resembling lemma, 3–7-nerved; keel central. Lodicules 2. Stamens usually 6, sometimes fewer. Stigmas 2, plumose. Caryopsis linear to ovoid.

1 Spikelets with ♂ flowers, strongly laterally compressed .... **Leersia**
  Spikelets unisexual, subterete or weakly flattened .... **Zizania**

LEERSIA Sw., 1788 *nom. cons.*

Perennials or annuals, rhizomatous, stoloniferous or tufted. Ligule membranous, entire or lacerate, usually glabrous. Leaf-blade flat, scabrid. Culm erect, or decumbent and straggling, terete or compressed. Panicle open or contracted, enclosed by leaf-sheaths or exserted, terminal or axillary. Spikelets 1-flowered, ♂, strongly laterally compressed; lower lemmas suppressed, rachilla modified to a short callus-like segment at base of deciduous floret; disarticulation above rudimentary glumes. Lemma chartaceous to coriaceous, keeled, 5-nerved with the outer lateral nerves ± at margins, usually awnless, keel and margins stiffly toothed. Palea similar in texture to lemma, much narrower, margins tightly enclosed by lemma margins. Stamens 6, or 3, 2, or 1.

Type species: *L. oryzoides* (L.) Sw.

c. 18 spp. of tropical and subtropical regions. Naturalised sp. 1.

1. L. oryzoides (L.) Sw., *Prodr. Veg. Ind. Occ.* 21 (1788).

rice cutgrass

Rhizomatous, yellow-green erect perennials. Leaf-sheath rounded, retrorsely scabrid between the fine nerves. Ligule *c.* 1 mm, chartaceous, truncate, or erose. Leaf-blade 9–20 cm × 7–10 mm, flat, finely prickle-toothed on both surfaces and on margins, finely pointed. Culm to 50 cm, internodes glabrous; nodes pubescent. Panicle 10–14 cm, enclosed at base by uppermost leaf-sheaths; branches filiform, flexuous, naked in lower ⅓. Spikelets *c.* 5 mm, elliptic-oblong, very pale green to later dark brown, shortly pedicelled, falling entire at maturity. Glumes reduced to a minute cupule at tip of pedicel. Lemma *c.* 5 mm, cymbiform, nerves 5, confluent at apex, firmly membranous, stiffly toothed on strong keel and scattered elsewhere, abruptly narrowed above to a short, blunt tip. Palea ≈ lemma, texture similar, hyaline, 3-nerved, keel stiffly long toothed. Stamens 3, sometimes 2; anthers *c.* 1.5 mm, 0.5 mm in cleistogamous flowers. Gynoecium: ovary glabrous; stigmas plumose, laterally inserted. Caryopsis *c.* 3 × 1.2 mm, laterally compressed, whitish to dark brown; embryo small; hilum linear, ≈ caryopsis.

Naturalised from North America and Europe.

N.: South Auckland (along Waikato R. and on margins of Lakes Waipapa, Arapuni, and Karapiro). Damp ground.

In all specimens seen the panicles had emerged from the leaf-sheaths and spikelets were chasmogamous. Cleistogamous spikelets with anthers 0.4–0.7 mm, in panicles enclosed by the leaf-sheaths, occur in CHR 467467 *P. J. de Lange* Waikato River, Mar 1990.

ZIZANIA L., 1753

Tall, aquatic, monoecious annuals or perennials, sometimes with creeping rhizomes. Leaves flat. Inflorescence a large, terminal panicle with lower branches spreading or ascending and bearing pendent, early deciduous ♂ spikelets, upper branches ascending to erect at maturity, bearing ascending, tardily deciduous ♀ spikelets. Spikelets 1-flowered, falling entire from the pedicel; lower lemmas suppressed, rachilla modified to a callus-like structure at base of floret. Glumes obsolete, a small, collar-like ridge at tip of pedicel. ♂ spikelets: lemma membranous, 5–7-nerved, acuminate to shortly awned; palea ≈ lemma, 3-nerved; stamens 6. ♀ spikelets: lemma chartaceous, 3-nerved, tapering to a long slender awn; palea 2-keeled, the keels tightly clasped by lemma margins. Lodicules 2. Styles fused, stigma narrowly plumose. Caryopsis linear; embryo ½ length caryopsis to ≈ caryopsis; hilum ≈ caryopsis. Fig. 3.

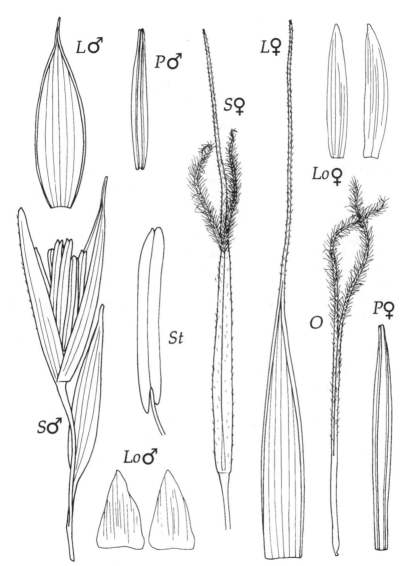

Fig. 3   *Zizania latifolia*
S♂, spikelet of ♂ floret, × 4.5; **S♀**, spikelet of ♀ floret, × 4.5; **L♂**, lemma of ♂ floret, × 4.5; **L♀**, lemma of ♀ floret, × 4.5; **P♂**, palea of ♂ floret dorsal view, × 4.5; **P♀**, palea of ♀ floret dorsal view, × 4.5; **Lo♂**, lodicules of ♂ flower, × 20; **Lo♀**, lodicules of ♀ flower, × 20; **O**, ovary, styles and stigmas of ♀ flower, × 4.5; **St**, anther of ♂ flower, × 9.

Type species: *Z. aquatica* L.

3 spp. of Asia and North America, in marshes and shallow water. Naturalised sp. 1; transient sp. 1.

1 Plants perennial, rhizomatous; usually ♂ flowers above, ♀ below, sometimes
    mixed .... **1. latifolia**
    Plants annual, tufted; ♀ flowers above, ♂ flowers below .... **palustris**ζ

1. Z. latifolia (Griseb.) Stapf, *Kew Bull. 1909*: 385 (1909).

Manchurian wild rice
Robust, stout perennials, to 3 m, with strong rhizomes and large, rather narrow, purplish panicle. Cataphylls broad. Leaf-sheath glabrous, many-nerved, spongy and honey-combed with numerous transverse septa beneath the stiff epidermis, pale greenish, straw-coloured or light brown to purplish; margins papery. Ligule 25–40 mm, nerved, smooth, acute, entire or later fimbriate. Collar conspicuous abaxially and separating sheath and blade. Leaf-blade 50–120 × 1–2.5 cm, linear-lanceolate, light green, tough, midnerve much thickened abaxially below, glabrous, margins scabrid; tip pungent. Culm to 2 cm diam., internodes glabrous. Panicle 30–70 cm, with sparsely scabrid pulvinate branches with long white hairs in axils, all branches and spikelets ascending, upper branches with ♀ spikelets, pedicels slender, capillary, tips cupule-like equally wide, sometimes short-ciliate; lower branches with ♂ spikelets and also ♂ spikelet at times among ♀ spikelets in middle branches. ♂ spikelets: lemma 12–15 mm, nerves 7, raised, purplish, linear-lanceolate, membranous, very finely scabrid on and between nerves, long-acuminate to shortly (*c.* 2 mm) awned. Palea *c.* 10.5 mm, 3-nerved, sparsely scabrid on keel and between nerves above, margins wide, hyaline. Lodicules 0.75 mm. Stamens 6; anthers (5)–6–7 mm, bright orange-yellow, tailed. ♀ spikelets: lemma 15–20 mm, 7-nerved, linear, chartaceous, greenish to sometimes slightly purplish, scabrid on nerves and very finely scabrid between nerves, awn scabrid, erect 12–30 mm. Palea 10–13 mm, 3-keeled, nerves scabrid, interkeels faintly scabrid. Gynoecium: ovary 1.5 mm, beaked; stigma-styles *c.* 20 mm, stigma white exserted beyond lemma apex. Caryopsis *c.* 6 × 1.5 mm. Fig. 3.

Naturalised from eastern Asia.

N.: North Auckland, Waikato, Hauraki (Turua), Wellington (Waikanae). Lagoons, river banks or tidal flats, roadside ditches and damp paddocks.

A notorious aquatic reed especially around Dargaville.

ζZ. palustris L.   Annual, erect tufts, *c.* 150 cm, rooting at nodes, with flat leaves and large rather narrow panicles. Leaf-blades 20–40 cm × 3–10 mm, with strong midrib abaxially; margins finely toothed. Panicle 30–40 cm, upper branches with ♀ spikelets, lower branches with ♂ spikelets. A North American species found once in a waterway CHR 156971A,B *F. Johansen* Tirohia–Rotokohu Drainage Board, Paeroa, 7.12.1966. It is now eradicated [Esler, A. E., Liefting, L. W. and Champion, P. D. *Biological Success and Weediness of the Noxious Plants of New Zealand*, M.A.F. (1993)]. First recorded for N.Z. under the name *Z. aquatica* L. by Healy, A. J. *Standard Common Names for Weeds in N.Z.* 141 (1969).

III SUBFAMILY POOIDEAE

Spikelets 1–many-flowered, laterally compressed. Lemma several-nerved, awn apical or dorsal, or 0. Lodicules 2–3, membranous or hyaline. Stamens 3, rarely fewer. Stigmas usually 2. Caryopsis usually ellipsoid to fusiform; embryo ⅓–⅙ length of caryopsis; hilum linear to round. Herbs, usually with linear leaf-blades. Ligule membranous. Inflorescence usually a panicle or raceme.

Tribes: 4. Nardeae; 5. Stipeae; 6. Poeae; 7. Hainardieae; 8. Meliceae; 9. Agrostideae; 10. Bromeae; 11. Brachypodieae; 12. Hordeeae

4. NARDEAE

Perennials. Ligule membranous. Leaf-blade setaceous, involute. Inflorescence a unilateral spike, the spikelets all alike and borne edgeways to rachis, with the lemma opposed to a shallow hollow in the tough rachis but not embedded. Spikelets 1-flowered, ♀, dorsally compressed; disarticulation above glumes; rachilla not prolonged. Glumes persistent, very reduced; lower very small, upper suppressed or almost so. Lemma 3-nerved; awn terminal. Palea 2-keeled. Lodicules 0. Stamens 3. Ovary glabrous; style 1. Caryopsis fusiform; embryo small; hilum linear.

NARDUS L., 1753

Monotypic Eurasian genus. The single species is naturalised in New Zealand, as well as in Tasmania and North America.

1. N. stricta L., *Sp. Pl.* 53 (1753).   mat grass

Stiff, dense, bright green or greyish green perennial tufts, 15–30–(40) cm, from shortly creeping rhizome, sometimes forming large swards; branching intravaginal. Leaf-sheath coriaceous, strongly ribbed, glabrous, shining, pale greyish brown or creamy. Ligule 0.5–1.4 mm, truncate, entire, abaxially with sparse prickle-teeth. Leaf-blade 6–18 cm × *c.* 0.5 mm diam., tightly involute, hard, pungent, spreading at right angles to sheath at maturity, abaxially smooth, or with minute prickle-teeth or hairs beside ribs, adaxially minutely papillose on ribs. Culm 10–32 cm, slender, erect, internodes glabrous, or minutely hairy or prickle-toothed. Spike 4.5–7.5 cm, secund, very slender; rachis concave-convex, smooth but margins finely scabrid, produced as a fine scabrid bristle 2–9 mm. Spikelets 6–10.5 mm, 1-flowered, in 2 rows along concave side of rachis, loosely to closely overlapping, sessile in notches on rachis, green or usually purplish. Glumes persistent, much reduced; lower a ± triangular, sometimes short-ciliate wedge, projecting upwards from rim of rachis-notch, *c.* 0.3–0.6 mm, upper usually 0. Lemma = spikelet, 3-nerved with central nerve inconspicuous and lateral nerves prominent, subcoriaceous with wide membranous margins, narrow-lanceolate or lanceolate-oblong, smooth to minutely scabrid above with some longer prickle-teeth on lateral nerves; awn 1.6–3 mm, terminal, straight. Palea slightly < lemma, membranous, apex rounded and very short-ciliate, keels 2. Callus ringed by minute hairs. Lodicules 0. Stamens 3; anthers 3.5–4 mm and fertile, or mostly 1–2 mm and pollenless. Gynoecium: ovary *c.* 1 mm, glabrous; stigma-style 5–6 mm. Caryopsis *c.* 3 × 0.5 mm, fusiform; embryo *c.* ⅙ caryopsis; hilum *c.* ⅓ caryopsis. Fig. 4.

Naturalised from Eurasia.

N.: Volcanic Plateau (Waimarino); S.: scattered and local, not recorded from Westland and Fiordland. Swampy pasture and damp ground.

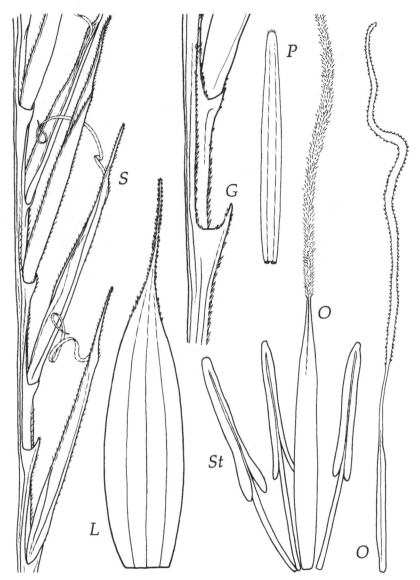

Fig. 4 *Nardus stricta*
S, portion of spike, showing secund spikelets, × 10; **G**, portion of spike showing lower glume, × 20; **L**, lemma, × 12; **P**, palea dorsal view, × 12; **St**, stamens × 20; **O**, ovary, style and stigma, × 12.

5. STIPEAE

Usually wiry, often bamboo-like perennials, or less frequently annuals. Ligule membranous. Leaf-blade setaceous to linear, usually convolute or folded. Inflorescence an open or contracted panicle. Spikelets 1-flowered, all ♀, terete to laterally compressed, or sometimes dorsally compressed; disarticulation above glumes; rachilla not prolonged. Glumes 2, persistent, usually > floret, hyaline to membranous, 1–7-nerved, acute to long-acuminate. Lemma (3)–5–9-nerved, membranous to crustaceous, often strongly indurate in fruit, terete to lenticular, usually enclosing palea; awn mostly once or twice geniculate, terminal from bidenticulate or entire lemma-tip. Palea usually ≈ lemma, hyaline to membranous, usually without keels, nerves 2 or 0, usually acute. Lodicules (2)–3. Stamens (1)–3. Ovary glabrous; styles 2, plumose. Caryopsis usually fusiform; embryo small; hilum linear, ½–≈ caryopsis, very rarely elliptic and < ½ caryopsis.

The disposition of genera as set out by Jacobs, S. W. L. and Everett, J. *Telopea* 6: 579–595 (1996) for Australia is accepted here; it follows from Barkworth, M. E. and Everett, J. *in* Soderstrom, T. R. et al. (Eds) *Grass Systematics and Evolution* 251–264 (1987) and realigns taxa discussed for New Zealand by Jacobs, S. W. L. et al. *N. Z. J. Bot.* 27: 569–582 (1989).

1 Margins of lemma overlapping, lemma 5–7-nerved 2
  Margins of lemma contiguous, lemma 3-nerved 3
2 Palea ≈ lemma, 2-nerved; lemma coriaceous .... **Austrostipa**
  Palea ⅓ lemma, nerveless; lemma silicified, tuberculate .... **Nassella**
3 Awn persistent; palea with long hairs .... **Achnatherum**
  Awn caducous; palea glabrous 4
4 Floret laterally compressed; stamen 1 .... **Anemanthele**
  Floret dorsally compressed; stamens 3 .... **Piptatherum**

ACHNATHERUM P.Beauv., 1812

Perennial with intra- or extravaginal branching. Inflorescence an open or narrow panicle; branches scabrid. Glumes > lemma, 1- and 3-nerved, hyaline. Lemma cylindrical, thickly membranous, margins meeting, hairy throughout or almost so, apex comate, not thickened. Palea ≤ lemma, nerves 2, hairy throughout. Callus short. Cleistogenes present or not.

Type species: *A. calamagrostis* (L.) P.Beauv.

c. 300 spp.; Eurasia, Africa, New World, Australasia. Endemic sp. 1; naturalised sp. 1.

1 Branching intravaginal; leaf-blade adaxially finely prickle-toothed; glume nerves conspicuous; lemma hairy on keel and outer margin; coma evident; awn to 20 mm; often cleistogamous .... **1. caudatum**
Branching extravaginal; leaf-blade adaxially clothed in hairs; glume nerves obscure; lemma hairy throughout; coma obscure; awn to 40 mm; always chasmogamous .... **2. petriei**

1. A. caudatum (Trin.) S.W.L.Jacobs et J.Everett, *Telopea 6*: 582 (1996).

Caespitose, tall, stout, bright green perennial with intravaginal branches with slender purpled inflorescences and cleistogenes in leaf-sheaths at the base. Leaf-sheath 10–20 cm, tough, glabrous, purpled below, margins long hairy above, apical tuft of hairs to 2 mm. Ligule *c.* 1 mm. Leaf-blade to 50 cm × 3 mm, ± rolled, glabrous, becoming very narrow and finely pointed, abaxially ribbed, glabrous, adaxially with many small prickles on ribs; margins with scattered long (1–1.5 mm) straggling hairs and prickle-teeth. Culm erect, to 1 m, nodes swollen, glabrous, internodes glabrous. Inflorescence to 30 cm, narrow, subtended by tufts of hairs, branches to 10 cm in fascicles similarly subtended, compound; rachis ridged, smooth; branches and pedicels short stiff hairy, naked below. Glumes ± equal, 7–8 mm, including lemma, conspicuously 3-nerved, awn tipped (1.3–1.5 mm), keel short stiff hairy, small prickles in internerves, centrally purple, hyaline elsewhere turning brown, margins ciliate below. Lemma 5–5.5 mm, fusiform, 5-nerved, purple towards margins, fading brown, lobes minute or 0, coma 0.5–1.0 mm, margins contiguous, keel and outer margins long (0.5–0.7 mm) hairy, minutely tuberculate elsewhere; awn to 18 mm, 1-geniculate, column 6–9 mm, arista to 10 mm. Palea 4.5–5 mm, lightly purpled, 2-nerved, internerve hairs reaching almost to ciliate apex. Callus to 1 mm, ± blunt, hairs to 1 mm. Lodicules 3, 1.2–2.0 mm, nerved. Anthers 2.5–3 mm in chasmogamous flowers, penicillate. Gynoecium: ovary 0.75–1.0 mm, trigonous; stigma-styles 1.75–2.0 mm, eccentric. Caryopsis 3 mm, obovate, irregularly ribbed, bases of old style eccentric, keeled opposite linear hilum to 2.5 mm, faintly rugose; embryo 1 mm. Chasmogamous in aerial inflorescences; cleistogamous in reduced inflorescences at culm nodes; cleistogenes in basal leaf-sheaths.

Naturalised from South America.

S.: Known only from Canterbury (Amberley) and Christchurch City.

In aerial inflorescences the caryopsis is so swollen that the lemma and palea are widely opened, and the caryopsis appears to be ready to fall free.

Ensheathed inflorescences at culm nodes bear florets as in aerial inflorescences but anthers are shorter (1.25 mm); caryopses as in aerial florets but appear more coriaceous.

Dark coloured, indurate cleistogenes are present on short inflorescences with glabrous or sparsely hairy branches of up to 8 spikelets on short hairy culms in leaf-sheaths of up to 4 successive culm nodes and up to 3 successive nodes of vegetative shoots. Spikelets are subtended by tufts of hairs to 1 mm; glumes *c.* 7 mm, 3-nerved, papery, becoming acute; lemma to 5 mm, apiculate, 1-nerved, brown, keel long hairy; palea 3–3.5 mm, 2-nerved, internerve hairy; anthers 3, 0.5–1.0 mm, barbed not penicillate, persisting on stigma-styles of caryopses; caryopsis 4–4.5 mm × 4–5 mm wide > aerial caryopsis, styles eccentric, pericarp dark, extremely coriaceous.

This taxon is distinguished by the criteria in Torres, M. A. *Monographia 12 Prov. Buenos Aires Comisión Invest. Cien.* (1993) and *Monographia 13* (1997) where it is treated as a species of *Stipa* L.

2. A. petriei (Buchanan) S.W.L.Jacobs et J.Everett, *Telopea 6*: 582 (1996) ≡ *Stipa petriei* Buchanan, *Indig. Grasses N.Z.* t. 17, 2 (1880); Holotype: WELT 59622! Buchanan's folio.

Erect, wiry perennial frequently branching at nodes; branching extravaginal; cataphylls short. Leaf-sheath to 3 cm, usually glabrous sometimes retrorsely pubescent. Ligule to 0.5 mm; auricular lobes to 1 mm, symmetrical or asymmetrical, often lightly pubescent. Collar thickened, occasionally with a very small tuft of hairs. Leaf-blade to 30 cm × 0.8 mm diam., narrow, involute, rigid, acicular, glabrous abaxially, adaxially clothed in short white hairs. Culm to 60 cm, wiry, internodes smooth, nodes purple, glabrous except for some very short hairs below.

Panicle to 25 cm, narrow; rachis smooth below, scabrid above, branches and pedicels scabrid. Glumes ± equal, to 7 mm, hyaline, shining, pink-suffused, produced into awn-like processes to 0.5 mm, or split at apex, < awn column; lower 1-nerved, upper 3-nerved. Lemma to 5.0 mm, cylindrical, 3- or 5-nerved, margins contiguous, fulvous, clothed in white ± appressed hairs; coma to 0.5 mm, lobes short and inconspicuous; awn to 40 mm, ± straight or weakly 1-geniculate, short stiff hairy, column loosely twisted to 10 mm, arista to 30 mm. Palea = lemma, clothed in long hairs, apex ciliate, 2-nerved. Callus short (to 0.3 mm), oblique, hairs white, to 1 mm. Lodicules 3, one usually emarginate, or entire, 1-nerved, to 1 mm. Anthers to 2.7 mm, weakly penicillate and shortly caudate. Caryopsis 2.5–3.5 mm; hilum linear.

Endemic.

S.: inland basins of Waitaki River, and Central Otago; up to 1000 m.

ANEMANTHELE Veldkamp, 1985

Monotypic and endemic to New Zealand as erected by Veldkamp, J. F. *Acta Bot. Neerl. 34*: 107 (1985) and followed by Jacobs and Everett (1996 op. cit.) and Jacobs et al. (1989 op. cit.).

1. A. lessoniana (Steud.) Veldkamp, *Acta Bot. Neerl. 34*: 108 (1985) ≡ *Agrostis lessoniana* Steud., *Nomencl. Bot.* ed. 2, *1*: 41, 42 (1840) ≡ *Oryzopsis lessoniana* (Steud.) Veldkamp, *Blumea* 22: 11 (1974) ≡ *Agrostis procera* A.Rich., *Ess. Fl. N.Z.* 125 (1832) non Retz. (1786) ≡ *Dichelachne procera* Steud., *Syn. Pl. Glum. 1*: 121 (1854); Holotype: P! *Lesson s.n.* = *D. rigida* Steud., *Syn. Pl. Glum. 1*: 120 (1854) ≡ *Oryzopsis rigida* (Steud.) Zotov, *T.R.S.N.Z. 73*: 235 (1943) ≡ *Agrostis rigida* A.Rich., *Ess. Fl. N.Z.* 124 (1832) non Spreng. (1825); Holotype: P! *Lesson s.n.* = *Apera arundinacea* Hook.f., *Fl. N.Z. 1*: 295, t. 67 (1853) ≡ *Stipa arundinacea* (Hook.f.) Benth., *J. Linn. Soc. Bot. 19*: 81 (1881); Lectotype: K *Colenso* (designated by Veldkamp 1985 op. cit. p. 108). = *Apera purpurascens* Colenso, *T.N.Z.I. 21*: 106 (1889); Lectotype: WELT 21978! *Colenso s.n.* Dannevirke (designated by Jacobs et al. 1989 op. cit. p. 580).

Densely tufted, wiry perennial with leaves drooping above, delicate nodding panicles, and short, creeping rhizome; branching extravaginal. Leaf-sheath to 15 cm, firm, outer margin usually ciliate. Ligule to 1.5 mm, usually asymmetrical, entire to fimbriate. Leaf-blade to 45 cm × 6 mm, stiff, involute or flat, abaxially shining, minutely scabrid towards filiform tip, adaxially smooth, dull, margins scabrid. Culm to 75 cm, simple, erect to nodding above, internodes smooth, rarely scaberulous below panicle, nodes glabrous. Panicle to 60 cm; branches capillary, spreading, in distant whorls; rachis slender, smooth, rarely scaberulous, branches and pedicels scaberulous. Spikelets laterally subcompressed, pale green to purplish. Glumes subequal, 2.5–3.5 mm, hyaline, acute to acuminate, keel scabrid; lower linear-lanceolate, 1-nerved, upper elliptic-lanceolate, 1–3-nerved. Flower ♀. Lemma 2 mm, 3-nerved, firmly membranous, elliptic-oblong, smooth below, scaberulous above, lobes minute, margins contiguous; awn to 8 mm, fine, scabrid, curved, caducous. Palea < lemma, weakly nerved, glabrous. Callus minute, rounded, ringed by very minute hairs. Lodicules 2, ligulate, glabrous, nerveless, 0.5 mm. Anther 1, 0.8–1.4 mm, apically thickened. Caryopsis *c.* 1.5 mm, very swollen, protruding from the anthoecium; hilum elliptic, < ½ length of caryopsis. $2n = 40$–44. Plate 6D.

Endemic.

N.: North Auckland and in southern half; S.: Nelson and Marlborough Sounds, Canterbury from Banks Peninsula southwards, near Dunedin, near Invercargill. Sea level to montane in forest, forest margins and scrub, also on bluffs. Occasionally found as a garden escape on roadsides.

Anemanthele lessoniana is not especially abundant in the wild but is much grown in New Zealand as an ornamental particularly in landscaped areas featuring native plants; there it has been confused with *Microlaena polynoda* (Ehrharteae). It seeds freely and spreads in shaded sites in parks, domains, and Botanic Gardens.

AUSTROSTIPA S.W.L.Jacobs et J.Everett, 1996

Perennial, caespitose or rhizomatous; branching intravaginal or extravaginal. Culm simple or bambusiform, sometimes branching at nodes. Inflorescence a narrow or open, much-branched panicle.

Spikelets 1-flowered; disarticulation oblique above persistent glumes. Glumes 2, equal or unequal, enclosing the floret, lower longer, 1–3–5-nerved, acute, acuminate, rarely mucronate, awned, or erose. Flower ♀. Lemma cylindrical, coriaceous, indurated, fusiform, pyriform or turbinate, pubescent, hairy or tubercular-scabrid or both, 5-nerved (7-in *A. flavescens*), margins overlapping, terminating in a rim with or without a coma of longer hairs; lobes small or 0. Awn 0–1–2-geniculate, often very long; column twisted, arista straight or curved. Callus sharply pointed (blunt in *A. verticillata*), hairs usually different in colour and density from those of lemma, extending over lemma base. Palea enclosed by lemma, membranous, hyaline or ± indurated, ≤ lemma, 2-nerved, internerve hairy or glabrous. Lodicules 2 or 3, hyaline, glabrous. Stamens 3, anthers penicillate or naked at apex and/or caudate at base; in cleistogamous flowers reduced to 1 small fertile anther and 2 sterile small anthers. Gynoecium: ovary glabrous, stigmas plumose. Caryopsis tightly enclosed and conforming to shape of lemma; embryo to about ⅓ of caryopsis; hilum linear, ≈ caryopsis. Flowering chasmogamous or cleistogamous in aerial inflorescences. Fig. 5.

Type species: *A. mollis* (R.Br.) S.W.L.Jacobs et J.Everett

61 spp. endemic to Australia, 1 sp. indigenous to New Zealand and Australia — *A. stipoides*; 9 spp. naturalised.

Segregated from *Stipa* L. by Jacobs and Everett (1996 op. cit.). This treatment of species substantially follows that of New Zealand stipoids in Jacobs et al. (1989 op. cit.) which was based on Vickery, J. W., Jacobs, S. W. L. and Everett, J. *Telopea 3*: 1–132 (1986). The nine naturalised species of *Austrostipa* are classified in six subgenera viz: three species in subgen. *Falcatae*, *A. nitida*, *A. nodosa*, *A. scabra*; two species in subgen. *Ceres*, *A. bigeniculata*, *A. blackii*. The remaining four are placed each in a separate subgenus; the Australasian indigenous species, *A. stipoides*, is included in subgen. *Lobatae*.

Connor, H. E., Edgar, E. and Bourdot, G. W. *N.Z. J. Agric. Res. 36*: 301–307 (1993) described the ecological history of naturalised species.

1 Culm nodes glabrous *2*
  Culm nodes pubescent... *6*

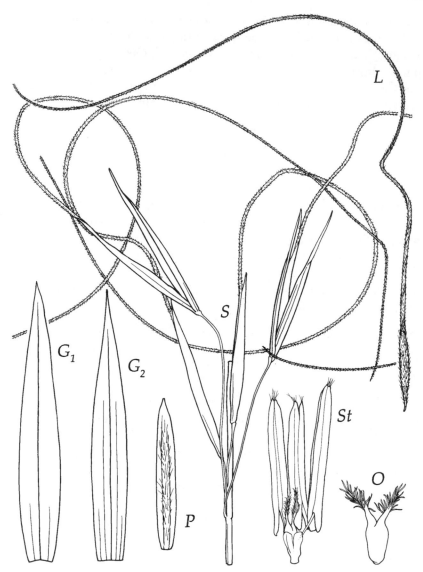

Fig. 5 *Austrostipa scabra*
S, spikelets, × 4; **G₁**, **G₂**, glumes, × 8; **L**, floret with lemma, awn column and arista, × 4;
P, palea dorsal view, × 12; **O**, ovary, styles and stigmas, × 16; **St**, stamens, lodicules and
gynoecium, × 16.

2 Leaf-blade glabrous abaxially; lemma lobes 1–2 mm, hairy .... **8. stipoides**
 Leaf-blade hairy or with prickles abaxially; lemma lobes 0–0.5 mm *3*

3 Branching extravaginal *4*
 Branching intravaginal *5*

4 Callus long sharp; awn falcate; glumes to 12 mm .... **5. nodosa**
 Callus short blunt; awn geniculate; glumes to 4 mm .... **10. verticillata**

5 Leaf-sheath hairy; innovations numerous; ligule to 1.5 mm .... **7. scabra**
  5a Panicle dense, spikelets close-set; longer auricular lobe of culm-leaf
  2–4 mm.. subsp. **scabra**
  Panicle open, spikelets spreading; longer auricular lobe of culm-leaf to
  1 mm... subsp. **falcata**
 Leaf-sheath glabrous; innovations few, ligule to 1 mm .... **4. nitida**

6 Palea internerve glabrous .... **3. flavescens**
 Palea internerve hairy *7*

7 Lower glume 15–20 mm; callus 1.75–2.5 mm.. *8*
 Lower glume 10–12 mm; callus 1–1.5 mm... *9*

8 Leaf-blade retrorsely scabrid abaxially; coma 2 mm .... **1. bigeniculata**
 Leaf-blade with long hairs abaxially; coma 3–3.5 mm................... **9. stuposa**

9 Lemma completely clothed in hairs; coma to 3.5 mm .... **2. blackii**
 Lemma lacking hairs above, tuberculate; coma 0 .... **6. rudis**

1. A. bigeniculata (Hughes) S.W.L.Jacobs et J.Everett, *Telopea* 6: 584 (1996).

Tall, erect, fine, perennial tussock, shoots very close set, slightly swollen at base; branching extravaginal; cataphylls hairy. Leaf-sheath to 10 cm, thin, very shortly retrorse pilose, occasionally with some long hairs. Ligule 0.2 mm, ciliate. Leaf-blade to 25 cm × 1 mm diam., stiff, weakly rolled, ribs few, abaxially finely retrorsely scabrid from columns of small prickle-teeth between ribs, abundant prickle-teeth adaxially and some few long hairs; margins with antrorse prickle-teeth and occasional long (1 mm) hairs. Culm to 60 cm, nodes retrorsely pilose, internodes below nodes densely retrorse-pilose, elsewhere sparsely scabrid. Panicle to 30 cm, narrow, subtended by tuft of hairs to 1.5 mm; branches short; rachis, branches and pedicels short stiff hairy. Glumes unequal, purple fading brown below, produced into hyaline awn-like processes to 2 mm, scabrid on nerves and above, > awn column; lower to 20 mm, 3-nerved, upper to 14 mm, 5-nerved. Lemma to 7 mm, clothed in dense white hairs, lobes minute

(0.1 mm) ciliate; coma to 2 mm; awn to 50 mm, 1-geniculate, column to 10 mm tightly twisted, short hairy, arista to 40 mm. Palea weakly 2-nerved, internerve hairs few, apex ciliate. Callus to 2 mm, hairs to 3 mm. Lodicules 3, ligulate or emarginate, posterior and anterior equal, to 1.3 mm, or posterior shorter. Anthers to 4 mm, penicillate.

Naturalised from eastern Australia.

S.: Canterbury (Port Hills, Christchurch, Taylors Mistake to Harris Bay; Weka Pass). In modified grasslands; sea level to 250 m.

2. A. blackii (C.E.Hubb.) S.W.L.Jacobs et J.Everett, *Telopea 6*: 584 (1996).

Tall, very coarse, erect perennial tussock with a stout rootstock; branching extravaginal; cataphylls numerous. Leaf-sheath to 5 cm, with retrorse long weak hairs, terminating in a small tuft of white hairs to 1.5 mm. Ligule to 0.5 mm, shortly ciliate. Leaf-blade to 30 cm × 5 mm, flat or weakly rolled, scabrid, dense long hairs abaxially, adaxially and on margins retrorse long hairs. Culm to 2 m, stout, smooth, compressed, nodes densely pubescent, internodes densely pubescent below nodes, glabrous elsewhere. Panicle to 40 cm, very open, verticels widely spaced; branches long, branches and pedicels shortly scabrid. Glumes unequal, 3-nerved, becoming purple-suffused, somewhat spreading, minutely scabrid, tip erose or apiculate, < awn column; lower to 14 mm, upper to 9 mm. Lemma to 6 mm, fuscous, clothed in white hairs, lobes short (0.2 mm); coma to 3.5 mm; awn to 35 mm, stiffly hairy, 2-geniculate, column strongly twisted to 10 mm, less strongly so to 7 mm above, arista to 20 mm. Palea firm, darker than lemma, internerve hairs long, apex erose. Callus 1 mm, off-white hairs to 2 mm. Lodicules 3, 2 anterior to 1.5 mm, posterior to 1 mm. Anthers to 5.2 mm in chasmogamous flowers, to 1.5 mm in cleistogamous flowers, penicillate.

Naturalised from eastern and southern Australia.

S.: Nelson (Richmond). In modified pasture.

3. A. flavescens (Labill.) S.W.L.Jacobs et J.Everett, *Telopea 6*: 585 (1996).

Tall, coarse, dense, erect or slightly decumbent perennial shortly rhizomatous tussock with close-set, white shoots, swollen at base; branching extravaginal; cataphylls numerous. Leaf-sheath to 2 cm, finely scabrid, terminating in a tuft of hairs to 1.5 mm. Ligule to 0.2 mm, ciliate. Leaf-blade to 10 cm × 1 mm diam., tightly rolled, rigid, abaxially scaberulous with short retrorse prickle-teeth, adaxially papillate and with occasional hairs and prickles, more so near collar; margins antrorsely scabrid. Culm to 50 cm, nodes pubescent, internodes below nodes pubescent, glabrous elsewhere. Panicle to 20 cm, contracted, verticels with many branches; rachis, branches and pedicels short stiff hairy. Glumes unequal, purple-suffused, green below, produced into fine hyaline awn-like processes 1 mm long; lower to 16.5 mm, 3-nerved, glabrous below ± reaching top of awn column, upper to 10 mm, 5-nerved, nerves with short stiff hairs below, becoming scabrid above except at tip. Lemma to 6 mm, 7-nerved, cylindrical, clothed in grey hairs becoming golden brown at maturity, lobes small (0.2 mm) or absent; coma to 2 mm; awn to 40 mm, 2-geniculate, column short hairy, tightly twisted, to 8 mm, loosely twisted to 8 mm above, arista to 25 mm. Palea weakly 2-nerved, internerve hairs few, apex ciliolate. Callus to 1.6 mm, hairs to 2.5 mm. Lodicules 3, ligulate, to 2 mm. Anthers to 2 mm, penicillate.

Naturalised from coastal southern states of Australia.

S.: Canterbury (Teddington). Modified grassland on steep headland.

4. A. nitida (Summerhayes et C.E.Hubb.) S.W.L.Jacobs et J.Everett, *Telopea 6*: 587 (1996).

Tall, erect, light-green, perennial tussock with fine crowded shoots and broad culm leaf-sheaths; branching intravaginal. Leaf-sheath to 10 cm, glabrous, auricular lobes to 2 mm, subtended by row of long hairs, symmetrical or asymmetrical, apex ciliate. Ligule 0.7 mm, fimbriate. Leaf-blade to 30 cm × 0.5 mm diam., inrolled, few-nerved, abaxially with antrorse prickle-teeth between veins and some short stiff hairs, adaxially with abundant short hairs, margins with antrorse prickle-teeth. Culm to 60 cm, nodes glabrous, internodes shortly pubescent below nodes, elsewhere glabrous. Panicle to 20 cm, contracted; rachis, branches and pedicels short stiff hairy. Glumes subequal, purple below, produced into long fine hyaline awn-like processes, 3-nerved, > awn column; lower to 15 mm, upper to 12 mm. Lemma to 4.5 mm, sparsely hairy becoming scabrid above, lobes to 0.5 mm; coma obscure or of few short hairs; awn to 40 mm, falcate, column 10 mm, arista to 30 mm, short hairy, lightly twisted. Palea internerve sparsely hairy, apex glabrous. Callus to 1.5 mm, hairs to 1.5 mm. Anthers 1.75 mm, penicillate. Our specimens are immature.

Naturalised from Australia.

S.: Marlborough (Dashwood and Weld passes). Depleted grassland and on roadsides.

Has not been collected since the gatherings of A. J. Healy in September 1941 and March 1942, although he found it on the Wither Hills, Marlborough, in 1947 (A. J. Healy *hic comm.*). There are no other reports and it has been searched for quite intensively.

5. A. nodosa (S.T.Blake) S.W.L.Jacobs et J.Everett, *Telopea 6*: 587 (1996).

Caespitose, open-based, perennial tussock with mostly floriferous, erect to nodding shoots; branching extravaginal; cataphylls hairy. Leaf-sheath to 4 cm, finely retrorse-scabrid, and densely short-hairy, terminating in a tuft of hair to 1.5 mm. Ligule to 0.6 mm, ciliate or sometimes glabrous. Leaf-blade to 20 cm × 1 mm diam., inrolled, rigid, ribs few, retrorse prickle-teeth in columns between ribs and some antrorse long hairs abaxially, abundant short hairs and long hairs especially near the margins adaxially; margins with antrorse prickle-teeth or glabrous. Culm to 40 cm, nodes glabrous, internodes pubescent below nodes, glabrous elsewhere. Panicle to 25 cm, subtended by hairy bract to 4 mm or hairs to 3 mm, erect; verticels close-set; branches short, widely spreading; rachis glabrous below, becoming scabrid above, pedicels and branches short stiff hairy. Glumes ± equal, purple fading brown, tapering to hyaline awn-like processes to 2 mm or acuminate, scabrid, short stiff hairs on midnerve, < awn column; lower to 14 mm, 3-nerved, upper to 13 mm, 3–5-nerved. Lemma 4.5 mm, fulvous, with dense, short hairs, usually scabrid above, lobes minute (0.1 mm); coma obscure or of few short hairs; awn to 85 mm, falcate, sometimes also 1-geniculate, column hairy, tightly twisted, to 10 mm, arista to 55 mm, usually intertwined with awns of adjacent florets. Palea internerve hairs long, apex glabrous, erose. Callus to 1.75 mm, hairs to 2.5 mm, white to brown.

Lodicules 3; anterior to 1.5 mm, ligulate; posterior to 0.4 mm. Anthers penicillate, to 3.75 mm in chasmogamous flowers, reduced to 1 fertile anther, 0.5–1.5 mm, and 2 sterile in cleistogamous flowers.

Naturalised from southern Australia.

N.: Bulls; S.: widespread in Marlborough, Canterbury and Otago in dry pastures, waste places, riverbeds and roadsides to 250 m.

Of recent years this was known as *Stipa variabilis* Hughes.

6. A. rudis (Spreng.) S.W.L.Jacobs et J.Everett, *Telopea 6*: 588 (1996)

subsp. rudis

Tall, erect, wiry, open perennial tussock, shoots slightly swollen at base; branching extravaginal; cataphylls glabrous shining. Leaf-sheath to 5 cm, with scattered retrorse long hairs, terminating in a small tuft of hairs to 1 mm. Ligule to 0.5 mm, apex shortly ciliate. Leaf-blade to 35 cm × 1 mm diam., usually inrolled, nerves few, abaxially finely antrorsely scabrid especially near margins, some few stiff long hairs towards base, adaxially with abundant long (0.5 mm) and short hairs, margins with long hairs at base becoming scabrid from small prickle-teeth above. Culm to 1 m, stout, nodes dark retrorsely pubescent, internodes pilose below nodes elsewhere smooth and shining, leaf-sheath with contra-ligule. Panicle to 35 cm, loosely contracted, subtended by hairy bract to 1 mm; rachis smooth becoming scabrid towards apex, branches and pedicels short stiff hairy. Glumes subequal, nerves conspicuously raised, straw coloured, scabrid, internerves purple-suffused, margins short hairy, tips membranous and erose, < awn column; lower to 12 mm, 3-nerved, much shorter than awn column, upper to 11 mm, 5-nerved. Lemma to 6.5 mm, very dark brown at maturity, tubercular scabrid, hairy below glabrous above, lobes 0; coma 0; awn to 50 mm, 2-geniculate, column short hairy, tightly twisted below less so above to 30 mm, arista to 20 mm. Palea weakly 2-nerved, internerve hairy sometimes slightly so, apex glabrous. Callus to 1.5 mm, golden hairs to 2 mm. Lodicules 2, ligulate, to 1.75 mm. Anthers in cleistogamous flowers, 2 small (0.5 mm) the third to 1.1 mm. Cleistogamous flowers only known.

Naturalised from higher altitudes in eastern Australia.

N.: North Auckland (near Henderson); S.: Marlborough (Wairau Valley). On roadsides and in paddock.

7. A. scabra (Lindl.) S.W.L.Jacobs et J.Everett, *Telopea* 6: 588 (1996).

Densely tufted, fine-leaved, perennial tussock with many innovation shoots; branching intravaginal. Leaf-sheath to 2 cm, with abundant retrorse short hairs, terminating in tuft of hairs to 1.5 mm subtending auricular lobes; auricular lobes symmetrical or asymmetrical to 0.5 mm. Ligule to 1.5 mm, shortly ciliate. Leaf-blade to 30 cm × 0.5 mm diam., inrolled, weakly acicular, abaxially very scabrid from short (0.25 mm) antrorse hairs and prickle-teeth in columns between ribs, adaxially with abundant short hairs; margins with antrorse prickle-teeth. Culm to 50 cm, pubescent below nodes, elsewhere glabrous. Panicle to 30 cm, sometimes subtended by long hairs or a bract; rachis, branches and pedicels short stiff hairy. Glumes subequal, purple below, tapering to hyaline awn-like processes, scabrid, > awn column; lower to 15 mm, 3-nerved, upper to 12 mm, 3–5-nerved. Lemma to 5 mm, clothed in white appressed hairs, less so above and becoming scabrid, lobes minute (0.25 mm); coma a few short hairs; awn to 65 mm, exceedingly falcate, ± 1-geniculate, column tightly twisted, short hairy, to 10–15 mm, arista to 50 mm. Palea internerve hairs long, apex glabrous. Callus to 1.5 mm, hairs to 2 mm. Lodicules 3, to 2 mm, ligulate. Anthers to 1.5 mm in cleistogamous, and to 3.5 mm in chasmogamous florets, penicillate. Fig. 5.

subsp. **scabra**

Panicle dense, spikelets close-set; longer auricular lobes of culm-leaf 2–4 mm, usually with some long hairs between leaf-blade and lobe.

Naturalised from Australia.

S.: Marlborough to Canterbury, and Waitaki Basin. In modified grassland to 250 m.

Late in the season a floriferous branch may develop at the uppermost culm node and emerge from the culm leaf-sheath below the earlier flowering terminal inflorescence; flowering there is cleistogamous.

subsp. **falcata** (Hughes) S.W.L.Jacobs et J.Everett, *Telopea* 6: 588 (1996).

Panicle open, spikelets spreading, longer auricular lobe of culm-leaf to 1 mm, usually subtended by a row of hairs ≥ lobe; culm more slender than in subsp. *scabra*.

Naturalised from eastern Australia.

S.: Marlborough and around Banks Peninsula, Canterbury. In modified grassland to 250 m.

The two subspecies co-occur in Marlborough and in Canterbury and the distinctions between them, as described for Australian plants, may become blurred, as is also true of parts of Australia.

8. A. stipoides (Hook.f.) S.W.L.Jacobs et J.Everett, *Telopea 6*: 589 (1996) ≡ *Dichelachne stipoides* Hook.f., *Fl. N.Z. 1*: 294, t. 66 (1853) ≡ *Stipa stipoides* (Hook.f.) Veldkamp, *Blumea 22*: 11 (1974); Lectotype: K! *Hooker* Bay of Islands [designated by Townrow, J. E. S. *Pap. Proc. Roy. Soc. Tasmania 112*: 261 (1978)]. = *S. teretifolia* Steud., *Syn. Pl. Glum. 1*: 128 (1854); Holotype: P, Australia (*fide* Vickery et al. 1986 op. cit., p. 116).

Tall, caespitose, stiff, erect perennial with swollen bases occasionally with elongating basal internodes and branching at nodes; branching intravaginal; cataphylls small. Leaf-sheath to 25 cm, smooth, very short hairs often between ridges. Ligule 3–7 mm, decurrent, symmetrical or asymmetrical, nerved, subacute. Leaf-blade to 60 cm × 0.7 mm diam., rigid, folded but appearing terete, abaxially glabrous, sharply acicular, adaxially with abundant short white hairs and some prickle-teeth towards base. Culm to 1 m, internodes glabrous, nodes concealed. Panicle to 20 cm, narrow; branches ascending; rachis, branches and pedicels glabrous. Glumes ± equal, to 20 mm, 3-nerved, produced into awn-like processes ± 1 mm, or acuminate, > awn column. Lemma 8–10 mm long, fuscous, hairs white or brown (± 2 mm), lobes hairy, 1–2 mm; coma to 3 mm; awn to 35 mm, short stiff hairy, 1-geniculate, column strongly twisted to 10 mm, arista to 25 mm. Palea with abundant long internerve hairs, few elsewhere, entire or shallowly lobed, apex shortly ciliate. Callus to 1.5 mm, hairs to 1.5 mm. Lodicules 3, usually 1–2-nerved, acute, ligulate, to 2.5 mm, often 1- or 2-fid, adhering to caryopsis. Anthers apically and basally caudate, to 7.0 mm. Plate 5C.

Indigenous.

N.: south to Mokau River, Taranaki, and Whakatane, Bay of Plenty, Wellington; S.: Nelson. Coastal rocks and mud flats.

Also indigenous on coast of south-east Australia, and Tasmania.

9. A. stuposa (Hughes) S.W.L.Jacobs et J.Everett, *Telopea* 6: 589 (1996).

Tall, erect, coarse, densely tufted perennial tussock with stout culms; branching extravaginal; cataphylls hairy. Leaf-sheath to 10 cm, with abundant long (0.5 mm) hairs from swollen bases, terminating in a tuft of long hairs to 2.5 mm. Ligule to 0.5 mm, conspicuously hair-fringed. Leaf-blade to 30 cm × 1 mm diam., loosely inrolled, stiff, harsh abaxially with dense erect or retrorse ± stiff long (0.5 mm) hairs from swollen bases in columns between ribs, adaxially with abundant short hairs or prickle-teeth; margins with long hairs or prickle-teeth. Collar curved, hair-fringed. Culm to 1 m, nodes and internodes densely retrorsely pilose. Panicle to 35 cm, violet or violet-suffused, narrow, subtended by hairs or hairy bracts; rachis scabrid, branches and pedicels short stiff hairy. Glumes unequal, 3-nerved, scabrid, produced into hyaline awn-like processes to 3 mm, < awn column; lower to 18 mm, upper to 15 mm. Lemma to 8 mm, fulvous, clothed in long white hairs, lobes minute or absent; coma to 3 mm; awn to 50 mm, 1- or weakly 2-geniculate, widely divergent, column twisted and hairy to 15 mm, less twisted for 8 mm, arista to 25 mm. Palea internerve hairs long, apex sparsely hairy. Callus to 2.5 mm, hairs to 3 mm. Lodicules 2, ligulate, to 2 mm. Anthers to 1.5 mm in cleistogamous flowers, to 3.5 mm in chasmogamous flowers, penicillate.

Naturalised from south-eastern Australia or Tasmania.

S.: Banks Peninsula (Pigeon Bay). In modified grassland and among short scrub to 200 m.

Collected by W. R. Boyce on 17 February 1937, from Bowen's Valley, Port Hills [Christchurch] CHR 92947 (duplicate CANU 834); it has not been found there since.

Flowering in spring is either chasmogamous or cleistogamous; cleistogamy occurs in both emerged or ensheathed inflorescences. In autumn a floriferous branch may develop at the uppermost culm node and emerge from the culm leaf-sheath; flowering there is cleistogamous.

Plate 3. Some weedy naturalised grasses. **A** *Lagurus ovatus*, harestail, found mainly on sand dunes (centre), and (arching above) *Bromus diandrus*, ripgut brome (Kaitorete, Canterbury); **B** *Agrostis capillaris*, browntop, one of the most widespread naturalised grasses (Queenstown); **C** *Critesion murinum*, barley grass, especially common on dry sunny slopes and livestock camps (Central Otago); **D** *Nassella trichotoma*, nassella tussock, a weedy tussock of hill-country in dry districts (North Canterbury).

Plate 4. Grasses in their habitats: **A** *Poa astonii* (on cliff) and *P. foliosa* (on talus slope) on a southern coast (Solander Island); **B** *Pennisetum clandestinum,* kikuyu grass, on a Northland coast; **C** *Parapholis incurva* in salt marsh (Lake Grassmere); **D** *Poa matthewsii* under kanuka woodland (Otago Peninsula); **E** *Anthoxanthum odoratum,* sweet vernal, among matagouri shrubland (Central Otago); **F** *Vulpia myuros,* hair grass, and *Aira caryophyllea,* silvery hair grass, annual grasses on an arid terrace (Central Otago).

10. A. verticillata (Nees) S.W.L.Jacobs et J.Everett, *Telopea 6*: 589 (1996).

Stout caespitose or shortly rhizomatous erect perennial, prolifically branching at nodes; branching extravaginal; cataphylls numerous. Leaf-sheath to 10 cm, short hairs below, glabrous above, terminating in a short tuft of hairs. Ligule to 2 mm, erose, truncate. Collar curved. Leaf-blade to 60 cm × 4 mm, flat, abaxially and on margin antrorsely scabrid, adaxially antrorsely scabrid or with long hairs. Culm to 2 m, bambusiform, internodes glabrous, nodes brown, rough. Panicle to 50 cm, subverticillate; branches ascending-spreading; rachis smooth below, scabrid above, branches and pedicels scabrid. Glumes equal, 3–4 mm, narrow, ≈ lemma, green-suffused, spreading at maturity, acuminate, sometimes shortly awned but mostly erose, 3-nerved, nerves scabrid, < awn column. Lemma 3 mm, becoming dark brown at maturity, clothed in long white appressed hairs but tubercular-scabrid beneath and at apex, lobes minute; awn to 35 mm, 1-geniculate or sometimes straight, short stiff hairy, column straight, scabrid, 7–12 mm, arista to 25 mm. Palea 1–1.5 mm, ligulate, *c.* ½ length of lemma, nerves with a few scattered hairs or glabrous, internerve darker, apex ciliolate. Callus to 0.2 mm, short, blunt and rounded, loosely clothed in short white hairs to 0.5 mm. Lodicules 2, to 1.3 mm, ligulate, ≈ palea. Anthers 3, penicillate and caudate, to 1.8 mm.

Naturalised from eastern Australia.

Has been in cultivation since the 1880s, and has escaped in Nelson, and Marlborough, and could be expected elsewhere. First recorded as *Streptachne ramosissima.*

NASSELLA Desv., 1854 *emend.* Barkworth, 1990

Perennial. Floret fusiform, laterally compressed. Lemma heavily silicified, with tightly overlapping margins, apex thickened, coronate or comate; awn persistent or caducous, centric or eccentric. Palea ≪ lemma, nerveless, glabrous. Callus obtuse to bluntly acute. Lodicules 2. Stamens 1–3. Caryopsis conforming to shape of, and as long as lemma; embryo small; hilum linear.

Type species: *N. pungens* Desv.

100 spp. of the New World, mainly South American. Naturalised spp. 3.

1 Floret gibbous; awn eccentric .... **3. trichotoma**
Floret fusiform or elliptic-oblong; awn subcentric *2*

2 Callus 3–4 mm, hairs to 4 mm; corona a prominent ridge with spines to 1 mm; ligule to 0.5 mm .... **1. neesiana**
Callus 0.2–0.4 mm, hairs to 1 mm; corona spines to 0.2 mm; ligule to 2.5 mm .... **2. tenuissima**

1. N. neesiana (Trin. et Rupr.) Barkworth, *Taxon 39*: 611 (1990)

var. **neesiana**   Chilean needle grass

Erect, strongly caespitose perennial with shoots swollen and close-set at base; branching intravaginal. Leaf-sheath to 5 cm, striate, densely pubescent at base, long hairs at apex, hairs sometimes forming a contra-ligule, elsewhere glabrous. Ligule ± 0.5 mm, flat-topped, finely and shortly ciliolate at apex or shallowly crenate. Leaf-blade to 40 cm × 5 mm, flat, scabrid, sometimes scattered long hairs abaxially, margins scabrid. Culm to 2 m, internodes smooth except for prickle-teeth below inflorescence, nodes appressed-pilose. Panicle to 30 cm, open; branches drooping, flexuous; rachis smooth to slightly scabrid, branches and pedicels stiff hairy. Glumes unequal, violet below, hyaline above, 3-nerved, nerves scabrid, produced into awn-like processes to 3 mm, < awn column; lower to 20 mm, margins long hairy, upper to 15 mm, glabrous. Lemma to 6 mm, tubercular-scabrid, median nerve long hairy, lobes minute; corona to 1 mm, violet or violet-suffused, glabrous, a conspicuous ring of spines to 1 mm at apex; awn to 70 mm, 1-geniculate; column tightly twisted and long hairy to 25 mm, then shallowly twisted and short stiff hairy to 15 mm above, arista scabrid, to 35 mm, usually intertwined with awns of adjacent florets. Palea 1.5 mm, membranous, glabrous. Callus to 4 mm, hairs to 4 mm. Lodicules 2, *c.* 1 mm, ≈ palea. Anthers penicillate, to 3.2 mm in chasmogamous flowers, reduced to 1 fertile, 0.5–0.7 mm, and 2 sterile, 0.1–0.2 mm, anthers in cleistogamous flowers.

Naturalised from South America.

N.: Waitakere Ranges, Waipawa; S.: Marlborough (near Blind River, Seddon, and Lake Grassmere). Roadsides and pastures.

Cleistogenes are formed at nodes towards the base of flowering culms. They are subtended by prophylls, and 1–3-flowered; glumes 2, awned, about half as long as in aerial florets; lower glume 1–3-nerved, upper to 7 nerves; lemma, corona, palea as in aerial florets; awn to 25 mm; callus ± 0.5 mm; anthers reduced to 1 fertile anther and 2 sterile small anthers; caryopsis round or planoconvex, to 4 mm. For frequencies see Connor et al. (1993 op. cit.).

Occasionally an aerial-type branch may be produced at the uppermost culm node, and so far as can be judged, remains ensheathed.

Locally troublesome [see Connor, H. E., Edgar, E. and Bourdôt, G. *N.Z. J. Agric. Res. 36*: 301–307 (1993)].

Specimens gathered at Auckland by M. Hodgkins (CHR 17437, CHR 18300) have long hairy leaf-blades, lemma nerves long hairy with the hairs on the main nerve reaching almost to the corona. These characters suggest a provenance different from the Marlborough plants.

2. N. tenuissima (Trin.) Barkworth, *Taxon 39*: 612 (1990).

Erect bright green perennial tussock with very narrow rolled leaves in slender shoots white at base, below a long narrow panicle with long fine intertwining awns; branching intravaginal and at nodes. Leaf-sheath to 10 cm, scabrid as leaf-blades, margins ciliate. Prophyll to 10 cm, hairy. Ligule 2.5 mm, becoming acute, minutely prickle-toothed. Leaf-blade to 60 cm × 0.3 mm diam., terete, abaxially beset with small antrorse prickles, adaxially and on margins very finely short hairy. Culm to 1 m, internodes hairy especially below nodes, antrorsely prickle-toothed elsewhere; nodes ± geniculate. Panicle to 30 cm, contracted, much branched; rachis, branches and pedicels finely stiff hairy often densely so. Glumes unequal, violet below, hyaline above, keels sparsely finely toothed, produced to awn-like processes to 5 mm; lower to 8 mm, upper to 6 mm. Lemma 2 mm, cylindrical, slightly compressed, finely tuberculate, median nerve with long hairs, lobes minute; corona 0.4 mm, spines to 0.2 mm; awn to 50 mm, very fine (0.1 mm), weakly 2-geniculate, column lacking. Palea 0.75–1 mm, membranous, glabrous. Callus 0.2 mm, hairs to 1 mm. Lodicules 2, 0.4 mm. Anthers to 2.5 mm in chasmogamous flowers,

1 fertile anther to 0.5 mm and 2 rudimentary anthers to 0.2 mm in cleistogamous flowers. Gynoecium: ovary 0.6 mm; stigma-styles to 1.25 mm.

Naturalised from horticultural importation; a New World sp.

N.: Auckland and Hamilton cities. A garden escape; cultivated sporadically in both Islands.

First recognised in 1994. See also Gardner, R. O., Champion, P. D. and de Lange, P. J. *Auck. Bot. Soc. J. 51*: 31–33 (1996).

The authority given this species by Barkworth, M. E. *Phytologia 74*: 17 (1993) as (Hitchcock) Barkworth is a *lapsus calami.*

3. N. trichotoma (Nees) Arechav., *An. Mus. Nac. Montevideo 1*: 276 (1895).   nassella tussock

Caespitose perennial with shoots swollen and white at base, and open ± purple panicle, branching intravaginal. Leaf-sheath to 5 cm, scabrid near ligule and on upper margins, becoming glabrous below. Ligule to 1.5 mm, decurrent, symmetrical or asymmetrical, rounded or apiculate, short-hairy or scabrid. Leaf-blade to 35 cm × 0.5 mm diam., inrolled appearing terete, very stiff, acicular, abaxially markedly antrorsely scabrid, adaxially clothed in short dense hairs; margins smooth. Culm to 70 cm, glabrous except for hairs above and below nodes. Panicle to 25 cm, open, much-branched, drooping, at maturity readily detaching with culm and blowing freely; rachis, branches and pedicels scabrid. Glumes ± equal, 5–8 mm, deeply purple-suffused, 3-nerved, produced into awn *c.* 2 mm, nerves and margins stiff hairy. Lemma to 2.5 mm, 5-nerved, gibbous, tubercular-scabrid except on margin, lobes short, apiculate; coma present; awn tardily caducous to 35 mm, weakly 1-geniculate, short stiff hairy, column loosely twisted, to 15 mm, arista to 20 mm. Palea *c.* 1 mm, completely enclosed by lemma, glabrous, hyaline, weakly bifid at apex. Callus to 0.3 mm, blunt, hairs to 1.5 mm. Lodicules 2, to 1 mm in chasmogamous flowers, to 0.6 mm in cleistogamous flowers. Anthers penicillate, in chasmogamous flowers to 2 mm; in cleistogamous flowers 1 fertile anther to 0.7 mm, and 2 aborted anthers to 0.3 mm. Plate 3D.

Naturalised from South America.

N.: scattered small infestations from Kaitaia to Coromandel Peninsula, and on east coast to Hastings; S.: east coast from Marlborough to Central Otago. Grasslands and scrub to 300 m.

An important pastoral weed especially in Marlborough and North Canterbury [Healy, A. J. *in* Knox, G. A. (Ed.) *Nat. Hist. Canty* 261–333 (1969)]; the Nassella Tussock Act, 1946, was passed to assist with its control; that was superseded by The Noxious Plants Act, 1978, which embodied most of the original provisions, but the Biosecurity Act 1993 repealed it.

Two species of *Austrostipa* are at times confused with *N. trichotoma*. The abundant *A. nodosa* is distinguishable by: absence of swollen white bases of shoots, with leaf-sheaths bearing long hairs on margins and terminating in a conspicuous tuft of abundant white hairs up to 1.5 mm long. The pinkness of the inflorescences and their shimmer in the distance may also mislead, but the heads do not freely detach. *Austrostipa scabra* lacks the conspicuously swollen bases of nassella tussock, its leaf-blade bears short stiff hairs, and the inflorescence does not readily detach. Differences in lemma shape and size easily separate the two genera.

PIPTATHERUM P.Beauv., 1812

Perennial. Leaf-blade flat or involute. Inflorescence a lax or contracted panicle. Spikelets ± compressed dorsally. Glumes subequal, 3–5-nerved. Lemma obscurely 3–5-nerved, coriaceous or membranous, with a caducous terminal awn. Palea ≈ lemma but narrower. Callus very small. Stamens 3.

Type species: *P. coerulescens* (Desf.) P.Beauv.

c. 25 spp. from eastern Mediterranean to south-western Asia. Naturalised sp. 1.

1. P. miliaceum (L.) Coss., *Notes Pl. Crit.* 129 (1851).

Loosely tufted, erect, perennial, branching at nodes, from short creeping rhizomes; branching extravaginal; cataphylls densely hairy. Leaf-sheath firmly membranous, smooth, sometimes terminating in a tuft of minute hairs. Ligule to 1.5 mm, truncate, entire. Leaf-blade to 40 cm × 5.5 mm, flat or involute, abaxially smooth, but scabrulous towards filiform tip and on margins, adaxially finely scabrid and with scattered soft hairs. Culm to 130 cm, wiry, glabrous. Panicle to 60 cm, lax; branches capillary, ascending to spreading, in dense distant whorls, rachis smooth, branches and pedicels scaberulous. Glumes subequal, 3–4 mm, 3-nerved, acute, often purple; lower slightly keeled and scabrid near tip. Lemma to 2.5 mm, 3-nerved, firmly membranous, smooth, shallowly lobed at apex; awn to 4.5 mm, straight, caducous. Palea ≈ lemma, 2-nerved, glabrous. Callus minute, blunt, glabrous. Lodicules 3. Anthers to 1.7 mm, penicillate.

Naturalised from Eurasia.

N.: scattered throughout; S.: vicinity of Nelson City, near Blenheim and Seddon in Marlborough, and also at Christchurch. In waste land, pasture, and on dry banks.

6. POEAE

Annuals or perennials. Leaf-sheath usually free, sometimes connate. Ligule membranous. Leaf-blade linear, filiform or setaceous, flat, plicate or convolute. Inflorescence a panicle, rarely a spike or raceme; florets all ♀, or some ♀ and others ∅, or unisexual in dioecious, gynomonoecious or gynodioecious spp. (*Poa*). Spikelets (1)–2–many-flowered, laterally compressed or terete; disarticulation above glumes and usually between florets. Glumes < spikelets, usually membranous. Lemma membranous to coriaceous, (3)–5–7–(13)-nerved, awnless or with a terminal awn, usually keeled. Palea 2-keeled. Lodicules 2, lanceolate-acute, entire or lobed. Stamens 3, rarely 2 or 1. Ovary glabrous or sometimes hairy at apex; styles 2. Caryopsis usually ellipsoid; hilum linear or punctiform.

1 ♀ spikelets ± concealed by persistent ∅ spikelets .... **Cynosurus**
  ♀ spikelets not so *2*

2 Inflorescence a spike, bearing spikelets edgeways to rachis; lower glume
   absent, except in terminal spikelet .... **Lolium**
   Inflorescence a dense or open panicle; both glumes present *3*

3 Lemma long-awned *4*
   Lemma awnless, or mucronate to shortly awned *5*

4 Glumes very unequal; plants annual .... **Vulpia**
  Glumes ± equal; plants perennial .... **Festuca**

5 Lemma keeled *6*
  Lemma rounded, rarely keeled above *8*

6 Lemma spinulose on keel .... **Dactylis**
  Lemma often scabrid on keel but not spinulose *7*

7 Rachilla smooth or scabrid .... **Poa**
  Rachilla hairy .... **Austrofestuca**

8 Leaf-blade with clasping auricles .... **Schedonorus**
  Leaf-blade lacking clasping auricles *9*

9 Lemma apex firm *10*
  Lemma apex thinly scarious to hyaline *11*

10 Panicle ± lax; spikelets on slender pedicels; plants perennial .... **Festuca**
   Panicle stiff; spikelets on short, stiff pedicels; plants annual .... **Catapodium**

11 Lemma orbicular, cordate at base .... **Briza**
   Lemma elliptic to ovate, not cordate at base *12*

12 Panicle open or racemose; lemma apex ± erose .... **Puccinellia**
   Panicle capitate; lemma apex 2–5-toothed .... **Sesleria**ζ (p. 211)

AUSTROFESTUCA (Tzvelev) E.B.Alexeev, 1976

Tussock-forming perennials, moderately tall, rhizomatous; branching intravaginal. Leaf-sheath open, glabrous. Ligule short, submembranous, truncate, shortly ciliate, abaxially glabrous. Leaf-blade stiff, rolled. Culm internodes glabrous. Panicle contracted, almost spike-like; branches short, stiff. Spikelets 3–6-flowered, pedicelled, laterally compressed; disarticulation above glumes; rachilla prolonged, ± hairy. Glumes 3–5-nerved, subequal, ≈ lemma of proximal floret, coriaceous, keeled, acute. Lemma 5–7-nerved, coriaceous, keeled; apex obtuse, or minutely notched with midnerve minutely excurrent. Palea subcoriaceous, keels 2, stiffly ciliate. Lodicules 2, bilobed, ciliate, and with a few minute hairs centrally. Callus ringed by short, stiff hairs. Stamens 3. Ovary glabrous, styles free to base, stigmas plumose, whitish. Caryopsis elliptic to cylindric, adaxially longitudinally grooved; embryo small; hilum oval, *c.* ¼ length of caryopsis.

Type species: *A. littoralis* (Labill.) E.B.Alexeev

4 spp. of Australasia. Indigenous sp. 1, shared with Australia.

Austrofestuca, based on *Festuca* subgenus *Austrofestuca* Tzvelev, was segregated by Alekseev, E. B. *Bjull. Moskovsk. Obšč. Isp. Prir., Otd. Biol. 81*: 55–60 (1976) from *Festuca*, to accommodate the Australasian tussock plants of coastal sand dunes originally described as *Festuca littoralis* Labill. The plants resemble *Poa* rather than *Festuca* in having a ± hairy callus, keeled glumes and lemmas, and short hilum, but the caryopsis has a deeply grooved ventral side as in *Festuca*. The lodicules are hairy — an unusual feature in tribe Poeae, but the significance of hairy lodicules in delimiting *Austrofestuca* from *Poa* is diminished by the presence of one or two hairs at the tips of lodicules of New Zealand species of *Poa* [Edgar, E. *N.Z. J. Bot. 24*: 425–503 (1986)] and most indigenous species of *Festuca* [Connor, H. E. *N.Z. J. Bot. 36*: 329–367 (1998)].

1. A. littoralis (Labill.) E.B.Alexeev, *Bjull. Moskovsk. Obšč. Isp. Prir., Otd. Biol. 81*: 55 (1976) ≡ *Festuca littoralis* Labill., *Nov. Holl. Pl. 1*: 22, t. 27 (1805) ≡ *Schedonorus littoralis* (Labill.) P.Beauv., *Ess. Agrost.* 99, 163, 177 (1812) ≡ *Triodia billardierei* Spreng., *Syst. Veg. 1*: 330 (1824) ≡ *Poa billardierei* (Spreng.) St.-Yves, *Candollea 3*: 284 (1927) non Benth. (1878) ≡ *Schedonorus billardiereanus* Nees, *Lond. J. Bot. 2*: 419 (1843) *nom. superfl.* ≡ *Arundo triodioides* Trin., *Sp. Gram.* t. 351 (1836) ≡ *Poa triodioides* (Trin.) Zotov, *T.R.S.N.Z. 73*: 236 (1943) *nom. superfl.*; Holotype: FI! *Labillardière* [Capite Van-Diemen]. = *Schedonorus littoralis* var. β *minor* Hook.f., *Fl. N.Z. 1*: 310 (1853); Lectotype: K! *Sinclair* Auckland (here designated).   sand tussock

Dense, very stiff, rhizomatous tussocks, to 100 cm, with pungent leaves usually overtopping culms. Leaf-sheath light brown to grey-brown, coriaceous, striate, glabrous. Ligule 0.7–1 mm, asymmetrical, stiffly membranous, ciliate. Leaf-blade 10–60 cm × *c*. 1 mm diam., inrolled, wiry, abaxially glabrous, adaxially densely minutely pubescent. Culm (8)–12–50 cm, erect, internodes glabrous. Panicle (3.5)–6–25 cm, stiff, erect, contracted, almost spike-like; rachis glabrous, branches erect, pedicels with dense stiff minute hairs and prickle-teeth. Spikelets 9–16 mm, 3–6-flowered, greenish brown. Glumes almost equal, coriaceous, elliptic-lanceolate, acute, minutely prickle-toothed above, densely prickle-toothed on keel; lower 5.5–10 mm, (3)–5-nerved, upper 6.5–11 mm, 3–(5)-nerved; margins finely prickle-toothed. Lemma 6–11.5 mm, 5–7–(9)-nerved, elliptic-lanceolate, obtuse or with shortly excurrent midnerve, ± covered with very dense, stiff prickle-teeth, often minutely hairy on nerves below. Palea 5–8 mm, subcoriaceous, keels very densely, short, stiff hairy, interkeel and flanks minutely prickle-toothed. Lodicules *c.* 1.5 mm, bilobed, ciliate. Callus ringed by short, stiff hairs. Rachilla sparsely to densely hairy especially near apex. Anthers (1.5)–2–3 mm. Gynoecium: ovary *c.* 2 mm; stigma-styles *c.* 3 mm. Caryopsis 2.5–4 × *c.* 1 mm. $2n = 28$. Plate 5B.

Indigenous.

N.: scattered throughout; S.: scattered, rare on east coast; St.; Ch. Coastal in sand dunes and on damp sandy or shingly flats.

Also indigenous to Australia.

Hair, J. B. *N.Z. J. Bot.* 6: 268 (1968) reported for *Austrofestuca littoralis* (as *Poa triodioides*) that at least four of the eight pairs of submedian chromosomes were distinctly more heterobrachial than elements of the same class in tetraploid species of *Poa.*

Buchanan, J. *Indig. Grasses N.Z.* t. 54 (1880) incorrectly recorded *Festuca littoralis* var. *triticoides* as common in New Zealand; this western Australian taxon is now known as *Austrofestuca pubinervis* (Vickery) B.K.Simon.

LECTOTYPIFICATION   In the protologue of *Schedonorus littoralis* var. *minor* Hooker (1853 op. cit. p. 310) cited 2 specimens "Auckland, *Sinclair* and Port William, *Lyall*". Only Sinclair's specimen is veritable *Austrofestuca littoralis,* Lyall's specimen is *Poa astonii.*

BRIZA L., 1753

Loosely tufted annuals or perennials. Leaf-sheath open at maturity. Ligule membranous. Leaf-blade flat or convolute, scaberulous. Culm erect, or ± geniculate at base. Inflorescence a lax or contracted panicle; branches and pedicels slender. Spikelets 3–many-flowered, ovoid or broadly triangular, compressed, pendent or erect; florets closely imbricate; disarticulation above glumes and between florets; rachilla prolonged, glabrous. Glumes ± equal, persistent, concave, ovate-orbicular or lanceolate, spreading, < adjacent lemmas, at least the upper cordate at base. Lemma 5–many-nerved, broad, faintly keeled or rounded, usually deeply concave, orbicular or suborbicular, gibbous, usually coriaceous and shining centrally, sometimes cordate at base, margins wide, scarious. Palea < lemma, orbicular, obovate or lanceolate; keels 2, well-separated, usually narrowly winged and minutely ciliolate. Callus glabrous. Lodicules 2, membranous. Stamens 3, or 1 (South American spp.). Ovary glabrous, styles free to base. Caryopsis dorsally rounded, ventrally flattened; embryo small; hilum variable.

Type species: *B. minor* L.

c. 16 spp. of north temperate regions and South America. Naturalised spp. 4.

Northern Hemisphere spp. are chasmogamous and either self-incompatible, e.g., *B. media,* or self-compatible, e.g., *B. minor, B. maxima* [Murray, B. G. *Heredity 33*: 285–292 (1974)]. Some South American spp. are cleistogamous.

1 Spikelets 10–25 mm, 1–12, or rarely to 18, in panicle .... **1. maxima**
  Spikelets 2–7 mm, numerous in panicle *2*

2 Panicle contracted; spikelets ± sessile .... **4. rufa**
  Panicle open; spikelets pedicelled *3*

3 Plants annual; ligule 2.5–5.5 mm, acute; lemma 1.5–2.5 mm; anthers 0.4–
  0.8 mm .... **3. minor**
  Plants perennial; ligule 0.5–1 mm, truncate; lemma 3–3.5 mm; anthers (1)–
  2–2.5 mm .... **2. media**

1. B. maxima L., *Sp. Pl.* 70 (1753).   large quaking grass

Annual loose tufts or solitary shoots, (18)–25–65–(90) cm. Leaf-sheath papery, glabrous, finely ribbed, light green, light brown or purplish. Ligule 2–4–(5) mm, glabrous, denticulate, centrally tapered. Leaf-blade (2.2)–5–23 cm × 2–8 mm, thin, flat, finely ribbed, gradually tapered to fine acicular tip, abaxially occasionally sparsely prickle-toothed on ribs, adaxially ribs usually finely scaberulous; margins finely scaberulous. Culm (12)–20–50 cm, erect or geniculate at base, internodes glabrous. Panicle (2.2)–4–13–(16) cm, secund, nodding, of 1–12, rarely to 18, papery spikelets; branches few, sparsely scabrid, pedicels filiform, 0.5–2–(3) cm. Spikelets 1–2–(2.5) × 0.6–1.5 cm, 8–15-flowered, silvery, light green to golden-brown, often dark purplish at base. Glumes subequal, 5–6–(6.6) mm, (5)–7–9-nerved, spreading, firmly membranous, concave, suborbicular, smooth, often purplish. Lemma 6.5–8.5 mm, 7–9–(11)-nerved, centrally hardened and concave, glabrous, base cordate, sometimes with minute glandular hairs especially on auricle-like extensions, and appressed fine hairs on broad, firmly membranous margins, apex subacute. Palea 3–4 mm, to ⅔ length of lemma, orbicular, keels finely winged and densely short-ciliate, interkeel glabrous. Lodicules (1)–2–3 mm, lanceolate, obtuse. Anthers 1.5–2 mm. Gynoecium: ovary *c.* 1 mm; stigma-styles 2 mm. Caryopsis 2.5–3 × 1.5–2 mm; embryo 0.7–1 mm; hilum linear, 1.2–1.5 mm.

Naturalised from south Europe.

N.: scattered; S.: near Nelson City, Canterbury (near Rangiora, near Christchurch and at Akaroa), Westland (Knights Point), Otago (near Dunedin); St.; Ch. Often near the coast on roadsides and sandy or shingly waste land.

2. B. media L., *Sp. Pl.* 70 (1753). quaking grass

Shortly rhizomatous perennial, 30–55 cm. Leaf-sheath sub-membranous, glabrous, distinctly ribbed. Ligule 0.5–1 mm, blunt, glabrous. Leaf-blade *c.* 10 cm × 2.5 mm, flat to folded, glabrous; margins finely scabrid, tip narrow, acute. Culm 25–45 cm, erect, internodes glabrous. Panicle 10–16 cm, erect, ± pyramidal with numerous spikelets; rachis smooth to sparsely scabrid above; branches erect to spreading, scabrid, pedicels very filiform, curved or flexuous, smooth or sparsely scabrid, 5–10 mm. Spikelets 4–6 × 3–5.5 mm, pendent, broadly ovate, laterally compressed, usually purplish. Glumes subequal, 2.5–3 mm, 3-nerved, spreading, concave, suborbicular, subcoriaceous, glabrous; margins wide, whitish, membranous, apex hooded. Lemma 3–3.5 mm, 7–9-nerved, similar to glumes in texture and colour, cordate at base. Palea ≈ lemma, elliptic, flat, narrowly winged. Lodicules *c.* 1 mm, linear-lanceolate, acute. Anthers (1)–2–2.5 mm. Gynoecium: ovary *c.* 1 mm; stigma-styles 2–2.5 mm. Caryopsis *c.* 1 × *c.* 0.8 mm; embryo *c.* 0.6 mm; hilum punctiform, *c.* 0.3 mm.

Naturalised from Eurasia.

S.: South Canterbury (Whitecliffs and on east side of Tasman Valley). In manuka scrub and tussock grassland.

3. B. minor L., *Sp. Pl.* 70 (1753).   shivery grass

Erect to sprawling annual tufts, (5)–15–90 cm. Leaf-sheath submembranous, glabrous, striate. Ligule 2.5–5.5 mm, lanceolate, acute, glabrous. Leaf-blade (2)–5–16 cm × 2–9 mm, flat, thin, striate, adaxially minutely scabrid on ribs; margins minutely scabrid, tip acute. Culm 10–70 cm, erect or slightly geniculate at base, internodes glabrous. Panicle (1)–5–20 × (0.4)–3–11 cm, lax, much branched, with numerous spikelets; rachis glabrous, branches erect to widely spreading, finely scabrid, pedicels very filiform, often flexuous or curved, 5–10 mm. Spikelets 3–4.5 × 2.5–3.5 mm, (3)–5–9-flowered, pendent, ± ovate to subtriangular, shining, usually pale green, sometimes purplish. Glumes ± equal, 1.7–2.6 mm, 3–(5)-nerved, patent, concave, centrally firmly membranous, margins wide, hyaline, apex hooded. Lemma 1.5–2.5 mm, 7-nerved, similar to glumes but hardened central portion bearing scale-like, readily caducous hairs, deeply concave, base cordate, margins membranous, very broad, apex inflexed at maturity. Palea 1.4–1.8 mm, somewhat < lemma, elliptic, flat, keels narrow, interkeel with scale-like, readily caducous hairs. Lodicules 0.5–0.7 mm, linear-lanceolate, acute, sometimes hair-tipped. Anthers 0.4–0.8 mm. Gynoecium: ovary 0.5–0.6 mm; stigma-styles *c.* 1.2 mm. Caryopsis *c.* 1 × 0.5 mm; embryo *c.* 0.4 mm; hilum punctiform, *c.* 0.2 mm.

Naturalised from Mediterranean.

N.; S.; St.: scattered; K. Usually on roadsides, sometimes on sandhills or swamp margins.

4. B. rufa (J.Presl) Steud., *Nomencl. Bot.* ed. 2, *1*: 225 (1840).

Tufted perennial to 35 cm with short ascending rhizomes; branching extravaginal. Leaf-sheath submembranous, distinctly ribbed, keeled above, glabrous. Ligule 1–1.2 mm, rounded, erose, abaxially minutely hairy. Leaf-blade 2–15 cm × 1.5–7 mm, flat, striate; ribs and margins minutely prickle-toothed, tip subacute, hooded. Culm 10–28 cm, erect, internodes glabrous, pubescent near panicle. Panicle 1.2–5 cm, contracted; rachis sparsely pubescent, branches hidden by numerous, dense spikelets, pedicels very short, with few minute prickle-teeth. Spikelets 2–2.5 × 1.5–2 mm, 3–4-flowered, almost sessile, ovate, pale green, brownish near base. Glumes ± equal, 1.5–2 mm, 3-nerved, spreading, firmly membranous, concave, keeled, suborbicular, minutely prickle-toothed above with a few longer hairs on midnerve. Lemma *c.* 2 mm, 7-nerved, centrally hardened, concave, brown, shining, glabrous, apex hyaline, minutely papillose, margins minutely ciliate. Palea *c.* 1.5 mm, elliptic-suborbicular, keels ciliate. Lodicules *c.* 1 mm. Stamen 1; anther *c.* 0.3 mm. Gynoecium: ovary 0.4–0.5 mm; stigma-styles 0.5–0.7 mm. Caryopsis not seen.

Naturalised from South America.

N.: Auckland City. Street verges.

Cultivated for genetic work at University of Auckland and now naturalised in Auckland City near the University.

CATAPODIUM Link, 1827

Annuals. Ligule membranous. Leaf-blade flat, convolute when dry. Inflorescence a panicle with short, stiff branches, or a spiciform raceme. Spikelets 5–14-flowered, laterally compressed, shortly pedicelled, disarticulating tardily, or above glumes. Glumes subequal, ±coriaceous, (1)–3–5-nerved, acute to obtuse. Lemma 5-nerved, with inner lateral nerves sometimes indistinct, glabrous, rounded, or keeled near firm, acute to obtuse, or sometimes emarginate and apiculate tip. Palea ≈ lemma, 2-keeled, apex sometimes shortly bifid. Lodicules 2, membranous, glabrous, entire or bidentate. Callus glabrous. Stamens 3. Ovary glabrous; styles free to base. Caryopsis narrowly ellipsoid-oblong; embryo small; hilum punctiform.

Type species: *C. loliaceum* (Huds.) Link

2 spp., of Eurasia and North Africa. Naturalised sp. 1.

1. C. rigidum (L.) C.E.Hubb., *in* Dony *Fl. Bedfordshire* 437 (1953).

hard grass

Annual, erect or spreading, rigid tufts, from narrow base, 5–40 cm; branching intravaginal. Leaf-sheath glabrous, ribbed; margins wide, hyaline. Ligule 1.5–3.8 mm, glabrous, truncate, erose. Leaf-blade (0.5)–4–17 cm × 0.6–2.4 mm, flat, finely ribbed and often scabrid on ribs; margins minutely scabrid, narrowed to long, acute tip. Culm (2)–5.5–25 cm, erect, or geniculate at base, internodes glabrous. Panicle (2.3)–4–11 cm, linear-lanceolate to ovate, secund, often rather narrow, sometimes more open, rigid, shortly branched; branches and pedicels angled, scabrid on angles. Spikelets 3.5–10 mm, 4–9-flowered, greenish or sometimes purplish. Glumes subequal, lanceolate, acute, almost hyaline but midnerve prominent, scabrid; lower 1.2–2 mm, 1–3-nerved, upper 1.5–2.5 mm, 3-nerved. Lemma 1.8–2.6 mm, faintly 5-nerved, rounded, obtuse, coriaceous with narrow membranous margins. Palea ≈ lemma, keels minutely scabrid. Anthers 0.4–0.5 mm. Caryopsis 1.5–2 × 0.4–0.6 mm.

Naturalised.

N.; S.: usually coastal, scattered in the east, rare in the west. On sandy and gravelly beaches, on limestone, in depleted pasture, tracksides and gardens and in shingle ballast; usually lowland, occasionally montane.

Indigenous to western and southern Europe, northern Africa and western Asia, now naturalised in temperate regions of both Hemispheres.

CYNOSURUS L., 1753

Annuals or perennials. Leaf-sheath rounded. Ligule membranous. Leaf-blade flat, ± lax. Inflorescence a ± secund, condensed panicle. Spikelets dimorphic, ♀ spikelets ± concealed by ∅ spikelets. ♀ spikelets

1–5-flowered, often laterally compressed, disarticulating above glumes at maturity, rachilla prolonged; Ø spikelets below ♀ spikelets, consisting of narrow rigid bract-like glumes and lemmas. Glumes of ♀ spikelets subequal, hyaline, 1-nerved, excurrently mucronate or shortly awned. Lemma 5-nerved, chartaceous, oblong to lanceolate, rounded, awned from just below minutely or obscurely 2-toothed apex. Palea ≈ lemma, 2-keeled, apex shortly bifid. Lodicules 2, oblong. Stamens 3. Ovary glabrous; styles distinct, short; stigmas loosely plumose, laterally exserted. Caryopsis oblong, ± adhering to lemma and palea; embryo small, < ⅓ caryopsis; hilum short.

Type species: *C. cristatus* L.

c. 4 spp. of Europe, western Asia and Africa. Naturalised spp. 2.

> *1* Panicle linear, usually < 10 mm wide; lemmas very shortly awned; plants perennial .... **1. cristatus**
> Panicle ovate-oblong, often lobed, > 10 mm wide; lemmas conspicuously awned; plants annual .... **2. echinatus**

1. **C. cristatus** L., *Sp. Pl.* 72 (1753). crested dogstail

Compactly tufted perennials, 10–75–(85) cm. Leaf-sheath subcoriaceous, smooth below, minutely scabrid and keeled above near ligule. Ligule 0.4–1.8 mm, truncate, erose to ciliate. Leaf-blade (2.5)–7–24 cm × 1–4 mm, glabrous, or adaxially with minute prickle-teeth or hairs; margins glabrous, narrowed to long, scabrid, acuminate tip. Culm 2–70 cm, erect, internodes glabrous. Panicle 2–10–(15) cm, to 12 mm wide but usually quite narrow, secund, linear, spike-like, stiff, narrow-oblong, erect or slightly curved; rachis and very short branches with minute prickle-teeth or hairs. ♀ and Ø spikelets mixed together in dense clusters, green to purplish; occasionally proliferous: Ø spikelets 3.3–5.5 mm, persistent, ovate, flattened, of 7–10 acute to acuminate narrow bracts, 2.5–3.8 mm, 1-nerved, scabrid on nerve, margins hyaline; ♀ spikelets 4–6 mm, 2–5-flowered, elliptic-oblong. Glumes persistent, ± equal, 3–4 mm, 1-nerved, hyaline, midnerve scabrid above. Lemma 3–4 mm, faintly 5-nerved, scabrid above, subcoriaceous with very narrow hyaline margin; awn minute,

0.4–1 mm. Palea 2.5–3.5 mm, hyaline, keels minutely scabrid, interkeel papillose. Anthers 1–2.5 mm, purple or yellow. Caryopsis 1.4–2 × 0.6–0.7 mm.

Naturalised from Eurasia.

N.: scattered throughout; S.: common in Canterbury, more scattered further south, rare in Nelson and not recorded from Marlborough; A., C. On grassy roadsides and in pasture, sometimes at swamp margins; lowland to subalpine.

Now widespread in temperate regions in both Hemispheres.

2. C. echinatus L., *Sp. Pl.* 72 (1753).   rough dogstail

Annuals, in tufts, or shoots solitary, 7–70–(110) cm. Leaf-sheath sub-membranous, smooth, or very minutely scabrid on ribs; upper sheaths ± inflated. Ligule (0.7)–2–6.5 mm, glabrous, rounded. Leaf-blade (2)–4–26 cm × 2–8 mm, abaxially mainly smooth, but finely scabrid on ribs above, adaxially finely scabrid on ribs, margins finely scabrid, tip finely tapered, acicular, scabrid. Culm 6–65–(100) cm, erect or geniculate at base, internodes glabrous. Panicle (1)–1.5–7.5 × (0.6)–1–4.5 cm, secund, spike-like, ovate to oblong or almost globose, bristly, shining; rachis and very short branches minutely scabrid-papillose on angles. ♀ and Ø spikelets mixed together in dense clusters, green to purplish; occasionally pro-liferous: Ø spikelets 6–11 mm, persistent, obovate, of 9–14 narrow-lanceolate, finely scabrid, rigid, finely awn-tipped bracts, 6–8.5 mm, spreading at maturity; ♀ spikelets 10–20 mm, 1–3-flowered, elliptic. Glumes persistent, ± equal and equalling spikelet, 1-nerved, hyaline, shining, narrow-lanceolate, midnerve and margins minutely scabrid, apex fine, acicular. Lemma 4–6.5 mm, faintly 5-nerved below, more obviously nerved above, submembranous, smooth in lower ½, scabrid above and on narrow hyaline margins; awn (5)–7–14.5 mm, strong, straight, scabrid. Palea ≈ lemma, scabrid near apex. Anthers 1.6–2.6 mm. Caryopsis 3–3.8 × 0.8–1.3 mm.

Naturalised from Mediterranean.

N.: scattered; S.: common in Canterbury, scattered in Marlborough, Otago and Southland. On roadsides, waste land, in pasture and crops, and on hillsides in grassland; lowland to montane.

A widespread weed in temperate regions of both Hemispheres.

DACTYLIS L., 1753

Perennials, with young shoots much flattened. Leaf-sheath closed, at least below. Ligule membranous. Leaf-blade flat or involute. Inflorescence a compound panicle, of dense, secund spikelet clusters terminating erect or spreading branches; lower branches often distant and longer. Spikelets 2–6-flowered, laterally compressed, shortly pedicelled; disarticulation above glumes; rachilla prolonged. Glumes subequal, keeled; lower 1–3-nerved, upper 3-nerved. Lemma > glumes, 5-nerved, stiffly chartaceous, keel spinulose, apex mucronate to shortly awned. Palea ≈ lemma, 2-keeled, keels finely ciliate. Lodicules 2, membranous, glabrous, denticulate. Stamens 3. Ovary glabrous; styles free, very short, terminal; stigmas plumose. Caryopsis dorsiventrally compressed; embryo small, < ⅓ length of caryopsis; hilum punctiform or shortly elliptic.

Type species: *D. glomerata* L.

c. 5 spp. of temperate Eurasia. Naturalised sp. 1.

1. D. glomerata L., *Sp. Pl.* 71 (1753).   cocksfoot

Dense, coarse, often greyish green tufts, 25–120 cm; branching intravaginal, shoots compressed. Leaf-sheath green to pale brown, subcoriaceous, keeled, ribs finely prickle-toothed or minutely hairy. Ligule 3–10.5 mm, acute, erose or sometimes ciliate, abaxially smooth, or with minute hairs or prickle-teeth. Leaf-blade (7)–15–55 cm × 1.5–9 mm, flat, midrib strong, lateral ribs fine, ribs, margins and acicular tip scabrid. Culm 15–95 cm, erect or spreading, internodes glabrous. Panicle (5)–7–20–(45) cm, secund, with spikelets in dense spike-like clusters at tips of few, strong, scabrid branches; lower branches often distant, naked below. Spikelets 6–8 mm, 3–7-flowered, stiff, almost sessile, oblong or cuneate, light green or purplish; often proliferous. Glumes persistent, unequal, membranous, ovate-lanceolate, strongly keeled; lower 3–5 mm, acute, upper 4–6 mm, acuminate, keel scabrid to long-ciliate. Lemma 4.5–6.5–(7) mm, 5-nerved, ovate-lanceolate, sparsely pubescent, scabrid, papillose or glabrous, keel scabrid to long-ciliate, margins hyaline, scabrid; awn terminal 0.2–1.2 mm. Palea ≤ lemma, keels short-ciliate, interkeel finely scabrid. Anthers 2–3.5 mm. Caryopsis *c.* 1 × 0.5 mm.

Naturalised.

N.: throughout, but rare in the north, and in the east from East Cape southwards; S.: common in the east in the upper half, rare in the west and in Otago and Southland; St.; K., Ch., C. Lowland in waste ground, pasture, gardens and dune hollows, to subalpine in short tussock grassland and scrub.

An important pasture grass; native to Europe, North Africa and temperate Asia and now naturalised in temperate regions in both Hemispheres.

FESTUCA L., 1753

Perennial caespitose, stoloniferous, or rhizomatous. Leaf-sheath split to base or margins fused, apical auricles small. Ligule ciliolate. Leaf-blade convolute, or folded; usually persistent; pungent in *F. ultramafica;* variable in TS. Culms unbranched; nodes glabrous. Inflorescence an open or contracted panicle of 2–many-flowered laterally compressed, pedicelled spikelets disarticulating above the glumes. Glumes unequal or subequal, persistent, ± keeled and ornamented, rarely exceeding lemmas. Lemma lanceolate-acute or awned, rounded on back or slightly carinate, membranous to thinly coriaceous, 5–7-nerved, awn terminal. Palea bidentate, keels scabrid. Flowers ♀, usually chasmogamous. Lodicules 2, unequally lobed, often hair-tipped. Stamens 3. Ovary glabrous or hispid hairy at apex. Caryopsis dorsiconvex, ventrally sulcate, free; embryo small; hilum long, linear. Fig. 6.

Type species: *F. ovina* L.

c. 450 spp. worldwide — "A large and variable genus for which no overall treatment is available..." (Clayton and Renvoize 1986 op. cit.). Endemic spp. 9, indigenous sp. 1; naturalised spp. 3.

The genus was revised for N.Z. indigenous species by Connor, H. E. *N.Z. J. Bot. 36*: 329–367 (1998); Howarth, W. O. *J. Linn. Soc. Lond. 48*: 57–77 (1928) — incorrectly cited as vol. *68* in Connor 1998 (op. cit.) — and Saint-Yves, A.

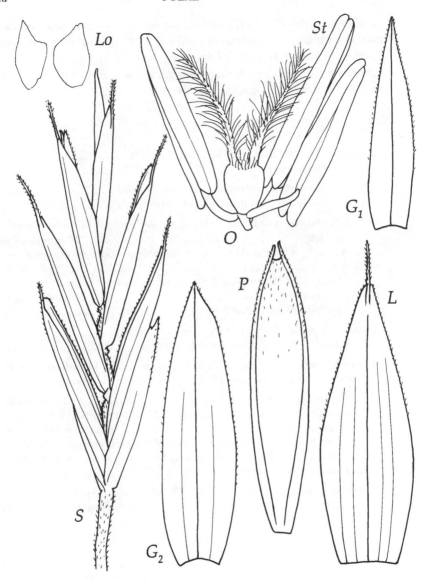

Fig. 6   *Festuca novae zelandiae*
S, spikelet, × 10; **G₁**, **G₂**, glumes, × 15; **L**, lemma, × 15; **P**, palea dorsal view, × 15; **Lo**, lodicules,
× 25; **O**, ovary, styles and stigmas, × 25; **St**, stamens, × 25.

Candollea 4: 293–307 (1931) had earlier treated endemic and naturalised taxa. Variation in leaf-blade anatomy for *F. matthewsii* and *F. novae-zelandiae* is detailed in Connor, H. E., *N.Z.J. Sci. 3*: 468–509 (1960). Experimental hybrids between *F. matthewsii* and *F. novae-zelandiae* are fertile [Connor, H. E. *N.Z.J. Bot. 6*: 293–308 (1968)]; none is known in the wild.

1 Branching intravaginal *2*
  Branching extravaginal *9*

2 Leaf-blades antrorsely scabrid *3*
  Leaf-blades smooth *5*

3 Lemma awnless or mucronate; spikelets 4–7 mm .... **5. filiformis**
  Lemma evidently awned; spikelets 8–20 mm *4*

4 Panicle with erect or lax branches not conspicuously borne above glaucous brown leaves .... **10. novae-zelandiae**
  Panicle with wide angled branches conspicuously borne above glaucous blue leaves .... **8. matthewsii**
    4a Rachis prickle-toothed on margins and abundantly minutely so elsewhere .... subsp. **aquilonia**
    Rachis prickle-toothed on angular margins .... subsp. **latifundii**

5 Panicle ± open; glumes not including lemmas *6*
  Panicle contracted; glumes including lemmas *8*

6 Lemma margins conspicuously hair-fringed .... **11. ovina** subsp. **hirtula**
  Lemma margins exceedingly finely prickle-toothed *7*

7 Leaf-blades persisting on sheath .... **8. matthewsii**
    7a Leaf-blades thick; stout stature; culm internodes glabrous; spikelets 14–20 mm .... subsp. **matthewsii**
    Leaf-blades thin; slender stature; culm internodes glabrous or prickle-toothed; spikelets 10–12 mm .... subsp. **pisamontis**
  Leaf-blades disarticulating at collar .... **1. actae**

8 Awns 6–13 mm, ≥ lemma; anthers 3.75–4.25 mm; chasmogamous ...... **3. coxii**
  Awns 1.5–2.5 mm, ≪ lemma; anthers 0.6–0.8 mm; cleistogamous ... **2. contracta**

9 Lemma glabrous *10*
  Lemma prickle-toothed and/or hairy *12*

10 Panicle branches wide-angled .... **4. deflexa**
  Panicle branches spreading-erect, occasionally otherwise *11*

11 Leaf-blades weakly hexagonal or terete, secund; awns 0–1.5 mm ....
    .... **9. multinodis**
  Leaf-blades conduplicate, pungent; awns 1.5–4.5 mm .... **13. ultramafica**

12 Glume apex conspicuously ciliate .... **6. luciarum**
    Glume apex pointed *13*
13 Anthers 0.5–1.0 mm; flowers cleistogamous .... **7. madida**
    Anthers 2–3 mm; flowers chasmogamous .... **12. rubra**
    13a Habit rhizomatous .... subsp. **rubra**
        Habit caespitose .... subsp. **commutata**

1. F. actae Connor, *N.Z. J. Bot. 36*: 335 (1998) ≡ *F. ovina* subsp. *matthewsii* var. *grandiflora* Howarth, *J. Linn. Soc. Lond. 48:* 66 (1928) ≡ *F. novae-zelandiae* var. *grandiflora* (Howarth) St.-Yves, *Candollea 4*: 298 (1931) a combination based on var. *grandiflora* is preoccupied by *F. grandiflora* Lam. (1791); Lectotype: K! *A. Wall* Lyttelton–Sumner Rd., sea level; close to Christchurch, Dec 1920 (Wall 7, page 5; Howarth I; Saint-Yves 1-I) (designated by Connor 1998 op. cit. p. 335). = *F. ovina* subsp. *novae-zelandiae* var. *grandiflora* Howarth, *J. Linn. Soc. Lond. 48*: 64 (1928); Lectotype: K! *D. Petrie* grown by Dr. Petrie in his garden at Auckland from plant at Sumner Road, Lyttelton (Wall 8, page 53; Howarth XXIII; Saint-Yves 8-XXIII) (designated by Connor 1998 op. cit. p. 335). = *F. ovina* subsp. *matthewsii* var. *eu-matthewsii* Howarth, *J. Linn. Soc. Lond. 48*: 65 (1928) non Hack. (1903), *nom. invalid.*; Lectotype: K photo! *A. Wall* Diamond Harbour, Lyttelton Harbour, near Christchurch among rocks at sea level (Wall 3, page 17; Howarth V; Saint-Yves 3-V) (designated by Connor 1998 op. cit. p. 335). = *F. petriei* forma *tenuifolia* Howarth, *J. Linn. Soc. Lond. 48*: 71 (1928); Holotype: K! *A. Wall* Lyttelton–Sumner Rd. 24 Dec. [1920]. "The more slender form of F. nov. zel." (Wall 8, page 35; Howarth XIV; Saint-Yves 8-XIV).

Slender tussock with intravaginal branches, with fine and long glaucous leaf-blades. Prophyll 3–5 cm, keels irregularly but mostly antrorsely ciliate. Leaf-sheath 5–10 cm, stramineous occasionally reddened, glabrous, striate, margins becoming membranous; apical auricles 0.4–0.7 mm, truncate or rounded, ciliate. Ligule as for auricles. Collar 0.6–1.7 mm, manifestly thickened, usually becoming brown coloured, adaxially with many small white hairs. Leaf-blade 20–60 cm × 0.4–0.7 mm diam., terete or hexagonal and ribbed, glaucous, glabrous except for prickle-teeth at pointed apex, disarticulating at collar, adaxially and on margin a multitude of small (0.1–0.15 mm) white hairs; TS: 5 vascular bundles,

7–11 sclerenchyma strands. Culm 25–60 cm, erect or geniculate at base, nodes 2–3 evident, internodes glabrous. Panicle 5–25 cm, with 7–9 nodes of 10–30 spikelets; basal branches 4–5 cm, binate, lax, of 2–6 spikelets and naked below, soon becoming single ascending branches, uppermost 3–5 spikelets solitary on pedicels; rachis glabrous or sparsely prickle-toothed below becoming more so, branches and pedicels prickle-toothed. Spikelets 10–17 mm × 6 mm, of 4–7–(12) florets. Glumes unequal, keeled, prickle-teeth on keel and at apex, variously elsewhere, usually green centrally lighter at margins, sometimes purpled, margin ciliate; lower 3–4.5 mm, 1-nerved, long triangular acute, upper 4.5–7.5 mm, 3-nerved occasionally 5-nerved, narrowly ovate, acute to obtuse. Lemma 6–9 mm, rounded, lobes small (0.1–0.2 mm), glaucous, prickle-teeth from outer nerve to ciliate margin, scattered elsewhere, denser near awn; awn 0.1–2.5 mm. Palea 6–8 mm, sometimes > lemma, deeply (0.4–0.6 mm) bifid, keels toothed to base or nearly so, flanks and interkeel hairy above. Callus 0.2–0.5 mm, margin very short stiff hairy; articulation flat. Rachilla 1–1.5 mm, very shortly antrorsely stiff hairy. Lodicules 0.7–1.5 mm, lobed or entire, hair-tipped. Anthers 3.5–4.4 mm, yellow. Gynoecium: ovary 0.5–1 mm, hispid hairs at apex (0.12 mm) in 2 lateral groups or occasionally surrounding apex and longer; stigma-styles 1.4–2.8 mm. Caryopsis 3–4.8 mm; embryo 1 mm; hilum 1.7–3 mm.

Endemic.

S.: Banks Peninsula. Rocks at sea level, rock outcrops and bluffs; sea level to 700 m.

Two distinct rachis vestures are present in *F. actae*: (i) abundant small prickles covering the whole rachis in the style of *F. novae-zelandiae*, (ii) prickles on the trigonous rachis margins. The former is found on plants from Port Hills, e.g., CHR 5016, CHR 202745–7, CHR 333045, WELT 68792, AK 215000, which are mostly "grandifloras". Prickle-toothed angles are found on specimens from the Peninsula proper, e.g., CHR 202748, CHR 221842, CHR 403939 but CHR 370394 *J. M. Hay* Whisky Bay, has some prickle-teeth on the rachis surfaces although not as densely as on those from the Port Hills.

Fine-leaved specimens, to which forma *tenuifolia* could apply, are for example: CHR 202753 *L. B. Moore* Lyttelton, Dec 1970; CHR 202746 *V. D. Zotov* Evans Pass, Dec 1970; CHR 357484 *E. Edgar* & *B. A. Matthews* Lyttelton, Dec 1979.

2. F. contracta Kirk, *T.N.Z.I. 27*: 353 (1895); Holotype: WELT 68607! *A. Hamilton* 1894. = *F. erecta* d'Urv., *Mém. Soc. Linn. Paris 4*: 601 (1825) non (Huds.) Wallr. 1822; Holotype: P! *J. D'Urville* I. Soledad, Falkland Islands.

Tufted tussock with pale, flabellate intravaginal branches of many smooth leaf-blades often exceeding the compact, short, stiffly erectly branched purple suffused inflorescences of many few-flowered closely packed spikelets with shortly prickle-toothed lemmas. Shoots dimorphic, narrow if vegetative, swollen if reproductive. Prophyll 4–5 cm, stramineous, pointed, keels antrorsely ciliate, interkeel antrorsely hairy. Branching intravaginal. Leaf-sheath 5–10 cm, glabrous, stramineous, strongly nerved, margins membranous; apical auricles 0.4–0.5 mm, ciliate, symmetrical. Ligule as for auricles. Leaf-blade 12–25 cm × 0.8–1.0 mm diam., glaucous, glabrous, stiff, pointed, terete or shallowly compressed-terete, adaxially and on margins a multitude of short white antrorse hairs; TS: 7–8 vascular bundles, sclerenchyma discontinuous of up to 10–12 strands, costal sclerenchyma present. Culm 12–16–20–(40) cm, nodes hidden, internodes finely antrorsely prickle-toothed and very short hairy, almost always hidden. Panicle (5)–7–13 cm, narrow, compact, spike-like, with 13–15 nodes all hidden by 20–50 spikelets; branches mostly binate, erect, stiff, very close-set, imbricate, > nearest internode, not naked below, basal branch appressed, 2–4 cm, of 5–7 spikelets, uppermost 5–7 solitary, on short (1 mm) pedicels; rachis, branches, and pedicels shortly and densely prickle-toothed on margins, elsewhere frequently with antrorse short fine hairs. Spikelet 9–10–12 mm × 1.5–1.8 mm, narrow, of 3 florets; glumes, lemmas, paleas purple suffused above; lemma included, awns exserted. Glumes ± equal, 8–11 mm, 3-nerved, upper prominently so, centrally green, prickle-teeth on keel and above, margin hyaline, finely ciliate. Lemma 7 mm, lobes small (0.25 mm), 5-nerved sometimes evident, abundant small uniform prickle-teeth present throughout; awn 1.5–2.5 mm. Palea 6.5–7 mm, ≤ lemma, apex bifid (0.2–0.4 mm), ciliate; keels toothed ± to base, interkeel hairs to base, flanks shortly prickle-toothed in upper ½. Callus 0.1–0.2 mm, very shortly bearded near rachilla; articulation flat. Rachilla 1–1.25 mm, very shortly stiff hairy.

Lodicules 0.6–0.8 mm, lobed, glabrous. Anthers 0.6–0.8 mm, purple; pollen shed from loculi. Gynoecium: ovary 1–1.25 mm, triangular turbinate, cap thickened, apex glabrous; stigma-styles 2–3 mm, stigmas sparingly branched. Caryopsis 2.7–3.0 mm, free but firmly enclosed by anthoecium; embryo 0.5–0.6 mm; hilum 2.0–2.5 mm. Cleistogamous. $2n = c.$ 170.

Indigenous. Circum-Antarctic.

M. Grassland, rocky places, peat; 20–400 m.

Also in Tierra del Fuego, Falkland Is, South Georgia, Îles Kerguélen.

3. F. coxii (Petrie) Hack., *in* Cheeseman *Man. N.Z. Fl.* 919 (1906) ≡ *Agropyrum coxii* Petrie, *T.N.Z.I. 34*: 395 (1902); Lectotype: AK 2009! *L. Cockayne & F. A. D. Cox* Herb. L. Cockayne 4024, rocks near the sea, Chatham Island, Jan 1901 (No 1515 to Hackel), (designated by Connor 1998 op. cit. p. 339).

Tall tufted tussock with intravaginal branching sometimes internodes elongating below and rooting at nodes; inflorescence short, compact, usually shorter than tall leaf-blades with long awned abundantly prickle-toothed florets in shortly pedicelled spikelets. Prophyll 4–6 cm, stramineous and dark-brown papery on margins, keels mostly retrorsely hairy. Leaf-sheath 8–12–(20) cm, thin, pale, much broader than leaf-blades, minutely retrorsely or antrorsely hairy between nerves becoming glabrous above, margins dark brown membranous below; apical auricles 0–0.5 mm, ciliate. Collar conspicuously thickened and curved. Ligule 0.3–0.5 mm, ciliate. Leaf-blade (11–15)–20–35–(40) cm × (0.5)–06–0.8–(0.9) mm diam., glaucous, softly sharp-pointed, terete to somewhat compressed, glabrous, adaxially and on margin a multitude of short (0.15 mm) antrorse or erect prickle-teeth becoming smaller above; TS: 5 vascular bundles, sclerenchyma continuous; costal sclerenchyma often present. Culm 25–45 cm, almost always included by leaf-blades; nodes 2–3 dark, glabrous, sometimes ± geniculate; internodes glabrous or densely antrorsely short hairy sometimes becoming less so below. Panicle 6–10–(15) cm, narrow, compact, with 8–11 nodes, of 12–18 close-set, usually imbricate, spikelets; branches short, erect-appressed, basal branch 1–2 cm of 3–5 spikelets, not naked below, uppermost 6–10 spikelets solitary on short pedicels; rachis, branches and pedicels prickle-toothed on margins, and frequently also densely antrorsely short hairy becoming less so above, or ± glabrous. Spikelets 15–25–(30) mm, 5–6 mm wide, of 5–7 florets. Glumes evidently unequal, usually green centrally, broad, narrowing and becoming awned, prominently keeled, glabrous except for prickle-teeth on keel above and near awn, margin hyaline sometimes short hairy or finely ciliate; usually ≈ lowermost lemma, twice as long as nearest proximate internodes or in upper panicle twice proximate internodes of solitary spikelets; lower 4–10 mm, 1-nerved, upper 6–12 mm, 3-nerved (both longer on The Sisters). Lemma 6–10 mm, becoming briefly canaliculate, lobes 0 or extremely minute, 5-nerved, slightly keeled, abundantly striately prickle-toothed (0.15 mm) throughout, longer teeth and/or hairs below, and on keel and at margins below; awn 6–13 mm, usually > lemma. Palea 6–9 mm, usually < lemma, apex deeply (0.5–2 mm) bifid, keels toothed to base, interkeel hairs above, margins of flanks very shortly toothed; sometimes folded. Callus 0.3–0.6 mm, abundantly long (0.2 mm) hairy on upper margin, shorter centrally; articulation acute. Rachilla 1–1.6 mm, densely antrorsely long hairy. Lodicules 0.7–1.5 mm, conspicuously hair-tipped, simple lozenge-shaped to slightly lobed, ≤ ovary. Anthers 3.7–4.2 mm, orange. Gynoecium: ovary 1–1.4 mm, triangular turbinate, apex with narrow rim enclosing base of styles and central tuft of hispid hairs (0.2 mm); stigma-styles 2.5–3 mm. Caryopsis 3.7–4.6 mm, free but firmly enclosed by anthoecium; embryo 0.7–1 mm; hilum ≈ caryopsis. $2n = 56$.

Endemic.

Ch.: known from all islands except The Forty Fours. Coastal cliffs and bluffs to 200 m.

The most remarkable endemic species of *Festuca* especially in the vesture of the lemma, awn ≥ lemma, the very short (5 mm) internodes of the upper inflorescence, and many-leaved shoots. It looks less like a *Festuca* than any other endemic. Petrie (1902 loc. cit.) not unreasonably interpreted it as a true *Agropyron* in spite of the short branches of the rachis and the shortly pedicelled spikelets.

4. F. deflexa Connor, *N.Z. J. Bot. 36*: 341 (1998); Holotype: CHR 389142! *A. P. Druce* Mt Baldy, Mt Arthur Range, NW Nelson, 4700 ft, tussockland, March 1982.

Elegant tufted slender or ± stout extravaginally branching grass rooting at nodes with tall inflorescences of deflexed branches of spikelets borne high above fine glaucous leaf-blades. Prophyll 3–6 cm, keels predominantly antrorsely ciliate. Branching extravaginal. Leaf-sheath 3–7–12 cm, keeled, red, purple or stramineous, glabrous or with short fine prickles sparse on ribs, margins membranous, glabrous; apical auricles 0.4–0.7 mm, rounded or truncate, symmetrical, ciliate. Ligule shorter centrally, erose, ciliate. Collar triangular, thickened. Leaf-blade 15–25 cm × 0.4–0.7 mm diam., hexagonal in outline and ribs evident or rounded, glaucous, usually persistent, abaxially smooth but occasionally prickle-toothed below and very occasionally with retrorse hairs, abaxially and on margins a multitude of short antrose hairs at base becoming fewer above; TS: 5 vascular bundles, 7 sclerenchyma strands. Culm (15)–30–65 cm, erect, much exceeding leaf-blades, internodes glabrous, nodes evident. Panicle 7–16 cm, with 4–10 nodes, 7–25 spikelets; branches widely spaced, pulvinate and spreading to deflexed, basal branch solitary or binate 5–8 cm with 4–8 spikelets, naked below, uppermost 3–9 spikelets solitary on 4–6 mm pedicels; rachis smooth below becoming prickle-toothed on edges, branches and pedicels prickle-toothed. Spikelets 8–14 mm × 3–4 mm wide, of 5–7 florets, ripening golden-brown, often very distant. Glumes unequal, keeled or faintly so, prickle-toothed on keel above, margins ciliate, apex truncate and sometimes shallowly cleft or obtuse; usually green or light purple centrally, lighter or hyaline at thin margins; lower 3–4.5 mm, 1-nerved, upper 3.5–6.5 mm, usually 3-nerved often conspicuous. Lemma 5–6.75 mm, lobes 0 or minute, lightly purpled, nerves evident, slightly keeled above, prickle-toothed from outermost nerve to very shortly ciliate chartaceous margin and near awn; awn 0.2–1.75 mm, sparsely prickle-toothed. Palea 5.5–7.5 mm, ≥ lemma, apex deeply (0.2–0.5 mm) bifid, keels toothed sometimes to base, interkeel hairs in upper ⅓, margins of flanks very shortly toothed. Callus 0.2–0.75 mm, bearded near rachilla; articulation ± flat. Rachilla 1–1.5 mm, short stiff hairy. Lodicules 0.7–1.1 mm, lobed or entire, hair-tipped. Anthers (2)–3–4.0 mm, yellow or purple tinged. Gynoecium: ovary 0.6–1 mm, hispid hairs at apex; stigma-styles 1.5–2.5 mm, stigmas almost to base. Caryopsis 3–3.5 mm; embryo 0.75 mm; hilum 2 mm.

Endemic.

S.: north-west Nelson. Predominantly in tussock grasslands but also on bluffs or in rocky places, often marble or calcareous; 600–1650 m.

Plants may be graceful, slender, and very finely leaved with inflorescences of deflexed branches borne high above them as in: CHR 249981 *A. P. Druce* Mt Luna, Jan 1971; CHR 197064 *A. P. Druce* Gouland Downs, Jan 1969. Stouter forms with longer leaves are represented by CHR 325755 *A. P. Druce* Burgoo Stream, Jan 1970. Inflorescences with few spikelets show very strikingly the pulvinately deflexed branching emphasised in the description but some inflorescences have less divergent branches, e.g., CHR 95718 *R. Mason & N. T. Moar 4671* Lake Sylvester, Feb 1957; CHR 202773 *I. M. Ritchie* Mt Arthur, Jan 1970; CHR 393736 *A. P. Druce* Owen Range, Jan 1983.

5. F. filiformis Pourr., *Mém. Acad. Roy. Sci. Toulouse 3*: 319 (1788).

Tufts of scabrid very fine leaves overtopped by darkly coloured narrow inflorescences with many spikelets of awnless florets. Caespitose or shortly stoloniferous; branching intravaginal. Leaf-sheath 1–5 cm, ≫ wider than leaf-blade, keeled, striate, open to base, glabrous or occasionally with scattered fine hairs; apical auricles *c.* 0.1 mm, finely ciliate. Collar thickened. Ligule 0.2–0.2 mm. Leaf-blade to 20 cm × 0.3–0.5 mm, elliptical to laterally compressed, capillary, abaxially scattered fine antrorse prickle-teeth, adaxially finely hairy; TS: 5–7 vascular bundles, 1 rib, sclerenchyma continuous. Culm to 30 cm, often a single internode, densely finely hairy or becoming glabrous, sometimes purpled. Panicle 2–4–8 cm, very narrow, of 10–12–14–(20) nodes, (10)–60 spikelets much overlapping; branches solitary, short, mostly appressed to rachis, scarcely naked below; basal branch 1.5–3 cm of 10–20 spikelets, uppermost 6–10 spikelets solitary on short pedicels; rachis prickled on angles below becoming dense above, branches and pedicels densely so. Spikelets 4–7 mm, 3–6 florets, widely opened at anthesis, sometimes bright violet. Glumes unequal, reflexed at anthesis, keeled, prickle-toothed above, margins shortly ciliate; lower 1.0–1.75 mm, 1-nerved, keeled, upper 1.75–2.5 mm, 3 evident or raised nerves, ± obtuse. Lemma 2.5–3 mm, 5-nerved, lobes minute or 0, rounded below, prickle-teeth above on midnerve and below on lateral nerves, margins ciliate below; awn 0, mucro, or 0.1–0.2–0.5 mm; golden to purple especially above. Palea 2.5–3 mm, ≥ lemma, broad, apex narrowed and bifid, keels toothed in upper ⅓, interkeel with some few hairs at apex. Rachilla 0.4–0.5 mm, short hairy. Callus 0.1–0.15 mm, shortly bearded laterally; articulation ± oblique. Lodicules 0.3–0.5 mm, entire or lobed. Anthers 1.4–1.8 mm, yellow or purple-suffused, or purple. Gynoecium: ovary 0.5–0.7 mm, glabrous; stigma-styles 1.0–1.25 mm. Caryopsis 1.4–1.75 mm, sulcate, free; embryo *c.* 0.3 mm; hilum linear.

Naturalised from Europe.

N.: central mountains, occasional elsewhere (Hawkes Bay, Manawatu); S.: Nelson (Cobb Valley, Takaka), in east from Marlborough to Otago especially in dry inland basins, and western Southland. Scrub, modified tussock-grasslands, and river terraces and beds; sea level to 1200 m.

6. F. luciarum Connor, *N.Z. J. Bot. 36*: 343 (1998); Holotype: AK 200090! *L. M. Cranwell* Maungapohatu, Urewera, 20–23 Jan. 1932.

Short extravaginally tufted to tall stoloniferous grass with long shoots bearing inflorescences of small panicles of few broad dark violet suffused usually patent spikelets on short prostrate to ascending culms above the shorter leaves. Branching extravaginal. Leaf-sheath 2–5–(8) cm, glabrous, striate, much wider than leaf-blade, becoming brown and fibrous below, margins membranous; apical auricles 0.3–0.4 mm, rounded, ciliate. Ligule as for auricles. Leaf-blade 3–12 cm × 0.6–1.0 mm diam.; ± hexagonal with evident ribs or folded, glabrous, glaucous, adaxially and on margins with a multitude of short antrorse hairs; TS: 5–7 vascular bundles, 7–9 sclerenchyma strands, rarely irregular. Culm 4–15 cm or 15–30–(50) cm, in swollen shoots, erect or erect-ascending, usually > leaf-blades; nodes visible, internodes glabrous. Panicle (2)–3–7–(10) cm, with 4–7 nodes, 5–11–20 spikelets; branches erect or weakly spreading usually solitary, basal branch 2–5 cm of 1–4 spikelets, not naked below, uppermost 4–5 spikelets solitary on short pedicels; rachis mostly glabrous and frequently tortuous below, branches and pedicels usually prickle-toothed. Spikelets 8–12 mm × (4)–5–7 mm wide, of 4–7–(10) florets, glaucous, dull violet suffused, imbricate, becoming evidently patent at anthesis and up to 10 mm wide. Glumes evidently unequal, keeled, narrowing to become acute or acuminate, smooth but occasionally prickle-toothed on keels, apex with long (0.2–0.5 mm) cilia, margins shortly or conspicuously long ciliate; lower 2.6–3–4 mm, l-nerved, upper 3–4.5–6 mm, 3-nerved, nerves sometimes evident. Lemma 5–6–(7) mm, lobes 0 or very short, 5-nerved, slightly keeled above, inrolled, ± prickle-toothed throughout or short stiff hairy and prickle-toothed; awn 0–0.5–1 mm; apex of lowest lemma usually awnless and often long (0.3–0.5 mm) ciliate. Palea 5.5–6.5–(7) mm, usually > lemma, apex deeply (0.3–1 mm) bifid, keels toothed in upper ⅓ occasionally more, interkeel hairs above, margins of flanks ciliate. Callus 0.2–0.3 mm, shortly stiffly bearded throughout; articulation oblique. Rachilla 0.75–1.5 mm, with short prickle-teeth or stiff hairs. Lodicules 1–1.4 mm, ≥ ovary, lobed, glabrous. Anthers 2–2.5–(3.5) mm. Gynoecium: ovary 1 mm, turbinate, hispid hairs at apex or glabrous; stigma-styles (1.75)–2–3 mm. Caryopsis 3 mm; embryo 1 mm; hilum linear, 2 mm.

Endemic.

N.: eastern and inland on Raukumara Range (Mt Hikurangi, Mt Wharekia), Maungaharuru Range, and Huiarau Range (Mt Maunga-pohatu). Limestone rocks and cliffs, tussock grasslands; 900–1500 m.

The four localities for this species are geographically separate; the morphological differences among plants at these sites are reflected chiefly in ornamentation of the lemma: prickle-toothed at all sites but hairs also present at Maungaharuru and Mt Maungapohatu; ovary apex glabrous at Mt Maungapohatu, hairy elsewhere. The range in size and the extent of stolon development varies quite considerably. An attempt to resolve the confusion among collecting dates of specimens from Mt Maungapohatu at CHR and AK is in Connor (1998 op. cit. p. 344); it is thought that *L. M. Cranwell* Jan. 1932 is the correct date for the only collections made there and that the specimens in CHR have incorrect labels.

7. F. madida Connor, *N.Z. J. Bot. 36*: 345 (1998); Holotype: CHR 74284! *A. P. Druce* margin of bog in tussock, Mokai Patea, Ruahine Mts *c.* 5000 ft, Feb. 1951.

Tufted to open extravaginally branched graceful, slender, upland grass of wet sites, with long-auricled fine leaves and smooth culms bearing short inflorescences of a few golden-violet spikelets, the lowest branch (es) usually widely divergent from the finely and minutely prickle-toothed rachis. Prophyll short (2–3 cm), keels glabrous. Branching extravaginal. Leaf-sheath 3–10 cm, striate, glabrous, much wider than leaf-blade, becoming red and fibrous; apical auricles (0.5)–0.7–1.7–(2) mm, lacerate, shortly ciliate. Ligule as for auricles. Leaf-blade 4–10–(20) cm × 0.3–0.4 mm diam., triangular or hexagonal, ridged, abaxially with long retrorse hairs below becoming glabrous, adaxially and on margins abundant short hairs; T.S.: 3–5 vascular bundles, 5–7 sclerenchyma strands. Culm (6)–10–30–(45) cm, slender, often greatly exceeding leaves, nodes hidden sometimes geniculate, internodes usually only one visible, glabrous. Panicle (1.5)–2–3.5–(6) cm, with 3–6–(9) nodes, (2)–3–6–(10) spikelets; basal branch pulvinate and widely divergent or erect ascending, (1)–1.5–2.5–(4) cm, solitary occasionally binate, with 1–2–(3) spikelets, not naked below; uppermost 3–6 spikelets solitary, imbricate, on short pedicels; rachis, branches and/or pedicels finely and minutely prickle-toothed on margins. Spikelets 6–9–(11) mm × 2–3 mm wide, golden-violet, of 3–5–(6) florets. Glumes unequal, keeled, glabrous, centrally golden-violet, smooth except for occasional prickles on keel, margins membranous and ciliate; lower 2.5–3.5 mm, 1-nerved, long triangular acute, upper 3–5 mm, 3-nerved, laterals often weakly developed, oblong, entire sometimes emarginate or slightly acute. Lemma 4–5–(6) mm, shortly (0.1 mm or less) lobed, upper margin broadly membranous, prickle-teeth throughout or sparse except on nerves at broad apex; narrowing abruptly to awn 0.5–1.5–(2) mm. Palea (3.5)–4–5.5 mm, ⩽ lemma, acute, very shortly (0.1–0.2 mm) bifid, keels usually toothed to base, interkeel hairs often to base but mostly in upper ⅓, margins ciliate above. Callus 0.1–0.2 mm, ± glabrous; articulation flat. Rachilla 0.8–1.1 mm, short stiff hairy. Lodicules 0.6–1.0 mm, lobed, occasionally hair-tipped. Anthers 0.5–0.9–(1.2) in cleistogamous florets, 1.2–1.6 mm in chasmogamous florets, purple. Gynoecium: ovary 0.6–0.8 mm, hispid hairs at apex, glabrous on Campbell Id; stigma-styles 1–1.25 mm in cleistogamous florets, 1.6 mm in chasmogamous florets. Caryopsis 2.5–3 mm; embryo 0.5 mm; hilum 2–2.25 mm.

Endemic.

N.: Kaweka and Ruahine Ranges; S.: mountain ranges in inland Otago and Southland, sporadically elsewhere; C. Wet places in subalpine and alpine tussock grasslands and shrub communities; 1100–1800 m.

Generally, North Id plants are slender and graceful, and South Id plants are often sturdier and form more compact clumps. CHR 320258 *W. G. Lee* talus slopes, Dana Peaks, Murchison Mts, Mar. 1979, is more robust than any other gathering. Campbell Id plants have glabrous leaf-blades and the apex of the ovary lacks hispid hairs.

North Id distribution is between latitudes 39–40°S. In South Id *F. madida* is abundant on mountain ranges in Otago and Southland, e.g., Rock and Pillar, Old Man, Garvie Mountains. There are sporadic South Id collections: Craigieburn Range, Canterbury; Danseys Pass, North Otago; Swampy Hill, Dunedin; Dana Peaks, Fiordland; Lake Monowai, Southland. It cannot be the absence of suitable habitats that limits the distribution because wet sites abound, but among other grasses *F. madida* may suffer from its insignificant stature.

Only *F. madida* among endemic species of *Festuca* occurs on mainland islands and on a subantarctic island, Campbell Id, latitude 52° 30' S.

8. F. matthewsii (Hack.) Cheeseman, *Man. N.Z. Fl.* 205 (1925) ≡ *F. ovina* subsp. *matthewsii* Hack., *T.N.Z.I. 35*: 385 (1903) ≡ *F. ovina* var. *matthewsii* (Hack.) Cheeseman, *Man. N.Z. Fl.* 918 (1906); Lectotype: W 8151 (centre)! *H. Matthews* Mount Bonpland, Otago (No 1496 to Hackel) (designated by Connor 1998 op. cit. p. 347). = *F. petriei* Howarth, *J. Linn. Soc. Lond. 48*: 68 (1928) ≡ *F. petriei* forma *petriei* (autonym, Howarth 1928 op. cit. p. 71); Lectotype: K! *D. Petrie No 131* plant from Tasman Valley near Mt Cook. Cultivated in my garden (Howarth XXXIII) (designated by Connor 1998 op. cit. p. 348).

Glaucous upland tussocks with intravaginal branching. Leaf-sheath 3–15 cm, wider than leaf-blade. Leaf-blade 10–30 cm, hexagonal or nearly so, smooth or scabrid at thickened collar; T.S.: 5 vascular bundles, 7 sclerenchyma strands. Culm (10)–40–100 cm, much taller than leaves; nodes evident. Panicle (2.5)–7–20 cm, with open, pulvinate branches; rachis prickle-toothed. Spikelets 10–20 mm. Lemmas usually glabrous, glaucous; awn 1–2–(4) mm. Lodicules usually hair-tipped. Anthers yellow to purple. Ovary commonly with hispid apical hairs.

S.: throughout except north-west Nelson and Buller, and east coast lowlands. Tussock grasslands; 500–1800 m.

subsp. **matthewsii**

Smooth soft to stiff green or glaucous tufted or occasionally shortly stoloniferous grass with long wide-angled inflorescences of evidently awned florets held high above hexagonal leaf-blades from swollen collars. Branching intravaginal. Prophyll 2–6 cm, brown, chartaceous, keels stiffly ciliate. Leaf-sheath (3)–9–11 cm, glabrous, keeled centrally, ridged elsewhere, much wider than leaf-blade, stramineous, margins becoming membranous; apical auricles 0.6–1.5 mm, rounded, shortly ciliate. Ligule as for auricles. Collar conspicuously thickened. Leaf-blade 10–20–(30) cm × 0.5–0.8 mm diam., hexagonal rarely circular, ribbed, abaxially glabrous except scabrid tip rarely minutely scabrid elsewhere, adaxially and on margins abundant short, white hairs; TS: 5 vascular bundles, 7 widely spaced sclerenchyma strands, rarely fewer, costal sclerenchyma very rare. Culm 25–45–(100) cm, usually ≫ leaves, nodes often geniculate below, uppermost node conspicuous, black, internodes smooth, often slightly scabrid below inflorescence node. Panicle (7)–10–20 cm, with (5)–7–10 nodes, of 15–30–(50) spikelets, open with pulvinate widely divergent often flexible branches which may contract, basal branches 2–6–(10) cm solitary or occasionally binate, with 3–4–(5) spikelets, naked below, branches becoming progressively shorter and uppermost 3–4 spikelets solitary on short pedicels; rachis, branches and pedicels prickle-toothed to sparsely so becoming smooth or almost so. Spikelets (9)–14–20 mm × 3–4–5 mm wide, stramineous or sometimes bronze or slightly purpled, of (4)–5–8 florets. Glumes unequal, keeled, centrally green to stramineous occasionally with purpled veins, long triangular acute to obtuse, apex ciliate; lower (3)–4.5–6 mm, 1-nerved, upper (3.5)–4–6–(8) mm, 3–(5)-nerved. Lemma 5–8–(9) mm, lobes 0 or very small, rounded on back becoming keeled above, prickle-teeth below on outer nerve and on margin, and above near awn; awn (1)–2.5–3–(4) mm. Palea 4.5–6.7–(9) mm, frequently < lemma, long acute, deeply (0.5–1.0 mm) bifid, keels prickle-toothed to base or almost so, interkeel hairs mostly in upper ⅓, margins of flanks ciliate above sometimes to base. Callus 0.2–0.4 mm, centrally glabrous very shortly bearded laterally; articulation oblique to ± flat. Rachilla 1–1.6 mm, short stiff hairy. Lodicules (0.6)–1.0–1.5 mm, ⩾ ovary, often lobed, frequently hair-tipped. Anthers (2.0)–3–4.0 mm, yellow or yellow and purpled. Gynoecium: ovary 0.6–1.0 mm, hispid hairs at apex; stigma-styles 1.5–2.5 mm. Caryopsis 3–4 mm; embryo 0.5–0.75 mm; hilum 2 mm.

Endemic.

S.: east and west of Main Divide from Waiau River, Canterbury, to Fiordland, occasional in Marlborough. Tussock grasslands; 500–1500 m.

The spikelets of the lectotype of *F. matthewsii*, and its duplicate, are 15–17 mm. Spikelets of these dimensions are found especially in Fiordland, occasionally in Otago and Westland, and uncommonly elsewhere. Some examples are: CHR 25033 *G. Simpson* McKinnon Pass; CHR 64090 *W. A. Thompson* Hollyford Valley; CHR 253921 *P. N. Johnson* Hopkins Valley; CHR 3926 *H. H. Allan* Fox Glacier; WELT 16101 *W. R. B. Oliver* Mt Moltke; WELT 68656 [*B.C. Aston*] George Sound.

Among specimens from Westland there are fine-leaved plants of slender stature, e.g., CHR 166832 *P. Wardle & I. R. Fryer* Mt Fox; CHR 185639 *P. Wardle & I. R. Fryer* Mt Arthur; CHR 195178 *P. Wardle* Pioneer Peak. Here the impression, too, is that spikelets are smaller than elsewhere but the range 10–17 mm is equal to that in Fiordland (11–18 mm), and Otago (10–19 mm), though the frequency may be higher in the modal range 11–14 mm.

    subsp. **aquilonia** Connor, *N.Z. J. Bot. 36*: 351 (1998); Holotype: CHR 51923! *A. P. Druce* Mt Travers, Nelson Prov. *c.* 4600 ft, tussock grassland, 5 April 1947.

Stiff scabrid leaved tussock with scabrid inflorescences of wide-angled branches of awned spikelets borne high above circular or hexagonal glaucous leaves. Prophyll 3–5 cm, stramineous, keels strongly ciliate. Leaf-sheath (5)–10–15 cm, glabrous or minutely scabrid, keeled, stramineous, margins membranous, brown below; apical auricles 0.3–0.7 mm, ciliate. Collar usually swollen. Leaf-blade 15–25–30 cm × 0.5–0.7 mm diam., glaucous, hexagonal and ribbed or terete, abaxially minutely or strongly antrorsely prickle-toothed sometimes glabrous, adaxially with abundant short white antrorse hairs on ribs and margins; TS: 5 vascular bundles, 7 sclerenchyma strands or continuous, costal sclerenchyma sometimes present. Culm 30–45–(70) cm, usually greatly exceeding leaf-blades, nodes sometimes geniculate, uppermost conspicuous, upper internodes sparsely prickle-toothed or glabrous. Panicle (7)–10–15 cm with 6–9 nodes, of 8–13–(25) spikelets; open wide-angled pulvinate branches sometimes contracted or becoming so, naked below, basal branch solitary with 2–4–(7) spikelets, uppermost 4–6 spikelets solitary on short (2–4 mm) pedicels; rachis, branches and pedicels with abundant small prickle-teeth especially on convex surfaces, longer prickle-teeth on margins. Spikelets 10–15 mm × 4.5 mm wide, of 4–6–(9) florets. Glumes unequal, keeled with prickle-teeth above, centrally green-bronze, margins pale membranous, ciliate; lower 3.4–5 mm, 1-nerved, triangular acute, upper 4.5–7 mm, 3-nerved, long triangular acute to obovate acute. Lemma (4.5)–6–7 mm, lobes small (0.05 mm) or 0, rounded below becoming keeled above, sparsely scabrid on margin below and near awn; awn (0.5)–1–2 mm. Palea (5)–6–7 mm, frequently > lemma, apex deeply (0.5–0.7 mm) bifid, ciliate, keels toothed ½–⅔ or more, interkeel hairs above, margins of flanks ciliate above. Callus 0.2–0.3 mm, margins sparsely bearded, articulation ± flat. Rachilla 1–1.2–1.5 mm, abundantly short stiff hairy. Lodicules 0.8–1.1 mm, > ovary, often hair-tipped. Anthers 2.5–3.75 mm, yellow or purpled. Gynoecium: ovary 0.5–0.75 mm, hispid hairs at apex; stigma-styles 1.8–2.5 mm. Caryopsis 3–3.5 mm; embryo 0.5 mm; hilum 2–3 mm.

Endemic.

S.: north-west Nelson, Mt Benson, central and southern Nelson from St Arnaud to Lewis Pass, Marlborough, Wairau Mountains, Richmond Range. Subalpine and alpine tussock grasslands; 900–1600 m.

Subsp. *aquilonia* has two characters mostly associated with *F. novae-zelandiae* viz scabrid leaves, and inflorescence rachis, branches, and pedicels with a multitude of small prickles especially on the ridged convex surface. Some TS leaf sections are also typical of *F. novae-zelandiae*.

subsp. **latifundii** Connor, *N.Z. J. Bot. 36*: 352 (1998); Holotype: CHR 98244! *H. E. Connor* Mt Longslip, Lindis Pass 4500 ft, 10.2.1958.

Glaucous tussock with scabrid leaves and tall culms bearing reflexed branches of large spikelets. Prophyll 2–3–(8) cm, keels antrorsely prickle-toothed, apex acute. Leaf-sheath 3–10 cm, glabrous, margins membranous; apical auricles 0.4–1.0 mm, ciliate, ≈ ligule. Leaf-blade 8–20–(30) cm × 0.5–0.9 mm diam., terete to ± hexagonal, abaxially with short (0.05 mm) antrorse prickle-teeth especially on ribs, adaxially and on margins with abundant longer (0.1 mm) white antrorse hairs; TS: 5 vascular bundles, 7 sclerenchyma strands or sometimes continuous. Culm 20–45–(80) cm, usually exserted high above leaves, nodes dark, evident, internodes smooth but sometimes antrorsely scabrid. Panicle (6)–8–13–(18) cm, erect with 6–8 nodes, 10–15–20 spikelets; branches solitary or often binate, naked below, pulvinate and reflexed, basal branch (1.5)–5–8 cm of 2–6 spikelets, uppermost 2–4 spikelets solitary on 2–4 mm pedicels; rachis margins prickle-toothed sometimes smooth below, branches and pedicels prickle-toothed on margins. Spikelets 10–16 mm × 4–6 mm, of 5–7 florets. Glumes unequal, keels sometimes prickle-toothed, centrally green, brown or reddish, margins ciliate; lower 3–5.5 mm, 1-nerved, upper 3.5–4.5–(6.5) mm, 3-nerved. Lemma (5)–6–7–(8) mm, 5-nerved, glaucous, keeled, smooth except for prickle-teeth at apex, margin antrorsely hairy; awn (1.0)–2–2.5–(3) mm. Palea 5–6–(7.5) mm, usually ≥ lemma, deeply (0.5 mm) bifid, keels toothed in upper ½ or more, interkeel hairs at apex, margins of flanks hairy above. Callus 0.2–0.3 mm, margin sparsely bearded; articulation ± oblique. Rachilla 1.0–1.5 mm, abundantly short stiff hairy. Lodicules 0.7–1.4 mm, usually hair-tipped. Anthers 3–4.2 mm, yellow to purple. Gynoecium: ovary 0.6–1.0 mm, hispid hairs at apex; stigma-styles 1.2–2.0 mm. Caryopsis 2–3 mm; embryo 0.5 mm; hilum ± 2.5 mm. $2n = 42$.

Endemic.

S.: Waitaki Basin and Central Otago. Tussock grasslands on hills and plains; 900–1500 m.

The leaf-blades of subsp. *latifundii* are as rough as those of subsp. *aquilonia* and of those of *F. novae-zelandiae*. The vesture of the rachis on CHR 98243 *H. E. Connor* Mt Longslip, Feb 1958, is primarily that of *F. novae-zelandiae*; the remainder of the specimen is typically *F. matthewsii* subsp. *latifundii*.

subsp. **pisamontis** Connor, *N.Z. J. Bot. 36*: 354 (1998); Holotype: CHR 74046! *V. D. Zotov* Mt Pisa, 4000–5700 ft, 15/1/50.

Tall to modestly short tussock with very smooth or minutely prickle-toothed fine leaf-blades, below laxly branched inflorescence of large spikelets with conspicuously awned florets. Prophyll 3–5 cm, stramineous, keels ± antrorsely toothed, apex ciliate. Leaf-sheath (2.5)–4.5–7 cm, glabrous, striate, stramineous, margin membranous; apical auricles (0.4)–0.5–0.8–(1.3) mm, rounded, ciliate. Ligule 0.3–0.6 mm, ciliate. Leaf-blade 10–15–(20) cm, occasionally 5–8 cm, × 0.4–0.6 mm diam., hexagonal and ribbed or ± terete, pointed, abaxially smooth or minutely or imperceptibly antrorsely prickle-toothed on ribs, adaxially and on margins abundantly short white antrorsely hairy; TS: 5 vascular bundles, 7 sclerenchyma strands. Culm (11)–30–40 cm, erect, well exserted above leaf-blades; nodes visible; internodes smooth but sometimes occasionally minutely prickle-toothed. Panicle (2.5)–6–10–(13) cm with 5–9 nodes, (4)–8–10–(16) spikelets; branches solitary sometimes binate, lax, naked below, basal branch (1.2)–2–4–(5.5) cm, of (1)–2–(5) spikelets, uppermost 3–5 spikelets solitary on short pedicels; rachis prickle-toothed on margins throughout or sometimes glabrous below becoming scabrid above, branches and pedicels prickle-toothed. Spikelets 10–12–(15) mm × 3–5 mm, of 5–6 florets. Glumes unequal, keeled or slightly so, prickle-teeth above, usually brown or purpled centrally, margins membranous, ciliate; lower 2.8–4.0 mm, 1-nerved, long triangular acute or blunt, upper 3.5–5.5 mm, 3-nerved, narrowly ovate, obtuse to acute. Lemma (3.7)–5–6.2 mm, lobes short (0.1 mm), rounded, glabrous except for prickle-teeth at apex and near awn, glaucous; awn 1.2–2.5 mm. Palea (4.5)–5.5–6.8 mm, often ≥ lemma, acute, deeply (0.6 mm) bifid, keels toothed in upper ½–⅔, interkeel hairs above, margins of flanks ciliate above. Callus 0.2–0.3 mm, margin very sparsely shortly bearded; articulation flat. Rachilla 1–1.5 mm, short stiff antrorsely hairy. Lodicules 0.8–1.2 mm, lobed, usually hair-tipped. Anthers 2.8–3.7 mm, yellow. Gynoecium: ovary 0.6–0.9 mm, obovate, apex usually with hispid hairs; stigma-styles 1.2–1.3 mm. Caryopsis 3 mm; embryo 1 mm; hilum ± 2 mm.

Endemic.

S.: mountains of Central Otago: Dunstan, Pisa, and Old Man Ranges; 600–1800 m.

9. F. multinodis Petrie et Hack., *T.N.Z.I. 44*: 186 (1912); Lectotype: WELT 68611! *B. C. Aston* Days Bay, Wellington, Feby 1906 (designated by Connor 1998 op. cit. p. 355).

Scrambling, prostrate ascending, extravaginally branching with buds bursting forth on many-noded stolons, or sometimes densely caespitose, glaucous grass with many-leaved vegetative shoots and inflorescences exceeding the usually secund leaves. Prophyll short, membranous throughout; keels usually with retrorse hairs. Branching extravaginal. Leaf-sheath 3–10 cm (longer on shoots with very many leaf-blades), glabrous, ribbed, manifestly broader than leaf-blade, becoming fibrous with evident white nerves; apical auricles 0.2–0.5–0.7 mm, rounded, ciliate. Ligule 0.2–0.5 mm, flat to ± triangular between auricles, ciliate. Collar scarcely or ± thickened. Leaf-blade (5)–10–25 cm × 0.3–0.5– (0.9) mm diam., weakly hexagonal and ribbed, often terete usually secund, glaucous, smooth except for prickle-teeth at apex, adaxially and on margins antrorsely short white hairy becoming less so above; TS: 5 vascular bundles and 7 sclerenchyma strands, occasionally 7 bundles and 9 strands. Culm 20–40–(50) cm, greatly exceeding leaf-blades, nodes brown to purple-brown usually geniculate, internodes glabrous. Panicle (2.5)–5–10–(20) cm, with 5–9 nodes, (6)–10–15–(25) spikelets; branches spreading erect or weakly so, occasionally ± divergent, binate or solitary, basal branch (1)–2–4–(10) cm of 3–6 spikelets, naked below or not naked below especially in Cook Strait, uppermost 5–6 spikelets, imbricate, solitary on short (0.5–1 mm) pedicels; rachis prickle-toothed often glabrous below, branches and pedicels prickle-toothed but glabrous throughout in Cook Strait; frequently tortuous below. Spikelets 7–15– (20) mm × 3–5 mm, of 4–6–(9) stramineous florets. Glumes unequal, evidently keeled, linear-oblong narrowing abruptly to an acute or mucronate apex, glabrous but occasionally prickle-toothed on keel above, apex sometimes shortly or evidently ciliate, margins membranous, ciliate above; lower 2.5–3–(4.5) mm, 1-nerved, upper 4–5.5–(6) mm, 3-nerved. Lemma 5–6 mm, glaucous, apex shortly lobed or 0, 5-nerved, keeled, smooth except for prickle-teeth at base and extending from callus to outer nerve below, and on keel above; awn 0 or 0.5–1.5 mm. Palea

(4.5)–5–5.7–(6.5) mm, ≥ lemma, acute, shortly (0.2–0.4 mm) bifid, keels toothed towards apex, interkeel hairs above but sometimes to base, flanks short ciliate above. Callus 0.3–0.4–(0.5) mm, upper margins shortly bearded, less so centrally; articulation ± oblique. Rachilla 1–1.3 mm, sparsely short stiff hairy. Lodicules 0.7–1.5 mm, bifid or lobed, usually glabrous but occasionally hair-tipped. Anthers 2.0–3.0 mm, yellow to orange. Gynoecium: ovary 0.6–1.0 mm, ± turbinate, apex glabrous or with hispid hairs; stigma-styles 1.5–2.5 mm. Caryopsis 3–3.5 mm; embryo 0.5 mm; hilum 3 mm. $2n = 56$.

Endemic.

N.: inland in south-west Kaimanawa Mts, north-west Ruahine Range, and Manawatu Gorge, coastal from Cape Turnagain to Cook Strait; S.: Marlborough Sounds, Kaikoura Ranges to Waipara, North Canterbury. Coastal rocks, cliffs and bluffs, limestone except in Kaimanawa Mts; sea level to 100 m on coasts, inland 300–1200 m.

The main habit is straggling, open, conspicuously stoloniferous with prostrate-ascending shoots bearing inflorescences, but plants may form dense clumps of many short stoloniferous shoots of very fine smooth leaves, e.g., CHR 197189 *A. P. Druce* Waipapa Pt, Kaikoura coast, from a raised gravel beach.

Panicle structure varies from erect, stiff branches, appressed or nearly so, to open, reflexed, sometimes flexuous branches, but this latter character is only sporadically distributed through the collections. Inflorescence length varies over a wide range; at the type locality, the Wellington Coast in Port Nicolson and nearby Cook Strait, inflorescences are shorter than anywhere else. Elsewhere in North Island there is a simple north–south gradient for panicle length. Inflorescences from Marlborough plants, mostly of mountainous limestone areas, are longer than elsewhere. Rachis, branches, and pedicels are glabrous throughout, or very infrequently prickle-toothed in specimens from Cape Palliser to Port Nicolson and Cook Strait. Elsewhere the rachis, branches, and pedicels are toothed to varying extents, commonly glabrous below but toothed above.

Plants referred to as octoploid *Festuca multinodis* by Connor, H. E. *N.Z. J. Bot. 6*: 295–308 (1968) do not belong to this taxon; they are a form of *F. rubra*.

10. F. novae-zelandiae (Hack.) Cockayne, *T.N.Z.I. 48*: 178 (1916) ≡ *F. ovina* subsp. *novae-zelandiae* Hack., *T.N.Z.I. 35*: 384 (1903) ≡ *F. ovina* var. *novae-zelandiae* (Hack.) Cheeseman, *Man. N.Z. Fl.* 917 (1906) ≡ *F. ovina* subsp. *novae-zelandiae* var. *eu-novae-zelandiae* Howarth, *J.*

Linn. Soc. Lond. 48: 63 (1928) *nom. invalid.* ≡ *F. ovina* subsp. *novae-zelandiae* var. *novae-zelandiae* subvar. *novae-zelandiae* (autonym, Howarth 1928 op. cit. p. 64); Holotype: W 8150! *T. F. Cheeseman* Slopes of Mt Torlesse, Canterbury, 3000 ft, Jan 1880 (No 1497 to Hackel). = *F. ovina* subsp. *novae-zelandiae* var. *novae-zelandiae* subvar. *pruinosa* Howarth, *J. Linn. Soc. Lond. 48:* 64 (1928); Lectotype: K! *A. Wall* Mt Herbert, 2000 ft, growing alongside No. 7 (designated by Connor 1998 op. cit. p. 361).
F. novae-zelandiae J.B.Armstr., *N.Z. Country J. 5*: 57 (1881); *nomen nudum.*   fescue tussock, hard tussock

Short tawny tussock with fine harshly scabrid rolled leaf-blades and inflorescences with small spikelets in ± appressed branches on rachides covered with small prickle-teeth. Prophyll 3–6 cm, red-brown, keels antrorsely toothed, scattered hairs on flanks; margin glabrous, membranous. Branching intravaginal. Leaf-sheath 5–25 cm, stramineous, striate, minutely scabrid or glabrous, margin membranous sometimes undulating; apical auricles 0.4–1.5 mm, unequal, rounded or acute, hairy, margins ciliate. Ligule = auricles, shortly triangular. Leaf-blade 10–40 cm × 0.4–0.7–(1.0) mm diam., terete, pointed, abaxially beset with small antrorse prickle-teeth, adaxially and on margins antrorsely short white hairy; TS: 5 vascular bundles, sclerenchyma usually continuous but frequently in 7 strands, occasionally various, costal sclerenchyma infrequent. Culm 20–100 cm, usually exceeding leaf-blades; nodes glabrous, concealed, brown; internodes antrorsely prickle-toothed, upper node often ensheathed. Panicle 5–25 cm, with (5)–6–9–(10) nodes, 15–25–30 spikelets, often included by leaves; branches solitary ascending, often naked below, sometimes binate, basal branch (2)–5–7–(10) cm of 3–6–(10) spikelets, uppermost 4–6 spikelets solitary on short pedicels; rachis, branches and pedicels abundantly shortly (0.05 mm) prickle-toothed on all surfaces. Spikelets 8–15 mm × 2.5–3 mm, 4–8 florets. Glumes unequal, keeled above, prickle-toothed on keel and above, centrally green-brown, pale stramineous elsewhere, thin, nerves evident, margin ciliate; lower 2.5–5.5 mm, 1-nerved, long triangular almost awned, upper 3.0–6.8 mm, 3-nerved, narrowly ovate, acute. Lemma 4–6 mm, lobes minute, rounded, prickle-toothed above and on margins below sometimes below on outermost nerve; awn 0–3 mm. Palea 4–6.5 mm, ≥ lemma, acute, shortly (0.1–0.3 mm) bifid, keels toothed in upper ½–⅔, interkeel hairs in upper ¼, margins of flanks with small hairs almost to base. Callus 0.1–0.4 mm, bearded on margins; articulation ± flat. Rachilla 0.75–1.5 mm, dense short stiff antrorsely hairy. Lodicules 0.5–1 mm, lobed, usually hair-tipped. Anthers 2–4 mm, yellow. Gynoecium: ovary 0.7 mm, turbinate, apex with hispid hairs; stigma-styles 1.0–1.75 mm. Caryopsis 2–3.8 mm; embryo 0.5 mm; hilum ± 2 mm. $2n = 42$. Fig. 6. Plate 1C.

Endemic.

N.: central Volcanic Plateau, Ruahine Range, Taranaki (Pouakai Range and Ahukawakawa Swamp); S.: Nelson uncommon, Marlborough to Southland east of Main Divide, occasional in Fiordland; St.: uncommon. Tussock grasslands and river beds; sea level to 1700 m.

Leaf-blades lacking antrorse prickles are uncommon, but in Taranaki in the Ahukawakawa Swamp, Pouakai Range, dominated by *Chionochloa rubra* var. *inermis*, *F. novae-zelandiae* usually has smooth leaf-blades, e.g., CHR 86390 *A. P. Druce* 1958, CHR 86511 *A. P. Druce* 1960, CHR 116688 *A. P. Druce* 1968; CHR 86127 *A. P. Druce* 1955 has more scabrid leaves than the others. The panicle rachis may appear to have smaller teeth than modally, but the pattern is as elsewhere.

On 8 Feb. 1910 D. Petrie collected *Festuca* at 4000 ft on Gordons Knob, Gordon Range, Nelson which is represented by WELT 68772, WELT 68798, CHR 1505 with a Petrie label, CHR 1525 and AK 152716. All appear to be parts of one specimen. Leaf-blades are smooth to very lightly scabrid. The culm is prickle-toothed sometimes slightly so; the rachis and branches bear some of the prickles characteristic of *F. novae-zelandiae* but never at the customary density; the lemma is typical of *F. novae-zelandiae*. The adaxial surface of the glumes is covered by very short white hairs — a condition unrecorded elsewhere.

No status is accorded to the high altitude ecotype discussed by Connor, H. E. *N.Z. J. Bot. 6*: 295–308 (1968), and Scott, D. *N.Z. J. Bot. 8*: 76–81 (1970).

This species in earlier times was discussed under the binomial *Festuca duriuscula* non L.

11. F. ovina subsp. **hirtula** (W.G.Travis) M.J.Wilk., *Bull. Soc. Ech. Pl. vasc. Eur. occ. Bass. médit. 20*: 72 (1985).

Tufted intravaginally branching short leaved clumps bearing tall ± erect culms with narrow inflorescences of spikelets whose lemma margins are conspicuously hair fringed. Leaf-sheath 3–10 cm, open to base, keeled, striate, densely hairy to less so below; apical auricles 0 or small, shortly ciliate. Collar curved and thickened. Ligule 0.1–0.3–0.5 mm, finely ciliate. Leaf-blade 4–6–10 cm, plano-compressed, narrowing to midrib, ± firm, abaxially hairy or irregularly so, adaxially finely hairy more so above ligule; margins finely antrorsely scabrid; TS: 7 vascular bundles, sclerenchyma in 3–5 strands, one always subtending median bundle, occasionally continuous. Culm 20–40 cm, ridged, internodes smooth or short hairy and becoming smooth, often only inflorescence internode visible. Panicle 4–7–10 cm, narrow, of 8–12 nodes, 15–25 spikelets, branches solitary, short, ± erect, > internodes; basal branch 1.5–3.5 cm of 4–7 spikelets naked below, uppermost 6–7 spikelets solitary on short pedicels; rachis glabrous below becoming toothed on margins, branches finely scabrid, pedicels with many short hairs. Spikelets 5–12 mm, of 4–7 florets. Glumes unequal, keels toothed above, margins ciliate or hair fringed but sparsely so on lower glume; lower 2–3 mm, 1-nerved, narrow triangular acute, upper 3–4.5 mm, shortly awned, 3 conspicuous nerves. Lemma (3.75)–4.5–5.5 mm, lobes minute or 0, 5-nerved, rounded, upper margins conspicuously hair fringed, apex prickle-toothed elsewhere smooth or clothed in stiff hairs throughout or hairs shorter and fewer towards base; awn (0.5)–2–(3) mm. Palea (3.5)–4.5–5.5 mm, ≥ lemma, apex shallowly bifid, keel toothed in upper ¼–⅔, interkeel hairy throughout or above. Callus 0.1–0.2 mm, very shortly bearded laterally or ± glabrous; articulation ± flat. Rachilla (0.6)–0.75–1.2 mm, with long stiff antrorse hairs. Lodicules 0.6–0.75 mm, lobed. Anthers (1.75)–2.5–2.75 mm, yellow occasionally purple. Gynoecium: ovary (0.5)–0.7–1 mm, glabrous, often with cap; stigma-styles (1)–1.5–2.5 mm, often sparsely branched. Caryopsis 2.5–3.5 mm; embryo 0.5 mm; hilum ≈ caryopsis.

Naturalised from Europe.

N.: Auckland City; S.: occasional in inland basins from Marlborough to Otago; sporadic elsewhere. Grassland, waste places.

Two forms may be found, and judged by herbarium representation they may co-occur. In one lemmas are hair-fringed on the upper margins and in the other lemmas are fully clothed in hairs; in an intermediate state very short hairs or potential hair sites occur on lower regions of the lemma. Once again based on herbarium representation, the form with densely hair-clothed lemmas is the more common. Hair-fringed glumes are coincident with hair-clothed lemmas.

The presence of any other naturalised subspp. of *F. ovina* is not indicated by herbarium specimens.

12. F. rubra L., *Sp. Pl.* 73 (1753).

Tufted or spreading from slender rhizomes. Branching extravaginal. Prophyll keeled, hairy. Leaf-sheath (1)–3–7–(10) cm, tubular, red-brown, hairs often retrorse, rarely glabrous, striate, becoming fibrous; apical auricles 0 to 0.25 mm, finely ciliate. Collar sometimes thickened. Ligule 0.25 mm, finely ciliate. Leaf-blade (2)–10–(25) cm × 0.3–0.7 mm, elliptic to rounded or regularly hexagonal, abaxially smooth rarely antrorsely scabrid or hairy below, adaxially finely hairy, margins finely hairy; TS: usually 5 vascular bundles, 7 sclerenchyma strands, occasionally fewer or continuous. Culm 10–20–(45) cm, internodes erect or ascending-erect, usually > leaf-blades, nodes usually glabrous but sometimes evidently hairy or prickle-toothed, often glaucous, sometimes violet-suffused and striate. Panicle (1)–5–10–(20) cm, of 6–10–13 nodes, of (3)–8–15–(30) spikelets, narrow and erect with short branches to open ascending with long branches naked below, often secund above; basal branch solitary or binate with 2–10 spikelets, uppermost 5–10 solitary on short pedicels; rachis toothed on angles sometimes hairy rather than toothed and occasionally ± pilose, usually tortuous below, branches toothed or hairy on angles more so on basal branch at internodes; pedicels toothed on angles. Spikelets (5)–8–10–(13) mm, of 3–6 florets, often glaucous and/or purple-suffused. Glumes unequal, nerves often raised, keeled, smooth except for prickle-teeth on keel and infrequently elsewhere; margins membranous and ciliate; lower 2.5–3.5 mm, 1-nerved, upper 3–5.5 mm, 3-nerved, sometimes shortly awned. Lemma 3.5–4.5–5 mm, 5-nerved, often glaucous, and violet-suffused, rounded, prickle-toothed above, margins ciliate below; awn 0.5–1.5–(2.5) mm. Palea 3.5–4.5–5 mm, apex pointed, shortly bifid, keels toothed to ½ or more, short interkeel hairs above. Callus 0.2–0.3 mm, bearded laterally; articulation ± oblique. Rachilla 0.8–1.25 mm, short stiff hairy, sometimes evident at anthesis. Lodicules 0.6–1.25 mm, usually lobed, often hair-tipped. Anthers (1.0)–

1.5–2–3 mm. Gynoecium: ovary 0.5–0.7 mm, sometimes hispid hairy at apex; stigma-styles *c.* 1.5 mm. Caryopsis 2.5–3.5 mm, free; embryo 0.5 mm; hilum in sulcus, = caryopsis.

Naturalised from Europe.

N.; S.; St.; Ch., A., Ant., C. Grasslands, scrublands, wasteplaces; sea level to 1850 m.

A very variable species to which European interpretations, in the absence of modern authenticated specimens in N.Z. herbaria, do not easily apply. Nor are there cytological data to assist in discriminating among subspecific taxa.

subsp. **rubra**   red-fescue

Rhizomatous; culms > leaves; spikelets 10–12 mm; anthers 1.5–2.5 mm.

Within subsp. *rubra* a morphological gradient can be detected from high alpine populations to those of lowland areas. To indicate some of that range (for to attempt to do so completely might be impractical) some morphological features of points on that gradient are offered.

i. Alpine: leaves 2–7 cm, glaucous; culm 3–15 cm, erect, violet; panicle 1–5 cm, contracted; spikelets 4–10 mm, usually solitary; anthers 1.5–2.5 mm. Mostly on rock outcrops and there freely rhizomatous; also in nearby damp grassland. $2n = 56$.
Mountains of Marlborough and Canterbury, 1000–1850 m.

Specimens gathered by A. Wall on the Richmond Range, Mackenzie Country, 4000–6000 ft, were interpreted by Howarth (1928 op. cit. p. 76) as var. *fallax* Hack., but Saint-Yves (1931 op. cit. p. 306) referred them to *F. briquetii* St.-Yves along with plants of similar stature.

ii. Alpine and subalpine: tall; leaves 4–10 cm, glaucous; culm 10–30–(40) cm, erect, often hairy, violet; panicle 2–7 cm, contracted, branches binate below; spikelets in groups, 7–11 mm; anthers 1.2–2 mm. Mostly in seepages, watercourses, broken ground and rock outcrops.
Mountains of central North Id and Nelson, Marlborough and Canterbury, less frequently in Otago and Southland, 750–1550 m.

Some specimens of this general appearance were referred to *F. rubra* var. *rubra* subvar. *vulgaris* Hack. by Howarth (1928 op. cit. p. 67), and by E. Hackel in his determinations for T. F. Cheeseman, and D. Petrie. Many other specimens were similarly treated as subvar. *vulgaris* including taller, open, widely rhizomatous plants.

Inflorescence length and culm length are highly correlated (r = 0.94) in alpine and subalpine specimens, many of which are considered indigenous by some collectors.

iii. Lowland: tall or often straggling; leaves 10–30 cm, glaucous or not; panicle to 20 cm, open to somewhat contracted, branches often lax, solitary or binate below; spikelets in groups, 10–12 mm; anthers 1.5–2.5 mm.
Throughout, usually in montane and lowland areas.

iv. grandiflora: leaves to 30 cm, hairy below; culm 40–60 cm, many prominent nodes; panicles 7–14 cm, open, branches long, naked below; spikelets 8–11 mm; anthers 2.5–3.5 mm.
Sown grassland on Campbell, Auckland and Antipodes Islands.

The specimen of *B. C. Aston* Antipodes, sent to Kew by D. Petrie is recorded by Saint-Yves (1931 op. cit. p. 303) as *F. rubra* var. *rubra* subvar. *grandiflora*, and by Zotov, V. D. *Rec. Dom. Mus.* 5: 124 (1965) as *F. rubra* var. *commutata* Gaudin. No special significance attaches to these subantarctic plants except Saint-Yves' and Zotov's determinations.

subsp. **commutata** Gaudin, *Fl. Helv. 1*: 287 (1828).

Chewings fescue
Caespitose, rhizomes not formed but aerial stems with few short internodes; culms ≫ leaves; panicle 10–30 cm, widely branched, branches naked below; spikelets 9–12 mm; anthers usually 3 mm.

N.; S.: throughout. Often in pastures and semi-natural grasslands but on riverbeds, in waste places and roadsides.

Imperfectly gathered specimens may be difficult to distinguish from subsp. *rubra*.

13. F. ultramafica Connor, *N.Z. J. Bot. 36*: 363 (1998); Holotype: AK 1993! *T. F. Cheeseman* Dun Mt, Nelson, 4000 ft, [1878] (No 1493 to Hackel).

Tufted shortly rhizomatous grass with pungent, stiff, thick leaf-blades much shorter than the tall smooth culmed inflorescence of ± violet suffused spikelets of long-awned, smooth, glaucous florets growing on ultramafic parent materials in South Island Dun Mountains mineral belt. Prophyll 2–3 cm, keels glabrous. Branching extravaginal. Leaf-sheath 3–4 cm, striate, glabrous or minutely antrorsely prickle-toothed, becoming red-brown and fibrous; apical auricle 0.5–1 mm, rounded, ciliate. Ligule 0.5–1 mm, erose, ciliate. Leaf-blade (6)–8–12 cm × 0.6–1.2 mm diam., conduplicate, somewhat compressed, pungent, stiff, strict or slightly curved, ribbed, glabrous, adaxially and on margins abundant short white hairs; TS: 5 vascular bundles, 7 sclerenchyma strands, exceptionally more. Culm 20–50 cm, ≫ leaves, nodes visible, internodes glabrous, sometimes violet suffused below. Panicle narrow, 6–12–(14) cm with 4–6–(10) nodes, (12)–15–25 spikelets; basal branch ascending, 3–8 cm, solitary or binate, with 3–5 spikelets, naked below, nodes 2–5 with branches of 2–4 spikelets, uppermost 3–5 spikelets solitary, imbricate on short pedicels; rachis glabrous or glabrous below becoming prickle-toothed, branches and pedicels prickle-toothed or glabrous, often tortuous below. Spikelets 8–12–(14) mm × 3–4 mm, sometimes violet tinged, of (3)–4–5 florets. Glumes unequal, margins long ciliate below shorter above, keeled, glabrous except for a few prickle-teeth on keel, usually violet suffused centrally or throughout; lower 2.5–4.5 mm, 1-nerved, long-triangular acute, awn 0.4–0.8 mm or absent, upper 3.5–5.3 mm, 3-nerved, oblong acute, awn 0.5–3 mm abrupt or tapering or absent. Lemma 5–7 mm, lobes 0 or minute, 5-nerved, glaucous, glabrous except below awn, scarcely keeled, margin long ciliate below shorter above; awn 1.5–4.2 mm. Palea 5.5–6 mm, ≥ lemma, acute, deeply (0.3–0.5 mm) bifid, keels toothed above, interkeel hairs to base denser at apex, flank margins ciliate. Callus 0.2–0.5 mm, sparsely bearded except near rachilla, articulation ± flat to oblique. Rachilla 1–1.6 mm, densely short stiff hairy. Lodicules (0.7)–1–1.2 mm, lobed, hair-tipped or glabrous. Anthers 2–3 mm, yellow or golden. Gynoecium: ovary 0.6 mm, hispid hairs at apex or absent; stigma-styles 2 mm. Caryopsis 3.5 mm; embryo 0.5 mm; hilum = caryopsis.

Endemic.

S.: Nelson (Dun Mt, near Mt Duppa, Motueka River). Soils developed from ultramafic parent material; 600–1100 m.

REPRODUCTIVE BIOLOGY   All species are chasmogamous except *F. contracta* and *F. madida* which are cleistogamous [Connor, H. E. *N.Z. J. Bot. 36*: 471–476 (1998)]. Inbreeding depression was measurable in progenies from self-pollination of *F. coxii, F. matthewsii* and *F. novae-zelandiae* [Connor, H. E., and Cook, A. B. *N.Z. J. Sci. Tech. A37*: 103–105 (1955); Connor, H. E. *N.Z. J. Agric. Res. 3*: 728–733 (1960), *N.Z. J. Bot. 26*: 163–167 (1988)]. There is some pseudo-self-incompatibility.

Although *F. madida* is predominantly cleistogamous, CHR 122577 *H. E. Connor* Danseys Pass, anthers 1.8 mm, and CHR 428895 *K. J. M. Dickinson* Whitcombe Creek, west of Umbrella Mts, anthers 1.3–1.5 mm, are chasmogamous. Cleistogamy persists in cultivation — see CHR 193320 *P. Wardle & I. R. Fryer* Mt Cockayne, Craigieburn Range, Canterbury.

The habit of *F. novae-zelandiae* is typically caespitose with leaves tightly packed together. After burning, or with time, the central part dies and a ± circular array of individual sets of shoots remains; many of these become large plants of identical self-incompatible genotype, but Lord, J. M. *N.Z. J. Bot. 31*: 133–138 (1993) concluded that population recruitment of that kind is not as great as that from seeds. For differentiation in seed germination see Lord, J. M. *N.Z. J. Bot. 32*: 227–235 (1994), and ovule failure and seed production see Lord, J. M. and Kelly, D. *N.Z. J. Bot. 37*: 503–509 (1999).

LOLIUM L., 1753

Annuals, or perennials, of variable habit. Leaf-sheath auriculate, or auricles 0. Ligule membranous, short, truncate or obtuse. Leaf-blade flat or ± folded, rolled or folded in young shoots. Culm ± decumbent to erect. Inflorescence a simple, stiff, terminal spike with mainly solitary spikelets alternating in two opposite rows and placed edgeways on and ± sunk in rachis. Spikelets 2–22-flowered, ± sessile, ± laterally compressed; disarticulation above glumes and between florets. Glumes both present and ± equal in terminal spikelet; lower absent in lateral spikelets, upper 3–7-nerved, coriaceous. Lemma 5–9-nerved, rounded, membranous to coriaceous; awn subterminal or 0. Palea ≈ lemma and similar in texture, 2-keeled, keels ciliolate. Callus glabrous. Lodicules 2, ± membranous, glabrous, ± lanceolate, entire. Stamens 3. Ovary glabrous; styles free. Caryopsis dorsiventrally compressed, longitudinally furrowed; embryo small; hilum linear, > ½ length of caryopsis.

Type species: *L. perenne* L.

8 spp. of temperate Eurasia. Naturalised spp. 4; transient sp. 1.

Two spp., *L. perenne*, perennial ryegrass and *L. multiflorum*, Italian ryegrass, are extremely important pasture grasses in N.Z., as in many other temperate regions. Because of a long history of breeding within these two spp., and introgression between them, cultivated and naturalised plants frequently do not conform to the classical taxonomic descriptions of the species.

Annual and biennial cultivars of Italian ryegrass as well as the hybrid *L. multiflorum* × *L. perenne* cultivars, are valued for their superior winter growth, faster establishment, and high acceptability to livestock. Though much grown in pasture, *L. multiflorum* does not occur so commonly in the wild as *L. perenne*.

This account is based on the revision of *Lolium* by Terrell, E. E. *Techn. Bull. U.S.D.A 1392*: 1–65 (1968).

1 Florets all awned, awns 3–13 mm 2
  Florets all awnless or some awned and awns < 3 mm 3

2 Upper glume usually ≪ spikelet; lemmas lanceolate or oblong; florets not
  turgid at maturity .... **1. multiflorum**
  Upper glume reaching or exceeding uppermost lemma; lemmas elliptic to
  ovate; florets turgid at maturity .... **4. temulentum**

3 Lemmas lanceolate or oblong; florets not turgid at maturity 4
  Lemmas elliptic to ovate; florets turgid at maturity 5

4 Perennial, dark green; spikelets evident .... **2. perenne**
  Annual, reddish brown; spikelets ± sunk in concavities of rachis and ±
  concealed by glumes .... **3. rigidum**

5 Glumes 10–22 mm; caryopsis 5–6 mm .... **4. temulentum**
  Glumes 7–9 mm; caryopsis *c.* 3.5 mm .... **remotum**ζ (p. 128)

1. L. multiflorum Lam., *Fl. Franc. 3*: 621 (1778).
                              Italian ryegrass, Westerwolds ryegrass
Annual to biennial, slender to rather stout tufts, (20)–40–100–(130) cm; branching intravaginal, leaf-blade rolled when young. Leaf-sheath glabrous, pink at base to light brown above, not becoming very fibrous; upper sheaths sometimes scabrid. Auricles usually present, to 1.6 mm. Ligule (0.6)–1–2.5 mm, truncate, glabrous. Leaf-blade (5.5)–8–30 cm × (0.7)–2.5–9–(11) mm, abaxially glabrous and shining, adaxially smooth or usually scaberulous; margins scaberulous, tapered to scabrid acute tip. Culm 10–60 cm, erect, or spreading or decumbent, often branched below, internodes smooth. Spikes (9.5)–15–35–(45) cm, erect or curved; rachis *c.* 1 mm wide, usually ± scabrid. Spikelets 14–25 mm, (4)– 9–14-flowered, green or purplish, overlapping, or more than their own length apart. Upper glume usually ≪ spikelet, 5–7-nerved, narrow-

lanceolate or narrow-oblong, obtuse or acute, glabrous. Lemma 6–8 mm, 5-nerved, oblong or oblong-lanceolate, smooth or minutely scaberulous especially on hyaline margins, not turgid at maturity, apex obtuse or shortly bifid; awn (2)–4–10 mm, fine, straight, subapical. Palea ≈ lemma, keels scaberulous. Anthers 3–4.5 mm. Caryopsis 2–4 × 0.5–1.5 mm.

Naturalised.

N.: scattered; S.: occasional; St.: local; C. On roadsides and waste land, found sometimes as a weed in pasture and in gardens.

An important pasture grass, indigenous to Europe, Asia and northern Africa. First recorded for N.Z. as *L. italicum* and known by that name until the 1930s.

Lolium multiflorum and *L. perenne* are very closely related. Both spp. are self-incompatible and are completely interfertile. Because they interbreed freely their separation into distinct species is often doubted and *L. multiflorum* regarded as a variety of *L. perenne*. The name *L. perenne* var. *multiflorum* was applied by Buchanan, J. *T.N.Z.I. 9*: 525 (1877), to plants on Kawau Id.

Westerwolds ryegrass originated in the Westerwolde area, Netherlands, as an unintentional selection of annual early-maturing plants from fields of Italian ryegrass (Terrell 1968 op. cit.). This type of Italian ryegrass is much grown in N.Z. and has been reported as naturalised, under the name *L. westwolticum* by Cockayne, A. H. *J. Agric. N.Z. Dept Agric. 13*: 212 (1916).

2. L. perenne L., *Sp. Pl.* 83 (1753).   perennial ryegrass

Perennials, forming loose to dense, dark green tufts (10)–20–75 cm; branching intravaginal, leaf-blade folded when young. Leaf-sheath glabrous, often pinkish at first, later light to dark brown and shredding into fibres. Auricles inconspicuous or 0. Ligule 0.5–1–(2) mm, truncate, glabrous. Leaf-blade (3)–6–20–(28) cm × 1.5–4–(5) mm, abaxially glabrous and shining, adaxially smooth or scaberulous, long-narrowed or abruptly narrowed to acute tip; margins scaberulous or smooth. Culm 10–50 cm, erect or spreading or decumbent, internodes glabrous. Spikes (4)–10–28 cm, erect or curved; rachis 0.5–1.2 mm wide, smooth, or scaberulous on angles. Spikelets 6–15 mm, (2)–5–9–(11)-flowered, green or purplish, overlapping, or more than their own length apart; sometimes proliferous. Upper glume usually < to ≥ spikelet, 5–7-nerved, narrow-lanceolate, subobtuse, glabrous. Lemma (4.5)–5–7–(8.5) mm, 5-nerved, oblong-lanceolate, obtuse to acute, smooth and firm below, scaberulous near margins and hyaline apex, not turgid at maturity, muticous, or occasionally minutely (to 1 mm) awned. Palea = lemma, keels scabrid. Anthers (1.5)–2.5–4 mm. Caryopsis 2–4 × 0.7–1.5 mm. Plate 2A.

Naturalised.

N.; S.: common throughout; St.: scattered, and on islands offshore; K., Three Kings Is, Ch., C. Roadsides and tracks, pasture, river flats and banks, waste land and sand dunes; lowland to montane.

An important pasture grass, commonly the main grass component of permanent pastures in New Zealand, indigenous to Europe, Asia and northern Africa, now widespread throughout temperate regions of both Hemispheres.

Aberrant or abnormal forms are common in *L. perenne* and plants with branched spikelets also occur; plants with branched inflorescences are probably hybrids with tall fescue, *Schedonorus phoenix* × *Lolium perenne*.

3. L. rigidum Gaudin, *Agrost. Helv. 1*: 334 (1811).

Stiff annuals, (7.5)–10–25–(45) cm, with shoots curving upwards, reddish brown, especially near base; branching extravaginal. Leaf-sheath glabrous, reddish purple. Auricles very short, *c.* 0.5–(1) mm, or 0. Ligule 0.4–1.2 mm, membranous, truncate. Leaf-blade (2)–6–18 cm × 1.5–4 mm, abaxially glabrous, adaxially scabrid on ribs; margins smooth, tip acute, scabrid. Culm (3)–6–30 cm, ensheathed by upper leaves, internodes glabrous. Spike (2.5)–5–20 cm, often curved; rachis 1–1.5–(2) mm wide, smooth, margins finely scabrid. Spikelets 5–10.5 mm, 2–5-flowered, less than their own length apart, usually ± concealed by glumes in concavities of rachis. Upper glume ≥ spikelet, 5–7–(9)-nerved, very stiff, lanceolate or narrowly oblong, obtuse or acute, glabrous; acutely angled to rachis at maturity. Lemma 3.5–6 mm, faintly 5-nerved, oblong or lanceolate-oblong, obtuse, smooth or finely scaberulous, not turgid at maturity, usually muticous; awn sometimes present, subterminal, short, fine, to 3 mm. Palea = lemma, keels scabrid. Anthers *c.* 1–1.5 mm. Caryopsis 2–3.5 × 0.7–1.3 mm.

Naturalised from Mediterranean.

N.: North and South Auckland, also in environs of Wellington City; K. Coastal on cliffs and sandy waste land.

4. L. temulentum L., *Sp. Pl.* 83 (1753). darnel

Stiff, stout, erect annuals, 35–100 cm; branching intravaginal. Leaf-sheath glabrous, light brown. Auricles to 1 mm, spreading, or 0. Ligule 0.8–1.5 mm, truncate. Leaf-blade 10–22 cm × 3–4.5–(6) mm, flat, abaxially glabrous, adaxially scaberulous or rarely smooth; margins smooth below, scabrid near tapered acute tip. Culm 20–60 cm, internodes closely scabrid above or smooth. Spike 10–20–(34) cm, erect; rachis 1–2 mm wide, smooth or finely scabrid on convex side, margins scaberulous. Spikelets 10–30 mm, 5–12-flowered, green; florets turgid at maturity. Upper glume 10–22 mm, < to > spikelet, 7–9-nerved, rigid, oblong-lanceolate, obtuse or subacute, smooth or scabrid. Lemma 5.5–9.5 mm, elliptic to ovate, smooth below, becoming hard at maturity, apex hyaline, scabrid, bifid, awned from sinus; awn fine, strongly scabrid, 1–13 mm, or 0. Palea = lemma. Anthers 2–3 mm. Caryopsis 5–6 × 2–2.5 mm.

Naturalised from Mediterranean.

N.: North Cape, Auckland City, Opotiki district, Wellington City and environs; S.: Canterbury (Akaroa and Ashburton), Otago (Hawea Flat and Matukituki R.). Roadsides, waste land, paddocks.

Lolium temulentum is widespread in temperate regions of both Hemispheres.

Slender specimens from Auckland — Beresford St., *T. Kirk* undated (WELT 71438, CHR 4102, and probably CHR 19265) and Hepburn St., Jan. 1864 (WELT 71444) — with culm 20–30 cm and spikes consisting of only 1–2 spikelets seem to be depauperate forms of *L. temulentum*.

Plants with very short awns or no awns, were recorded as *L. arvense* With., by Hooker, J. D. *Handbk N.Z. Fl.* 343 (1864). Terrell (1968 op. cit.) regarded the degree of development or absence of awns as a minor character in *L. temulentum*, and considered that the rank of forma was better applied to muticous or short-awned plants i.e., *L. temulentum* forma *arvense* (With.) Junge.

No specimens have been found to substantiate J. F. Armstrong's record of *L. temulentum* var. *ramosum* Guss. for Canterbury [*T.N.Z.I. 4*: 309 (1872)]. This name has been applied in Europe to plants with branched spikes.

ζ**L. remotum** Schrank Annual, 50–95 cm. Leaf-blade flat, long-narrowed to fine tip. Spike 12–24 cm, rachis < 1 mm wide. Spikelets 3–7-flowered, rather distant along rachis; perfect florets turgid at maturity. Upper glume 7–9 mm, ≈

spikelet. Lemma 4–5 mm, elliptic to ovate. Caryopsis 3.5 × 1.5 mm. Weed of flax-crops in Europe, now rare; in N.Z. known from one specimen, CHR 35181 *Woodcock* Dunedin, in flax-crops, 1942; recorded as a seed impurity in 1944 (CHR 25545); no subsequent records.

HYBRIDS   All spp. of *Lolium* are ± interfertile. They also cross readily with *Schedonorus phoenix*, tall fescue, and its allies, forming ± sterile hybrids.

Lolium multiflorum × *L. perenne* hybrids (*L.* ×*hybridum* Hausskn.) often occur in N.Z. and were first recorded as naturalised by Allan, H. H. *N.Z. J. Sci. Tech.* 9: 260 (1929). Hybrid cultivars have also been bred in N.Z. to combine the persistence of *L. perenne* with the higher productivity of *L. multiflorum*. These hybrids when growing in the wild tend to segregate and it is impossible to give a definitive description since many combinations of the parental characters occur. An earlier name, *L.* ×*boucheanum* Kunth, is applied to these hybrids in New Zealand. Terrell (1968 op. cit.) placed *L.* ×*boucheanum* in synonymy under *L. perenne*. He was uncertain whether the name referred to an awned form of *L. perenne* or to a *L. multiflorum* × *L. perenne* hybrid.

Plants of *L. multiflorum* with branched inflorescences are probably hybrids with tall fescue, *Schedonorus phoenix* × *Lolium multiflorum*. Forms with abnormal, branched spikelets are also found occasionally.

POA L., 1753

Tufted annuals or perennials, sometimes forming large tussocks, often rhizomatous or stoloniferous; branching intra- or extravaginal. Leaf-sheath open or closed, membranous to coriaceous. Ligule membranous, entire or rarely lacerate, sometimes ciliate, sometimes reduced to a truncate, minutely ciliate rim. Leaf-blade persistent or disarticulating at ligule, flat, folded, sometimes setaceous or filiform, margins sometimes inrolled, coriaceous, or soft and flaccid, tip hooded, or finely acute to ± pungent. Culm unbranched above. Inflorescence an open or contracted panicle. Spikelets 2–several-flowered, pedicelled, laterally compressed, uppermost florets reduced, disarticulation above glumes and between florets; rachilla prolonged; ♀, dioecious, gynomonoecious. Glumes subequal or unequal, usually < adjacent lemmas, keeled, acute to subobtuse; lower 1–3-nerved, upper 3–(5)-nerved. Lemma (3)–5–(7)-nerved, keeled, obtuse to acute to acuminate, awnless but midnerve occasionally slightly excurrent, nerves shortly hairy in lower half, or short scabrid, or glabrous, internerves variously short hairy, scabrid, glabrous, or occasionally papillose. Palea ≈ lemma, keels and interkeel scabrid to ciliate, or glabrous. Callus short, with loose woolly hairs or glabrous. Lodicules 2, glabrous, occasionally hair-tipped. Stamens 3. Gynoecium: ovary glabrous; styles 2, free; stigmata plumose. Caryopsis ellipsoid; embryo < ⅓ length of caryopsis; hilum punctiform or shortly elliptic, < ½ length of caryopsis.

Type species: *P. pratensis* L.

Cosmopolitan, *c.* 500 spp. Endemic spp. 37, indigenous sp. 1 — *P. cookii*; naturalised spp. 10; transient spp. 2.

Poa occurs throughout the N.Z. botanical region, but has a marked southern and subantarctic centre of speciation. New Zealand spp. were revised by Edgar, E. *N.Z.J. Bot. 24*: 425–503 (1986); two new endemic spp. were added by Edgar, E. and Connor, H. E. *N.Z.J. Bot. 37*: 63–70 (1999) and one by Edgar and Molloy *in* Molloy, B. P. J. et al. *N.Z.J. Bot. 37*: 41–50 (1999). Zotov, V. D. *Rec. Dom. Mus. 5*: 101–146 (1965), presented a synoptic treatment of the grasses of the Subantarctic Islands including 12 endemic, 1 indigenous, and 3 naturalised spp. of *Poa*.

Endemic spp. of *Poa* are found from sea level to high alpine areas, usually in grassland or on open rocky ground, cliffs, and scree, occasionally in open forest, or in seepages. Naturalised spp. from Europe occur mainly on roadsides and in waste places. Cosmopolitan *P. annua* is a common weed throughout. *Poa pratensis*, Kentucky bluegrass, is grown as a pasture grass in both North and South Is, and is locally common in tussock grassland, sometimes at high altitudes. Two Australian tussock-forming spp. occur in modified tussock grassland.

Edgar (1986 op. cit. p. 427) discussed the apparent relationship of subantarctic *P. ramosissima* and *P. cookii* to the large tussock sp. *P. flabellata* of southern South America, South Georgia, Falkland and Gough Is. The monotypic genus *Parodiochloa* C.E.Hubb. based on *Poa flabellata*, is distinct from *Poa* only in the elongated unbranched sparsely papillate stigmas and Clayton and Renvoize (1986 op. cit.) treat *Parodiochloa* as a synonym of *Poa*.

Poa triodioides (Trin.) Zotov, formerly referred to *Festuca* as *F. littoralis* Labill., is now included within *Austrofestuca* (Tzvelev) E.B.Alexeev.

Synopsis

A. Massive tussocks, or large or sometimes small tufts; leaf-blades persistent, ± tough, wide and flat, or folded, tips often semi-pungent; ligule entire or lacerate; panicle branches usually smooth; plants sexually dimorphic, dioecious or gynomonoecious, rarely (*P. tennantiana*) ♀; anthers > 0.6 mm

  1. Large tussocks or sward-forming (*P. ramosissima*); panicle contracted; southern and subantarctic islands: 16. *cookii*, 18. *foliosa*, 37. *ramosissima*, 45. *tennantiana*

  2. Tufted, sometimes long-rhizomatous; panicle open or ± contracted; subalpine to alpine South Id (and North Id *P. novae-zelandiae*): 32. *novae-zelandiae*, 38. *schistacea*, 43. *subvestita*, 44. *sudicola*

B. Short tussocks, or small or sometimes large tufts; leaf-blades disarticulating at ligule, wiry, folded with inrolled margins, tips acuminate or naviculate; ligule apically glabrous; panicle open or contracted, branches scabrid or smooth; anthers > 0.6 mm, rarely shorter (*P. maniototo*): 1. *acicularifolia*, 5. *astonii*, 14. *colensoi*, 19. *hesperia*, 29. *maniototo*

C. Large tussocks, clumps, or tufts, erect or trailing down banks, rarely slender (*P. pusilla*); leaf-blades persistent, usually folded, sometimes flat, tips often semi-pungent, or naviculate; ligule a truncate rim, minutely ciliate; panicle open, sometimes contracted, branches usually scabrid; anthers > 0.6 mm

  1. Branching intravaginal: 12. *cita*, 13. *cockayneana*, 25. *labillardierei*, 27. *litorosa*, 40. *sieberiana*

  2. Branching extravaginal: 2. *anceps*, 11. *chathamica*, 35. *pusilla*, 48. *xenica*

D. Tufted, or rhizomatous or stoloniferous (rarely in compact tufts, *P. bulbosa*, with culms swollen at base), perennial or annual; leaf-blades persistent, usually flat, sometimes folded; ligule apically glabrous or ciliate; panicle open, sometimes compressed, branches scabrid or smooth; anthers > 0.6 mm (except *P. infirma*): 3. *annua*, 4. *antipoda*, 9. *bulbosa*, 15. *compressa*, 17. *dipsacea*, 22. *infirma*, 31. *nemoralis*, 33. *palustris*, 34. *pratensis*, 47. *trivialis*, *alpina*

E. Tufted, sometimes shortly rhizomatous or stoloniferous; leaf-blades persistent, flat or folded, sometimes filiform; ligule apically glabrous (ciliate in *P. intrusa*); panicle open, branches scabrid or smooth

  1. Panicle branches slender but firm, spikelets relatively large and conspicuous; anthers usually > 0.6 mm: 6. *aucklandica*, 10. *celsa*, 23. *intrusa*, 24. *kirkii*, 28. *maia*, 42. *sublimis*

  2. Panicle branches and spikelets delicate; anthers 0.2–0.6 mm: 7. *breviglumis*, 20. *imbecilla*, 30. *matthewsii*, *remota*

F. Tufts small, compact (rarely stoloniferous *P. senex*) with culms often almost entirely included within leaf-sheaths; leaf-blades persistent, often stiff and inrolled, or flat; ligule apically glabrous (ciliate in *P. spania*); panicle open or contracted, branches scabrid, or smooth (*P. pygmaea*); anthers 0.2–0.6 mm (> 0.6 mm in *P. pygmaea* and *P. spania*)

1. Internerves of lemma silky haired for ½ length or more: 26. *lindsayi*, 36. *pygmaea*, 41. *spania*, 46. *tonsa*

2. Internerves of lemma smooth, scabrid or papillose, rarely a few hairs near base: 8. *buchananii*, 21. *incrassata*, 39. *senex*

1 Leaf-blades disarticulating at ligule *2*
  Leaf-blades persistent *7*

2 Lemma internerves smooth or scabrid, occasionally with a few hairs near base *3*
  Lemma internerves short-pubescent in lower ½ or almost throughout *6*

3 Ligule > 2.5 mm .... **14. colensoi**
  Ligule < 2.5 mm *4*

4 Panicle narrow, with stiff ± erect branches; lemma acute to acuminate, usually shortly mucronate .... **5. astonii**
  Panicle open, with lax, spreading branches; lemma subobtuse to subacute, entire *5*

5 Tussock-forming; leaf-blades stiff, usually blue-green and 0.5 mm diam.; lemma scabrid or smooth on nerves and internerves, occasionally with a few hairs near base .... **14. colensoi**
  Rhizomatous, turf-forming; leaf-blades lax, light green, 0.5–1 mm wide; lemma with fine silky hairs on lower ⅓ to ½ of nerves, glabrous above on nerves and internerves .... **19. hesperia**

6 Panicle open; spikelets 4–8 mm; anthers 1.8–3 mm .... **1. acicularifolia**
    6a Mat forming; on limestone .... subsp. **acicularifolia**
        Tufted; on ultramafic soils .... subsp. **ophitalis**
  Panicle spike-like; spikelets 2–3.5 mm; anthers 0.2–0.4 mm .. **29. maniototo**

7 Ligule ciliate, usually truncate, usually *c.* 0.5 mm *8*
  Ligule apically glabrous, rounded or tapered or truncate, sometimes deeply lacerate, sometimes erose, 0.2–16 mm *23*

8 Branching intravaginal; tussocks, or forming dense swards *9*
  Branching extravaginal; tufted, often trailing, rarely sward-forming *15*

9 Leaf-blades folded or inrolled *10*
  Leaf-blades flat *13*

10 Leaf-blades abaxially smooth *11*
  Leaf-blades abaxially scabrid *12*

11 Panicle open; spikelets (3.5)–6–9 mm .... **12. cita**
   Panicle contracted; spikelets (7)–11–14 mm .... **27. litorosa**

12 Leaf-blades folded, *c.* 1.5–2 mm wide; callus with long tuft of soft tangled hairs ....**25. labillardierei**
   Leaf-blades inrolled, *c.* 0.5 mm diam.; callus glabrous or with a few hairs ...
   .... **40. sieberiana**

13 Plants sward-forming or clump-forming; lemma 5–6.5 mm .... **13. cockayneana**
   Plants tussock-forming; lemma 2–5–(6) mm *14*

14 Leaf-blades always smooth abaxially; tussocks not packed with dead dry leaves at base .... **12. cita**
   Leaf-blades scabrid or smooth abaxially; tussocks with large bulk of dead dry leaves at base .... **25. labillardierei**

15 Leaf-blade inrolled, *c.* 0.5 mm diam.; leaf-sheath scabrid throughout or just inside margins .... **35. pusilla**
   Leaf-blade flat or folded, (1)–2–6.5 mm wide; leaf-sheath smooth or finely retrorsely hairy, occasionally scabrid near ligule *16*

16 Leaf-tip stiff, ± pungent, acuminate or abruptly acute; leaf-blade and sheath equally stiffly coriaceous *17*
   Leaf-tip either firm and blunt or curved, or tip weak and finely acuminate; leaf-blade subcoriaceous, sometimes soft and weak, and sheath usually more membranous *19*

17 Leaf-blades adaxially smooth between inconspicuous nerves; spikelets 3–7.5 mm .... **2. anceps**
   17a Leaf-blades 1–6.5 mm wide; panicle 10–25 cm; anthers 1.5–2.5 mm; North and South Is .... subsp. **anceps**
   Leaf-blades 1–2.5 mm wide; panicle (3.5)–5–15 cm; anthers *c.* 1.5 mm; Kermadec Is .... subsp. **polyphylla**
   Leaf-blades adaxially ribbed and entirely covered with short prickle-teeth; spikelets 6.5–14.5 mm *18*

18 Panicle to 12 cm; callus with loose crinkled hairs; plants ♂ ..... **11. chathamica**
   Panicle to 25 cm; callus glabrous; plants dioecious .... **48. xenica**

19 Lemma distinctly 5–(7)-nerved *20*
   Lemma appearing 3-nerved, with inner lateral nerves obscure *22*

20 Lemma internerves silky for ¾ length .... **41. spania**
   Lemma internerves glabrous or scabrid in lower ½ *21*

21 Lemma minutely scabrid on nerves, sometimes with a few hairs near base of midnerve; callus glabrous, rarely with a few sparse hairs .... **23. intrusa**
   Lemma sparsely to densely hairy on nerves to about ½ length, scabrid above; callus with tuft of long crinkled hairs .... **34. pratensis**

22 Plants rhizomatous; leaf-tip curved; culms elliptic in transverse section .... **15. compressa**
   Plants not rhizomatous; leaf-tip finely to abruptly acute; culms rounded in transverse section .... **31. nemoralis**

23 Ligule entire, or erose *24*
   Ligule deeply and sharply lacerate *58*

24 Lemma nerves hairy, or sometimes ciliate-scabrid in lower ¼ to ½; callus with obvious tuft of crinkled hairs to glabrous... *25*
   Lemma nerves glabrous or rarely with a few sparse hairs near base; callus glabrous, sometimes with a few wispy hairs *49*

25 Midnerve of lemma smooth above basal hairs; midnerve of glumes smooth, or rarely (*P. subvestita*) with a few prickle-teeth near tip......................... *26*
   Midnerve of lemma scabrid above basal hairs; midnerve of glumes scabrid, especially towards tip, rarely smooth *30*

26 Leaves stiff, coriaceous; plants perennial; lemma 3-nerved, or 5-nerved with the inner lateral nerves often faint *27*
   Leaves soft, thin; plants annual, sometimes perennial; lemma 5-nerved....... *29*

27 Plants long-rhizomatous; leaf-blades inrolled, to 1 mm diam., adaxially minutely pubescent.. **38. schistacea**
   Plants tufted; leaf-blades flat or folded, 1–4–(6.5) mm wide, adaxially ± prickle-toothed *28*

28 Lemma acute to usually acuminate; anthers 0.8–1.2–(1.4) mm; plants gynomonoecious .... **32. novae-zelandiae**
   Lemma usually subobtuse; anthers (1.5)–2–3 mm, or 0.3–0.7 mm and pollen-sterile; plants dioecious .... **43. subvestita**

29 Spikelets with crowded florets; anthers 0.6–1 mm; lower panicle-branches spreading or deflexed after anthesis... **3. annua**
   Spikelets with rather distant florets; anthers 0.1–0.4 mm; lower panicle-branches erect to spreading after anthesis .... **22. infirma**

30 Panicle contracted with erect branches ± concealed by numerous (> 100) spikelets; plants forming massive tufts or tussocks................................... *31*
   Panicle open with spreading branches obvious among spikelets, occasionally contracted and spikelets fewer than 30; plants not forming massive tufts or tussocks *32*

31 Lower leaf-sheaths scabrid between nerves; callus glabrous .... **45. tennantiana**
   Lower leaf-sheaths smooth; callus with tuft of crinkled hairs .... **18. foliosa**

32 Plants rhizomatous *33*
   Plants caespitose, occasionally stoloniferous *38*

33 Panicle branches and spikelet pedicels scabrid *34*
Panicle branches smooth, spikelet pedicels usually smooth, occasionally scabrid *36*

34 Anthers 0.3–0.7 mm .... **6. aucklandica**
34a Close-packed tufts with stiff inrolled leaves; ligule 0.3–1 mm; anthers 0.3–0.4 mm .... subsp. **campbellensis**
Loose tufts with lax, flat to folded leaves; ligule (0.5)–1–2 mm; anthers 0.4–0.7 mm *34b*
34b Culms (4.5)–6.5–12–(30) cm; lemmas acute .... subsp. **aucklandica**
Culms 15–40 cm; lemmas obtuse .... subsp. **rakiura**
Anthers 1–2 mm *35*

35 Leaf-blade (1.5)–2–4 mm wide, flat or folded; glumes unequal, lower 1–3-nerved .... **34. pratensis**
Leaf-blade *c.* 0.5 mm wide, inrolled; glumes ± equal, lower 3-nerved .... **35. pusilla**

36 Leaf-blade inrolled, *c.* 1 mm diam., tip acicular .... **44. sudicola**
Leaf-blade flat, 1.5–4 mm wide, tip blunt, curved *37*

37 Glumes acute; spikelets 4–6 mm; anthers 1.2–1.7 mm .... **34. pratensis**
Glumes subacute to subobtuse; spikelets 5–8.5 mm; anthers 1.6–2.7 mm .... **17. dipsacea**

38 Lemma acute to acuminate *39*
Lemma obtuse *43*

39 Leaf-blade soft, flat *40*
Leaf-blade coriaceous, usually folded, often with inrolled margins *41*

40 Anthers 1.5–2 mm; leaf-sheath usually scabrid above .... **47. trivialis**
Anthers 0.5–1 mm; leaf-sheath smooth .... **4. antipoda**

41 Spikelets 6–12 mm .... **32. novae-zelandiae**
Spikelets 3–3.5–(4) mm *42*

42 Panicle contracted; shoots swollen at base; anthers 1–1.5 mm .... **9. bulbosa**
Panicle lax; shoots slender at base; anthers 2 mm .... **alpinaζ** (p. 191)

43 Lemma internerves silky for ½ length or more *44*
Lemma internerves glabrous except for occasional hairs at base, or sometimes scabrid or very sparsely hairy in lower ½ *45*

44 Leaf-blade grey-green or blue-green, rarely reddish, usually folded; lemma 1–2 mm, covered almost throughout or to *c.* ⅔ length with appressed silky hairs .... **26. lindsayi**
Leaf-blade light green, flat; lemma 2–2.5 mm, covered with short hairs in lower ½ .... **46. tonsa**

45 Plants cushion-forming, to 2.5 cm; culms scarcely overtopping leaves ....
.... **36. pygmaea**
Plants tufted, 5–85 cm; culms obviously exceeding leaves *46*

46 Spikelets 2–3 mm; leaf-blades smooth adaxially .... **39. senex**
Spikelets 3–8 mm; leaf-blades scabrid adaxially, at least near base *47*

47 Leaf-blade flaccid, minutely scabrid throughout .... **33. palustris**
Leaf-blade subcoriaceous, abaxially smooth or scabrid only on nerves, adaxially scabrid in lower ½ and on nerves... *48*

48 Leaf-sheath minutely but densely pubescent-scabrid especially below ligule and on midnerve; spikelets 6–8 mm; anthers 1–1.5 mm .... **10. celsa**
Leaf-sheath smooth, rarely sparsely scabrid; spikelets 3–6.5 mm; anthers 0.6–1–(1.2) mm .... **24. kirkii**

49 Panicle contracted, stiff; plants glaucous *50*
Panicle lax, soft and delicate; plants light or bright green, sometimes brownish or purplish green, very rarely glaucous *51*

50 Leaf-blade adaxially densely scabrid throughout .... **8. buchananii**
Leaf-blade adaxially smooth, sparsely hairy near ligule .... **21. incrassata**

51 Lower glume 0.3–0.8–(1) mm, ≪ upper glume; palea keels smooth, sometimes with a few prickle-teeth near tip .... **7. breviglumis**
Lower glume 0.8–4 mm, ≤ upper glume; palea keels scabrid throughout or in upper ½ *52*

52 Leaf-sheath pubescent-scabrid near hyaline margin *53*
Leaf-sheath glabrous near hyaline margin *54*

53 Leaf-blade 1–2 mm wide; spikelets (2)–3–4–(5)-flowered, ± clustered at branchlet tips .... **30. matthewsii**
Leaf-blade 3–4 mm wide; spikelets 2-flowered, solitary ...... **remota**ζ (p. 192)

54 Anthers 0.2–0.5 mm *55*
Anthers 0.6–1.2–(1.5) mm *57*

55 Panicle branches smooth; spikelet pedicels smooth to occasionally scabrid .... **42. sublimis**
Panicle branches scabrid; spikelet pedicels scabrid *56*

56 Panicle (3.5)–8–15–(25) cm, with delicate, elongate branches .... **20. imbecilla**
Panicle 0.5–2.5 cm, with stiff short branches .... **21. incrassata**

57 Lemma nerves scabrid, at least in part, and internerves usually minutely ± sparsely scabrid; leaf-blades 1–3 mm wide, usually flat .... **24. kirkii**
Lemma completely smooth apart from a few prickle-teeth on midnerve near tip; leaf-blades 0.5–1.5 mm wide, usually folded .... **28. maia**

58 Leaf-blade scabrid-papillose on both surfaces; spikelets scabrid-papillose
     throughout; (Auckland and Campbell Is) .... **37. ramosissima**
     Leaf-blade scabrid-papillose adaxially, smooth abaxially; spikelets glabrous;
     (Macquarie Id) .... **16. cookii**

1. P. acicularifolia Buchanan, *Indig. Grasses N.Z.* t. 49A (1880);
Lectotype: WELT 59604 (Buchanan's folio)! without locality, collector,
or date (designated by Edgar 1986 op. cit. p. 442).

Small, blue-green perennial, 10–20 cm at flowering with culms far
overtopping leaves, from a woody, much-branched rhizome, with wiry,
very long-creeping roots at nodes and numerous fine rootlets; branching
intravaginal; leaf-blades disarticulating at ligule. Leaf-sheath light cream
to later greyish brown, much wider than leaf-blade, glabrous,
membranous, sparsely ribbed; margins very wide hyaline. Ligule 1–5 mm,
apically glabrous, tapered, entire, abaxially scabrid near base or smooth,
occasionally extending as a rim-like membranous contra-ligule. Leaf-
blade 0.5–2.5 cm, rolled and *c.* 0.5 mm diam., glabrous; margins sparsely
prickle-toothed, slightly narrowed to firm abruptly shortly curved,
sometimes pungent tip. Culm (3)–5–15 cm, with 1–2 small cauline leaves,
internodes glabrous. Panicle 1.5–3.5 cm, lax; rachis glabrous, branches
capillary, smooth or very finely scabrid, with 1–2 spikelets at branchlet
tips. Spikelets 4–8 mm, (2)–3–5-flowered, light grey-green. Glumes
subequal, 2–3.5 mm, 3-nerved, elliptic-ovate, midnerve scabrid near
subobtuse tip, margins often finely scabrid. Lemma 3–4 mm, 5-nerved,
oblong-elliptic, obtuse, short-pubescent throughout lower ⅓ to ½ but
central internerves sometimes glabrous, scabrid above on midnerve and
occasionally towards tip; margins scabrid above. Palea 2.5–3.5 mm, keels
ciliate-scabrid, interkeel minutely hairy on lower ½. Callus ringed by
short soft hairs. Rachilla 0.5–1 mm, usually ciliate; prolongation twice as
long. Lodicules 0.5 mm. Anthers 1.8–3 mm. Caryopsis *c.* 1.5–2×0.5 mm.

   subsp. **acicularifolia**

Low tight mats of interlacing, stiff, short, hard, curved leaves. $2n = 28$.

Endemic.

S.: Marlborough, North and Central Canterbury. Lowland to alpine
on limestone.

subsp. **ophitalis** Edgar, *N.Z. J. Bot. 24*: 444 (1986); Holotype: CHR 389678! *A. P. Druce* Dun Mt, Nelson, 3500 ft, tussock land (on mineral belt), Jan. 1980.

Small tufts with erect leaves; tufts sometimes aggregated or rhizomes interconnecting to form loose mats.

Endemic.

S.: eastern Nelson and western Marlborough. Subalpine on ultramafic soil.

Poa acicularifolia subsp. *ophitalis* in ultramafic sites at Dun Mt and Wooded Peak in Nelson, and on Red Hills in western Marlborough, may be mistaken for *P. colensoi*, but can be distinguished by the adaxially smooth, not scabrid, leaf-blades and the short-hairy lemmas; the lemma of *P. colensoi* from eastern Nelson and the Mineral Belt is almost glabrous or shortly scabrid.

2. P. anceps G.Forst., *Prodr.* 8 (1786) ≡ *P. anceps* G.Forst. var. *anceps* (autonym, J. D. Hooker 1853 op. cit. p. 306); Lectotype: K! *Forster* New Zealand (designated by Edgar 1986 op. cit. p. 450). = *P. anceps* var. α *elata* Hook.f., *Fl. N.Z. 1*: 306 (1853); Lectotype: K! *Colenso 1150* New Zealand [Aropauanui River, Hawke's Bay], grass (designated by Edgar 1986 op. cit. p. 450). = *P. anceps* var. β *foliosa* Hook.f., *Fl. N.Z. 1*: 306 (1853); Lectotype: K! *J. D. H[ooker]* Bay of Islands (designated by Edgar 1986 op. cit. p. 450). = *P. anceps* var. δ *densiflora* Hook.f., *Handbk N.Z. Fl.* 339 (1864) ≡ *P. anceps* var. *condensata* Cheeseman, *Man. N.Z. Fl.* 904 (1906) *nom. superfl.*; Lectotype: K! *Colenso 3809* [cliffs of Parimaha, near Turnagain Cape], grass (designated by Edgar 1986 op. cit. p. 451). = *P. affinis* var. α *multiflora* Hook.f., *Fl. N.Z. 1*: 307 (1853); Lectotype: K! *Colenso 52* N. Zealand [Hawke's Bay] (designated by Edgar 1986 op. cit. p. 450). = *P. affinis* var. β *agrostoidea* Hook.f., *Fl. N.Z. 1*: 307 (1853); Lectotype: K! *Colenso 798* N. Zealand [Ahuriri] 1847 (designated by Edgar 1986 op. cit. p. 451).

Very variable, coarse, light green to greenish brown to bluish green perennial tufts to *c.* 70 cm, with stiff erect leaves < stems, or often scrambling and trailing to 2 m, with hanging leaves and stems drooping from thick stolons, rooting at nodes below tufts; branching extravaginal, with up to three, short, glabrous, obtuse, bract-like sheaths at base; leaf-blades persistent. Leaf-sheath light green to light brown, coriaceous, folded and strongly keeled, lateral ribs conspicuous, smooth or slightly scabrid above, rarely minutely scabrid throughout. Ligule *c.* 0.5 mm, a truncate usually long-ciliate rim, scabrid abaxially. Leaf-blade coriaceous, folded to flat, abaxially with prominent, thickened midrib, and numerous, distinct lateral ribs, smooth apart from prickle-teeth near tip, adaxially smooth between inconspicuous ribs; margins thickened, tip acuminate or abruptly acute, often pungent, scabrid. Culm often not far exserted beyond uppermost leaf-sheath, internodes glabrous. Panicle much-branched, branches whorled, very slender; rachis and primary branches often smooth, secondary branchlets finely, sharply, densely or sparsely scabrid or smooth, often spikelet-bearing ± throughout. Spikelets numerous, light green. Glumes subequal, narrow- to elliptic-lanceolate, acute to subobtuse, occasionally smooth throughout, or upper ⅔ scabrid; lower slightly shorter, upper (2.5)–3–4–(5) mm, 3-nerved. Lemma 3–4.5 mm, elliptic-oblong, with short crinkled hairs on lower ½ of midnerve and near base of outer lateral nerves; margins minutely scabrid. Palea 2.5–4 mm, keels finely scabrid, interkeel and flanks smooth or minutely scabrid. Callus with thick tuft of soft crinkled hairs. Rachilla *c.* 0.5 mm, smooth or minutely, sparsely scabrid; prolongation *c.* twice as long. Lodicules *c.* 0.5 mm, occasionally hair-tipped.

subsp. **anceps**

Very variable in habit, stiff and erect or scrambling and trailing. Leaf-sheath smooth or variably scabrid. Leaf-blade 10–30–(40) cm × 1–6.5 mm, adaxially smooth, or scabrid on ribs, rarely papillose-scabrid, occasionally with fringe of stiff short hairs above ligule; margins smooth or scabrid. Culm (15)–25–50–(70) cm. Panicle *c.* 10–25 cm, usually open with spreading branches, sometimes contracted. Spikelets (3)–4–7.5 mm, (2)–3–5–(8)-flowered. Lower glume (2)–2.5–3.5–(4.5) mm, 3-nerved. Lemma 5–7-nerved, acute to subobtuse, internerves finely scaberulous throughout, occasionally only minutely papillose. Anthers 1.5–2.5 mm. Caryopsis *c.* 2 × 0.5 mm. $2n = 28$.

Endemic.

N.: throughout; S.: northern and western coasts to George Sound and on the east at Banks Peninsula; Three Kings Is. Lowland, frequently coastal, to subalpine, in open forest, scrub or grassland, often hanging down rocks or banks.

On rocky coastal sites on the eastern coast of North Id, from islands off the Auckland Coast to the Wellington Heads, plants often have numerous, thick, coarse, erect leaves and a very condensed panicle with branches almost concealed by closely clustered spikelets. This form was described by Hooker as *P. anceps* var. *densiflora* (= *P. anceps* var. *condensata* Cheeseman) but cannot be clearly separated from other plants growing on cliffs with slightly less condensed heads.

Specimens collected on Three Kings Is in 1983 and cited by Edgar (1986 op. cit. p. 447) as *P. cita*, viz AKU 15553, CHR 407574 A, B *E. K. Cameron 2409*, are an unstable, habitat-induced form of *P. anceps* which gradually loses its tussock form in cultivation (P. J. de Lange *in litt.*).

   subsp. **polyphylla** (Hack.) Edgar, *N.Z. J. Bot. 24*: 452 (1986) ≡ *P. polyphylla* Hack., *T.N.Z.I. 35*: 383 (1903) ≡ *P. anceps* var. *polyphylla* (Hack.) Zotov, *T.R.S.N.Z. 73*: 236 (1943) ≡ *P. polyphylla* Hack. forma *polyphylla* (autonym, Hackel 1903 op. cit. p. 383); Lectotype: W 9646! *T. F. C[heeseman]* Kermadec Islands, N.E. from New Zealand [Aug. 1887] (No 1444 to Hackel) (designated by Edgar 1986 op. cit. p. 452). = *P. polyphylla* forma *compacta* Hack., *T.N.Z.I. 35*: 383 (1903); Holotype: W 9648! *R. H. Shakespear* Kermadec Islands, N.E. from N.Z. (No 1445 to Hackel).

Loose tufts, with trailing culms, equalled or overtopped by numerous dull green leaves. Leaf-sheath glabrous. Leaf-blade 6.5–20 cm × 1–2.5 mm, adaxially densely minutely papillose or glabrous, minutely pubescent-scabrid just above ligule; margins sparsely ciliate near ligule and scabrid near tip. Culm (7)–20–30–(70) cm. Panicle (3.5)–5–15 cm, contracted, ± oblong, with few, stiff, erect branches, or slightly more open with finer branches. Spikelets 4.5–5.5 mm, 3–4-flowered. Lower glume 2–3.5 mm, (1)–3-nerved. Lemma 5-nerved, acute or occasionally apiculate, internerves glabrous. Anthers *c.* 1.5 mm. Caryopsis *c.* 1.5 × 0.5 mm. $2n = 28$.

Endemic.

K.: On coastal cliffs and roadside banks.

3. P. annua L., *Sp. Pl.* 68 (1753).   annual poa

Loosely to compactly tufted, usually bright green annual, to 35 cm, or short-lived perennial; branching intravaginal; leaf-blades persistent. Leaf-sheath pale green, hyaline, glabrous, keeled. Ligule 0.5–2 mm, entire, rounded, glabrous throughout. Leaf-blade 1.5–9 cm × 3–4 mm, flat or folded, soft, glabrous, often crinkled when young; margins sparsely, minutely scabrid, tip rounded to blunt point. Culm 3–20 cm, erect, spreading or prostrate, internodes glabrous. Panicle 2–10 cm, usually open, erect; branches smooth to sparsely scabrid, spreading to deflexed after anthesis. Spikelets 4–6 mm, with 3–6 crowded florets, light green or sometimes purplish. Glumes ± unequal, glabrous; lower 1.5–2.5 mm, 1-nerved, lanceolate to ovate, acute, upper 2–3 mm, 3-nerved, elliptic to oblong, subobtuse. Lemma 3–3.5 mm, 5-nerved, nerves densely hairy for about ½ length and glabrous above hairs, internerves glabrous, rarely lemma entirely glabrous, purplish just below hyaline band at obtuse tip. Palea 2.5–3 mm, keels long-ciliate, or rarely glabrous, interkeel glabrous. Callus glabrous. Rachilla 0.5–1 mm, glabrous; prolongation much shorter. Lodicules 0.3 mm. Gynomonoecious: each spikelet with 2–3 lower flowers ♂, anthers (0.6)–0.7–1 mm, gynoecium 1–1.5 mm; upper flowers ♀, pollen-sterile anthers to 0.2 mm or 0, gynoecium *c.* 1.5 mm. Caryopsis *c.* 1.5 × 0.5 mm, tightly enclosed by anthoecium.

Naturalised from Europe.

N.; S.: throughout; St.; K., Ch., Sn., Ant., A., C., M. A weed of cultivated land and waste places throughout. FL Jan–Dec.

Cosmopolitan and weedy.

4. P. antipoda Petrie, *in* Chilton *Subantarct. Is N.Z.* 2: 478 (1909); Lectotype: WELT 66428! *T. Kirk* Antipodes Island, Jan. 1890 (No 1474 to Hackel) (designated by Edgar 1986 op. cit. p. 454).

Soft, ± drooping, light green stoloniferous perennial tufts, *c.* 20–60 cm, rooting at nodes; branching extravaginal; leaf-blades persistent. Leaf-sheath light green to light brown, submembranous, distinctly ribbed, glabrous. Ligule 1–4.5 mm, entire, apically glabrous, gradually narrowed and subacute, abaxially slightly scabrid. Leaf-blade 7.5–25 cm

× 2–4.5 mm, flat, soft, smooth almost throughout but minutely scabrid abaxially near straight-sided narrow acute tip and adaxially just above ligule. Culm (8)–18–50 cm, internodes glabrous. Panicle 5–15 cm, ± lax and usually open with spreading branches; rachis usually ± smooth, branches ± smooth to sparsely scabrid. Spikelets 4–6 mm, (2)–3–4-flowered, brownish green. Glumes ± unequal, narrow-lanceolate, acute to acuminate, glabrous, apart from a few prickle-teeth on midrib in upper ½; lower 1.5–3 mm, 1-nerved, upper 2–3.5 mm, 3-nerved. Lemma 2.5–5 mm, 3–(5)-nerved, ± elliptic, acute, with long hairs on lower ½ of midnerve and at base of lateral nerves and fine prickle-teeth on midnerve above, internerves glabrous or sometimes slightly scabrid near tip. Palea 2–4 mm, keels shortly ciliate-scabrid, interkeel glabrous. Callus with narrow tuft of long fine hairs. Rachilla 0.5 mm, glabrous; prolongation twice as long. Lodicules 0.4–0.5 mm. Anthers 0.5–1 mm, mainly pollen-sterile. Caryopsis not seen.

Endemic.

St.: Herekopere Id; Ant., A., C. Coastal on cliffs and rock outcrops; inland on damp banks, and in herbfield.

No mature caryopses were found among the many specimens of *P. antipoda* examined at CHR, AK, and WELT (Edgar 1986 op. cit. 455).

5. P. astonii Petrie, *T.N.Z.I. 38*: 423 (1906) ≡ *P. astonii* Petrie var. *astonii* (autonym, Zotov 1943 op. cit. p. 236); Lectotype: WELT 66186! *D. P[etrie]* Brighton, near Dunedin (designated by Edgar 1986 op. cit. p. 439). = *P. oraria* Petrie, *T.N.Z.I. 42*: 196 (1910) ≡ *P. astonii* var. *oraria* (Petrie) Zotov, *T.R.S.N.Z. 73*: 236 (1943); Lectotype: WELT 66041! *B. C. Aston* Deep Cove, West Coast Sounds, Otago, Jan. 1909 (designated by Edgar 1986 op. cit. p. 439).

Tightly packed, wiry-leaved small glaucous tussocks, 10–40 cm, spreading in narrow fans from a slender rhizome; branching intravaginal; leaf-blades disarticulating at ligule. Leaf-sheath light brownish to stramineous, sometimes purple, glabrous, shining, ± coriaceous with broad membranous margins, ribs inconspicuous. Ligule 1–2 mm, apically glabrous, entire, rounded and slightly tapered, abaxially slightly scabrid near base, or densely scabrid throughout. Leaf-blade (5)–10–30–(40) cm

× 0.5–1 mm diam., inrolled, not keeled, usually erect, occasionally curled above, abaxially glabrous, adaxially pubescent, tip long-acicular. Culm (5)–20–30–(45) cm, ≤ leaves, internodes glabrous. Panicle 5–10–(12) cm, narrow, stiff; branches short, ± erect, glabrous to finely pubescent-scabrid, tipped by few, large spikelets. Spikelets 5.5–12 mm, 3–6–(8)-flowered, light greenish brown. Glumes subequal, subacute to acuminate, ovate to elliptic-lanceolate, ± smooth, occasionally a few prickle-teeth on midnerve near tip; lower 2–5–(6) mm, 1–3-nerved, upper 2–5.5–(7) mm, 3–(5)-nerved; margin finely ciliate. Lemma 3.5–7 mm, 3–5-nerved, elliptic-lanceolate, acute to acuminate, usually shortly mucronate, outer lateral nerves very distinct, internerves minutely scabrid or very rarely smooth, or prickle-teeth fewer between lateral nerves and keel, but always very dense and in several rows along keel, with longer crinkled hairs on lower ½ of midnerve, and near base of lateral nerves, often also on internerves near base; margins finely fimbriate. Palea 3–6 mm, keels and flanks with dense short prickle-teeth or minute hairs, the teeth on keels arranged in several rows, interkeel with minute prickle-teeth or hairs, often almost glabrous centrally; rarely palea glabrous except at fimbriate tip. Callus with short tuft of crinkled hairs. Rachilla 0.5– *c.* 1 mm, with sparse, short, fine hairs, or densely to minutely finely scabrid; prolongation twice as long. Lodicules 0.5–1 mm, occasionally hair-tipped. Anthers *c.* 2–3 mm. Caryopsis *c.* 1.5 × 0.5 mm. $2n = 28$. Plate 4A.

Endemic.

S.: Canterbury on Banks Peninsula near Akaroa Heads, Otago south from Dunedin and in Fiordland; St.: and on islands nearby; Solander Id, Sn., A., C. Coastal, on beaches or rocky cliffs.

Specimens from coastal cliffs near Dunedin have small spikelets (5.5–9 mm), short, ovate-elliptic glumes (2–3.5 mm) and shortly mucronate lemmas (3.5–4.5 mm), whereas many specimens from Fiordland, those from the Snares, and some from Southland, Stewart Id and the Subantarctic Is, have longer spikelets (8–12 mm), longer, narrower glumes (3–5.5–(7) mm), and longer very acuminate lemmas (5–7 mm). Both forms are present in some localities and there is no obvious definite geographical or ecological separation between the two kinds.

A specimen of *P. astonii* from the south-western Mutton Bird Is (CHR 261888) has proliferation in its spikelets.

The only known specimen from Canterbury was collected from Palm Gully, near Akaroa Heads, *H. D. Wilson* 4.12.1984 (CHR 417312). There are also some specimens collected by Buchanan from Nelson (WELT 65918, WELT 66313 and CHR 29296) but with no further details of locality or date of collection.

No recent collections of *P. astonii* have been made from Auckland and Campbell Is.

6. P. aucklandica Petrie, *in* Chilton *Subantarctic Is N.Z.* *2*: 478 (1909) ≡ *P. kirkii* var. *aucklandica* (Petrie) Zotov, *T.R.S.N.Z.* *73*: 236 (1943); Lectotype: WELT 66441! *B. C. Aston* Mt top behind Camp Cove, Carnley H[ar]b[ou]r., Auckland Is, Nov. 1907 (designated by Edgar 1986 op. cit. p. 465).

Tufted, dull green, slenderly rhizomatous perennial with narrow, ± erect leaves, ± reaching top of culms; branching extravaginal at plant base, intravaginal above; leaf-blades persistent. Leaf-sheath very light green or pale brown, membranous, shredding into fibres. Ligule apically glabrous, ± obtuse and slightly erose, abaxially slightly scabrid. Leaf-blade 0.5– *c.* 1.5 mm wide, abaxially smooth, adaxially minutely ciliate-scabrid on ribs; margins and midrib smooth, tip curved. Culm slender, internodes usually slightly scabrid below panicle. Panicle lax, with few large spikelets borne singly at tips of filiform, finely scabrid branches. Spikelets 3–5-flowered, brownish green or greenish purple. Glumes smooth, but occasionally scabrid on midnerve near tip; lower 3-nerved, upper 3–(5)-nerved. Lemma 5–7-nerved, nerves minutely scabrid with a few short hairs near base of midnerve and outer lateral nerves, internerves glabrous, often scabrid towards tip. Palea minutely closely scabrid on keels, smooth elsewhere. Callus with very small tuft of long crinkled hairs. Rachilla *c.* 1 mm, sparsely scabrid; prolongation to twice as long. Lodicules 0.5–0.7 mm. Caryopsis *c.* 1.5 × 0.7 mm.

subsp. **aucklandica**

Slender lax tufts with culms (4.5)–6.5–12–(30) cm. Leaf-sheath glabrous. Ligule 1–1.3 mm. Leaf-blade 4–12–(16) cm, flaccid, soft, flat to folded. Panicle 3–5.5–(8) cm. Spikelets 4.5–8 mm. Glumes subequal, oblong-elliptic, acute; lower 3.5–5 mm, upper 4–5.5 mm. Lemma 4–5.5 mm, oblong-elliptic, acute; midnerve occasionally very shortly excurrent. Palea 3–4.5 mm. Anthers 0.4–0.7 mm. $2n = 28$.

Endemic.

A. Upland herbfield and rock crevices.

subsp. **campbellensis** (Petrie) Edgar, *N.Z. J. Bot. 24*: 467 (1986) ≡ *P. campbellensis* Petrie, *T.N.Z.I. 50*: 211 (1918) ≡ *P. kirkii* var. *campbellensis* (Petrie) Zotov, *T.R.S.N.Z. 73*: 236 (1943); Lectotype: WELT 66443! *B. C. Aston* Campbell Island, 12.1.1909 (designated by Edgar 1986 op. cit. p. 467).

Small tufts with close-packed culms 2–10 cm. Leaf-sheath glabrous. Ligule 0.3–1 mm. Leaf-blade 2–5 cm, stiff, inrolled. Panicle 1.5–5 cm. Spikelets 4.5–6.5 mm. Glumes unequal, elliptic-ovate, acute to subacute; lower 3–4 mm, upper 4–4.5 mm. Lemma 4–4.5 mm, elliptic-ovate, subacute to acute; midnerve occasionally very slightly excurrent. Palea 3–4 mm. Anthers 0.3–0.4 mm. $2n = 28$.

Endemic.

C. Upland rocky herbfield.

subsp. **rakiura** Edgar, *N.Z. J. Bot. 24*: 467 (1986); Holotype: CHR 362702! *H. D. Wilson* Mt Anglem, Stewart Id, summit area, 4.2.1980.

Slender lax tufts with culms 15–40 cm. Leaf-sheath glabrous or sometimes shortly pubescent-scabrid near margins. Ligule (0.5)–1–2 mm. Leaf-blade (4)–8–20 cm, flaccid, soft, flat to folded, occasionally entirely smooth or with finely scabrid midrib and margins. Panicle 4–7.5 cm. Spikelets 5.5–7.5 mm. Glumes ± unequal, elliptic, acute; lower 3–4.5 mm, upper 4–5 mm. Lemma 3.5–4.5 mm, oblong-elliptic, obtuse. Palea 3–4 mm. Anthers 0.5–0.7 mm.

Endemic.

St.: Mt Anglem, summit. In *Chionochloa crassiuscula* grassland among rocks.

7. P. breviglumis Hook.f., *Fl. Antarct. 1*: 101 (1845) ≡ *P. imbecilla* var. *breviglumis* (Hook.f.) Cheeseman, *Man. N.Z. Fl.* 201 (1925) ≡ *P. breviglumis* Hook.f. var. *breviglumis* (autonym, Zotov 1965 op. cit. p. 131); Holotype: K! *J. D. Hooker* Campbell's Island, moist banks near the sea, not uncommon, Dec. 1840. = *P. breviglumis* var. *brockiei* Zotov, *Rec. Dom. Mus. 5*: 131 (1965); Holotype: CHR 119414! *V. D. Zotov* Campbell Is., Tucker Cove, 8/1/1961. = *P. breviglumis* var. *moarii* Zotov, *Rec. Dom. Mus. 5*: 132 (1965); Holotype: CHR 88823A! *N. T. Moar & J. B. Hair 1394* Lord Auckland Islands, Port Ross, Ranui Cove, 11.11.1954.

Soft, weak, loosely tufted, stoloniferous or rhizomatous perennial, *c.* 5–40 cm, rooting at lower nodes, smaller tufts sometimes with close-packed shoots; leaves < culms, bright green, leaf-blades persistent; branching extravaginal near plant base, intravaginal above. Leaf-sheath light green, sometimes purplish, later very light brown, membranous, distinctly ribbed, glabrous. Ligule 0.5–1.5–(3) mm, entire, glabrous throughout. Leaf-blade *c.* 2–9 cm × 0.5–1.5 mm, flat, abaxially slightly scabrid near acute, ± incurved tip, slightly scabrid adaxially and on margins. Culm (3)–10–25–(35) cm, internodes glabrous. Panicle 3–10–(16) cm, lax and delicate; branches often paired, very fine, scabrid, spreading, bearing few spikelets, pedicels scabrid on angles, often purplish. Spikelets 2–3–(5) mm, 2–3–(5)-flowered, light green, often purple-tinged. Glumes usually noticeably unequal, often purplish, smooth; lower 0.3–0.8–(1) mm, 1-nerved, triangular, subacute, upper (0.8)–1–1.5 mm, 3-nerved, more ovate, subobtuse, rarely with a few prickle-teeth on midnerve; margins entire. Lemma 1.2–2 mm, 3-nerved, subobtuse, smooth apart from a few prickle-teeth on midnerve; margins entire, usually purplish towards tip. Palea 1.0–1.5 mm, keels smooth, or with a few prickle-teeth near tip, interkeel smooth. Callus glabrous. Rachilla *c.* 0.5 mm, glabrous. Lodicules *c.* 0.4 mm, rarely hair-tipped. Anthers 0.2–0.3 mm. Gynoecium: ovary 0.4 mm; stigma-styles 0.7–0.8 mm. Caryopsis 0.9–1 × 0.3–0.5 mm. $2n = 28$.

Endemic.

N.: central and southern; S.: throughout; St.; Ant., A., C. Sea level to alpine in open forest, scrub, tussock grassland or herbfield.

Poa breviglumis var. *brockiei* was based on a plant from Campbell Id which was rather densely tufted and rather well-branched with short narrow leaves. Plants ± tufted at base and with well-branched stems seem to occur naturally throughout the range of the species; plants grown in pots often take this form also.

Only two collections have been made from Auckland Is (Edgar 1986 op. cit. p. 474). In these the lemma is *c.* 2.5 mm and sharply acute.

8. P. buchananii Zotov, *T.R.S.N.Z. 73*: 236 (1943) ≡ *P. anceps* var. ε *alpina* Hook.f., *Handbk N.Z. Fl.* 339 (1864) ≡ *P. sclerophylla* Berggr., *Minneskr. Fïsiogr. Sällsk. Lund* Art. *8*: 30 (1878) non Kunth (1833) ≡ *P. albida* Buchanan, *Indig. Grasses N.Z.* t. 50C (1880) non Trin. (1831); Holotype: K! *J. Haast 629* Mt Darwin and Mt Dobson, Canterbury, New Zealand, 4–6000 ft, 1862.

Glaucous, small, stiff, dense perennial tufts, 7–25 cm; leaf-blades persistent; branching intravaginal, with a few extravaginal shoots at plant base. Basal leaf-sheaths very light brown, upper sheaths purple-green, coriaceous, distinctly ribbed, glabrous, very much wider than leaf-blade. Ligule 0.5–1.5 mm, apically glabrous, rounded, ± erose, abaxially scabrid. Leaf-blade (0.5)–1.5–5 cm long, folded, *c.* 3 mm wide if flattened, rigid, coriaceous, prominently nerved, abaxially usually smooth, adaxially densely, minutely scabrid, midrib scabrid near curved scabrid tip; margins scabrid, incurved. Culm 6–18–(25) cm, often entirely enclosed by sheaths of uppermost leaves, internodes minutely scabrid below panicle. Panicle 1.5–5.5 cm, contracted, oblong, narrow and spike-like, with numerous, densely packed spikelets; rachis and short stiff branches minutely scabrid. Spikelets 2.5–3.5 mm, (2)–3–4-flowered, light green, often purple-tinged. Glumes ± equal, almost overtopping spikelets, (1.5)–2–2.5–(3) mm, 3-nerved, midnerve sparsely scabrid; lower elliptic-lanceolate, subacute, upper oblong-elliptic, subobtuse, often scabrid near tip; margins sparsely scabrid. Lemma 2–2.5 mm, 5-nerved, ovate-elliptic, obtuse, midnerve scabrid, internerves minutely papillose and often minutely scabrid; margins membranous, wide, edged with scattered prickle-teeth. Palea *c.* 1.5 mm, keels scabrid, interkeel minutely papillose with a few prickle-teeth. Callus glabrous. Rachilla *c.* 0.5 mm, minutely papillose; prolongation occasionally twice as long. Lodicules 0.3–0.4 mm. Anthers 0.4–0.6 mm. Gynoecium: ovary 0.4–0.5 mm; stigma-styles 0.6–1.2 mm. Caryopsis *c.* 1 × 0.5 mm. $2n = 28$.

Endemic.

S.: east of Main Divide. Subalpine to alpine in scree, stony ground on moraines and loose debris in fellfield.

This glaucous scree plant is very distinct from other N.Z. spp. of *Poa* because of its stiff, scabrid, close-packed leaves and compact, spike-like panicle.

9. P. bulbosa L., *Sp. Pl.* 70 (1753).   bulbous poa

Tightly packed, swollen-based perennial tufts to *c.* 10 cm; branching intravaginal; leaf-blades persistent. Leaf-sheath cream, membranous, glabrous, later shredding into fibres; inner basal sheaths thick, swollen and hard at base; sheaths of cauline leaves light green, some purplish. Ligule 0.5–2.5 mm, entire, tapered, apically glabrous or very sparingly ciliate, abaxially glabrous. Leaf-blade 2.5–5 cm long, rather stiff, folded, grey-green, glabrous; margins minutely scabrid, midrib also scabrid near abruptly pointed tip. Culm 5–10 cm, internodes glabrous. Panicle 2–4 cm, contracted, ± oblong, with densely packed spikelets; rachis smooth, branches few, ± erect, sparsely scabrid. Spikelets 3–3.5 mm, 2–3-flowered, light green to purplish. Glumes ± equal, 2–3 mm, ovate, acute, keel distinct, green, scabrid; lower 1-nerved, upper 3-nerved; margins membranous, wide, purplish. Lemma *c.* 3 mm, 5-nerved, ovate, acute, membranous, glabrous, nerves densely ciliate on lower ½, midnerve scabrid above hairs. Palea *c.* 2.5 mm, keels minutely ciliate, interkeel glabrous. Callus with tuft of tangled hairs. Rachilla *c.* 0.5 mm, glabrous; prolongation much shorter. Lodicules *c.* 0.4 mm. Anthers 1–1.5 mm. Caryopsis not seen.

Naturalised from Europe.

S.: Central Otago (near Alexandra). Roadside.

No specimens have been found to support the record of Kirk, T. W. *Ann. Rep. Dept Agric. 13*: 332 (1905) from Pleasant Point. *Poa bulbosa* was collected from the cemetery at Alexandra in 1966 and became well established in the vicinity (A. J. Healy, *in litt.*).

The proliferous form, *Poa bulbosa* var. *vivipara* Koeler, in which the spikelet is replaced above the glumes by a miniature shoot, has also been collected near Alexandra.

10. P. celsa Edgar, *N.Z. J. Bot. 24*: 463 (1986); Holotype: CHR 275245! *A. P. Druce* Luna L., NW Nelson, 4500 ft, rocky ground at foot of cliff, Jan. 1974.

Erect, light purplish green perennial tufts to 50 cm, culms overtopping leaves; branching extravaginal; leaf-blades persistent. Leaf-sheath purplish, to later greyish, subcoriaceous, ribs inconspicuous, abaxially with minute prickle-teeth or hairs especially on midrib near ligule. Ligule *c.* 1.5–3 mm, apically glabrous, rounded, sometimes erose, abaxially with minute prickle-teeth or hairs. Leaf-blade 6–20 cm × (2)–3–6 mm, folded or flat, subcoriaceous, abaxially ± minutely scabrid on ribs, adaxially scabrid in lower ½ and on ribs above; margins scabrid, with some stiff short hairs at collar, tip ± curved, acuminate to often apiculate. Culm 10–40 cm, erect, internodes scabrid below panicle. Panicle 8–15 cm, lax, with slender spreading scabrid branches. Spikelets 6–8 mm, (2)–3–4–(5)-flowered, green or purplish. Glumes ± equal, 3-nerved, ovate-elliptic, nerves and internerves near tip ± scabrid; lower 3.5–5.5 mm, subacute, upper 4–5.5–(6) mm, subobtuse; margins scabrid. Lemma 4.5–6 mm, 5-nerved, oblong, obtuse, pubescent-scabrid on nerves, margins, and often on internerves. Palea 3.5–4.5 mm, keels ciliate-scabrid, interkeel scabrid. Callus with tuft of a few long crinkled hairs. Rachilla *c.* 1 mm, with a few long hairs; prolongation twice as long. Lodicules 0.5–0.6 mm. Anthers 1–1.5 mm. Caryopsis *c.* 1–2 × 0.5 mm.

Endemic.

S.: Nelson and in Lewis Pass area, and north Westland. Subalpine to alpine in grassland, often on rock.

Both *Poa celsa* and *P. intrusa* are closely related to *P. kirkii* but plants are larger, with longer anthers. *Poa celsa* particularly differs from other spp. in the *P. kirkii* group in its pubescent-scabrid leaf-sheaths.

11. P. chathamica Petrie, *T.N.Z.I. 34*: 394 (1902) ≡ *P. anceps* var. *chathamica* (Petrie) Zotov, *T.R.S.N.Z. 73*: 236 (1943); Holotype: WELT 66386a! *L. Cockayne 6575 & F. A. D. Cox* growing on sphagnum or very boggy ground, S. end of Chatham Island (No 1473 to Hackel).

Dense, drooping perennial swards from long, narrow rhizomes, or stiff tufts to 90 cm, with light green leaves ≈ culms; branching extravaginal near plant base, intravaginal above; leaf-blades persistent. Leaf-sheath light green to light brown or straw-coloured, coriaceous, ribbed, smooth to finely scabrid, keeled. Ligule (0.2)–0.5–1 mm, a truncate stiff ciliate rim, abaxially with matted stiff minute hairs. Leaf-blade 5–30 cm × 2.5–4.5 mm, folded to flat, or inrolled and *c.* 1 mm diam., coriaceous, abaxially smooth, adaxially ribbed, covered with short prickle-teeth and sometimes short stiff hairs; margins ± thickened, smooth to sparsely scabrid, midrib scabrid near straight-sided, pungent tip. Culm 10–80 cm, internodes smooth, densely scabrid below panicle. Panicle 5.5–12 cm, lax or contracted; branches spreading or erect, sparsely to densely scabrid. Spikelets 6.5–14.5 mm, (2)–3–5-flowered, greyish green to light greenish brown. Glumes subequal, 3-nerved, elliptic-lanceolate, acute to acuminate, often with minute fine hairs near tip, occasionally scabrid throughout, midnerve ciliate-scabrid especially on upper ½; lower 4.5–7.5 mm, upper 4.5–8 mm; margins ciliate. Lemma 4.5–9 mm, 5-nerved, elliptic- to oblong-lanceolate, subobtuse to subacute, scabrid above or occasionally throughout, midnerve with long fine hairs to ½ length, lateral nerves hairy near base; margins minutely ciliate. Palea 3.5–7.5 mm, keels rather stiffly ciliate-scabrid, interkeel with sparse minute hairs. Callus with loose web of long fine crinkled hairs. Rachilla 0.5–1 mm, glabrous to sparsely minutely pubescent; prolongation twice as long. Lodicules 0.5–2 mm, occasionally hair-tipped. Anthers 2–3.5 mm. Caryopsis 2 × 0.5 mm. $2n = 112$.

Endemic.

Ch. Rocky sites on coast and inland, sand dunes, and open sites in peat.

Although plants of *P. chathamica* vary considerably in habit, it is impossible to subdivide the species. Plants growing on the sandy foreshore at Kaingaroa (e.g., CHR 95402, CHR 96577) are distinct in having the following features in combination: a very contracted panicle with densely scabrid branches and glumes, long spikelets, and short ligules. However, there is no clear-cut separation between plants from sandy ground and others from cliffs or peat.

12. P. cita Edgar, *N.Z. J. Bot. 24*: 446 (1986) ≡ *P. caespitosa* Spreng., *in* Biehler *Pl. Nov. Herb. Spreng.* 7, no. 10 (1807) non Poir., *in* Lamarck (1804); Holotype: B (Herb. Willdenow no. 1894)! *Forster* habitat in Nova Zeelandia. = *P. caespitosa* var. *leioclada* Hack., *in* Cheeseman *Man. N.Z. Fl.* 908 (1906); Lectotype: W 10641! *D. Petrie* Mount Egmont, Taranaki, 4000 ft [6.1.1901] (No 1364 to Hackel) (designated by Edgar 1986 op. cit. p. 446). = *P. caespitosa* var. *planifolia* Petrie, *T.N.Z.I. 47*: 58 (1915); Lectotype: WELT 66405! *D. P[etrie] ex H. J. Matthews* grown partly in my [Petrie's] garden Epsom and partly in a Dunedin garden, ...[originally] from the Auckland Islands, end Novr. 1912 (designated by Edgar 1986 op. cit. p. 447). = *P. laevis* var. β *filifolia* Hook.f., *Fl. N.Z. 1*: 307 (1853); Lectotype: K! *Colenso 2393* [woods, nr. Tarawera] grass (designated by Edgar 1986 op. cit. p. 446).

silver tussock

Dense rather shining tussock, 30–100 cm, sometimes hanging down steep banks and up to 2 m, with light brownish green sometimes glaucous leaves smooth to the touch; branching intravaginal; leaf-blades persistent. Leaf-sheath pale green, later creamy brown, shining, ± membranous, smooth, or very minutely scabrid throughout, ribs inconspicuous. Ligule *c.* 0.5 mm, truncate, short-ciliate, abaxially minutely hairy. Leaf-blade (10)–20–60 cm × *c.* 1–1.5–(2.5) mm, folded or ± flat, coriaceous, abaxially glabrous, midrib and lateral ribs not very prominent, adaxially closely minutely pubescent throughout; margins inrolled, scabrid, midnerve scabrid near acicular, often pungent tip. Culm (10)–20–50–(70) cm, ≤ leaves, internodes smooth to scabrid below panicle. Panicle (5)–10–20–(25) cm, open; branches slender, ± erect to spreading, sparsely to very closely scabrid, naked below, spikelets many on shorter secondary branchlets. Spikelets (3.5)–6–9 mm, (2)–3–5-flowered, light green to later light brown. Glumes subequal, (2)–3.5–5–(6) mm, acute to subacuminate, sparsely minutely scabrid to smooth, midnerve with sparse prickle-teeth in upper ½; lower 1–(3)-nerved, narrow-lanceolate, upper 3–(5)-nerved, elliptic-lanceolate; margins membranous with scattered fine prickle-teeth above or almost throughout. Lemma 3–5–(6) mm, 5-nerved, elliptic-oblong, obtuse, minutely scabrid, midnerve with long crinkled hairs in lower ½, scabrid above, outer lateral nerves with crinkled hairs near base; margins membranous, sparsely minutely scabrid. Palea

(2)–3–4 mm, keels densely ciliate-scabrid, interkeel minutely scabrid to pubescent-scabrid, flanks short-scabrid. Callus with tuft of crinkled hairs. Rachilla *c.* 1 mm, scabrid or with occasional hairs. Lodicules *c.* 0.5 mm, occasionally hair-tipped. Anthers (1)–2–3 mm. Caryopsis *c.* 1 × 0.5 mm. $2n = 84$. Plate 7D.

Endemic.

N.: throughout, but rare in the west from Raglan to Manawatu except on Mt Egmont; S.: throughout; St.; K., Ch. Lowland to subalpine in grassland, and pasture, open scrub and forest, coastal cliffs; on relatively fertile soils.

This widespread tussock grass varies throughout its range. In central North Id plants are small, with narrow, needle-like, abaxially rounded leaves, scabrid or smooth culm internodes and drooping lower panicle branches bearing few, rather small spikelets. Plants from Mt Egmont have wide leaves, ± ribbed abaxially, usually smooth culm internodes, and few, large spikelets. Plants from the Wellington Coast and from coastal Nelson and northern Marlborough also have wide abaxially ribbed leaves and smooth culm internodes but the panicle is rather contracted and bears numerous spikelets; in northern North Id there are some even wider-leaved, larger plants. Most South Id plants are rather uniform with scabrid leaf-sheaths, abaxially ± ribbed leaves, usually scabrid culm internodes, and very scabrid panicle branches with numerous spikelets.

Specimens of *P. cita* with proliferous spikelets have been collected, e.g., WELT 65969 from Lake Lyndon, and WELT 66022 from Coleridge Pass.

Specimens on which Edgar (1986 op. cit. p. 447) based a record for Three Kings Is are referred to *P. anceps.*

Varietal combinations made under *P. caespitosa* Spreng. are illegitimate but the varietal epithets are legitimate. *P. caespitosa* var. *planifolia* was described from plants grown at Dunedin and Auckland from seed collected from Antipodes Is by H. J. Matthews in 1895, but labelled by Petrie as from Auckland Is. Edgar (1986 op. cit.) recorded an MS comment from V. D. Zotov that the plants were probably a local contaminant from Dunedin. Godley, E. J. *N.Z. J. Bot.* 27: 531–563 (1989) correctly excluded *P. cita* from the flora of Antipodes Id.

13. P. cockayneana Petrie, *T.N.Z.I. 45*: 274 (1913); Lectotype: WELT 66379a! *L. Cockayne* Rolleston R., Westland [7.4.1911] (designated by Edgar 1986 op. cit. p. 448).   avalanche grass

Large green perennial clumps or extensive swards, to *c.* 40–(80) cm, not tussock-forming, stoloniferous between tufts, and shoots ± decumbent at base; branching intravaginal; leaf-blades persistent. Leaf-sheath straw-brown to reddish brown or grey-brown, glabrous, membranous, distinctly ribbed. Ligule *c.* 0.5 mm, truncate, ciliate, abaxially minutely hairy. Leaf-blade (6)–10–35 cm, flat, to 4.5 mm wide, coriaceous, abaxially glabrous, ribs prominent, adaxially minutely pubescent; margins scabrid or smooth, midnerve scabrid near acicular tip. Culm 20–40–(60) cm, usually < leaves, internodes smooth to slightly scabrid below panicle. Panicle 10–20 cm, very lax with few rather large spikelets in clusters of 2–3 at tips of very slender, widely spreading, pubescent-scabrid branches. Spikelets 7–9.5 mm, 3–5–(6)-flowered, light green. Glumes ± equal, (4)–5–6.5–(7) mm, subacuminate, sparsely short scabrid to smooth, midnerve scabrid in upper ½; lower (1)–3-nerved, narrow-lanceolate, upper 3–(5)-nerved, elliptic-lanceolate; margins membranous with scattered fine prickle-teeth above or almost throughout. Lemma 5–6.5 mm, prominently 5-nerved, elliptic-oblong, obtuse, short-scabrid to pubescent-scabrid throughout, midnerve with long fine crinkled hairs in lower ½, lateral nerves with hairs on lower ⅓; margins membranous sparsely scabrid above or almost throughout. Palea 4–4.5 mm, keels ciliate-dentate, interkeel with minute hairs and prickle-teeth in lower ⅔, flanks smooth to slightly scabrid; margin prickle-toothed near tip. Callus with long tuft of crinkled hairs. Rachilla 1–1.5 mm, almost glabrous with a few prickle-teeth or scattered hairs; prolongation twice as long. Lodicules *c.* 0.5 mm. Anthers *c.* 2 mm. Caryopsis *c.* 2 × 1 mm. $2n = 112$.

Endemic.

S.: along Main Divide, occasionally to east in Marlborough and Canterbury, more common to the west in Westland and also in Fiordland. Alpine to subalpine in grassland, shrubland and damp disturbed stony ground affected by spring avalanches; rarely on coastal cliffs in the south.

Poa cockayneana is very similar to *P. cita* but is sward-forming not tussock-forming, it has wider flat leaves, larger fewer spikelets, and longer glumes and lemmas. Both spp. frequently grow together but *P. cockayneana* extends to higher altitudes than *P. cita.*

14. P. colensoi Hook.f., *Handbk N.Z. Fl.* 340 (1864) ≡ *P. colensoi* Hook.f. var. *colensoi* (autonym, Cheeseman 1906 op. cit. p. 908); Lectotype: K! *Colenso 1589* [top Ruahine Mt] grass (designated by Edgar 1986 op.

cit. p. 440). = *P. colensoi* var. *intermedia* Cheeseman, *Man. N.Z. Fl.* 908 (1906) ≡ *P. intermedia* Buchanan, *Indig. Grasses N.Z.* t. 48A (1880) non Koeler (1802); Lectotype: WELT 59602 (Buchanan's folio)! without locality, collector or date (designated by Edgar 1986 op. cit. p. 440). = *P. guthrie-smithiana* Petrie, *T.N.Z.I.* 45: 275 (1913) ≡ *P. colensoi* var. *guthrie-smithiana* (Petrie) Zotov, *T.R.S.N.Z.* 73: 236 (1943); Lectotype: WELT 66433a! *D. P[etrie] ex H. G. S. [H. Guthrie-Smith]* from Herekopere Island near Stewart Island grown in my [Petrie's] garden, Epsom, Auckland (designated by Edgar 1986 op. cit. p. 440). = *P. colensoi* var. *breviligulata* Petrie, *T.N.Z.I.* 47: 57 (1915); Lectotype: CHR 1468! *D. Petrie* Mt Egmont about 4000 ft, early Feb. 1912 (designated by Edgar 1986 op. cit. p. 440).   blue tussock

Small, blue-green or light green, narrow-leaved, stiff tussocks, 5–30–(70) cm, with tightly-packed shoots; branching intravaginal; leaf-blades disarticulating at ligule. Leaf-sheath light straw-coloured, later greyish brown, somewhat coriaceous, glabrous, indistinctly ribbed, tightly inrolled with narrow hyaline margins; upper sheaths sometimes purplish. Ligule 0.5–5.5 mm, entire, apically glabrous, rounded to tapered, abaxially minutely scabrid or smooth. Leaf-blade 5–15–(30) cm, usually tightly inrolled and *c.* 0.5 mm diam., abaxially glabrous, adaxially slightly to densely short pubescent-scabrid; margins finely scabrid, tip truncate, almost smooth with a few minute prickle-teeth, or fine-acicular and finely scabrid. Culm 5–20–(40) cm, erect, usually overtopping but often = leaves, internodes glabrous. Panicle 1–10–(15) cm; branches ± spreading, few to many, often short, almost filiform, smooth to ± finely scabrid, tipped by 1–3 rather large spikelets. Spikelets (3.5)–5–7–(10.5) mm, 2–5–(7)-flowered, light green to brownish. Glumes subequal, subacute to acute, glabrous, midnerve with a few minute prickle-teeth near tip; lower 2–3–(4) mm, 1–3-nerved, narrow elliptic-lanceolate, upper (2)–2.5–3.5–(4) mm, 3-nerved, ovate- to elliptic-lanceolate; margins minutely fimbriate. Lemma (2.5)–3–4–(5) mm, 5-nerved, ovate- to elliptic-lanceolate, subobtuse, in some long-liguled plants with fine hairs on lower ½–⅓ of midnerve and short hairs near base of lateral nerves, otherwise nerves scabrid or smooth, internerves densely scabrid to smooth, rarely with a few scattered hairs near base; margins glabrous to finely ciliate-scabrid. Palea 2–4–(5) mm, keels finely ciliate-scabrid, interkeel minutely pubescent-scabrid. Callus in long-liguled forms sometimes with a tuft of tangled hairs, otherwise glabrous or very rarely with a few short wispy hairs. Rachilla 0.5–1 mm, smooth to strongly densely scabrid. Lodicules *c.* 0.5 mm. Anthers 1–2 mm. Caryopsis *c.* 1–2 × 0.5 mm. $2n = 28$. Plate 7E.

Endemic.

N.: south from Coromandel Range; S.: throughout, but very rare in coastal North Canterbury, and rarely extending into Westland and Fiordland except near Main Divide; St.: throughout, and Herekopere Id. Lowland to alpine, widespread in tussock grassland and herbfield, often on rock.

A long-liguled (> 2 mm) form is dominant in North Id and also in northern South Id in Nelson, Marlborough and in Canterbury about as far south as Banks Peninsula; on Mt Egmont, North Id and in South Id southwards from Banks Peninsula plants are predominantly short-liguled (0.5–1.5 mm). In plants with long ligules the leaf tips are blunt and the lemma nerves may be silky haired towards the base. In some long-liguled plants from South Id, however, the lemmas are completely glabrous or occasionally scabrid. In contrast, short-liguled plants have acicular leaf-tips and vary in scabridity of the lemma; lemmas may be densely minutely prickle-toothed, occasionally with scattered hairs near the base, especially on the nerves; or lemmas may be completely glabrous, or minutely prickle-toothed only on the nerves.

On Mt Egmont, North Id, and in western Otago, Southland, and Stewart Id, some plants may have lemmas with almost glabrous internerves and a few short hairs near the base of the nerves and between the outer lateral nerves and margins, thus resembling *P. hesperia* in features of the lemma. However, the stiff, short, blue-green, tussock-habit and shorter spikelets distinguish these western and southern plants of *P. colensoi* from the lax, rhizomatous, light green, turf-forming *P. hesperia*.

Very fine-leaved plants are also found, e.g., CHR 387358 A,B, *A. P. Druce*, Pelorus Valley, Marlborough, Dec. 1981, a long-liguled plant; in marked contrast are the stiff, very pungent-leaved plants on the Remarkables, e.g., OTA 34687 *C. D. Meurk* 2.5.1971.

15. P. compressa L., *Sp. Pl.* 69 (1753).

Stiff, loosely tufted, bluish or greyish green perennial to *c.* 40 cm, with erect, or geniculate and ascending culms from wiry rhizomes; branching extravaginal; leaf-blades persistent. Leaf-sheath light yellowish green to light brown, subcoriaceous, glabrous, keeled, otherwise indistinctly ribbed. Ligule *c.* 0.5–1 mm, ± truncate to somewhat rounded, very minutely ciliate, abaxially scabrid. Leaf-blade 2–7 cm × 1.5–2.5 mm, subcoriaceous, folded or flat, abaxially smooth, but scabrid on nerves and keel near curved tip, adaxially finely scabrid throughout; margins short-scabrid. Culm (8)–12–40 cm, flattened, nodes usually purplish, internodes glabrous. Panicle (1)–3–8 cm, usually contracted, stiff, with numerous, densely clustered spikelets, rarely more open and somewhat branched; rachis, branches and pedicels sparsely scabrid. Spikelets 3–6 mm, 3–6-flowered, green or purplish. Glumes ± equal, 3-nerved, ovate-elliptic, acute, smooth, but slightly scabrid on midnerve near tip; lower 2–2.3 mm, upper 2.2–2.5 mm. Lemma 2.2–2.7 mm, 5-nerved but lateral nerves faint, oblong, obtuse, with broad, purple, thin hyaline band above, internerves glabrous, midnerve glabrous or softly hairy in lower ½, lateral nerves glabrous or softly hairy near base. Palea ≈ lemma, keels short-scabrid, interkeel smooth. Callus usually with small tuft of soft hairs. Rachilla *c.* 0.4 mm, very minutely papillose. Lodicules 0.3–0.4 mm. Anthers 1–1.3 mm. Caryopsis *c.* 1 × 0.5 mm.

Naturalised from Europe.

N.: near Auckland City, Makino near Feilding, Wellington City; S.: Marlborough (Pelorus Sound), Canterbury (near Christchurch), Otago (Lauder, and near Queenstown). Not common; roadsides and shingle banks.

16. P. cookii (Hook.f.) Hook.f., *Phil. Trans. Lond. 168*: 22 (1879) ≡ *Festuca cookii* Hook.f., *Fl. Antarct. 2*: 382, t. 139 (1846); Lectotype: K! *J. D. H[ooker]* Kerguelen's Land, Christmas Harbour, on rocks and in moist places always near the sea, May–August 1840 (designated by Edgar 1986 op. cit. p. 433). = *Poa hamiltonii* Kirk, *T.N.Z.I. 27*: 353 (1895); Holotype: K! *A. Hamilton* Macquarie Island, 1894.

Large dense perennial tufts to 50 cm, with flat, spreading, light green leaves overtopping erect panicles, fibres from older sheaths in a tangled mass at plant base; branching intravaginal; leaf-blades persistent. Leaf-sheath greenish brown to dark brown, hyaline, glabrous; ribs prominent. Ligule 4–7.5 mm, deeply and sharply lacerate, glabrous throughout. Leaf-blade (9)–15–30 cm × 3.5–5.5 mm, strongly ribbed, subcoriaceous, flat, abaxially glabrous, obviously keeled, adaxially deeply furrowed, ribs densely, minutely papillose-scabrid; margins and curved tip smooth. Culm 9–25 cm, simple, erect, leafy almost throughout, internodes glabrous. Panicle 5–25 cm, oblong, cylindrical to ± clavate, contracted, with lowermost branches sometimes slightly distant; rachis glabrous, branches erect, short, only slightly spreading, ± papillose, bearing close-set spikelets. Spikelets 4.5–8.5 mm, 2–4-flowered, light green. Glumes unequal, membranous, glabrous; lower 2–3–(3.5) mm, 1–(3)-nerved, very narrow oblong-lanceolate, acute, upper 3–4 mm, 3-nerved, narrow ovate-elliptic, acuminate. Lemma 4.8–6 mm, 5-nerved, oblong-elliptic, narrowed to long-acuminate tip, almost glabrous but nerves and internerves with short scattered hairs near base, and midnerve finely scabrid above hairs. Palea *c.* 4 mm, very narrow, keels ciliate-scabrid in upper ½. Callus glabrous. Rachilla *c.* 1 mm, glabrous. Lodicules slightly < 1 mm, rarely hair-tipped. Gynomonoecious: each spikelet with 1–(2) lower flowers ♂, anthers 2–3 mm, gynoecium *c.* 1.5 mm; upper flowers ♀ with minute colourless anthers 0.2–0.4 mm, on long filaments, gynoecium *c.* 1.5 mm. Caryopsis *c.* 1 × 0.5 mm.

Indigenous.

M.: Among rocks and peat mounds near the sea, and to 200 m on slopes; often near penguin colonies.

Also on Îles Kerguélen, Heard, and Prince Edward Is.

Poa cookii is the only sp. of *Poa* which is indigenous and not endemic to the N.Z. Botanical Region.

17. P. dipsacea Petrie, *T.N.Z.I. 26*: 271 (1894) ≡ *P. kirkii* var. *dipsacea* (Petrie) Zotov, *T.R.S.N.Z. 73*: 236 (1943); Holotype: WELT 68267a! *D. P[etrie]* source of Broken River, 3500–4000 ft. = *P. cheesemanii* Hack., *T.N.Z.I. 35*: 383 (1903) ≡ *P. kirkii* var. *cheesemanii* (Hack.) Zotov, *T.R.S.N.Z. 73*: 236 (1943); Holotype: W 11052! *T. F. C[heeseman]* Lake Tennyson, Nelson Mts, 3000 ft. [Jan. 1893] (No 1416 to Hackel). = *P. wallii* Petrie, *T.N.Z.I. 54*: 571 (1923); Lectotype: WELT 66662a! *A. Wall* Craigieburn Mts, Canterbury Alps, *c.* 6000 ft, mid. Feb., 1921 (designated by Edgar 1986 op. cit. p. 460).

Erect, open, rather stiff, shortly rhizomatous perennial tufts, to *c.* 30 cm, with leaves blue-green abaxially and bright green adaxially; branching extravaginal; leaf-blades persistent. Leaf-sheath light-coloured, green to creamy brown or purplish, later grey, membranous, shining, glabrous, obviously ribbed. Ligule *c.* 0.5–1 mm, apically glabrous, truncate to obtuse, abaxially smooth or with a few minute prickle-teeth. Leaf-blade (1.5)–5–20–(25) cm × 1.5–3–(4) mm, flat or folded, subcoriaceous, abaxially smooth, adaxially usually minutely scabrid on midrib and lateral ribs near blunt, curved tip; margins smooth or scabrid, sometimes with short hairs just above ligule. Culm (4.5)–10–20–(35) cm, erect, often only slightly overtopping leaves but sometimes much longer, internodes glabrous. Panicle (2.5)–4–6–(9.5) cm, open, with 1–(3) large spikelets borne at tips of slender smooth branches; pedicels usually smooth, rarely slightly scabrid. Spikelets 5–8.5 mm, (3)–4–5–(7)-flowered, light green to usually purplish. Glumes rather unequal, elliptic, subacute to subobtuse, submembranous, smooth except occasionally for some minute prickle-teeth on midnerve; lower 2.5–5 mm, (1)–3-nerved, upper 3.5–5–(6) mm, 3–(5)-nerved. Lemma 3.5–5.5 mm, 5–(7)-nerved, elliptic-oblong, subobtuse, long-ciliate on midnerve in lower ½ and on lateral nerves near base, scabrid on midnerve above hairs and usually on lateral nerves, internerves usually glabrous. Palea 3–4.5 mm, keels ciliate-scabrid, interkeel with sparse minute hairs, or glabrous. Callus with web of soft tangled hairs. Rachilla *c.* 0.5 mm, very minutely papillose. Lodicules 0.5–0.7 mm. Anthers 1.6–2.7 mm. Caryopsis *c.* 2 × 0.7 mm. $2n = 28$.

Endemic.

S.: Nelson, Marlborough and Canterbury. Subalpine to alpine, along streams or seepages in tussock grassland or herbfield, sometimes on rock.

Poa dipsacea is very variable in size and different names have been applied to small (*P. wallii*), medium (*P. dipsacea*), and large (*P. cheesemanii*) plants.

Poa dipsacea may be confused with *P. pratensis* but differs in the larger spikelets with all floral parts larger, and the submembranous, subacute to subobtuse glumes.

18. P. foliosa (Hook.f.) Hook.f., *Handbk N.Z. Fl.* 338 (1864)
≡ *Festuca foliosa* Hook.f., *Fl. Antarct. 1*: 99, t. 55 (1845) ≡ *Poa foliosa*
(Hook.f.) Hook.f. var. *foliosa* (autonym, Cheeseman 1925 op. cit.
p. 188); Holotype: K! *J. D. Hooker* Lord Auckland Isld., on the
ground forming large green tufts on the cliffs never far from the
sea, 2–3 ft, Dec. 1840; ♂. = *Festuca foliosa* var. β Hook.f., *Fl. Antarct.
1*: 100, t. 55 (1845); Lectotype: K (central specimen)! *J. D. Hooker
1633* Campbell's Island, in elevated dense tufts on banks near
the sea, Dec. 1840; ♂ (designated by Edgar 1986 op. cit. p. 434).

Massive, lush green dioecious tussocks to 1.5 m, from short,
narrow, woody stolons, with shoots covered at base by abundant
fibrous remnants of sheaths; branching extravaginal; leaf-blades
persistent. Leaf-sheath light brown, coriaceous, glabrous, closely
striate, keel prominent above. Ligule 1–3 mm, apically glabrous,
entire, rounded, abaxially finely scabrid. Leaf-blade 15–40–
(50) cm × (1)–3–6 mm, coriaceous, tough, flat, abaxially smooth
with prominent midrib and many lateral ribs, adaxially short-
scabrid, two prominent ridges along centre; margins thickened,
smooth, tip straight-sided, smooth, semi-pungent. Culm 20–40–
(60) cm, internodes glabrous. Panicle 10–25 cm, dense, with all
branches, except the longer ones, bearing spikelets almost to base;
rachis and branches smooth or with very occasional prickle-teeth.
Spikelets (5.5)–7–9 mm, (3)–4–6-flowered, light greenish brown.
Glumes subequal, long-acuminate, membranous, except for
thickened nerves, smooth, but with a few prickle-teeth on nerves
above and occasionally on margins; lower 3–5–(6) mm, 1–(3)-
nerved, narrow-lanceolate, upper 4–5.5–(6.5) mm, 3-nerved,
narrow elliptic-lanceolate. Lemma 5–6–(7) mm, 5-nerved, acute
or with midnerve very shortly excurrent, scabrid except near base,
midnerve ciliate to more than halfway, outer lateral nerves,
internerves, and margins with minute hairs in lower ⅓. Palea 3.5–
4.5 mm, keel rather densely ciliate-scabrid, interkeel and flanks
with sparse minute hairs and prickle-teeth. Callus with large tuft
of crinkled hairs just below midnerve of lemma and a few hairs
below lateral nerves. Rachilla *c.* 0.5 mm, glabrous. Lodicules 0.4–
0.7 mm, occasionally hair-tipped. Dioecious: ♂ with anthers (2)–

2.5–3.3 mm, gynoecium 0; ♀ with pollen-sterile anthers *c.* 0.6–1 mm, often on long filaments; stigma-styles *c.* 2 mm; caryopsis *c.* 2 mm; rarely ♂. 2*n* = 28. Plate 4A.

Endemic.

St.: islands to north-east; Solander Id, Ant., A., C., M. Coastal, on steep slopes, sometimes in turf near shore; often forming extensive swards.

Poa foliosa has not been seen recently on Stewart Id though it occurs locally on islands nearby, Wilson, H. D. *Field Guide: Stewart Id Plts* 380 (1982). *Poa foliosa* was recorded from the Snares but Fineran, B. A. *T.R.S.N.Z. (Bot.) 3*: 237–270 (1969) regarded the records as doubtful.

In habit and habitat *Poa foliosa* resembles *Poa flabellata* of the southern South American region. Both form massive, very palatable tussocks close to the sea on cliffs.

19. P. hesperia Edgar, *N.Z. J. Bot. 24*: 442 (1986); Holotype: CHR 223875! *P. Wardle* Lower Otoko Pass, at head of Clarke tributary of Landsborough R., dominant in stony grassland, 4500 ft, 21.2.1972.

Slender, fine-leaved, often rhizomatous perennial, in low lax clumps or forming a light green turf, to 40 cm; branching intravaginal; leaf-blades disarticulating at ligule. Leaf-sheath pale creamy brown, later greyish, somewhat coriaceous with wide hyaline margins, folded, glabrous. Ligule 0.5–*c.* 1 mm, apically glabrous, entire, rounded, abaxially minutely scabrid. Leaf-blade 5–10–(20) cm × 0.5–1 mm, coriaceous, ± folded with inrolled margins, abaxially glabrous, adaxially glabrous or with scattered short fine hairs; margins and long-acicular tip finely scabrid. Culm (5)–15–30–(40) cm, erect, usually overtopping but sometimes = leaves, internodes glabrous. Panicle 2–5.5–(10) cm; branches ± spreading, few to many, often short, almost filiform, smooth to finely scabrid, tipped by 1–3 rather large spikelets. Spikelets 4.5–10–(12) mm, (2)–4–8-flowered, light green to purplish. Glumes ± equal, lanceolate, acute to acuminate, glabrous; lower (2)–2.5–3.5 mm, (1)–3-nerved, upper (2.5)–3–4 mm, 3-nerved; margins smooth, occasionally finely fimbriate. Lemma (3)–3.5–4.5–(5) mm, 5-nerved, elliptic-lanceolate, subacute to subobtuse, midnerve with short fine silky hairs on lower ⅓ to ½, glabrous above, lateral nerves with very short fine hairs near base, internerves glabrous apart from a few, occasional scattered hairs near base usually between lateral nerve and margin; margins finely ciliate above, glabrous or longer ciliate near base. Palea 2.5–4–(4.5) mm, keels with short hairs near base gradually decreasing in length to short prickle-teeth near tip, interkeel sometimes with a few hairs in lower ⅓ or glabrous. Callus with tuft of fine tangled hairs. Rachilla *c.* 0.5 mm, smooth, or with a few fine to very minute prickle-teeth; prolongation twice as long. Lodicules *c.* 0.5 mm. Anthers (1)–1.5–2.5–(3) mm. Caryopsis *c.* 1 × 0.5 mm. $2n = 28$.

Endemic.

S.: along Main Divide and to the west, occasionally to the east in Canterbury, Otago, and Southland, also in Fiordland; St. Alpine to subalpine in tussock grassland, herbfield or scrub, often on rock; at lower altitudes in the south.

Turf-forming, light green *P. hesperia* is distinct in habit from tussock-forming blue-green *P. colensoi* and also in spikelet characters from most short-liguled plants of *P. colensoi* other than those of Mt Egmont, western Otago, Southland and Stewart Id, see above under *P. colensoi*. The lemmas are mostly longer than those of *P. colensoi* and less ovate, with midnerve completely glabrous above and silky hairy in the lower ½, and internerves completely glabrous except for a few basal hairs.

20. P. imbecilla Spreng., *in* Biehler *Pl. Nov. Herb. Spreng.* 9, no. 14 (1807) ≡ *P. sprengelii* Kunth, *Révis. Gram. 1*: 116 (1829) *nom. superfl.* ≡ *Eragrostis imbecilla* (Spreng.) Benth., *Fl. Austral. 7*: 643 (1878) non (R.Br.) Steud. (1854) ≡ *Poa imbecilla* Spreng. var. *imbecilla* (autonym, Cheeseman 1925 op. cit. p. 201); Holotype: B (Herb. Willdenow no. 1896)! *Forster* Nova Zelandia. = *P. matthewsii* var. *minor* Petrie, *T.N.Z.I. 34*: 393 (1902); Lectotype: WELT 67003a! *D. Petrie* Manuherikia plain, Vincent Co. & Blacks, same station, Nov. 1899 (No 1463 to Hackel) (designated by Edgar 1986 op. cit. p. 471).

Very fine-leaved, light green, shortly rhizomatous, small, lax, perennial tufts 10–50 cm, with conspicuous light-coloured leaf-sheaths, and slender, erect culms ≫ drooping leaves, topped by extremely delicate, green panicles; branching extravaginal; leaf-blades persistent. Leaf-sheath light green, later creamy brown sometimes purplish, membranous, ribbed, glabrous. Ligule 0.2–0.4–(1) mm, apically glabrous, entire, ± rounded, abaxially slightly scabrid. Leaf-blade (2)–5–12–(16) cm × 0.5–1 mm, folded, usually filiform, soft, culm-leaves 1–3 mm wide, flat, minutely scabrid abaxially on midrib towards long, fine, acute, triangular tip, and adaxially just above ligule and on margins. Culm (4)–10–32–(50) cm, nodes swollen, internodes smooth, rarely scabrid below panicle. Panicle (3.5)–8–15–(25) cm, lax and delicate; rachis smooth to finely scabrid above, elongate branches and pedicels filiform, finely scabrid, spreading, with few spikelets. Spikelets 3–4 mm, 3–4–(5)-flowered, light green, often purplish. Glumes ± unequal, obtuse, midnerve often with a few prickle-teeth; lower 0.8–1.5 mm, 1-nerved, linear-lanceolate, upper 1.3–1.8 mm, 3-nerved, ovate-lanceolate; margins finely scabrid. Lemma (1.6)–1.8–2.2 mm, 5-nerved but inner lateral nerves often rather faint, glabrous, margin and nerves near obtuse tip minutely scabrid. Palea 1.2–2 mm, keels obviously ciliate, interkeel and often margins sparsely prickle-toothed. Callus glabrous, very rarely with a few fine hairs. Rachilla 0.5–1 mm, glabrous. Lodicules *c.* 0.2 mm. Anthers 0.2–0.4 mm. Gynoecium: ovary 0.3 mm; stigma-styles 0.5–0.9 mm. Caryopsis *c.* 1 × 0.3 mm, tightly enclosed by lemma and palea. $2n = 28$. Plate 7A.

Endemic.

N.: throughout except in Bay of Plenty and Taranaki; S.: Nelson, and east of Main Divide; St.; Three Kings Is, Ch. Open forest and scrub, often on or near rock; occasionally a weed; from sea level to subalpine zone.

Specimens may have either wide, folded or flat leaves, and more obviously 5-nerved lemmas, or fine, filiform leaves, with the lateral lemma nerves scarcely visible. The two forms do not appear to be separated geographically, and plants growing as garden weeds in exposed dry situations have filiform leaves whereas those growing in shade have wider leaves. At sites of high fertility plants are very densely tufted and tussock-like with yellow-green leaves, erect to suberect inflorescences, and 4–5-flowered spikelets (B.P.J. Molloy, pers.comm.).

Specimens from Stewart Id (CHR 370996 *Wilson* 1981 and CHR 399109 *Wilson & Meurk* 1980) have very fine leaves, lemmas *c.* 3 mm and anthers *c.* 0.6 mm.

21. P. incrassata Petrie, *T.N.Z.I. 34*: 394 (1902) ≡ *P. kirkii* var. *incrassata* (Petrie) Zotov, *T.R.S.N.Z. 73*: 236 (1943); Holotype: WELT

66452! *F. R. Chapman* plant taken from the Auckland Islands (Jan. 1890) and grown in Mr F. R. Chapman's garden at Dunedin, Jan. 1891. = *P. exigua* Hook.f., *Handbk N.Z. Fl.* 338 (1864) non Dumort. (1823); Holotype: K! *Hector and Buchanan no. 15* Otago, lake district, alpine.

Small perennial tufts, 5–15 cm, narrow, glaucous; branching extravaginal at plant base, intravaginal above; leaf-blades persistent. Leaf-sheath shining, greenish or reddish purple, later light grey-brown, membranous, glabrous, ribs conspicuous. Ligule 0.2–0.7 mm, apically glabrous, entire, rounded or sometimes centrally narrowed to a point, abaxially minutely papillose. Leaf-blade 1–6–(9) cm × 0.6–2 mm, usually folded, abaxially smooth, but scabrid on midrib near curved tip, adaxially glabrous, but shortly hairy above ligule and occasionally with a few prickle-teeth on midrib, lateral ribs indistinct; margins inrolled, shortly prickle-toothed. Culm 2–9–(16) cm, internodes often closely short-scabrid below panicle or smooth. Panicle 0.5–2.5–(3) cm, compact, spike-like or racemose, very rarely more open with few, very short branches; rachis and branches stiff, angular, ± densely scabrid, spikelets few on scabrid pedicels. Spikelets 2.5–3.5 mm, 2–3–(4)-flowered, light green tinged purple. Glumes subequal, usually 1.5–2.5 mm, glabrous, with a few prickle-teeth on nerves in upper ½; lower occasionally smaller, 1–1.5 mm, 1–(3)-nerved, narrow- to ovate-elliptic, subobtuse to obtuse, upper 3-nerved, ovate, obtuse. Lemma *c.* 2–2.5 mm, 5–(7)-nerved, ovate, strongly folded about midnerve, obtuse, sometimes almost hooded, minutely papillose or occasionally minutely scabrid near distinct membranous margin, midnerve scabrid towards tip. Palea 1.5–2.3 mm, keels scabrid on upper ½, interkeel and flanks glabrous. Callus glabrous. Rachilla *c.* 0.5 mm, glabrous; prolongation usually twice as long. Lodicules 0.4–0.5 mm. Anthers 0.2–0.5 mm. Gynoecium: ovary 0.4–0.5 mm; stigma-styles 0.75–1.1 mm. Caryopsis 1–1.3 × *c.* 0.5 mm.

Endemic.

S.: western and Central Otago, north-eastern Southland; St.; A. Subalpine and alpine in damp ground, and among windswept short tussock grassland.

No plants resembling *P. incrassata* have been collected in recent subantarctic expeditions, but it may easily be overlooked.

22. P. infirma Kunth, *in* Humboldt, Bonpland and Kunth *Nov. Gen. Sp. 1*: 158 (1816).

Loosely tufted, yellow-green, very short-lived, free-seeding annual, to 25 cm; branching intravaginal; leaf-blades persistent. Leaf-sheath very pale green to light brown, hyaline, glabrous, keeled. Ligule 0.5–3 mm, entire, rounded or tapering, glabrous throughout. Leaf-blade 1–5 cm × 1.5–4 mm, flat or folded, thin, glabrous; margins sparsely and minutely scabrid, tip rounded to blunt point. Culm 0.5–20 cm, erect, spreading or prostrate, internodes glabrous. Panicle 0.5–6 cm, lax, erect, with very fine, glabrous branches spreading after anthesis. Spikelets 3–5 mm, with 2–5 rather distant florets, light green. Glumes subequal to unequal, glabrous; lower 1–2 mm, 1-nerved, elliptic, subobtuse, upper 1.5–2.5 mm, 3-nerved, ovate-oblong, obtuse. Lemma 2–2.5 mm, 5-nerved, ovate-oblong, obtuse, strongly silky haired on nerves for more than ½ length, nerves glabrous above hairs, internerves glabrous, often purplish on hyaline band at tip; margins hyaline. Palea 2–2.5 mm, = lemma, keels strongly ciliate, often purplish, interkeel glabrous. Callus glabrous. Rachilla *c.* 0.5 mm, glabrous; prolongation minute. Lodicules *c.* 0.2 mm. Anthers (0.1)–0.2–0.4 mm. Caryopsis *c.* 1 × 0.5 mm, tightly enclosed by lemma and palea.

Naturalised from Europe.

N.: Hawkes Bay (Pukeora), near Wellington; Cook Strait (Brothers Id); S.: Marlborough (Blenheim), Canterbury (from near Christchurch to Ashburton), Central Otago (Alexandra). Stony roadsides, shingle paths and waste places. FL Aug–Sep.

23. P. intrusa Edgar, *N.Z. J. Bot. 24*: 463 (1986); Holotype: CHR 187790! *I. M. Ritchie* Craigieburn Range, head of Craigieburn [Valley], tussock grassland among *Chionochloa pallens* and *C. flavescens, c.* 4500 ft, 19.2.1968.

Loose, green to purple-green perennial tufts, to 60 cm, culms usually only slightly overtopping leaves at flowering, later elongating; branching extravaginal; leaf-blades persistent. Leaf-sheath light green, often slightly purplish, later greyish, shining, submembranous, glabrous, keeled, distinctly ribbed, occasionally scabrid just below ligule especially on keel.

Ligule 0.5–1.5–(2.5) mm, evenly narrowed to a short point, strongly finely ciliate, abaxially minutely hairy. Leaf-blade 7–15.5 cm × 2–3.5 mm, flat or folded, subcoriaceous, abaxially glabrous except near curved, shortly apiculate scabrid tip, adaxially minutely scabrid; margins finely scabrid, with a few hairs just above ligule. Culm 15–45 cm, erect, internodes glabrous below panicle. Panicle 6.5–18 cm; branches slender, spreading, minutely scabrid, tipped by 2–4 large spikelets. Spikelets 6.5–9.5 mm, 3–5-flowered, usually purplish. Glumes ± equal, *c.* 4–5 mm, 3-nerved, elliptic- to ovate-lanceolate, subobtuse, midrib scabrid, tip minutely ciliate. Lemma 4.5–5 mm, 5-nerved, ± oblong, obtuse, minutely scabrid on nerves and internerves especially in lower ½, rarely with a few sparse short hairs near base of keel, tip minutely scabrid. Palea 4.5–5 mm, keels ciliate-scabrid, interkeel and flanks with minute hairs and prickle-teeth. Callus glabrous, rarely with a few sparse hairs. Rachilla *c.* 1 mm, sparsely, minutely scabrid; prolongation twice as long. Lodicules 0.7–0.9 mm. Anthers 1.2–1.7–(2) mm. Caryopsis *c.* 2 × 0.7 mm.

Endemic.

S.: Canterbury and Otago, rare in southern Marlborough and Southland. Montane and subalpine grassland and open forest.

Poa intrusa is similar in overall size and anther length to *P. celsa*, but rarely has hair on the lemma, and the leaf-sheaths are not scabrid abaxially except occasionally just below the ligule. The short, finely-ciliate ligule distinguishes *P. intrusa* from both *P. celsa* and *P. kirkii* in which the ligule is longer and entire or occasionally erose, but not ciliate.

24. P. kirkii Buchanan, *Indig. Grasses N.Z.* t. 51B (1880) ≡ *P. kirkii* Buchanan var. *kirkii* (autonym, Hackel *in* Cheeseman 1906 op. cit. p. 910); Holotype: WELT 59610 (Buchanan's folio)! *A. Mackay* Mt Arthur, 4200 ft [1874]. = *P. mackayi* Buchanan, *Indig. Grasses N.Z.* t. 51A (1880) ≡ *P. kirkii* var. *mackayi* (Buchanan) Hack., *in* Cheeseman *Man. N.Z. Fl.* 910 (1906); Lectotype: WELT 59609 (left specimen, Buchanan's folio)! *H. H. Travers* Tararua Mts, 5000 ft (designated by Edgar 1986 op. cit. p. 461). = *P. kirkii* var. *collinsii* Hack., *in* Cheeseman *Man. N.Z. Fl.* 910 (1906); Lectotype: W 14314! *T. Kirk* Mt Fyffe, Marlborough (No 1411 to Hackel) (designated by Edgar 1986 op. cit. p. 461).

Usually slender perennial tufts, *c.* 20–50 cm, occasionally stoloniferous, with brownish green leaves much overtopped by erect culms; branching extravaginal; leaf-blades persistent. Leaf-sheath often reddish purple, later grey-brown, submembranous, smooth or rarely slightly scabrid, ribs conspicuous. Ligule 1–4 mm, apically glabrous, entire, tapered, abaxially glabrous to finely pubescent-scabrid. Leaf-blade (3.5)–7.5–20 cm × *c.* 1–3 mm, subcoriaceous, often flat, abaxially smooth, or scabrid on ribs, adaxially scabrid just above ligule and on ribs for some distance; margins finely scabrid with a few hairs just above ligule, midrib scabrid near curved tip. Culm (10)–15–35–(65) cm, internodes smooth or occasionally finely scabrid below panicle. Panicle 5–10–(15) cm, loosely spreading, with rather large spikelets at tips of fine, smooth to scabrid branches. Spikelets 3–6.5 mm, (2)–3–4–(6)-flowered, light green, often purplish. Glumes ± equal, usually purplish, subacute to acute, midnerve usually finely prickle-toothed near tip; lower 1.5–4 mm, 1–3-nerved, narrow elliptic-lanceolate, upper 2–5 mm, 3-nerved, elliptic-ovate; margins entire or finely ciliate near tip. Lemma (1.5)–2–5 mm, 5–(7)-nerved, ± ovate to oblong, obtuse to almost truncate, tip hyaline, usually purplish, nerves finely scabrid throughout or only in upper or lower part; in North Id plants lower ⅓ to ½ of midnerve and lower ¼ of outer lateral nerves with short hairs, internerves glabrous; in plants from South and Stewart Is nerves usually glabrous, rarely with a few wispy hairs on midnerve, internerves usually minutely ± sparsely scabrid, or smooth. Palea 1.5–4 mm, keels ciliate-scabrid, interkeel glabrous to shortly pubescent. Callus glabrous, occasionally with a few wispy hairs. Rachilla 0.5–1 mm, glabrous; prolongation twice as long. Lodicules 0.3–0.6 mm, rarely hair-tipped. Anthers 0.6–1–(1.2) mm. Caryopsis 1–1.5 × *c.* 0.5 mm. $2n = 28$.

Endemic.

N.: Ruahine and Tararua Ranges; S.: almost throughout; St. Montane to alpine in grassland, scrub and open forest, sometimes on rock.

Most North Id plants of *Poa kirkii* are distinct from most South Id plants in lemma vesture, having a definite fringe of short hairs on the lower ½ of the midnerve and also near the base of the outer lateral nerves. The internerves are entirely glabrous in North Id plants and the callus always glabrous. Lemmas from South Id plants generally have a glabrous keel and callus, and the nerves and internerves may be scabrid or smooth. However, occasional South Id plants have a few wisps of short hair near the base of the midnerve or on the callus and a few have more obvious hair near the base of the midnerve and lateral nerves. Plants with hair on lemma nerves and callus were treated as *P. mackayi* but this morphological difference is insufficient for species recognition and the geographical separation is not sufficiently clear-cut to warrant subspecific rank.

25. P. labillardierei Steud., *Syn. Pl. Glum. 1*: 262 (1854).

rough poa tussock

Coarse, dense, green or blue-green often very scabrid tussocks, to *c.* 120 cm, with a large bulk of dead dry leaves densely packed at the narrow base; branching intravaginal; leaf-blades persistent. Leaf-sheath light greenish brown, later grey-brown, rarely slightly purplish, subcoriaceous, ± distinctly ribbed, ± minutely scabrid. Ligule 0.2–0.6 mm, truncate, minutely ciliate, abaxially scabrid. Leaf-blade 35–65 cm × 1.8–3 mm, coriaceous, flat, to folded and ± inrolled, abaxially usually scabrid, occasionally almost smooth, adaxially finely ribbed, finely scabrid on ribs; margins obviously scabrid, tip ± acicular, straight-sided. Culm 60–110 cm, erect or drooping, internodes scabrid below panicle. Panicle 10–25 cm, open; branches fine, not drooping, slightly spreading, finely scabrid, in clusters of 3–7, naked and undivided below, with numerous spikelets clustered at tips. Spikelets 3–7 mm, 3–5-flowered, light green, often purplish. Glumes usually ± equal, 1.5–3.5 mm, 3-nerved, often scabrid near tip and on midnerve and margins, narrow- to ovate-lanceolate, acute. Lemma *c.* 2–4 mm, 5-nerved, oblong-elliptic, obtuse, ± short scabrid but usually softly hairy on midnerve and lateral nerves near base. Palea 2–3.5 mm, keels scabrid in upper ⅔ and smooth below, interkeel scabrid. Callus with long tuft of soft tangled hairs. Rachilla slightly > 0.5 mm, minutely scabrid with some longer hairs; prolongation sometimes twice as long. Lodicules 0.5–0.6 mm. Anthers 1–2 mm. Caryopsis 1–1.5 × 0.5–1 mm.

Naturalised from Australia.

N.: scattered and local from North Auckland to Bay of Plenty and Mt Egmont, and in the Wairarapa and Wellington; S.: scattered in east near Blenheim, Waipara, and on Banks Peninsula; more common in Central and South Otago and in Southland. Lowland to montane in modified tussock grassland and waste ground.

Poa labillardierei is very variable and at least 3 forms are present in N.Z.: a) large plants with broad green leaves in Hutt Valley, on the Port Hills near Christchurch, in Central Otago and also in Marlborough, where the leaves are barely scabrid; b) very large, very scabrid plants with broad blue-green leaves also on the Port Hills and in Central Otago; c) plants with narrow green leaves in Central Otago.

Poa labillardierei was first recorded for N.Z. under the name *P. australis* agg. which was applied for a long time in Australia to large tussock-forming spp. of *Poa*.

26. P. lindsayi Hook.f., *Handbk N.Z. Fl.* 340 (1864); Lectotype: K! *L. Lindsay* northern slopes of Saddle Hill, near Dunedin, Otago, 20 Nov. 1861 (designated by Edgar 1986 op. cit. p. 474).

Small, delicate grey-green or blue-green, rarely reddish perennial tufts, 5–40 cm, with densely packed wiry, short and curved, or longer and erect leaves usually ≪ culms; branching intravaginal, with a few extravaginal shoots at plant base; leaf-blades persistent. Leaf-sheath very pale brown, often purplish, membranous, distinctly ribbed, minutely scabrid abaxially, occasionally smooth. Ligule 0.2–0.7 mm, truncate, erose, glabrous throughout. Leaf-blade 0.5–1.5–(5) cm × 1–1.5–(3) mm, usually fine and folded, adaxially with minute scattered prickle-teeth, abaxially smooth, but midrib scabrid near naviculate tip; margins inrolled, sparingly scabrid. Culm (1.5)–5–15–(40) cm, very slender, purplish, usually bearing 2 small cauline leaves, internodes glabrous. Panicle 1.5–6.5–(8) cm, open; rachis glabrous, branches capillary, sometimes flexuous, very minutely scabrid with prickle-teeth scarcely visible except at high magnification, with rather few spikelets, clustered 2–3 at branch tips. Spikelets 1.5–4.5 mm, (1)–3–5–(6)-flowered, silvery purple-green. Glumes subequal, 1–2 mm, with wide hyaline margins, midnerve slightly scabrid in upper ½ or near tip; lower narrower, ovate-lanceolate, acute, 1–(3)-nerved, upper ovate, obtuse, 3-nerved, sometimes with a few prickle-teeth on margins and on lateral nerves above. Lemma 1–2 mm, 5-nerved, ovate, obtuse, covered almost throughout or for *c.* ⅔ length with long, appressed, silky hairs, *c.* 0.2 mm, midnerve scabrid and internerves glabrous near membranous tip. Palea 1–2 mm, keels long-ciliate, interkeel with appressed hairs, flanks with some hairs. Callus glabrous. Rachilla to 0.5 mm, with scattered long hairs or almost glabrous. Lodicules 0.2–0.4 mm. Anthers 0.2–0.5 mm. Gynoecium: ovary 0.3–0.5 mm; stigma-styles 0.7–1 mm. Caryopsis *c.* 1 × 0.5 mm. $2n = 28$. Plate 7F.

Endemic.

N.: near Lake Taupo, and Kaweka and Ruahine Ranges; S.: east of Main Divide from Marlborough to Southland, occasionally extending into Westland and Fiordland. Dry stony places in riverbeds, moraines, and depleted short tussock grassland; lowland to alpine.

A specimen with red anthers *c.* 1 mm long, Gooses Neck, Lake Benmore, depleted short tussock grassland, CHR 221915 *D. Kelly* 19.10.1971, agrees in other respects with *P. lindsayi*. It is tall (to 30 cm) with reddish leaves, and spikelets *c.* 4 mm.

27. P. litorosa Cheeseman, *Man. N.Z. Fl.* 1156 (1906) ≡ *Festuca scoparia* Hook.f., *Fl. Antarct. 1*: 98 (1845) a combination in *Poa* is preoccupied by *P. scoparia* Kunth (1832); Holotype: K! *J. D. Hooker* Dea's head on rocks near the sea in large tufts Ld Auckland's Islands, Nov. 1840. = *F. scoparia* var. β Hook.f., *Fl. Antarct. 1*: 99 (1845); Holotype: K! *J. D. Hooker* on ledges of rock, 1000 ft, Campbell's Island, Dec. 1840.

Stiff, wiry, straw-coloured tussock, to 1.8 m, with acicular-tipped leaves overtopping culms; rhizome slender, from base of tufts; branching intravaginal; leaf-blades persistent. Leaf-sheath light green to straw-coloured to grey-brown, ±shining, coriaceous, distinctly ribbed, smooth or short-scabrid between ribs. Ligule 0.3–1 mm, a truncate ciliate rim, abaxially with a dense mat of minute hairs. Leaf-blade 10–65 cm, inrolled *c.* 1 mm diam., coriaceous, abaxially glabrous, adaxially with stiff short hairs; tip straight-sided, pungent. Culm 5–50 cm, erect, internodes usually very slightly scabrid just below panicle. Panicle (4.5)–6–15 cm, contracted, with few, short, erect, scabrid branches. Spikelets (7)–11–14 mm, 4–6–(7)-flowered, light green, glumes and lemmas brown-tipped. Glumes unequal, 3-nerved, narrow-lanceolate, subobtuse, ±scabrid above; lower (4)–5.5–7 mm, upper (5)–6–7.5 mm. Lemma (5.5)–6–7.5 mm, 5-nerved, elliptic-oblong, obtuse, short-scabrid, with hairs on lower ⅓ of mid-nerve and outer lateral nerves. Palea 5.5–6.5 mm, keels ciliate-scabrid, interkeel with minute hairs and prickle-teeth, tip bifid. Callus with narrow tuft of long crinkled hairs.

Rachilla 1–1.5 mm, stiff-ciliate; prolongation twice as long. Lodicules 0.8–1.5 mm, occasionally hair-tipped. Anthers (2.5)–3–4 mm. Caryopsis *c.* 3 × 1 mm. 2*n* = *c.* 266.

Endemic.

Ant., A., C., M. Coastal and upland in tussock grassland, on peat, often in enriched sites.

More stunted tussocks are generally found on the inland plains in Antipodes Is (E. J. Godley *in* Edgar 1986 op. cit. p. 450) and are more slender with smaller spikelets than those from coastal Antipodes, Auckland and Campbell Is.

28. P. maia Edgar, *N.Z. J. Bot. 24*: 470 (1986); Holotype: CHR 324197! *A. P. Druce* saddle betw. Mt Owen & Lookout Ra., NW Nelson, 3800 ft, silver beech forest (wet ground), Jan. 1972.

Erect, shortly rhizomatous, perennial light green to purplish tufts, to 40 cm, with loosely packed shoots, and culms overtopping leaves; branching extravaginal; leaf-blades persistent. Leaf-sheath green or purplish, later light brown, membranous, glabrous, distinctly ribbed. Ligule (0.4)–0.7–1.5 mm, apically glabrous, entire, obtuse, abaxially slightly scabrid. Leaf-blade 1.5–16 cm × 0.5–1.5 mm, usually folded, rather wiry, often filiform rarely flat, smooth, but scabrid adaxially near ligule and abaxially on midrib near acute, ± incurved tip; margins scabrid. Culm (4)–12–35–(50) cm, internodes glabrous. Panicle (1.5)–3–9 cm, very lax; rachis glabrous, branches often in pairs, finely, sparingly scabrid, spreading, often purplish, each bearing few, conspicuous spikelets. Spikelets (2.3)–3–6 mm, (1)–2–3–(5)-flowered, green, often purplish. Glumes rather unequal, usually purplish; lower 1–2.5 mm, 1-nerved, linear-lanceolate, subacute, upper *c.* 2–3 mm, 3-nerved, elliptic-lanceolate, obtuse, often scabrid on nerves. Lemma (2)–2.5–3.5 mm, 5-nerved, oblong-lanceolate, glabrous, apart from minute prickle-teeth on midnerve near obtuse, usually purplish tip. Palea 2–3 mm, keels shortly scabrid in upper ½, interkeel glabrous. Callus glabrous. Rachilla *c.* 0.5 mm, glabrous; prolongation almost twice as long. Lodicules 0.3–0.5 mm. Anthers (0.6)–0.8–1.2–(1.5) mm. Caryopsis *c.* 1.5 × 0.5 mm.

Endemic.

N.: Mt Egmont; S.: Nelson, Marlborough Sounds. Montane to subalpine in wet open forest and occasionally in damp sites in tussock grassland.

29. P. maniototo Petrie, *T.N.Z.I. 22*: 443 (1890); Lectotype: AK 1940! *D. Petrie* Maniototo Plain, Upper Clutha, Otago, 1000–3000 ft (designated by Edgar 1986 op. cit. p. 446).   desert poa

Small, compact, greyish perennial tufts, < 10 cm, with densely crowded rigid leaves overtopped by slender flowering culms; branching intravaginal; leaf-blades disarticulating at ligule. Leaf-sheath much wider than leaf-blade, membranous, glabrous, ribs distinct, lower sheaths whitish straw-coloured, shining, upper greenish. Ligule consisting of 2 lobes at either side of leaf-blade, 0.5–1 mm, much shorter or 0 across leaf-blade, finely ciliate, abaxially with sparse minute hairs; usually extending as a rim-like minutely ciliate contra-ligule, to 0.2 mm. Leaf-blade 0.25–1.5 cm, inrolled, to 0.5 mm diam., stiff, often curved, grey-green, glabrous, ribs distinct; margins and curved tip scabrid. Culm 1.5–8 cm, longer culms often bearing a single cauline leaf *c.* midway, internodes glabrous. Panicle 0.5–1.5 cm, compact, oblong, spike-like; branches and pedicels very short, sparsely scabrid. Spikelets *c.* 2–3.5 mm, 3–5-flowered, silky greyish green to greyish purple. Glumes ± equal, 1.5–2–(2.5) mm, 3-nerved, glabrous; lower narrow-elliptic, subobtuse, upper ovate-elliptic, obtuse. Lemma *c.* 1.5–2–(2.5) mm, faintly 3–5-nerved, elliptic, subobtuse to obtuse, shortly finely pubescent almost throughout, ± glabrous near tip or rarely throughout, margins wide membranous above. Palea *c.* 1.5–(2) mm, ± tangled pubescent throughout. Callus with small tuft of short crinkled hairs. Rachilla minute, *c.* 0.2–0.3 mm, minutely hairy; prolongation twice as long. Lodicules 0.2–0.3 mm. Anthers 0.2–0.4 mm. Gynoecium: ovary 0.4 mm; stigma-styles 0.75 mm. Caryopsis *c.* 0.5–1 × 0.3 mm. $2n = 28$. Plate 7F.

Endemic.

S.: eastern areas, rare in North Canterbury and not recorded from Southland. Lowland to alpine in short-tussock grassland, on rocky, stony, or depleted ground.

Poa maniototo is not only distinct because of its silky spikelets, but the ligule, largely consisting of 2 lateral membranous lobes and extending as a contra-ligule, is unusual in *Poa*; a contra-ligule occurs also occasionally in *P. acicularifolia.*

30. P. matthewsii Petrie, *T.N.Z.I. 34*: 392 (1902) ≡ *P. imbecilla* var. *matthewsii* (Petrie) Hack., *in* Cheeseman *Man. N.Z. Fl.* 913 (1906) ≡ *P. matthewsii* Petrie var. *matthewsii* (autonym, Petrie 1902 op. cit. p. 393); Lectotype: WELT 66983! *D. P[etrie]* Waipahi, S. Otago, by banks of river (designated by Edgar 1986 op. cit. p. 470). = *P. matthewsii* var. *tenuis* Petrie, *T.N.Z.I. 34*: 393 (1902); Lectotype: WELT 66929! *D. Petrie* Catlin's River, Otago, sea level, Dec. 1891 (designated by Edgar 1986 op. cit. p. 470).

Markedly stoloniferous perennial, with bright green, rarely glaucous leaves, tending to lie along the ground, and very slender culms erect or geniculate and ascending, to 50 cm; branching extravaginal; leaf-blades persistent. Leaf-sheath pale green to light brown, membranous, mainly glabrous but with minute hairs and prickle-teeth near hyaline margin, occasionally near ligules and on keel. Ligule 0.7–1.5 mm, apically glabrous, entire, rounded, abaxially with dense, minute hairs. Leaf-blade (5.5)–10–17 cm × 1–2 mm, flat, soft, abaxially glabrous, adaxially with very minute fine hairs; margins ciliate near ligule, midrib minutely scabrid especially near filiform, acuminate tip. Culm 10–25–(36) cm, internodes very minutely scabrid just below panicle or rarely smooth. Panicle 10–25 cm, lax and delicate; rachis very minutely, sparsely scabrid, branches finely scabrid, spreading, with few spikelets at tip. Spikelets 3.5–5–(6) mm, (2)–3–4–(5)-flowered, green, rarely purplish. Glumes unequal; lower 0.8–1.5 mm, 1-nerved, narrow-lanceolate, subobtuse, upper 1.5–2 mm, 3-nerved, elliptic-lanceolate, obtuse, midnerve scabrid. Lemma 2–3 mm, 5-nerved, narrow-lanceolate, subobtuse, nerves minutely scabrid above, lateral nerves occasionally with long hairs near base, internerves glabrous. Palea 1.5–2 mm, keels ciliate-scabrid, interkeel ± minutely hairy. Callus glabrous, or often with a few fine long hairs. Rachilla 0.5–1 mm, glabrous or with a few hairs. Lodicules 0.2–0.4 mm. Anthers 0.3–0.5 mm. Gynoecium: ovary 0.3–0.5 mm; stigma-styles 1.0–1.25 mm. Caryopsis *c.* 1 × 0.3 mm. $2n = 28$. Plates 4D, 7B.

Endemic.

N.: South Auckland (Mangapu Caves), and central and southern regions; S.: east of Main Divide and in north-west Nelson. Lowland to montane in open forest, often not far from the coast.

Poa matthewsii is morphologically close to *P. imbecilla* in its delicate leaves and panicle, almost glabrous spikelets and minute anthers, but can be distinguished by its stoloniferous habit and by the scabrid hyaline margins of the leaf-sheath.

31. P. nemoralis L., *Sp. Pl.* 69 (1753).

Loose, bright green perennial tufts, to 50 cm; branching extravaginal; leaf-blades persistent. Leaf-sheath light green to light brown, membranous, smooth, or lower sheaths sparsely scabrid, striate. Ligule 0.4–0.5–(0.8) mm, truncate, minutely ciliate, abaxially minutely scabrid. Leaf-blade 5–12–(15) cm × 1.5–2 mm, weak, flat, abaxially glabrous, adaxially minutely pubescent-scabrid, often densely scabrid just above ligule; margins minutely scabrid, tip finely to abruptly acute. Culm 20–30–(40) cm, erect or spreading, terete, internodes glabrous. Panicle 5–20 cm, lax and open; branches few to many, filiform, finely scabrid, naked below with few small spikelets near tips. Spikelets 3–4.5 mm, 2–4-flowered, light green. Glumes equal or subequal, acute to acuminate, 3-nerved, midnerve finely scabrid above; lower 2.2–3 mm, narrow-lanceolate, upper (2.2)–2.5–3.5 mm, lanceolate to ovate-lanceolate; margins glabrous. Lemma *c.* 3.0 mm, finely 5-nerved but inner lateral nerves not always distinct, narrow oblong-lanceolate, midnerve and outer lateral nerves with fine hairs to ½ length; margins, midnerve and internerves minutely scabrid near subacute tip. Palea ≈ lemma, keels densely, minutely scabrid, interkeel glabrous. Callus with small tuft of long soft hairs or a few wispy hairs. Rachilla *c.* 1 mm, softly short-ciliate; prolongation usually twice as long. Lodicules *c.* 0.4 mm. Anthers *c.* 1.5 mm. Caryopsis 1.2–1.3 × 0.3–0.4 mm.

Naturalised from Europe.

N.: Kawau Id, Auckland City, Egmont National Park; S.: near Christchurch, Dunedin at Kaikorai. Rare; roadsides.

The only collections made this century are from Kaikorai (CHR 75024) and from N. Egmont Rd, Egmont National Park (CHR 129983).

32. P. novae-zelandiae Hack., *T.N.Z.I. 35*: 381 (1903) ≡ *P. novae-zelandiae* Hack. var. *novae-zelandiae* (autonym, Hackel 1903 op. cit. p. 382) ≡ *P. novae-zelandiae* Hack. forma *novae-zelandiae* (autonym, Hackel 1903 op. cit. p. 382); Lectotype: W 7886! *T. F. C[heeseman]* Arthurs Pass, Canterbury Alps, alt. 3000 ft (No 1339 to Hackel) (designated by Edgar 1986 op. cit. p. 435). = *P. foliosa* var. β Hook.f., *Handbk N.Z. Fl.* 338 (1864); Lectotype: K! *Dr Lyall* Milford Sound (designated by Edgar 1986 op. cit. p. 435). = *P. foliosa* var. γ Buchanan, *Indig. Grasses N.Z.* t. 43B (1880); Lectotype: WELT 59594c (Buchanan's folio)! *J. Morton* Mt Eglinton, 3000 ft (here designated). = *P. novae-zelandiae* forma *laxiuscula* Hack., *T.N.Z.I. 35*: 382 (1903); Holotype: W 9513! *T. Kirk* Bealey Gorge, Canterbury Alps, alt. 3000 ft (No 1343 to Hackel). = *P. novae-zelandiae* forma *humilior* Hack., *T.N.Z.I. 35*: 382 (1903) ≡ *P. novae-zelandiae* var. *wallii* Cheeseman, *Man. N.Z. Fl.* 189 (1925); Lectotype: W 9512! *T. F. C[heeseman]* Nelson Alps, Raglan Range, Wairau Valley, alt. 5500 ft (No 1342 to Hackel) (designated by Edgar 1986 op. cit. p. 435).

Stout perennial, bright green tufts, variable in size, to 30 cm, with leaves usually erect and much narrowed above; branching intravaginal; leaf-blades persistent. Leaf-sheath light green, sometimes purpled, membranous, ribbed, glabrous, or with a few prickle-teeth on margin near ligule. Ligule 1–2–(6) mm, apically glabrous, entire, smooth, somewhat narrowed above to a point, abaxially pubescent-scabrid. Leaf-blade 5–15 cm × 1–3–(5) mm, folded, often with inrolled margins, coriaceous, mainly smooth, but adaxially sparsely scabrid near ligule, rarely finely scabrid throughout; margins with a few scattered prickle-teeth and midrib occasionally scabrid near the curved, often shortly apiculate tip. Culm 5–15–(30) cm, internodes glabrous. Panicle (3)–5–10 cm, dense with erect branches or more spreading and often drooping above; rachis glabrous, branches numerous, slender, glabrous, tipped by large, elliptic, acute spikelets. Spikelets 6–12 mm, 5–8-flowered, light green to purplish. Glumes ± equal, glabrous, smooth, rarely scabrid on midnerve, acuminate; lower 3–6 mm, 1–(3)-nerved, narrow-lanceolate, upper 3.5–7 mm, 3-nerved, narrow elliptic-lanceolate. Lemma 3.5–8 mm, 3–(5)-nerved, elliptic-lanceolate, acute to usually acuminate, glabrous except for

Plate 5. Some native grasses of coastal habitats. **A** *Spinifex sericeus*, a foredune grass of mainly northern New Zealand (Aotea Harbour); **B** *Austrofestuca littoralis*, a grass of sand dunes mainly of southern New Zealand (Stewart Island); **C** *Austrostipa stipoides*, a tussock of coastal rocks and mudflats mainly in Northland (Rangitoto Island); **D** *Hierochloe redolens*, a coastal but also inland grass (Stewart Island).

Plate 6. Some native grasses of rocky and forest habitats. **A** *Rytidosperma setifolium*, a small tussock of mainly alpine rock (Fiordland); **B** *Lachnagrostis lyallii*, a "wind grass" of open sites including rocky lake shores (Manapouri, Fiordland); **C** *Microlaena avenacea*, bush rice-grass, found in humid forests (Dunedin); **D** *Anemanthele lessoniana*, a tussock of dry forest, now commonly cultivated as an ornamental (Dunedin Botanic Garden).

long hairs near base of nerves, rarely scabrid on midnerve above hairs; margins narrow-membranous, entire. Palea 2.5–5 mm, keels sparsely short-scabrid to almost smooth, interkeel and flanks glabrous. Callus with thick tuft of long hairs just below midnerve of lemma and some hairs below lateral nerves. Rachilla 0.5–1 mm, glabrous. Lodicules *c.* 0.7 mm. Gynomonoecious: each spikelet with 1–2–(3) lower flowers ♂, anthers 0.8–1.2–(1.4) mm, gynoecium *c.* 2 mm; upper flowers ♀ with small to minute colourless pollen-sterile anthers 0.1–0.3–(0.6) mm, gynoecium *c.* 1.5 mm. Caryopsis 1.5–2.5 × *c.* 0.5 mm. $2n = 28$. Plate 7C.

Endemic.

N.: mountains from Raukumara Range southwards; S.: mountains throughout, especially along and west of Main Divide; St.: on Mt Anglem. Montane to alpine in scree and on rocks, in grassland, riverbeds and fellfield.

Poa novae-zelandiae is frequently found growing with dioecious *P. subvestita.*

LECTOTYPIFICATION   Buchanan (1880 op. cit.) cited 2 specimens in the protologue of *P. foliosa* var. γ: Mt Eglinton, 3000 ft *J. Morton* and Mt Arthur 4–200 ft *A. Mackay*; the specimen from Mt Eglinton designated lectotype was illustrated in t. 43B.

33. P. palustris L., *Syst. Nat.* ed. 10, *2*: 874 (1759).

Loosely tufted, bright green perennial, to 80 cm, with flaccid leaves overtopped by culms; branching extravaginal at plant base, intravaginal above; leaf-blades persistent. Leaf-sheath light green to light brown, membranous, smooth, or scabrid above, distinctly ribbed, keeled near ligule. Ligule 1.5–3 mm, entire, oblong, slightly rounded and slightly short-ciliate, abaxially with sparse minute hairs. Leaf-blade 10–20 cm × 2–4 mm, flat, minutely scabrid throughout; tip acute. Culm 30–70 cm, erect, internodes glabrous, or rarely papillose and minutely pubescent just below nodes. Panicle 10–25 cm, very loose and open; rachis glabrous, branches delicate, finely scabrid, spreading and flexuous, naked below, branched above with few spikelets on each branchlet. Spikelets 3.5–5 mm, 3–4-flowered, light green tinged purple and brown. Glumes equal or subequal, 3-nerved, acute to acuminate, midnerve scabrid above; lower 2.3–2.8 mm, lanceolate, upper 2.5–3.2 mm, narrowly-ovate. Lemma 2.6–2.8 mm, 5-nerved but inner lateral nerves often obscure, oblong, obtuse, glabrous apart from fringe of short hairs to ½ length of midnerve and outer lateral nerves, scabrid on midnerve above hairs, sometimes purplish, with shining yellow-brown band above. Palea 2.3–2.6 mm, keels ciliate, interkeel glabrous. Callus with long tuft of soft crinkled hairs. Rachilla *c.* 0.5 mm, minutely scabrid; prolongation twice as long. Lodicules *c.* 0.4 mm. Anthers 0.7–1 mm. Caryopsis *c.* 1.3 × 0.5 mm.

Naturalised from Europe.

N.: Moutoa Swamp near Foxton; S.: local in Nelson, Marlborough and Otago. Lowland in swamps and boggy ground.

34. P. pratensis L., *Sp. Pl.* 67 (1753).   Kentucky bluegrass

Rather narrow, loose to compact perennial tufts *c.* 10–50 cm from slender rhizomes with ± soft, bright green to greyish green leaves < culms; branching extravaginal at plant base, sometimes intravaginal above; leaf-blades persistent. Leaf-sheath light green to very light brown, membranous to ± coriaceous, distinctly ribbed, glabrous, or slightly short-scabrid to shortly hairy near ligule. Ligule in lower leaves 0.2–0.5 mm, truncate, often very short-ciliate, and abaxially minutely hairy, in upper leaves 1–2 mm, apically glabrous, entire and rounded. Leaf-blade (1.5)–10–20–(35) cm × (1.5)–2–4 mm, flat or folded, subcoriaceous, glabrous or sparsely finely hairy especially near ligule or only adaxially; margins sparsely scabrid to smooth, sometimes hairy near ligule; midrib scabrid near blunt, curved tip. Culm (5)–20–55 cm, internodes glabrous. Panicle (2)–5–10–(15) cm, ovate to pyramidal, or oblong, erect or nodding, loose and open to somewhat dense and contracted; rachis glabrous, branches in clusters of 3–5, spreading, filiform, flexuous, smooth or finely scabrid with spikelets clustered at tips. Spikelets 4–6 mm, (2)–3–6-flowered, light green or purplish. Glumes ± unequal, acute, membranous, midnerve scabrid; lower 1.5–3 mm, 1–3-nerved, narrow-ovate, upper 2–3.5–(4) mm, 3-nerved, ovate-elliptic. Lemma 2.5–4 mm, 5–(7)-nerved, ovate-oblong, subobtuse to acute, midnerve and marginal nerves thinly to densely hairy in lower ½, midnerve finely scabrid above hairs, internerves glabrous. Palea 2–3 mm, keels finely scabrid, interkeel glabrous. Callus with tuft of long crinkled hairs. Rachilla *c.* 0.5 mm, glabrous or very minutely sparsely papillose; prolongation twice as long. Lodicules 0.6–0.8 mm. Anthers 1.2–1.7 mm. Caryopsis *c.* 1.5 × 0.5 mm, tightly enclosed by anthoecium.

Naturalised from Eurasia.

N.; S.: throughout; St.; Ch., A., C. Lowland to alpine in grassland and pasture, cultivated or waste ground.

Poa pratensis is treated here in the broad sense, including *P. angustifolia* L. and *P. subcaerulea* Sm. In N.Z. the name *P. angustifolia* may be applied to greyish green, densely caespitose, narrow-leaved plants. *Poa subcaerulea* which Allan, H. H. *T.R.S.N.Z. 65*: 2 (1935) suggested was probably widely distributed in North Id, is distinguished from *P. pratensis* and *P. angustifolia* in having scattered, ± solitary stems, and equal, acuminate, 3-nerved glumes.

Hooker formally named a New Zealand plant of Kentucky bluegrass as *P. anceps* var. γ *breviculmis* Hook.f., *Fl. N.Z. 1*: 306 (1853); Lectotype: K (lowest specimen)! *J. D. H[ooker]* Bay of Islands [1841] (designated by Edgar 1986 op. cit. p. 459).

35. P. pusilla Berggr., *Minneskr. Fisiogr. Sällsk. Lund* Art. *8*: 31 (1878) ≡ *P. pusilla* Berggr. var. *pusilla* (autonym, Cockayne 1911 op. cit. p. 34); Holotype: LD! *S. Berggren* in alpibus ad flum. Bealey, Ins. austr. Novae Zelandiae, Feb. 1874. = *P. seticulmis* Petrie, *T.N.Z.I. 34*: 391 (1902) ≡ *P. pusilla* var. *seticulmis* (Petrie) Cockayne, *Dept Lands Rep. Dune Areas N.Z.* 34 (1911); Lectotype: WELT 66866! *D. P[etrie]* Maunganui Bluff, N. of Kaihu, Jan. 1896 (designated by Edgar 1986 op. cit. p. 453). = *P. anceps* var. ε *debilis* Buchanan, *Indig. Grasses N.Z.* t. 46E (1880); Holotype: WELT 59600 (Buchanan's folio)! *Kirk* Auckland district, hot springs [Feb. 1872]. = *P. anceps* var. ζ *minima* Buchanan, *Indig. Grasses N.Z.* t. 46F (1880); Holotype: WELT 59599 (Buchanan's folio)! *A. Mackay* Mt Arthur, 4200 ft [1874]. = *P. anceps* var. *gracilis* Cheeseman, *Man. N.Z. Fl.* 904 (1906); Lectotype: AK 1820! *T. F. C[heeseman]* Hawera, Taranaki, Jan. 1885 (No 1443 to Hackel) (designated by Edgar 1986 op. cit. p. 453).

Slender, often delicate, open, perennial tufts, 5–*c.* 35 cm, with few, very narrow, flaccid, bright green leaves, and slender, long-creeping rhizome; branching extravaginal; leaf-blades persistent. Leaf-sheath greenish purple becoming light brown, ± membranous, distinctly keeled, ribs strong, whitish, shaggy pubescent-scabrid, or minutely scabrid throughout or only near hyaline margins, occasionally with a few longer hairs on margins near ligule, rarely completely glabrous. Ligule 0.2–0.5 mm, truncate, ciliate, sometimes erose, abaxially with minute hairs. Leaf-blade (1)–5–15–(25) cm × *c*. 0.5 mm diam., folded, rarely flat and wider, margins inrolled, abaxially distinctly ribbed, glabrous, adaxially with scattered short hairs especially on midrib, tip subacute to curved, scabrid on midrib. Culm 3–30–(50) cm, very slender, erect or drooping above, internodes glabrous. Panicle 1.5–8.5 cm; branches spreading, few, filiform, scabrid, bearing 1–4 spikelets at tips. Spikelets 3–7.5 mm, 2–5–(7)-flowered, yellow-green, faintly purplish. Glumes subequal, 2–4–(4.5) mm, 3-nerved, narrow elliptic- to ovate-lanceolate, acute or subacute, smooth, midnerve and sometimes lateral nerves sparsely scabrid. Lemma 2–4–(5) mm, 5-nerved, oblong- to ovate-elliptic, obtuse, midnerve with long silky crinkled hairs in lower ½, sometimes scabrid above hairs, lateral nerves usually with conspicuous silky hairs near base or in lower ½, internerves usually smooth, rarely finely scabrid, with hyaline band at slightly erose tip. Palea 2–3.5 mm, keels short ciliate-scabrid, interkeel short pubescent-scabrid, flanks smooth to scabrid. Callus with thick tuft of long, silky crinkled hairs. Rachilla (0.3)–0.6–0.8 mm, with scattered long hairs. Lodicules *c*. 0.5 mm. Anthers 1–2 mm. Caryopsis *c*. 1 × 0.5 mm. $2n = 28$.

Endemic.

N.: throughout; S.: throughout but rarely recorded from eastern plains and dry inland basins; St.; Three Kings Is. Lowland to subalpine, in open forest, scrub, tussock grassland, in stony places or on sand dunes.

An extremely slender fine-leaved form of *P. pusilla* was collected on the west coast of Wellington Province at Himatangi (1931), Foxton (1932), and at Hokio (1967), where it formed dense mats in damp sand flats among dunes, from densely interwoven rhizomes; A. P. Druce (*in litt.*) considered that all sites where the plants grew are so modified that this capillary-like form of *P. pusilla* may no longer exist.

36. P. pygmaea Buchanan, *Indig. Grasses N.Z.* t. 50A (1880); Holotype: WELT 59606 (Buchanan's folio)! *D. Petrie* Mt Pisa, 4–6000 ft (No 1356 to Hackel).

Small, dull green, much-branched, very densely packed perennial, rooting from short prostrate branches and forming compact, rigid cushions to 10 cm across and 2.5 cm high, slightly overtopped by filiform flowering culms with few spikelets; branching intravaginal, new shoots with densely imbricating leaves; leaf-blades persistent. Leaf-sheath much wider than leaf-blade, pale greyish brown, membranous, glabrous, shining, ± indistinctly ribbed. Ligule 0.5–1 mm, apically glabrous, entire, narrowed centrally to a short point, abaxially sparsely to densely minutely pubescent. Leaf-blade 0.3–0.7 cm × 1–2 mm, stiff, inrolled, abaxially glabrous, ribs prominent, adaxially strongly ribbed, ribs whitish with fine minute hairs or prickle-teeth; midrib scabrid near very curved, subobtuse or occasionally acute tip. Culm 1.5–2.5 cm, internodes mainly smooth with a few minute prickle-teeth above. Panicle 0.5–1 cm, contracted, racemose, with 1–3 spikelets; branches and pedicels glabrous. Spikelets 4.5–5.5 mm, 2–5-flowered, light green to purplish. Glumes equal, 2.5–3.5 mm, 3-nerved, ovate-elliptic, subobtuse, smooth or rarely with a few prickle-teeth on midnerve near tip; margins entire. Lemma 3–3.5 mm, 5-nerved, ovate, obtuse, lower ½ with fine, often crinkled hairs on nerves and sparse somewhat shorter hairs on internerves, upper ½ sparsely scabrid on midnerve and near wide membranous margin; midnerve occasionally minutely excurrent. Palea 2.5–3 mm, keels ciliate, interkeel with sparse minute hairs. Callus with few long hairs. Rachilla *c.* 0.5 mm, sparsely ciliate; prolongation twice as long. Lodicules 0.4–0.6 mm. Anthers 1.2–1.6 mm. Caryopsis *c.* 1.5–2 × 0.5 mm. $2n = 28$.

Endemic.

S.: Pisa Range, and Mount St Bathans, Otago. In subalpine to alpine snow hollows and on lake margins.

37. P. ramosissima Hook.f., *Fl. Antarct. 1*: 101 (1845); Holotype: K! *J. D. Hooker* Lord Aucklands Islands, hangs down from the cliffs and rocks near the sea, common, very stoloniferous, Nov. 1840. = *P. ramosissima* var. β Hook.f., *Fl. Antarct. 1*: 101 (1845); Holotype: K!

J. D. Hooker in rigid naked stemmed tufts on sloping ground from the top of the hills to the sea (1000 ft) on the weather side of Campbell's Island, Dec. 1840; (the sheet bears a second label "1625, covering the ground at an elevat. of 700 ft on the windward side of the Isld, Campbell's Isld").

Bright green, soft, turfy perennial patches, from long, bare prostrate culms, distally becoming erect, leafy and much-branched; branching intravaginal; leaf-blades persistent. Leaf-sheath greenish brown to purplish, glabrous, hyaline, ribs prominent. Ligule (1.5)–2.5–4 mm, deeply and sharply lacerate, glabrous throughout. Leaf-blade (4)–9–15 cm × 1–2 mm, thin, weak, flat, ribs many, strong, minutely papillose-scabrid, adaxially furrowed, evenly narrowed to very finely obtuse or subobtuse tip; margins glabrous. Culm (1)–10–30–(40) cm, internodes glabrous. Panicle (2)–4–5–(10) cm, ± oblong, contracted, usually overtopped by leaves; rachis glabrous, branches erect, short, scarcely spreading, finely papillose-scabrid, bearing few spikelets. Spikelets 4.5–7.5 mm, 3–5-flowered, greenish brown, very minutely papillose-scabrid. Glumes subequal or the lower obviously shorter, both narrow-lanceolate, acute or acuminate; lower 2.5–4 mm, 1–3-nerved, upper 3–4.5 mm, 3-nerved. Lemma 3.8–5 mm, 5–(7)-nerved, elliptic, drawn out to acute or acuminate tip, mid- and lateral nerves with a few short hairs near base. Palea 2.5–4.5 mm, very narrow, keels with a few short hair-like prickle-teeth. Callus with small tufts of long, twisted hairs below midnerve of lemma and occasionally below lateral nerves. Rachilla 0.5–1 mm, glabrous. Lodicules 0.3–1 mm, occasionally hair-tipped. Gynomonoecious: each spikelet with 1–2 lower flowers ♂, anthers 1.5–2.5 mm, gynoecium *c.* 1.5 mm; upper flowers ♀ with minute colourless anthers 0.1–0.7 mm, gynoecium *c.* 1.5 mm. Caryopsis *c.* 1 × 0.5 mm. $2n = 28$.

Endemic.

A., C. Coastal cliffs, usually associated with bird colonies.

Poa ramosissima is closely related to *P. cookii* of Macquarie, Heard, and Kerguelen Is. Both spp. are characterised by deeply lacerate ligules, papillose adaxial leaf surfaces and panicle branches, and gynomonoecious flowers. *Poa ramosissima* is more slender than the large, tufted *P. cookii,* and both leaf surfaces and all parts of the spikelet are scabrid-papillose in *P. ramosissima,* whereas the leaf is glabrous abaxially in *P. cookii* and the spikelets are glabrous.

38. P. schistacea Edgar et Connor, *N.Z. J. Bot. 37:* 63 (1999); Holotype: CHR 395536A! *A. P. Druce* Two-mile V[alley], Hector Mts, 5300 ft, rocks at foot of cliff, Mar 1985; ♀.

Long-rhizomatous dioecious perennial, 10–30–(60) cm, with narrow, involute, greyish green, stiffly erect to curved leaves, often equalling the culms; branching extravaginal; leaf-blades persistent. Leaf-sheath light greenish to brownish grey or stramineous, glabrous, ± shining, submembranous, keeled above, with several prominent ribs. Ligule 0.5–1.0–(1.5) mm, apically glabrous, entire, truncate or sometimes rather abruptly tapered centrally, abaxially minutely pubescent. Leaf-blade (2.5)–4.5–18.5–(28) cm × 0.5–1 mm diam., involute, abaxially glabrous, adaxially minutely pubescent; tip acicular. Culm 4.5–25–(53) cm, internodes smooth, sparsely minutely scabrid below panicle. Panicle (2)–3–6.5–(8) cm, with short ascending branches and close-set spikelets; rachis and branches smooth or with sparse prickle-teeth. Spikelets 4.5–8.5–(10) mm, (2)–3–6–(7)-flowered, green to stramineous, often purple-suffused. Glumes subequal, glabrous; lower 2.5–4 mm, 1-nerved, elliptic-lanceolate, acute to acuminate, upper 3–5.5 mm, 3-nerved, elliptic-ovate, acute to subobtuse. Lemma 4–5 mm, 5-nerved, inner lateral nerves sometimes faint, elliptic-ovate, acute, glabrous apart from a few hairs at base of keel and outer lateral nerves. Palea 3–3.5 mm, narrower than lemma; keels with sparse minute prickle-teeth in upper ½–¾ or rarely smooth, occasionally keels and lower ½ of interkeel with sparse minute hairs. Callus with tufts of long fine hairs below lemma nerves. Rachilla < 1 mm, glabrous. Dioecious: ♂ with anthers 2–3 mm; ♀ with lodicules 0.3–0.6 mm, sterile anthers 0.2–1 mm, ovary 0.5–0.9 mm, stigma-styles 1.2–2.6 mm, caryopsis 1.5–2.5 × 0.5–0.8 mm.

Endemic.

S.: mountains of western Otago and north-eastern Southland. Usually in scree derived from schist, 550–1980 m.

39. P. senex Edgar, *N.Z. J. Bot. 24*: 477 (1986); Holotype: CHR 133878!
V. D. Zotov Old Man Range, Otago, water-course, 5200', 13.2.1963.

Small, stoloniferous, almost completely glabrous, brownish green
perennial tufts, *c.* 5–15–(25) cm, culms overtopping leaves; branching
extravaginal; leaf-blades persistent. Leaf-sheath membranous, glabrous,
ribbed, keeled. Ligule 0.5–1.5 mm, entire, tapered, glabrous
throughout. Leaf-blade 1–3.5 cm × 1–2 mm, flat or folded,
subcoriaceous, smooth, but midrib scabrid near curved tip; margins
finely scabrid. Culm 3–15–(20) cm, very slender, erect or geniculate at
base, internodes glabrous. Panicle 1–3 cm, ± open or contracted, with
few, ovate spikelets; rachis, branches and pedicels slender with sparse,
scattered prickle-teeth. Spikelets 2–3–(3.5) mm, 2–3-flowered, light
green, tinged purple. Glumes unequal, submembranous with hyaline
margins, a few prickle-teeth on midnerve near tip; lower 1.5–2 mm,
1-nerved, narrow-lanceolate, acute, upper 2–2.5 mm, (1)–3-nerved,
elliptic-oblong, subobtuse to obtuse. Lemma 2–2.5 mm, 5-nerved,
elliptic-ovate, obtuse, glabrous, but midnerve with short crinkled hairs
to *c.* ½ length and sparsely prickle-toothed near tip, lateral nerves with
a few hairs near base. Palea 1.5–1.8 mm, keels minutely scabrid,
interkeel glabrous. Callus with a few wispy hairs. Rachilla *c.* 0.5 mm,
glabrous. Lodicules *c.* 0.1 mm. Anthers 0.3–0.4 mm. Gynoecium: ovary
0.4–0.5 mm; stigma-styles 0.8–1 mm. Caryopsis *c.* 1 × 0.5 mm.

Endemic.

S.: mountains of Central and western Otago, and Eyre Mts, Southland.
Alpine in damp ground.

Known only from the Old Man and Pisa Ranges until late 1980s when A. P.
Druce collected it from Mt Cardrona in the Crown Range, Treble Cone in the
Harris Mts, Old Woman Range, and Jane Peak in the Eyre Mts. In most of these
specimens the culms are taller than in those specimens on which the original
description was based.

Poa senex is one of three low-growing, small-anthered, alpine spp. found in damp
situations; it differs from both *P. sublimis* and *P. incrassata* in having hairs on the
lemma nerves; from *P. sublimis* it is also distinguished by the more contracted
panicle with firmer branchlets, and from *P. incrassata* by the wider leaves, and by
the flat, not hooded, lemma tip and glabrous, not papillose, lemma internerves.

40. P. sieberiana Spreng., *Syst. Veg. 4, Cur. Post.* 35 (1827).

rough poa tussock

Stiff, fine-leaved, greyish green, small, wiry, tussocks to *c.* 80 cm, from a narrow base packed with dead dry leaves; branching intravaginal; leaf-blades persistent. Leaf-sheath light brown, becoming dull grey-brown, coriaceous, smooth or slightly scabrid. Ligule *c.* 0.3 mm, a truncate minutely ciliate rim, abaxially with minute prickle-teeth. Leaf-blade 15–25–(35) cm, inrolled, *c.* 0.5 mm diam., abaxially usually finely scabrid, adaxially densely short-scabrid; margins finely scabrid, tip long, fine, acicular. Culm 30–75 cm, erect, internodes smooth to finely scabrid. Panicle 10–20 cm, contracted at first, later lax with spreading, slender, finely scabrid branches tipped by numerous spikelets. Spikelets (3.5)–4–5–(6.5) mm, 3–6–(8)-flowered, light green often purpled. Glumes ± unequal, narrow-lanceolate to narrow-ovate, acute to subacuminate, keels and sometimes internerves scabrid; lower 1.5–2 mm, 1–3-nerved, upper 2–2.5 mm, 3-nerved; margins minutely scabrid. Lemma *c.* 2.5 mm, 5-nerved, elliptic-oblong, ± closely pubescent on lower ½, usually with slightly longer hairs on keel and marginal nerves, tip obtuse, short-ciliate. Palea 2–2.5 mm, keels scabrid above and finely ciliolate in lower ½, interkeel pubescent especially in lower ½. Callus glabrous or with a few long hairs. Rachilla *c.* 0.5 mm, with a few short hairs. Lodicules 0.4–0.6 mm. Anthers 1.2–1.5 mm. Caryopsis *c.* 1.5 × 0.5 mm.

Naturalised from Australia.

S.: Nelson (Lake Rotoiti), Marlborough (between Blenheim and Cape Campbell). Lowland to montane grassland.

41. P. spania Edgar et Molloy, *N.Z. J. Bot. 37*: 43 (1999); Holotype: CHR 511252! *B. P. J. Molloy* North Otago, Waitaki Valley, Awahokomo Creek, true left, 14 Nov 1996.

Small, greyish green or dull green, sometimes purple tinged, short-lived perennial, forming slender tufts to *c.* 22 cm with wiry leaves < culms; branching extravaginal and occasionally intravaginal above; leaf-blades persistent. Leaf-sheath pale brown to purplish, membranous, distinctly ribbed, almost smooth with a few minute prickle-teeth just below ligule especially near margins. Ligule

1–1.5 mm, ± lacerate with at least one deep cleft, ciliate, tip dentate, abaxially minutely pubescent-scabrid. Leaf-blade to 6 cm × 1 mm, folded, abaxially with sparse minute prickle-teeth on midrib, elsewhere smooth, adaxially with minute stiff hairs; margins ± inrolled, with very minute prickle-teeth. Culm to 15 cm, very slender, purplish, with 1–2 cauline leaves spreading at right angles, internodes glabrous. Panicle 2–6 cm, open, sparingly branched; rachis glabrous, branches capillary, sometimes flexuous, smooth, with a few minute prickle-teeth above, and with 1–2 spikelets at tips. Spikelets 2.2–3 mm, 2–3-flowered, silvery purplish green. Glumes subequal, 2–2.5 mm, with wide, hyaline margins, midnerve with sparse prickle-teeth near tip; lower narrower, elliptic-lanceolate, acute to acuminate, 1–3-nerved, upper ovate, obtuse, 3-nerved, sometimes with minute prickle-teeth on margins and on lateral nerves above. Lemma 2–2.4 mm, 5-nerved, ovate, obtuse, covered for ¾ length with appressed, silky hairs *c.* 0.1 mm, and with sparse minute prickle-teeth near membranous tip. Palea *c.* 2 mm, keels and interkeel with minute appressed hairs in lower ¾, keels minutely prickle-toothed above hairs. Callus with minute fine hairs. Rachilla to 0.5 mm with scattered long hairs or almost glabrous. Lodicules *c.* 0.4 mm. Anthers 0.7–1.1 mm, sometimes reddish. Stigma-styles 1–1.4 mm. Caryopsis *c.* 1.5 × 0.7 mm.

Endemic.

S.: Otago (Awahokomo Bluff). In remnant scrub/grassland on weathered limestone.

Poa spania resembles *P. lindsayi* in the silky haired lemmas, inrolled greyish leaves and sparingly branched slender panicle but differs in extravaginal branching, and in the ± lacerate, abaxially scabrid ligule, rather than truncate and glabrous as in *P. lindsayi*. Anthers in *P. spania* (0.7–1.1 mm) are ≫ anthers of *P. lindsayi* (0.2–0.5 mm).

42. P. sublimis Edgar, *N.Z.J. Bot. 24*: 465 (1986); Holotype: CHR 25232! *V. D. Zotov* Arthurs Pass National Park, [Mt] Blimit, 6000 ft, 21.2.1943.

Low-growing perennial, forming small tufts or cushions to *c.* 15 cm, with culms usually somewhat overtopping the rather stiff, green or purplish leaves; branching extravaginal; leaf-blades persistent. Leaf-sheath creamy brown or purple, glabrous, hyaline, ribs distinct. Ligule *c.* 0.5–(1) mm, entire, narrowed to a subobtuse point, glabrous throughout. Leaf-blade 1–5 cm × (0.5)–1–1.5 mm, folded, smooth but with a few prickle-teeth abaxially on midrib near incurved tip, and on margins near ligule and tip. Culm 1–10–(15) cm, internodes glabrous. Panicle 1–4.5 cm, lax; rachis and slender spreading branches and pedicels filiform, smooth, rarely sparsely scabrid below the very few spikelets. Spikelets 3–4 mm, 3–4-flowered, light green, usually purplish. Glumes subequal, smooth, but midnerve prickle-toothed in upper ½; lower 1.5–2.5 mm, 1-nerved, ovate-lanceolate, acute, upper 1.8–3.2 mm, 3-nerved, ovate, subobtuse. Lemma 2–3.5 mm, 5–7-nerved, ovate, obtuse, smooth, but upper ⅓ of midnerve sparsely scabrid. Palea 1.5–2.5 mm, keels short-scabrid, interkeel glabrous. Callus glabrous. Rachilla *c.* 0.5 mm, glabrous. Lodicules 0.3–0.6 mm. Anthers (0.2)–0.3–(0.5) mm. Caryopsis *c.* 1 × 0.5 mm.

Endemic.

S.: along and west of Main Divide and in Fiordland, to the east in Southland. Subalpine to alpine, in snow hollows and on stream margins among rocks.

Poa sublimis differs from *P. kirkii* and other related spp. in being almost smooth in all parts and in having very minute anthers.

43. P. subvestita (Hack.) Edgar, *N.Z. J. Bot. 24*: 436 (1986) ≡ *P. novae-zelandiae* var. *subvestita* Hack., *T.N.Z.I. 35*: 382 (1903); Lectotype: W 9510! *L. Cockayne* Arthurs Pass, Canterbury Alps [1898] (No 1346 to Hackel); ♀ (designated by Edgar 1986 op. cit. p. 436).

Often robust, bright green perennial tufts to 40 cm, with erect, wide, rather stiff leaves overtopped by culms; branching extravaginal; leaf-blades persistent. Leaf-sheath green to light brown, membranous, distinctly ribbed, glabrous; midrib thickened above. Ligule 1–2.5–(6.5) mm, apically glabrous, entire, sometimes ± erose, shortly tapered, often narrowed to a long point, abaxially short-pubescent or rarely glabrous. Leaf-blade 5–20 cm × 2–4–(6.5) mm, flat, abaxially smooth, adaxially with minute, ± appressed prickle-teeth; margins thickened, smooth, rarely with a few prickle-teeth near base, abruptly narrowed and curved to ± thickened, ovate to acute tip. Culm 10–20–(40) cm, internodes glabrous. Panicle (3)–6–10–(15) cm, lax and often drooping above or sometimes dense; branches short, fine, ± crowded, smooth, very rarely sparsely prickle-toothed, tipped by large spikelets. Spikelets (6)–8–10 mm, 4–6-flowered, light green to purple. Glumes subequal, smooth throughout or midnerve sparsely scabrid above; lower (2.5)–3–4.5 mm, 1-nerved, narrow-lanceolate, usually acuminate, upper 3–5.5 mm, 3-nerved, elliptic-lanceolate, acute or subobtuse. Lemma (3.5)–4.5–6 mm, 3-nerved, elliptic-lanceolate, acute to obtuse, glabrous except for soft tangled hairs on lower ½ of nerves; margins membranous wide, occasionally with a few sparse prickle-teeth. Palea (3)–3.5–4–(4.5) mm, keels sparsely ciliate in upper ½, interkeel and flanks glabrous. Callus with thick tufts of long hairs below lemma nerves. Rachilla *c.* 1 mm, glabrous, or with occasional hairs. Lodicules 0.5–1 mm. Dioecious: ♂ with anthers (1.5)–2–3 mm, gynoecium *c.* 0.2 mm; ♀ with ovary *c.* 1 mm, stigma-styles *c.* 1.5 mm, pollen-sterile anthers 0.3–0.7 mm, caryopsis 0.5–0.8 × 0.2–0.3 mm. $2n = 28$.

Endemic.

S.: along and west of Main Divide and in Fiordland, occasionally to the east. Montane to alpine on shaded rocks and cliffs, particularly on limestone and marble; often covered by spray from waterfalls.

Poa subvestita can usually be distinguished from *P. novae-zelandiae* without dissecting the florets to determine whether plants are dioecious (*P. subvestita*) or gynomonoecious (*P. novae-zelandiae*). Plants are usually robust, often wider-leaved than in *P. novae-zelandiae* and the lemmas are obtuse-tipped rather than long drawn out to an acuminate point as in many plants of *P. novae-zelandiae*.

44. P. sudicola Edgar, *N.Z. J. Bot. 24*: 437 (1986); Holotype: CHR 369894A! *A. P. Druce* Pike P[ea]k, Allen Ra., NW Nelson, 4900 ft, limestone scree, Feb. 1982; ♀.

Long-rhizomatous, narrow-leaved, greyish green perennial tufts, to 25 cm; branching intravaginal; leaf-blades persistent. Leaf-sheath light greenish brown, faintly purpled above, smooth or occasionally scabrid between ribs, membranous, keeled, with a few prominent ribs. Ligule 0.5–1–(1.5) mm, apically glabrous, entire, truncate with ± abrupt central peak, abaxially minutely pubescent-scabrid. Leaf-blade (2)–5–9–(12) cm, inrolled, *c.* 1 mm diam., wiry, abaxially glabrous, adaxially glabrous or with scattered short hairs especially near margins; tip acicular. Culm 10–25 cm, slender, internodes glabrous. Panicle 2–4.5–(6) cm, with few, ± spreading branches; rachis glabrous, branches very slender, glabrous, tipped by very few, narrow spikelets. Spikelets 6–9 mm, 3–4-flowered, greenish brown. Glumes subequal, glabrous, rarely with a few minute prickle-teeth on midnerve near tip, acute to subobtuse; lower (3)–3.5–4 mm, 1–3-nerved, narrow-elliptic, upper (3.5)–4–4.5 mm, 3–5-nerved, more broadly ovate. Lemma 4.5–5–(6) mm, 5–7–(9)-nerved, elliptic-lanceolate, acute, glabrous apart from short hairs on lower ⅔ of midnerve and on lateral nerves near base, with some longer hairs at very base, smooth apart from minute prickle-teeth on nerves near tip, and near narrow-membranous, entire margins. Palea 3–4–(5) mm, much narrower and ≪ lemma, keels sparsely, minutely fine-ciliate, interkeel finely scabrid, flanks glabrous. Callus with tufts of long fine hairs below lemma nerves. Rachilla < 1 mm, smooth, or occasionally with a few microscopic prickle-teeth. Lodicules 0.3–0.5 mm. Dioecious: ♂ with anthers 2.9–3.3 mm, gynoecium 0.7–1.3 mm; ♀ with pollen-sterile anthers 0.6–1–(1.2) mm, ovary 0.3–0.5 mm, stigma-styles 0.75–2.2 mm, mature caryopses not seen.

Endemic.

S.: north-west Nelson, Matiri Range, Pike Peak on Allen Range and Turks Cap Range. Subalpine, on limestone scree and steep mudstone slopes.

Edgar (1986 op. cit. p. 439) cited two specimens from Matiri Range (CHR 354914, CHR 354915) as seed-bearing female plants of *P. sudicola*; these specimens are now referred to *P. novae-zelandiae.*

Poa sudicola resembles *P. novae-zelandiae* and *P. subvestita* in having comparatively large spikelets on slender, glabrous panicle branches; it is distinguished from both species by the very long rhizomes. The type, from Pike Peak, was sympatric with *P. novae-zelandiae* and *P. subvestita.*

45. P. tennantiana Petrie, *in* Chilton *Subantarctic Is N.Z. 2*: 476 (1909) ≡ *P. foliosa* var. *tennantiana* (Petrie) Cheeseman, *Man. N.Z. Fl.* 188 (1925); Lectotype: WELT 36063! *T. Kirk* The Snares, 9 Jan. 1890 (designated by Edgar 1986 op. cit. p. 434).

Stout, stiff-leaved, yellow-green perennial tufts to *c.* 1 m, from strong rhizomatous base covered by abundant fibrous remnants of leaf-sheaths; branching intravaginal; leaf-blades persistent. Leaf-sheath submembranous, striate, densely, retrorsely, minutely pubescent-scabrid between ribs in basal leaves, glabrous in cauline leaves, very light brown, later becoming darker and shredding into fibres. Ligule 6–16 mm, apically glabrous, entire, narrowed to a long fine point, abaxially short-pubescent. Leaf-blade 16–38 cm × 4.5–9 mm, flat, finely striate, abaxially smooth, adaxially minutely papillose, very rarely minutely ciliate-scabrid on ribs; margins ciliate-fimbriate for a short distance above ligule, otherwise glabrous, gradually narrowed to straight-sided acute tip. Culm 15–33 cm, internodes glabrous. Panicle 9–16 cm, broad, dense but much-branched, upper branches almost completely hidden by numerous, rather small spikelets, lower branches naked towards base; rachis smooth, branches and pedicels ±scabrid to occasionally smooth. Spikelets 3.5–4.5 mm, 2–3-flowered, light brown. Glumes subequal, ovate-lanceolate, acute, glabrous; lower 2–2.5 mm, 1-nerved, upper 2.5–3 mm, 3-nerved; midnerve near tip and margins with sparse prickle-teeth. Lemma 3–3.5 mm, 5-nerved, ovate-elliptic, minutely pubescent-scabrid, nerves longer ciliate on lower ⅓, acute, or midnerve very shortly excurrent. Palea 2.5–3 mm, keels scabrid in upper ½, interkeel smooth or with a few prickle-teeth. Callus glabrous. Rachilla *c.* 0.5 mm, glabrous; prolongation twice as long. Lodicules 0.5–0.8 mm. Anthers 1–1.5 mm. Caryopsis *c.* 1.5 × 0.5 mm. $2n = 56$.

Endemic.

S.: Otago, at Taieri R. mouth; St.: islands offshore except to north-east; Sn., A. Coastal, in forest margins, and clearings in scrub and on banks.

No recent collections have been made from Auckland Is, and earlier records of *Poa tennantiana* on Antipodes Is are erroneous [Godley, E. J. *N.Z. J. Bot. 27*: 531–563 (1989)].

46. P. tonsa Edgar, *N.Z. J. Bot. 24*: 477 (1986); Holotype: CHR 175630! *J. Wells & A. F. Mark* Omarama Saddle, Central Otago, 5500', occasional in snow tussock grassland, 9.1.1967.

Small, compact, light green perennial tufts, to *c.* 15 cm; branching intravaginal, sometimes extravaginal at plant base; leaf-blades persistent. Leaf-sheath light creamy brown to reddish purple, membranous, ribbed, smooth or papillose above, especially on keel. Ligule 0.5–1 mm, apically glabrous, centrally tapered, erose, abaxially ciliate. Leaf-blade 1–3.5 cm × 1–2 mm, flat, subcoriaceous, abaxially smooth, adaxially minutely scabrid; margins densely, finely scabrid, tip curved. Culm 2–10 cm, slender, purplish, usually with one short cauline leaf; internodes smooth or slightly scabrid below panicle. Panicle (1)–2–4.5–(7) cm, open; rachis slender, finely, sparsely scabrid, branches few, filiform, finely scabrid, tipped by 1–2 relatively large, ovate spikelets. Spikelets 3–4.5 mm, 3–5-flowered, purplish green. Glumes subequal, 1.5–2 mm, with wide hyaline margins, scabrid on midnerve; lower 1–3-nerved, ovate-elliptic, acute, upper 3-nerved, ovate, obtuse, often scabrid on lateral nerves and on internerves near tip and margin. Lemma 2–2.5 mm, 5–7-nerved, ovate, obtuse, minutely hairy in lower ½, scabrid above with midnerve closely distinctly scabrid; margins hyaline, very wide. Palea 1.5–2 mm, keels scabrid, interkeel with minute hairs and flanks with a few hairs. Callus with a few wispy hairs. Rachilla *c.* 0.5 mm, glabrous. Lodicules *c.* 0.2 mm. Anthers 0.3–0.6 mm. Gynoecium: ovary 0.5 mm; stigma-styles *c.* 1 mm. Caryopsis *c.* 1–1.5 × 0.5 mm.

Endemic.

S.: Shingle Range, Upper Awatere, Marlborough and in mountains of Central and west Otago. Subalpine to alpine in snow tussock grassland, on rock, and river flats.

The lemmas are devoid of hair in the upper part, distinguishing this sp. from the closely related *P. lindsayi.*

47. P. trivialis L., *Sp. Pl.* 67 (1753).　　rough-stalked meadow grass

Open, often wide-leaved, light to darkish green, stoloniferous perennial tufts, *c.* 20–90 cm; branching extravaginal; leaf-blades persistent. Leaf-sheath light green, sometimes purplish, ± membranous, usually finely scabrid on ribs, especially near ligule, keeled. Ligule 2.5–10 mm, apically glabrous, entire, tapered to a point, abaxially with scattered very minute prickle-teeth. Leaf-blade 10–20 cm × 1.5–7 mm, soft, flat, finely scabrid on ribs or smooth; margins and fine pointed tip finely scabrid. Culm 40–70 cm, erect, internodes smooth, or finely scabrid below panicle. Panicle (5)–10–20 cm, usually open and very lax; rachis smooth to finely scabrid above, branches fine, spreading, finely, densely scabrid, with numerous spikelets toward tips. Spikelets 3–4 mm, 2–3-flowered, green or purplish. Glumes ± unequal, finely scabrid on nerves; lower 1.5–2.5 mm, 1-nerved, narrow-lanceolate, upper 2–3 mm, 3-nerved, elliptic- to ovate-lanceolate; tips fine, acute. Lemma 2–3 mm, 5-nerved, narrow-oblong, glabrous, but midnerve with short fine hairs on lower ½ and scabrid above hairs, marginal nerves with a few short hairs near base; margins and acute tip membranous. Palea 2–2.5 mm, keels extremely minutely scabrid, interkeel glabrous. Callus with long slender tuft of fine crinkled hairs. Rachilla ± 0.5 mm, glabrous. Lodicules 0.2–0.4 mm. Anthers 1.5–2 mm. Caryopsis *c.* 1.5 × 0.5 mm, tightly enclosed by anthoecium.

Naturalised from Europe.

N.; S.: throughout; St.; Ch., A., C. Lowland to montane in grassland and pasture, on roadsides and in waste places.

48. P. xenica Edgar et Connor, *N.Z. J. Bot. 37*: 65 (1999); Holotype: CHR 514884a! *G. Jane* Riwaka River, bluffs in South Branch, 600 m, 3 Dec 1997; ♂.

Coarse long-leaved, dioecious, extravaginally branching often pendulous grass with elongate internodes and rooting at nodes; leaf-blades persistent. Leaf-sheath 10 cm, keeled, ribbed, open to base, becoming dull brown and fragile, finely retrorsely hairy, margin membranous, darker brown. Ligule 1 mm, ciliate, abaxially finely hairy. Collar conspicuous, margin short hairy. Leaf-blade to 80 cm × 3–4 mm, coriaceous, folded below and at apex, ± flat elsewhere, abaxially glabrous with some long hairs near collar, adaxially clothed with many small antrorse hairs, denser near ligule; margins very sparsely prickle-toothed below becoming almost smooth but apex prickle-toothed and sharp-pointed. Culm to 85 cm, many noded, erect or with some geniculate nodes, nodes swollen, coloured, glabrous; internodes glabrous. Panicle to 25 cm, open, violet-suffused, subtended by bract to 2.5 mm; several to many solitary rarely binate branches at internodes, naked below with solitary spikelets, branches > internodes; rachis glabrous, branches and pedicels glabrous to sparsely shortly prickle-toothed. Spikelets 10–12 mm × 2 mm, gaping at anthesis, (1)–3–4–6 widely separated florets. Glumes unequal, centrally green, violet elsewhere, nerves elevated; lower 2.5–3.5 mm, 1–3-nerved, upper 3.5–5 mm, 5-nerved (3 long, 2 short) adaxially shortly hairy at apex, margins ciliate. Lemma 4.5–6 mm, 5-nerved, centrally green, violet elsewhere, abundantly finely pubescent throughout, keel prickle-toothed above, margins ciliate, membranous above, apex recurved, shortly lobed and mucronate (0.05 mm or 0) becoming erose. Palea 4.5–5.2 mm, ≤ lemma, apex bifid, keels minutely stiff hairy. Callus 0.25 mm, short, blunt, glabrous or with a few prickle-teeth. Rachilla 1.5 mm, glabrous or with a few prickle-teeth at base; prolonged. Lodicules 0.5–1.0 mm, acutely lobed, tip ciliate. Anthers (a) male flowers: 2.75–3–3.75 mm, yellow or violet-suffused; (b) female flowers: 2–2.2 mm, white, pollenless. Gynoecium (a) male flowers: 0.8–1.0 mm; (b) female flowers: ovary 0.8–1.2 mm, glabrous; stigma-styles 2–3–3.5 mm, widely disposed, stigmatic hairs almost to base. Caryopsis not seen.

Endemic.

S.: Nelson (South Branch Riwaka River). Scree at foot of steep marble bluffs and in shrublands; known only from this area; 500–700 m.

Ovaries, and these resemble caryopses, are infected by the smut fungus *Tilletia cathcartae*, also known in New Zealand on three specimens of *Poa pusilla* (E. H. C. McKenzie *in* Edgar and Connor 1999 op. cit. p. 67).

G. Jane (*in litt.*) remarks that this species is known only from four particular sites on very steep marble bluffs and the scree below. A preference for damp shaded sites is detected. The populations are quite large, inaccessible, and not imminently endangered.

ζ**P. alpina** L.   Tufts 10–30 cm, with mostly basal leaves and thickened at base with fibrous remains of leaf-sheaths; branching intravaginal; leaf-blades persistent. Ligule 3 mm, erose. Leaf-blade to 3.5 cm × 3 mm, stiff, flat or folded, smooth.

Panicle 3–5 cm, lax; spikelets ovate, often proliferous, 3.5–4 mm, on slender, almost smooth branches. Glumes subequal, acute, upper ½ of midnerve sparsely prickle-toothed; margins wide, membranous. Lemma 5-nerved, hyaline, acute, nerves densely long hairy in lower ½, midnerve scabrid above hairs. Callus with a few wispy hairs. Anthers 2 mm. A European species only known from two early, but undated collections (CHR 5098 and CHR 5103 *J. B. Armstrong* fields near Christchurch, "rare"; and WELT 68265 *A. C. Purdie* head of Lake Wakatipu, 2000 ft).

ζ**P. remota** Forselles   Tufted. Leaf-sheath scabrid above. Ligule *c.* 2 mm, obtuse, entire. Leaf-blade 3–4 mm wide, soft, flat, persistent. Panicle 16 cm; branches few, paired, long, drooping, filiform, scabrid, each tipped by a single, pale greenish brown spikelet. Spikelets 4–5 mm, 2-flowered. Glumes unequal; lower *c.* 2 mm, upper *c.* 3 mm. Lemma 5-nerved, obtuse, nerves finely scabrid; margin wide, hyaline. Palea ciliate-scabrid on keels. Callus glabrous. Anthers *c.* 0.6 mm. A European species only known from one collection, WELT 68720 *D. Petrie* Nevis Valley, 2000 ft, Feb. 1890.

REPRODUCTIVE BIOLOGY   Anton, A. M. and Connor, H. E.   *Aust. J. Bot. 43*: 577–599 (1995) is a comprehensive survey of floral biology and reproduction in the genus. The majority of N.Z. spp. of *Poa* are ♀, but a few dioecious spp., without floret dimorphism, and some gynomonoecious spp. are also present. Dioecious spp. are: *P. foliosa, P. schistacea, P. subvestita, P. sudicola,* and *P. xenica.* Gynomonoecious spp. are: *P. cookii, P. novae-zelandiae, P. ramosissima,* and naturalised *P. annua.*

Five collections of *P. foliosa* from the Stewart Id area were ♀ rather than dioecious. Anthers in ♀ plants were 2–2.5 mm, stigma-styles *c.* 1.5 mm, and seed was set. There seemed to be no vegetative or inflorescence differences between ♀, ♂ or ♀ specimens.

For dioecious *P. sudicola* there is only one collection of ♂ plants (CHR 369893 from Pike Peak). Collections from Matiri Range and Turks Cap Range are ♀ plants with developing but immature caryopses.

Connor, H. E.   *N.Z. J. Sci. Tech. 38A*: 742–751 (1957), demonstrated that *P. colensoi* was self-incompatible, *P. cita* (as *P. caespitosa*) inefficiently self-compatible, and *P. breviglumis* in all probability self-compatible.

All spp. in N.Z. flower chasmogamously. Species with short anthers, *P. breviglumis, P. buchananii, P. imbecilla, P. incrassata, P. lindsayi, P. maniototo, P. matthewsii, P. senex, P. tonsa,* have tendencies to cleistogamy and all seem very self-fertile.

CYTOLOGY   Chromosome counts for indigenous spp. were reported by Hair, J. B. *N.Z. J. Bot. 6*: 267–276 (1968). Most are tetraploid, $2n = 28$, but species are also 6x, 8x, 12x, 16x and *c.* 38x. Connor, H. E. and Edgar, E. in Barlow, B. A.

(Ed.) 1986 (*Flora and Fauna Alpine Australasia* 413–434) reported two evident polyploid series within related spp., one within the wide-leaved tussock spp. of southern and subantarctic islands, i.e., *P. foliosa*, 4*x* (2*n* = 28), and *P. tennantiana*, 8*x* (2*n* = 56), the second series occurring in the folded- or narrower-leaved tussock spp., i.e., *P. anceps*, 4*x* (2*n* = 28), *P. cita*, 12*x* (2*n* = 84), *P. cockayneana* and *P. chathamica*, 16*x* (2*n* = 112) and finally *P. litorosa*, *c.* 38*x* (2*n* = *c.* 266). So far as is known, the count for *P. litorosa* is the highest chromosome number yet recorded in Gramineae or in the Monocotyledons as a whole [Hair, J. B. and Beuzenberg, E. J. *Nature 189*: 160 (1961)].

The chromosome count, 2*n* = 28, recorded for *P. exigua* by Hair, J. B. (1968 op. cit.) was based on a plant collected from Mt Potts, S. Canterbury, CHR 102397. The specimen has not been traced, and, as Mt Potts is outside the known range of *P. incrassata* (= *P. exigua*) the chromosome count must be regarded as doubtful. A count of 2*n* = 29, recorded for *P. foliosa* on Macquarie Id by Moore, D. M. *Bot. Not. 113*: 187 (1960) is also regarded as doubtful.

HYBRIDS   Nine hybrid combinations between indigenous spp. of *Poa* were recorded by Edgar (1986 op. cit. pp. 487–495): *P. astonii* × *P. foliosa*, Stewart Id region (≡ *P.* ×*poppelwellii* Petrie, *T.N.Z.I. 46*: 38 (1914); Holotype: WELT 66222! *D. Petrie ex D. L. Poppelwell* grown on [from] a live plant from Herekopere Islet, early Dec. 1912); *P. astonii* × *P. tennantiana*, Stewart Id region, 2*n* = 42; *P. aucklandica* subsp. *campbellensis* × *P. foliosa*, Campbell Id (≡ *P. novae-zelandiae* var. *desiliens* Zotov, *Rec. Dom. Mus. 5 (15)*: 127 (1965); Holotype: CHR 119433! *V. D. Zotov* Campbell Is., Mt Lyall, 6.1.1961); *P. breviglumis* × *P. colensoi*, Ruahine Mts, Mt Stokes, Garvie Mts; *P. colensoi* × *P. novae-zelandiae*, Tararua Ra., South Id mountains; *P. hesperia* × *P. subvestita*, Fiordland; *P. imbecilla* × *P. lindsayi*, Canterbury, North Otago; *P. lindsayi* × *P. novae-zelandiae*, Mt Cook; *P. maniototo* × *P. pygmaea*, Pisa Ra.

All are sterile; the anthers were either pollenless, or pollen-sterile, or only slightly pollen-fertile, the stigmas were usually imperfect and no caryopses were seen.

Edgar (1986 op. cit. p. 493) cited the distribution of *P. hesperia* × *P. subvestita* as Fiordland and western Otago; specimens from western Otago are now referred to *P. schistacea.*

Two further putative hybrids are reported here:

P. colensoi × *P. pygmaea*
Small dense tufts, 3–4 cm, with short, stiff, rolled leaves; some leaf-blades disarticulating. Panicle 1–1.5 cm, of 3–4 spikelets. Lemma with very short fine hairs on lower ½. Anthers 1.2–1.3 mm, pollen-sterile. Gynoecium with well-developed stigmas.
Only two specimens seen: CHR 395475 *A. P. Druce* Mt Pisa, 6300 ft, herbfield, 1988; CHR 394597 *A. P. Druce* Mt Pisa, 6400 ft, rocky ground, 1989.

P. kirkii × *P. subvestita*

Rhizomatous tufts 20–40 cm, with narrow, folded, rather stiff leaves. Ligule 1–4 mm. Panicles erect, ± pyramidal, 4–6 cm; branches glabrous. Spikelets 5–6 mm, 3–4-flowered, purplish. Lemma 5-nerved, glabrous apart from short hairs on lower ⅓ of keel and near base of outer lateral nerves. Callus with tuft of crinkled hairs. Anthers 0.5–0.7 mm, pollen-sterile. Gynoecium 1–2 mm. Upper florets sometimes with anthers *c.* 0.3 mm and gynoecium 0.5 mm.

Specimens seen: CHR 394596, CHR 395184 *A.P. Druce* Pisa Range, wet seepages, 1989, and CHR 395204 *A.P. Druce* Harris Mts, wet seepage, 1989.

A sterile *Poa*, *c.* 70 cm, CHR 223383 *G. C. Kelly* hanging down almost vertical cliffs at Dan Rogers Creek, Akaroa Harbour, 1971, may be a hybrid, *P. anceps* × *P. astonii*, or *P. anceps* × *P. colensoi*; branching extravaginal; leaf-blades narrow, inrolled; panicle 6 cm, shortly branched; spikelets *c.* 7 mm; glumes and lemmas acute to acuminate and glabrous, callus with copious long crinkled hairs; anthers 1.5–2.5 mm, somewhat pollen-fertile; gynoecium 0.1 mm, without stigmas.

UNASSIGNED OR EXCLUDED NAMES

Poa laevis var. α *brevifolia* Hook.f., *Fl. N.Z. 1*: 307 (1853): no specimens so labelled were found at Kew; the name may refer to *P. colensoi*.

Poa hypopsila Steud., *Syn. Pl. Glum. 1*: 263 (1854): no specimen could be found at P; the original locality cited is N.Z. and no collector is given.

Eragrostis eximia Steud., *Syn. Pl. Glum. 1*: 279 (1854): Steudel cited *Urville* as the original collector, from "N. Holl. N. Zeelandia"; the type specimen at P is a single panicle of *Poa*, labelled in Steudel's hand "Bromus *Urville* N. Zeelandia"; glumes and lemmas are longer, narrower, more acute and more prominently nerved than in N.Z. spp. and the name probably refers to an Australian species.

PUCCINELLIA Parl., 1850, *nom. cons.*

Perennials, biennials, or annuals. Leaf-sheath open. Ligule membranous. Leaf-blade flat, folded, or involute. Inflorescence an open or contracted panicle. Spikelets 2–several-flowered, pedicelled, laterally compressed or almost cylindric; disarticulation above glumes and between florets; rachilla prolonged, glabrous. Glumes usually unequal and < adjacent lemmas, obtuse or acute, rounded, awnless, herbaceous to membranous, margins hyaline; lower 1–(3)-nerved, upper 3–(5)-nerved. Lemma 5–(7)-nerved, rounded, usually oblong, obtuse or sometimes acute, awnless, often pubescent near base especially on nerves, or rarely completely glabrous; apex usually scarious or hyaline and ± erose. Palea ≈ lemma, keels scaberulous to ciliate, rarely slightly excurrent, apex shortly bifid. Callus glabrous. Lodicules 2, membranous, lanceolate, acute. Stamens 3. Ovary glabrous; styles free. Caryopsis oblong; embryo small; hilum subbasal, punctiform or shortly elliptic. Chasmogamous or cleistogamous.

Type species: *P. distans* (L.) Parl.

c. 80 species of temperate regions throughout the world, but principally in Asia; usually in coastal salt marshes, or inland on saline or alkaline soils. Endemic spp. 3, indigenous spp. 1 shared with Australia; naturalised spp. 3.

The N.Z. spp. were revised by Edgar, E. *N.Z. J. Bot. 34*: 17–32 (1996), and earlier by Allan, H. H. and Jansen, P. *T.R.S.N.Z. 69*: 265–269 (1939).

1 Leaf-blade involute; lemma glabrous or with minute hairs near base (visible only at high magnification *c.* ×40); palea keels finely scabrid *2*
Leaf-blade flat or folded, sometimes with involute margins; *either* lemma with basal hairs visible at ×10 *or* lemma ± glabrous but palea keels conspicuously ciliate at midway *3*

2 Branching intravaginal; panicle overtopping leaves; upper glume usually 2–3.5 mm .... **6. stricta**
Branching extravaginal; panicle rarely formed, but if present then overtopped by leaves and upper glume < 1.5 mm .... **4. raroflorens**

3 Lemma 2.7–5 mm; upper glume 2–5 mm *4*
Lemma 1.5–2.6 mm; upper glume 1–2 mm *6*

4 Panicle ovate or oblong, strongly secund .... **5. rupestris**
Panicle linear-lanceolate, fusiform, or rarely with spreading branches *5*

5 Leaf-sheath firmly membranous, straw-coloured to greyish or greenish brown or purplish; leaf-blade scabrid adaxially and on margins; palea keels not conspicuously ciliate at midway .... **7. walkeri**
    5a Palea keels shortly excurrent .... subsp. **antipoda**
        Palea keels reaching only to apex *5b*
    5b Panicle branches scabrid .... subsp. **walkeri**
        Panicle branches smooth .... subsp. **chathamica**
Leaf-sheath hyaline, whitish; leaf-blade completely smooth; palea keels conspicuously ciliate at midway .... **3. macquariensis**

6 Lemma subobtuse with midnerve reaching apex or often slightly excurrent;
　panicle branches erect, sometimes spreading at maturity .... **2. fasciculata**
　Lemma truncate with midnerve not reaching broad hyaline upper margin;
　panicle branches deflexed at maturity .... **1. distans**

1. P. distans (L.) Parl., *Fl. Ital. 1*: 367 (1850).　　　reflexed salt grass

Perennial, erect or prostrate tufts, 8–80 cm; branching intravaginal.
Leaf-sheath glabrous, firmly membranous, light brown, sometimes
purplish near base. Ligule 1–1.5–(2) mm, entire and rounded, rarely
tapered and subacute. Leaf-blade 3–6.5–(12) × 1–3 mm, flat, or folded
with inrolled margins, abaxially glabrous, adaxially shallowly ribbed,
ribs scabrid; margins sparsely scabrid, tip scabrid, acute or hooded.
Culm (5)–10–60 cm, enclosed by uppermost leaf-sheath at flowering,
later visible, internodes glabrous. Panicle (3)–10–24 × 2–18 cm, ovate
to ± triangular, very lax and open; branches finely scabrid, naked below,
later spreading to deflexed. Spikelets 3–6 mm, 3–7-flowered, greenish
or purplish. Glumes ± unequal, minutely scabrid at obtuse apex; lower
0.7–1.2 mm, 1-nerved, elliptic-lanceolate, upper 1.1–1.8 mm, 3-nerved,
ovate. Lemma 1.5–2.4 mm, indistinctly 5-nerved, with short hairs at
base and on nerves near base, broad-elliptic or ovate, ± membranous
with nerves not reaching wide hyaline upper margin; apex minutely
ciliate, broadly obtuse to almost truncate. Palea ≈ lemma, keels ciliate-
scabrid in upper ⅔, apex shortly bifid. Rachilla 0.6–0.8 mm. Anthers
0.6–1 mm. Caryopsis 1–1.7 × 0.3–0.6 mm.

Naturalised.

N.: Auckland City, Bay of Plenty (Ohiwa Harbour), Hawkes Bay
(Porangahau), inland at Te Aroha; S.: Nelson city and environs, North
Canterbury and Banks Peninsula, near Dunedin, inland in Central
Otago. Mud flats and salt marsh at sea level; inland in salt pans, 100–
400 m.

Puccinellia distans is indigenous to temperate regions of the Northern Hemisphere.
It is well distinct from other spp. of *Puccinellia* in N.Z. by the truncate lemmas with
broad hyaline band at apex into which the nerves do not extend.

Sterile and fertile plants of *P. distans* grow together at Okains Bay, Banks Peninsula
A. J. *Healy* 20 Feb 1945 (CHR 48699a,b, CHR 48805, CHR 48806 — sterile; CHR

48739 — fertile). Sterile plants are 35–50 cm with flat to folded leaves, panicles with fine, widely spreading branches, lemmas *c.* 3 mm, anthers 0.7–1 mm, mature caryopses absent. In CHR 48806 sterile and fertile shoots were entwined.

2. P. fasciculata (Torr.) E.P.Bicknell, *Bull. Torrey Bot. Club 35:* 197 (1908).   salt grass

Perennial, yellowish green or glaucous tufts, (3)–10–65 cm, with culms often much overtopping leaves; branching intravaginal. Leaf-sheath glabrous, subcoriaceous, greyish to straw-coloured. Ligule 0.5–1.5–(2) mm, truncate to rounded, or shortly tapered. Leaf-blade (3)–5–12 cm × (1.5)–2–3.5 mm, flat or folded, abaxially glabrous, adaxially shallowly ribbed, ribs scaberulous; margins scaberulous, tip hooded. Culm 15–40 cm, erect or spreading, internodes glabrous. Panicle (1.5)–4–15 × 0.4–5–(7) cm, lanceolate, or narrow-oblong to ovate, usually contracted, sometimes more open; branches scabrid, all, or at least the shortest, bearing densely clustered spikelets to the base. Spikelets 3.5–5.5 mm, (3)–4–5–(8)-flowered, greyish green, often purple tinged. Glumes ± unequal, elliptic to ovate, obtuse to subacute, midnerve often finely scabrid above, apex minutely ciliate; lower 0.7–1.5–(1.8) mm, 1-nerved, upper 1.2–2 mm, 3-nerved. Lemma 1.7–2.6 mm, 5-nerved, with short hairs at base and on lateral nerves for a short distance above base or sometimes on lower half, elliptic-oblong, apex subobtuse, minutely ciliate, midnerve minutely scabrid near lemma apex, often minutely excurrent Palea ≈ lemma, apex shallowly bifid, keels ciliate-scabrid in upper ½–⅔. Rachilla 0.4–0.8 mm. Anthers 0.5–0.9 mm. Caryopsis 1.2–1.5 × 0.4–0.6 mm.

Naturalised.

N.: Gisborne (Wherowhero Lagoon), Hawkes Bay (lagoons near Wairoa, near Napier); S.: scattered on the eastern coast from Christchurch to Invercargill; inland in North and Central Otago; St. Mud flats and salt marsh at sea level; inland in salt pans, 100–600 m.

Indigenous to North America and Europe.

Allan and Jansen (1939 op. cit.) treated all N.Z. plants of *P. fasciculata* as indigenous; either their var. *novozelandica*, from the Otago coast between Waikouaiti and Dunedin, or var. *caespitosa* from "alkali patches" in Central Otago. Further collecting reveals that plants growing inland in Central Otago cannot be delimited from *P. fasciculata* growing elsewhere.

Allan and Jansen (1939 op. cit. p. 268) described two varieties of *P. fasciculata* based on New Zealand plants; their species *P. scott-thomsonii* was also described from New Zealand plants of *P. fasciculata* (Edgar 1996 op. cit. p. 29). No status is accorded to the following taxa; all are regarded as synonyms of *P. fasciculata*.

P. fasciculata var. *novozelandica* Allan et Jansen, *T.R.S.N.Z. 69*: 268 (1939); Lectotype: L 956.180 097 (Herb. Jansen et Wachter 44765)! *J. Scott Thomson* Tomahawk Lagoon, near Dunedin, South Island, Jan 1936 (designated by Edgar 1996 op. cit. p. 29).

P. fasciculata var. *caespitosa* Allan et Jansen, *T.R.S.N.Z. 69*: 268 (1939); Lectotype: WELT 68537! *D. Petrie* Chatto Creek, Manuherikia Plain, Vincent Co., Otago, 30 Dec 1910 (designated by Edgar 1996 op. cit. p. 29).

P. scott-thomsonii Allan et Jansen, *T.R.S.N.Z. 69*: 267 (1939) ≡ *P. fasciculata* var. *scott-thomsonii* (Allan et Jansen) Zotov, *T.R.S.N.Z. 73*: 236 (1943); Lectotype: CHR 17120! *J. S. Thomson* Waikouaiti, 29 Dec 1935 (designated by Edgar 1996 op. cit. p. 29).

3. P. macquariensis (Cheeseman) Allan et Jansen, *T.R.S.N.Z. 69*: 268 (1939) ≡ *Triodia macquariensis* Cheeseman, *Aust. Antarct. Exped. 1911–14, Sci. Rep.* Ser. C, 7 (3): 34 (1919); Lectotype: AK 1732! *H. Hamilton* Macquarie Island, coastal form only found near sea [designated by Edgar, E. *in* George, A. S. et al. (Eds) *Fl. Australia 50*: 572 (1993)].

Perennial tufts, 4–25 cm, with culms ± hidden by leaf-sheaths and panicles overtopped by the soft, dull green leaves; branching intravaginal. Leaf-sheath glabrous, hyaline, whitish green, much wider than leaf-blade. Ligule 0.7–1.5 mm, erose. Leaf-blade 2–8 cm × *c.* 1 mm, flat or folded, sometimes with inrolled margins, glabrous throughout, tip obtuse. Culm erect, or geniculate at base, internodes glabrous. Panicle 1.5–6 × 0.5–1 cm, lanceolate; branches few, short, erect, sharply angled, smooth to sparsely scabrid. Spikelets 4–6.5–(8) mm, 3–5-flowered, pale green, sometimes purplish. Glumes ± unequal, elliptic-ovate to elliptic-oblong, obtuse; lower 1.6–2.6 mm, 1–3-nerved, upper 2–3.5 mm, 3–(5)-nerved. Lemma 3–4 mm, 5-nerved, elliptic-ovate to elliptic-oblong, glabrous or with a few hairs at base and on nerves near base, midnerve almost reaching obtuse apex. Palea ≈ lemma, keels conspicuously long-ciliate at midway with shorter prickle-teeth in upper ¼, and glabrous in lower ¼, apex truncate to bifid, keels not excurrent. Rachilla 0.7–1.2 mm. Anthers 0.4–0.9 mm. Caryopsis 1.5–1.7 × 0.6–0.8 mm.

Endemic.

M. Common in dense patches on coastal rock stacks and cliffs.

Meurk, C. D. *N.Z. J. Bot. 13*: 721–742 (1975) recorded *P. macquariensis* from near Courrejolles Point, Campbell Id *D. V. Merton* 27 Feb 1971 (CHR 202761, CANU 15446, OTA 30790). A flowering specimen *D. R. Given 9373* 5 Feb 1976 (CHR 284730) confirms *P. walkeri* subsp. *chathamica* rather than *P. macquariensis.*

4. P. raroflorens Edgar, *N.Z. J. Bot. 34*: 22 (1996); Holotype: CHR 402693! *B. Patrick 3* Otago Land District, Alexandra, Conroys Road, salty soil patch on side of small valley in dry rolling country, 280 m, 19 Nov 1993.

Low-growing perennial to 4.5 cm, sometimes forming a loosely woven mat to 2 m diam. or more, usually almost entirely covered by soil, rarely flowering and only the short very narrow dull green curved leaves visible; branching extravaginal. Leaf-sheath glabrous, submembranous, much wider than leaf-blade, ribs few, distinct. Ligule 0.2–0.6 mm, subhyaline, obtuse or truncate. Leaf-blade 1–3 cm × 0.2–0.5 mm, setaceous, involute, glabrous, subacute, margins minutely scaberulous. Culm entirely hidden by leaf-sheaths. Panicle rarely present, 10–16 mm, overtopped by leaf-blades, bearing up to 12 spikelets, lowermost branches enclosed by leaf-sheaths, smooth, uppermost branches scaberulous. Spikelets 3–4–(4.8) mm, 4–6-flowered, green to brownish green. Glumes unequal, ovate, obtuse, submembranous; lower 0.6–1 mm, 1-nerved, upper 1.1–1.4 mm, 3-nerved. Lemma 1.8–2.5 mm, 5-nerved, ovate-elliptic, entirely glabrous, midnerve ± reaching subacute apex. Palea ≈ lemma, keels scabrid in upper ⅔. Rachilla 0.4–0.5 mm. Anthers 0.4–0.6 mm. Caryopsis 1.2–1.6 × 0.6–0.7 mm.

Endemic.

S.: Central Otago; St. (Paterson Inlet). In saltpans, salty slicks and scarps from 100–600 m, in the barest, saltiest ground, in Otago; in stony low turf on tidal bank at sea level on St.

Flowering specimens are not often seen, those that have been collected may bear panicles in profusion, e.g., CHR 190047 *A. J. Healy 68/389* Conroy's Gully, near Alexandra, Central Otago, 10/11/1968; CHR 508522 *B. H. Patrick* Conroy's Rd, near Alexandra, Central Otago, 20 Oct 1995.

5. P. rupestris (With.) Fernald et Weath., *Rhodora 18*: 10 (1916).

Annual or biennial tufts, (6)–10–35 cm, with procumbent to ascending culms and wide leaves; branching intravaginal. Leaf-sheath glabrous, firmly membranous, closely ribbed, light green to creamish brown. Ligule (1.2)–1.5–2 mm, truncate to ± acute centrally. Leaf-blade 2.5–6 cm × 2–5 mm, flat or folded, abaxially finely scabrid on midrib, adaxially scabrid on ribs; margins scabrid, tip hooded. Culm (5)–15–30 cm, overtopping leaves, internodes glabrous. Panicle (2.5)–4–8 × (0.7)–1–5 cm, ovate or oblong, stiff, strongly secund; branches short, scabrid on angles, bearing close-set, very shortly pedicelled spikelets along one side almost to base. Spikelets 4.5–5.5–(7) mm, 3–4-flowered, green. Glumes unequal, strongly nerved, ovate-elliptic, apex minutely scabrid, obtuse, or midnerve sometimes slightly excurrent; lower 1.3–2.2 mm, 3-nerved, upper 2–3 mm, 3–5-nerved. Lemma 2.7–3.5 mm, strongly 5-nerved, with short hairs at base and on lateral nerves near base, elliptic, firm, margin hyaline, apex minutely scabrid, obtuse, or midnerve sometimes very slightly excurrent. Palea = lemma, keels ciliate-scabrid in upper ¾, apex ± entire, shallowly bifid. Rachilla ⩾ 1 mm. Anthers 0.8–1.2 mm. Caryopsis 1.6–2 × 0.5–0.7 mm.

Naturalised from western Europe.

S.: Otago, known only from Tomahawk Lagoon and Company Bay, Otago Peninsula. Damp muddy coastal lagoon margin and on harbour-edge road embankment at sea level.

6. P. stricta (Hook.f.) Blom, *Acta Hort. Gothob. 5*: 89 (1930) ≡ *Glyceria stricta* Hook.f., *Fl. N.Z. 1*: 304 (1853) ≡ *Atropis stricta* (Hook.f.) Hack., *in* Cheeseman *Man. N.Z. Fl.* 914 (1906) ≡ *Puccinellia stricta* (Hook.f.) Blom forma *stricta* (autonym, Allan and Jansen 1939 op. cit. p. 266); Lectotype: K! *Gunn 1463* [Tasmania] Launceston, marsh, 24.12.1844 (designated by Edgar 1996 op. cit. p. 23). = *Atropis stricta*

var. *suborbicularis* Hack., *in* Cheeseman *Man. N.Z. Fl.* 915 (1906)
≡ *Puccinellia stricta* var. *suborbicularis* (Hack.) Allan et Jansen,
T.R.S.N.Z. 69: 266 (1939); Holotype: W 10913! *D. Petrie* Oamaru,
North Otago (No 1483 to Hackel). = *P. stricta* forma *luxurians* Allan
et Jansen, *T.R.S.N.Z. 69*: 266 (1939); Lectotype: L 956.179 578
(Herb. Jansen et Wachter no. 44268)! *D. Petrie* salt meadow at
Thames, Auckland Province, North Island [Nov 1904] (designated
by Edgar 1996 op. cit. p. 23). = *P. stricta* forma *pumila* Allan et Jansen,
T.R.S.N.Z. 69: 266 (1939); Lectotype: L 956.179 583 (Herb. Jansen
et Wachter no. 44270)! *V. D. Zotov* Island Bay, Wellington, North
Island, in soil filled pockets of rock near high-tide mark, [10] Dec
1928 (designated by Edgar 1996 op. cit. p. 23).

Light bluish green or rarely pale yellow-green perennial tufts, (2.5)–10–
50–(65) cm, with stiff culms and leaves, or with finer and less rigid leaves;
branching intravaginal. Leaf-sheath glabrous, submembranous to
subcoriaceous, ± distinctly ribbed, light brownish to purplish. Ligule
0.7–1.5–(2) mm, glabrous, rounded to truncate, or ± tapered centrally
and acute. Leaf-blade 1–12 cm × *c.* 0.5 mm diam., involute, rigid and
erect, or finer and softer, abaxially glabrous, adaxially sparsely scabrid
on ribs, or sometimes densely scabrid throughout; margins scabrid,
narrowed to fine acute tip. Culm (2)–5–20–(40) cm, erect, hidden by
uppermost leaf-sheath at flowering, later visible, internodes glabrous.
Panicle (2)–4.5–14.5–(20) × 0.2–5.5 cm, at first narrow-linear, racemose
above, with few, erect, usually scabrid branches below, later more open
with ± spreading branches naked below. Spikelets 3.5–9–(10.5) mm, 2–
10-flowered, narrow, almost terete, light green to purplish. Glumes often
quite unequal, elliptic-oblong, margins and sometimes midnerve
minutely ciliate near apex; lower (0.6)–0.8–2.3 mm, 1-nerved, subacute
to subobtuse, upper (1.3)–2–3.5 mm, 3-nerved, subobtuse to obtuse.
Lemma (2)–2.5–3–(4) mm, 5-nerved, broad-elliptic, with very minute
hairs at base not usually visible at ×10, and occasionally a few minute
hairs on nerves near base, midnerve usually not quite reaching finely
ciliate, obtuse apex. Palea ≤ lemma, apex bifid, keels scabrid in upper
⅔. Rachilla 0.5–1 mm. Anthers 0.4–1 mm. Caryopsis 1–1.8 × 0.4–0.7 mm.

Indigenous.

N.: southwards from Auckland City, but not recorded from Bay of Plenty, Gisborne, or Taranaki; S.: coastal Nelson, and on eastern to south-eastern coasts throughout, inland in Otago near Sutton, and in Central Otago; St. Salt marsh, and sandy or stony ground at high tide level; inland in salt pans, 100–600 m.

Also indigenous to Australia.

Puccinellia stricta is very variable in height and appearance. Allan and Jansen (1939 op. cit. p. 266) separated N.Z. plants into two forms of widespread distribution: forma *luxurians*, with culms 30–40 cm and leaf-blades to 15 cm, and forma *pumila* with culms ± 10 cm and shorter leaf-blades. It is impossible to delimit two distinct groups in N.Z. on that basis.

Allan and Jansen (1939 op. cit. p. 265) indicated for *P. stricta* that "Gunn's specimens [at K]... must be taken as the type". Although Edgar (1996 op. cit. pp. 23–24) stated that Allan and Jansen chose the lectotype, they did not, in fact, cite the actual specimen.

7. P. walkeri (Kirk) Allan, *N.Z. DSIR Bull. 49*: 157 (1936) ≡ *Poa walkeri* Kirk, *T.N.Z.I. 17*: 224 (1885) ≡ *Atropis walkeri* (Kirk) Cheeseman, *Man. N.Z. Fl.* 203 (1925); Lectotype: WELT 66495! *T. Kirk* The Old Neck, Stewart Island, 31 Dec 1883 (designated by Edgar 1996 op. cit. p. 25). = *Glyceria novae-zealandiae* Petrie, *T.N.Z.I. 33*: 329 (1901) ≡ *Atropis novae-zealandiae* (Petrie) Hack., *in* Cheeseman *Man. N.Z. Fl.* 915 (1906) ≡ *Puccinellia novae-zealandiae* (Petrie) Allan et Jansen, *T.R.S.N.Z. 69*: 266 (1939); Lectotype: WELT 68550! *T. Kirk* Riverton, Jan 1887 (No 1477 to Hackel) (designated by Edgar 1996 op. cit. p. 25).

Erect, bluish green or light green to pale brownish green, stiff, dense, very leafy tufts, (6)–10–50 cm, with uppermost leaves usually overtopping culms; branching intravaginal. Leaf-sheath glabrous, firmly membranous, striate. Ligule 0.6–3 mm, obtuse or truncate, or sometimes tapered centrally and acute, glabrous, entire. Leaf-blade 5–25 cm × 5–10 mm, folded, or sometimes almost flat and wider, abaxially glabrous, adaxially shallowly ribbed, ribs sparsely scabrid; margins scabrid, tip smooth, ± firmly acute, sometimes subobtuse and ± apiculate. Culm usually hidden by sheaths, internodes glabrous. Panicle ± overtopped by leaves, linear-lanceolate, erect, contracted, dense; branches stiff, erect, sometimes ± spreading at maturity. Spikelets (3)–5–9 mm, 2–5-flowered, bluish green or sometimes purplish. Glumes ± unequal, narrow-lanceolate to elliptic-oblong, subacute; lower 1.5–3.6–(4.2) mm, 1–3-nerved, upper 2–4.5–(5) mm, 3–5-nerved. Lemma 3–5 mm, 5–7-nerved, elliptic-oblong, usually with a few hairs at base and on nerves near base or to *c.* midway. Palea ≤ lemma. Rachilla 0.8–1.5–(1.8) mm. Anthers (0.6)–0.8–1.5 mm. Caryopsis (1.5)–1.8–2.6 × 0.4–0.8 mm.

subsp. **walkeri**

Plants (9.5)–15–50 cm. Leaf-sheath straw-coloured to light greenish brown or grey-brown, sometimes purplish. Panicle (3)–6–17 × 0.5–2.5–(11) cm; branches scabrid, acute-angled. Lemma apex minutely ciliate-scabrid, subobtuse to obtuse, midnerve not excurrent. Palea keels scabrid in upper ½ – ⅔, reaching only to apex.

Endemic.

Cook Strait (The Brothers); S.: at scattered localities on the eastern and southern coasts from Banks Peninsula to Riverton; St. In salt marshes and on saline ground in estuaries; also on sandy and stony ground at high tide level.

Allan and Jansen (1939 op. cit. p. 266) treated *Poa walkeri* Kirk as a *nomen nudum*; Kirk's diagnosis distinguishes this taxon from other puccinelliae in N.Z.

subsp. **antipoda** (Petrie) Edgar, *N.Z. J. Bot. 34*: 26 (1996) ≡ *Atropis antipoda* Petrie, *in* Chilton *Subantarct. Is N.Z.* 2: 480 (1909) ≡ *Puccinellia antipoda* (Petrie) Allan et Jansen, *T.R.S.N.Z. 69*: 266 (1939); Lectotype: WELT 68542a! *B. C. Aston* Antipodes Island, Jan 1909 (designated by Edgar 1996 op. cit. p. 26).

Plants 6.5–25 cm. Leaf-sheath straw-coloured to light grey-brown or greenish brown. Panicle 2.5–8 × 0.4–1.5 cm; branches smooth to sparsely scabrid, sharply acute-angled. Lemma apex ciliate-scabrid, subobtuse, midnerve slightly excurrent. Palea keels scabrid in upper ½, excurrent.

Endemic.

Ant.: coastal and common.

subsp. **chathamica** (Cheeseman) Edgar, *N.Z. J. Bot. 34*: 26 (1996) ≡ *Atropis chathamica* Cheeseman, *Man. N.Z. Fl.* 203 (1925) ≡ *Puccinellia chathamica* (Cheeseman) Allan et Jansen, *T.R.S.N.Z. 69*: 266 (1939); Lectotype: AK 1978! *F. A. D. Cox* Chatham Islands (designated by Edgar 1996 op. cit. p. 27).

Plants (6)–10–40–(50) cm. Leaf-sheath light greenish brown, sometimes dark brown or purplish. Panicle 2–10–(13) × 0.5–1.5 cm; branches smooth, subacute-angled. Lemma apex minutely ciliate-scabrid, obtuse with slightly excurrent midnerve, very rarely entire; lemma usually 5-nerved, rarely 7-nerved. Palea keels scabrid in upper ⅓, very rarely slightly ciliate below, reaching only to apex.

Endemic.

Ch., A., C. In salt marsh or on stony shores and hill slopes exposed to the sea.

REPRODUCTIVE BIOLOGY   Species of *Puccinellia* are self-compatible [Connor, H. E. *N.Z. J. Sci. Tech. 38A*: 742–751 (1957); *N.Z. J. Bot. 17*: 547–574 (1979)]. New Zealand spp. are chasmogamous but in *P. raroflorens* some florets remain enclosed in the uppermost leaf-sheath and probably flower cleistogamically.

In collections determined as *P. scott-thomsonii*, and now regarded as *P. fasciculata*, the lemmas are *c.* 3 mm, and anthers indehiscent, either pollenless, or pollen sterile. It is possible that the sterility is caused by some disease especially since a similar phenomenon has been found in *P. distans* from Okains Bay.

SCHEDONORUS P.Beauv., 1812

Perennial and caespitose. Leaf-sheath split. Auricles falcate, clasping, hairy or glabrous. Ligule shortly truncate. Leaf-blade flat with many ribs. Culm unbranched, nodes glabrous. Inflorescence an open panicle of many-flowered laterally-compressed spikelets, disarticulation above glumes and between florets. Lemma lanceolate or oblong-lanceolate, rounded, variously prickle-toothed; awn from between membranous apical lobes, or awnless. Rachilla prolonged. Stamens 3; anthers caudate. Ovary apex glabrous; styles subterminal. Caryopsis adherent to palea.

Type species: *S. elatior* (L.) P.Beauv. *nom. specific. rejic.*

c. 5 spp. of temperate Eurasia. Naturalised spp. 2.

Schedonorus as a separate genus as treated by Holub, J. *Preslia 70*: 112–113 (1998) is preferred to retention in *Festuca* subgenus *Schedonorus*, the common style, or placement in *Lolium* as favoured by Darbyshire, S. J. *Novon 3*: 239–243 (1993). Modern molecular data are summarised by Soreng, R. J. and Terrell, E. E. *Phytologia 83*: 85–88 (1997) who opted for *Schedonorus*; the actual date of publication of their combinations is August 1998 (although *Phytologia* is dated 1997). Holub's combinations predate it, and priority lies with Holub including that for the nothogenus ×*Schedololium* Holub, *Preslia 70*: 111 (1998).

1. S. phoenix (Scop.) Holub, *Preslia 70*: 113 (1998).   tall fescue

Tall robust tussock forming perennial, commonly of river beds. Prophyll 2.5 cm or more, keels antrorsely prickle-toothed and with mixed short hairs. Branching intravaginal and extravaginal when shortly rhizomatous. Leaf-sheath 5–20 cm, striate, finely scabrid or smooth becoming scabrid near ligule. Ligule 0.5–1–3 mm, firm. Auricles clasping 0.7–1.5 mm, hairs *c.* 0.3 mm. Leaf-blade 10–100 cm × 5–10 mm wide, many-ribbed, midrib prominent, finely antrorsely scabrid below becoming very scabrid above, margins finely prickle-toothed. Culm to 1.5 m, stout, nodes conspicuous, dark, constricted, internodes glabrous. Panicle to 40 cm, erect or nodding, with 10 or more nodes of many spikelets; basal branches to 20 cm, binate, one long, naked below, other shorter with spikelets to base, other nodes with similar binate branches becoming shorter and eventually solitary near apex or solitary throughout and spikelets to base of branches; rachis smooth below becoming scabrid above pedicels and branches with prominent prickle-teeth. Spikelets 10–15 mm, narrow, of 4–7 florets. Glumes unequal, shortly awned, keel scabrid above otherwise smooth, margins membranous; lower 3–8 mm, 1-nerved, upper 4–9 mm, 3-nerved. Lemma 7–10 mm, 5-nerved, rounded, fine prickle-teeth on central and lateral nerves, margins membranous shortly toothed throughout broadly membranous above terminating in small (0.3–0.6 mm) hyaline prickled lobes; awn 1–2.5 mm, visible through apical lobes. Palea ≥ lemma, 7–10 × 1.5 mm wide, apex bifid ciliate, keels toothed ± to base, interkeel glabrous. Callus 0.25 mm, margin hairs 0 or few; articulation flat. Rachilla 1.5 –1.75 mm, stiff hairs scattered throughout. Lodicules 0.75–1.6 mm, > ovary, deeply to ± irregularly lobed, glabrous. Anthers 3–4 mm, caudate. Gynoecium: ovary 0.6–1.0 mm, apex glabrous; stigma-styles subterminal, 1.8–3.0 mm, hairs almost to base. Caryopsis 3 mm, obovate, adherent to palea; embryo 1 mm; hilum 2 mm.

Naturalised from Europe.

N.; S.; St.; Ch. River beds and banks, and waste places; grown in pastures for summer herbage.

Traditionally *S. phoenix* was treated here, and elsewhere, as *Festuca arundinacea* Schreb. The inapplicability of the epithet *arundinacea* in *Schedonorus* is illuminated by Holub (1998 loc. cit. p. 112) where the argument in favour of the name *S. phoenix* is set out.

In N.Z. herbaria there are many specimens labelled *Festuca pratensis* Huds.; nearly all fail the test of the simplest of characters for that taxon i.e. auricles glabrous, and the shorter of the binate inflorescence branches with 1–2 spikelets. Intuitively, it is easy to believe that *F. pratensis* is or was sporadic in N.Z.; we would include it in *Schedonorus* as *S. pratensis* (Huds.) P.Beauv.

Should a key be required, the following will satisfy:

1 Auricles hairy; shorter binate basal inflorescence branch with 3 or more spikelets .... **1. phoenix**
  Auricles glabrous; shorter binate basal inflorescence branch with 1–2 spikelets .... **2. pratensis**

×**Schedololium holmbergii** (Dörfl.) Holub, *Preslia 70*: 112 (1998). *Lolium perenne* × *Schedonorus phoenix*

Tall, openly caespitose sometimes ± rhizomatous naturally occurring hybrid. Leaf-sheath glabrous. Branching intravaginal and extravaginal. Ligule 1–1.5 mm. Auricles stiff, 1 or 2, clasping, hairs few or glabrous. Leaf-blade *c.* 30 cm × 4 mm, multiribbed, midrib conspicuous, adaxially finely prickle-toothed, abaxially shining and smooth, margins prickle-toothed. Culm 50–70 cm, ribbed, nodes and internodes glabrous. Panicle 12–20–(30) cm, strict of 10–20 nodes, branches binate, the longer with up to 5 spikelets or more the shorter with 1 spikelet though sometimes > 1, uppermost 5–8 with solitary spikelets; rachis and branches prickle-toothed on margins. Spikelets 10–20 mm of 4–8 florets. Glumes unequal; lower 4–6 mm, 3-nerved, upper 5–8 mm, 5-nerved of which 1–3 are

Plate 7. Some native species of *Poa*, a large genus of many habitats. **A** *P. imbecilla*, often below rocks in open scrub and forest (Otago); **B** *P. matthewsii*, in open dry forest (Otago); **C** *P. novae-zelandiae*, in alpine wet stony sites (Eyre Mts); **D** *P. cita*, silver tussock, on open sites with fertile soils (Central Otago); **E** *P. colensoi*, blue tussock, on rock and in mountain grasslands (Fiordland); **F** *P. lindsayi* (centre) and *P. maniototo* (right), both mainly in stony or bared ground in dry districts (Central Otago).

Plate 8. *Cortaderia* and *Chionochloa*. **A** *Cortaderia fulvida* (Kaikoura); **B** *Cortaderia richardii* (Fiordland); **C** *Chionochloa conspicua* subsp. *conspicua* (Fiordland); **D** *Chionochloa rubra* subsp. *cuprea* (mid-Canterbury); **E** *Chionochloa teretifolia* (left) and *Chionochloa crassiuscula* subsp. *directa* (Longwood Range); **F** *Chionochloa acicularis* (South Westland); **G** *Chionochloa bromoides* (Northland); **H** *Chionochloa australis* (north-west Nelson).

usually prominent. Lemma 6–8.5 mm, 5-nerved, rounded, small prickle-teeth scattered throughout, denser laterally below and on membranous margins; apex with hyaline lobes, awnless or awn to 3 mm. Palea 6–7.5 mm, apex bifid, ciliate, keels finely toothed almost to base, interkeel hairs few at apex. Callus 0.25 mm, glabrous. Rachilla 1–1.5 mm, short finely stiff hairy. Lodicules 1–2 mm, deeply lobed, sometimes twice-lobed; glabrous. Anthers 2.5–3 mm, caudate, sometimes malformed. Gynoecium: ovary *c.* 1 mm, apex glabrous; stigma-styles *c.* 2 mm.

N.; S. Occasionally where parent species co-occur.

Somewhat variable especially in the inflorescence and its components; the genetic composition of the parents and the ecological conditions both influence appearance. Although *Schedonorus phoenix* itself may be awned, *Lolium multiflorum* is evidently awned, and longer awned florets in hybrids may reflect the incorporation of genes from *L. multiflorum*; the intergeneric *S. phoenix* × *L. multiflorum* lacks a formal name according to Holub (1998 op. cit. p. 112). The nothogenus ×*Schedololium* Holub in N.Z. appears to involve *S. phoenix* only, but the hybrid may combine genomes from commercial cultivars of *Lolium* in which both *L. multiflorum* and *L. perenne* genes are present (*vide supra Lolium*). It may be simpler to refer all hybrids to ×*Schedololium holmbergii*; where greater precision is warranted the alternative form of hybrid nomenclature may be desirable viz *Lolium multiflorum* × *Schedonorus phoenix* if certainty is established.

Teratological flowers may occur e.g., twin gynoecia, three stigma-styles.

VULPIA C.C.Gmel., 1805

Short-lived annuals, rarely perennials; branching intravaginal. Leaf-sheath rounded. Ligule membranous. Leaf-blade flat, convolute when dry. Culm erect or decumbent, branching mainly at base. Inflorescence a narrow, usually secund panicle, usually sparsely branched, or sometimes a spiciform or compound raceme. Spikelets laterally compressed, (1)–3–12-flowered, ♀, the upper 1–2–(3) florets ♂ or ∅; disarticulation above and sometimes below glumes, and below each ♀ floret. Glumes usually very unequal; lower sometimes minute, 1-nerved, or nerves 0, upper 1–3-nerved, acute to acuminate. Lemma (3)–5-nerved, rounded or occasionally keeled, subcoriaceous, acuminate, mucronate or tapering to a long straight awn. Palea ≈ lemma, 2-keeled, apex bifid. Callus rounded or pointed, glabrous or pubescent. Lodicules 2, membranous, bilobed or bidentate, narrow or sometimes ovate. Stamens 1–(3), often small. Ovary apex glabrous or pubescent; styles free. Caryopsis linear to narrowly ellipsoid, dorsiventrally compressed, longitudinally grooved; embryo small; hilum linear, at least ¾ length of caryopsis.

Type species: *V. myuros* (L.) C.C.Gmel.

22 spp. of temperate and subtropical regions of Northern Hemisphere and South America. Naturalised spp. 2.

The genus *Vulpia* is very variable and the morphological and anatomical variation within it was reviewed by Cotton, R. and Stace, C. A. *Bot. Not. 130*: 173–187 (1977). The two spp. present in N.Z. belong to sect. *Vulpia* which is characterised by annual plants, with inflorescence a panicle or rarely subracemose, with most florets ♀ or with a group of ∅ but not smaller florets at the apex, and with disarticulation below each ♀ floret. The florets are usually cleistogamous with 1 small anther, 0.3–0.8 mm, but there may be up to 3 anthers to 1.8 mm, and the larger anthers may be slightly exserted at anthesis. The variation in flowering behaviour in *Vulpia* is discussed by Auquier, P. and Stace, C. A. *Pl. Syst. Evol. 136*: 47–52 (1980).

Stace, C. A., *in* Davis, P. H. (Ed.) *Fl. Turkey 9*: 451 (1985), noted that "Glume-ratio is an important character in this genus, but care must be taken to avoid spikelets which are terminal on the inflorescence or on its branches...; on such terminal spikelets the lower glume is often relatively long."

1 Lower glume ½–¾ length of upper glume; panicle lanceolate to narrow-oblong, erect or sometimes nodding .... **1. bromoides**
   Lower glume <¼–almost ½ length of upper glume; panicle usually linear, mostly curved or nodding .... **2. myuros**
      1a Lemma margins glabrous near apex .... var. **myuros**
        Lemma margins ciliate near apex .... var. **megalura**

1. V. bromoides (L.) Gray, *Nat. Arr. Brit. Pl. 2*: 124 (1821).

vulpia hair grass

Annual, slender, rather stiff tufts or solitary shoots, 3–60–(80) cm, with panicles usually conspicuously exserted above uppermost leaf-sheath. Leaf-sheath hyaline, glabrous, green to purplish, lower sheaths darker brown with paler ribs. Ligule 0.2–0.5 mm, a membranous, truncate rim. Leaf-blade (1.5)–3–10 cm × 0.8–1.5 mm, flat or involute, abaxially glabrous, adaxially short-ciliate on ribs; margins and acute tip scabrid. Culm 3–40 cm, erect or decumbent at base, internodes glabrous, or sparsely to more densely hairy. Panicle 1.5–12–(17) cm, secund, erect or slightly nodding, lanceolate to narrow-oblong, often rather lax below and with upper branches erect-appressed, sometimes reduced to a single spikelet; rachis, branches and stiff pedicels scabrid on angles. Spikelets (8)–10–20–(25) mm, 5–7-flowered, uppermost 1–2–(3) florets ∅, cuneate or oblong; flowering usually cleistogamous. Glumes very unequal with broad green nerves and hyaline margin, acute to apiculate; lower 2.5–4.2 mm, ½–¾ length of upper, 1-nerved, subulate, smooth, upper 4.7–7.8 mm, 3-nerved, oblong-lanceolate, midnerve scabrid near tip. Lemma 5.5–7.5–(10.5) mm, 5-nerved, firmly membranous, rounded, elliptic-lanceolate, scabrid; awn 4.5–11 mm, fine, minutely scabrid. Palea keels scabrid, interkeel scabrid near apex. Callus glabrous. Rachilla glabrous. Anthers usually 1, rarely 2, (0.2)–0.4–0.6 mm, or 0.8–1.5 mm. Caryopsis (2)–3–4.2 × 0.5 mm.

Naturalised from Eurasia.

N.; S.: throughout; St.; K., Three Kings Is, Ch., C., M. Waste land and disturbed ground, stony river beds and dry depleted grassland; sea level to subalpine.

A weed of temperate regions in both Hemispheres.

The record for Macquarie Id is based on ANU 11748 *R. Hnatiuk* Macquarie Island, Met. Station, 16.1.1972 (duplicate CHR 276460). This record was not included in the account of the grasses of Macquarie Id by Edgar. E., *in* George, A. S. et al. (Eds) *Fl. Australia 50*: 461–470 (1993).

A slender form from the Bay of Islands was described by J. D. Hooker [*Fl. N.Z. 1*: 309 (1853)] as *Festuca bromoides* var. *tenella* Hook.f. No specimens were found at K.

2. V. myuros (L.) C.C.Gmel., *Fl. Bad. 1*: 8 (1805). vulpia hair grass

Annuals, (3)–9–90 cm, in dense or loose tufts, or as solitary shoots, with base of panicle usually hidden by leaf-sheath. Leaf-sheath firmly membranous, glabrous, green to purplish, or light brown. Ligule 0.2–0.5–(0.7) mm, a truncate membranous ciliate rim, sometimes asymmetrical. Leaf-blade (1.5)–3.5–20 cm × 0.3–0.5 mm diam., inrolled, or flat and 0.5–2.5 mm wide, abaxially glabrous, adaxially covered with minute silky hairs; margins and fine acute tip scabrid. Culm 5–40–(60) cm, erect, or curved upwards from decumbent base, internodes glabrous. Panicle linear, secund, curved or nodding, dense, usually ± racemose above, shortly branched below; rachis, stiff ± appressed branches and short stiff pedicels scabrid on angles. Spikelets 15–25 mm, 5–6–(7)-flowered, uppermost 1–(2) florets ⊘, oblong or cuneate, green to purplish. Glumes extremely unequal, glabrous, acute; lower 0.5–1–(2) mm, ⅙ to almost ½ length of upper, 1-nerved, subulate, upper (2.7)–3–4.5 mm, 3-nerved, narrow-lanceolate. Lemma 5.5–8.5 mm, faintly 5-nerved, firm, rounded, linear-lanceolate, scabrid, especially above near awn; awn 9–14.5 mm, fine, scabrid. Palea keels scabrid near apex. Anthers usually 1, rarely 2, (0.4)–0.5–0.6 mm, or 0.8–1.5–(2.5) mm. Caryopsis 3.5–4.5 × 0.4–0.6 mm.

var. **myuros**

Panicle (2.5)–4–24 cm. Lemma entirely glabrous. Plate 4F.

Naturalised from Eurasia.

N.: common; S.: common, except Fiordland and Southland; St.; K., Three Kings Is. Waste land, shingly river flats, and in rough pasture; sea level to montane.

var. **megalura** (Nutt.) Auquier, *Bull. Jard. Bot. Nat. Belg. 47*: 123 (1977).

Panicle (5)–10–28 cm. Lemmas ciliate on margins above; sometimes almost entirely glabrous in some spikelets.

Naturalised from Mediterranean.

N.; S.: scattered; St.; K. Waste land, often stony.

Auquier, P. *Bull. Jard. Bot. Nat. Belg. 47*: 117–137 (1977), in his revision of Belgian spp. of *Vulpia*, discussed *V. megalura* (Nutt.) Rydb. This taxon, described from North America, differs from Eurasian *V. myuros* only in the lemmas being ciliate on the margins near the apex. Lonard, R. I. and Gould F. W. *Madroño 22*: 217–230 (1974), considered these plants were introduced to North America. Auquier (1977 op. cit.)

recognised 3 infraspecific units within *V. myuros*: var. *myuros* (lemmas glabrous), var. *megalura* (lemmas ciliate on margins near apex), and var. *hirsuta* Hack. of Portugal (lemmas hairy throughout), which does not occur in N.Z.; Stace, C. A. and Cotton, R. *Watsonia 11*: 72–73 (1976), also recognised these 3 units with the rank of forma. Auquier (1977 op. cit.) considered that var. *megalura* was native to the Mediterranean and had been introduced further north in Europe and also to North America.

ζ**Sesleria** Scop., from Europe, is a genus of ± caespitose perennial spp. distinguished by spiciform, cylindrical to globose, bluish to silvery panicles subtended by 2 well-developed scarious bracts. Allan, H. H. *N.Z. DSIR Bull. 49*: 64 (1936) recorded that two species had been collected in N.Z. but neither appeared to have become established. Both were collected by J. F. and J. B. Armstrong, near Christchurch, undated [late 19th Century] and labelled "rare". They have not been seen since. *S. albicans* Schult. (CHR 5064) has the uppermost leaf < 1 cm and a dense, ovate panicle; in *S. autumnalis* (Scop.) F.W.Schultz (CHR 5065) the uppermost leaf is > 4 cm, and the panicle lax, ± linear-oblong.

7. HAINARDIEAE

Ligule membranous. Leaf-blade linear to linear-lanceolate. Inflorescence a cylindrical spike. Spikelets 1–(6)-flowered, florets all ♀, sessile, alternate, sunk in cavities of rachis. Glumes 2, or 1, > lemma, coriaceous, 3–9-nerved, appressed to and covering the cavity of the rachis except at maturity. Lemma usually hyaline and awnless. Palea 2-keeled. Lodicules 2, lanceolate, entire. Stamens 3. Ovary glabrous; styles 2. Caryopsis narrowly oblong; hilum linear to elliptic.

1 Spikelets edgeways to rachis; glume 1 (2 in apical floret) .... **Hainardia**
  Spikelets broadside to rachis; glumes 2 .... **Parapholis**

HAINARDIA Greuter, 1967

Monotypic; Mediterranean to western Asia; the sole species is naturalised in New Zealand and also in Australia.

The genus was formerly known as *Monerma* P.Beauv. which is considered to be a superfluous name for *Lepturus* R.Br. [Greuter, W., *in* Greuter, W. and Rechinger, K. H. *Boissiera 13*: 177–178 (1967)]. However, see Scholz, H. *Feddes Repert. 106*: 169–171 (1995) on the status of *Monerma* relative to *Lepturus*.

1. H. cylindrica (Willd.) Greuter, *Boissiera 13*: 177 (1967).

barb grass

Small annual tufts 10–30–(45) cm, often growing closely together to form large patches, plants usually rigidly erect but sometimes with shoots appressed to ground. Leaf-sheath submembranous, glabrous, straw-coloured, striate. Ligule 0.2–0.7 mm, membranous, glabrous, truncate, erose, projecting slightly upwards on either edge. Leaf-blade 2.5–9–(11.5) cm × 1.2–1.7 mm, flat, or inrolled and *c.* 0.5 mm diam., abaxially glabrous, adaxially ribbed, ribs minutely scabrid. Culm 5–30 cm, sometimes purplish, usually branched near base, internodes glabrous. Inflorescence a single simple spike, breaking up at maturity by disarticulation of rachis below each spikelet. Spikes 6.5–15 cm, cylindric, rather thick, straight or slightly curved, light green; rachis internodes ≤ spikelets, hollowed on one side. Spikelets 5–7.5 cm, 1-flowered, sessile, solitary, edgewise to rachis, alternating on opposite sides of rachis, ± sunk in cavities within it; each floret ± concealed by one glume, uppermost floret with 2 glumes. Glumes 9-nerved, very hard, thick and rigid, oblong-subulate, acute, outer nerves slightly scabrid near apex; glume concealing floret at first, but later sometimes recurved and projecting stiffly at 45° from the rachis. Lemma 4.5–6.2 mm, with back appressed to rachis, membranous, finely 3-nerved with lateral nerves indistinct, lanceolate, acute, glabrous. Palea ≈ lemma, 4–5.5 mm, hyaline, glabrous, keels 2, indistinct. Lodicules 2, glabrous, 1–1.5 mm. Stamens 3; anthers 1.5–2.5 mm. Caryopsis *c.* 3–3.5 × 0.8–1.1 mm, longitudinally grooved; embryo small; hilum short, linear to oblong.

Naturalised.

N.: Auckland City, near Kawhia, Tauranga; S.: North Canterbury (Saltwater Creek), near Banks Peninsula (head of Lyttelton Harbour, Kaituna near Lake Ellesmere, Kaitorete Spit). On salt flats.

Indigenous to western Europe, the Mediterranean and western Asia and now naturalised in temperate regions of both Hemispheres; formerly known as *Lepturus cylindricus* and *Monerma cylindrica.*

PARAPHOLIS C.E.Hubb., 1946

Short, caespitose annuals or biennials. Leaf-sheath open. Ligule membranous. Leaf-blade flat to rolled, narrow. Culm glabrous. Inflorescence a single, simple spike, breaking up at maturity by disarticulation of rachis below each spikelet. Spikelets 1-flowered, sessile, solitary and alternating on opposite sides of rachis, broadside to rachis and ± sunk in cavities within it, falling away with rachis internodes at maturity. Glumes 2, adjacent, strongly nerved, coriaceous, ≥ lemma, tip acute. Lemma 3–5-nerved, membranous, at right angles to rachis. Palea ≤ lemma, hyaline, keels 2, rather indistinct. Lodicules 2, glabrous. Stamens 3. Ovary glabrous; styles free to base. Caryopsis longitudinally grooved; embryo small; hilum punctiform or shortly elliptic. Fig. 7.

Type species: *P. incurva* (L.) C.E.Hubb.

6 spp. of western Europe, Mediterranean, to eastern Asia. Naturalised spp. 2.

Mediterranean spp. were revised by Runemark, H. *Bot. Not. 115*: 1–17 (1962); his treatment is followed here.

1 Anthers < 1 mm; culms and spikes often strongly curved .... **1. incurva**
   Anthers 2–3 mm; culms and spikes straight to slightly curved ..... **2. strigosa**

1. P. incurva (L.) C.E.Hubb., *Blumea Suppl. 3*: 14 (1946).

sickle grass

Small annual tufts to *c.* 35 cm. Leaf-sheath submembranous, glabrous, purplish, rounded, ribs ± inconspicuous. Ligule 0.3–1 mm, truncate, erose, projecting slightly upwards at either margin. Leaf-blade (1)–2.5–7.5 cm, flat, or incurved, *c.* 0.5 mm wide when dry, abaxially glabrous, adaxially with fine, hair-like prickle-teeth on ribs; margins and acute tip scabrid. Culm (1)–5–25 cm, loosely to densely tufted, sometimes solitary, usually much-branched below, prostrate, curved and ascending, or erect, few- to many-noded, internodes glabrous. Spikes (3)–7–10–(12) cm, curved or rarely straight, cylindric, rigid, green or purplish, rachis internodes ≤ spikelets, hollowed on one side, disarticulating horizontally below each spikelet at maturity.

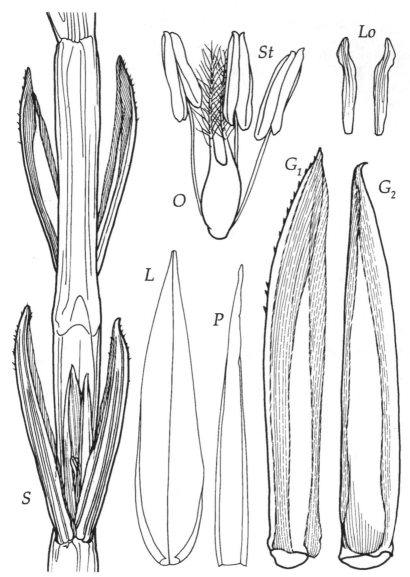

Fig. 7 *Parapholis incurva*
S, spikelet on rachis segment, × 12; **G₁, G₂**, glumes, × 24; **L**, lemma ventral view, × 24; **P**, palea dorsal view, × 24; **Lo**, lodicules, × 30; **O**, ovary, styles and stigmas, × 30; **St**, stamens, × 30.

Spikelets 5–6.5 mm, embedded in cavities in rachis, solitary and alternating on opposite sides of rachis. Glumes ± equal and equalling spikelets, inserted side by side on rachis and appressed to it obscuring cavity, 4–5-nerved, rigid, much thickened, narrowly oblong-subulate, acute, hyaline margins minutely hairy. Lemma 3.5–4.5 mm, lanceolate, membranous, finely 3-nerved, lateral nerves short and very indistinct. Palea ≈ lemma. Lodicules *c.* 1 mm. Anthers 0.6–0.9 mm. Caryopsis 2.5–3 × 0.6–0.8 mm, enclosed between hardened glumes and segment of rachis. Fig. 7. Plate 4C.

Naturalised from Mediterranean.

N.: North and South Auckland, scattered further south; S.: Nelson, Marlborough, Canterbury to Banks Peninsula, Otago (near Dunedin, and inland at Conroys Gully). Usually coastal, on sand dunes, on shingle and rocks near water's edge, and in saline estuaries, mud flats, and reclamation areas.

Originally recorded for New Zealand as *Lepturus incurvatus* and also known as *Lepturus incurvus* and *Pholiurus incurvus*. *Parapholis incurva* resembles littoral species of *Lepturus* (Chlorideae) in which spikelets are also deeply embedded within fragile racemes, but the glumes in *Lepturus* are very unequal and only the upper glume covers the embedded spikelet, except in the topmost floret.

2. P. strigosa (Dumort.) C.E.Hubb., *Blumea Suppl. 3*: 14 (1946).

Annual, stiff, erect, ± open tufts, (12)–20–35 cm. Leaf-sheath subcoriaceous, glabrous, rounded, ribs ± inconspicuous. Ligule 0.3–0.6 mm, truncate, erose. Leaf-blade 2–6 cm × *c.* 2 mm, flat, rolled or folded and < 1 mm diam., abaxially glabrous, adaxially ribbed, with scattered minute hairs on ribs, tip acute. Culm 15–25 cm, tufted or solitary, rigid, usually loosely branched, few- to many-noded, internodes glabrous. Spikes (2.5)–5–12 cm, cylindric, stiffly erect, light green, rachis internodes ≤ spikelets, hollowed on one side, disarticulating horizontally below each spikelet at maturity. Spikelets 4.5–5.5 mm, embedded in cavities in rachis, solitary and alternating on opposite sides of rachis. Glumes ± equal and equalling spikelets, inserted side by side on rachis and appressed to it, 3–5-nerved, rigid, much thickened, narrowly oblong-subulate, acute, hyaline margins minutely papillose. Lemma 4–4.5 mm, ovate-oblong, membranous, 3-nerved, lateral nerves very faint and short. Palea ≈ lemma. Lodicules *c*. 1 mm. Anthers 2–3 mm. Caryopsis 2.7–3 × 0.6–0.8 mm, enclosed between hardened glumes and segments of rachis.

Naturalised from Europe.

N.: North Auckland (Hokianga Harbour, Waitemata Harbour), Auckland City and to the west, eastern Coromandel. Coastal salt marsh and grass flats, dry rocky slopes and shell banks on mud flats.

An earlier record for Canterbury, as plentiful in salt meadows on Lyttelton Harbour and on Lake Ellesmere, Mason, R. *in* Knox, G. A. (Ed.) *Nat. Hist. Canty* 100 (1969), was based on specimens of *Hainardia cylindrica*.

8. MELICEAE

Perennials. Leaf-sheath usually closed. Ligule membranous. Inflorescence a panicle or raceme. Spikelets laterally compressed, or terete, with few to many ♀ florets and imperfect florets above; disarticulation above glumes and usually between florets. Glumes unequal. Lemma herbaceous to coriaceous, rounded, usually scarious at apex, distinctly nerved, awned or awnless. Palea ≈ lemma, membranous, 2-keeled. Lodicules 2, truncate, usually connate, fleshy, glabrous. Stamens 3, rarely 2. Ovary glabrous; styles 2. Caryopsis ellipsoid to terete; embryo small; hilum linear, = caryopsis.

1 Glumes < spikelet, ≈ lowest floret; upper glume 1-nerved .... **Glyceria**
  Glumes ≥ florets; upper glume 3-nerved .... **Melica**ζ (p. 223)

GLYCERIA R.Br., 1810 *nom. cons.*

Glabrous, often tall, aquatic perennials, with creeping and rooting bases or with creeping rhizomes. Leaf-sheath often closed. Ligule membranous. Leaf-blade flat. Culm robust or slender, erect, or often decumbent and ascending. Inflorescence a ± lax panicle; rachis trigonous. Spikelets ovate to linear, florets 3–many, often closely imbricate; disarticulation above glumes and between florets. Glumes unequal or subequal, persistent, usually hyaline,

1–(3)-nerved, < lowest lemma. Lemma 5–11-nerved, broad, rounded, usually obtuse, awnless, nerves parallel, usually prominent. Palea ≈ lemma. Lodicules 2, small, ± connate, fleshy, glabrous. Stamens 3. Ovary apex glabrous; styles free. Caryopsis dorsiventrally compressed, longitudinally grooved; embryo small; hilum ± linear. Fig. 8.

Type species: *G. fluitans* (L.) R.Br.

c. 40 spp., mainly in Northern Hemisphere, especially North America. Naturalised spp. 5, and one probable natural hybrid.

Species are either aquatic or grow in damp or swampy ground, and are collectively known in N.Z. as floating sweet grass, except for *G. maxima* and *G. striata.*

1 Spikelets 8–30–(40) mm; palea keels winged above *2*
　Spikelets 2–10–(12) mm; palea keels not winged *5*

2 Lemma apex with 3–5 distinct teeth; palea deeply bifid, with aristate teeth, distinctly exceeding lemma apex .... **1. declinata**
　Lemma apex without distinct teeth; palea rounded, bidenticulate or shallowly bifid, with teeth not, or slightly exceeding lemma apex *3*

3 Lemma 6–7.5 mm; anthers 1.5–2.5 mm .... **2. fluitans**
　Lemma 3.5–5.5–(6) mm; anthers < 1.5 mm, or up to 2 mm but sterile *4*

4 Spikelets fragile, florets readily disarticulating at maturity; lemma usually 3.5–5 mm; anthers dehiscent, pollen-fertile .... **5. plicata**
　Spikelets ± persistent; lemma usually 5–5.5 mm; anthers indehiscent, pollen-sterile .... **4. ×pedicellata**

5 Leaf-blade 10–20 mm wide, cross-veinlets numerous, especially on leaf-sheath; spikelets 5–12 mm .... **3. maxima**
　Leaf-blade 1–4 mm wide, cross-veinlets few; spikelets to 3 mm .... **6. striata**

1. G. declinata Bréb., *Fl. Normandie* ed. 3, 354 (1859).

Loosely tufted, rather flaccid; branching extravaginal. Leaf-sheath submembranous, closed, keeled, striate, glabrous, often purplish, scattered cross-veinlets inconspicuous. Ligule 3.5–7 mm, hyaline, tapered to a point. Leaf-blade (4)–8–17 cm × 2.5–7 mm, folded at first, becoming flat, abaxially minutely papillose, striate, minutely

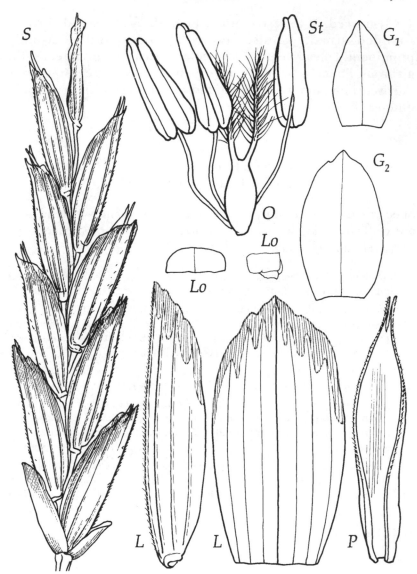

Fig. 8 *Glyceria declinata*
S, spikelet, × 10; **G₁**, **G₂**, glumes, × 16; **L**, lemma lateral and dorsal view, × 16; **P**, palea dorsal view, × 16; **Lo**, lodicule dorsal and lateral view, × 32; **O**, ovary, styles and stigmas, × 32; **St**, stamens, × 32.

scabrid on midrib and lateral ribs, adaxially glabrous, channelled on either side of midrib; margins scabrid, abruptly narrowed to stiff point. Culm (15)–25–45 cm, erect or ascending from curved, bent, or prostrate base, internodes glabrous. Panicle 20–30–(40) cm, narrow, with few spikelets, spiciform, ± secund, with 1–3 branches at lower panicle-nodes; rachis smooth, branches with a few minute prickle-teeth below spikelets. Spikelets 15–25–(30) mm, 8–14-flowered, narrowly oblong, greyish green or purplish. Glumes noticeably unequal, 1-nerved, hyaline, ovate-oblong, obtuse; lower 1.5–2.5 mm, upper 2.5–3.5 mm. Lemma 4–5.5 mm, 7-nerved, ovate- or elliptic-oblong, 3-lobed or 3–5-toothed, minutely papillose to short-scabrid, green and firm, nerves purple tipped extending into wide hyaline upper margin. Palea oblong-lanceolate or elliptic, deeply bifid into often purplish, aristate teeth, usually projecting slightly beyond lemma apex, keels narrowly winged, very minutely scabrid. Rachilla *c.* 1.5 mm, glabrous. Anthers (0.9)–1–1.4 mm. Caryopsis (1.7)–2–2.5 × 0.8–1 mm. Fig. 8.

Naturalised from Europe.

N.; S.: throughout; St.; Ch. In damp ground in swamps, on stream margins, along drains and in damp pasture; lowland to montane.

Early collections in N.Z. were not distinguished from *G. fluitans.*

2. G. fluitans (L.) R.Br., *Prodr.* 179 (1810).

Loosely tufted or forming loose masses in shallow water. Leaf-sheath closed, papery, light brown or purplish, glabrous, striate, keeled. Ligule (4)–8–10 mm, membranous, tapered. Leaf-blade 10–23 cm × (2)– 4.5–7.5 mm, folded or flat, striate, minutely papillose to tubercular-scabrid especially on ribs; margins finely scabrid, rather abruptly narrowed to stiff tip. Culm (20)–45–75 cm, erect or spreading, sometimes prostrate or floating at base, few-noded, internodes glabrous. Panicle (20)–30–55 cm, secund, lax, linear, open at anthesis, later contracted; rachis smooth, branches sparingly scabrid, paired at lower panicle nodes, longer branch with 1–4 spikelets, shorter branch with 1–2 spikelets. Spikelets (15)–20–30 mm, 7–13-flowered, narrowly oblong, green or purple. Glumes unequal, elliptic-oblong; lower 2–3–(4) mm, 1-nerved, acute, upper 3–4–(5.5) mm, 1–(3)-nerved, subobtuse. Lemma 6–7.5 mm, 7-nerved, elliptic-oblong or oblong, subobtuse to subacute, minutely scabrid, margins later incurved; nerves not extending into wide hyaline upper margin. Palea = lemma, oblong-lanceolate, apex sharply, shortly bidentate, keel very minutely scabrid, scarcely winged. Rachilla 1.5–1.8 mm, glabrous. Anthers 1.5–2.5 mm. Caryopsis 2–3 × 0.7–1.2 mm.

Naturalised.

N.: Auckland City, Bay of Plenty, Manawatu, Wairarapa, Wellington City; S.: Nelson, Canterbury, Otago, Southland, rare in Westland and Fiordland, not recorded from Marlborough. In still and slow waters, or in soft mud on margins; lowland to montane.

Indigenous to temperate regions of the Northern Hemisphere.

3. G. maxima (Hartm.) Holmb., *Bot. Not. 1919*: 97 (1919).

reed sweet grass

Tall, luxuriant, wide-leaved perennial sending up numerous long leafy shoots from a stout creeping rhizome. Leaf-sheath entire at first, later splitting, light green to pale brown, often purplish, paper-like, shining, very finely scabrid to almost smooth, with numerous conspicuous cross-veinlets. Ligule (4)–4.5–6.5 mm, firmly membranous, glabrous, blunt, but tapered centrally to fine point. Leaf-blade (20)–30–50 cm × 10–20 mm, flat, finely striate, very finely scabrid to almost smooth, abaxially scabrid above on conspicuous midrib; margins finely scabrid, tapering abruptly to short ± rigid tip. Culm (5)–10–180 cm, robust, internodes glabrous. Panicle 20–30–(40) cm, lax and open, or contracted and rather dense, broadly ovate to oblong, with numerous spikelets; rachis, branches and pedicels finely scabrid, longer branches naked below. Spikelets (5)–6–7.5–(10) mm, (3)–4–8-flowered, green, or tinged purplish brown. Glumes subequal, hyaline, 1-nerved; lower 1.7–2.5 mm, elliptic-oblong, subacute, upper 2–3.5 mm, ovate-oblong, obtuse. Lemma 2.5–3.5 mm, 7-nerved, elliptic to ovate-elliptic, broadly obtuse, subcoriaceous; nerves minutely prickle-toothed, not reaching upper hyaline minutely prickle-toothed margins. Palea oblong, keels tubercular-scabrid. Rachilla 0.7–1 mm, glabrous. Anthers (1)–1.5–1.8 mm. Caryopsis *c.* 1.5 × 1 mm.

Naturalised from Europe.

N.: common in North and South Auckland, scattered further south; S.: Canterbury (near Christchurch, near Hororata), South Otago and Southland, one record from Nelson (Matakitaki R.), and from Westland (Kaniere). An aquatic sp. found in lowland swamps, pools, edges of slow-flowing streams; grows in water to *c.* 1 m deep.

4. G. ×pedicellata F.Towns., *Ann. Mag. Nat. Hist.* Ser. 2, **5**: 105 (1850). *G. fluitans* × *G. plicata*

Forming large patches with long floating stolons. Leaf-sheath papery, light brown or purplish, minutely scabrid or smooth. Ligule (3)–5–10 mm, membranous. Leaf-blade (9)–17–25 cm × 3–6 mm, folded or flat, minutely scabrid throughout, or only abaxially, or entirely smooth; margins finely scabrid, tip narrowed abruptly to stiff point. Culm (25)–45–75 cm, branched at base, internodes spongy, glabrous, smooth. Panicle (18)–25–45 cm, branches erect, or finally spreading, below in clusters of 2–3 at nodes, borne singly above or sometimes throughout, unequal, longer branches with up to 9 spikelets, shorter branches with 1–2 spikelets; rachis and branches glabrous. Spikelets 20–27 mm, 8–14-flowered, linear-oblong, green, rarely purplish, ± persistent. Glumes unequal, broadly oblong to broadly elliptic, 1-nerved, obtuse, membranous; lower 2–3.5 mm, upper 3–5.3 mm. Lemma (4.5)–5.5–6 mm, 7-nerved, elliptic-oblong, obtuse; nerves finely scabrid, not extending into wide hyaline upper margin. Palea oblong, shortly bidentate with teeth not projecting beyond lemma apex, keels narrowly winged and minutely scabrid. Rachilla *c.* 1.5 mm, glabrous. Anthers 1.2–2 mm, indehiscent, pollen-sterile. Caryopsis 0.

Of probable natural hybrid origin in New Zealand.

N.: scattered localities in Auckland, Hawke's Bay and Wellington provinces; S.: Marlborough (Tuamarina), Canterbury (near Christchurch), scattered localities in South Otago and Southland. Lowland in swampy ground, on edges of creeks and streams and on drain banks.

This hybrid occurs sporadically in the British Isles and Europe [Stace, C. A. (Ed.) *Hybridization and the Flora of the British Isles* (1975)].

5. G. plicata (Fr.) Fr., *Novit. Fl. Suec., Mant. 3*: 176 (1842).

Stout, with leafy shoots ascending from a prostrate rooting base. Leaf-sheath papery, light brown, closed, keeled, finely striate (ribs becoming undulate), with very fine prickle-teeth or hairs above and on ribs, cross-veinlets few, inconspicuous. Ligule 3–10 mm, membranous, tapered above, abaxially minutely, finely prickle-toothed. Leaf-blade (5)–15–25 cm × (2)–3.5–7 mm, folded to flat, finely ribbed with minute fine prickle-teeth on ribs, or adaxially smooth; margins finely tubercular-scabrid. Culm 30–40–(65) cm, finely ribbed, branched near base, several-noded, internodes spongy, glabrous. Panicle 18–40 cm, very lax, branches at lower nodes in clusters of 2–5 with one longer branch finally spreading, the rest much shorter with very few spikelets; rachis, branches and pedicels sparsely finely scabrid. Spikelets 12–22 mm, 8–12-flowered, linear-oblong, light green or purplish. Glumes unequal, membranous, 1-nerved, broadly elliptic-oblong, obtuse; lower 1.5–2–(2.5) mm, upper 2.5–3–(4.5) mm. Lemma 4–4.5–(5.5) mm, 7-nerved, broadly oblong-obovate, obtuse or slightly 3-lobed; nerves minutely scabrid, not extending into wide hyaline upper margin. Palea 3.5–4–(4.5) mm, ⩽ lemma, elliptic, rather firm, shortly bidenticulate, keels minutely scabrid, narrowly winged. Rachilla 1–1.5 mm, glabrous. Anthers 1–1.5 mm. Caryopsis 1.5–2.2 × 0.9–1.1 mm.

Naturalised.

N.: scattered; S.: Canterbury, Otago, Southland; St.; Ch. In swampy creeks, streams or ponds, on the margin of roadside drains or water races; lowland to montane.

Indigenous to Europe, south-western to central Asia and north-western Africa.

6. G. striata (Lam.) Hitchc., *Proc. Biol. Soc. Wash. 41*: 157 (1928).

fowl manna-grass

Stoloniferous clumps. Leaf-sheath light brown, membranous, closed, striate, ribs minutely scabrid, cross-veinlets few, inconspicuous. Ligule 1.5–3 mm, membranous, glabrous, tapered to a point. Leaf-blade 10–25 cm × 1–4 mm, flat, midrib minutely scabrid; margins minutely scabrid, narrowed to long fine tip.

Panicle 15–25 cm, slender, with numerous small spikelets on slender branches naked below; rachis, branches and pedicels finely scabrid. Spikelets 2–3 mm, 3–4-flowered, light green to purplish. Glumes subequal, 1-nerved, membranous with conspicuous midnerve, ovate, subobtuse, margins ciliate; lower < 1 mm, upper *c*. 1 mm. Lemma 1.3–1.8 mm, 7-nerved, ovate, obtuse; nerves minutely tubercular-scabrid, extending into wide upper hyaline margin. Palea 1–1.5 mm, ≈ lemma, obovate, obtuse, keels minutely tubercular-scabrid. Rachilla 0.3–0.4 mm, minutely tubercular-scabrid. Anthers 0.3–0.4 mm. Caryopsis 0.8 × 0.6 mm.

Naturalised from North America.

N.: North and South Auckland, Taranaki (between Te Kiri and Egmont National Park); S.: Nelson (Matiri Scenic Reserve). Damp ground, dairy pasture.

ζ**Melica minuta** L.   The single specimen consists of one shoot, 30 cm; ligule to 6 mm, lacerate; leaf-blade 1.5 mm diam., convolute; panicle 10 cm, with 2 short branches below and ± simple above; spikelets 8–9.5 mm, with 2 fertile florets and a clavate, pedicelled mass of ∅ florets; glumes unequal, > florets, upper glume 3-nerved; lemmas covered with minute prickle-teeth, 9-nerved. A Mediterranean species only known from one 19th Century collection, CHR 5090 *J. F. & J. B. Armstrong* in fields near Christchurch, "rare", undated.

9. AGROSTIDEAE

Annual or perennial herbs. Ligule membranous. Leaf-blade linear, flat or convolute. Inflorescence a panicle. Spikelets 1–several-flowered, laterally compressed; usually disarticulating below each floret. Glumes usually persistent, (except *Holcus*), usually > adjacent lemmas, often as long as spikelet, commonly membranous and shining, margins thin, hyaline. Lemma hyaline to coriaceous and often with thin shiny margins, (3)–5–11-nerved, often with a geniculate, dorsal awn. Lodicules 2, free. Stamens 3. Stigmas 2. Caryopsis mostly ellipsoid; hilum usually round or oval; endosperm sometimes soft, occasionally liquid.

AGROSTIDINAE

Spikelets with 1 ♀ floret; glumes commonly enclosing floret.

1 Glumes villous throughout, or ciliate on keel *2*
  Glumes glabrous, or finely scabrid, especially on keel, sometimes finely ciliate
  on margins *5*

2 Lemma awnless or shortly mucronate .... **Phleum**
  Lemma awned *3*

3 Leaves pubescent throughout; lemma with 2 awn-tipped lateral teeth ....
  .... **Lagurus**
  Leaves mainly glabrous but adaxially with hairs near base and sometimes
  scattered hairs; lemma truncate to acute, or with entire lateral lobes *4*

4 Panicle soft, narrow-cylindric; awns slender; glumes united near base or for
  c. ¼ length .... **Alopecurus**
  Panicle tough, ovate-globose to oblong; awns stout, bristly; glumes free to
  base .... **Echinopogon**

5 Glumes swollen near base .... **Gastridium**
  Glumes never swollen *6*

6 Glumes slightly < ½ length of lemma .... **Simplicia**
  Glumes, both or one, > lemma or ≈ lemma *7*

7 Lemma with glabrous callus *8*
  Lemma with hairy callus, hairs sometimes minute *10*

8 Lemma coriaceous, much harder than glumes .... **Milium**ζ (p. 292)
  Lemma ± membranous, of similar texture to glumes or softer *9*

9 Spikelets disarticulating below glumes and falling entire with piece of pedicel
  attached .... **Polypogon**
  Spikelets disarticulating above the persistent glumes .... **Agrostis**

10 Lemma ± translucent, soft, membranous or very thin at maturity, smooth
  and ± shining, sometimes ± covered by hairs *11*
  Lemma ± opaque, ± hard and rigid at maturity, ± scaberulous or scabrid at
  least above, never completely smooth and shining, hairs 0 .... *13*

11 Callus hairs = lemma .... **Calamagrostis**ζ (p. 292)
  Callus hairs < lemma *12*

12 Culm with panicle rachis persistent; lemma usually glabrous, very rarely with
  a few hairs near margins in lower ½ .... **Agrostis**
  Culm fragile below panicle, breaking off with panicle attached and readily
  wind-blown; lemma usually softly, ± sparsely hairy .... **Lachnagrostis**

13 Lemma sharply keeled ....**Ammophila**
　　 Lemma rounded or lightly keeled *14*

14 Lemma with long central awn and four shorter lateral awns .... **Pentapogon**
　　 Lemma with a single awn *15*

15 Awn ⩾ 10 mm; lemma entire or minutely bifid at apex .... **Dichelachne**
　　 Awn < 10 mm; lemma minutely 4-toothed at apex .... **Deyeuxia**

AGROSTIS L., 1753

Annuals or perennials, tufted, or widely creeping and rhizomatous or stoloniferous. Leaf-sheath rounded. Ligule membranous. Leaf-blade flat or involute. Culm of various height, erect, or decumbent and geniculately ascending; nodes glabrous. Inflorescence a diffuse and often delicately branched or contracted panicle, sometimes spike-like, rarely much reduced and hidden among leaves; rachis persistent at maturity. Spikelets small, numerous to few, 1-flowered; disarticulation above glumes; rachilla usually not prolonged. Glumes equal to subequal, lanceolate, acute to acuminate, sometimes subobtuse, membranous, shining, 1-nerved, keel ± scabrid. Lemma usually < glumes, thinner in texture than glumes, hyaline to firmly membranous, broadly ovate, usually truncate to dentate, rounded, 3–5–(7)-nerved with nerves sometimes excurrent, awnless or with a dorsal awn. Palea hyaline, 2-keeled or nerveless, usually < lemma, sometimes very small or obsolete, almost covered by inrolled margins of lemma. Callus short or minute, blunt, glabrous, or with ± minute lateral tufts of hairs. Lodicules 2, lanceolate, hyaline. Stamens 3. Ovary glabrous; styles distinct, free to base, short; stigmas 2, plumose. Caryopsis ovoid or fusiform, ± dorsally compressed, longitudinally grooved or rarely terete; embryo small; hilum punctiform or elongate; endosperm sometimes liquid. Chasmogamous (N.Z. spp.).

Type species: *Agrostis canina* L.

c. 220 spp. of temperate regions throughout the world and in the tropics at high altitudes. Endemic spp. 8, indigenous spp. 2; naturalised spp. 4.

N.Z. spp. of *Agrostis* were revised by Edgar, E. and Forde, M. B. *N.Z. J. Bot. 29*: 139–161 (1991), with an assessment by M. B. Forde of the extent and taxonomic significance of hybridism among naturalised bent grasses.

1 Palea *c.* ⅓ length of lemma or shorter (indigenous) *2*
  Palea ½–⅔ length of lemma (naturalised) *13*

2 Panicle contracted with branches hidden, narrow-linear, or densely oblong
  to lanceolate, or a minute (< 7 mm) dense ovate head *3*
  Panicle loose and often spreading with obvious branches, ovate to pyramidal
  or laxly oblong to lanceolate, never contracted to a dense head *9*

3 Lemma awn 3 mm or longer; spikelets 3.5–5 mm .... **6. magellanica**
  Lemma awnless, or with awn 0.5–2–(2.5) mm; spikelets < 3.3 mm, rarely to
  4 mm *4*

4 Panicles all < 4.5 cm *5*
  Panicles all, or at least some, > 4.5 cm *7*

5 Plant forming rounded cushions or mat-like patches; panicle < 0.7 cm; leaves
  usually reflexed .... **8. muscosa**
  Plant forming ± stiff tufts; panicle 1–4.5 cm; leaves ± erect *6*

6 Glumes finely scabrid on keel and sometimes near tip, otherwise smooth or
  papillose, usually tinged deep reddish purple, occasionally entirely green
  .... **7. muelleriana**
  Glumes minutely pubescent-scabrid except near base, light green, very rarely
  light purple-tinged .... **14. subulata**

7 Panicle 2–3 mm wide .... **5. imbecilla**
  Panicle > 5 mm wide *8*

8 Panicle oblong; lemma 1.4–2 mm .... **7. muelleriana**
  Panicle lanceolate; lemma 2.3–3.3 mm .... **3. dyeri**

9 Leaf-blade strictly involute, wiry, abaxially very densely scabrid-papillose
  throughout; anthers 1.3–1.8 mm, rarely *c.* 1 mm .... **12. petriei**
  Leaf-blade flat or folded, or filiform-involute, not wiry, abaxially smooth, but
  scabrid near tip, or on nerves or very rarely throughout if flat; anthers 0.4–
  1.2 mm *10*

10 Lemma 2.3–3.3 mm; spikelets usually 3–3.5 mm, rarely to 4 mm ...... **3. dyeri**
  Lemma 1.2–2–(2.2) mm; spikelets usually 1.5–2.5 mm, rarely to 3 mm ... *11*

11 Branching intravaginal at base; lemmas all awned; ligule abaxially smooth
  .... **9. oresbia**
  Branching extravaginal at base; lemmas awnless, or occasionally awned and
  then ligule abaxially finely scabrid *12*

12 Tufts 6.5–30–(60) cm; panicle (2)–4.5–8.5–(16) cm; spikelets usually
2–2.5 mm; glumes acute, > lemma .... **11. personata**
Tufts 3–5–(20) cm; panicle 0.6–2–(3) cm; spikelets usually 1.5–2 mm; glumes
subobtuse, usually ≈ lemma .... **10. pallescens**

13 Ligule 0.6–2 mm, shorter than wide; plants tufted or very shortly rhizomatous
or stoloniferous; lemma usually 3-nerved .... **1. capillaris**
Ligule 0.7–6 mm, taller than wide; plants long-rhizomatous or with long
stolons; lemma usually 5-nerved *14*

14 Plant with long stolons; panicle contracted after flowering, 0.5–2.5 cm wide
.... **13. stolonifera**
Plant long-rhizomatous; panicle remaining open and lax, or if contracted
after flowering, then 4–8 cm wide *15*

15 Leaf-blade 3–8 mm wide, bright green; panicle usually pyramidal, always very
lax; spikelet pedicels always closely scabrid .... **4. gigantea**
Leaf-blade 1.5–4 mm wide, grey-green; panicle usually linear-lanceolate, often
contracted after flowering; spikelet pedicels glabrous to ± scabrid ....
.... **2. castellana**

1. A. capillaris L., *Sp. Pl.* 62 (1753).   browntop

Loose to dense perennial tufts to *c.* 100 cm, often forming a dense
turf or loose sward, spreading by very short rhizomes; branching intra-
and extravaginal, the intravaginal shoots erect or sometimes
decumbent or forming trailing stolons, the extravaginal shoots
subterranean, bearing pale scale-leaves. Leaf-sheath glabrous,
greenish to purplish. Ligule 0.6–2 mm, truncate to rounded, in non-
flowering shoots shorter than wide. Leaf-blade (1.5)–2–13–(15) cm
× 1–2.5–(5) mm, flat or ± involute, ± scabrid, acute. Culm erect or
geniculate, often procumbent and branching then erect or ascending,
sometimes remaining ± horizontal, internodes usually smooth, rarely
scabrid below panicle. Panicle 2–15–(28) cm, oblong to ovate or
pyramidal, usually very lax with many spreading branches; rachis
smooth or sometimes scabrid above, branches and pedicels smooth
or slightly scabrid. Spikelets 1.5–3.5 mm, purplish brown to greenish.
Glumes subequal, lanceolate, acute; lower slightly scabrid above on
midnerve, upper smooth. Lemma 1.5–2.5 mm, smooth or sparsely
scabrid, ovate or elliptic, obtuse or truncate, 3-nerved with marginal
nerves sometimes slightly excurrent and midnerve not usually
reaching lemma tip, or rarely 5-nerved with midnerve produced to

an awn arising middorsally and occasionally geniculate and projecting slightly beyond tip of lemma. Palea usually ½–⅔ length of lemma, apex usually bifid. Callus glabrous or with very short hairs. Anthers 1–1.5 mm. Caryopsis *c.* 1 × 0.4 mm. Plate 3B.

Naturalised from Eurasia.

N.; S.; St.: common throughout; Ch., A., C. Sea level to subalpine in disturbed ground and in modified grassland.

Throughout the world and for many years *Agrostis tenuis* Sibth. was accepted as the correct name for browntop.

Browntop is a notoriously variable sp. with plants differing in size, habit, presence or absence of stolons or rhizomes, type of inflorescence, and in spikelet structure. It is widespread in pastures of several kinds in N.Z. and a standard component of lawn mixtures. Some of the variation may arise as a result of hybridisation with related spp.

Forde, M. B. *in* Edgar and Forde (1991 op. cit.), confirmed that where *A. capillaris* and *A. castellana* coexist a blurring of the boundaries between the spp. has already occurred and will continue, with characteristics of *A. castellana* remaining in populations of *A. capillaris*. However, the difference of two weeks in flowering time between the two spp. may restrict the formation of F_1 hybrids and, once formed, these hybrids will most likely backcross repeatedly to *A. capillaris*, the dominant sp.

Lee, W. G., Mark, A. F. and Wilson, J. B. *N.Z. J. Bot. 21*: 141–156 (1983) reported their experiments on an ultramafic ecotype of *A. capillaris* (as *A. tenuis*) at Black Ridge, Southland. Rapson, G. L. and Wilson, J. B. *Functional Ecol. 2*: 479–490 (1988) found very little evidence of adaptation of populations of *A. capillaris* sampled from a wide range of habitats in southern South Id and transplanted back into their own and each other's sites.

Plants of browntop infected by smut fungus, *Tilletia sphaerococca*, are dwarfed and tufted, with small compact panicles, undulating panicle branches, and glumes often shorter and broader than usual. Plants in which the eelworm *Anguina agrostis* had formed galls in the spikelets and caused elongation of glumes and lemmas were recorded by Healy, A. J. *N.Z. J. Agric. Res. 1*: 265–266 (1958).

2. A. castellana Boiss. et Reut., *Diagn. Pl. Nov. Hisp.* 26 (1842).

dryland browntop

Grey-green, rather fine-leaved perennials, forming tufts 30–90 cm, with long slender rhizomes; branching intravaginal. Leaf-sheath glabrous, brownish green or purplish. Ligule 0.7–2 mm, or up to 3 mm in culm leaves, taller than wide, somewhat tapered or rounded, minutely ciliate. Leaf-blade 4–16 cm × 1.5–4 mm, flat, finely scabrid, adaxially finely ribbed; margins scabrid, tip acute. Culm erect, internodes glabrous. Panicle 10–24 cm, linear-lanceolate to spreading, later more contracted; rachis, branches and pedicels sparsely scabrid to almost smooth. Spikelets 2–3 mm, grey-green to purplish; terminal spikelets usually with awned, 5-nerved lemmas, and lateral spikelets with awnless, 3–5-nerved lemmas. Glumes subequal, elliptic-lanceolate, acute or acuminate; lower with keel scabrid in upper ½, upper slightly shorter, smooth to scabrid on keel near tip. Lemma 1.6–2.2 mm, (3)–5-nerved, ovate-oblong, truncate-lobulate, the outer lateral nerves usually minutely excurrent, awned lemmas often with a few hairs in lower ½ near margin, awnless lemmas glabrous; awn, when present, from near base of lemma and ≈ lemma or up to twice length of lemma, occasionally awn very minute, middorsal or subapical and not projecting beyond lemma. Palea ½–⅔ length of lemma, apex shallowly bifid. Callus with minute tufts of hairs on either side, or sometimes glabrous in spikelets with awnless, 3-nerved, glabrous lemmas. Rachilla sometimes prolonged and tipped with a small tuft of hairs in awned spikelets. Anthers 1–1.5 mm. Caryopsis *c.* 1 × 0.4 mm.

Naturalised from Mediterranean.

N.: throughout; S.: throughout, but most common in Canterbury and Otago; C. Lowland to montane on light soils and dry, often stony waste ground, roadsides, track margins, and river terraces; sometimes found in dry hill-country grassland with *A. capillaris.*

Dryland browntop is now used as a specialist lawn grass; because it is a frequent contaminant of browntop lawn seed it has become a common garden weed.

Dryland browntop was first recorded for N.Z. as a native ecotype of *A. capillaris* originating in Canterbury and differing from normal browntop in having finer, grey-green leaves, a stronger rhizome, longer ligule, and more compact, coarser panicles.

In typical *A. castellana*, all spikelets are awned from near the base of the lemma and the lemma is pubescent on the margins of the lower half, and 5-nerved, the outer lateral nerves being conspicuously excurrent (to 0.5 mm), with the rachilla shortly prolonged and bearded, and the callus laterally bearded with hairs to 0.6 mm. However, variants of *A. castellana* occur lacking pubescence on the back of the lemma, or the awn, or prolonged rachilla, but still recognised by the bearded callus. Some of the awnless variants have 3-nerved lemmas. From these variants of *A. castellana* there is a series of intermediates to florets which would be recognised as *A. capillaris*. Plants of N.Z. dryland browntop are often found with spikelets of two kinds in the same panicle. The terminal spikelets of the branches and branchlets have basally awned lemmas, excurrent lateral nerves, and a hairy callus, whereas the lateral spikelets are awnless with a glabrous callus and have 3–4-nerved glabrous lemmas with only minutely excurrent lateral nerves. The description of *A. castellana* encompasses all these variants, which probably result from hybridism with *A. capillaris*, and the two spp. cannot be clearly separated by spikelet characters.

3. A. dyeri Petrie, *T.N.Z.I. 22*: 441 (1890) ≡ *A. dyeri* Petrie var. *dyeri* (autonym, Hackel *in* Cheeseman 1906 op. cit. p. 865); Lectotype: WELT 69294! *D. Petrie* Ruahine Mountains (E. side) 4500 ft, Jan. 1889 (No 1112 to Hackel) (designated by Edgar and Forde 1991 op. cit. p. 148). = *A. dyeri* var. *aristata* Hack., *in* Cheeseman *Man. N.Z. Fl.* 865 (1906); Lectotype: W 36259 (specimen on right)! *D. Petrie* Clinton Valley, Lake Te Anau [Jan. 1892] (No 1091 to Hackel) (designated by Edgar and Forde 1991 op. cit. p. 149).

Tufts usually stiff, occasionally laxer with wide leaves, (10)–15–55 cm, with culms usually much overtopping leaves; branching intravaginal. Leaf-sheath firmly membranous, light green to light greyish brown, distinctly ribbed, smooth, or scabrid above near ligule, or scabrid throughout, or only basal sheaths scabrid, later shredding into fibres. Ligule 0.8–2.2 mm, truncate, erose to lacerate, abaxially smooth or scabrid. Leaf-blade 3.5–10–(17.5) cm × 1–2.5–(4) mm, usually flat, abaxially smooth or finely scabrid on ribs, very rarely densely scabrid throughout, adaxially finely scabrid, tip acute to acuminate. Culm erect or geniculate at base, internodes glabrous. Panicle (4)–5–12–(20) cm, narrowly branched, lanceolate, either contracted after flowering but > 5 mm wide, or remaining open; rachis smooth, branches and pedicels filiform, sparsely scabrid. Spikelets *c.* 3–3.5–(4) mm, green to brownish purple. Glumes ± equal, smooth, elliptic-lanceolate, acute, keel finely scabrid above, margins scabrid near tip. Lemma 2.3–3.3 mm, glabrous, 5-nerved, ovate-elliptic, truncate, minutely denticulate; awn 0, or rarely fine, straight, scabrid, to *c.* 2 mm, arising from midpoint to upper ⅓ of lemma, scarcely or not projecting beyond glumes. Palea 0.5–0.8 mm, ovate. Lodicules ≤ palea. Callus with small tuft of hairs or only a few minute hairs below each lemma margin. Anthers 0.6–1.2 mm. Caryopsis 1–1.8 × 0.3–0.5 mm. $2n = 42$.

Endemic.

N.: in southern mountains; S.: mountains along and west of Main Divide and in Fiordland, on the east in Marlborough and Canterbury and occasionally further south. Tussock grassland, scrub, herbfield or scree; subalpine to alpine.

Two forms have been recognised within *A. dyeri*, see Druce, A. P. *in* Druce, A. P. et al. *N.Z. J. Bot. 25:* 41–78 (1987). Throughout its range in North Id and in wetter localities in South Id the panicle is contracted after flowering, all leaf-sheaths are smooth and anthers usually 0.6–0.8 mm. In South Id on drier sites the panicle remains open after flowering, all leaf-sheaths or at least the basal sheaths are scabrid and the anthers usually 0.8–1.2 mm. Druce (1987 op. cit.) noted "Occasionally plants of the two [forms] may be found growing side by side in the same community".

4. A. gigantea Roth, *Tent. Fl. Germ. 1*: 31 (1788).   redtop

Tufted, bright green, wide-leaved perennials, 60–120 cm, with strong rhizomes; branching intra- or extravaginal, with the extravaginal shoots spreading underground as long rhizomes. Leaf-sheath smooth, or very minutely scabrid, especially near margins. Ligule 1.5–6 mm, denticulate, taller than wide. Leaf-blade 6–15 cm × 3–8 mm, flat, finely scabrid on ribs and on margins, tip acute. Culm erect or geniculate at base, sometimes trailing, internodes glabrous. Panicle 18–30 cm, pyramidal, erect, branches spreading almost horizontally; rachis usually scabrid above, branches and pedicels closely scabrid. Spikelets 1.8–2.2 mm, green or purplish. Glumes subequal, lanceolate, acute; lower = spikelet, scabrid on upper ½ of midnerve, upper usually slightly shorter, scabrid towards tip with less noticeable prickle-teeth. Lemma 1.5–2 mm, glabrous, ovate, truncate, 3–5-nerved, lateral nerves usually very shortly excurrent, midnerve sometimes produced to very short awn subapically, or dorsally ± at midpoint, usually not projecting beyond lemma tip. Palea ½–⅔ length of lemma, bifid or ± truncate. Callus usually with very short hairs. Anthers 1.3–1.6 mm. Caryopsis *c.* 1 × 0.5 mm.

Naturalised from Eurasia.

N.: southern Waikato; S.: Southland. Damp waste ground.

Agrostis gigantea is the most robust of the bent grasses with the widest leaves, longest ligules and largest panicle. Levy, E. B. *N.Z. J. Agric. 28*: 73–91 (1924), noted that it forms a coarse, open turf, as tufts appear at fairly widely spaced intervals on the long-creeping, strong rhizome. Though redtop was extensively sown in N.Z. in the latter years of the 19th Century it has not persisted in pastures or in the wild except in southern Waikato and in Southland.

5. A. imbecilla Zotov, *T.R.S.N.Z. 73*: 233 (1943) ≡ *A. tenella* Petrie, *T.N.Z.I. 22*: 442 (1890) non Hoffm. (1800); Holotype: WELT 69601! *D. P[etrie]* Macraes, *c.* 1800 ft, Waihemo Co, Otago, Feby 1889.

Very slender, lax, perennial tufts, 15–35 cm, with long, very slender, spike-like panicles on slender culms much overtopping leaves; branching extravaginal. Leaf-sheath narrow, membranous, light creamish brown to reddish, glabrous, distinctly ribbed. Ligule 0.8–1.6 mm, obtuse, often erose, abaxially glabrous. Leaf-blade 2–13 cm × 0.2–0.4 mm diam., involute, filiform, flaccid, abaxially finely striate and, in young leaves, finely scabrid, later smooth, adaxially scabrid on ribs; margins scabrid, tip fine, blunt. Culm almost filiform, erect or geniculate at base, internodes glabrous. Panicle (2.5)–5–9 cm × 2–3 mm, contracted, extremely narrow-linear; rachis and few short erect branches and pedicels sparsely, finely scabrid, often pale purplish. Spikelets 1.5–2.4 mm, pale greenish purple to straw coloured. Glumes ± equal, lanceolate, minutely scabrid on keel and on margins near acute tip. Lemma 1.5–1.9 mm, at times ≈ glumes, glabrous, faintly 5-nerved, ovate-oblong, apex denticulate, awnless, or rarely a few lemmas subapically awned, awn straight, fine, up to 1 mm, scarcely projecting beyond glumes. Palea 0.1–0.2 mm, ovate. Lodicules 0.3–0.4 mm. Callus with few very minute hairs. Anthers 0.6–1 mm. Caryopsis 1–1.4 × 0.3 mm.

Endemic.

N.: one record from Upper Moawhango R., Kaimanawa Mts; S.: scattered in Canterbury and Otago. Tussock grassland in inland, montane to subalpine regions, often in damp places.

6. A. magellanica Lam., *Tabl. Encycl. 1*: 160 (1791); Holotype: P! *Commerson* Magellanes. = *A. multiculmis* Hook.f., *Fl. Antarct. 1*: 95 (1845); Holotype: K! *J. D. Hooker 1628* on sloping banks about 500 ft above the sea on the west side of Campbell's Island, prostrate, Decr. 1840.

Tufted leafy perennials, very variable in size, usually ± erect, but occasionally with prostrate culms, 5–45 cm, leaves overtopping panicles, or panicles projecting beyond leaves; branching extravaginal. Leaf-sheath firmly membranous, very pale creamy brown, closely ribbed, glabrous, later shredding into fibres. Ligule 1–5 mm, truncate, irregularly denticulate, abaxially scabrid. Leaf-blade 2–16 cm × 0.5–1 mm diam., folded and inrolled, sometimes flattish and up to 2 mm wide, tough, abaxially smooth with midrib prominent only near base and scabrid near acute tip, adaxially strongly, closely ribbed, ribs finely prickle-toothed; margins scabrid. Culm erect or slightly geniculate at base, internodes finely scabrid below panicle or occasionally smooth. Panicle 2–10–(12) cm, contracted, erect, densely elliptic- or linear-oblong; rachis, branches, and pedicels erect and closely scabrid. Spikelets 3.5–5 mm, greenish purple. Glumes subequal, usually finely scabrid especially near acuminate tip, keel strongly scabrid throughout or only above. Lemma (1.5)–1.8–2.2–(2.5) mm, glabrous, 5-nerved with nerves thickened and sometimes minutely scabrid above, ovate, truncate and shortly 4-toothed, always awned; awn 3–4–(5) mm, middorsal, scabrid, very fine, usually flexuous and twisted near base, often projecting well beyond glumes. Palea 0.4–0.7 mm, ovate-elliptic. Lodicules 0.5–0.6 mm. Callus with minute tufts of hair on either side of lemma base. Anthers 0.5–0.9 mm. Caryopsis 1.2–1.6 × 0.4–0.5 mm. $2n = 84$.

Indigenous.

S.: in south-western regions; Ant., A., C., M. On stony or rocky ground, and also in dry to wet peat and moss cushions in the Subantarctic Islands; subalpine to alpine, and at lower altitudes in the Subantarctic.

Also indigenous to South America (from Chile to Tierra del Fuego), and on Falkland, Kerguelen, Crozet, and Marion Is.

In southern South America much taller, broader-leaved plants of damp ground are referred to as *A. magellanica*, whereas *A. meyenii* Trin. is an alpine grass with stature and habit similar to *A. magellanica* from South Id (P. Wardle pers. comm.).

7. A. muelleriana Vickery, *Contrib. N.S.W. Natl Herb. 1*: 103 (1941) ≡ *A. canina* var. β Hook.f., *Handbk N.Z. Fl.* 328 (1864) ≡ *A. gelida* F.Muell., *Trans. Vict. Inst.* 43 (1855) non Trin. (1845) ≡ *A. canina* var. β *gelida* (F.Muell.) Buchanan, *Indig. Grasses N.Z.* t. 20A (1878) ≡ *A. muelleri* Benth., *Fl. Austral. 7*: 576 (1878) non C.Presl (1844); Holotype: K! *F. Mueller* [Australia, Victoria], Cobboras [Cobberas] Mountains, 6000 feet.

Dense, often small, dull green, usually strict, perennial tufts, (3)–5–30 cm, with narrow, usually dark purplish, sometimes green spike-like panicles on slender culms much overtopping to scarcely overtopping leaves; branching intravaginal. Leaf-sheath ± membranous, light green or pale brownish, glabrous, ribs few distinct, rarely minutely scabrid above. Ligule 1–2–(3) mm, acute to truncate, usually fimbriate to lacerate, abaxially scabrid. Leaf-blade 1–7 cm × 0.4–0.7 mm diam., folded with involute margins, often filiform, or strict and firm, ± erect, sometimes more flattened and wider, to 2 mm, abaxially smooth or finely scabrid, adaxially finely scabrid on ribs; margins finely scabrid, narrowed to fine subobtuse tip; uppermost culm leaves abaxially scabrid. Culm erect, internodes glabrous. Panicle 1–4.5–(10) cm, contracted, erect, densely narrow-linear, longer panicles densely oblong, > 5 mm wide; rachis smooth or obscurely scabrid, branches and pedicels short, erect, capillary, finely scabrid. Spikelets (1.5)–2–3–(3.5) mm. Glumes ± unequal, lanceolate to ovate-lanceolate, acute to subobtuse, usually tinged deep reddish purple, occasionally entirely green, margins colourless or purplish, usually finely scabrid on keel and at times near tip, otherwise smooth or papillose; lower = spikelet, upper (1.4)–2–2.5–(3) mm. Lemma 1.4–2 mm, glabrous, 5-nerved, lateral nerves often faint, oblong-ovate, obtuse or truncate, usually minutely denticulate; awn 0, or present and dorsal, from midpoint or from just below upper ⅓, 0.5–2–(2.5) mm, delicate, usually projecting beyond glumes. Palea *c.* 0.4 mm, ovate. Lodicules 0.3–0.4 mm. Callus with very few minute hairs or glabrous. Anthers 0.4–1 mm. Caryopsis *c.* 1 × 0.3–0.5 mm.

Indigenous.

N.: central mountains; S.: mountains throughout. Montane to alpine in rocky ground or in seepages.

Also indigenous to Australia; New South Wales, Victoria, and Tasmania.

Specimens from Mt Cook, western Otago, and Fiordland are often more luxuriant than usual, with wider, flat leaves, and wider panicles to 10 cm long. Sometimes the lower branches of the panicle may be slightly spreading but the spikelets are always densely crowded. These plants are often awned and resemble *A. magellanica* but the spikelets are smaller and the awns shorter and finer.

Although *A. muelleriana* resembles *A. subulata* in the short, contracted panicle, the glumes, panicle branches, and leaves of *A. muelleriana* are distinctly less scabrid than those of *A. subulata* which seems confined to Auckland and Campbell Is.

8. A. muscosa Kirk, *T.N.Z.I. 13*: 385 (1881); Lectotype: WELT 69300! *T. Kirk* Lake Wanaka, 8 Jan., 1877 (designated by Edgar and Forde 1991 op. cit. p. 147). = *A. aemula* subsp. β *spathacea* Berggr., *Minneskr. Fisiog. Sällsk. Lund* Art. *8*: 32, t. 7, figs 41–47 (1878); Lectotype: LD! *S. Berggren* Omatangi, Jan. 1875 (designated by Edgar and Forde 1991 op. cit. p. 147). = *A. parviflora* var. *perpusilla* Hook.f., *Fl. N.Z. 1*: 296 (1853) ≡ *A. canina* var. γ Hook.f., *Handbk N.Z. Fl.* 328 (1864); Holotype: K! *Colenso 1731* Summit of Ruahine Mountains, N. Zealand.

pincushion grass

Small, tight, pale green or glaucous perennials, forming rounded cushions or mats 0.6–2–(6) cm high and up to 10 cm diam., with many culms, much-branched at base; branching intravaginal. Leaf-sheath hyaline, whitish, glabrous, conspicuously, sparsely ribbed, usually completely concealing culms, much wider than and sometimes longer than leaf-blade. Ligule 0.5–1.5–(4) mm, truncate to acute, usually lacerate or fimbriate, abaxially with scattered prickle-teeth. Leaf-blade (0.1)–0.4–1.2–(4) cm × 0.2–0.4 mm diam., involute, spreading and recurved, abaxially smooth, adaxially ribbed, finely scabrid on ribs; margins minutely scabrid, tip obtuse. Culm erect, internodes glabrous. Panicle usually almost hidden among leaf-sheaths, often recurved, contracted to a minute, dense, ovate head 3.5–7 mm, with few or many spikelets; rachis, branches and pedicels very short, minutely, sparsely scabrid. Spikelets 1.5–2 mm, greenish. Glumes ± equal, ovate-lanceolate, acute, keel green, scabrid, margins wide, glabrous, hyaline, minutely ciliate near tip. Lemma 1.2–1.4 mm, glabrous, faintly 5-nerved, ovate-oblong, truncate, awnless. Palea 0.2–0.4 mm, orbicular or ovate. Lodicules 0.2–0.3 mm. Callus glabrous. Anthers 0.4–0.6 mm. Caryopsis 0.7–1 × 0.3–0.4 mm. $2n = 42$.

Endemic.

N.: eastern Bay of Plenty, Volcanic Plateau, Ruahine and Tararua Ranges, Mt Egmont and on Taranaki coast; S.: throughout except in Westland and Fiordland, more common in South Canterbury, Otago, and Southland; St. Montane to subalpine, rarely alpine, on stony bare ground in tussock grassland or on margins of tarns and lakes; lowland in turf above coastal cliffs.

9. A. oresbia Edgar, *N.Z. J. Bot. 29:* 143 (1991); Holotype: CHR 132825! *A. P. Druce* Park V[alley] cirque, Tararua Ra, watercourse, Jan. 1965.

Bright green perennial tufts 5–25 cm, with culms overtopping leaves, occasionally stoloniferous; branching intravaginal. Leaf-sheath hyaline, light green to light brown, prominently ribbed, usually smooth, rarely with sparse prickle-teeth. Ligule 1–4.5 mm, truncate to obtuse, denticulate or entire, glabrous. Leaf-blade 2–7 cm × 1.5–2 mm, flat, or folded and *c.* 0.5 mm diam., abaxially smooth, rarely uppermost leaf scabrid, adaxially prominently ribbed, ribs usually scabrid; margins often entirely smooth, sometimes sparsely, or rarely densely, minutely scabrid, tip fine, blunt, scabrid. Culm erect, internodes glabrous. Panicle (1.5)–2–6 cm, open, laxly oblong to pyramidal, with spreading sometimes flexuous branches; rachis smooth, branches and pedicels smooth or scabrid. Spikelets 2–2.5–(3) mm, usually purplish. Glumes ± unequal, acute to acuminate, keel finely scabrid near tip, margins smooth, rarely scabrid at tip; lower = spikelet, ovate-lanceolate, upper 1.7–2.5–(2.9) mm, elliptic-lanceolate. Lemma 1.5–2 mm, glabrous, faintly 5-nerved, ovate, obtuse; awn (1.5)–2–3 mm, usually middorsal, usually geniculate, sometimes ± straight, slightly projecting beyond glumes. Palea *c.* 0.5 mm, ovate. Lodicules *c.* 0.3 mm. Callus with minute hairs. Anthers (0.4)–0.6–0.9 mm. Caryopsis *c.* 1.5 × 0.7 mm.

Endemic.

N.: Raukumara, Ruahine, and Tararua Ranges; S.: north-west Nelson at Lake Aorere and Mt Domett. Tussock grassland, often in watercourses and seepages, and on rocky ground, cliffs, screes, river flats, usually in shade; subalpine to alpine.

Specimens from north-west Nelson have narrow glumes and small anthers, 0.4–0.5 mm.

Agrostis oresbia resembles *A. magellanica* in the consistently awned lemmas but is smaller in all respects with a laxer panicle. It seems nearest to *A. muelleriana* especially in the size and colour of the spikelets, but again *A. oresbia* has a more lax panicle, and the glumes are less scabrid and less papillose than in *A. muelleriana.*

10. A. pallescens Cheeseman, *T.N.Z.I. 53*: 423 (1921) ≡ *A. muelleri* var. *paludosa* Hack., *in* Cheeseman *Man. N.Z. Fl.* 864 (1906); Lectotype: W 36804! *T. F. C[heeseman]* swamps in the Tasman Valley, Mt. Cook, alt. 2000 ft [Jan. 1897] (No 1110 to Hackel) (designated by Edgar and Forde 1991 op. cit. p. 152).

Small, finely rhizomatous tufts 3–5–(20) cm, with fine, pale green or blue-green, to light brown leaves, and narrowly branched, delicate panicles usually scarcely overtopping leaves; branching extravaginal at base, intravaginal above. Leaf-sheath ± hyaline, light green to light brown, glabrous, ribs few, prominent. Ligule 0.1–0.6 mm, truncate, minutely ciliate or erose, abaxially glabrous. Leaf-blade 0.8–2.5 cm × 0.2–0.4 mm diam., narrow-linear, involute, filiform, abaxially smooth except near blunt, minutely scabrid tip, adaxially usually finely scabrid on ribs. Culm slender, erect, internodes glabrous. Panicle 0.6–2–(3) cm, open, ovate to pyramidal, with few, minutely, sparsely ciliate branches each tipped by a single spikelet. Spikelets (1.2)–1.5–2 mm, pale green to light creamy brown, sometimes purplish. Glumes ± equal, and usually ≈ lemma, ovate-lanceolate, subobtuse, keel scabrid near tip, margins sometimes finely ciliate near tip. Lemma 1.2–1.8 mm, glabrous, faintly 5-nerved, ovate, obtuse or almost truncate, awnless. Palea 0.4–0.6 mm, ovate. Lodicules 0.2–0.4 mm. Callus glabrous. Anthers 0.4–0.7–(1) mm. Caryopsis *c.* 1 × 0.5 mm.

Endemic.

N.: central mountains; S.: mountains in Nelson, along Main Divide, and to the east, rare in Fiordland; St. Boggy ground, seepages in tussock grassland; montane to alpine, lowland in the south.

Larger plants of *A. pallescens* resemble very small forms of *A. personata* in shape of panicle, and number and size of spikelets; *A. pallescens*, however, has smooth lower leaf-sheaths, and subobtuse glumes slightly > lemma whereas in *A. personata* leaf-sheaths are often scabrid and the acute glumes are obviously > lemma.

11. A. personata Edgar, *N.Z. J. Bot. 29*: 149 (1991) ≡ *A. dyeri* var. *delicatior* Hack., *in* Cheeseman *Man. N.Z. Fl.* 865 (1906); Lectotype: W 36260 (specimen on left)! *D. Petrie* Lake Te Anau, Otago [Jan. 1892] (No 1089 to Hackel) (designated by Edgar and Forde 1991 op. cit. p. 149).

Lax, slender tufts 6.5–30–(60) cm, with delicate panicles equalling or overtopping the usually soft flaccid leaves; branching extravaginal. Leaf-sheath membranous, light green to light brown; lower sheaths often with minute retrorse prickle-teeth or hairs, upper sheaths glabrous. Ligule 0.8–2 mm, truncate to ± obtuse, slightly erose, abaxially finely scabrid. Leaf-blade 2–8 cm × 0.5–1.5–(3) mm, flat, or folded and inrolled at margins, abaxially smooth, or sometimes scabrid, especially towards tip and on uppermost culm-leaves, adaxially ciliate-scabrid on ribs; margins sparsely scabrid, occasionally almost smooth, tip acute, scabrid. Culm usually geniculate at base or erect, internodes glabrous. Panicle (2)–4.5–8.5–(16) cm, ovate-lanceolate to pyramidal with filiform branches naked below and spreading widely at maturity; rachis, branches and pedicels sparsely scabrid. Spikelets *c.* 2–2.5–(3) mm, purplish green. Glumes subequal, acute, with greenish central portion and wide, colourless to purplish margins scabrid near tip; lower narrow-lanceolate, keel scabrid in upper ⅔, upper ovate-lanceolate, keel scabrid in upper ⅓ to ½. Lemma 1.4–2–(2.2) mm, glabrous, 5-nerved, ovate-oblong, truncate, ± denticulate, awnless or with delicate awn 1–3 mm, inserted at *c.* midpoint to *c.* upper ⅓ of lemma. Palea 0.3–0.6 mm, orbicular. Lodicules 0.3–0.6 mm, = or usually > palea. Callus with tufts of minute hairs. Anthers 0.4–0.7–(0.9) mm. Caryopsis 1–1.5 × 0.4–0.6 mm. 2*n* = 42.

Endemic.

N.: south from East Cape, Volcanic Plateau, and Mt Egmont; S.: common
west of Main Divide and in Fiordland, to the east in scattered localities;
St. Open ground in tussock grassland, in scrub, or forest margins,
sometimes in seepage; lowland to alpine.

Specimens collected from Teneriffe, Wairarapa are very slender, with spikelets
and anthers smaller than usual, but still match *A. personata* in extravaginal
branching and ± ascending panicle branches.

This sp. has been recorded under many different names. Early collections were
referred to European *A. canina* L., which differs from *A. personata* in the longer
anthers (1–1.5 mm), and acute ligules. Hackel (*in* Cheeseman 1906 op. cit.)
described Fiordland plants as *A. dyeri* var. *delicatior*, but *A. dyeri* is well distinct in
its longer spikelets, longer anthers, and intravaginal branching. Plants of
A. personata were also equated with Australian *A. parviflora* R.Br., and North
American *A. scabra* Willd., and *A. perennans* (Walter) Tuck.; the differences
between *A. personata* and these three spp. are outlined by Edgar and Forde (1991
op. cit. pp. 151–152).

12. A. petriei Hack., *T.N.Z.I. 35*: 379 (1903) ≡ *A. petriei* Hack. var.
petriei (autonym, Hackel 1903 op. cit. p. 379); Holotype: W 36494!
D. Petrie Nevis Valley, Hector Mts, Central Otago [Feb. 1890] (No
1092 to Hackel). = *A. petriei* var. *mutica* Hack., *T.N.Z.I. 35*: 379 (1903);
Holotype: W 7926! *D. Petrie* Cromwell, Central Otago (No 1085 to
Hackel).

Perennial, loose, tussocky, bluish green or greyish green clumps (15)–
30–55 cm, with slender, strictly involute, finely scabrid leaves ≪ culms;
branching extravaginal. Leaf-sheath firm, light brown; lower sheaths
minutely scabrid, upper sheaths smooth, sparsely ribbed. Ligule
1–4.5 mm, obtuse to truncate, denticulate, abaxially scabrid. Leaf-
blade 4–10.5–(18) cm × 0.3–1 mm, stiff and wiry, densely minutely
papillose-scabrid throughout, tip obtuse. Culm erect or geniculate at
base, internodes glabrous. Panicle 4–16 cm, laxly oblong; rachis smooth,
branches and pedicels ± spreading, very delicate, reddish brown, finely,
sparsely scabrid. Spikelets 2.5–3.4 mm, pale greenish to brownish,
sometimes purplish. Glumes ± equal, lanceolate, keel and margins usually
finely scabrid near acute tip. Lemma 2–2.6 mm, glabrous, 5-nerved, ovate,
obtuse to truncate, minutely denticulate; awn (1.5)–2–3 mm, middorsal,

straight or slightly flexuous, sometimes 0. Palea 0.2–0.4 mm, orbicular. Lodicules 0.2–0.4 mm. Callus with very few short hairs. Anthers (*c.* 1)–1.3–1.8 mm. Caryopsis *c.* 1.5 × 0.5 mm.

Endemic.

S.: Mid and South Canterbury, mostly inland; Central Otago. Montane to subalpine on dry stony ground on river flats or in tussock grassland, on rock outcrops.

13. A. stolonifera L., *Sp. Pl.* 62 (1753).   creeping bent

Perennial tufts of variable habit, to *c.* 100 cm, spreading by long stolons, sometimes low-growing and trailing, or turf-forming, sometimes loosely or densely tufted; branching intra- and/or extravaginal, the intravaginal shoots often trailing and elongating as stolons or sometimes erect, the extravaginal shoots at once ascending with 2–3 scale leaves at base. Leaf-sheath smooth or slightly scabrid, green or purple-tinged. Ligule (1)–2–6 mm, rounded, fimbriate, taller than wide. Leaf-blade 1–20 cm × *c.* 1–8 mm, flat, minutely scabrid on ribs, tip acute. Culm geniculate to rarely erect, or trailing and branching, rooting at lower nodes, internodes glabrous. Panicle 3–28 × 0.5–2.5–(6) cm, lanceolate to ovate, lax at flowering, later ± contracted with branches ± erect and appressed to rachis; rachis smooth, branches and pedicels scabrid. Spikelets 1.5–2.5–(3) mm, greenish or yellowish, becoming brownish or purplish. Glumes ± unequal, elliptic-lanceolate, sometimes slightly scabrid above, acute to acuminate, midnerve scabrid above. Lemma 1.3–2 mm, ± truncate, smooth to ± scabrid below, 5-nerved, nerves sometimes scabrid, outer lateral nerves usually slightly excurrent, apex minutely lobed above inner lateral nerves, midnerve not reaching lemma apex, usually terminating as an extremely minute projection in upper ⅓ of lemma, or midnerve rarely produced as a short, subterminal awn. Palea usually ½ to ⅔ length of lemma, apex usually obtuse, rarely shallowly notched. Callus glabrous or with very few minute hairs. Anthers 0.9–1.5 mm. Caryopsis *c.* 1 × 0.4 mm. Plate 2B.

Naturalised.

N.; S.; St.: throughout; Ch., A., C. Sea level to subalpine, often in damp or boggy sites, in disturbed ground, modified grassland or open scrub.

Indigenous to Europe, temperate Asia, North America.

14. A. subulata Hook.f., *Fl. Antarct. 1*: 95, t. 53 (1845); Holotype: K! *J. D. Hooker 1627* Alpine grass; grows rather prostrate on the highest rocks on the mts of Campbell's Island, Dec. 1840.

Small, dense perennials, forming tufts 3–9 cm, with strict leaves often overtopping the light green spike-like panicles; branching intravaginal. Leaf-sheath ± hyaline, light brown, strongly ribbed, finely scabrid especially near margins above. Ligule 0.8–1–(1.6) mm, denticulate, abaxially scabrid. Leaf-blade 1.5–4 cm × 0.3–0.6 mm, firm, inrolled, strongly ribbed, minutely, closely scabrid, tip blunt. Culm hidden among leaves, erect, internodes scabrid. Panicle 1–2 cm, contracted, densely oblong; rachis, branches, and pedicels erect and closely short-scabrid. Spikelets 2–3.3 mm. Glumes subequal, lanceolate, pale green, rarely tinged light purple, with antrorse prickle-teeth or minute hairs, but glabrous near base. Lemma 1.3–1.6 mm, glabrous, distinctly 5-nerved, ovate, truncate, minutely denticulate, margins minutely scabrid near apex, usually awnless, but occasionally with delicate awn to 2 mm from upper ⅓ of midnerve. Palea 0.3–0.4 mm, ovate. Lodicules 0.4 mm. Callus glabrous. Anthers (0.4)–0.5–0.7 mm. Caryopsis *c.* 1 × 0.5 mm.

Endemic.

A., C. On peat-covered rock-ledges, and in *Chionochloa* tussock or in herbfields.

CYTOLOGY   Beuzenberg, E. J. and Hair, J. B. *N.Z. J. Bot. 21*: 13–20 (1983), reported chromosome counts for some indigenous spp.; all were hexaploid. For *A. magellanica* from Campbell Id, Beuzenberg and Hair recorded *n* = 42. For *A. magellanica* from Macquarie Id, Moore, D. M. *Bot. Not. 113:* 185–191 (1960), recorded 2*n* = 72, which does not fit well with other numbers for *Agrostis*.

×Agropogon littoralis (Sm.) C.E.Hubb., *J. Ecol. 33*: 333 (1946).

Agrostis stolonifera × *Polypogon monspeliensis*　　　perennial beard grass

Loose tufts, 15–40 cm, sometimes stoloniferous. Leaf-sheath distinctly ribbed. Ligule 2–5.5 mm, truncate or rounded, becoming lacerate. Leaf-blade 1.5–7 cm × 1–3 mm, flat, scabrid. Culm geniculate at base, erect or ascending above, internodes glabrous. Panicle 2–8 × 0.5–2.2 cm, dense, oblong or lanceolate, often lobed, brownish green; rachis smooth, branches and short pedicels scabrid. Spikelets persistent, 3–5 mm, green to purplish brown. Glumes = spikelet, firmly membranous, very finely scabrid; awn apical, 1–2.3 mm, straight, fine, finely scabrid. Lemma < glumes, 1.2–1.8 mm, glabrous, elliptic-oblong, truncate, apex irregularly and very shortly denticulate; awn very fine, subapical, 0.7–2 mm. Palea < lemma, glabrous, finely 2-nerved. Callus glabrous. Stamens 3; anthers 0.7–1 mm, pollenless. Caryopsis usually not developed, very rarely present and 1–1.5 × *c.* 0.5 mm.

Of probable natural hybrid origin in New Zealand rarely occurring in Buller (Orowaiti estuary near Westport) and Canterbury (Kairaki, Le Bons Bay, Lake Ellesmere) on salt flats and damp sandy or silty ground near the coast, growing with *Polypogon monspeliensis* and *Agrostis stolonifera*.

×*Agropogon littoralis* occurs sporadically in coastal areas in the Northern Hemisphere and Australia; the hybrids are intermediate between the parents in floral characters, with conspicuous awns and persistent spikelets; vegetatively they resemble *Agrostis stolonifera*.

Excluded Species　　*Agrostis canina* L., velvet bent, was first recorded for N.Z. by Hooker, J. D. *Fl. N.Z. 1*: 296 (1853), but this was a misidentification of endemic *A. personata*. Velvet bent was recorded as naturalised from Europe by Kirk, T. *T.N.Z.I. 3*: 160 (1871), but Allan, H. H. *N.Z. DSIR Bull. 49*: 108 (1936), considered it "very rare in the naturalised state". *Agrostis canina* has occasionally been cultivated in N.Z. as a lawn grass but no specimens collected from the wild have been seen. It is a loosely tufted, often stoloniferous perennial, rather similar to *A. capillaris* and *A. castellana* but the ligules are acute and the palea is minute.

ALOPECURUS L., 1753

Low growing to moderately tall perennials or annuals. Leaf-sheath open. Ligule membranous. Leaf-blade flat, striate, midrib inconspicuous. Culm few- to many-noded, sometimes geniculate at base. Panicle cylindric, spike-like; rachis stout, culm-like, branches and pedicels very short and slender. Spikelets 1-flowered, protogynous, strongly laterally compressed, disarticulating below glumes, falling entire at maturity. Glumes similar, ≈ lemma, keeled, 3-nerved, margins often united below. Lemma membranous, 4–5-nerved, margins often united below, usually awned dorsally from near midpoint or below, awn included within glumes, or often projecting beyond spikelet. Palea 0. Lodicules 0. Callus glabrous. Rachilla not prolonged. Stamens 3. Ovary glabrous; styles fused near base. Caryopsis lenticular; embryo small; hilum punctiform to very short; endosperm liquid or solid.

Type species: *A. pratensis* L.

36 spp. of temperate Eurasia and South America. Naturalised spp. 3; transient sp. 1.

1 Glumes obtuse, united only at base *2*
  Glumes acute, united for up to ½ length *3*
2 Awn inserted near lemma base and projecting *c.* 2–3 mm beyond spikelet
  .... **2. geniculatus**
  Awn inserted just below midpoint of lemma keel and included within spikelet,
  or projecting < 1.5 mm beyond spikelet .... **1. aequalis**
3 Lemma keel ciliate .... **3. pratensis**
  Lemma keel smooth .... **myosuroides**ζ (p. 245)

1. A. aequalis Sobol., *Fl. Petrop.* 16 (1799).   orange foxtail

Annuals or short-lived perennials, with culms usually many-noded and ascending from a geniculate or prostrate base, often rooting at nodes. Leaf-sheath very light brown, glabrous, striate, subhyaline, lower sheaths usually purplish. Ligule (1)–2.5–5 mm, membranous, obtuse, entire, glabrous, but abaxially with short fine hairs near base. Leaf-blade 3.5–8 cm × 1–2.5 mm, flat, finely ribbed, minutely scabrid on ribs and margins, tapering to acute, long-triangular tip. Culm 15–25–(35) cm, internodes ridged, glabrous. Panicle 2–5.5 cm, spike-like, with densely crowded silver-green spikelets. Glumes 1.5–2.5 mm, ± equal, 3-nerved, narrowly elliptic-oblong, blunt, membranous, with scattered hairs, keeled, keel silky-ciliate, margins united near base. Lemma ≈ glumes, 4-nerved, broadly elliptic, very blunt, thinly membranous, glabrous, keeled, margins united below for *c.* ½ length; awn 0.8–1.5 mm, delicate, inserted near midpoint of keel, scarcely reaching top of lemma or projecting *c.* 0.5 mm beyond glumes. Palea 0. Anthers 0.6–1.2 mm, bright orange or golden-yellow. Caryopsis 1–1.6 × 0.7–0.8 mm.

Naturalised from Eurasia.

N.: Pukekohe and near Gisborne; S.: to east of Main Divide and at Fiordland lakes. In swampy ground, lowland and coastal or in inland basins.

2. A. geniculatus L., *Sp. Pl.* 60 (1753).   kneed foxtail

Perennials; culms spreading from geniculate base and rooting at nodes. Leaf-sheath light green, glabrous, striate, lower sheaths purplish. Ligule (1.5)–2–4.5 mm, membranous, blunt, or tapering above to a point, entire, glabrous, but abaxially with short fine hairs near base. Leaf-blade 4–12 cm × 1.5–2.5 mm, flat, dark green to occasionally glaucous, finely ribbed, finely scabrid on ribs and margins, prickle-teeth not so evident abaxially, tapering to acute tip. Culm (12)–20–30 cm, internodes ridged, glabrous. Panicle 2–4 cm, spike-like, with densely crowded light green to purplish spikelets. Glumes (2)–2.5–3 mm, ± equal, 3-nerved, narrowly elliptic-oblong, blunt, membranous, short hairy, keeled, keel silky-ciliate, margins united near base. Lemma 2–3 mm, ≈ glumes, 4-nerved, broadly oblong or ovate, thinly membranous, glabrous except for a few hairs on margins at very blunt apex, keeled, margins united for short distance below; awn fine, 2.5–4.5 mm, inserted near base, projecting 1.4–2.5 mm beyond glumes. Palea 0. Anthers 1–1.7 mm, yellow or purple. Caryopsis 1.2–1.5 × 0.5–0.8 mm.

Naturalised from Eurasia.

N.; S.: throughout; St.; Ch., A., C. Widespread in damp, waste ground, along drains, swamp edges, on river flats and in brackish lagoon margins, in drained pools and shingle pits, the stems often floating in water, rarely on dry shingly roadsides; coastal and inland.

3. A. pratensis L., *Sp. Pl.* 60 (1753).   meadow foxtail

Loosely or compactly tufted tall perennials, often shortly rhizomatous. Leaf-sheath green, glabrous, striate, lower sheaths dark brown, sometimes purplish. Ligule (0.5)–1–2–(2.8) mm, membranous, blunt, sparsely ciliate, abaxially very minutely hairy. Leaf-blade 10–20–(30) cm × 2–5 mm, flat, finely ribbed, finely scabrid on ribs and margins, or almost glabrous, tapered to fine pointed tip. Culm 30–100 cm, erect or geniculate at base and rooting at lower nodes, nodes few, internodes ridged, glabrous. Panicle 3.5–9.5 cm, spike-like, with densely packed silvery green to purplish spikelets. Glumes 4–6 mm, ± equal, 3-nerved, margins united for ¼ length, acute, firm, finely hairy below, keeled, keel and lateral nerves fine-ciliate. Lemma 4–5.5 mm, 4-nerved, ovate or elliptic, blunt, membranous, keeled, keel finely hairy, margins united below midpoint; awn fine, 4.5–8–(9.5) mm, inserted near base, projecting 3–4 mm beyond glumes. Palea 0. Anthers 2–3.5 mm, yellow to purple. Caryopsis 1–1.2 × 0.6–0.8 mm.

Naturalised from Eurasia.

N.: scattered, more common near Auckland City; S.: scattered in eastern areas, more common about Christchurch. Damp and swampy pasture, grassland, roadsides, often beside drains, and in damp, grassy waste land.

Spikelets may be proliferous, e.g., CHR 96192 *V. D. Zotov* Palmerston North, May 1935.

ζ**A. myosuroides** Huds.   Culm to 15 cm. Panicle *c.* 5.5 cm. Glumes *c.* 4.5 mm, acute, margins united for ⅓ length, lateral nerves with short hairs near base. Lemma *c.* 5 mm, ≥ glumes, ovate, blunt, membranous, keeled, glabrous, margins united below for *c.* ½ length; awn fine, *c.* 9 mm, inserted near base, projecting to *c.* 5 mm beyond lemma. Palea 0. Anthers and caryopsis not seen. An annual sp., indigenous to Eurasia and North Africa, recorded for N.Z. in 1844 but only one comparatively recent specimen has been found, CHR 215133, Mid Canterbury, 16.1.1941.

AMMOPHILA Host, 1809

Tough, rather coarse perennials with long-creeping rhizomes. Leaves tightly involute, pungent, glaucous. Inflorescence a dense, cylindric, spike-like panicle. Spikelets 1-flowered, strongly compressed laterally; disarticulation above glumes. Glumes ≥ lemma, firm with wide hyaline margin, keeled, persistent; lower 1-nerved, upper 3-nerved. Lemma 3–5-nerved, coriaceous, apex shortly bifid and shortly mucronate-awned. Palea ≈ lemma, subcoriaceous, keels close together. Lodicules 2, nerveless. Rachilla prolonged, hairy. Caryopsis enclosed by hardened anthoecium, longitudinally grooved; embryo small; hilum ± linear, *c.* ⅔ length of caryopsis; endosperm solid.

Type species: *A. arundinacea* Host *nom. illeg.*

2 spp., of Europe, North Africa and North America. Naturalised sp. 1.

1. A. arenaria (L.) Link, *Hort. Berol. 1*: 105 (1827).   marram grass

Long-rhizomatous, wiry-leaved perennials forming compact, glaucous-grey tufts to 170 cm; with strong woody rhizomes and numerous fibrous roots; branching intravaginal. Leaf-sheath glabrous, light straw-brown or brownish purple. Ligule 15–30 mm, membranous, narrowed to a point, abaxially covered with very minute hairs. Leaf-blade 35–60 cm × *c.* 1.5 mm diam., tightly involute, abaxially glabrous, adaxially ribbed, ribs densely short stiff hairy. Culm 70–145 cm, internodes usually scabrid, sometimes smooth to slightly scabrid. Panicle 15–25–(35.5) cm, spike-like, cylindric, tapering above, with numerous crowded light green spikelets; rachis minutely scabrid, erect branches and pedicels hidden among spikelets. Glumes 9.5–14 mm, ± equal, persistent, subhyaline, narrow-lanceolate, glabrous, keel and margins minutely scabrid, tip blunt or obtuse; lower usually 1-nerved, upper usually 3-nerved. Lemma 8.5–13 mm, < glumes, 5-nerved, narrow-lanceolate, obtuse, minutely apiculate, keeled, keel minutely scabrid. Palea 2–4-nerved, subhyaline, central nerves minutely scabrid. Lodicules 1.5–2.5 mm. Callus ringed by long straight silky hairs. Rachilla prolongation bearing long, straight, silky hairs. Anthers 3.5–5.5 mm. Caryopsis *c.* 3 mm.

Naturalised from Europe.

N.; S.: throughout; St.; Ch. Coastal sand dunes and occasionally inland — in Volcanic Plateau and at Erewhon Park in Rangitata Valley.

Ammophila arenaria is an effective binder of drifting sand and has become naturalised in most temperate countries. In N.Z. marram grass is much planted to stabilise sand dunes where it spreads rapidly vegetatively and occasionally from seed. In many areas it has replaced the native *Spinifex sericeus* and the golden sand-sedge or pingao, *Desmoschoenus spiralis*.

DEYEUXIA P.Beauv., 1812

Tufted perennials, sometimes rhizomatous; branching extravaginal. Leaf-sheath rounded. Ligule membranous. Leaf-blade flat, involute, convolute or conduplicate. Culm erect, or decumbent at base, persistent at maturity; nodes glabrous. Inflorescence paniculate, contracted and ± spicate, often lobed below, or sometimes more lax, persistent. Spikelets 1-flowered, rarely 2-flowered, laterally compressed; disarticulation above glumes; rachilla usually prolonged and tipped by a tuft of silky hairs or rarely by a floret, sometimes rachilla prolongation glabrous and very short. Glumes ± equal and equalling the spikelet, lanceolate, acute or acuminate, membranous to submembranous, 1–(3)-nerved, keel scabrid. Lemma usually < glumes and firmer, submembranous to subcoriaceous, lanceolate, usually minutely denticulate, rounded, ± scabrid, ± obscurely 5-nerved, lateral nerves occasionally excurrent; awn dorsal, scabrid, geniculate or straight, inserted from near base of lemma to slightly above midpoint, to subterminal, rarely 0. Palea ≈ lemma, hyaline, 2-keeled. Lodicules 2, lanceolate or linear, acute to obtuse, membranous, glabrous. Callus short, blunt, with minute hairs, or the hairs to ⅔ or rarely ≈ lemma, rarely glabrous. Stamens 3; anthers rarely penicillate. Ovary glabrous; styles distinct, free to base, short; stigmas 2, plumose. Caryopsis compressed, fusiform to oblong or obovoid; embryo small; endosperm doughy or dry. Chasmogamous or cleistogamous.

Type species: *D. montana* (Gaudich.) P.Beauv. *nom. illeg.*

c. 110 spp., Australasia, Malaysia, New Guinea, South America. Endemic spp. 4, indigenous sp. 1, shared with Australia.

Deyeuxia is often included in *Calamagrostis* Adans., e.g., by Clayton and Renvoize (1986 op. cit.). However, Watson, L. and Dallwitz, M. J. *Grass Genera of the World* (1992) indicate that *Deyeuxia* is distinct from *Calamagrostis* in having callus hairs < lemma, or 0, and in the lemma being at least ¾ length of the glumes and firmer than them. In *Calamagrostis* callus hairs are = or ≫ lemma, and the lemma is *c.* ½–¾ length of the glumes, usually hyaline, and less firm, or of similar texture to the glumes. Watson and Dallwitz also report significant differences between the two genera in leaf anatomy.

New Zealand spp. were revised by Edgar, E. *N.Z. J. Bot. 33*: 1–33 (1995); a new endemic sp. was added by Edgar, E. and Connor, H. E. *N.Z. J. Bot. 37*: 63–70 (1999).

1 Callus hairs *c.* ½ length of lemma or longer; hair tuft of rachilla prolongation reaching top of lemma or almost to top and clearly visible *2*
  Callus hairs to ⅓ length of lemma; hair tuft of rachilla prolongation reaching to ⅔ length of lemma, often not visible without dissection, or sometimes 0 .... *3*

2 Lemma (2)–2.4–3.8 mm; awn curved, projecting from between glumes; culm nodes usually inconspicuous .... **1. aucklandica**
  Lemma (4)–5–6 mm; awn straight, usually not projecting beyond glumes, or 0; culm nodes conspicuous .... **5. youngii**

3 Rachilla prolongation 0 or ≤ 0.3 mm .... **4. quadriseta**
  Rachilla prolongation always present, (0.5)–0.8–1.5–(2) mm *4*

4 Awn from near base of lemma; rachilla tipped by an almost equally long hair tuft .... **2. avenoides**
  Awn from upper ¼ of lemma; rachilla glabrous at tip .... **3. lacustris**

1. D. aucklandica (Hook.f.) Zotov, *Rec. Dom. Mus. 5*: 139 (1965) ≡ *Agrostis aucklandica* Hook.f., *Fl. Antarct. 1*: 96 (1845) ≡ *Deyeuxia filiformis* var. *aucklandica* (Hook.f.) Zotov, *T.R.S.N.Z. 73*: 235 (1943) *comb. illeg., varietal epithet legit.*; Neotype: WELT 76210! *B. C. A[ston]* [Auckland Is], Port Ross, 8.1.1909 (designated by Zotov 1965 op. cit. p. 139). = *D. setifolia* Hook.f., *Fl. N.Z. 1*: 299, t. 65B (1853) ≡ *Agrostis setifolia* (Hook.f.) Hook.f., *Handbk N.Z. Fl.* 329 (1864) non Brot. (1804) ≡ *Calamagrostis setifolia* (Hook.f.) Cockayne, *N.Z. Dept Lands Rep. Bot. Surv. Tongariro Natl Park* 35 (1908); Lectotype: K! *Colenso 922* N. Zealand, grass [small grass from near summit (Ruahines)] (designated by Edgar 1995 op. cit. p. 6).

Open to dense, stiff tufts or small tussocks, 6–35–(55) cm; branching extravaginal. Leaf-sheath subcoriaceous, distinctly ribbed, smooth, or sparsely scabrid to densely scaberulous between ribs, light green to light brown, rarely purplish. Ligule 0.5–2 mm, truncate, abaxially smooth or rarely sparsely minutely hairy or scabrid. Leaf-blade 2.5–15 cm × 0.3–1.5 mm, folded with inrolled margins, setaceous to filiform or wiry, abaxially smooth, or rarely scabrid, adaxially scabrid on ribs, margins sparsely scabrid, narrowed to blunt tip. Culm 4–35 cm, erect or curved, sometimes geniculate at base, nodes usually inconspicuous, internodes either distinctly scabrid below panicle (most North Id plants), to slightly scabrid near panicle or completely smooth. Panicle 2–5.5–(11) cm × (2)–3–18–(22) mm, ± loosely spiciform, narrowly branched, widest at anthesis, later becoming contracted; rachis, branches and pedicels densely scabrid, or rachis smooth and branches scabrid or smooth. Spikelets 3–5.5 mm, greenish or purplish, becoming pale brown. Glumes 1–3-nerved, linear- to elliptic-lanceolate, acute, membranous, often scabrid near tip and on midnerve above. Lemma (2)–2.4–3.8 mm, usually ± ¾ length of glumes, submembranous, smooth to finely papillose, sometimes finely scabrid above, ovate-lanceolate, apex hyaline, truncate, denticulate; awn 2–4.5 mm, usually stout and reflexed, projecting from between glumes, occasionally slightly twisted at base, from upper ⅓ of lemma or *c.* middorsal. Palea ≈ lemma, keels scabrid in upper ½, interkeel sometimes sparsely scabrid, apex obtuse or bifid. Callus hairs rather dense, 1.5–2.5 mm, *c.* ½–⅔ length of lemma, very rarely = lemma. Rachilla prolongation 1–1.5 mm, tipped by a strong tuft of hairs 1–2 mm reaching lemma apex or slightly overtopping lemma. Lodicules 0.6–1 mm, lanceolate, subacute. Anthers 0.6–1.4 mm. Caryopsis 1.6–2.2 × 0.4–0.9 mm.

Endemic.

N.: mountains south from East Cape and on Mt Egmont; S.: mountains along and west of Main Divide, and in Fiordland, Otago and Southland, scattered to the east in Marlborough and Canterbury; St.; A. Tussock grassland, open scrub, and open rocky sites; montane to alpine.

Edgar (1995 op. cit. p. 7) noted that CHR 86368 *A. P. Druce* between Ruapehu and Mt Hauhungatahi, *c.* 3500', bog, March 1958, appeared to be a small form of *D. aucklandica* with panicles < 2 cm, spikelets *c.* 2.5 mm, minutely (*c.* 0.2 mm) awned lemma and glabrous palea.

The record of ?*D. setifolia* Hook.f. for Australia, from Mt Kosciusko and the Bogong High Plains [Willis, J. H. *Handbk Pl. Vict.* ed. 2, *1*: 432 (1970)] referred to the closely related Australian endemic *D. affinis* M.Gray.

2. D. avenoides (Hook.f.) Buchanan, *Indig. Grasses N.Z.* Add. et Corrig. 11 (1880) ≡ *Agrostis avenoides* Hook.f., *Handbk N.Z. Fl.* 330 (1864) ≡ *Calamagrostis avenoides* (Hook.f.) Cockayne, *N.Z. Dept Lands Rep. Bot. Surv. Tongariro Natl Park* 35 (1908) ≡ *Deyeuxia avenoides* (Hook.f.) Buchanan var. *avenoides* (autonym, Hackel *in* Cheeseman 1906 op. cit. p. 871); Lectotype: K! *Sinclair & Haast* Prov. Canterbury, New Zealand, 1860–1 (designated by Edgar 1995 op. cit. p. 8). = *D. avenoides* var. *brachyantha* Hack., *in* Cheeseman *Man. N.Z. Fl.* 871 (1906); Lectotype: W 37901! *T. F. C[heeseman]* Tokoroa Plains, Taupo [Jan. 1884] (No 1189 to Hackel) (designated by Edgar 1995 op. cit. p. 8).   mountain oat grass

Dark green to grey-green usually slender tufts, (12)–20–110 cm; branching extravaginal. Leaf-sheath firmly membranous, obviously ribbed, smooth, or rarely scabrid near ligule, brownish, sometimes purplish. Ligule 0.5–4.5 mm, truncate or tapered, ciliate, abaxially with very minute hairs. Leaf-blade (2)–4–15–(24) cm × 0.3–1.5 mm diam., inrolled, abaxially smooth, very rarely scabrid, adaxially scabrid, margins scabrid, tip acute, scabrid. Culm (7.5)–15–75 cm, varying in diameter, usually 0.3–0.5 mm diam. in plants with cleistogamous flowers and 0.8–2.5–(3) mm diam. in plants with chasmogamous flowers, erect or curved upwards from decumbent base, internodes glabrous. Panicle (1)–3.5–20–(28) cm × 1.5–17 mm, erect or ± nodding, shining, lanceolate, cylindric above, narrowly branched with spikelets densely crowded throughout, lower branches ± distant; rachis smooth, branches and pedicels sparsely scaberulous. Spikelets 3–7 mm, 1–(2)-flowered, light green to purplish green. Glumes 1-nerved, lanceolate, acute to acuminate, smooth, midnerve minutely scabrid in upper ⅔. Lemma 2.8–5.3 mm, ≈ glumes, firmly membranous, minutely scabrid-papillose, narrow elliptic-lanceolate, involute, apex acute or occasionally finely bifid; awn 4.5–8.5 mm, from lower ⅓ of lemma and overtopping glumes. Palea ≈ lemma, keels scabrid in upper ⅔, tip acute to subacute. Callus hairs few, to *c.* 1 mm, ⅕–⅓ length of lemma or shorter. Rachilla prolongation (0.5)–0.8–1.5–(2) mm, tipped by an almost equally long, slender tuft of fine hairs reaching to ⅔ length of lemma. Lodicules 0.5–1 mm, ovate-lanceolate, subacute. Anthers 0.2–0.7–(1.2) mm in cleistogamous flowers, (0.5)–0.8–1.4–(2) mm in chasmogamous flowers. Caryopsis 1.5–2.3 × 0.4–0.9 mm.

Endemic.

N.: throughout; S.: throughout; St.; Ch. Open forest, scrub, and tussock grassland; lowland to alpine.

3. D. lacustris Edgar et Connor, *N.Z. J. Bot. 37*: 68 (1999); Holotype: CHR 313063! *A. P. Druce* near Lake Tennyson, North Canterbury, tarn margin, 3700 ft, Jan 1976.

Reddish green tufts 18–30 cm, with erect culms topped by dense cylindric panicles and much exceeding the leaves at maturity; branching extravaginal. Leaf-sheath chartaceous, glabrous, light green to light reddish brown, ribs more prominent above. Ligule 2–3 mm, erose, often tapered centrally to a fine tip, abaxially glabrous. Leaf-blade 2–9 cm × 1.5–2 mm, folded, abaxially smooth, adaxially slightly ribbed, sparsely scabrid on ribs; margins finely scabrid, tip hooded. Culm 10–24 cm, erect or geniculate at base, nodes hidden by leaf-sheaths, internodes glabrous. Panicle 1.5–4.5 cm × 5–8 mm, cylindric, dense; rachis smooth, branches and pedicels rather sparsely finely scabrid. Spikelets 5–7 mm, purplish green. Glumes subequal, 1-nerved, submembranous, elliptic-lanceolate, acute, smooth, keel finely scabrid on upper ⅔. Lemma 4–5 mm, ⅔–⅘ length of glumes, subcoriaceous, smooth below, finely papillose to prickle-toothed above, ± elliptic, apex shortly bifid; awn 3.5–4.5 mm, from upper ¼ of lemma, twisted about twice at base, ± recurved above. Palea *c.* ⅔ length of lemma, hyaline, keels thickened, faintly prickle-toothed above. Callus hairs 0.3–1 mm. Rachilla prolongation 0.5–1 mm, bearing minute hairs throughout except at glabrous tip. Lodicules 0.4–0.5 mm. Anthers 0.3–0.65 mm. Gynoecium: ovary 0.6 mm; stigma-styles 1.0–1.5 mm. Caryopsis *c.* 2 × 0.6 mm, with dehisced anthers entangled in stigmas. Cleistogamous.

Endemic.

S.: north-west Nelson at Lake Sylvester, and North Canterbury at Lake Tennyson. Lake margins on rocky ground or at tarn margins; 1120–1340 m.

In the dense cylindrical panicle *D. lacustris* resembles *D. quadriseta* and *D. avenoides*; in *D. lacustris* all specimens seen were cleistogamous, but *D. quadriseta* and *D. avenoides* may be chasmogamous or cleistogamous. The awn in *D. lacustris* arises much higher on the lemma than in the other two species; the rachilla is longer than that of *D. quadriseta*, and lacks the strong hair tuft of *D. avenoides*.

4. D. quadriseta (Labill.) Benth., *Fl. Austral.* 7: 581 (1878) ≡ *Avena quadriseta* Labill., *Nov. Holl. Pl. 1*: 25, t. 32 (1805) ≡ *Agrostis quadriseta* (Labill.) R.Br., *Prodr.* 171 (1810) ≡ *Calamagrostis quadriseta* (Labill.) Spreng., *Syst. Veg. 1*: 253 (1824); Holotype: FI! *Labillardière* N. Holl.

Variable, rather slender to robust, dense or lax tufts, 15–125 cm, with culms erect or curved upwards from base, much overtopping leaves; branching extravaginal. Leaf-sheath subcoriaceous, greenish brown, keeled above, distinctly ribbed, smooth or finely scabrid, or ribs rarely short-ciliate. Ligule 1.5–3.5–(6) mm, oblong, truncate, erose or finely ciliate, abaxially glabrous. Leaf-blade 2.5–15–(20) cm × 0.5–3.5–(6) mm, flat or inrolled, abaxially smooth or ribs finely scabrid, adaxially strongly ribbed, margins finely scabrid, tip long-tapered, acute to acuminate, finely scabrid. Culm (10)–15–105 cm, internodes smooth or faintly scabrid. Panicle 3–25–(35) × 0.4–2 cm, narrow-linear, cylindric, or up to 3.5 cm wide and lobed below; rachis smooth to scaberulous, branches finely scabrid, usually ± appressed to rachis or slightly spreading in larger panicles, densely covered with spikelets almost to base, pedicels scabrid. Spikelets (3)–3.5–4–(5) mm, greenish to purplish. Glumes 1-nerved, hyaline, elliptic-lanceolate, acute or acuminate, smooth or sometimes scabrid, midnerve scabrid, margins scaberulous. Lemma 2–3.5–(4) mm, *c.* ¾ length of glumes, submembranous to firm, smooth and shining below, and finely scabrid near apex, or lightly scabrid to distinctly scabrid-papillose throughout, narrow-lanceolate, apex denticulate with lateral nerves excurrent as 4 minute awns; central awn 1.6–6 mm, geniculate, usually arising from lower ⅓ of lemma, often subbasal, occasionally almost middorsal, ± enclosed within or projecting beyond glumes. Palea ≈ lemma, hyaline, keels faintly scabrid in upper ½. Callus hairs wispy, 0.4–1 mm, *c.* ⅓ length of lemma. Rachilla prolongation 0, or 0.05–0.3 mm, including hairs if any. Lodicules *c.* 0.3 mm, linear, obtuse. Anthers 0.3–1.3–(1.8) mm. Caryopsis 1.5–2 × 0.4–0.7 mm.

Indigenous.

N.: from North Auckland south to Volcanic Plateau and Taranaki, further south near Wellington; S.: very scattered; St.; Three Kings Is. Lowland form in gumland, scrub, and on roadsides, montane to subalpine form in boggy ground; lowland to subalpine.

Also indigenous to Australia.

Deyeuxia quadriseta in Australia is very variable and may comprise several distinct taxa [Vickery, J. W. *Contrib. N.S.W. Natl Herb. 1*: 43–82 (1940)]. For New Zealand Edgar (1995 op. cit. p. 11) distinguished an inland, montane to subalpine form of boggy ground, with narrow, involute leaves, and short, narrow-cylindric panicles with scabrid lemmas bearing awns ± included within the glumes, and a wide-leaved coastal form of disturbed sites, with large ± lobed panicles, long awns, and lemmas which are smooth and shining at the base and only slightly scabrid above. This second form may have been introduced from Australia because it is found in very disjunct localities—gumlands of North Auckland, Taita near Wellington, north-west Nelson, Christchurch, and Bluff.

5.　D. youngii (Hook.f.) Buchanan, *Indig. Grasses N.Z.* Add. et Corrig. 11 (1880) ≡ *Agrostis youngii* Hook.f., *Handbk N.Z. Fl.* 330 (1864) ≡ *Calamagrostis youngii* (Hook.f.) Petrie, *T.N.Z.I. 47*: 57 (1915) ≡ *Deyeuxia youngii* (Hook.f.) Buchanan var. *youngii* (autonym, Hackel *in* Cheeseman 1925 op. cit. p. 162); Holotype: K! *J. Haast* Canterbury, New Zealand, 1862. = *Calamagrostis petriei* Hack., *T.N.Z.I. 35*: 380 (1903) ≡ *Deyeuxia petriei* (Hack.) Cheeseman, *Man. N.Z. Fl.* 872 (1906) ≡ *Calamagrostis youngii* var. *petriei* (Hack.) Petrie, *T.N.Z.I. 47*: 57 (1915) ≡ *Deyeuxia youngii* var. *petriei* (Hack.) Cheeseman, *Man. N.Z. Fl.* 162 (1925); Holotype: W 29192! *D. Petrie* Swampy Hill, 1500 ft, Dunedin (No 1190 to Hackel).

Tufts rather narrow, 30–130 cm; branching extravaginal. Leaf-sheath chartaceous with membranous margins, distinctly ribbed, glabrous or finely hairy, light green to light brown. Ligule 0.8–2.5 mm, truncate, ciliate to lacerate, abaxially smooth, or densely minutely pubescent-scabrid. Leaf-blade 12–35–(60) cm × 1–4 mm, stiff, flat to frequently rolled, especially towards tip, abaxially smooth, adaxially distinctly ribbed and finely scabrid on ribs; margins finely scabrid, tip filiform, acute. Culm 40–115 cm, nodes conspicuous, internodes mostly smooth, but sparsely scabrid below panicle. Panicle 7–16–(22) cm × 5–23 mm, linear-lanceolate, shortly and narrowly branched, or more open below; rachis smooth, branches slightly scabrid, with ± densely crowded spikelets on slightly scabrid pedicels. Spikelets 4.5–7.5 mm, light green or purplish. Glumes 1-nerved, submembranous, elliptic-lanceolate, acute, scabrid near apex, keel scabrid. Lemma (4)–5–6 mm, ≈ glumes, subcoriaceous, scabrid, elliptic-lanceolate, apex denticulate; awn (0.5)–1–3 mm, straight, from upper ⅓ of lemma or rarely middorsal, usually barely reaching but occasionally overtopping lemma apex, rarely awn 0. Palea ≈ lemma, keels and interkeel scabrid. Callus hairs fine, ± dense, *c.* ½ length of lemma. Rachilla prolongation 1.5–2.5 mm, tipped by dense tuft of fine hairs 2–3–(3.5) mm, almost reaching lemma apex. Lodicules 0.6–0.8 mm, linear, acute. Anthers 1.2–1.9 mm, penicillate. Caryopsis 2.5–3 × 0.7–1 mm.

Endemic.

S.: north-west Nelson at Lake Sylvester, Marlborough on Chalk Range and on Mt Fyffe, scattered localities in Mid and South Canterbury and Otago. Tussock grassland, shrubland and open sites, and on rocky sites; lowland to alpine.

Deyeuxia youngii has only been collected from scattered localities.

Deyeuxia youngii is the only sp. of *Deyeuxia* with penicillate anthers. In tribe Agrostideae this type of anther is otherwise found only in *Dichelachne* Endl. which has spp. with penicillate and spp. with non-penicillate anthers.

REPRODUCTIVE BIOLOGY   Cleistogamous flowers occur throughout the range of *D. avenoides*. Plants with cleistogamous flowers usually have more slender culms, much shorter anthers and narrower panicles than plants with chasmogamous flowers. There is no geographical separation and plants with cleistogamous flowers may be found growing close to plants with chasmogamous flowers.

A plant of *D. quadriseta* growing in the wild at Taita had cleistogamous flowers and anthers 0.4 mm (CHR 179594 *A. P. Druce* 1967) and another from nearby had chasmogamous flowers and anthers 0.6–0.7 mm (CHR 179595 *A. P. Druce* 1967).

All known specimens of *D. lacustris* are cleistogamous.

DICHELACHNE Endl., 1833

Slender, erect perennials, tufted at base, with wiry ± nodding culms. Leaf-sheath densely, minutely puberulous to glabrous. Ligule short, membranous, truncate. Leaf-blade rather stiff, flat; margins often inrolled, gradually tapered to acute tip. Culm much overtopping leaves at maturity, few-noded. Panicle contracted, densely branched, to open, with few lax branches. Spikelets 1–(rarely 3)-flowered; disarticulation above glumes. Glumes 1–(3)-nerved, unequal to subequal, <, or ≥ lemma, hyaline except for scaberulous midnerve, acute to acuminate to shortly aristate. Lemma 5-nerved, fusiform; awn long, flexuous, inserted below entire or minutely bifid tip, ± geniculate, column scarcely, or much twisted. Palea < lemma, hyaline, narrow, with 2 close-set scabrid keels. Callus very short, ringed by short hairs. Rachilla prolongation minute. Lodicules 2, oblong, nerveless, ± bidentate, glabrous or sparsely ciliate. Stamens 1, 3, or variably 1–3, very small in cleistogamous flowers; anthers sometimes penicillate. Ovary glabrous; styles apical, free to base; stigmas plumose. Caryopsis cylindric, beaked, adaxially longitudinally grooved; embryo minute; hilum subbasal, punctiform; endosperm liquid. Chasmogamous or cleistogamous.

Type species: *D. montana* Endl.

c. 8 spp. of Australasia, New Guinea, Timor, Easter Id. Endemic sp. 1, indigenous spp. 3, shared with Australia; naturalised spp. 2 (from Australia). They are collectively known as plume grass.

N.Z. spp. were revised by Edgar, E. and Connor, H. E. *N.Z. J. Bot. 20*: 303–309 (1982); a new endemic species was added by Edgar, E. and Connor, H. E. *N.Z. J. Bot. 37*: 63–70 (1999). The genus was revised for Malesia by Veldkamp, J. F. *Blumea 22*: 5–12 (1974), and for Australia by Jacobs, S. W. L., McClay, K. L., and Simon, B. K. *Telopea 5*: 325–328, for *Fl. N.S.W. 4*: 582–584 (1993).

1 Awn inserted 1–3 mm below lemma tip 2
  Awn inserted < 1 mm below lemma tip 3

2 Callus hairs < 1 mm; awn 20–30 mm, column straight .... **1. crinita**
  Callus hairs ≥ 3 mm; awn 8–12 mm, column twisted .... **3. lautumia**

3 Longer panicle branches with florets ± to base; lower glume ≥ lemma..... *4*
Longer panicle branches naked below; lower glume < lemma *5*

4 Lower glume usually < 4 mm, ≈ lemma .... **4. micrantha**
Lower glume > 4 mm, noticeably > lemma .... **5. rara**

5 Awn column slightly twisted; culms glabrous or minutely scabrid below nodes
....**2. inaequiglumis**
Awn column very tightly twisted; culms usually shortly puberulous, especially
at nodes .... **6. sieberiana**

1. D. crinita (L.f.) Hook.f., *Fl. N.Z. 1*: 293 (1853) ≡ *Anthoxanthum crinitum* L.f., *Suppl. Pl. 13*: 90 (1782) ≡ *Dichelachne forsteriana* Trin. et Rupr., *Mém. Acad. Imp. Sci. Saint-Pétersbourg VI*, Sect. Nat. *5*: 4 (1843) *nom. superfl.* ≡ *Deyeuxia crinita* (L.f.) Zotov, *T.R.S.N.Z. 73*: 234 (1943) ≡ *Dichelachne crinita* (L.f.) Hook.f. var. *crinita* (autonym, Hackel *in* Cheeseman 1906 op. cit. p. 874); Holotype: S (Herb. Bäck) *n.v.*, *Forster* (*fide* Veldkamp 1974 op. cit. p. 10).

Light green, ± stout, moderately tall, extravaginal tufts or rosettes, leaves < ± nodding, narrow-plumed culms. Leaf-sheath light brown, with minute, soft, appressed, retrorse hairs. Ligule 0.5–1.5 mm, membranous, abaxially minutely scabrid, ± truncate, minutely ciliate, occasionally asymmetric. Leaf-blade to 40 cm × 1.5–5 mm, somewhat stiff, flat or slightly inrolled, gradually tapered, strongly ribbed, abaxially scabrid near tip, adaxially minutely scabrid on margins and ribs. Culm (30)–50–100–(120) cm, internodes glabrous or minutely scaberulous below panicle. Panicle 10–25 cm, erect, spike-like, light green to stramineous, densely branched, close-set erect branches hidden by spikelets pulled together by entwining awns; rachis, branchlets and pedicels closely short-scabrid; spikelets numerous, shining. Glumes very narrow, linear-lanceolate, silvery; lower 4.5–8–(9) mm, ≈ lemma, often shortly aristate, upper 5.5–8.5–(10) mm, ≥ lemma, tip acuminate. Lemma (4.5)–5–7–(8) mm, minutely scabrid, tip scarcely bifid; awn 20–30 mm, light green to purple, inserted 1.5–3 mm below lemma tip, column straight, awn ± falcate and twisted about once. Palea 3–5 mm, very narrow, keels minutely scabrid near ciliate tip. Callus hairs to 0.7 mm. Rachilla prolongation to 0.1 mm. Lodicules 0.5–0.7 mm, membranous, elliptic, acute, apically ciliate. Anthers 1–3, (0.7)–1.0–2.0 mm in chasmogamous flowers, 0.2–0.9 mm in cleistogamous flowers. Caryopsis 2.0–2.5 × 0.3–0.4 mm; embryo 0.3–0.4 mm; hilum 1.9–2.3 mm. $2n = 70$.

Indigenous.

N.: throughout; S.: throughout but less common in Fiordland; St.; K., Three Kings Is, Ch. In open forest, scrub, grassland, and modified sites, in lowland and montane zones, and colonising depleted areas.

Also in Australia and Pacific Is.

Many Australian plants referred to *D. crinita* have fine awns 40–50 mm, and slender lemmas 4–5 mm, compared to the stiffer awns, *c.* 30 mm, and stouter lemmas (4.5)–5–7–(8) mm, in N.Z. plants and in some Tasmanian specimens. The holotype of *D. crinita* (L.f.) Hook.f. is a N.Z. specimen; the longer-awned Australian plants which are found also in Tasmania, Norfolk Id, and Easter Id have received no further recognition by Jacobs et al. (1993 op. cit.). One plant with long awns is CHR 310050 *S. Bowman, W. Versluys & E. McKenzie* Waikumete Cemetery, Auckland, 16.12.1976; this may be a recent introduction from Australia.

CHR 402411 *A. P. Druce* Moawhango Valley, Kaimanawa Mts, 800 m, grassland, Feb. 1985, appears to be a form of *D. crinita* with narrow rolled leaves, florets with awn shorter than usual as also are callus hairs and rachilla.

2. D. inaequiglumis (Hack.) Edgar et Connor *N.Z. J. Bot.* **20**: 307 (1982) ≡ *D. sciurea* var. *inaequiglumis* Hack., *in* Cheeseman *Man. N.Z. Fl.* 874 (1906) ≡ *D. micrantha* var. *inaequiglumis* (Hack.) Domin, *Biblioth. Bot.* **85**: 353 (1915); Holotype: W 26408! *D. Petrie* vicinity of Auckland [Western Park, Dec. 1895] (No 1068 to Hackel).

Rather small, slender extravaginal tufts, leaves ≪ culms. Leaf-sheath dull or light brown, glabrous or minutely scabrid below collar, or lowermost sheaths scabrid throughout, sometimes with scattered hairs. Ligule 0.1–0.5 mm, membranous, truncate, ciliate, abaxially minutely scabrid. Leaf-blade to 15 cm × 1–2.5 mm, flat, tapered; margins minutely scabrid towards tip, occasionally both, or only adaxial, surface with short hairs, adaxially ribs scabrid. Culm 25–70 cm, internodes entirely glabrous, or slightly and minutely scabrid below purplish nodes. Panicle 10–30 cm, lax; branches few, slender, in ± distant whorls, naked below; rachis, branchlets and pedicels short-scabrid, visible between spikelets. Spikelets purple, few. Glumes elliptic-lanceolate, acute; lower (2.5)–3–3.5–(4.5) mm, < lemma, upper (3.5)–4–5–(5.5) mm, ≤ or rarely > lemma. Lemma 4–6 mm, minutely scabrid, tip shortly bifid; awn 13.5–18.5 mm, purple, inserted *c.* 0.5 mm below lemma tip, column ± straight, awn twisting 2–3 times. Palea 3.5–4 mm, narrow-linear, keels scabrid near ciliate tip. Callus hairs (0.5)– *c.* 1 mm. Rachilla prolongation to 0.2 mm. Lodicules 0.4–0.7 mm, cuneate, somewhat bilobed with a few apical hairs. Anthers 3, (0.9)–1.0–1.5–(2.5) mm in chasmogamous flowers, (0.1)–0.2–0.7 mm in cleistogamous flowers. Caryopsis 2–2.5×0.3–0.4 mm; embryo 0.2–0.4 mm; hilum 1.8–2.3 mm.

Indigenous.

N.: common from North Cape to Auckland, one collection in Taranaki and common in Wellington Province especially in the Wairarapa and Hutt Valley; S.: common in Nelson, Marlborough, and Westland. Scrub, grassland, and modified sites in lowland and montane zones, common in low hill country in "danthonia" grassland.

Also in eastern Australian states and Tasmania.

3. D. lautumia Edgar et Connor, *N.Z. J. Bot. 37*: 67 (1999); Holotype: CHR 514885C! *G. Jane* Flaxbourne River, Marlborough, north side, near mouth, disused limestone quarry, 100 m, 4 Dec 1997.

Robust grey-green tufts, 45–65 cm; branching extravaginal. Leaf-sheath chartaceous with membranous margins, ± distinctly ribbed, glabrous, stramineous, or reddish purple, especially in culm leaves. Ligule 0.7–1.0 mm, truncate, erose, sparsely ciliate, abaxially scabrid. Leaf-blade 6–20 cm × 1.5–3.0 mm, greyish green to later reddish, abaxially smooth, adaxially strongly ribbed, prickle-toothed on ribs; margins finely prickle-toothed. Culm 30–45 cm, nodes green to purple with an upper fringe of dense appressed hairs, internodes glabrous. Panicle 10–16 × 1–1.5 cm, linear-lanceolate, dense, contracted above, longer lower branches in slightly more distant clusters, scarcely spreading, spikelet-bearing to base; rachis, branches, and pedicels with short hair-like prickle-teeth. Spikelets 6–8 mm, green to stramineous or purple-suffused. Glumes 1-nerved, equal or upper very slightly longer, *c.* 6 mm, submembranous, elliptic-lanceolate, acuminate, keels finely prickle-toothed. Lemma ≤ glumes, 6–7 mm, subcoriaceous, papillose, minutely prickle-toothed above, apex bifid with hyaline finely acuminate lobes 0.5–1 mm; awn 8–12 mm, middorsal, or from slightly above midway, geniculate, twisted, very hairy below. Palea < lemma, folded, keels prickle-toothed above, apex ciliate. Callus *c.* 0.2 mm, hairs to 4 mm. Rachilla 0.2–0.7 mm, glabrous or with a few hairs; prolongation 0.5 mm. Lodicules *c.* 0.5 mm, cuneate, ciliate. Stamens 3; anthers 1.0–1.5 mm. Gynoecium: ovary 1–1.3 mm; stigma-styles 0.7–1.5 mm, hairs almost to base. Caryopsis *c.* 3.5 × 1 mm, beaked; embryo 0.5 mm; hilum 2.0 mm. $2n = 70$ (+ 2f).

Endemic.

S.: Marlborough, near mouth of Flaxbourne River; confined to a disused limestone quarry, 100 m.

In CHR there are two specimens originally collected from the Waima Valley, Marlborough, but grown on at Pinehaven, Hutt Valley; CHR 252333 *A. P. Druce* tributary of Waima River, Jan 1974, and CHR 387531 *A. P. Druce* Isolation Creek, Jan 1982. Flowers in both are cleistogamous, and anthers are 0.6–0.9 mm long in florets of smaller dimension than in *D. lautumia* from Flaxbourne River, *c.* 20 km to the north-east, where all plants are regarded as chasmogamous. Of the Isolation Creek site A. P. Druce (*in litt.*) noted that these plants were abundant on the stream bank there. Edgar, E. *N.Z. J. Bot. 33*: 11 (1995) had included these specimens in her discussion of forms of *Deyeuxia quadriseta.*

4. D. micrantha (Cav.) Domin, *Biblioth. Bot. 85*: 353 (1915) ≡ *Stipa micrantha* Cav., *Icon. 5*: 42, t. 467, fig. 2 (1799); Holotype: MA *n.v.*, photo! *Née* Nova-Hollandia. = *Agrostis sciurea* R.Br., *Prodr.* 171 (1810) ≡ *Dichelachne sciurea* (R.Br.) Hook.f., *Fl. N.Z. 1*: 294 (1853); Holotype: BM! *R. Brown No. 6211* Port Jackson. = *D. crinita* var. *intermedia* Hack., *in* Cheeseman *Man. N.Z. Fl.* 874 (1906); Lectotype: W 26415! *D. Petrie* Bay of Islands (No 1063 to Hackel) (designated by Edgar and Connor 1982 op. cit. p. 307).

Stout, rather rigid, extravaginal tufts, leaves < stiff, erect culms. Leaf-sheath stramineous to dull brown, with minute, appressed, scattered hairs. Ligule 0.3–0.5–(1) mm, membranous, truncate, minutely ciliate, abaxially scabrid, often asymmetric. Leaf-blade to 20 cm × 1.5–2.5 mm, rather stiff, flat, tapered towards tip, abaxially sparingly minutely scabrid, adaxially scabrid on ribs towards tip, minutely scabrid on margins. Culm 40–100 cm, internodes minutely scaberulous throughout, or glabrous but minutely scaberulous below panicle, variously purplish. Panicle 10–25 cm, erect, spike-like, often purplish, branches spreading at first; rachis, branchlets and pedicels closely short-scabrid. Spikelets numerous, close-set on branchlets, rather delicate. Glumes narrow-lanceolate, acute to acuminate, often suffused with purple; lower 3–4 mm, ≈ lemma, upper 3.5–5 mm, > lemma. Lemma 2.5–4 mm, sometimes purplish; awn 12–18 mm, very fine, column straight, awn curving above and twisted 2–3 times along whole length, inserted 0.6–0.9 mm below minutely bifid lemma-tip. Palea 2–3 mm, narrow-linear, keels scabrid above, tip ciliate. Callus hairs 0.3–0.5 mm. Rachilla prolongation *c.* 0.05 mm. Lodicules (0.4)– 0.6–0.8 mm, hyaline, elliptic-oblong, unequally bilobed, occasionally minutely ciliate. Anthers 1, 1.2–1.4 mm in chasmogamous flowers, 0.6–0.8 mm in cleistogamous flowers. Caryopsis 2.0–2.3 × 0.3–0.4 mm; embryo 0.2–0.4 mm; hilum 1.8–1.9 mm.

Indigenous.

N.: North Cape to Auckland. Scrub, roadsides, banks, usually not far from the coast, in lowland zone.

Also indigenous to Norfolk Id, Australia, Easter Id and New Guinea.

5. D. rara (R.Br.) Vickery *Contrib. N. S. W. Natl Herb. 1*: 337 (1950).

Stout or slender, moderately tall, extravaginal tufts, leaves < culms. Leaf-sheath light brown, sometimes purplish, usually glabrous below and scaberulous above, rarely with projecting hairs. Ligule (0.5)–1– 1.5 mm, membranous, truncate, ciliate, abaxially scabrid to ciliate, often asymmetric. Leaf-blade to 10 cm × 1.5–2.5 mm, rather stiff, flat, minutely scabrid on ribs and margins, tapered towards tip. Culm 35– 70 cm, internodes glabrous, often minutely scaberulous near nodes. Panicle (5.5)–10–16 cm, ± lax, with short, ± erect, few-flowered branches; rachis, branchlets and pedicels short-scabrid, partly hidden among spikelets. Spikelets pale, shining. Glumes ± equal, > lemma, elliptic-lanceolate, aristate; lower 4–5.5–(6) mm, upper 4.5–6 mm. Lemma 3.5–5 mm, surface minutely scabrid, tip opaque, scabrid, bifid; awn 10–20 mm, once-geniculate, brownish, inserted 0.3–0.5–(0.9) mm below lemma tip, column tightly twisted several times. Palea 3–4 mm, narrow-linear, keels scabrid near ciliate tip. Callus hairs 0.5–1 mm. Rachilla prolongation usually obvious, up to *c.* 1 mm. Lodicules 0.5–0.7 mm, elliptic-oblong, bifid at shortly ciliate tip. Anthers 1–3, 0.8–1.9 mm in chasmogamous flowers, 0.4–0.9 mm in cleistogamous flowers. Caryopsis (1.4)–1.8–2 × 0.3–0.4 mm; embryo 0.2–0.4 mm; hilum (1.2)–1.6–1.8 mm.

Naturalised from eastern Australia.

N.: North Cape to Auckland and Coromandel, southern Wellington Province; S.: Marlborough Sounds and Bluff; Three Kings Is. Manuka or kanuka scrub and disturbed sites in lowland zone.

Although first recorded in 1974 it has been here since 1878 at least (Edgar and Connor 1982 op. cit. p. 303).

6. D. sieberiana Trin. et Rupr., *Mém. Acad. Imp. Sci. Saint-Pétersbourg VI*, Sect. Nat. *5*: 2 (1843).

Stout or slender, rather tall, extravaginal tufts, leaves < culms. Leaf-sheath light brown, sometimes purplish, with numerous short projecting hairs. Ligule 1–1.5 mm, membranous, truncate, ciliate, abaxially scabrid to ciliate, often asymmetric. Leaf-blade to 15 cm × 1–1.5–(3) mm, rather stiff, flat, both surfaces with short stiff projecting hairs; margins minutely scabrid, adaxially ribs minutely scabrid, tapered towards tip. Culm 40–90 cm, nodes ringed above and below with short appressed hairs, internodes usually shortly puberulous, especially below, sometimes scabrid. Panicle (7.5)–12–25 cm, very lax, branches spreading and visible among spikelets; rachis, branchlets and pedicels short-scabrid. Spikelets few, often purplish. Glumes unequal, elliptic-lanceolate, acute; lower 3.5–5 mm, ≤ lemma, upper 4–6 mm, ≤ or > lemma. Lemma 4–6 mm, densely scaberulous, tip hyaline and less so, shortly bifid; awn 10–15 mm, inserted 0.3–0.5 mm below lemma tip, column very tightly twisted, awn usually geniculate at base of column and again above. Palea 3.5–5.5 mm, narrow-linear, keels scabrid near ciliate tip. Callus hairs 1.0–1.5 mm. Rachilla prolongation up to 0.5 mm. Lodicules 0.5–0.6 mm, elliptic-oblong, bifid, glabrous. Anthers 3, 0.9–2.8 mm in chasmogamous flowers, 0.4–0.6 mm in cleistogamous flowers. Caryopsis (2.3)–2.5–2.9 × 0.4–0.6 mm; embryo 0.4–0.5 mm; hilum 2.1–2.7 mm.

Naturalised from eastern Australia.

N.: Whangaroa Harbour and Bay of Islands; S.: Nelson (Moutere Hill), Canterbury (Christchurch). Roadsides and waste ground in lowland zone.

FLORAL BIOLOGY   Stamen numbers in *Dichelachne* vary: 1–3 in *D. crinita* and *D. rara*; 1 in *D. micrantha*; 3 in *D. inaequiglumis, D. lautumia* and *D. sieberiana*; there are no differences in stamen number between chasmogamous or cleistogamous flowers. There is no inbreeding depression in S_1 and S_2 families of *D. crinita* [Connor, H. E. and Cook, A. B. *N.Z. J. Sci. Tech. 34A*: 369–371 (1952); Connor, H. E. *N.Z. J. Sci. Tech. 38A*: 742–751 (1957)].

ECHINOPOGON P.Beauv., 1812

Tufted perennials. Leaf-sheath rounded. Leaf-blade flat, linear or linear-lanceolate. Panicle spike-like, oblong-cylindric or ovate, with densely crowded florets; branches short. Spikelets 1-flowered, shortly pedicelled or sessile; disarticulation above glumes; rachilla prolonged to a ± short bristle. Glumes subequal, 1-nerved, membranous, keeled, ≤ floret, acute to acuminate. Lemma 5–7–(11)-nerved, thinly coriaceous, apex bidentate, or setaceously bilobed, or rarely entire, stiffly awned apically or subapically, rarely mucronate. Palea ≈ lemma, 2-keeled. Lodicules 2, membranous, ciliate or glabrous. Callus ringed by short hairs. Stamens 3. Ovary apex hairy or glabrous; styles free. Caryopsis longitudinally grooved; embryo small; hilum linear; endosperm liquid, or solid.

Type species: *E. ovatus* (G.Forst.) P.Beauv.

7 spp. of New Guinea and Australasia. Indigenous sp. 1, shared with Australia.

The genus was last revised by Hubbard, C. E. *Icon. Pl. 33*: t. 3261 (1935).

1. E. ovatus (G.Forst.) P.Beauv., *Ess. Agrost.* 42, 148, t. 9, fig. 5 (1812) ≡ *Agrostis ovata* G.Forst., *Prodr.* 40 (1786) ≡ *Cinna ovata* (G.Forst.) Kunth, *Révis. Gram. 1*: 67 (1829) ≡ *Echinopogon asper* Trin., *Fund. Agrost.* 126 (1820) *nom. superfl.*; Holotype: GOET! *Forster* "Nova Zeelandia". = *E. purpurascens* Gand., *Bull. Soc. Bot. France 66*: 300 (1919); Holotype: LY! *G. M. T[homson]* Dunedin, 1896. = *E. novae-zelandiae* Gand., *Bull. Soc. Bot. France 66*: 300 (1919); Holotype: LY! *Petrie 10019* Aratapu, Kaipara Harbour, Auckland, N. Zelandia, 1880 (No 1050 to Hackel).

Lax, bluish-green tufts to 140 cm, or solitary culms, from a slender creeping rhizome. Leaf-sheath closely appressed to culms, firmly membranous, striate, retrorsely scabrid or smooth, sometimes purplish. Ligule 0.5–2.7 mm, finely denticulate, abaxially glabrous. Leaf-blade (2.5)–5–12–(25) cm × (1.5)–2–7–(8.5) mm, slightly constricted at base and tapering gradually to acute tip, abaxially minutely retrorsely scabrid on ribs, adaxially scabrid or smooth, sometimes with scattered hairs; margins rather closely and finely scabrid. Culm decumbent and sometimes geniculate at base, erect above, upper internodes with scattered minute prickle-teeth and retrorsely scabrid below panicle, or all internodes smooth. Panicle 0.7–5.5 × 0.4–4.0 cm, ovate-globose to narrow-oblong with conspicuous bristling awns; rachis smooth, branches almost smooth, bearing a few scattered prickle-teeth. Spikelets 4–12–(20) mm, green or occasionally purplish. Glumes ≤ spikelet, lanceolate, acute, ± membranous, keel conspicuous, thickened, green-margined, ciliate. Lemma 3–4–(4.5) mm, 5-nerved, linear-lanceolate, finely scabrid above, apex minutely bilobed; awn (2.5)–3.5–9–(14) mm, stout, straight, finely scabrid. Palea ≤ lemma, keels and apex ciliate. Callus hairs 0.8–1.4 mm. Rachilla prolongation ⅓–½ length of palea, usually > 1 mm and hairy, or < 1 mm and glabrous. Lodicules minutely ciliate or glabrous. Anthers 0.9–1.5 mm. Ovary apex hairy. Caryopsis 1.5–2 × 0.6–0.8 mm. $n = 21$.

Indigenous.

N.; S.: scattered throughout, but rare in the west; Three Kings Is, Ch. Sea level to montane; in forest or scrub, or among stones and boulders in grassy shrubby areas, on dry banks or in waste places. FL Oct–Jan. FT Jan–Mar.

Also indigenous to Australia.

Purple-coloured forms are not uncommon.

The name *Hystericina alopecuroides* Steud., *Syn. Pl. Glum. 1*: 35 (1853) was included in the synonymy of *E. ovatus* by Buchanan, J. *Indig. Grasses. N.Z.* t. 13B (1878), by Cheeseman (1925 op. cit. p. 150) and also by Hubbard, C. E. (1935 op. cit. p. 7). No specimen could be found at P; the original locality cited is "N. Zeeland" and no collector is given.

GASTRIDIUM P.Beauv., 1812

Annuals. Leaf-sheath rounded, slightly keeled above. Ligule membranous. Leaf-blade flat. Culm internodes glabrous. Panicle dense, spike-like, narrow-cylindric, sometimes lobed. Spikelets 1-flowered, laterally compressed, swollen and globular at base; disarticulation above glumes; rachilla very shortly prolonged. Glumes unequal, swollen, subcoriaceous and shining below, membranous and keeled above, 1-nerved. Lemma ≪ glumes, 5-nerved, broadly elliptic, rounded, truncate, denticulate, with slender dorsal awn from upper ⅓, or awn subapical, or 0. Palea = lemma, 2-nerved. Lodicules 2, glabrous. Stamens 3. Ovary glabrous; styles free. Caryopsis slightly dorsiventrally compressed; embryo small; hilum punctiform; endosperm liquid.

Type species: *G. australe* P.Beauv. *nom. illeg.*

3 spp., of Eurasia and N.E. Africa. Naturalised sp. 1.

1. G. ventricosum (Gouan) Schinz et Thell., *Vierteljahrsschr. Naturf. Ges. Zürich 58*: 39 (1913). nit grass

Erect tufts (9)–18–65 cm. Leaf-sheath submembranous, striate, smooth or sometimes scabrid, sometimes ± inflated. Ligule 1–3 mm, striate, truncate or rounded, becoming lacerate, abaxially often minutely scabrid. Leaf-blade 2–10 cm × 1–3 mm, scabrid, or sometimes smooth abaxially; margins scabrid, tip very finely acuminate. Culm erect or geniculate at base, internodes glabrous. Panicle 2–12–(14.5) × 0.6–1.5 cm, shining, lanceolate or cylindric, usually tapered above; rachis and branches minutely scabrid, hidden. Spikelets 3.5–5 mm, light green. Glumes smooth, swollen and rounded below, narrowed above and tapering to long-acuminate tip, keel scabrid; lower = spikelet, upper *c.* ¾ length of spikelet. Lemma 0.9–1.2 mm, hyaline, sparsely short-hairy near margins and in upper ⅓; awn 3–4.2 mm, very fine, geniculate, or 0 within the same spikelet. Palea hyaline, glabrous. Anthers 0.6–0.8 mm. Caryopsis 0.7–0.9 × 0.4–0.6 mm.

Naturalised.

N.: North Cape southwards to Waikato and at Wellington; S.: Marlborough (Blenheim), Canterbury (near Ashburton). Open pasture, roadsides and waste places.

Indigenous to Europe, North Africa and western Asia.

LACHNAGROSTIS Trin., 1820

Tufted, sometimes short-lived perennials, or annuals; branching intra- or extravaginal. Leaf-sheath rounded. Ligule membranous. Leaf-blade flat, or folded with inrolled margins. Culm erect or geniculately ascending, fragile at maturity; nodes glabrous. Inflorescence a lax, often delicately branched panicle; at maturity completely detached from the plant together with a fragment of the uppermost culm internode, or, in annual plants, soon withering and persisting. Spikelets small, 1- or occasionally 2-flowered, laterally compressed; disarticulation above glumes; rachilla often prolonged, tipped with soft hairs and almost equalling the palea, or prolongation minute and glabrous. Glumes ± equal and equalling the spikelet, ovate-elliptic to lanceolate, acute, membranous, often shining, 1–(3)-nerved, lateral nerves, if present, very short; keel ± scabrid. Lemma <, and thinner in texture than glumes, rarely (*L. tenuis*) ≈ glumes and firmer, elliptic-oblong, truncate, rounded, (3)–5-nerved, often denticulate and with lateral nerves minutely excurrent, usually with dense or sparse, soft, fine hairs, or sometimes glabrous, rarely (*L. tenuis*) finely scabrid; awn, finely scabrid, geniculate or straight, ± middorsal or very short and subapical, rarely 0. Palea from ½ to ≈ lemma, hyaline, keels 2, usually faint, close together at centre. Callus minute, blunt, with short, soft hairs, to ½ or rarely to ⅔ length of lemma, sometimes 0. Lodicules 2, linear to lanceolate, hyaline, glabrous. Stamens 3; anthers not penicillate. Ovary glabrous;

styles distinct, free to base, short; stigmas 2, plumose. Caryopsis fusiform; embryo small; endosperm doughy or dry. Chasmogamous (N.Z. spp.). Fig. 9.

Type species: *L. filiformis* (G.Forst.) Trin.

c. 20 spp. of Australasia, New Guinea and Easter Id, with the type sp. occurring throughout the range of the genus and naturalised in South America, South Africa and Malesia. Endemic spp. 10, indigenous spp. 2, 1 shared with Australia, 1 shared with Australia, New Guinea and Easter Id.

Lachnagrostis is generally included within *Agrostis*, e.g., Clayton and Renvoize (1986 op. cit. p. 134). However, in *Lachnagrostis* the whole inflorescence together with a fragment of the uppermost culm internode becomes detached at maturity and blown by the wind, whereas the rachis is persistent in *Agrostis*. Two features which distinguish *Lachnagrostis* from many, though not all, spp. of *Agrostis* are the well-developed callus hairs and the soft hairs on the lemma (though in a few spp. of *Lachnagrostis* the lemma is glabrous). Spp. of *Lachnagrostis* can be further distinguished from endemic and indigenous spp. of *Agrostis* by the well-developed palea and well-developed hairy rachilla.

The N.Z. spp. were revised by Edgar, E. *N.Z. J. Bot. 33*: 1–33 (1995), who briefly compared *Lachnagrostis* with *Deyeuxia* and *Calamagrostis*.

1 Awn geniculate 2
  Awn straight or ± curved, or awn 0 7
2 Branching extravaginal .... **8. lyallii**
  Branching intravaginal *3*
3 Lemma with scattered to dense hairs *4*
  Lemma glabrous 6
4 Perennial with firm panicle and leaves usually 2–6.5–(10) mm wide ....
  .... **9. pilosa**
    4a Awn geniculate, (3)–4.5–8.5 mm; palea ½–⅗ lemma ...... subsp. **pilosa**
      Awn straight, (0.5)–1–3 mm; palea ≈ lemma .... subsp. **nubifera**
    Annual or short-lived perennial with delicate panicle and leaves usually 0.5–
    3 mm wide 5

5 Panicle with naked primary and secondary branches mainly very unequal in length; lemma usually 1.3–2 mm; anthers 0.2–0.5 mm .... **4. filiformis**
Panicle with naked primary and secondary branches ± equal in length; lemma usually 1.8–3.0 mm; anthers usually 0.4–0.7 mm .... **7. littoralis**
   5a Lemma (1.5)–1.8–2.5 mm, awn geniculate; plants usually 5–20 cm; glumes equal, or the upper slightly shorter, both 1-nerved ....
.... subsp. **littoralis**
   Lemma 2.5–3 mm, awn straight or ± curved; plants 20–60 cm; glumes dissimilar, the lower slightly shorter, 1-nerved, the upper 3-nerved ..
.... subsp. **salaria**

6 Leaf-blades flat, 2.5–6–(10) mm wide; lemma smooth below, often scabrid above on nerves; spikelets (4)–5–6 mm .... **2. billardierei**
Leaf-blades inrolled, 0.3–0.9 mm diam.; lemma papillose-scabrid throughout; spikelets 3–5 mm .... **11. tenuis**

7 Branching intravaginal ....**6. leptostachys**
Branching extravaginal *8*

8 Lemma 2.5–3.8 mm; panicle very lax and ultimate branches bearing single spikelets at tip *9*
Lemma 1.2–2.5 mm; panicle rather contracted at first, later more open, the ultimate branches often bearing two or three spikelets towards tip *10*

9 Lemma glabrous or sparsely hairy; plants 30–80 cm .... **3. elata**
Lemma very densely hairy; plants 15–35 cm .... **1. ammobia**

10 Lemma densely hairy; anthers usually 0.2–0.5 mm .... **10. striata**
Lemma glabrous, or with scattered hairs; anthers 0.5–1.3 mm *11*

11 Callus hairs conspicuous, to 1 mm; palea ½ to ⅔ lemma .... **12. uda**
Callus hairs few, *c.* 0.1 mm or 0; palea ≈ lemma .... **5. glabra**

1. L. ammobia Edgar, *N.Z. J. Bot. 33*: 14 (1995); Holotype: CHR 467998! *B. D. Rance* Southland, Omaui, Three Sisters Dune, sand dune, 25.1.1990.

Lax, bluish green, perennial tufts, 15–35 cm, usually with narrow, involute leaves and very lax panicles; branching extravaginal, each shoot with 1–2 blunt-tipped sheaths at base and with the lowest leaf-blade often much reduced. Leaf-sheath submembranous, distinctly ribbed, smooth or sometimes scabrid above. Ligule (1.5)–2.5–4.5 mm, tapered above to acute tip or denticulate, abaxially scabrid. Leaf-blade 6–16 cm × 0.4–0.8 mm diam., inrolled, to 1.2 mm wide if flat, abaxially smooth to scabrid (especially near tip), adaxially finely scabrid; margins finely scabrid, tip fine, obtuse. Culm 10–15 cm, slender, usually included

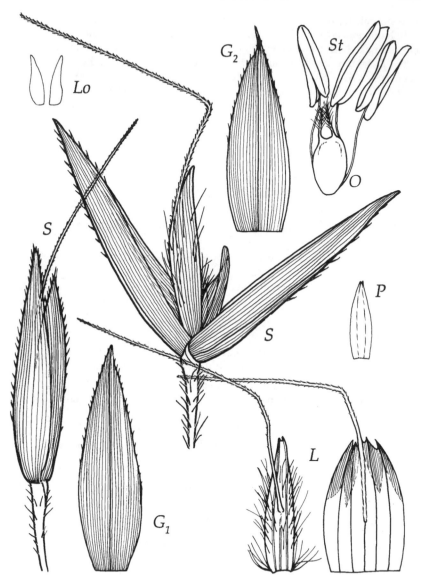

Fig. 9 *Lachnagrostis pilosa* subsp. *pilosa*
S, spikelet: closed, × 18; open, × 24; **G₁**, **G₂**, glumes, × 18; **L**, lemma dorsal view and flattened with hairs removed to show nervation, × 18; **P**, palea dorsal view, × 18; **Lo**, lodicules, × 18; **O**, ovary, styles and stigmas, × 30; **St**, stamens, × 30.

within uppermost leaf-sheath, internodes glabrous. Panicle (7)–10–16 × 1.5–12–(15) cm; branches few, long, filiform, ± closely, finely scabrid, each tipped by a single spikelet. Spikelets (4.5)–5–7 mm, light green. Glumes ± equal, narrowly elliptic-lanceolate, scabrid above and on midnerve and hyaline margins, or sometimes sparsely scabrid throughout. Lemma (2.8)–3.2–3.7 mm, ± ½ length of glumes, 5-nerved, densely covered throughout by long, silky hairs, oblong, truncate, lateral nerves minutely excurrent, rarely lemma less hairy near apex and scabrid nerves clearly visible; awn 4–5.5–(6) mm, very fine, straight or ± curved, ± middorsal. Palea ⅔–¾ length of lemma, keels *c.* 0.1 mm apart, apex bifid. Callus hairs dense, 1.5–2 mm, to ⅔ length of lemma. Rachilla prolongation 0.5–0.8 mm, tipped by a dense tuft of hairs 1–2 mm. Lodicules 0.9–1.2 mm, linear. Anthers 0.9–1.3 mm. Caryopsis *c.* 2 × 0.7–1 mm.

Endemic.

S.: southern Westland (Jacksons Bay), Fiordland (Milford Sound), and southern coast between Catlins and Bluff; St. Coastal in damp sand; near sea level.

Lachnagrostis ammobia resembles *L. lyallii* and *L. elata* in its lax habit with extravaginal branching. It may be distinguished from *L. lyallii*, though not from *L. elata*, by the very lax panicle, long lemma and ± straight awn.

2. L. billardierei (R.Br.) Trin., *Fund. Agrost.* 128, t. 10 (1820) ≡ *Agrostis billardierei* R.Br., *Prodr.* 171 (1810) ≡ *Deyeuxia billardierei* (R.Br.) Kunth, *Révis. Gram. 1*: 77 (1829) ≡ *Calamagrostis billardierei* (R.Br.) Steud., *Nomencl. Bot.* ed. 2, *1*: 249 (1840); Holotype: BM! *R. Brown no. 6218* Port Jackson, [New South Wales, Australia].   sand wind grass

Stiff, bluish green perennial tufts, 10–60 cm, with capillary-branched panicles sometimes overtopped by leaves; branching intravaginal. Leaf-sheath chartaceous, with wide membranous margins, closely striate, smooth but sometimes scaberulous above on nerves, light brown. Ligule (1)–2–4.5 mm, tapered above, entire to erose, abaxially scabrid. Leaf-blade (5)–10–24 cm × 2.5–6–(10) mm, flat, rather harsh, scaberulous on ribs and on margins throughout, ± abruptly narrowed to firm, ± blunt, ± hooded tip. Culm (4)–7–40 cm, erect, or decumbent at base, included within uppermost leaf-sheath, internodes densely finely scabrid. Panicle 6–24 × 10–24 cm, lax, with long, whorled, ascending branches, later spreading and panicle becoming as broad as long; rachis scaberulous, branches scaberulous, spikelets borne singly at tips of ultimate panicle branchlets, on pedicels thickened above. Spikelets (4)–5–6 mm, pale green to purplish. Glumes 1–(3)-nerved, narrow-lanceolate, acuminate, usually smooth, sometimes sparsely scabrid, margins wide, hyaline, midnerve scabrid. Lemma 3–4 mm, ± ⅔ length of glumes, smooth, or often scabrid above especially on nerves, firmly membranous, shining, elliptic-lanceolate, lateral nerves excurrent to short awns 0.5–1 mm; central awn (4.5)–5–9 mm, fine, geniculate, from lower ⅓ of lemma or rarely middorsal. Palea *c.* ⅔ length of lemma, hyaline with 2 faint keels near centre, keels finely scabrid at minutely bifid apex. Callus hairs ± dense, very short, 0.3–0.7 mm, *c.* ⅒ length of lemma. Rachilla prolongation 0.5–1 mm, tipped by a thick tuft of hairs 1–1.5 mm and ≈ palea. Lodicules slightly > 0.5 mm, lanceolate, acute. Anthers (0.5)–0.7–1 mm. Caryopsis 1.3–1.8 × 0.5–0.8 mm.

Indigenous.

N.: throughout; S.: northern half and near Dunedin and Invercargill; St.; Three Kings Is, Ch. Coastal cliffs, sandy flats and dunes, occasionally inland on limestone cliffs and at hot springs; lowland.

Also indigenous to Australia.

3. L. elata Edgar, *N.Z. J. Bot. 33*: 18 (1995); Holotype: CHR 389251! *A. P. Druce* Burgoo St[rea]m, N.W. Nelson, 3400 ft, periodically flooded riverflat, March 1982.

Rather slender, perennial tufts, 30–80 cm, usually with narrow leaves and very lax panicles with few branches tipped by few spikelets; branching extravaginal. Leaf-sheath submembranous, with few, distinct ribs, very sparsely, finely scabrid throughout or only above. Ligule 1.2–5–(8) mm, truncate and erose in lower leaves, tapering in upper leaves, abaxially scabrid. Leaf-blade 3–25 cm × 0.4–1.5–(2.5) mm, flat, or folded and ± involute, narrow-linear, abaxially almost smooth near base to closely scabrid near acute tip, adaxially ribbed and scabrid on ribs, margins scabrid. Culm 6–50 cm, ± geniculate at base, internodes minutely retrorsely scabrid above. Panicle 4–30 × 3–20–(30) cm, very lax at maturity, often wider than long; primary branches naked, capillary, finely scabrid, becoming horizontal or reflexed, secondary branchlets much shorter, each tipped by a single spikelet. Spikelets 3–8 mm, light green to purplish. Glumes ± equal, elliptic-lanceolate, scabrid above on keel and near hyaline margin, tip acute to sometimes acuminate. Lemma (2.5)–3–3.8 mm, *c.* ¾ length of glumes, 5-nerved, with scattered short hairs usually near margins on lower ⅔, or rarely glabrous, elliptic-oblong, truncate, denticulate, lateral nerves usually very shortly excurrent, margins and nerves finely scabrid near apex; awn (0.3)–1–5 mm, usually straight, sometimes curved, arising dorsally from just above midpoint of keel to just below lemma apex, rarely awn 0. Palea *c.* ½ length of lemma or shorter, keels ± distinct, *c.* 0.2 mm apart, apex shallowly bifid. Callus hairs to 1 mm, to ⅓ length of lemma. Rachilla prolongation 0.1–0.8 mm, with tuft of hairs to 1 mm; rarely 0. Lodicules 0.6–0.9 mm, linear, acute. Anthers (0.6)–0.8–1.2 mm. Caryopsis 1.5–1.8 × 0.5–0.7 mm.

Endemic.

N.: Urewera National Park, Volcanic Plateau and Mt Egmont; S.: to west of Main Divide and in Fiordland, to the east in scattered localities; St. Usually in damp ground in tussock grassland and open forest; lowland to subalpine.

Lachnagrostis elata resembles taller plants of *L. lyallii* in being a lax-growing perennial with extravaginal branching. However, panicles in *L. elata* are more lax, often with fewer branches and spikelets than in *L. lyallii* and lemmas in *L. elata* are longer, usually 3–3.8 mm, compared to 2–2.5 mm in *L. lyallii*.

The length of the awn and the position where it arises on the lemma vary considerably. Very minute awns (< 2 mm) arise subapically whereas longer awns (to 5 mm) arise ± middorsally. In some plants the awn is completely lacking.

Rarely spikelets proliferous, e.g., CHR 324489 *A. P. Druce* L. Cobb, north-west Nelson, Apr 1969.

4. L. filiformis (G.Forst.) Trin., *Fund. Agrost.* 128, t. 10 (1820) ≡ *Avena filiformis* G.Forst., *Prodr.* 9 (1786) ≡ *Agrostis filiformis* (G.Forst.) Spreng., *in* Biehler *Pl. Nov. Herb. Spreng.* 7 (1807) non Vill. (1787) ≡ *Calamagrostis filiformis* (G.Forst.) Cockayne, *N.Z. Dept Lands Rep. Bot. Surv. Tongariro Natl Park* 35 (1908) non Griseb. (1868) ≡ *Deyeuxia* *filiformis* (G.Forst.) Petrie, *in* Chilton *Subantarct. Is N.Z. 2*: 474 (1909) non Hook.f. (1896) ≡ *Agrostis avenacea* J.F.Gmel., *Syst. Nat. 2, 1*: 171 (1791) ≡ *Lachnagrostis avenacea* (J.F.Gmel.) Veldkamp, *Blumea 37*: 230 (1992) *nom. superfl.* ≡ *Agrostis forsteri* Roem. et Schult., *Syst. Veg. 2*: 359 (1817) *nom. superfl.* ≡ *Lachnagrostis forsteri* (Roem. et Schult.) Trin., *Gram. Unifl.* 217 (1824) *nom. superfl.* ≡ *Deyeuxia forsteri* (Roem. et Schult.) Kunth, *Révis. Gram. 1*: 77 (1829) *nom. superfl.* ≡ *Agrostis solandri* F.Muell., *Veg. Chatham Is* 60 (1864) *nom. superfl.* ≡ *Lachnagrostis filiformis* (G.Forst.) Trin. var. *filiformis* [autonym, Zotov *Rec. Dom. Mus. 5*: 142 (1965)]; Lectotype: B (Herb. Willdenow no. 2208)! *Forster* [Herb. Forster ex Sprengel], Habitat in Nova Zeelandia (designated by Edgar 1995 op. cit. p. 20).   New Zealand wind grass

Open, bluish green or light green, annual or short-lived perennial tufts, (12)–20–70 cm, whole plant often withering early and culms not breaking up below panicle; branching intravaginal. Leaf-sheath firmly membranous, distinctly ribbed, glabrous below, very finely scabrid above, light green, later light brown. Ligule (1)–2–5 mm, oblong, rounded or tapered, later lacerate, abaxially with sparse prickle-teeth. Leaf-blade 2.5–8–(18) cm × 1.5–3 mm, usually flat, sometimes involute and 0.5–1 mm diam., glabrous, or ribs scabrid; margins very finely scabrid, tip fine, acute. Culm 10–35 cm, erect to spreading, internodes usually densely, minutely scabrid, occasionally smooth. Panicle 9–30 × (0.5)–2–25 cm, delicate, enclosed at base by the uppermost leaf-sheath, at first contracted, later very lax; branches filiform, numerous, unequal, very finely scabrid, primary branches naked for much of their length, with spikelets in clusters of 2–several, towards tips of the much shorter capillary secondary branchlets. Spikelets 2.5–3.5–(4.2) mm, pale silvery green to purplish. Glumes subequal, acute to acuminate, usually glabrous, membranous, very narrow linear-lanceolate, lower glume usually slightly longer and more acuminate; keel scabrid almost to base. Lemma 1.3–2–(2.3) mm, ½–⅔ length of glumes, 5-nerved, moderately covered with very short hairs, oblong-ovate, glabrous near hyaline, truncate, erose apex, lateral nerves very shortly excurrent; awn 3–6 mm, geniculate, ± middorsal or from *c.* upper ⅓. Palea ¾–⅘ length of lemma, keels *c.* 0.1 mm apart, apex subobtuse. Callus ringed by minute hairs 0.3–0.4 mm, to ¼ length of lemma. Rachilla prolongation 0, or to *c.* 0.3 mm tipped with hairs to *c.* 0.8 mm. Lodicules 0.5–0.9 mm, linear, acute. Anthers 0.2–0.5 mm. Caryopsis 0.8–1.3 × 0.3–0.5 mm.

Indigenous.

N.: throughout; S.: scattered throughout near coast, and on lake margins inland, once recorded from the margin of a saline area near Alexandra; St.; K., Three Kings Is, Ch. In damp ground, often on lake margins, also in disturbed ground and on roadsides; usually lowland, occasionally montane.

Also indigenous to Australia, New Guinea and Easter Id; naturalised in South America, South Africa, and Malesia.

The name *Agrostis aemula* R.Br. was applied by Buchanan. J. *Indig. Grasses N.Z.* t. 21 (1879) to New Zealand plants of *L. filiformis* but *L. aemula* (R.Br.) Trin. of Australia is a coarser grass than *L. filiformis* with larger spikelets and a more spreading panicle.

5. L. glabra (Petrie) Edgar, *N.Z. J. Bot. 33*: 20 (1995) ≡ *Deyeuxia glabra* Petrie, *T.N.Z.I. 46*: 36 (1914); Lectotype: WELT 77017a! *D. P[etrie]* "Ocean Beach", Bluff, Southland, 21.1.1913 (designated by Edgar 1995 op. cit. p. 20).

Lax, partly sprawling and stoloniferous perennials, 18–35 cm, with sparse, narrow, soft, bright green leaves; branching extravaginal. Leaf-sheath membranous, distinctly ribbed, glabrous. Ligule 3–5 mm, oblong, erose, abaxially finely scabrid. Leaf-blade 4–15 cm × 1.5–2.5 mm, flat, thin, abaxially smooth, adaxially finely scabrid on ribs; margins finely scabrid, tip fine, subacute. Culm 7–25 cm, usually included within uppermost leaf-sheath, internodes finely scabrid below panicle. Panicle 5–20 × 1–6 cm, contracted, later spreading; branches filiform, finely scabrid, bearing few spikelets towards tip on long delicate pedicels. Spikelets 3–3.8 mm, light green. Glumes subequal, ovate-elliptic, rarely with scattered prickle-teeth near scabrid midnerve; margins hyaline, scabrid near acute tip. Lemma 1.8–2.5 mm, ±⅔ length of glumes, faintly 3–5-nerved, glabrous, apex ± truncate and erose, lateral nerves hardly excurrent; awn 0–1.2 mm, straight, from just below lemma apex. Palea ≈ lemma, nerves very close set, apex hardly bifid.

Callus hairs few, *c.* 0.1 mm or 0. Rachilla prolongation 0.3–0.5 mm, tipped by a few hairs to *c.* 0.8 mm. Lodicules linear, acute. Anthers 0.5–0.9 mm. Caryopsis not seen.

Endemic.

S.: Southland (Weydon Burn, Bluff, Fortrose); St. Moist ground by the sea and on flood plains near sea level.

6. L. leptostachys (Hook.f.) Zotov, *Rec. Dom. Mus.* 5: 143 (1965) ≡ *Agrostis leptostachys* Hook.f., *Fl. Antarct. 1*: 94 (1845); Neotype: CHR 119486! *V. D. Zotov* Beeman, Campbell Island, 3.1.1961 (designated by Edgar 1995 op. cit. p. 23). = *Deyeuxia forsteri* var. *micrathera* Hack., *in* Cheeseman *Man. N.Z. Fl.* 869 (1906) *comb. illeg., varietal epithet legit.*; Holotype: AK 1471! *T. Kirk* Antipodes Id, 17.1.1890 (No 1142 to Hackel).

Pale green, erect tufts, 15–35–(65) cm, with numerous, narrow, involute leaves often overtopping the lax, delicately branched panicles; branching intravaginal. Leaf-sheath membranous with few, distinct ribs, smooth or scabrid above, pale green to very light brown. Ligule 2–4.5 mm, oblong, very slightly tapered above, truncate, denticulate, abaxially ± scabrid, especially near base. Leaf-blade 3.5–10–(15) cm × 1–2 mm, often involute and 0.5–1 mm diam., abaxially sparsely to closely finely scabrid, adaxially ribbed and closely ciliate-scabrid on ribs; margins finely scabrid, tip acute. Culm 3–18 cm, often included within leaf-sheaths, internodes finely scabrid. Panicle (3)–7–16 × 2–12 cm, very lax; rachis finely scabrid, branches capillary, finely scabrid, tipped by few spikelets. Spikelets 4.5–5.5–(6.5) mm, light green. Glumes ± equal, lanceolate, acute to acuminate, green and submembranous, usually sparsely scabrid especially above, midnerve with sparse prickle-teeth except near base, margins hyaline, sparsely scabrid near tip. Lemma 2.5–3 mm, *c.* ½ length of glumes, 5-nerved, with soft hairs scattered almost throughout except at tip, elliptic, truncate, lateral nerves shortly excurrent; awn 4–6 mm, straight or ± curved, ± middorsal, or from upper ⅓ of lemma. Palea *c.* ½–⅔ length of lemma, elliptic, nerves not evident, margins with 1–2 minute hairs, apex very shallowly notched. Callus hairs ± dense, especially below lemma margins, fine, to *c.* 1 mm, ¼–⅓ length of lemma. Rachilla prolongation to 0.5 mm with fine hair tuft to 1 mm; or 0. Lodicules *c.* 1 mm, narrow-lanceolate, acute. Anthers (0.6)–0.8–1.2 mm. Caryopsis *c.* 1.5 × 0.5 mm.

Endemic.

Ant., A., C. Rocky places, especially near the sea, and on open ground in grassland.

7. L. littoralis (Hack.) Edgar, *N.Z. J. Bot. 33*: 23 (1995) ≡ *Deyeuxia forsteri* var. *littoralis* Hack., *in* Cheeseman *Man. N.Z. Fl.* 869 (1906) *comb. illeg.*, *varietal epithet legit.* ≡ *Lachnagrostis filiformis* var. *littoralis* (Hack.) Zotov, *Rec. Dom. Mus. 5*: 142 (1965); Lectotype: W 7916! *R. H. Shakespear* Kermadec Islands, N. E. from New Zealand (No 1136 to Hackel) (designated by Edgar 1995 op. cit. p. 23).

Usually dense, light or greyish green, or glaucous, annual tufts; branching intravaginal. Leaf-sheath finely striate. Ligule oblong, tapered, subobtuse or denticulate, or later lacerate. Leaf-blade firm, abaxially smooth or minutely papillose, sometimes scabrid above, adaxially minutely scabrid on ribs; margins minutely scabrid, tip acute to subobtuse. Culm included within leaf-sheaths, internodes very finely scabrid below panicle, rarely visible until culm breaks up at maturity. Panicle delicate, ± contracted, later spreading, enclosed at base by sheath of uppermost culm-leaf; branches and branchlets ± erect, all ± equal in length, scabrid, naked for much of their length, the ultimate branchlets tipped by 1–2 spikelets. Spikelets light green or greenish brown. Glumes smooth, keel scabrid, margins hyaline, finely scabrid above. Lemma to ¾ length of glumes, 5-nerved, with scattered to rather dense short silky hairs, often glabrous above, lateral nerves shortly excurrent, faintly scabrid; awn ± middorsal. Palea slightly < lemma, keels minutely excurrent, faintly scabrid at tip. Callus hairs dense, very short, to ⅕ length of lemma. Rachilla prolongation almost 0, or very short and tipped by a longer hair tuft. Lodicules linear, acute.

subsp. **littoralis**

Tufts (3)–5–20–(40) cm, whole plant often withering early and culms not breaking up below panicle. Leaf-sheath subhyaline, smooth, or often minutely scabrid above. Ligule (0.2)–0.5–3 mm, abaxially sparsely scabrid. Leaf-blade 1–8 cm × 1.5–3–(5) mm, flat, or sometimes involute and *c.* 0.5 mm diam. Culm (1)–4–14 cm. Panicle (2)–3–12 × 0.5–8.5 cm; branches slender, sparsely scabrid. Spikelets (2.5)–3–4–(6) mm. Glumes narrow elliptic-lanceolate, usually equal or the upper slightly shorter, acute to acuminate to shortly mucronate, 1-nerved. Lemma (1.5)–1.8–2.5 mm, oblong-ovate; awn 3–6 mm, geniculate, slightly twisted near base. Palea nerves 0.1–0.2 mm apart. Callus hairs to 0.5 mm. Rachilla prolongation 0.2–1 mm, with hair tuft 0.5–1.5 mm. Lodicules *c.* 0.7 mm. Anthers (0.3)–0.4–0.7–(1) mm. Caryopsis 1–1.5 × 0.4–0.7 mm.

Endemic.

N.: North and South Auckland on coast and offshore islands, scattered further south; S.: Nelson on western coast, Marlborough at Marfell Beach.; K., Three Kings Is, Ch. Coastal, in rock crevices.

Sykes, W. R. *N.Z. DSIR Bull. 219*: 170 (1977) noted that plants of *Lachnagrostis* from L'Esperance Id in Kermadec Is were larger and had wider leaves than plants from elsewhere in Kermadec Is. He referred the L'Esperance plants to *L. pilosa* (as *L. richardii*) but Edgar (1995 op. cit. p. 24) considered that all their floral characters were consistent with *L. littoralis*.

subsp. **salaria** Edgar, *N.Z. J. Bot. 33*: 24 (1995); Holotype: CHR 468642! *H.D. Wilson BP 164* Canterbury, Banks Peninsula near shore of Lake Forsyth, *c.* 2 km above lake outlet, growing in ± upright tufts from turf of *Selliera* and *Scirpus* between tall clumps of *Juncus maritimus*, 5.2.1984.

Rather coarse, stiff tufts, 20–60 cm. Leaf-sheath firmly membranous, glabrous. Ligule 1.5–3 mm, abaxially smooth, rarely sparsely scabrid. Leaf-blade 4–15 cm × *c.* 0.5 mm diam., often involute, or flat and up to 2 mm wide. Culm 7–30 cm. Panicle 4–20 × 1–18 cm; branches closely scabrid. Spikelets 3–4 mm. Glumes dissimilar; lower slightly shorter, elliptic-lanceolate, acute, 1-nerved, upper elliptic-ovate, ± acuminate, 3-nerved. Lemma 2.5–3 mm, elliptic-lanceolate; awn 2.5–4 mm, straight or ± curved. Palea keels *c.* 0.3 mm apart. Callus hairs 0.3–0.4 mm. Rachilla prolongation almost 0, or to 0.9 mm, with hairs to 1.2 mm. Lodicules 0.8–0.9 mm. Anthers (0.5)–0.6–0.7 mm. Caryopsis 1.3–1.6 × 0.4 mm.

Endemic.

S.: eastern coast from North Canterbury to Bluff; St. Coastal, often in salt meadow; one record from the margin of an inland salt lake.

Plants of subsp. *salaria* are taller and coarser than plants of subsp. *littoralis.* Occasional plants of subsp. *littoralis* may reach 40 cm but usually have leaves to 5 mm wide and geniculate awns, whereas in subsp. *salaria* leaves are always narrow (to 2 mm wide) and the awns straight or curved.

Lachnagrostis littoralis subsp. *salaria* was once collected inland, CHR 402571 *A. P. Druce* Sutton Salt Lake Scenic Reserve, south of Middlemarch, Otago, *c.* 250 m, Jan. 1994.

8. L. lyallii (Hook.f.) Zotov, *Rec. Dom. Mus. 5*: 142 (1965) ≡ *Agrostis lyallii* Hook.f., *Fl. N.Z. 1*: 297 (1853) ≡ *Deyeuxia forsteri* var. *lyallii* (Hook.f.) Hack., *in* Cheeseman *Man. N.Z. Fl.* 869 (1906) *comb. illeg., varietal epithet legit.*; Holotype: K! *Dr Lyall* Milford Sound, New Zealand. = *D. forsteri* var. *semiglabra* Hack., *in* Cheeseman *Man. N.Z. Fl.* 869 (1906) *comb. illeg., varietal epithet legit.* ≡ *Lachnagrostis filiformis* var. *semiglabra* (Hack.) Zotov, *Rec. Dom. Mus. 5*: 142 (1965); Holotype: AK 1449! *T. F. C[heeseman]* Broken River, Canterbury Alps, 2000 ft (No 1131 to Hackel).

Lax, sometimes stoloniferous, perennial tufts, (5)–10–50 cm, with wide, flat, or narrow and folded, light or dull green leaves, and lax panicles often large in proportion to the plant; branching extravaginal; short plants with narrow involute leaves may be more densely tufted. Leaf-sheath submembranous, distinctly ribbed, smooth, or minutely scabrid above, green to creamy brown, sometimes purplish. Ligule 0.6–4 mm, ovate-oblong, ± truncate, denticulate to later ± lacerate, abaxially minutely scabrid. Leaf-blade 2–15 cm × 0.3–3.5–(6) mm, flat or folded, abaxially smooth to finely scabrid on ribs throughout or near tip, adaxially finely scabrid on ribs; margins minutely scabrid, tip subacute. Culm 3–25–(45) cm, erect or geniculate at base, longer culms projecting beyond uppermost sheaths, internodes usually finely scabrid below panicle. Panicle 4–25 × 1–16 cm, very lax or sometimes contracted after flowering, with widely spreading, finally horizontal branches; branches rather few, very finely, ± sparsely scabrid, with 1–2–several spikelets at branchlet-tips. Spikelets 2.3–4.5–(5) mm, pale green to light straw-coloured, greenish brown or purplish. Glumes subequal, ovate-elliptic or narrowly elliptic-lanceolate, acute, midnerve and hyaline margins scabrid near tip or in upper ½. Lemma (1.5)–2–2.5–(2.8) mm, ½–¾ length of glumes, 5-nerved, with numerous to scattered short fine hairs, sometimes almost completely glabrous, or the upper ⅓ ± glabrous, elliptic-oblong, truncate, lateral nerves slightly to obviously excurrent; awn (1.7)–2–6 mm, geniculate, ± middorsal. Palea *c*. ½ length to ≈ lemma, keels very faint, 0.1–0.2 mm apart, apex truncate, or obtuse, or shallowly notched. Callus hairs moderately dense, to 1 mm, to ⅓ length of lemma. Rachilla prolongation 0, or up to 1 mm, tipped by a few equally long, or longer hairs. Lodicules 0.7–0.9 mm, linear, acute. Anthers (0.3)–0.6–1.2–(1.5) mm. Caryopsis 1.2–2 × 0.4–0.8 mm. Plate 6B.

Endemic.

N.: one collection from Kaihu (north of Dargaville) and south from East Cape, Rotorua and Kawhia Harbour; S.: throughout; Ch. Open ground in tussock grassland and in other open sites; lowland to alpine.

Two forms can be distinguished within *L. lyallii*. Plants throughout North Id and in northern South Id are lax, with flat leaves, spikelets *c*. 4–5 mm, the lemma often almost glabrous *c*. 2.5 mm, the palea *c*. ½ the length of the lemma and the anthers usually 0.8–1.2 mm. The type of *Deyeuxia forsteri* var. *semiglabra* belongs here. In central North Id and in South Id, especially in southern and western regions, plants have inrolled leaves, smaller spikelets *c*. 2.5–3.5 mm, smaller, more densely hairy lemmas *c*. 1.8–2.2 mm, with the palea *c*. ¾ to ≈ lemma and anthers often rather short (0.3)–0.6–0.8 mm. In inland central to South Canterbury and in Otago, plants are small and densely tufted, but in western areas plants with inrolled leaves are taller and more lax, as in the type of *L. lyallii*. However no satisfactory geographical separation can be found, and it is even more difficult to separate the forms morphologically because so many intermediates occur.

In many specimens of *L. lyallii* there is no rachilla prolongation but it can be well-developed in others; in CHR 285393 *M. J. A. Simpson 7326* The Monument, Banks Peninsula, some rachillas were tipped by a rudimentary second floret.

9. L. pilosa (Buchanan) Edgar, *N.Z. J. Bot. 33*: 27 (1995) ≡ *Deyeuxia pilosa* Buchanan, *Indig. Grasses N.Z.* Add. et Corrig. 11 (1880) ≡ *Lachnagrostis richardii* Zotov, *Rec. Dom. Mus. 5*: 143 (1965) *nom. superfl.* ≡ *Agrostis pilosa* A.Rich., *Ess. Fl. N.Z.* 134, t. 23 (1832) non Retz. (1791); Holotype: P (Herb. Richard)! [*D'Urville*] Nlle Zélande, Hâvre de l'Astrolabe.

Often robust, wide-leaved, perennial tufts, (5)–30–90 cm, with dull green somewhat harsh leaves drying dark green, and large, firm, pale green, many-flowered panicles; branching intravaginal. Leaf-sheath similar in texture to blade, firm, striate, smooth, or finely scabrid above, green to light brown. Ligule 1.3–6 mm, tapered, rounded, becoming denticulate, abaxially finely scabrid. Leaf-blade (2)–6–28 cm × (0.5)–2–6.5–(10) mm, abaxially with numerous fine, finely scabrid ribs, midrib prominent near base, adaxially finely ribbed and finely scabrid on ribs; margins finely scabrid, tip subobtuse. Culm 20–40–(66) cm, erect. Panicle (2)–12–28 × (1)–6–20 cm, at first stiff and contracted with erect branches, at maturity more lax and ovate-elliptic; rachis smooth below, scabrid above, primary branches sparsely scabrid, filiform, very long, erect, secondary branchlets much shorter, very numerous, capillary, scaberulous, tipped by several clustered spikelets. Spikelets (3)–3.5–5.5–(6) mm, light green. Glumes ± equal, smooth and membranous, rarely sparsely scabrid in upper half, acute to acuminate; lower elliptic-lanceolate, scabrid on upper ⅔ of keel, upper linear-lanceolate, scabrid on upper ½ of keel. Lemma < ⅔ length of glumes, elliptic-oblong, truncate, lateral nerves shortly excurrent. Palea elliptic-oblong, nerves *c.* 0.2 mm apart, apex shallowly bifid. Lodicules *c.* 0.5 mm, lanceolate, acute. Caryopsis 1.3–1.8 × 0.4–0.6 mm. Fig. 9.

subsp. **pilosa**

Leaf-blade usually flat, sometimes harsh and involute above. Culm internodes usually entirely smooth, sometimes with a few prickle-teeth just below panicle, or rarely scabrid. Lemma 2–2.8 mm, with long hairs in lower ⅔; awn (3)–4.5–8.5 mm, geniculate, middorsal. Palea ½–⅗ length of lemma, nerves scarcely evident. Callus hairs conspicuous, to 1.5 mm, to ½ length of lemma. Rachilla prolongation < 0.5 mm with hairs to *c.* 1 mm, or often 0. Anthers 0.6–1.2–(1.8) mm.

Endemic.

N.: South Auckland (Cuvier Id), uncommon in mountains further south; locally common on Wellington and Wairarapa coasts; S.: locally common in Nelson and Marlborough and on Banks Peninsula, near Dunedin, and near Invercargill, more common in south Westland and

Fiordland, south-western Canterbury and western Otago; Ch. On rock outcrops and cliffs, both coastal and inland, also on river flats and in scrub; lowland to alpine.

Lachnagrostis pilosa subsp. *pilosa* is characterised by the wide, flat leaves, and robust, usually smooth culms. However, plants with scabrid culms occasionally occur and are more common on Banks Peninsula and the Chatham Is. Though usually very robust it can be very variable in size. Coastal plants from hills near Wellington and from cliff tops in the Marlborough Sounds are sometimes quite small, 5–15 cm, with rather stiff, usually involute leaves, 2–6 cm × 0.5–2 mm; panicles 2–6 × 1–4 cm, ovate-elliptic; spikelets *c.* 3 mm; lemma *c.* 2 mm; awn 3–4 mm.

Although the name *Agrostis pilosa* A.Rich. was illegitimate, being a later homonym of *A. pilosa* Retz., the homotypic name *Deyeuxia pilosa* has priority from 1880 and should be cited as *D. pilosa* Buchanan not *D. pilosa* (A.Rich.) Buchanan.

subsp. **nubifera** Edgar, *N.Z. J. Bot. 33*: 28 (1995); Holotype: CHR 134107! *R. A. Falla* Webling Bay, Auckland Islands, early Jan. 1963.

Leaf-blade flat, or rather harsh and involute above in plants from Antipodes Id. Culm usually included within uppermost leaf-sheath, internodes finely scabrid. Lemma 1.5–2 mm, with dense short hairs throughout; awn (0.5)–1–3.5 mm, straight, from upper ⅓ of lemma. Palea ≈ lemma, nerves distinct. Callus hairs dense, very short, to 0.5 mm, ⅕–¼ length of lemma. Rachilla prolongation to 0.5 mm, tipped by hairs to *c.* 1 mm. Anthers 0.5–0.8 mm.

Endemic.

Ant., A. Coastal, on margins of forest or in open scrub, or in damp ground at margins of tussock grassland or among sedges.

Plants of *L. pilosa* subsp. *nubifera* differ from those of subsp. *pilosa*, and also from subantarctic *L. leptostachys*, in the shorter, more densely hairy lemma with a short, straight (not geniculate) awn and palea = lemma rather than *c.* ½ length of lemma. Leaves of plants of *L. pilosa* subsp. *nubifera* on the Auckland Is are much wider than those of *L. leptostachys* but on Antipodes Id the leaves of subsp. *nubifera* may be almost as narrow as those of *L. leptostachys*.

In specimens of *L. pilosa* subsp. *nubifera* from Auckland Is the rachilla is often terminated by a rudimentary second floret consisting of a short lemma only.

10. L. striata (Colenso) Zotov, *Rec. Dom. Mus.* 5: 142 (1965)
≡ *Agrostis striata* Colenso, *T.N.Z.I.* 21: 107 (1889) ≡ *Deyeuxia forsteri* var.
humilior Hack., *in* Cheeseman *Man. N.Z. Fl.* 869 (1906) *comb. illeg.,*
varietal epithet legit.; Holotype: AK 199712! *H. Hill* (ex Herb. Colenso),
"small mountainous Agrostis", Lake Waikaremoana, 3/[18]88.

Tufts perennial, usually fine-leaved and low-growing, 5–15–(40) cm;
branching extravaginal. Leaf-sheath membranous, sparsely ribbed,
smooth, rarely minutely scabrid, green to light brown. Ligule 1–3–
(5.5) mm, ± oblong, sometimes slightly tapered, denticulate, abaxially
sparsely prickle-toothed. Leaf-blade (1)–2–10 cm × 0.8–2 mm wide,
usually folded and appearing narrower, abaxially smooth except near
tip, adaxially scabrid on prominent ribs; margins scabrid, tip subobtuse.
Culm 1.5–15–(30) cm, erect or ascending, longer culms often
projecting beyond uppermost sheaths, internodes minutely, retrorsely
scabrid. Panicle 2–10–(20) × 1.5–7–(12) cm, delicate, rather contracted
at first, later lax; branches few, filiform, finely scabrid, tipped by 1–few
spikelets. Spikelets (2)–2.2–3.5 mm, green to purplish. Glumes
subequal, acute, scabrid on midnerve in upper ½ or almost throughout,
hyaline margins with a very few prickle-teeth near tip; lower slightly
longer, linear-lanceolate, upper elliptic-lanceolate. Lemma 1.2–2–
(2.3) mm, ⅔–¾ length of glumes, usually densely covered by stiff,
straight hairs, ovate-oblong, truncate, lateral nerves scarcely excurrent;
awn 0.2–2–(3) mm, straight, very fine, often purple, from upper ⅓ of
lemma. Palea *c.* ⅔ to ≈ lemma, keels *c.* 0.1 mm apart, slightly excurrent,
apex minutely bifid, sparsely prickle-toothed. Callus ringed with short
hairs to *c.* 0.3 mm, *c.* ¼ length of lemma. Rachilla *c.* 0.5 mm, tipped by
equally long, or slightly longer hairs. Lodicules *c.* 0.3 mm, linear, acute.
Anthers 0.2–0.5–(0.6) mm. Caryopsis 0.8–1.4 × 0.4–0.6 mm.

Endemic.

N.: scattered from Auckland south, common on Volcanic Plateau; S.:
scattered to east of Main Divide, more common in Fiordland. Damp
ground in tussock grassland, often on lake margins; lowland to subalpine.

In cultivation plants reach 30–55 cm with leaves to 2 mm wide; their florets match
those of plants growing in the wild in all dimensions with awns remaining minute.
The panicle, however, is longer with more numerous spikelets. Plants from central

North Id growing on lake margins are similar in size to those in cultivation. They generally have a short lemma (slightly < 2 mm and covered with stiff hairs), a short, fine, straight awn, and minute anthers. However, the awns may be elongate and geniculate in mature panicles.

Although Hackel applied the name *Deyeuxia forsteri* var. *humilior* to small, densely tufted specimens of *L. lyallii* from Canterbury, the type specimen of *Agrostis striata* Colenso was cited in the protologue, hence var. *humilior* must be treated as a synonym of *L. striata* rather than *L. lyallii*.

11. L. tenuis (Cheeseman) Edgar, *N.Z.J. Bot. 33*: 30 (1995) ≡ *Deyeuxia billardierei* var. *tenuis* Cheeseman, *Man. N.Z. Fl.* 870 (1906) ≡ *D. tenuis* (Cheeseman) Zotov, *Rec. Dom. Mus. 5:* 139 (1965); Holotype: WELT 77014 a! *H. J. Matthews* Catlins River, Clutha Co., Otago, on coast, March 1896 (No 1148 to Hackel).

Stiff, leafy tufts 18–36 cm, with narrow ± inrolled leaves and lax panicles; branching intravaginal. Leaf-sheath subcoriaceous, ± obviously ribbed, glabrous, light brown. Ligule 1–3.2 mm, truncate, erose, minutely ciliate, abaxially scabrid. Leaf-blade 2–12 cm × 0.3–0.9 mm diam., inrolled, abaxially smooth, sometimes scabrid near tip, adaxially ribbed and short-scabrid on ribs, margins smooth, tip obtuse. Culm 5–20 cm, usually included within uppermost leaf-sheath, internodes densely, finely scabrid below panicle. Panicle (3)–6–12 × 2–8–(12) cm, delicate; branches fine, scabrid, erect to spreading, bearing spikelets at tips. Spikelets 3–5 mm, light green to light brown. Glumes 1–(3)-nerved, elliptic-lanceolate, ± acuminate, strongly keeled, keel finely scabrid in upper ½–⅔. Lemma 3–4.2 mm, ≈ glumes, firmly membranous, glabrous, papillose-scabrid, elliptic-lanceolate, lateral nerves sharply excurrent to *c.* 1 mm; central awn 3.5–6.5 mm, geniculate, from lower ⅓ of lemma. Palea 2.5–3 mm, *c.* ⅔ length of lemma, lanceolate, apex bifid, keels scabrid, margins submembranous, scabrid near apex. Callus hairs ± dense, very short, 0.3–0.8 mm, *c.* ¼ length of lemma. Rachilla prolongation 1.5–2.5 mm, with very short hairs throughout except sometimes near tip, rarely completely glabrous. Lodicules 0.7–1 mm, lanceolate, acute. Anthers (0.5)–0.7–0.9 mm. Caryopsis 1.4–1.6 × 0.5–0.7 mm.

Endemic.

S.: Canterbury near Christchurch (South Brighton, Teddington), and on south eastern coast from Otago Peninsula to Bluff; St. Coastal slopes, and at sea level in salt marsh or tidal ground.

12. L. uda Edgar, *N.Z. J. Bot. 33*: 32 (1995); Holotype: CHR 395485! *A. P. Druce* nr. Skeleton Lakes, Garvie Mts, 4900 ft, bog, March 1988.

Rather lax, perennial tufts, 9–35 cm; branching extravaginal. Leaf-sheath firmly membranous, inconspicuously ribbed, glabrous, light green to dull brown. Ligule 1–2.5 mm, truncate, erose or denticulate, abaxially sparsely, finely scabrid. Leaf-blade 3–10–(15) cm × 1–2 mm, flat or folded, abaxially smooth, adaxially sparsely, finely scabrid on ribs; margins minutely, sparsely scabrid, tip obtuse. Culm 4–20 cm, often included within uppermost leaf-sheath, internodes finely scabrid below nodes and below panicle. Panicle 3–14 × 1.5–8 cm, rather contracted at first, later lax; rachis finely scabrid, branches filiform, finely scabrid, tipped by rather few spikelets. Spikelets (2.5)–3–3.5–(4) mm, pale green or tinged faint purple. Glumes ± equal, elliptic-lanceolate, midnerve and margins scabrid in upper ¼, occasionally in upper ½ or ⅔. Lemma 2–2.5 mm, *c.* ¾ length of glumes, with scattered soft hairs, very rarely glabrous, ovate-oblong, truncate, lateral nerves not, or scarcely, excurrent, midnerve produced to a mucro, or sometimes a straight subapical awn, to 2.5 mm. Palea ½–⅔ length of lemma, apex shallowly bifid, keels *c.* 0.3 mm apart, very faint. Callus hairs conspicuous, to 1 mm, to ½ length of lemma. Rachilla prolongation *c.* 0.2 mm, tipped by hairs to *c.* 0.8 mm. Lodicules 0.5–0.8 mm, lanceolate, acute. Anthers 0.7–1.3 mm. Caryopsis 1.2–1.8 × 0.5–0.6 mm.

Endemic.

S.: south-western Canterbury (lower Godley River valley), Central and western Otago, Southland (Eyre Mts, Garvie Mts). In seepages, creek margins and boggy ground; subalpine to alpine.

Lachnagrostis uda appears close to both *L. striata* and *L. lyallii*; *L. uda* matches *L. striata* in the short, straight awns, but has much longer anthers, 0.7–1.3 mm, compared to 0.2–0.5 mm in *L. striata*; *L. uda* matches *L. lyallii* in anther length, but the straight awns of *L. uda* are much shorter (to 2.5 mm), than the geniculate awns of *L. lyallii* (2–6 mm). In its preference for a very damp habitat *L. uda* resembles *L. striata* rather than *L. lyallii*.

Two-flowered spikelets are occasionally found in plants of *Lachnagrostis*. Many 2-flowered spikelets were present in panicles of *L. uda*, OTA 43900 *K. J. M. Dickinson & B. D. Rance* Leithen Burn headwaters, Black Umbrella Range, 31.1.1986.

LAGURUS L., 1753

Monotypic genus of the Mediterranean region; the one sp. naturalised in N.Z.

1. L. ovatus L., *Sp. Pl.* 81 (1753).   harestail

Softly hairy, erect, tufted annuals, 6–80 cm, with velvety greyish green leaves and erect to nodding, soft, silky, rounded panicles. Leaf-sheath densely pubescent, rounded, firmly membranous, distinctly ribbed; uppermost sheath ± inflated. Ligule 1.4–2.6 mm, membranous, truncate, abaxially densely pubescent. Leaf-blade (2)–4–17 cm × (2)–4–9 mm, flat, soft, densely pubescent, linear-lanceolate, tapered to acuminate tip. Culm slender, erect or ascending, simple or branched near base, internodes minutely pubescent. Panicle 1.5–7.5 × 1.6–4.5 cm, pale silvery green to cream, sometimes purplish, dense, ovoid to cylindric or globose, with numerous, soft, projecting bristles; rachis and very short branches hidden among spikelets, sparsely hairy. Spikelets 7–10 mm, 1-flowered, laterally compressed, subsessile; disarticulation above glumes; rachilla prolonged, hairy. Glumes equalling spikelet, 1-nerved, membranous, narrow-lanceolate, tapering to a fine bristle, covered throughout with long, fine silky hairs. Lemma ≈ glumes, 5-nerved, membranous, rounded, 4–5.6 mm, elliptic, hairy below, much narrowed above and narrowly diverging into 2 apical, fine, straight, minutely scabrid bristles 2–6 mm; awn dorsal, arising from upper ⅓ of lemma, 13–22 mm, somewhat geniculate near base, often purplish. Palea ≈ lemma and much narrower, 2-nerved. Lodicules 2, hyaline. Stamens 3; anthers 1.4–2.7 mm. Ovary glabrous, styles short. Caryopsis 1.6–2.6 × (0.3)–0.5–0.7 mm; embryo small; hilum short; endosperm liquid. Plate 3A.

Naturalised from Mediterranean.

N.; S.; St.; Ch. Usually near the coast in sandy ground or in shingle, often locally abundant.

PENTAPOGON R.Br., 1810

Monotypic genus indigenous to south-eastern Australia and Tasmania; the one sp. naturalised in N.Z.

1. P. quadrifidus (Labill.) Baill., *Hist. Plant. Monogr. Cypér. 12*: 280 (1893).   five-awned spear-grass

Erect annual or short-lived perennial tufts, 15–60 cm; branching extravaginal. Leaf-sheath rounded, membranous, light green to cream, or light brown with strong, paler ribs, ± glabrous and shining to densely pubescent, shredding into fibres at maturity. Ligule 0.3–0.6 mm, membranous, truncate, erose, abaxially ciliate. Leaf-blade 3–6 cm × 0.5–0.9 mm diam., strongly involute, ± densely pubescent, tip subacute. Culm erect, slender, nodes pubescent, internodes ± villous especially near nodes. Panicle 4–8 cm, rather dense to ± lax, shortly and narrowly branched; rachis glabrous to pubescent, branches pubescent. Spikelets 1-flowered, 12–20 mm, light greenish brown to purplish; disarticulation above glumes; rachilla very shortly prolonged. Glumes subequal, 1–3-nerved, submembranous, acute to shortly mucronate, keeled, keel scabrid; lower 1-nerved, upper 3-nerved. Lemma 3.2–4.5 mm, 5-nerved, chartaceous, firmer than glumes, tightly involute, finely, closely scabrid above, apex 2-lobed with stout geniculate awn from sinus and each lobe tipped by 2 short fine awns; central awn with twisted column 2–6 mm and inclined ± straight arista 6–12 mm, lateral awns straight, 2–7.5 mm. Palea hyaline, shorter and narrower than lemma, 2-keeled, keels short-ciliate near apex. Callus hair-fringed, hairs 1–1.3 mm. Lodicules 2, glabrous. Stamens 3; anthers 0.4–0.7 mm. Ovary glabrous, styles free. Caryopsis *c.* 2.3 × 0.5–0.8 mm; embryo small; hilum punctiform; endosperm liquid.

Naturalised from Australia; south-eastern Australia and Tasmania.

S.: Marlborough (Ure Valley), North Canterbury (Upper Glenmark), Otago (Waitati). In grassland.

In Victoria, Australia, "Wide variation exists in the dimensions of the panicle and the awns of the lemma, and the extremes appear vastly different, but these are linked by a complete range of intermediates." [Walsh, N. G. and Entwisle, T. J. (Eds) *Fl. Victoria 2*: 491 (1994)]. In Tasmania var. *parviflorus* (Benth.) D.I.Morris is recognised.

PHLEUM L., 1753

Annuals or perennials. Leaf-sheath open, or closed near base, glabrous or puberulent. Ligule membranous. Leaf-blade flat. Culm erect. Inflorescence a dense, spike-like, ovoid or cylindric panicle. Spikelets 1-flowered, strongly laterally compressed; disarticulation above glumes. Glumes subequal, membranous, > lemma, keeled, shortly awned, acuminate or acute, margins overlapping for much of their length. Lemma hyaline, broadly truncate or obtuse, awnless, 1–7-nerved, glabrous to densely pubescent. Palea ≈ lemma, ovate-lanceolate, obtuse, 2-keeled. Lodicules 2, or 0. Stamens 2–3. Ovary with glabrous apex. Caryopsis terete or laterally compressed, not grooved; embryo small; hilum punctiform or shortly elliptic; endosperm liquid.

Type species: *P. pratense* L.

c. 15 spp. of the northern temperate zone and temperate South America. Naturalised sp. 1.

1. P. pratense L., *Sp. Pl.* 59 (1753).   timothy

Loose or dense tufts to 1 m. Leaf-sheath subcoriaceous, rounded, open, striate, glabrous, green to dark brown at base. Ligule 1–2.5–(4) mm, membranous, truncate or slightly rounded. Leaf-blade 9–17.5–(25) cm × 3–4.5 mm, flat, narrowed to finely pointed tip, striate, minutely scabrid throughout or only towards tip; margins minutely scabrid. Culm (30)–60–85 cm, erect or ascending from geniculate base, usually stout, internodes smooth, or finely scabrid below panicle; basal nodes sometimes swollen and tuberous. Panicle (4)–12–23 cm × 6–9 mm, dense, spike-like, cylindric, green, greyish green or purplish; rachis very stout, with scattered fine hairs concealed beneath spikelets, pedicels very short. Spikelets 2.5–3.5 mm, tightly packed. Glumes (3)–3.5–5 mm, oblong, truncate, 3-nerved, membranous, covered with short fine hairs, margin of lower glume softly hairy; keel stout, with large stiff hairs, and produced to a short scabrid awn (0.8)–1–1.5 mm. Lemma 1.5–2.5–(3) mm, 5–7-nerved, membranous, broadly oblong, obtuse, minutely hairy. Palea (1.2)–1.5–2–(2.5) mm. Lodicules 0.4–0.6 mm. Anthers 1.3–2.2 mm. Caryopsis slightly > 1 mm.

Naturalised from Eurasia.

N.: scattered; S.: scattered throughout, more common in Canterbury; St.; Ch., C. Pastures, waste land, roadsides and track margins.

Plants with proliferous inflorescences sometimes occur.

POLYPOGON Desf., 1798

Annuals or perennials. Leaf-sheath rounded. Ligule membranous. Leaf-blade flat. Culm nodes glabrous. Panicle dense, ± spike-like, often lobed. Spikelets falling entire at maturity with the pedicel disarticulating a short distance below the glumes, 1-flowered, somewhat laterally compressed; rachilla not prolonged. Glumes submembranous, subequal, 1-nerved, > lemma, apically awned or awnless. Lemma membranous or hyaline, 5-nerved, truncate, irregularly denticulate at apex; awn short, caducous, subapical or dorsal, or 0. Palea hyaline, 2-keeled. Callus glabrous. Lodicules 2, small. Stamens 3. Ovary glabrous; styles short, free. Caryopsis longitudinally grooved; embryo small, *c.* ¼ length of caryopsis; hilum punctiform or shortly elliptic; endosperm liquid or solid.

Type species: *P. monspeliensis* (L.) Desf.

18 spp. in warm temperate regions and on mountains in the tropics. Naturalised spp. 3.

Clayton and Renvoize (1986 op. cit. p. 140) noted that *Polypogon* is close to *Agrostis* and hybridises with it, but that *Polypogon* is distinguished by its deciduous spikelets with stipitate bases.

1 Glumes entire, awnless ....**3. viridis**
  Glumes with a fine awn arising in slight notch *2*

2 Awn of glume usually ≤ glume, rarely very slightly longer .... **1. fugax**
  Awn of glume 2–3 times > glume .... **2. monspeliensis**

1. P. fugax Nees ex Steud., *Syn. Pl. Glum. 1*: 184 (1854).

Loosely tufted annuals, very variable in size, 4–70–(90) cm. Leaf-sheath ± loose, chartaceous, striate, usually finely scabrid above. Ligule (1.5)–4–9 mm, oblong, rounded, lacerate, abaxially finely scabrid.

Leaf-blade (1)–6–15 cm × (1.5)–3–4.5–(8) mm, flat, minutely scabrid throughout or sometimes only on ribs near tip and on margins. Culm (3)–10–60 cm, ± erect, or ascending from geniculate base, internodes glabrous. Panicle 1–6 × 0.5–4–(7) cm, dense, cylindric to ovate-lanceolate, often lobed to ± interrupted at base; rachis smooth to scabrid above, with fascicled, finely scabrid branches and short, finely scabrid pedicels disarticulating near base. Spikelets 3–5 mm, oblong, light green, sometimes purplish, falling entire at maturity with most of pedicel attached. Glumes equalling spikelet, ± evenly scabrid, margins usually sparsely ciliate in lower ½, narrowed to shallowly notched apex with very fine, straight, terminal scabrid awn 1.5–2.5–(3) mm. Lemma 1–1.4 mm, hyaline, glabrous, elliptic-oblong, denticulate-truncate or minutely 4-toothed, usually with a fine, caducous apical awn (1.4–2.6 mm). Palea slightly shorter and narrower than lemma, keels ± excurrent. Anthers 0.5–0.7 mm. Caryopsis 0.9–1.2 × 0.4–0.5 mm.

Naturalised.

N.: common in North Auckland, scattered further south; S.: Canterbury (near Christchurch); K.: Raoul Id. Roadsides, sand flats, damp waste land, steep banks, rocky and swampy ground.

Indigenous to Asia and tropical Africa.

Because of its short awns *P. fugax* is sometimes confused with ×*Agropogon littoralis*, but that hybrid between *P. monspeliensis* and *Agrostis stolonifera* persists for more than 1 year, has ∅ anthers, and only very rarely produces a mature caryopsis.

2. P. monspeliensis (L.) Desf., *Fl. Atlant. 1*: 67 (1798).

beard grass

Annual tufts, very variable in size, 5–70–(100) cm. Leaf-sheath chartaceous, striate, smooth or scabrid above, green to light brown; uppermost sheath ± inflated. Ligule 1.5–10–(18) mm, oblong, tapered, ciliate-lacerate, abaxially scabrid. Leaf-blade 1–28 cm × 1–10 mm, linear-lanceolate, finely striate, scabrid, tip acute. Culm 3–60 cm, in tufts or solitary, erect, or geniculate and sometimes rooting at lower nodes, usually branched near base, internodes smooth or occasionally scabrid near panicle. Panicle 0.7–17 × 0.5–4 cm, very dense, cylindric or lobed, bearing numerous fine bristles; rachis smooth to scabrid, branches very short, closely scabrid, pedicels scabrid, very short, disarticulating. Spikelets 4–12 mm, oblong, silvery or yellowish green, rarely purple-tinged, falling entire at maturity with a minute piece of pedicel. Glumes equalling spikelet, scabrid, with longer prickle-teeth in lower ½, margins finely ciliate, apex emarginate, with very fine, straight terminal scabrid awn (3–9 mm) from the short notch. Lemma 1–1.5 mm, hyaline, glabrous, elliptic, truncate and minutely denticulate, awnless or with extremely fine, terminal awn (0.8–1.8 mm). Palea slightly shorter and narrower than lemma, keels slightly excurrent. Anthers 0.4–0.7 mm. Caryopsis 1–1.3 × 0.4–0.6 mm.

Naturalised.

N.: throughout, but rare in Taranaki; S.: Nelson, Westland (Hokitika), on eastern coast and inland in Central Otago; K.: Macauley Id; Three Kings Is. Usually coastal, on mud flats, salt marsh, lagoon or swamp margins, damp sand flats and sand dunes, roadsides, waste ground, coastal rocks and cliff edges; inland in Central Otago in moist ground along ditches or creeks or on river terraces.

Indigenous to Europe, North Africa and Asia; now naturalised in most warm temperate regions.

3. P. viridis (Gouan) Breistr., *Bull. Soc. Bot. France 110* (Sess. Extra. 89): 56 (1966).   water bent

Perennials or annuals, 15–50–(90) cm, loosely tufted, often with creeping stolons rooting at nodes. Leaf-sheath chartaceous, striate, glabrous, greyish green. Ligule 1.5–5.5 mm, truncate, erose, ciliate, abaxially minutely scabrid. Leaf-blade 2.5–18 cm × 2–9 mm, linear-lanceolate, finely striate, scabrid, tip acute. Culm 9–50 cm, decumbent and rooting at lower nodes, internodes glabrous. Panicle 3–14 × 1–3–(5) cm, ±dense, oblong but much lobed, sometimes interrupted near base; rachis smooth, branches scabrid, with spikelets almost to base, pedicels densely scabrid, very short, disarticulating. Spikelets 1.5–1.9 mm, pale green or purplish, falling entire at maturity with portion of pedicel attached. Glumes equalling spikelet, 1-nerved, elliptic, entire, firmly membranous, scabrid, keel more prominent in upper ½, awnless. Lemma *c.* 1 mm, hyaline, broadly elliptic, glabrous, awnless, apex denticulate, minutely ciliate. Palea narrower than lemma, apex bifid, ciliate. Anthers 0.4–0.6 mm. Caryopsis $0.8-1 \times 0.3-0.6$ mm.

Naturalised.

N.: scattered; S.: Nelson. Lowland in waste ground, roadsides and edges of ditches, often in damp places, occasionally a garden weed.

Indigenous to Mediterranean and North Africa.

Though *P. viridis* has often been referred to *Agrostis*, as *A. semiverticillata*, because it lacks awns, its spikelets are deciduous at maturity, falling away with a portion of the pedicel attached, a feature regarded as diagnostic for *Polypogon*.

SIMPLICIA Kirk, 1897

Tufted, lax-leaved perennials. Leaf-sheath rounded, uppermost sheaths ± keeled above. Ligule membranous. Leaf-blade narrow linear-lanceolate, flat. Culm often decumbent, to erect. Panicle open or contracted. Spikelets small, 1–(2)-flowered; disarticulation above glumes; rachilla prolonged. Glumes unequal, ≪ florets, membranous, keeled; lower 1-nerved, upper 1–3-nerved. Lemma 3-nerved, sometimes with 2 additional weak nerves, chartaceous, keeled, with or without subapical mucro or awnlet. Palea similar to lemma in size and texture, keeled, 1–2-nerved. Callus glabrous. Lodicules membranous, ovate-lanceolate, often asymmetrically lobed. Stamens 3. Ovary glabrous; styles apical; stigmas plumose. Caryopsis laterally compressed, ovoid; hilum punctiform, basal.

Type species: *S. laxa* Kirk

2 spp., endemic to N.Z.

Simplicia was revised by Zotov, V. D. *N.Z. J. Bot. 9*: 539–544 (1971); his treatment is followed here. Both spp. are chasmogamous and self-compatible. *S. laxa* remains a rare and endangered sp.

The spp. appear to resemble *Poa*, but, as Clayton and Renvoize (1986 op. cit.) point out, the palea in *Simplicia* is keeled, unlike that of *Poa*.

1 Panicle ± linear, branches erect, ± appressed; lemma scabrid ..... **1. buchananii**
Panicle ± pyramidal, branches spreading to reflexed; lemma shortly
pubescent .... **2. laxa**

1. S. buchananii (Zotov) Zotov, *N.Z. J. Bot.* 9: 542 (1971) ≡ *S. laxa*
var. *buchananii* Zotov, *T.R.S.N.Z.* 73: 236 (1943) ≡ *Poa uniflora*
Buchanan, *Indig. Grasses N.Z.* t. 49B (1880) non Muhl. (1817);
Holotype: WELT 59605 (Buchanan's folio)! *A. Mackay* Mt Arthur
[1874].

Tufts usually slender, (25)–40–60–(105) cm. Leaf-sheath membranous,
strongly ribbed, glabrous, sometimes scabrid on ribs, or the dark brown
basal sheaths pubescent. Ligule 2–4 mm, erose, abaxially glabrous. Leaf-
blade 10–20 cm × 1.5–4 mm, smooth, or finely scabrid on ribs; margins
finely scabrid, tip acuminate. Culm internodes glabrous. Panicle 4–18 cm,
± linear; rachis glabrous, branches short, erect, glabrous, bearing spikelets
almost to base, pedicels short, glabrous, ± appressed to branchlets.
Spikelets 2.8–3 mm, 1–(2)-flowered, lanceolate, light green. Glumes
glabrous, ovate-lanceolate to ovate, acute to subacute, margins ciliate;
lower 0.7–1 mm, upper 1–1.5 mm. Lemma ≈ spikelet, scabrid, 3-nerved,
or with 2 additional fainter lateral nerves, ovate-lanceolate, acute to
mucronate, or with a subapical awnlet. Palea 2.3–2.7 mm, 1–2-nerved,
scabrid. Rachilla prolongation *c.* 0.5 mm, glabrous. Anthers (0.7)–
1–1.3 mm, purplish or yellow. Caryopsis *c.* 1.5 mm long. $2n = 28$.

Endemic.

S.: Nelson, especially in north-west. Stony or rocky ground in forest
or in open places in the shade of taller plants or rocks; montane to
subalpine.

2. S. laxa Kirk, *T.N.Z.I.* 29: 497, t. 44 (1897) ≡ *S. laxa* Kirk var. *laxa*
(autonym, Zotov 1943 op. cit. p. 236); Lectotype: WELT 43017!
D. Petrie Waikouaiti, Otago, (designated by Zotov 1971 op. cit. p. 541).

Plants weak, 25–50 cm. Leaf-sheath membranous, strongly ribbed,
pubescent, especially the light brown basal sheaths. Ligule 1–3 mm,
erose, abaxially pubescent. Leaf-blade 6–10 cm × 1.5–2 mm, scabrid
on ribs or smooth, adaxially sometimes shortly and sparsely
pubescent; margins minutely scabrid. Culm internodes glabrous.

Panicle 10–15 cm, ± pyramidal; rachis glabrous, branches ± scabrid, binate, spreading to reflexed, naked in lower ½, pedicels short, pubescent, ± appressed to branchlets. Spikelets 2.5–3 mm, 1-flowered, lanceolate, light green. Glumes glabrous, ovate-lanceolate to ovate, acute, margins ciliate; lower 0.5–0.7 mm, upper 0.8–1 mm. Lemma ≈ spikelet, shortly pubescent, 3-nerved or with 2 additional fainter lateral nerves, ovate-lanceolate, acute to mucronate. Palea 2–2.5 mm, 1–2-nerved, shortly pubescent, especially on nerves. Rachilla prolongation *c.* 0.5 mm, very minutely ciliate. Anthers *c.* 1 mm. Caryopsis *c.* 1.5 mm long. $2n = 28$.

Endemic.

N.: Wairarapa at Ruamahanga Valley; S.: Otago at Ngapara, on Rock and Pillar and Old Man Ranges, and at Waikouaiti. Stony ground in shade of rocks or other shelter; lowland to montane.

ζ**Calamagrostis epigejos** (L.) Roth   Plants to 160 cm. Leaf-blade stiff, harsh, flat, to later inrolled, to 80 cm × 7 mm, abaxially closely short-scabrid, adaxially closely ribbed. Panicles dense-flowered, lanceolate, purplish, to 30 × 3 cm. Spikelets *c.* 6 mm. Glumes stiff, narrow, lanceolate. Lemma *c.* 3 mm, hyaline, surrounded and exceeded by a ring of white callus hairs concealing weak middorsal awn. Anthers *c.* 2 mm. A species indigenous to Eurasia and Africa, once collected on the Kermadec Is, Raoul Id, CHR 55415 *J. H. Sorensen* 1944, and known from one North Id locality, CHR 498081 *C. C. Ogle 2798* Milsons Line, Palmerston North, at road edge in ungrazed rough pasture, 19.5.1994.

ζ**Milium effusum** L., wood millet   Plant loosely tufted, to 75 cm, with dull green, soft, flat leaves. Leaf-sheath glabrous. Ligule 4–5 mm, glabrous. Leaf-blade 15–20 cm × 10 mm, abaxially sparsely, finely scabrid on ribs, adaxially smooth; margins finely scabrid, tip acute. Culm erect, or geniculate at base, internodes glabrous. Panicle 15–20 cm, very lax, nodding; branches filiform, spreading to deflexed, bearing few spikelets near tip. Spikelets *c.* 2.5 mm, awnless, pale green, becoming whitish, elliptic or ovate-elliptic. Glumes equalling spikelets, membranous, 3-nerved, glabrous, acute. Lemma *c.* 2 mm, finely 5-nerved, rounded, coriaceous, glabrous, shining, becoming brown. Palea ≈ lemma and similar in texture. Anthers not seen. Caryopsis *c.* 1.5 × 1 mm. A species indigenous to Eurasia and North America, known from one collection CHR 143507A,B *A. J. Healy & I. Fryer 63/115* "Otahuna" Tai Tapu, Canterbury, 26.2.1963.

AVENINAE

Spikelets with 2–several ♀ florets, or with 1 ♀ and 1 ♂ floret. Glumes equal or unequal, the upper sometimes slightly < lemma and seldom enclosing spikelet.

1 Spikelets with 1 ♀ floret, or rarely a few of the spikelets with 2 ♀ florets .. *2*
   Spikelets with 2 or more ♀ florets *3*

2 Spikelets disarticulating above glumes; lower lemma awned ..... **Arrhenatherum**
   Spikelets disarticulating below glumes; lower lemma awnless .... **Holcus**

3 Lemma conspicuously awned; awn reflexed, or very stout and geniculate or
   straight *4*
   Lemma awnless, or shortly mucronate or inconspicuously awned; awn, if
   present, very fine, straight or geniculate *7*

4 Ovary glabrous *5*
   Ovary hairy *6*

5 Awns recurved .... **Trisetum**
   Awns geniculate or straight...................................... **Amphibromus**

6 Plants annual; glumes rounded ....**Avena**
   Plants perennial; glumes keeled .... **Helictotrichon**ζ (p. 338)

7 Lemma keeled *8*
   Lemma rounded *9*

8 Plants perennial; lemma awnless or with mucro to 1.5–(3.2) mm .... **Koeleria**
   Plants annual; lemma usually with awnlet to 3 mm .... **Rostraria**ζ (p. 338)

9 Plants perennial; florets well separated; lemma apex irregularly broadly
   toothed .... **Deschampsia**
   Plants annual; florets ± adjacent; lemma apex finely bidenticulate .... **Aira**

AIRA L., 1753

Slender, annual or biennial, glabrous, herbaceous tufts. Leaf-sheath rounded. Ligule membranous. Leaf-blade convolute to flat, narrow. Panicle with delicate branches. Spikelets ± laterally compressed, small, 2-flowered; disarticulation above glumes and between florets; rachilla not, or occasionally, prolonged. Glumes subequal, membranous, delicate, ≈ or > lemmas, 1–3-nerved. Lemma 5-nerved, lanceolate, becoming ± chartaceous below, hyaline above, apex shortly bifid, with a fine, dorsal, geniculate awn, or the lower or both lemmas awnless. Palea < lemma, 2-keeled. Callus very short, hairy. Lodicules 2, lanceolate, acute, glabrous. Stamens 3. Ovary glabrous; styles free. Caryopsis dorsiventrally compressed, longitudinally grooved; embryo small; hilum punctiform or shortly elliptic; endosperm solid.

Type species: *A. praecox* L.

8 spp., from temperate Europe, western Asia and North Africa. Naturalised spp. 3; transient sp. 1.

1 Panicle linear, dense, with erect branches .... **3. praecox**
   Panicle ± ovate, lax, with spreading branches *2*

2 Pedicels of most spikelets at least twice length of glumes.............. **2. elegans**
   Pedicels of most spikelets less than twice length of glumes........................ *3*

3 Glumes 2–3.5 mm, entire, acute; both lemmas awned .... **1. caryophyllea**
   3a Glumes 2.5–3.5 mm; plants usually 4–35 cm .... subsp. **caryophyllea**
      Glumes 2–2.5 mm; plants usually 20–50 cm .... subsp. **multiculmis**
   Glumes *c.* 2 mm, mostly denticulate, often mucronate; lower lemma mostly
   awnless .... **cupaniana**ζ (p. 297)

1. A. caryophyllea L., *Sp. Pl.* 66 (1753).   silvery hair grass

Grey-green to reddish green, erect, slender, variable in height, consisting of a single delicate culm or a tuft of numerous culms; branching intravaginal. Leaf-sheath membranous to submembranous, ribs well-spaced, minutely scabrid above. Ligule 3–4.5–(6) mm, tapered, shortly denticulate, abaxially very sparsely scabrid. Leaf-blade 0.5–6–(12) cm × 0.5–1 mm diam., inrolled, ribs minutely scabrid; margins and subobtuse tip minutely scabrid. Culm internodes glabrous. Panicle 2–11 × 1–6 cm, ± ovate, lax; branches spreading, filiform, often flexuous, sparsely scabrid, naked at base with pedicelled spikelets clustered at tips. Spikelets silvery green or purplish. Glumes 1–(3)-nerved, shining, ovate-lanceolate, acute, keel sparsely scabrid, margins very finely ciliate-scabrid. Lemma firmly membranous and brownish below, colourless, hyaline, scabrid above; awn from lower ⅓ of lemma, projecting beyond glumes, column dark brown, twisted. Palea narrower than lemma, keels rounded, glabrous. Callus ringed by minute hairs. Rachilla minute, glabrous, not prolonged. Anthers 0.4–0.5–(0.6) mm. Caryopsis *c.* 1.5 × 0.4–0.5 mm.

subsp. **caryophyllea**

Plants usually 4–35 cm. Longer spikelet pedicels usually > 5 mm. Glumes 2.5–3.5 mm. Lemma 2–2.5 mm; awn 2.5–3.5 mm. Plate 4F.

Naturalised from Eurasia.

N.; S.: throughout; St.; Three Kings Is, Ch. Usually in dry, stony, or sandy ground, or in disturbed sites; rarely in damp ground at swamp or lagoon margins; sea level to subalpine.

subsp. **multiculmis** (Dumort.) Bonnier et Layens, *Fl. France* 358 (1984).

Plants usually 20–50 cm. Longer spikelet pedicels < 5 mm. Glumes 2–2.5 mm. Lemma 1.5–2 mm; awn 2–2.5 mm.

Naturalised from south-western Europe.

N.: North Auckland, Auckland City, southern Coromandel, Manawatu (Bainesse). Sandy waste land, cliffs and banks, rocky grassland, roadsides.

2. A. elegans Kunth, *Enum. Pl. 1*: 289 (1833).

Slender, erect, reddish green tufts 5–25–(30) cm, with very delicate loose panicles; branching intravaginal. Leaf-sheath submembranous, rather closely distinctly ribbed, ribs minutely scabrid almost to base. Ligule 1.5–4 mm, smooth, tapered, minutely denticulate, abaxially occasionally sparsely scabrid. Leaf-blade 1–6.5 cm × 0.2–0.3 mm diam., inrolled, abaxially minutely scabrid on ribs, adaxially with minute hairs; margins minutely scabrid, tip subacute to obtuse. Culm internodes glabrous. Panicle 2–10 × 2–8 cm, ovate, very lax; branches smooth to sparsely scabrid, capillary, spreading, pedicels ≫ spikelets. Spikelets silvery green. Glumes 1.5–2.5 mm, 1-nerved, shining, ovate-lanceolate, acute, keel sparsely minutely scabrid, margins very finely ciliate-scabrid. Lemma 1.2–1.5 mm, firmly membranous, brownish, hyaline near apex, upper ½ minutely scabrid; lower lemma sometimes awnless, upper lemma, or both lemmas, with awn 1.8–2.5 mm, from lower ⅓ of lemma, projecting beyond glumes. Palea narrower than lemma, keels rounded, glabrous. Callus ringed by minute hairs. Rachilla minute, glabrous; prolongation vestigial. Anthers *c.* 0.3 mm. Caryopsis *c.* 1 × 0.3 mm.

Naturalised from Mediterranean.

N.: North Auckland (near Kaitaia, Kawau Id), Bay of Plenty (near Te Puke); S.: Marlborough (Vernon), Canterbury (near Christchurch). Lowland in open ground in scrub or depleted tussock grassland, and on swamp margins.

Kartesz, J. T. and Gandhi, K. N. *Phytologia 74*: 49 (1993) found that Kunth (1833) validated the name *A. elegans.* Previously Kartesz, J. T. and Gandhi, K. N. *Phytologia 69*: 301–302 (1990) considered that *A. elegans* Kunth was invalid and some authors, including Edgar, E., O'Brien, M.-A. and Connor, H. E. *N.Z. J. Bot. 29*: 101–116 (1991) in a checklist, used the later name *A. elegantissima* Schur (1853) for this taxon.

3. A. praecox L., *Sp. Pl.* 65 (1753).   early hair grass

Small, erect, reddish green or greyish green tufts, 3–20–(25) cm, with spike-like panicles. Leaf-sheath firmly membranous, sparsely ribbed, very minutely scabrid below ligule or entirely smooth. Ligule (0.6)–1.5–3.2 mm, tapered, shortly denticulate, abaxially usually sparsely scabrid. Leaf-blade 0.5–4 cm × 0.3–0.5 mm diam., inrolled, smooth, or abaxially sparsely scabrid on ribs; margins minutely scabrid, tip obtuse. Culm internodes usually densely scabrid below nodes, or smooth throughout. Panicle (0.5)–1–4.5 cm × 3–9 mm, oblong or lanceolate; rachis, short erect branches and pedicels rather sparsely, finely scabrid. Spikelets green, tinged purple. Glumes *c.* 3 mm, 1–3-nerved, elliptic-lanceolate, acute, often scabrid near tip, keel scabrid, margins minutely ciliate-scabrid. Lemma 2.5–3 mm, ≈ glumes, smooth and firmly membranous in lower ½, densely minutely scabrid towards bifid hyaline apex; awn 3.2–4.2 mm, from lower ⅓ of lemma, column brown, finely twisted. Palea narrower than lemma, keels glabrous, rounded. Callus ringed by minute hairs. Rachilla very short, glabrous, prolonged. Anthers *c.* 0.3 mm. Caryopsis 1.5–2 × *c.* 0.5 mm.

Naturalised from Europe.

N.: scattered, but abundant in pumice land near Taupo; S.: Nelson, Westland, Fiordland and on the east from Banks Peninsula southwards; St.; Three Kings Is, Ch. Dry sandy or stony waste ground, on roadsides, in depleted pasture or in scrub; usually lowland, occasionally montane to subalpine.

ζ**A. cupaniana** Guss.   Slender, 10–15 cm. Panicle 3.5–8 cm, delicate, open. Spikelets silvery green. Glumes 2 mm, mostly denticulate and often mucronate. Lemma ½ to ⅔ length of glumes, awned from lower ⅓; lower lemma mostly awnless. Callus hairs minute, *c.* 1 mm. Anthers 0.3 mm. A species of southern Europe twice collected from one locality in the Wither Hills, Marlborough, in danthonia grassland *A. J. Healy* 13.10.1944 (CHR 45385) and 26.10.1966 (CHR 172970). The first collection was a depauperate plant.

AMPHIBROMUS Nees, 1843

Slender, tufted perennials, sometimes stoloniferous, with culms decumbent and rooting at nodes, or rhizomatous. Leaf-sheath rounded. Ligule membranous. Leaf-blade linear, narrow, flat or inrolled. Culm erect or geniculate at base, occasionally with corm-like swellings at lower nodes. Panicle narrow, loose, elongated; rachis terete and smooth below, becoming angled and scabrid towards apex, branches erect, slender, often flexuous. Spikelets 3–10-flowered, ♀, or uppermost ♂, solitary at branchlet tips, pedicelled; disarticulation above glumes and between lemmas; rachilla prolonged. Glumes unequal to subequal, submembranous or hyaline, ± keeled; lower shorter, (1)–3–(5)-nerved, upper broader, 3–7-nerved. Lemma > glumes and firmer, rounded, 5–7-nerved, 2–4-toothed or -lobed, lobes aristate to obtuse; awn dorsal, ± geniculate, ± recurved, arising from near midpoint to just below lemma apex. Palea hyaline, 2-keeled, apex bifid. Callus prominent, hairy. Lodicules membranous, glabrous, spathulate to narrow-triangular, sometimes with a lateral lobe. Ovary glabrous, or slightly hairy at apex; styles free. Caryopsis terete; embryo small, to ⅓ length of caryopsis; hilum linear, to ½ length of caryopsis; endosperm solid. Chasmogamous or cleistogamous.

Type species: *A. neesii* Steud.

12 spp. of Australia, N.Z. and South America, with 9 spp. endemic to Australia. Indigenous sp. 1, also occurring in Australia.

The Australian spp. were revised by Jacobs, S. W. L. and Lapinpuro, L. *Telopea 2*: 715–729 (1986), and their treatment is the basis for this account. They referred *A. gracilis* P.Morris of New South Wales, Victoria and Tasmania to *A. fluitans*.

Clayton and Renvoize (1986 op. cit. p. 123) treated *Amphibromus* as a synonym of Northern Hemisphere *Helictotrichon*, noting that the glabrous ovary in *Amphibromus* seemed an inadequate basis for generic separation.

1. A. fluitans Kirk, *T.N.Z.I. 16*: 374 (1884); Lectotype: WELT 68389! *T. K[irk]* Waihi Lake, Waikato (here designated).

Weak tufts, (7)–20–40 cm, stoloniferous, with decumbent culms rooting at lower nodes, erect or floating above; branching extravaginal. Leaf-sheath chartaceous, smooth, or scabrid, especially towards ligule, uppermost sheaths often scabrid throughout. Ligule 1.5–4.5–(5) mm, long-tapered, acute, entire, becoming lacerate, abaxially smooth or sparsely, minutely scabrid. Leaf-blade 5.5–12.5 cm × (0.6)–1–3 mm, flat, or inrolled, papillose-scabrid, abaxially shallowly ribbed, adaxially more obviously ribbed; margins minutely scabrid, tip acute. Culm internodes smooth, sometimes scabrid just below nodes. Panicle 6.5–13 cm, ± erect, enclosed below by uppermost leaf-sheath; branches and pedicels filiform, scabrid. Spikelets 15–25 mm, 3–6-flowered, pale green; florets chasmogamous and cleistogamous within the same spikelet. Glumes unequal, glabrous, obtuse, margins ciliate-scabrid; lower 2–3.3 mm, 1-nerved, narrow-lanceolate, upper (2.2)–3.5–4 mm, 3-nerved, ± ovate-lanceolate. Lemma (4)–4.5–5.5 mm, 7-nerved, firm, green central portion sparsely scabrid, smooth near base, hyaline margin very wide, finely minutely hairy or scabrid; lemma lobes 2, obtuse; awn 7.5–18 mm, stout, straight, arising from near midpoint of lemma. Palea < lemma, keels densely stiff-ciliate, interkeel glabrous. Callus with ring of fine hairs < 1 mm. Rachilla prolongation 1.5–1.8 mm, hairy above. Anthers in chasmogamous florets 1.4–1.7 mm; in cleistogamous florets 0.4–0.5–(0.8) mm. Caryopsis 1.5–2 × 0.5–0.7 mm.

Indigenous.

N.: Rare and scattered, recently collected only in Kaweka Range and Kaimanawa Mts, and at Paekakariki and L. Wairarapa; S.: Westland (Mahers Swamp) and Canterbury (L. Tekapo). Lowland to montane in moderately fertile wetland.

Also indigenous to Australia; southern New South Wales, Victoria and Tasmania.

Though *A. fluitans* was earlier recorded from the Waikato and in Taranaki, no recent collections have been made from these areas. Ogle, C. *Wellington Bot. Soc. Bull. 43*: 29–32 (1987), discussed the future of the sp. in N.Z. and considered it threatened because the type of wetlands in which it occurs have either been lost to intensive pastoral farming or overrun by introduced grasses and other plants. Jacobs and Lapinpuro (1986 op. cit. p. 729) state that *A. fluitans* is now apparently uncommon in Australia.

LECTOTYPIFICATION   Of 3 specimens at WELT collected by Kirk at Lake Waihi, WELT 68389 was designated lectotype because the florets still remained on the panicle; the other two specimens were overmature.

REPRODUCTIVE BIOLOGY   Flowering in *Amphibromus* may be chasmogamous or cleistogamous. The cleistogamous florets, which have shorter anthers, may be mixed with chasmogamous florets in the same spikelet, or the plant may bear reduced entirely cleistogamous inflorescences which remain enclosed within the leaf-sheaths.

ARRHENATHERUM P.Beauv., 1812

Perennials. Leaf-sheath rounded. Ligule membranous. Leaf-blade flat, convolute when young. Panicle moderately dense. Spikelets somewhat laterally compressed, usually 2-flowered, the two florets falling together at maturity; lower floret ♂ with long, geniculate awn from lower ⅓ of lemma, upper floret ♀ or ♀ with ± straight awn from upper ⅓ of lemma or awnless, rarely both florets ♀ and long-awned; rachilla prolonged. Glumes persistent, hyaline, unequal, sometimes keeled, the lower shorter. Lemma submembranous. Palea hyaline, slightly < lemma, 2-keeled. Lodicules distally membranous, lanceolate, acute, glabrous. Callus hairy. Ovary apex hairy; styles free. Caryopsis ± dorsiventrally compressed, not grooved; embryo small; hilum linear; endosperm solid.

Type species: *A. avenaceum* P.Beauv.

c. 6 spp. of Europe, Mediterranean and western Asia. Naturalised sp. 1.

1. A. elatius (L.) J.Presl et C.Presl, *Fl. Cechica* 17 (1819).

tall oat grass

Loosely tufted, coarse perennials, (40)–60–150 cm. Leaf-sheath ± chartaceous, smooth, or rarely minutely scabrid, sometimes with scattered long fine hairs. Ligule 1–3–(4) mm, truncate to ± obtuse, erose, ± ciliate, abaxially minutely scabrid to minutely hairy. Leaf-blade 15–30–(50) cm × 2–12 mm, minutely scabrid on ribs, to ± smooth, adaxially often with scattered long fine hairs; margins minutely, closely scabrid, tip acute, scabrid. Culm erect or ± spreading, stout. Panicle (4)–14–32 cm, erect or nodding, dense or lax, slender and lanceolate or with lower branches spreading; rachis smooth below, scabrid above, branches and pedicels slender, scabrid. Spikelets 7.5–9.5–(11) mm, shining, green or purplish. Glumes acute, sometimes finely scabrid, nerves finely scabrid; lower (3.5)–4.5–6–(6.8) mm, lanceolate, 1-nerved, upper ≤ spikelet, ovate-lanceolate, 3-nerved. Lemma (7)–8–9 mm, 7-nerved, ovate-lanceolate, hyaline near acute tip, nerves distinct, finely scabrid, upper or both lemmas with scattered fine hairs in lower ½; awn of lower lemma 8–16.5 mm; awn of upper lemma 0.5–10 mm, or 0. Palea narrower than lemma, keels finely ciliate, interkeel minutely scabrid, apex shortly bifid. Callus usually ringed by short stiff hairs. Rachilla prolongation 1.5–2.5 mm, delicate, glabrous. Anthers 3.5–5 mm. Gynoecium: ovary 0.6–0.8 mm; stigma-styles 1.5–1.8 mm. Caryopsis 2.5–3.8 × 0.8–1.2 mm.

1   Culm with basal internodes of uniform width .... subsp. **elatius**
     Culm with basal internodes swollen and bulbous .... subsp. **bulbosum**

subsp. **elatius**

Culm nodes glabrous, or occasionally hairy; internodes glabrous, basal internodes of uniform width throughout, sometimes with sparse, long, retrorse hairs.

Naturalised.

N.: western areas; S.: common to locally abundant throughout; C. Along roadsides and railway lines, on clay banks, in waste ground, paddocks and sometimes on dune margins, or a garden weed; sea level to montane.

Indigenous to Europe, western Asia and North Africa; now naturalised and widespread in temperate regions of both Hemispheres.

Spikelets rarely proliferate, e.g., CHR 96196 *V. D. Zotov* Wellington, Jan. 1943, in which many spikelets are 3-flowered.

subsp. **bulbosum** (Willd.) Schübl. et G.Martens, *Fl. Würtemberg* 70 (1834).   onion twitch

Culm nodes densely hairy; internodes sometimes with scattered long fine hairs above and below nodes, basal internodes swollen, bulbous or pear-shaped.

Naturalised from south-western Europe.

N.: North Auckland and further south in western areas; S.: eastern areas; Ch. A weed of gardens and pasture, in waste ground and often in river beds; lowland to montane.

AVENA L., 1753

Annuals or biennials, with solitary culms or tufted. Leaf-sheath chartaceous, striate, rounded. Ligule membranous. Auricles 0. Leaf-blade flat, rarely convolute. Panicle loose to sometimes contracted or secund with large nodding spikelets. Spikelets with 1–6 ♀ florets and 1–(2) ♂ or rudimentary florets; disarticulation below each floret, or only above glumes, or not at all in cultivated spp.; rachilla prolonged. Glumes usually equal and equalling spikelet, 3–11-nerved, rounded, lanceolate, acuminate, herbaceous to thinly chartaceous with thick scarious margins. Lemma usually coriaceous at maturity, bidentate to biaristulate and then rarely with 2 additional setae, rarely subentire; awn middorsal, stout and geniculate, column usually twisted, cultivated spp. often with reduced awns or awnless. Palea tough, ≤ lemma, 2-keeled, keels ciliate. Lodicules 2, linear-lanceolate, or unequally bilobed or with a rudimentary side-lobe. Callus hairy or glabrous. Stamens 3. Ovary hairy; styles free. Caryopsis terete, grooved; embryo relatively small to very large; hilum linear; endosperm solid.

Type species: *A. sativa* L.

c. 30 spp. of Europe, western and central Asia, North and north-eastern Africa, naturalised in many temperate regions. Naturalised spp. 5; transient sp. 1.

Edgar, E. *Proc. 33rd N.Z. Weed and Pest Control Conf.* 230–236 (1980) discussed characters distinguishing species of *Avena* growing wild in N.Z.

The genus was monographed by Baum, B. R. *Canada Dept Agric. Monogr. 14*: 1–463 (1977) and his treatment is followed here.

1 Lemma lobes toothed beside central dorsal awn *2*
  Lemma lobes drawn into fine awns or bristles beside central dorsal awn *5*

2 Lowest floret separating easily at maturity from rachilla leaving a regular scar; callus with ring of stiff hairs *3*
  Lowest floret detached from rachilla only by fracture; callus glabrous or with only a few hairs *4*

3 All florets of spikelet easily detached from rachilla leaving a regular scar at base .... **2. fatua**
  Only lowest floret of spikelet separating readily leaving a scar, all upper florets detaching from rachilla only by fracture .... **4. sterilis**
    3a Lemma 25–30 mm; awn stout, 60–70 mm .... subsp. **sterilis**
    Lemma 18–20 mm; awn slender, 25–50 mm .... subsp. **ludoviciana**

4 Lowest floret breaking from rachilla with an almost horizontal fracture; spikelets 17–28 mm; mature culms yellowish .... **3. sativa**
  Lowest floret breaking from rachilla with an oblique (*c.* 45°) fracture; spikelets 25–30 mm; mature culms reddish .... **byzantina**ζ (p.306)

5 Lemma stiffly long hairy below awn insertion; all florets easily detached from rachilla and falling away early .... **1. barbata**
  Lemma ± glabrous, with a few hairs below awn insertion; all florets persistent on rachilla, finally falling after fracture of rachis .... **5. strigosa**

1. A. barbata Link, *J. Bot. (Schrader)* 2: 315 (1800).   slender oat

Tufts green to dark green, often very slender, 65–100–(170) cm, erect or drooping above in taller plants. Leaf-sheath smooth, or slightly scabrid above, occasionally red-tinged. Ligule 2.7–5–(9) mm, obtuse, denticulate, abaxially finely scabrid. Leaf-blade 12.5–30 cm × 2–8 mm, finely scabrid on ribs and margins. Culm internodes glabrous. Panicle (12)–18–40 cm, very lax, equilateral; rachis smooth, branches and pedicels very filiform, finely scabrid. Spikelets 40–50 mm, 2–(3)-flowered; disarticulation below each floret at maturity. Glumes 7–9-nerved. Lemma 12–18 mm, narrow-lanceolate, light brown, with long stiff reddish hairs below level of awn insertion, green and scabrid above, each lobe-tip produced to a fine bristle (2–3 mm); lemmas awned, awns 30–36 mm, geniculate, column very dark brown, twisted. Palea keels with 1–2 rows of cilia, interkeel glabrous, flanks sometimes sparsely scabrid. Callus long hairy, with horse-shoe shaped scar. Rachilla long hairy. Anthers 2–3 mm. Caryopsis 5–7 × 0.8–1.3 mm, light brown, silky hairy.

Naturalised from Mediterranean.

N.: North Auckland, South Auckland, Bay of Plenty (Opotoki), Hawke's Bay (near Napier), Wellington (Wanganui, Paekakariki and Wellington City). Near the sea, especially on cliffs, dry banks, and steep grassy slopes, occasionally inland on disturbed ground, roadsides and in pasture.

2. A. fatua L., *Sp. Pl.* 80 (1753).　　　　　　　　　　wild oat

Tall, green or light green tufts to 1.5 m, robust, or sometimes with rather slender culms. Leaf-sheath slightly hairy, upper sheaths glabrous. Ligule 2–7 mm, truncate to oblong and obtuse, denticulate, abaxially scabrid. Leaf-blade 15–50 cm × 4–15 mm, scabrid on ribs and margins, often with scattered hairs, especially on margins. Culm erect or occasionally geniculate at base, internodes glabrous. Panicle (15)–20–38 cm, lax, equilateral, ± nodding; branches widely spreading, slender, drooping, scabrid. Spikelets 35–50 mm, 2–3-flowered, pendulous on fine unequal pedicels; disarticulation below each floret at maturity. Glumes 9–11-nerved. Lemma 15–17.5 mm, narrow-lanceolate, light brown at first becoming tough and darker brown at maturity, stiffly hairy below level of awn insertion, or sometimes glabrous, scabrid on nerves above, narrowed to bidentate scarious apex; awn 30–40 mm, geniculate, stout, column dark brown, twisted. Palea keels with one row of cilia, interkeel scabrid. Callus with horse-shoe shaped scar densely bordered by long hairs. Anthers 2.5–3.5–(4) mm. Caryopsis 6–8 × 1.8–2.2 mm, light grey, densely hairy.

Naturalised.

N.: Auckland City, Bay of Plenty (Opotiki), Hawkes Bay (Takapau), more common in the Manawatu and Wairarapa, also from near Wellington City; S.: common in Canterbury, North and Central Otago, less so in Nelson, Marlborough, South Otago and Southland; Ch. A problem weed of crops, on roadsides and waste land and also in brackish soil and reclamation areas.

Indigenous to Europe, Central Asia and North Africa; now naturalised and a widespread weed in temperate regions of both Hemispheres.

Occasional hybrids *A. fatua* × *A. sativa* may occur within an oat crop, or where oats have been grown. The two spp. are strongly self-fertilising and few hybrids are found.

3. A. sativa L., *Sp. Pl.* 79 (1753).   oat

Erect, green to glaucous tufts, 50–150 cm. Leaf-sheath glabrous or slightly hairy. Ligule (2)–3–5 mm, ± rounded, finely denticulate, abaxially finely scabrid. Leaf-blade 15–40 cm × (2.5)–7–15 mm, ± finely scabrid on ribs; margins finely scabrid, sometimes with a few scattered hairs. Culm erect or decumbent and rooting at base, internodes glabrous. Panicle 14–32 cm, very variable in shape, loose or contracted, equilateral to secund; branches horizontal to ± erect, slender, finely scabrid. Spikelets (15)–20–30–(40) mm, 2–(3)-flowered, pendulous on fine unequal pedicels; falling intact at maturity. Glumes 9–11-nerved. Lemma 10–18 mm, ovate-lanceolate, glabrous, or with a few hairs towards base, apex obtuse or shallowly notched, red-brown, grey, yellow, white or blackish, usually awnless, but lowest or sometimes all florets in certain cultivars bearing a weak, straight or curved awn, not geniculate and scarcely twisted at base. Palea keels with 1 row of cilia, interkeel scabrid or glabrous. Anthers 2–4.2 mm. Caryopsis 4–9 × 1.5–3 mm.

Naturalised from Eurasia.

N.; S.; Foveaux Strait on Pig Id; St.; C. On roadsides, on railway embankments, and near racecourses and in farmland, occasionally in damp ground.

Fatuoid oats, a mutant form of oat, were first recorded for N.Z. by Wright, G. M. and Shillito, N. L. *N.Z. J. Agric. 128*: 45 (1974). Fatuoids, or false wild oats, occur spontaneously in oat crops. Their florets resemble those in *A. fatua*; the awn is geniculate with twisted column, and the callus horse-shoe shaped and bordered by hairs. The spikelet disarticulates below all florets at maturity. In other respects fatuoids resemble the oat cultivar within which they appear. Fatuoids may seem taller than other plants in the crop because they shed their florets early and the empty glumes stand out above the drooping grain-filled spikelets of normal plants.

4. A. sterilis L., *Sp. Pl.* ed. 2, 118 (1762).

Strong, rather stiff, erect green tufts to 150 cm, with heavy panicles. Leaf-sheath glabrous, or rarely with scattered soft hairs. Ligule 3–8 mm, truncate, or more tapered centrally, shortly denticulate, finely ciliate. Leaf-blade 15–45 cm × 3–15 mm, very finely scabrid on ribs, adaxially sometimes smooth, rarely with scattered hairs; margins minutely scabrid. Culm internodes glabrous. Panicle 20–40 cm, equilateral, lax; rachis smooth, branches few, widely spreading, filiform, scabrid. Spikelets often drooping; disarticulation above glumes; rachilla tough and continuous between florets. Glumes 9–11-nerved. Lemma elliptic-lanceolate, tough and hairy below level of awn insertion, greenish above and scabrid on nerves, narrowed to scarious bidentate apex; lower 1–2 lemmas awned; awn geniculate, column red-brown, twisted. Palea keels with 1–3 rows of cilia, interkeel usually minutely hairy above. Callus bordered by dense short hairs.

subsp. **sterilis**

Spikelets 6–8 cm, 3–5-flowered, often drooping. Lemma 25–30 mm, light reddish brown, with dense stiff reddish brown hairs below awn insertion; awn very stout, 60–70 mm. Anthers 3–4 mm.

Naturalised from Mediterranean.

N.: North Auckland (near Whangarei), Auckland City, Waikato (Waerenga); Cook Strait (Stephens Id); S.: Dunedin. In waste land.

subsp. **ludoviciana** (Durieu) Gillet et Magne, *Fl. Franç.* ed. 3, 532 (1873). winter wild oat

Spikelets 3.5–6 cm, 2–3-flowered, spreading. Lemma 18–20 mm, light brown or greyish, with ± dense light brown or whitish hairs below awn insertion; awn ± slender, 25–50 mm. Anthers 2–5 mm.

Naturalised from Eurasia.

N.: Feilding, Wellington City; S.: Nelson City, Blenheim, more common in Canterbury, North Otago (near Oamaru). A weed in crops.

Avena sterilis subsp. *ludoviciana* was formerly known as *A. persica.*

5. A. strigosa Schreb., *Spicil. Fl. Lips.* 52 (1771).   sand oat

Rather slender, erect, glaucous tufts, 80–120 cm. Leaf-sheath glabrous or finely hairy. Ligule 2.5–5.5 mm, obtuse, denticulate, abaxially minutely scabrid. Leaf-blade 15–30 cm × 4.5–13.5 mm, finely scabrid on ribs and margins, occasionally with scattered sparse hairs. Culm internodes glabrous. Panicle (12)–18–26 cm, erect or drooping at tip, rather dense, equilateral, rather narrow; rachis smooth, branches and pedicels fine, minutely scabrid. Spikelets 35–40 mm, 2-flowered, falling at maturity only after fracture of rachis. Glumes 7–(9)-nerved. Lemma 11–15 mm, narrow-lanceolate, light brown or straw-coloured, later becoming darker, smooth and shining below level of awn insertion, green and scabrid above, often with a very few fine hairs near base of awn, each lobe-tip produced to a fine purplish bristle 5–7.5 mm, and often with lateral setae to 1 mm; both lemmas awned, awn 25–35 mm, geniculate, column mid-brown, twisted. Palea keel with one row of cilia, interkeel finely scabrid to minutely hairy above. Callus glabrous or with a few short hairs. Rachilla prolongation glabrous, tipped by few to numerous short hairs. Anthers 2.2–4 mm.

Naturalised from Europe.

N.: Waverley, Wanganui, Palmerston North, Wellington; S.: Canterbury (Weedons), Otago (Balclutha). A weed in crops in light soil and on roadsides; also recorded in ballast and as a seed impurity (in 1883) in imported Russian wheat.

ζ**A. byzantina** K.Koch, Algerian oat   Tufts rather slender, rather stiff, 60–100 cm. Culm reddish at maturity. Panicle *c.* 20 cm, erect, equilateral, with rather few spikelets. Spikelets 30–50 mm, 3–(4)-flowered, lowest floret breaking from rachilla with an oblique (*c.* 45°) fracture. Lemmas reddish brown, glabrous except for tuft of long hairs at base of lowest lemma; awn ± straight, or geniculate at base, usually weak or lacking in upper lemma. Caryopsis *c.* 10 × 2.5 mm, light brown, very hairy. Algerian oats have been grown in N.Z. for many years, especially in North Id. Collected as an escape from cultivation only in North Canterbury, in waste places near Christchurch, and at Motunau Id.

DESCHAMPSIA P.Beauv., 1812

Perennials, usually caespitose; branching intravaginal. Leaves flat, involute or setaceous. Ligule membranous, acute or obtuse. Inflorescence a lax or contracted panicle. Spikelets rather small, membranous, laterally compressed, pedicelled, 2- or sometimes 3-flowered, chasmogamous or cleistogamous; disarticulation above glumes; rachilla prolonged. Glumes subequal, ≤ florets, 1–3-nerved. Lemma rounded, truncate, irregularly toothed at apex, obscurely 5-nerved; awn dorsal or subapical, or 0. Palea ≈ lemma, 2-keeled. Lodicules 2. Stamens 3, sometimes 1–2. Caryopsis oblong; embryo small; hilum punctiform or shortly elliptic; endosperm solid.

Type species: *D. cespitosa* (L.) P.Beauv.

c. 40 spp. from temperate and circumpolar regions of both Hemispheres, in the tropics at high altitudes. Endemic spp. 4; indigenous sp. 1; naturalised sp. 1.

1 Lemma subapically awned, or awnless 2
  Lemma dorsally awned, awn inserted near base to *c.* upper ⅓ of lemma .. 5

2 Rachilla glabrous or sometimes with a few minute hairs *c.* 0.2 mm *3*
  Rachilla hairs 0.5–1 mm *4*

3 Tufts 5–35 cm, often stoloniferous or sward-forming; panicle lax, (1.5)–2–12 cm ....**2. chapmanii**
  Tufts 2–4.5 cm, compact; panicle ± compact, (0.5)–0.8–1.8 cm ..... **5. pusilla**

4 Leaves filiform, flaccid; culms usually overtopping leaves; lower glume 0.5–2 mm, 1-nerved ....**6. tenella**
  Leaves strict; culms often hidden by leaf-sheaths; lower glume 2–3.3–(4) mm, (1)–3-nerved .... **4. gracillima**

5 Awn straight; leaf-blade flat, sometimes inrolled; ligule acute ... **1. cespitosa**
  Awn geniculate; leaf-blade tightly inrolled; ligule obtuse .... **3. flexuosa**

1. D. cespitosa (L.) P.Beauv., *Ess. Agrost.* 91, t. 18, fig. 3 (1812) ≡ *Aira cespitosa* L., *Sp. Pl.* 64 (1753); Holotype: LINN *n.v.* (microfiche!), Europe. = *A. australis* Raoul, *Choix Pl. Nouv.-Zél.* 12 (1846); Holotype: P *n.v.*, *Raoul* Akaroa. = *Deschampsia penicillata* Kirk, *T.N.Z.I.* 27: 354 (1895); Holotype: WELT 69438! *A. Hamilton* Macquarie Id., [1894?]. = *D. cespitosa* var. *macrantha* Hack., *in* Cheeseman *Man. N.Z. Fl.* 876 (1906); Lectotype: W 26864! *T. F. C[heeseman]* Clarence Valley, Nelson Alps, 2000 ft (No 1192 to Hackel) (here designated). tufted hair grass

Stiff, yellow-green, sometimes bluish green leafy tufts or strong tussocks, 30–120 cm, with rigid panicles on erect culms often well overtopping leaves. Leaf-sheath subcoriaceous, keeled, glabrous, faintly ribbed. Ligule 2.5–7 mm, striate, sometimes almost keeled, acute. Leaf-blade (3)–5–45 cm × 2–3.5 mm, flat, stiff, sometimes folded or inrolled, abaxially smooth and shining, adaxially ribbed, ribs scabrid, tip acute, subpungent. Culm erect, or slightly geniculate at base, internodes glabrous. Panicle (3)–10–30 cm, ± erect, lax at anthesis, bearing spikelets near branchlet tips, later more contracted; rachis and branches sparsely to moderately, finely scabrid. Spikelets (3.5)–4–7 mm, yellow-green to brownish or purplish. Glumes ± equal, slightly overtopping or equalling florets, rarely overtopped by florets in 3-flowered spikelets, acute, keel often very sparsely scabrid; lower 1-nerved, upper 3-nerved. Lemma (2.3)–3–5 mm, oblong, often sparsely scabrid above, apex 4-toothed, margins in upper ½ finely scabrid; awn (0.5)–1–4 mm, fine, straight, minutely scabrid, insertion middorsal or ± basal or rarely in upper ⅓, slightly overtopping or ± reaching lemma apex, occasionally ± appressed to lemma and not reaching apex. Palea narrower than lemma, apex bifid, keels and margins in upper ½ finely scabrid. Callus hairs to 1.5 mm. Rachilla hairs to 2 mm, dense, silky. Anthers (0.8)–1.2–2.2 mm. Gynoecium: ovary *c.* 0.8 mm; stigma-styles *c.* 1.5 mm. Caryopsis 1.2–1.6 × 0.4–0.7 mm. $2n = 26$.

Indigenous.

N.: Volcanic Plateau; S.: scattered in southern half, especially along Main Divide and to west, rare in Marlborough and Nelson; St.; Ch., A., M. Lowland to subalpine in damp ground in grassland, margins of lakes or tarns, or in coastal swamps, sometimes on rock.

Cosmopolitan, from temperate and cold (including arctic and subantarctic) regions; a very polymorphic sp.

Cheeseman's (1906 op. cit. p. 876) comment that N.Z. plants differ from Northern Hemisphere specimens in the larger spikelets and higher awn insertion cannot be upheld; the range in spikelet length in N.Z. specimens falls within those given for *D. cespitosa* in European flora treatments and awn insertion varies from near basal to middorsal or above as in *D. cespitosa sens. lat.* from the Northern Hemisphere. Cheeseman merely noted that the New Zealand form was distinguished by Hackel as var. *macrantha*. He did not give a translation of Hackel's Latin diagnosis of this variety (*in litt.* to Cheeseman 30-7-1902) as he customarily did for those of Hackel's new varieties published in *Man. N.Z. Fl.*, see Edgar, E. and Connor, H. E. *N.Z. J. Bot. 25*: 455–457 (1987).

The features on which Hackel based var. *macrantha* are insufficiently distinct and the variety is here treated as a synonym of *D. cespitosa*.

Plants on Macquarie Id form small stoloniferous tufts rather than coarse tussocks, their spikelets resemble those of plants of *D. cespitosa* from southern South Id and South Georgia except in size of all parts, and no polliniferous specimens have been seen [Edgar, E. Poaceae, *Fl. Australia 50*: 468 (1993)].

Deschampsia cespitosa is now exceedingly rare in North and South Is because of browsing pressure (P. J. de Lange *in litt.*).

The name *Triodia splendida* Steud., *Syn. Pl. Glum. 1*: 249 (1854) based on a specimen from "N. Zeelandia" was referred to *D. cespitosa* by Buchanan, J. *Indig. Grasses. N.Z.* t. 37 (1879). No specimens have been seen.

LECTOTYPIFICATION   The three specimens of *D. cespitosa* (Nos 1192–1194) sent by Cheeseman to Hackel are at W and determined in Hackel's hand as var. *macrantha*. W 26864 (No 1192 to Hackel) is designated lectotype because it most closely matches Hackel's Latin diagnosis (*in litt.* to Cheeseman, 30.7.1902). The panicle is ± contracted and spikelets 5–6 mm; in the other two specimens the panicle is slightly more open and spikelets 4–5 mm.

2. D. chapmanii Petrie, *T.N.Z.I. 23*: 401 (1891 [May]); Lectotype: WELT 69437! *F. R. Chapman* Auckland Islands, Jan. 1890 (here designated). = *Catabrosa antarctica* Hook.f., *Fl. Antarct. 1*: 102, t. 56 (1845) a combination in *Deschampsia* is preoccupied by *D. antarctica* E.Desv. (1853) ≡ *Triodia antarctica* (Hook.f.) Benth., *J. Linn. Soc. Bot. 19*: 111 (1881) non Hook.f. (1846) ≡ *Sieglingia antarctica* (Hook.f.) Kuntze, *Rev. Gen. Pl. 2*: 789 (1891) ≡ *Deschampsia hookeri* Kirk, *J. Bot. 29*: 238 (1891 [August]) *nom. superfl.*; Holotype: K! *J. D. Hooker* Campbell's Islands, apparently very scarce on rocky ledges of the mts, Dec. 1840. = *D. novae-zelandiae* Petrie, *T.N.Z.I. 23*: 402 (1891 [May]); Lectotype: WELT 69475a! *D. P[etrie]* Hector Mts, *c.* 5000 ft. [Feb. 1890] (here designated).

Slender, erect, smooth, glabrous, leafy tufts, often stoloniferous and forming swards, 5–35–(55) cm, with delicate shining panicles. Leaf-sheath membranous, broader than leaf-blade, distinctly ribbed, midrib more prominent, lower sheaths usually reddish brown, sometimes light greyish, upper sheaths light green. Ligule (0.5)–2–4.5–(7) mm, tapering to acuminate tip. Leaf-blade 1–8 cm × 0.3–0.7–(1) mm, involute to ± flat, often filiform, sometimes disarticulating at ligule, glabrous; margins minutely scabrid, tip acute. Culm erect or ascending, branched at base. Panicle (1.5)–2–14 cm, slender, lax; branches filiform, erect to later spreading, smooth to minutely scaberulous, tipped by few spikelets. Spikelets 2.2–4.5 mm, green to brown, sometimes purplish. Glumes unequal, < spikelet, keel sometimes minutely scabrid near tip, and margins minutely ciliate-scabrid; lower 1–2 mm, 1-nerved, narrow-lanceolate to narrow-elliptic, acute, upper 1.5–2.7 mm, 3-nerved, elliptic-ovate to ovate, subacute to subobtuse. Lemma 1.4–2.2 mm, ovate-oblong, hyaline, irregularly 3–5-toothed, sometimes minutely mucronate or with a minute subterminal awn to 1 mm. Palea narrower than lemma, keels sparsely ciliate-scabrid or glabrous. Callus hairs *c.* 0.1 mm. Rachilla glabrous, rarely with one or two minute (*c.* 0.1 mm) hairs. Anthers 0.2–0.6–(0.7) mm. Caryopsis 0.8–1 × 0.3–0.4 mm. $2n = 26$.

Endemic.

N.: central mountains; S.: throughout; St.; A., C., M. Moist to very wet places in grassland, herbfield and forest margins; sea level to alpine.

Cheeseman (1906 op. cit. p. 877; 1925 op. cit. p. 166) regarded *D. novae-zelandiae* as distinct from subantarctic *D. chapmanii* in characters of the lemma apex: in *D. novae-zelandiae* irregularly minutely denticulate with awn wanting; in *D. chapmanii* ± irregularly 3–5-toothed with awn usually present. It seems impossible to separate two taxa on the basis of this variable character. In fact, some lemmas on the type specimen of *D. novae-zelandiae* have minute awns and many are 3-toothed. Although plants from the Subantarctic Islands are stoloniferous with dark brown basal leaf-sheaths, leaf-blades hardly filiform and lemmas definitely, though very shortly, awned, many plants from North and South Is also share these characters and no clear cut distinction can be made between these plants and others, often growing in the same locality, which are hardly stoloniferous, have lighter greyish brown basal sheaths, more filiform leaves, and awnless, to shortly mucronate, to shortly awned lemmas.

The names *D. chapmanii* and *D. novae-zelandiae* published at the same time by Petrie (1891 op. cit.) have equal priority at species rank (ICBN Art. 11.5). We adopt the name *D. chapmanii* because the type specimen is more representative of the united taxon throughout its whole range.

LECTOTYPIFICATION   The lectotype of *D. chapmanii*, WELT 69437, is from Petrie's herbarium whereas a similar specimen collected by F. R. Chapman, WELT 69436 in Kirk's herbarium, was annotated by Kirk as from the Antipodes Is; Godley, E. J. *N.Z. J. Bot. 27*: 553 (1989) drew attention to Kirk's mislabelling of specimens from the Auckland Is.

Of the 4 specimens of *D. novae-zelandiae* in WELT collected by Petrie on the Hector Mts, the lectotype WELT 69475a consists of 2 similar-sized good tufts from a single altitude *c.* 5000 ft, whereas the other sheets bear many smaller pieces of different sizes collected from altitudes between 4000 and 6000 ft.

3. D. flexuosa (L.) Trin., *Bull. Sci. Acad. Imp. Sci. Saint-Pétersbourg 1*: 66 (1836).   wavy hair grass

Tufts 30–75 cm, with numerous, short, very narrow, tightly rolled leaves much overtopped by culms; occasionally shortly rhizomatous. Leaf-sheath submembranous, rounded, finely scabrid, striate. Ligule (0.2)–0.5–1–(3) mm, truncate, or broadly and shallowly bilobed, abaxially scabrid, tip ciliate-scabrid. Leaf-blade 4.5–15 cm × 0.3–0.6 mm diam., strongly involute, smooth, scabrid near acute tip. Culm erect, or geniculate at base, internodes glabrous. Panicle 6–13.5 cm, ± erect, lax; branches few, very slender, flexuous, scarcely scabrid, naked for some distance towards base, tipped by few spikelets. Spikelets 4.5–6 mm, silvery greenish brown or light brown, purplish near base. Glumes subequal, hyaline, elliptic, sometimes sparsely scabrid above, margins very finely scabrid; lower 4–5 mm, 1-nerved, upper ≈ spikelet, 3-nerved but lateral nerves short. Lemma 4–5 mm, elliptic-oblong, tip ± truncate, irregularly minutely denticulate, minutely scabrid in upper ⅔; awn (3.5)–5–6.5–(7) mm, fine, geniculate, column dark brown, twisted, inserted near base of lemma. Palea narrower than lemma, apex slightly notched, keels minutely scabrid, interkeel minutely scabrid above. Callus hairs 0.7–1.2 mm. Rachilla hairs to 1 mm, fine, silky. Anthers 2–3.5 mm. Caryopsis 1.7–2.5 × 0.5–0.9 mm.

Naturalised.

N.: Auckland City; S.: Nelson (Canaan, Denniston), Canterbury (Arthurs Pass, Craigieburn Range), Otago (Naseby, Dunedin, Tapanui); St.: Ulva Id. Forest margins, damp ground and waste land.

Indigenous to Europe, northern Asia, North and South America.

4. D. gracillima Kirk, *J. Bot. 29*: 237 (1891); Lectotype: WELT 69440! *T. Kirk* above Carnley Harbour, Lord Auckland's Group, 12 Jan. 1890 (here designated).

Slender densely leafy tufts (2.5)–4–20–(25) cm, with stiff panicles often scarcely overtopping leaves; drying dull greenish brown. Leaf-sheaths submembranous, rounded, glabrous, light to dark brown, ribs few, ± prominent. Ligule 1.5–2 mm, glabrous, long tapered to acuminate tip. Leaf-blade (2)–3–7–(9) cm × 0.3–0.6 mm diam., involute, strict, glabrous, ribs few; tip hooded, very minutely sparsely prickle-toothed. Culm ± erect, often hidden by leaf-sheaths, internodes glabrous. Panicle 1.5–6 cm, erect, open, often deltoid; rachis glabrous, branches ascending to spreading, glabrous or sparsely minutely prickle-toothed, tipped by 1–2 spikelets. Spikelets 3.5–4.5 mm, light green, not very shining, becoming dull brown. Glumes subequal or unequal, < spikelet, ± keeled, minutely prickle-toothed on midnerve and margins near tip; lower 2–3.3–(4) mm, (1)–3-nerved, narrow- to oblong-lanceolate, acute to acuminate, upper 2.5–4–(4.5) mm, elliptic-lanceolate, 3-nerved, obtuse. Lemma 2–3 mm, ovate-oblong, ± firmly membranous, apex irregularly 4-toothed; awn subapical, 0.2–1–(1.5) mm. Palea ≤ lemma, apex shortly bifid, keels minutely scaberulous. Callus hairs 0.6–1.3 mm, dense. Rachilla hairs 0.6–1 mm, ± dense. Lodicules 0.3 mm. Anthers 0.4–0.8 mm. Gynoecium: ovary 0.7–0.8 mm; stigma-styles 1.0–1.2 mm. Caryopsis *c.* 1 × 0.4 mm.

Endemic.

S.: Fiordland, rare; St.; A., C. Subalpine to alpine in exposed rocky ground or herbfield, sometimes in boggy ground, 300–1300 m, rarely to sea level in Subantarctic Islands.

Glumes in plants from Fiordland and Stewart Id (2–3 mm) are shorter than in plants from Subantarctic Islands (2.5–4 mm) and the lower glume is 1-nerved rather than 3-nerved.

LECTOTYPIFICATION   There are 5 specimens of *D. gracillima* at WELT collected by Kirk from Carnley Harbour, Auckland Is, Jan 1890; the lectotype WELT 69440 was labelled "*Deschampsia gracillima* MS" in Kirk's hand, and Kirk's handwritten description is mounted on the sheet.

5. D. pusilla Petrie, *T.N.Z.I. 23*: 403 (1891); Lectotype: WELT 69433! *D. P[etrie]* Hector Mts, 6000 ft [Feb 1890] (here designated).

Compact, leafy tufts, 2–4.5 cm, leaves often reaching to base of panicle. Leaf-sheath wider than leaf-blade, firmly membranous, light brownish grey, rounded, ribs few, inconspicuous. Ligule (1.5)–2–3.5 mm, tapered to acute tip. Leaf-blade 5.5–17 × 0.3–0.5 mm, involute, smooth; margins finely scabrid, tip subobtuse. Culm 1.2–2.5–(4) cm, branched at base, often almost completely hidden among leaves, internodes glabrous. Panicle (5)–8.5–18 mm, ± compact; branches few, sparsely, minutely scabrid. Spikelets 2.2–3 mm, greenish, shining. Glumes ± equal and equalling spikelets, keels sparsely scabrid above, tip subacute to acute to almost mucronate; lower 1-nerved, narrow-lanceolate, upper 3-nerved, elliptic-lanceolate. Lemma 1.5–2.2 mm, faintly 5-nerved, oblong, 4–5-toothed with short terminal mucro, 0.3–0.5 mm, in notch. Palea ≤ lemma, apex deeply bifid, keels finely scabrid. Callus hairs *c.* 0.2 mm. Rachilla hairs *c.* 0.2 mm, fine, or 0. Anthers (0.3)–0.5–0.6 mm. Caryopsis *c.* 1 × 0.4 mm. Cleistogamus.

Endemic.

S.: Central Otago on Hector and Humboldt Mts, Pisa Range, Remarkables, Garvie and Umbrella Mts. Alpine.

Mature culms had elongated to 10 cm in CHR 394440 *A. P. Druce* Gem L[ake] area, Umbrella Mts, 4700 ft, March 1986.

LECTOTYPIFICATION   Of the 4 specimens of *D. pusilla* at WELT collected by Petrie from the Hector Mts, WELT 69433 was designated lectotype because it was from Petrie's private herbarium and annotated "type".

6. D. tenella Petrie, *T.N.Z.I. 23*: 402 (1891) ≡ *D. tenella* Petrie var. *tenella* (autonym, Petrie 1901 op. cit. p. 329); Lectotype: WELT 69304a! *D. P[etrie]* Catlins' River, Clutha Co., Otago (here designated). = *D. tenella* var. *procera* Petrie, *T.N.Z.I. 33*: 329 (1901); Lectotype: WELT 69469a! *D. Petrie* Clinton Valley, Te Anau, Jan. 1892 (here designated).

Slender, smooth, leafy tufts, sometimes sward-forming, 9–35–(50) cm, with delicately branched shining panicles. Leaf-sheath membranous, broader than leaf-blade, rounded, sparsely strongly ribbed, light brown, sometimes purplish. Ligule 1–4 mm, tapered to acuminate tip. Leaf-blade (3)–5–15–(20) cm × 0.3–0.6 mm diam., involute, filiform, flaccid, smooth, tip minutely scabrid. Culm erect, or geniculate at base, green or purplish, usually overtopping leaves, internodes glabrous. Panicle (4)–10–(12) cm, very lax, branches few, filiform, erect to spreading, very sparsely minutely scabrid, naked below, tipped by 1–few spikelets. Spikelets 2–3–(4) mm, shining, hyaline to light green to brownish. Glumes unequal, < spikelet, ± keeled, sparsely minutely prickle-toothed on midnerve and margins near tip; lower 0.5–2 mm, 1-nerved, narrow- to oblong-lanceolate, acute, upper 1.2–2.5 mm, (1)–3-nerved, elliptic-lanceolate or elliptic-oblong, acute to obtuse. Lemma 1.5–2.5 mm, ovate-oblong, hyaline, shining, irregularly 3–4-toothed, awnless or awn subapical to 1–(1.5) mm. Palea ≤ lemma, apex shortly bifid, keels minutely scaberulous. Callus hairs 0.2–0.7–(1) mm, ± sparse to dense. Rachilla hairs *c.* 0.5 mm, silky, ± dense. Anthers 0.3–0.7–(1) mm. Caryopsis 1–1.5 × 0.3–0.5 mm. $2n = 26$.

Endemic.

N.: southern half; S.: common in Nelson and Fiordland, scattered in eastern areas; St. Sea level to alpine in forest, scrub or tussock grassland, often on stream banks or in boggy ground.

CHR 402550 *A. P. Druce* West Dome, Southland, 2300 ft, March 1993, matches *D. tenella* in the silky hairy rachilla, but the plant is tall, 50 cm, and stouter than usual, and the glumes almost equal the florets.

LECTOTYPIFICATION   From a number of sheets at WELT of Petrie's gathering of *D. tenella* at Catlins's River, WELT 69304a was designated lectotype because it is composed of two good large tufts; the other material is similar but has been divided into smaller pieces.

Petrie (1901 op. cit.) described *D. tenella* var. *procera* as larger in all its parts with spikelets twice the usual size. WELT 69469a was designated lectotype because it is a single good tuft. The specimen is similar in size to other Fiordland plants of *D. tenella* and the spikelets are 3–3.5 mm, only slightly longer than in other specimens of *D. tenella*; a duplicate, CHR 25070, has spikelets to 4 mm.

HOLCUS L., 1753 *nom. cons.*

Tufted to long-rhizomatous perennials or tufted annuals. Leaf-sheath rounded. Ligule membranous. Leaf-blade flat. Panicle moderately dense. Spikelets 2–(3)-flowered, ± laterally compressed, falling entire at maturity, lower floret ♀ and usually raised on a curved rachilla, upper floret(s) usually ♂, sometimes ♀; rachilla often shortly prolonged. Glumes subequal, thinly chartaceous, ± enclosing florets, keeled from tip to midpoint or below, rarely awned; lower 1-nerved, upper 3-nerved. Lemma often polished cartilaginous, rounded, or keeled above, indistinctly 3–5-nerved, obtuse to bidentate; lemma of lower floret usually awnless, lemma of upper floret with geniculate or hooked or straight dorsal awn from upper ⅓. Palea membranous, ≤ lemma, 2-keeled. Callus hairy or glabrous. Lodicules membranous, glabrous, sometimes denticulate. Ovary apex glabrous; styles free. Caryopsis laterally compressed, rarely grooved; embryo small; hilum punctiform to shortly elliptic, rarely linear; endosperm liquid or solid.

Type species: *H. lanatus* L.

6 spp., from Canary Is, North Africa, and Eurasia. Naturalised spp. 2.

1 Plant tufted, sometimes stoloniferous; upper glume usually shortly awn-tipped; awn of upper lemma uncinate, ± hidden by glumes .... **1. lanatus**
Plant strongly rhizomatous; upper glume not awned; awn of upper lemma slightly bent but not hooked, protruding from glumes .... **2. mollis**

1. H. lanatus L., *Sp. Pl.* 1048 (1753). Yorkshire fog

Tufted, sometimes stoloniferous, softly hairy, greyish green perennials, 35–110 cm. Leaf-sheath firmly membranous, striate, ± keeled above near ligule, villous with long, fine, silky hairs, basal sheaths often purplish. Ligule (0.6)–1–3 mm, ± truncate, denticulate, apex finely ciliate and occasionally with a few long hairs, abaxially villous. Leaf-blade (2.5)–4.5–15–(20) cm × (1.5)–3–10 mm, pubescent and with longer soft hairs, narrowed to acuminate tip; margins finely ciliate, and with longer hairs. Culm erect, or ascending from geniculate base, nodes and internodes pubescent. Panicle 4–18 × 1.5–7 cm, lanceolate, ovoid or sometimes oblong, dense to lax, erect or nodding; rachis and branches pubescent. Spikelets 3.8–5.5 mm, whitish, pale green, or purplish. Glumes stiff-ciliate on keels and lateral nerves of upper glume, elsewhere minutely scabrid to shortly pubescent; lower sometimes slightly shorter, narrow-lanceolate, acute to mucronate, upper broader, elliptic, mucronate, or awned to 0.8 mm. Lemma 2–2.7 mm, elliptic-lanceolate, glabrous, shining, keel sparsely finely prickle-toothed, apex minutely ciliate-scabrid; lower lemma awnless, upper lemma narrower, awn stout, 1.4–2.8 mm, inserted *c.* 0.4 mm below tip of lemma, becoming recurved and uncinate. Palea keels sparsely short-ciliate above, apex obtusely lobed, minutely ciliate. Callus hairs few, to 1 mm. Rachilla glabrous. Anthers 1–2.2 mm. Caryopsis 1–1.5 × 0.4–0.5 mm.

Naturalised from Europe.

N.; S.: throughout; St.; K., Ch., A., C. Roadsides, waste land, pasture and grassland, along tracks or streams in forest, often in damp or swampy ground; sea level to montane, rarely subalpine.

2. H. mollis L., *Syst. Nat.* ed. 10, *2*: 1305 (1759).   creeping fog

Strongly rhizomatous, dull bluish green perennials, 15–100 cm, forming loose swards. Leaf-sheath chartaceous, striate, glabrous or shortly pubescent with some longer hairs. Ligule 1–3 mm, ± truncate, denticulate, abaxially shortly pubescent. Leaf-blade 5–17 cm × 2–9 mm, rather stiff, shortly pubescent-scabrid, sometimes with scattered longer hairs or almost glabrous; margins minutely scabrid, tip acute. Culm erect or spreading, often decumbent, nodes loosely to densely bearded, internodes glabrous. Panicle (1)–4–10–(16) × (0.5)–1–2 cm, narrow-oblong or lanceolate, dense to rather lax, erect; rachis with ± sparse minute hairs, branches and very slender pedicels with slightly longer hairs. Spikelets 4–5.8 mm, whitish green, sometimes purplish. Glumes shortly stiff-ciliate on keel, minutely scabrid elsewhere, acute to acuminate; lower slightly shorter, narrow-lanceolate, upper elliptic-lanceolate with strong lateral nerves. Lemma 1–2.5 mm, elliptic- to ovate-lanceolate, with a few prickle-teeth on and near keel or scattered above, otherwise glabrous, ± shining, margins minutely ciliate-scabrid; lower lemma awnless, upper lemma with slightly curved awn, 3–5 mm, inserted *c.* 0.5 mm below lemma apex, sometimes twisted near base. Palea keels short-ciliate, apex obtusely lobed, short-ciliate. Callus hairs 0.2–0.3 mm, or callus of lower floret glabrous. Rachilla sparsely soft-hairy, hairs to 0.6 mm. Anthers (1)–1.5–2.2 mm. Caryopsis *c.* 1 × 0.7 mm.

Naturalised from Europe.

S.: eastern areas and to the west from Arthur's Pass southwards, rare in Marlborough; St.; Ch. Roadsides, waste land and pasture; sea level to montane.

HYBRIDS   Carroll, C. P. and Jones, K. *New Phytol. 61*: 72–84 (1962) gave a tentative list of some diagnostic features for identification of hybrids *H. lanatus* × *H. mollis.* For the most part the hybrids resemble *H. mollis* but generally have shrunken indehiscent anthers, and awns somewhat shorter than is usual in *H. mollis*; glumes may be ± pilose, somewhat obtuse with a slightly exserted awn. Occasional hybrid plants may resemble *H. lanatus* but be weakly rhizomatous. Specimens which are apparently sterile and may be the hybrid *H. lanatus* × *H. mollis* have been noted among N.Z. collections, e.g., CHR 107032 *R. Mason* & *N. T. Moar 5287* West of Rimu, Westland, 15.2.1958; CHR 174277 *A. J. Healy 67/11* Ashley State Forest, near airfield, 23.1.1967; CHR 183917 *W. R. B[oyce]* Port Hills [Christchurch], 16.2.1939; CHR 234610 *A. J. Healy 73/37* near Morven, South Canterbury, 10.2.1973.

KOELERIA Pers., 1805

Tufted or rhizomatous perennials; branching extravaginal and intravaginal. Leaf-sheath glabrous to pubescent, or scabrid; margins hyaline. Ligule membranous, truncate to subacute, entire, erose or somewhat ciliate. Leaf-blade flat, folded, involute or convolute. Culm erect, sometimes geniculate at base. Inflorescence a contracted panicle, spike-like, cylindrical, or sometimes lanceolate, occasionally interrupted, especially near base; branches and pedicels glabrous, scabrid or pubescent. Spikelets 2–3–(5)-flowered, laterally compressed, third and upper florets reduced, or very reduced and ∅; disarticulation above glumes and between florets. Glumes subequal, membranous, with hyaline margins above, keeled, not awned; lower 1-nerved, shorter and narrower than upper, upper 3-nerved. Lemma membranous with hyaline margins above, (1)–3–5-nerved, keeled, entire, or mucronate, or with a short apical or subapical, inconspicuous, usually straight awn. Palea hyaline, shining, gaping when mature, 2-keeled, apex ± bifid.

Callus short, hairy. Rachilla hairy. Flowers ♂, chasmogamous. Lodicules 2, hyaline, bifid. Stamens 3. Ovary glabrous; styles free to base; stigmas plumose. Caryopsis oblong-fusiform, slightly laterally compressed; hilum short; endosperm liquid.

Type species: *K. gracilis* Pers. *nom. illeg.*

c. 60 spp., of temperate regions throughout the world. N.Z. spp. 3, all endemic.

N.Z. spp. were revised by Edgar, E. and Gibb, E. S. *N.Z. J. Bot. 37*: 51–61 (1999), who commented on their similarity to N.Z. spp. of *Trisetum*.

1 Culms usually villous or pubescent ± throughout; leaves firm, semi-pungent, dense; panicle oblong, very dense, occasionally lobed at base ....
.... **1. cheesemanii**
Culms glabrous or often minutely scabrid-pubescent below panicle and/or hairy near nodes; leaves flexible, soft-tipped, lax, or sometimes dense; panicle spike-like or occasionally lanceolate, ± interrupted 2
2 Plant caespitose or very shortly rhizomatous; lemma of second floret usually > palea; plants of well drained habitat .... **2. novozelandica**
Plant finely rhizomatous; lemma of second floret = palea; plants of poorly drained habitat .... **3. riguorum**

Very rarely culm vesture differs from that indicated in the key, and *K. cheesemanii* with glabrous culms, *K. novozelandica* with culms pubescent throughout, and *K. riguorum* with pubescent or villous culms may be found.

1. K. cheesemanii (Hack.) Petrie, *T.N.Z.I. 48*: 192 (1916) ≡ *Trisetum cheesemanii* Hack., *T.N.Z.I. 35*: 281 (1903); Holotype: W 31707! *T. F. C[heeseman]* Hooker Glacier, Mount Cook District, alt. 4000 ft (No 1221 to Hackel). = *Koeleria novozelandica* var. *pubiculmis* Domin, *Biblioth. Bot. 65*: 117 (1907); Holotype: W 22048! *L. Cockayne* Otira Gorge, Westland (No 1333 to Hackel).

Densely caespitose, of low to medium stature (7)–9–17–(43) cm; with stiff, semi-pungent, pale greyish green leaves, their sheaths often imbricate and of ± equal length, usually overtopped by densely villous culms bearing dense cylindrical panicles; branching intravaginal, sometimes extravaginal at base. Leaf-sheath abaxially glabrous, scabrid, pubescent or rarely ± villous. Ligule 0.3–0.8–(1.6) mm, rim-like or truncate, sometimes erose, ciliate, abaxially glabrous or hairy. Leaf-blade 1.2–9–(21) cm × (0.5)–0.6–3.2–(5.5) mm, usually folded, involute, or convolute, or sometimes flat, linear-lanceolate or narrower, abaxially minutely scabrid to densely pubescent or sometimes villous, midrib often prominent below, adaxially scabrid or pubescent on ribs; margin with minute prickle-teeth or longer hairs, apex hooded, acute to acuminate. Culm (1)–4–27–(35) cm, pale green to purple, internodes sparsely to densely pubescent to villous, very rarely glabrous. Panicle densely contracted, usually cylindrical (to spike-like) occasionally interrupted especially at base, (1.3)–2.6–5.0–(7.3) × (0.3)–0.8–1.5–(2.2) cm; branches erect, pubescent, appressed to rachis. Spikelets (3.8)–4.3–5.2–(6.9) mm, 2–3-flowered, pale green, purplish or golden. Glumes subequal, sometimes scabrid above and on keel; lower 3.5–3.8–(4.8) mm, 1–(3)-nerved, oblong, with prickle-teeth on upper ½ to ⅔ of keel, upper (3.2)–4.0–4.5–(6.1) mm, 3-nerved, ovate, with prickle-teeth on upper ⅔ of keel. Lemma 3.5–6.0 mm, 3–5-nerved, oblong, narrow obovate, or lanceolate, acute, acuminate, or obtuse, entire to mucronate or shortly awned and sometimes minutely bidentate; awn apical or subapical, to 1.5–(3.2) mm. Palea 3.0–5.2 mm. Callus hairs 0.3–0.6 mm. Rachilla 0.8–1.0–(1.4) mm with hairs to 0.5 mm; prolongation *c.* 1.5 mm with hairs to 0.3 mm. Lodicules (0.7)–0.8–1.2 mm, ± bifid, erose. Anthers 0.6–1.1 mm. Gynoecium: ovary *c.* 0.7 mm; stigma-styles *c.* 1 mm. Caryopsis 2–2.5 × 0.6–0.7 mm.

Endemic.

N.: Mt Hikurangi, Kaweka Range, north-east Ruahine Mts; S.: mountains throughout. Rocky outcrops, scree, cliffs in high mountains; usually above 1000 m to 2800 m.

Occasional rhizomatous forms have been collected, and also specimens with culm internodes which are largely glabrous or are either glabrous or sparsely to densely pubescent (Edgar and Gibb 1999 op. cit. p. 55).

Koeleria cheesemanii is often confused with *Trisetum spicatum*, a similarly compact, tufted grass with densely villous culms and dense, ± oblong panicles found at high altitudes on rocky exposed ground in the same localities as *K. cheesemanii*. Usually the long, 2–6 mm, recurved, dorsal awns of *T. spicatum* separate this species from plants of *K. cheesemanii* in which the lemma is entire, to shortly apically awned.

2. K. novozelandica Domin, *Biblioth. Bot. 65*: 116 (1907) ≡ *K. novozelandica* Domin var. *novozelandica* (autonym, Domin 1907 op. cit. p. 117) ≡ *K. novozelandica* var. *typica* Domin, *Biblioth. Bot. 65*: 116 (1907) *nom. invalid.*; Holotype: Z! *L. C[ockayne]* stony ground, Waterfall Creek, [Craigieburn Range], Southern Alps, 2500 ft, 18.I.[18]96. = *K. novozelandica* var. *parvula* Domin, *Biblioth. Bot. 65*: 117 (1907); Holotype: W 22046! *L. Cockayne* Broken River, Canterbury Alps (No 1334 to Hackel). = *K. superba* Domin, *Biblioth. Bot. 65*: 118 (1907); Holotype: W 22047! *T. F. C[heeseman]* Broken River, Canterbury Alps, 2500 ft (No 1331 to Hackel). = *K. gintlii* Domin, *Biblioth. Bot. 65*: 126 (1907); Holotype: W 22045! *T. F. C[heeseman]* Hooker Valley, Mount Cook District, 2500 ft (No 1329 to Hackel).

Extremely variable, slender, occasionally lush, tufted or very shortly rhizomatous perennial, of low to medium stature, 4–46–(82) cm, with leaves varying from greyish green to green and from tightly inrolled to flat, with erect culms and usually spike-like panicles, sometimes greatly overtopping leaves; branching extravaginal. Leaf-sheath abaxially glabrous, scabrid, pubescent or villous. Ligule membranous, truncate, often erose and/or somewhat ciliate, (0.2)–0.3–1.3 mm. Collar often thickened and paler near margins, often with long hairs on upper margins. Leaf-blade flat, folded or involute, linear to ± filiform, 3–15–(30) cm × 1–1.5 mm diam., or up to 2–(4) mm wide, abaxially glabrous, scabrid, pubescent or villous, adaxially ribbed, scabrid to densely villous, especially on ribs; margins ± scabrid, often with long hairs below, apex acute, hooded. Culm to 41–(66) cm, internodes entirely glabrous or frequently minutely pubescent below panicle, often scabrid to villous above and/or below nodes, very occasionally pubescent throughout. Panicle (2.0)–3.0–16 × 0.5–1.2 cm, spike-like or occasionally lanceolate, sometimes interrupted, with appressed-ascending branches; branches and pedicels glabrous, scabrid or pubescent to villous; bract subtending panicle glabrous, or often tipped by long hairs, or absent. Spikelets (3.8)–4.5–5.8–(7) mm, 2–3–(5)-flowered, light green, often purplish to brownish. Glumes subequal, membranous with wide hyaline margins above, oblong-lanceolate to elliptic; lower 3.0–4.8 mm, 1–3-nerved, upper 3.2–5.2 mm, 3–4-nerved. Lemma (3.0)–3.8–4.7–(5.5) mm, 3–5-nerved, oblong, narrow obovate, or lanceolate, acute, acuminate or occasionally obtuse, entire, mucronate or shortly awned (sometimes within one panicle), apex sometimes minutely bidentate; awn apical or subapical, to 1.5–(3.2) mm. Palea 3.5–4.3 mm, recurved from base, keel prickle-toothed above. Callus hairs (0.1)–0.2–0.5–(0.7) mm. Rachilla 0.6–1.5–(1.9) mm, hairs 0.2–1.4 mm; prolongation *c.* 1 mm, hairs short (0.3 mm). Lodicules 0.7–1.1 mm. Anthers 0.8–2.5 mm. Gynoecium: ovary 0.5–0.8 mm; stigma-styles 1–1.4 mm. Caryopsis *c.* 2 × 0.4 mm.

Endemic.

N.: Kaimanawa Mts; S.: throughout although rare in Westland and Fiordland. Rock outcrops, tussock grassland, on wide range of rock substrates from limestone to ultramafic; (*c.* 100)–500–2000 m.

Koeleria novozelandica is extremely variable. No clear morphological, ecological or geographical distinctions were found between the wide-leaved tall, or involute-leaved small forms, nor between plants with glabrous culms and those with pubescent culms (Edgar and Gibb 1999 op. cit. p. 58). Many plants from Mt Cook and some from Broken River are large and lax with flat leaves, but this form is found occasionally elsewhere; plants from Banks Peninsula and North Otago are often smaller with involute leaves, but again these forms occurred throughout the range of the species; plants with mainly pubescent culms are occasionally found in South Id.

Koeleria novozelandica was originally referred in New Zealand [Hooker, J. D. *Handbk N.Z. Fl.* 334 (1864); Buchanan, J. *Indig. Grasses. N.Z.* t. 38 (1879)] to *K. cristata* of Europe and North America; Cheeseman (1906 op. cit. p. 897; 1925 op. cit. p. 184) equated New Zealand plants with *K. kurtzii* Hack. of Argentina.

Connor, H. E. *N.Z. J. Agric. Res. 3*: 728–733 (1960) recorded that *K. novozelandica* (as *K. kurtzii*) was self-fertile with normal selfed progeny.

3. K. riguorum Edgar et Gibb, *N.Z. J. Bot. 37*: 59 (1999); Holotype: CHR 401013! *A. P. Druce* nr Lonely L[ake], Douglas Ra., N.W. Nelson, 1400 m, seepage in tussock land, Feb 1984.

Slender, low-growing, to 25–(36) cm, with fine rhizomes, leaves often reddish purple, diverging at an angle of *c.* 30°, and erect panicles at maturity well overtopping leaves; branching extravaginal, often intravaginal above; sheaths persistent and becoming dark brown.

Leaf-sheath strongly ribbed, abaxially puberulous between ribs, pale straw to purple-suffused. Ligule 0.6–1.1 mm, truncate to subacute, ciliate. Leaf-blade 3.2–9.5–(17) cm × 0.3–1.2 mm, usually folded or involute, or flat, linear to filiform, abaxially glabrous, very rarely pubescent, adaxially strongly ribbed, ribs puberulous; margin finely scabrid. Culm 5.8–20–(30) cm, often geniculate at base, erect, stramineous to reddish purple, internodes glabrous, or rarely pubescent or villous. Panicle 2–5–(7) × 0.3–1.0 cm, spike-like, ± interrupted; rachis, branches, and pedicels pubescent, rarely glabrous. Spikelets (3.6)–4.2–6.5 mm, 2–(3)-flowered, bright green, noticeably banded with purple and golden brown, later stramineous. Glumes subequal, lanceolate to elliptic-ovate, acute or obtuse, margin often minutely toothed above; lower 3.5–4.1 mm, 1–3-nerved, narrower than upper, keel scabrid in upper ⅓, upper 4.0–4.2 mm, 3–5–(7)-nerved. Lemma 4.0–4.6 mm, elliptic, often minutely scabrid above, entire or mucronate or shortly apically awned or with a short, slender, straight, subapical awn to 2 mm, apex entire or minutely bidentate. Palea 4.0–4.2 mm, keels prickle-toothed above; palea of lower floret < lemma, palea of upper floret = lemma. Callus hairs 0.2–0.4 mm. Rachilla 0.8–1.2 mm, hairs to 0.9 mm; prolongation to 1.5 mm, hairs to 1.0 mm. Lodicules (0.5)–1.0–1.2 mm. Anthers (0.5)–0.9–1.7 mm. Gynoecium: ovary 0.8–1 mm; stigma-styles *c.* 1 mm. Caryopsis *c.* 2.0 × 0.7 mm.

Endemic.

S.: north-west Nelson, St Arnaud Ra., Spenser Mts. Wet areas in tussock grassland; 1480–1670 m.

In its lax habit and narrow interrupted panicles *K. riguorum* resembles some forms of *K. novozelandica* but differs from that species in the longer more slender rhizomes, leaf-blades diverging at an angle of *c.* 30° rather than ± erect, reddish leaves and culms, the conspicuously purplish and golden-brown colour banding in the spikelets, and the longer palea, relative to the lemma, in the second floret.

A preference for damp habitats is unusual in *Koeleria* but Moore, D. M. *Fl. Tierra del Fuego* 299 (1983) reported the habitat of *K. fueguina* Calderón as moist grassland and bogs.

TRISETUM Pers., 1805

Perennials, tufted or rhizomatous; branching extravaginal or rarely intravaginal. Ligule membranous, truncate or rounded, often ciliolate. Leaf-blades linear, narrow, flat or involute. Inflorescence a contracted to lax panicle. Spikelets 2–several-flowered, laterally compressed; disarticulation above glumes and between lemmas; rachilla usually hairy, prolonged. Glumes unequal to subequal, < to ≈ spikelets, keeled, not awned; lower 1–3-nerved, upper 3–5-nerved. Lemma membranous to thinly coriaceous, 3–7-nerved, keeled, bidentate or bicuspid, dorsally awned from upper ½; awn reflexed, or geniculate (not N.Z. spp.). Palea gaping, silvery, thinner than lemma, 2-keeled. Callus short, hairy. Flowers ♂, chasmogamous or cleistogamous. Lodicules 2, membranous, glabrous or ciliate. Stamens 3. Ovary glabrous; styles free. Caryopsis slightly compressed laterally; embryo small; hilum short; endosperm liquid. Fig. 10.

Type species: *T. pratense* Pers. *nom. illeg.*

c. 85 spp., of temperate regions throughout the world. Meadows, mountain slopes, upland grasslands, weedy places. Endemic spp. 7, indigenous spp. 2.

N.Z. spp. of *Trisetum* were revised by Edgar, E. *N.Z. J. Bot. 36*: 539–564 (1998), incorporating the intensive field observations of A. P. Druce. Most species are variable, especially in form of leaf-blade and in culm vesture. In general *Trisetum* is found in N.Z. on rocky sites: on limestone or marble cliffs, or on bluffs; in alpine herbfield, fellfield or scree; and on ultramafic sites. Three species, *T. lasiorhachis*, *T. lepidum* and *T. youngii* occur in scrub, open forest or tussock grassland.

All species are very variable. Features normally characteristic of a species are not invariably present and it was not feasible to include every variant in the key.

1 Glumes markedly dissimilar in shape, the lower linear-lanceolate or narrow-oblong, the upper elliptic-lanceolate to ovate, unequal in length, the lower < upper 2
  Glumes ± similar in shape, the lower narrow-elliptic, the upper elliptic-ovate, ± equal in length 8

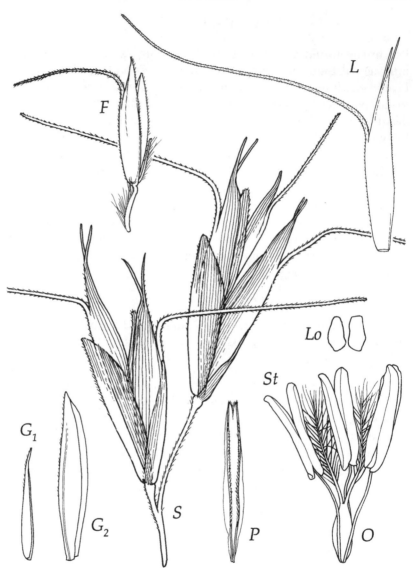

Fig. 10 *Trisetum lepidum*
S, spikelets, × 9; **G$_1$, G$_2$**, glumes, × 9; **F**, floret and rachilla, × 9; **L**, lemma lateral view, × 9; **P**, palea dorsal view, × 9; **Lo**, lodicules, × 24; **O**, ovary, styles and stigmas, × 24; **St**, stamens, × 24

2 Panicle very lax, with individual spikelets mostly conspicuous ... **5. lepidum**
  Panicle compact or interrupted at base, or somewhat open but with spikelets
  clustered and not individually conspicuous *3*

3 Culms with long straight soft hairs above nodes and few elsewhere, or densely
  pilose or pubescent throughout *4*
  Culms glabrous throughout, or sometimes minutely pubescent above basal
  nodes or below panicle *6*

4 Panicle compact or interrupted at base, rachis hidden except near base ....
  .... **7. spicatum**
  Panicle somewhat open, rachis clearly visible *5*

5 Lemma 5–7.5 mm; awn ≥ lemma; leaf-blade flat or rarely inrolled, 1.3–3 mm
  wide .... **4. lasiorhachis**
  Lemma 3–4.5 mm; awn ≤ lemma; leaf-blade inrolled, < 1 mm diam., rarely
  flat and up to 2 mm wide .... **6. serpentinum**

6 Tufts open; panicle whitish or pale straw-coloured, sometimes purple-tinged
  .... **2. arduanum**
  Tufts dense; panicle grey- or light-green to brownish red or amber *7*

7 Panicle compact, dense, oblong, sometimes with spreading lower branches;
  leaf-blade abaxially glabrous, scabrid above .... **1. antarcticum**
  Panicle ± interrupted, not very dense, lanceolate, shortly branched
  throughout; leaf-blade abaxially densely minutely scabrid or sometimes
  smooth throughout or sparsely scabrid above .... **3. drucei**

8 Panicle compact, or interrupted at base, rachis hidden except near base;
  culms often villous to shortly sparsely pubescent, usually not much
  overtopping leaves .... **7. spicatum**
  Panicle ± spike-like but rachis visible, lower branches sometimes slightly
  distant; culms glabrous, occasionally villous or pubescent, usually far
  overtopping leaves *9*

9 Spikelets 5.5–8 mm; panicle 3–15–(24) cm; leaf-blade 1–7 mm wide ....
  .... **9. youngii**
  Spikelets (3.5)–4–5.5 mm; panicle 0.5–8.5 cm; leaf-blade 0.3–1.5–(2) mm
  wide .... **8. tenellum**

1. T. antarcticum (G.Forst.) Trin., *Mém. Acad. Imp. Sci. Saint-Pétersbourg
VI 1:* 61 (1831) non (Thunb.) Nees (1832) ≡ *Aira antarctica* G.Forst.,
Prodr. 41 (1786) ≡ *Avena antarctica* (G.Forst.) Roem. et Schult., *Syst.
Veg. 2:* 676 (1817) non Thunb. (1794) ≡ *Trisetum antarcticum* (G.Forst.)
Trin. subsp. *antarcticum* [autonym, Petrie *T.N.Z.I. 44:* 187 (1912)]

= *T. antarcticum* (G.Forst.) Trin. var. *antarcticum* (autonym, Hackel *in* Cheeseman 1906 op. cit. p. 880); Holotype: GOET! *G. Forster* Nova Zeelandia [Queen Charlotte Sound]. = *T. saxeticolum* Cockayne et Allan, *T.N.Z.I. 57*: 60 (1926); Lectotype: CHR 1389B! *H. H. Allan* Island Bay, Wellington, on rocks (designated by Edgar 1998 op. cit. p. 543).

Tufts usually dense, to 40 cm, with dull green rather rigid leaves usually reaching or sometimes overtopping the dense, spike-like panicles; branching extravaginal at plant base, sometimes intravaginal above. Leaf-sheath to 4 cm, very minutely pubescent or with extremely minute appressed hairs between ribs. Ligule 0.2–0.6 mm, truncate, erose, often sparsely minutely ciliate, abaxially sometimes minutely prickle-toothed. Leaf-blade 3.5–22 cm × 1.5–3–(4) mm, flat or inrolled, abaxially smooth but scabrid near long-narrowed tip, adaxially ribbed with sparse to dense minute hairs and prickle-teeth on ribs, hairs slightly longer near ligule; margins minutely prickle-toothed. Culm 4–30 cm, internodes glabrous, occasionally a few minute prickle-teeth below panicle. Panicle (2)–4–10–(15) × 0.6–2.5 cm, compact, oblong, very dense, or with some lower branches more obvious and slightly spreading; rachis glabrous, branches and pedicels smooth or with sparse minute prickle-teeth or sometimes minute hairs. Spikelets 4.5–7 mm, greyish green or brownish amber. Glumes unequal, hyaline, keel thickened with sparse long prickle-teeth on upper ½; lower ⅔–⅘ length of upper, linear-lanceolate, upper slightly < spikelet, elliptic; margins with prickle-teeth near acuminate to almost mucronate tip. Lemma 4–6.2 mm, bidentate to shortly bicuspid, papillose, prickle-toothed above and on keel; awn 3.5–6 mm, straight to later recurved, insertion in upper ¼ of lemma. Palea with minute prickle-teeth on keels and usually on margins. Callus hairs to 0.5 mm. Rachilla hairs to 2.5 mm. Lodicules *c.* 1 mm, glabrous. Anthers 1–1.7 mm. Gynoecium: ovary *c.* 0.8 mm; stigma-styles to 1.3 mm. Caryopsis 2.5–3 × *c.* 0.6 mm.

Endemic.

N.: south Egmont coast, Kapiti Id, and western and southern Wellington coasts to Cape Palliser; S.: Nelson near Cape Farewell, Marlborough Sounds and islands. Coastal, on gravel, sand, and bluffs; sea level to 30 m. FL Nov–Feb.

Schweickerdt, H. G. *Bothalia 3*: 185–203 (1937) clarified the nomenclatural confusion which arose for some authors between *Aira antarctica* G.Forst., *Avena antarctica* (G.Forst.) Roem. et Schult. and *Avena antarctica* Thunb. from South Africa for which the type specimen is missing.

2. T. arduanum Edgar et A.P.Druce, *N.Z. J. Bot. 36*: 545 (1998); Holotype: CHR 225566! *A. E. Esler 3597* Motuorahi, Coromandel Islands, Hauraki Gulf, coastal rocks, 16.10.1971.

Open tufts to 60 cm, erect or often drooping, with slender pale yellow-green to glaucous leaves and whitish open panicles; branching extravaginal. Leaf-sheath to 6 cm, glabrous, or with fine, very minute prickle-teeth, to minutely puberulous. Collar hairs few, long. Ligule 0.2–0.3 mm, truncate, slightly erose, glabrous, or rarely sparsely minutely ciliate. Leaf-blade 10–20–(30) cm × 0.5–1.5–(4) mm, usually narrow and inrolled, sometimes flat, abaxially smooth or with sparse to dense prickle-teeth towards tip, rarely throughout, adaxially ± shallowly ribbed, sparsely minutely prickle-toothed or sparsely pubescent on ribs; margins minutely prickle-toothed. Culm 6–35 cm, internodes glabrous, occasionally with a few short hairs above basal nodes, sometimes with a few minute prickle-teeth below panicle. Panicle (3)–8–18–(27) × 0.7–3.5–(5) cm, oblong-lanceolate, becoming lax with spreading branches but spikelets clustered and individually inconspicuous; rachis, branches, and pedicels with sparse to dense fine prickle-teeth, or rachis smooth below. Spikelets (4.5)–5–7–(7.5) mm, whitish or pale straw-coloured, sometimes purple-tinged. Glumes unequal, hyaline, keels minutely prickle-toothed in upper ½; lower ⅔–⅘ length of upper, linear-lanceolate, upper slightly < spikelet, elliptic-lanceolate; margins entire, or with a few minute prickle-teeth near acuminate tip. Lemma 4.5–6 mm, bicuspid, papillose; awn 3.5–8 mm, straight to later recurved, insertion in upper ¼ of lemma. Palea with minute prickle-teeth on keels and rarely on margins. Callus hairs to 0.5 mm. Rachilla hairs to 2 mm. Lodicules to 1 mm, glabrous. Anthers 1.2–2 mm. Gynoecium: ovary to 1 mm; stigma-styles to 1.3 mm. Caryopsis *c.* 2.6 × 0.6 mm.

Indigenous.

N.: scattered localities, usually coastal but frequently inland in Waikato and Wellington Province; S.: Marlborough, few records in eastern mountains; Three Kings Is. Coastal and inland on bluffs, usually on calcareous rocks, sometimes on basalt, on serpentinite at North Cape, and near bird roosts on offshore islands in the Hauraki Gulf; 3–800 m. FL Oct–Jan.

Also indigenous to Norfolk Id.

Trisetum arduanum was collected by Banks and Solander in N.Z. on Cook's first voyage, e.g, WELT 63935, and described in Solander's unpublished MS under the name *Avena flavescens*. It was formerly more common in the Auckland region but is now collected only occasionally north of Auckland [de Lange, P. J. and Crowcroft, G. M. *Auck. Bot. Soc. J. 51*: 38–49 (1996)].

Trisetum was not previously recorded for Norfolk Id, see Green, P. S. *Fl. Australia 49*: 442–499 (1994); *T. arduanum* was recently collected there, CHR 529149 *P. J. de Lange NF 223* Norfolk Id, Mt Pitt National Park, Bridle Track near Cooks Monument, 13.11.1998.

3. T. drucei Edgar, *N.Z. J. Bot. 36*: 548 (1998); Holotype: CHR 260294! *A. P. Druce* SE of Imjin Camp, Moawhango R., Kaimanawa Mts, 2700 ft, cliff in gorge, Jan 1974.

Dense tufts to 60 cm with culms often much overtopping the densely, very minutely scabrid leaves; branching intravaginal. Leaf-sheath 2–5–(8) cm, firmly membranous, grey-brown to light greenish brown, sometimes reddish, very densely minutely papillose or minutely scabrid to pubescent, sometimes glabrous. Ligule 0.8–1 mm, truncate, erose, sparsely finely ciliate. Collar sometimes with a few stiff hairs to 1.5 mm. Leaf-blade (3)–12–25–(30) cm × (0.5)–1–4 mm, folded with inrolled margins or flat, hard, dull green or reddish, abaxially scabrid with very dense minute prickle-teeth, or smooth, or sparsely scabrid above, adaxially ribbed with dense fine minute prickle-teeth or minutely densely pubescent on ribs; margins finely scabrid. Culm 10–40 cm, internodes glabrous or finely pubescent above, rarely pilose, or minutely prickle-toothed below panicle. Panicle 5–20 × 1–1.5–(3) cm, lanceolate, with short, inconspicuous branches, ± interrupted, not very dense, rarely more open, but spikelets clustered and individually inconspicuous; rachis, branches, and pedicels very densely minutely strigose or rarely puberulous. Spikelets 6–8 mm, light green often tinged reddish brown. Glumes unequal, membranous, keels with minute prickle-teeth in upper ½; lower *c.* ¾ length of upper, narrow oblong to narrow ovate, upper ¾ to ≈ spikelet, elliptic-oblong; margins broadly hyaline, sparsely prickle-toothed near acute to acuminate sometimes mucronate tip. Lemma 3.5–6 mm, bidentate or sometimes bicuspid, papillose, minutely prickle-toothed near midnerve; awn 3.5–7.5 mm, recurved, insertion in upper ¼ of lemma. Palea minutely prickle-toothed on keels almost throughout and on margins near tip. Callus hairs to 1 mm. Rachilla hairs to 2 mm. Lodicules 0.8–1.2 mm, glabrous. Anthers 0.8–1.5–(2) mm. Gynoecium: ovary *c.* 0.8 mm; stigma-styles *c.* 1.2 mm. Caryopsis 2.5–3 × 0.7–0.8 mm.

Endemic.

N.: eastern mountains, from Mt Wharekia, East Cape, to near Taihape, S.: north-west Nelson, eastern Marlborough from Waima River south to North Canterbury near Parnassus. On limestone and marble cliffs, occasionally on greywacke; 30–1500 m. FL Oct–Feb.

Trisetum drucei and sometimes *T. antarcticum* have intravaginal innovation shoots, a feature unusual for the genus.

In South Id the leaf-blades are often reddish and in both North and South Is the spikelets may be reddish tinged. Specimens from North Id often have long hairs at the leaf collar margins but in specimens from South Id collars are usually glabrous. In North Id leaf-blades have dense very minute prickle-teeth throughout; in South Id leaves are abaxially smooth to slightly scabrid near the tip with rather sparse prickle-teeth, and adaxially densely prickle-toothed on the ribs or very densely shortly pubescent.

4. T. lasiorhachis (Hack.) Edgar, *N.Z. J. Bot. 36*: 549 (1998) ≡ *T. antarcticum* var. *lasiorhachis* Hack., *in* Cheeseman *Man. N.Z. Fl.* 880 (1906); Lectotype: W 27977! *D. Petrie 10144* Mt Hikurangi, East Cape, [5000 ft] (No 1217 to Hackel) (designated by Edgar 1998 op. cit. p. 549).

Strong, dense but narrow tufts, (5)–12–60–(85) cm, sometimes rhizomatous, with pale straw-coloured leaf-sheaths and dull green to grey-green leaf-blades usually with scattered long hairs, with pilose culms and often pubescent panicle-rachis and branches; branching extravaginal. Leaf-sheath to 6 cm, pubescent. Collar hairs few, very long, fine. Ligule 0.3–0.5–(0.8) mm, truncate, erose, glabrous or ciliate. Leaf-blade 5–25 cm × 1.3–3 mm, flat, rarely inrolled and narrower, often with scattered long hairs, abaxially prickle-toothed towards tip, adaxially ribbed with ± scattered minute prickle-teeth or short hairs on ribs; margins minutely prickle-toothed, and often with scattered long hairs. Culm 10–50 cm, internodes pilose above and below nodes, densely pilose, pubescent, or sometimes glabrous towards panicle. Panicle (3)–7–14–(21) × 1–3–(5) cm, lanceolate, somewhat open with visible rachis but spikelets clustered and individually inconspicuous; rachis, branches, and pedicels densely pilose to sparsely, minutely hairy. Spikelets 5–8 mm, pale green, often purple-tinged. Glumes unequal, keels often strong, prickle-toothed in upper ½ or almost throughout; lower ⅔–⅘ length of upper, narrow-oblong, tapered to often long-acuminate tip, upper < spikelet, elliptic, acute to shortly acuminate; margins almost entire with very few prickle-teeth near tip. Lemma 5–7.5 mm, bicuspid, minutely prickle-toothed or papillate throughout; awn 5–9.5 mm, straight to later recurved, insertion in upper ½ to upper ⅓ of lemma. Palea minutely prickle-toothed on keels and rarely on margins. Callus hairs to 0.8 mm. Rachilla hairs to 2 mm. Lodicules *c.* 1.3 mm, glabrous. Anthers 1.8–2.2 mm. Gynoecium: ovary to 1 mm; stigma-styles to 2.4 mm. Caryopsis *c.* 2.5 × 0.8 mm.

Endemic.

N.: Auckland (Mt Wellington) and to south-west, mountains of East Cape, Hawke's Bay, the Central Plateau, and Mt Egmont; S.: Marlborough, eastern mountains. Often on cliffs or riverbanks, also in scrub, open forest, and tussock grassland; usually montane to subalpine, 600–1500 m, sometimes lower towards the limit of its range. FL (Oct)–Dec–Feb.

The amount of hair on the culm internodes varies from densely pilose throughout with the panicle, rachis, and branches also pubescent, as in the lectotype from Mt Hikurangi, to slightly pilose on the upper culm with panicle-rachis and branches only sparsely minutely hairy. In all plants, however, the long hairs above and below the culm nodes are conspicuous.

5. T. lepidum Edgar et A.P.Druce, *N.Z. J. Bot. 36*: 553 (1998); Holotype: CHR 227700! *M. J. A. Simpson 6735* Mt Owen, Nelson, marble bluffs nth of Granity Pass, 22.1.1972.

Erect, often slender, usually reddish tufts, 12–65–(100) cm, with very lax, sometimes large panicles well overtopping the leaves; branching extravaginal. Leaf-sheath to 6 cm, sometimes reddish, glabrous to pubescent, margins often with longer hairs above. Collar with some long hairs. Ligule 0.2–0.6 mm, truncate, erose, glabrous to minutely or sometimes obviously ciliate. Leaf-blade 3–16 cm × 0.5–2–(6) mm, dull green to later reddish, flat, distinctly ribbed, abaxially smooth, often scabrid towards tip, adaxially minutely pubescent or minutely prickle-toothed on ribs; margins smooth or with very minute prickle-teeth. Culm 6–45–(70) cm, often geniculate at base, internodes glabrous. Panicle 3–20–(28) × 1–10 cm, lax, with very fine, spreading branches bearing individually evident or scarcely overlapping spikelets; rachis smooth below or with few prickle-teeth, more scabrid above, branches and pedicels with slightly denser long fine prickle-teeth. Spikelets 4.5–9 mm, light green to later brownish, usually red-tinged. Glumes very unequal in shape and size, ± hyaline, keels with minute close-set prickle-teeth in upper ½; lower *c.* ½, or rarely to ¾ or ≈ upper, very narrowly ovate almost subulate, subacute to acuminate, upper < spikelet, broadly ovate, subobtuse to obtuse, rarely acuminate or mucronate; margins entire. Lemma 4.5–7 mm, bicuspid, lower lemma smooth near base, papillose above and on keel, upper lemma(s) almost entirely papillose or minutely prickle-toothed, sometimes all lemmas equally smooth or equally scabrid; awn 5–9 mm, insertion in upper ¼ of lemma. Palea minutely prickle-toothed on upper ⅔ of keels, otherwise smooth. Callus hairs to 0.5 mm. Rachilla hairs to 1.3 mm. Lodicules to 1.3 mm, glabrous or ciliate. Anthers 0.8–1.5 mm. Gynoecium: ovary to 1 mm; stigma-styles to 1 mm. Caryopsis 2–3 × 0.5–0.6 mm. $2n = 28$. Fig. 10.

Endemic.

N.: south from Mt Hikurangi, the Volcanic Plateau, and Mt Egmont; S.: throughout.; St.; Ch. In open forest, scrub or tussock grassland, often on river or stream banks, and on calcareous rock; from near sea level to 1500 m but usually montane. FL Oct–Feb.

A plant with unusually contracted panicles but with glumes and lemmas matching those of *T. lepidum* was collected from Popotunoa, Otago by *J. Buchanan*, WELT 69040; other specimens from the same locality have the customary lax panicle, i.e. WELT 69042, WELT 69061.

The very lax panicles distinguish *T. lepidum* from other indigenous spp. with unequal glumes. The leaves and spikelets in *T. lepidum* are often tinged red, the lower lemma is usually smooth in the lower ½, and the glumes are usually very unequal in size and shape. Even when the two glumes are less disparate in length the difference in shape remains, with the lower glume very narrow-ovate and the upper broadly ovate.

Beuzenberg, E. J. and Hair, J. B. *N.Z. J. Bot. 21*: 14 (1983) made a single count from New Zealand *Trisetum*; they published the record under the name *T. antarcticum* but their voucher specimen, CHR 100118 [*H. E. Connor*] Tara Hills, Omarama, is *T. lepidum.*

6. T. serpentinum Edgar et A.P.Druce, *N.Z. J. Bot. 36*: 554 (1998); Holotype: CHR 401623! *A. P. Druce* Motueka R., (L[eft] Branch) Gordon Ra., 740 m, river terrace, mineral belt (dry site), Jan 1985.

Rather open tufts to 50 cm, with dull green, narrow, inrolled leaves overtopped by narrow to ± lax, usually purplish panicles; branching extravaginal. Leaf-sheath 1–3 cm, densely softly pubescent, with longer scattered hairs on sheaths of culm-leaves, and sometimes on margins. Ligule 0.4–0.8 mm, erose, minutely ciliate. Leaf-blade 2–16 cm, usually inrolled and < 1 mm diam., rarely flat and up to 2 mm wide, sometimes with scattered long hairs, abaxially smooth or with very minute prickle-teeth, adaxially shallowly ribbed with scattered very fine prickle-teeth on ribs; margins minutely prickle-toothed, sometimes with scattered long hairs. Culm 8–34–(40) cm, slender, internodes with long fine hairs above nodes and a small band of shorter hairs below nodes, uppermost internode glabrous. Panicle 2–10 × 0.5–2 cm, narrow-lanceolate to somewhat open, with visible rachis and short, ascending to spreading branches bearing clustered spikelets; rachis and branches with moderately dense, fine prickle-teeth and often a few longer hairs at lower nodes of rachis and at base of panicle. Spikelets 4–6.5 mm, often purplish. Glumes unequal, with sparse short prickle-teeth on keel; lower ⅔–¾ length of upper, oblong-lanceolate, upper < to ≈ spikelet, elliptic-lanceolate; margins with very few minute prickle-teeth near acute to acuminate tip. Lemma 3–4.5 mm, bicuspid, minutely papillose; awn 3–4 mm, ± recurved, insertion in upper ¼ of lemma. Palea minutely prickle-toothed on keels and margins. Callus hairs *c.* 0.2 mm. Rachilla hairs to 0.8 mm. Lodicules to 1 mm, glabrous. Anthers to 1.5 mm. Gynoecium: ovary to 0.8 mm; stigma-styles to 1 mm. Caryopsis not seen.

Endemic.

N.: North Cape at Surville Cliffs; S.: north-west Nelson, Cobb and Takaka Valleys, east Nelson, Dun Mt and Richmond Forest Park; Marlborough, Wairau Valley, Upper Motueka Valley. Open ground in short-tussock grassland or on river terraces, on ultramafic soil or rock; 600–1200 m. FL Dec–Jan.

One specimen from Red Hills, Wairau Valley, CHR 387495 *A. P. Druce* Dec 1980, is much taller than the rest, the single culm being 57 cm and the panicle 16.5 cm. The spikelets are 6–6.5 mm, lemmas 5–5.5 mm, awns to 5 mm, and anthers *c.* 2 mm.

7. T. spicatum (L.) K.Richt., *Pl. Eur. 1*: 59 (1890) ≡ *Aira spicata* L., *Sp. Pl. 1*: 64 (1753) non L., *Sp. Pl. 1*: 63 (1753) ≡ *A. subspicata* L., *Syst. Nat.* ed. 10, *2*: 873 (1759) *nom. superfl.* ≡ *Trisetum subspicatum* (L.) P.Beauv., *Ess. Agrost.* 88, 149, 180 (1812) *nom. superfl.*; Holotype: LINN 85.7 *n.v., Linné* Sweden, Lapland, 1732 [*fide* Veldkamp, J. F. and Van der Have, J. C. *Gard. Bull. Singapore 36*: 131 (1983)].

Compact tufts 3–20–(55) cm, with stiff leaves, densely pubescent to villous culms usually not much overtopping leaves, and dense ± oblong panicles usually from two to four times as long as wide; branching extravaginal. Leaf-sheath to 3 cm, glabrous to shortly hairy above, to densely pubescent throughout; margins glabrous to ciliate. Ligule 0.5–2–(4) mm, ± erose, truncate to tapered, short-ciliate, abaxially glabrous or with minute hairs. Leaf-blade 1.5–14 cm × 0.5–4.5 mm, rolled or folded, often with inrolled margins, or flat (mainly on Subantarctic Islands), abaxially glabrous except near semipungent minutely prickle-toothed tip, or with sparse prickle-teeth in upper ½, to densely pubescent throughout, adaxially ribbed, sparsely minutely hairy or prickle-toothed, to densely pubescent on ribs; margins with scattered long hairs especially near base, or with short hairs or prickle-teeth. Culm 2–17–(40) cm, internodes either densely villous with soft retrorse hairs, and antrorse hairs just below panicle, or with all hairs antrorse to spreading, or with very short, sparse hairs, exceptionally glabrous on Chatham Is. Panicle 1–6.5–(15) × 0.3–2–(2.5) cm, dense, spike-like, sometimes interrupted near base, oblong, to lanceolate- or ovate-oblong; rachis and branches pubescent. Spikelets 4.3–6.5–(7.4) mm, brownish to greenish, often with purplish bands on glumes and lemmas, later stramineous. Glumes subequal or sometimes unequal, with minute to hair-like prickle-teeth on keel and margins in upper ½, and in Chatham Is, Auckland Is and Campbell Id with minute hairs or prickle-teeth scattered on surface; lower ≤ upper, narrowly elliptic, upper ≤ spikelet, more broadly elliptic; apex subacute to acuminate, sometimes shortly mucronate. Lemma 3.5–6.2 mm, bidentate to conspicuously bicuspid, papillose, minutely prickle-toothed only near base of awn to throughout lemma; awn 2–6.5 mm, straight to recurved, insertion in upper ¼ of lemma. Palea minutely prickle-toothed on upper ¾ of keel or almost throughout, in Subantarctic Is with prickle-teeth also on flanks. Callus hairs 0.2–0.6 mm. Rachilla hairs 0.4–0.8 mm, very sparse to moderately dense. Lodicules *c.* 1 mm, glabrous or ciliate, dentate or bilobed. Anthers 0.6–1–(1.5) mm. Gynoecium: ovary *c.* 0.8 mm; stigma-styles *c.* 1.2 mm. Caryopsis 1.5–1.8 × 0.5–0.7 mm.

Indigenous.

S.: east of Main Divide, in Nelson and Marlborough and south from Arthur's Pass; Ch., A., C. Subalpine to alpine in open exposed rocky ground in fellfield, herbfield, tussock grassland or on scree, 1000–2000 m, in South Id; on fellfield, peaty ledges, and rock crevices, descending to sea level in Chatham and Subantarctic Is. FL Jan–Feb.

A widespread circumpolar species; indigenous to Eurasia, North and South America and Australasia.

The villous or pubescent culm is a distinctive feature of *T. spicatum* but culms were glabrous in a specimen from the Chatham Is, AK 210702 *E. H. McKenzie 90* Nov 1992.

Randall, J. L., and Hilu, K. W. *Syst. Bot. 11*: 567–578 (1986) succinctly define *T. spicatum* as a circumpolar, bipolar species with diploid, tetraploid, and hexaploid cytotypes. Hultén, E. *Svensk Bot. Tidskr. 53*: 203–228 (1959), considered the *T. spicatum* complex over its whole range and distinguished 14 subspecies and 9 varieties within it. For the N.Z. Botanical Region Edgar (1998 op. cit. p. 557) tabulated three forms: Otago/Southland with short panicles, villous culms bearing mainly retrorse hairs, and leaves abaxially glabrous; Marlborough/ Canterbury with long panicles, villous or shortly pubescent culms bearing antrorse to tangled hairs, and leaves abaxially glabrous; Chatham/Subantarctic Is with short to long panicles, villous culms bearing antrorse to tangled hairs, and leaves abaxially densely pubescent. Edgar (1998 op. cit. p. 557) deemed it unwise to distinguish any infraspecific taxa in the N.Z. Botanical Region without an extensive biosystematic and cytological investigation as well as comparison with European and North and South American material.

Hultén (1959 op. cit. p. 221) listed four specimens from N.Z. as *T. spicatum* subsp. *australiense*. Edgar (1998 op. cit. p. 557) commented that specimens at CHR from Mt Kosciusko, New South Wales, agreed well with Hultén's description of subsp. *australiense* but matched none of the N.Z. plants.

8. T. tenellum (Petrie) A.W.Hill, *Index Kewensis, Suppl. 9*: 289 (1938) ≡ *T. antarcticum* subsp. *tenellum* Petrie, *T.N.Z.I. 44*: 187 (1912), "tenella" ≡ *T. antarcticum* var. *tenellum* (Petrie) Cheeseman, *Man. N.Z. Fl.* 169 (1925); Lectotype: WELT 68724! *D. P[etrie]* Hooker River, Mt Cook, 13.2.1911 (designated by Edgar 1998 op. cit. p. 558).

Small tufts 2–35–(55) cm, with slender, often almost filiform culms usually ≫ grey-green often involute leaves at maturity; branching extravaginal. Leaf-sheath 0.5–2 cm, glabrous, or sparsely finely prickle-toothed on ribs to densely pubescent. Ligule 0.2–1 mm, truncate or rounded, ciliate. Leaf-blade 1–12–(18) cm × 0.3–1.5–(2) mm, abaxially glabrous to finely scabrid to pubescent, sometimes with scattered long hairs, adaxially ribbed, minutely scabrid to shortly pubescent; margins minutely prickle-toothed, occasionally with scattered long hairs. Culm (1)–2–20–(45) cm, internodes glabrous, or sometimes densely pubescent with ascending or tangled spreading hairs, occasionally hairs retrorse below. Panicle 0.5–8.5 × 0.5–1.2 cm, ± spike-like with very short close-set branches but rachis visible, or more open with short ascending branches; rachis glabrous and branches with a few minute prickle-teeth, or rachis and branches pubescent. Spikelets (3.5)–4–5.5 mm, pale green or red-tinged.

Glumes subequal, membranous, keels with prickle-teeth in upper ⅓ to upper ½; lower ≤ upper, elliptic, very rarely oblong-lanceolate, upper ≤ spikelet, broadly elliptic; margins with very minute prickle-teeth near obtuse to acute rarely almost mucronate tip. Lemma 3.5–5 mm, minutely bidentate, papillose, minutely scabrid near keel; awn 2–3–(4) mm, insertion in upper ¼ of lemma. Palea minutely prickle-toothed on keels except near base and sparsely minutely prickle-toothed on margins. Callus hairs 0.2 mm. Rachilla hairs few, to 0.8 mm. Lodicules *c*. 0.7 mm, glabrous. Anthers 0.6–1.3 mm. Gynoecium: ovary *c*. 0.7 mm; stigma-styles *c*. 1.2 mm. Caryopsis *c*. 2 × 0.5 mm.

Endemic.

S.: north-west Nelson and to east of Main Divide in Marlborough and from south of Arthur's Pass and Banks Peninsula, extending slightly to the west in Fiordland. Rocky or stony ground on moraines, scree, river beds, and eroded sites, on limestone, marble, and ultramafic sites in tussock grassland, fellfield and herbfield, sometimes in swampy sites; (300)–600–1860 m. FL Dec–Feb.

Plants with pubescent culms, panicle-rachis and branches are commonly found in north-west Nelson, and occasionally elsewhere throughout the range of the species, sometimes growing with plants with glabrous culms.

Trisetum tenellum occupies ultramafic sites in Nelson, Marlborough, and Southland, but plants there are indistinguishable from those elsewhere.

Laing, R. M. and Gourlay, H. W. *T.R.S.N.Z. 64*: 3 (1934) listed "*Trisetum tenellum* (Petrie) Allan and Zotov (Sp. ined.)" in anticipation of a taxonomic revision by Allan and Zotov which was never written. Laing and Gourlay therefore made no new combination but A. W. Hill who cited the basionym validly published the new combination as *T. tenellum*.

9. T. youngii Hook.f., *Handbk N.Z. Fl.* 335 (1864); Lectotype: K! *J. Haast 672* Canterbury, New Zealand, 1862 (designated by Edgar 1998 op. cit. p. 560).

Erect ± open tufts to 100 cm, with flat dull green leaves usually ≪ culms and long, narrow-lanceolate, silvery panicles; branching extravaginal. Leaf-sheath to 5 cm, often pubescent or puberulous, sometimes minutely prickle-toothed or glabrous, upper sheaths usually with long fine soft scattered retrorse hairs. Ligule 1–1.5 mm, erose or lacerate, shortly stiff-ciliate, abaxially often with minute hairs. Leaf-blade 3–15 cm × 1–7 mm, sometimes with scattered long hairs, abaxially glabrous to minutely scabrid above, adaxially finely ribbed, ribs minutely prickle-toothed; margins scabrid, sometimes with scattered long hairs. Culm 9–60 cm, internodes mainly glabrous but usually with soft hairs above and below nodes, scabrid to puberulous below panicle and often with a few long hairs, occasionally pubescent to villous throughout. Panicle 3–15–(24) × 0.5–2 cm, lanceolate-oblong, ± spike-like and rachis visible, or lower branches sometimes slightly distant, narrowly branched, each branch bearing few spikelets crowded to base; rachis usually scabrid, sometimes smooth or pubescent, lower nodes often with tufts of long hairs, branches scabrid or sometimes pubescent. Spikelets 5.5–8 mm, light green, purplish or brownish tinged. Glumes subequal, membranous, keels with prickle-teeth in upper ½; lower ≤ upper, elliptic-oblong to elliptic, upper ≈ spikelet, wider elliptic; margins very minutely prickle-toothed near acute or sometimes finely mucronate tip. Lemma 5–6.5 mm, bidentate to shortly bicuspid, closely finely scabrid with prickle-teeth more prominent near keel; awn 4–7 mm, slightly to strongly recurved, insertion in upper ¼ of lemma. Palea minutely prickle-toothed on keels throughout, on flanks, and on margins above. Callus hairs to 0.5 mm. Rachilla hairs to 1 mm. Lodicules *c.* 1 mm, glabrous. Anthers 0.8–1.3 mm. Gynoecium: ovary to 1 mm; stigma-styles to 2 mm. Caryopsis *c.* 2.7 × 0.8 mm.

Endemic.

N.: Mt Hikurangi, Ruahine and Tararua Ranges; S.: along Main Divide and to the west, also in Fiordland, extending east to the Two Thumb Range in South Canterbury and to the Old Man Range in Otago. Subalpine to alpine grassland, shrubland and forest margins, on shadier slopes, in rocky or damp ground, sometimes on limestone or marble cliffs, 760–1980 m. FL Dec–Feb.

Some spikelets may have unequal glumes, e.g., AK 1587 *T. F. Cheeseman* Arthur's Pass, Jan 1883 (No. 1227 to Hackel). Specimens from Mt Mytton, Cobb Valley, and Mt Owen, north-west Nelson, show similar variation.

Plants with villous culms are uncommon but occur throughout the range of the species. Plants in which the culms are only shortly pubescent occur more frequently in north-west Nelson and also in Fiordland.

REPRODUCTIVE BIOLOGY   All N.Z. species appear to be chasmogamous. Connor, H. E. *N.Z. J. Sci. Tech. 38A*: 742–751 (1957) recorded *T. lepidum* from Omarama (as *T. antarctium*), and *T. youngii,* as self-sterile.

DISJUNCT DISTRIBUTION   Disjunct distribution is marked: *T. serpentinum* is recorded only at North Cape and on the Mineral Belt in northern South Id; *T. drucei* occurs in mountains of eastern North Id, and in South Id in north-west Nelson and on the Chalk Range of eastern Marlborough; *T. lasiorhachis* of the central mountains of North Id has outliers in north-west Nelson and on the Chalk Range; even in *T. spicatum* and *T. tenellum* which spread eastwards from the Main Divide in South Id, there is a noticeable gap in distribution in North Canterbury from the boundary with Marlborough southwards to Arthur's Pass and Banks Peninsula.

EXCLUDED SPECIES   Eurasian *T. flavescens* (L.) P.Beauv. was recorded several times for New Zealand (Edgar 1998 op. cit. p. 562) but only one specimen, AK 99562 *D. Petrie* Ranfurly Road, Epsom, at roadside, Jan 1919, has been found. This plant probably escaped from Petrie's garden at Epsom; the geniculate awns distinguish *T. flavescens* from N.Z. species.

ζ**Helictotrichon pubescens** (Huds.) Pilg.   Plants *c.* 65 cm. Leaf-blade softly long-hairy. Panicle 10–12 cm; branches slender, flexuous, sparsely scabrid. Spikelets *c.* 25 mm, 2-flowered and with a rudimentary floret at tip of densely villous rachilla. Glumes membranous, keeled. Lemma chartaceous below, hyaline above near denticulate tip, minutely scabrid; awn 15–20 mm, twisted below. This Eurasian species was collected in N.Z. at the end of 19th Century. Only two collections have been seen: CHR 183865 *G. M. Thomson* Waikari, and CHR 3422 *D. Petrie* Dunedin, [also WELT 71121 (dated 1886), WELT 71122].

ζ**Rostraria cristata** (L.) Tzvelev   Tufts 20–40 cm. Leaf-blade flat, 4–16 cm × 1.5–5 mm, narrowly tapered, pubescent. Culm geniculate or erect, internodes glabrous. Panicle dense, spike-like, 1.5–9 cm, cylindric or somewhat lobed. Spikelets *c.* 3.5 mm, 4-flowered, laterally compressed. Glumes pubescent, unequal; lower *c.* ¾ length of upper and much narrower. Lemma 5-nerved, keeled, with subapical straight awn (*c.* 0.5 mm). This Eurasian species has been collected in N.Z. twice; WELT 71125, 71127 *T. Kirk* Great Barrier Id, Dec. 1867 and CHR 33183 *A. J. Healy* Railway Yards, Wellington, Dec. 1940.

PHALARIDINAE

Spikelets 3-flowered, the 2 lower florets ♂ or ∅, the uppermost ⚥ or ♀, disarticulating above glumes; glumes, or at least the upper, usually enclosing florets; lemma of ⚥ or ♀ floret rounded.

1 Spikelets apparently 1-flowered, with (1)–2, lower, greatly reduced Ø lemmas
....**Phalaris**

  Spikelets 2–3-flowered *2*

2 Glumes unequal; lower florets reduced to empty lemmas (rarely ♂); lodicules
  0 .... **Anthoxanthum**

  Glumes ± equal; lower florets ♂; lodicules 2 .... **Hierochloe**

ANTHOXANTHUM L., 1753

Perennials or annuals, ± fragrant with coumarin. Leaf-sheath rounded. Ligule membranous. Leaf-blade flat, usually ± flaccid. Culm slender. Panicle narrow, spike-like. Spikelets with 3 florets, the lower 2 Ø or rarely ♂, the third much shorter, ♀, protogynous; disarticulation above glumes; rachilla not prolonged. Glumes unequal with the lower shorter, membranous, 1–3-nerved, keeled. Lemmas of Ø florets equal, keeled, hairy, dorsally awned; lemma of ♀ floret hyaline, 1–7-nerved. Palea 1-nerved. Lodicules 0. Stamens 2 in ♀ florets, sometimes 3 in ♂ florets. Ovary glabrous; styles free. Caryopsis terete; embryo ¼ length of caryopsis; hilum punctiform or shortly elliptic; endosperm solid.

Type species: *A. odoratum* L.

c. 20 spp., of temperate and tropical regions. Naturalised spp. 2.

Schouten, Y. and Veldkamp, J. F. *Blumea 30*: 319–351 (1985) considered that it was impossible to uphold *Anthoxanthum* and *Hierochloe* as separate genera and transferred spp. of *Hierochloe* to *Anthoxanthum*, the older name. However, Clayton and Renvoize (1986 op. cit. p. 133) found *Anthoxanthum* sufficiently distinct from *Hierochloe* to justify generic status.

1 Glumes glabrous; plants annual; culm branched .... **1. aristatum**

  Glumes ± hairy; plants perennial; culm simple .... **2. odoratum**

1. A. aristatum Boiss., *Voy. Bot. Espagne 2*: 638 (1842).

Annuals, 9–20 cm, with culms often much-branched near base, faintly coumarin-scented. Leaf-sheath submembranous, strongly ribbed, sometimes sparsely villous above. Collar hairs long, in small lateral tufts. Ligule 0.7–2 mm, glabrous, tapering, becoming erose. Leaf-blade

1–5.5 cm × 1–2.5 mm, thin, scabrid, sometimes sparsely hairy, tip narrow, subacute. Culm erect or spreading, branched, almost filiform, internodes glabrous. Panicle 1.5–2.4×0.5–1.2 cm, ovate- or lanceolate-oblong, rather lax or more dense, with few to numerous spikelets; rachis glabrous, branches and pedicels very short, scabrid or minutely hairy. Spikelets 5.5–7 mm, green or brownish green. Glumes glabrous, acute to acuminate, keels green, thickened, sparsely scabrid; lower ½–⅔ length of upper, 1-nerved, ovate, upper > florets and enclosing them, 3-nerved, elliptic-ovate. Ø florets: lemma 2.6–3.5 mm, 4–5-nerved, oblong, firmer and brown below with rather stiff silky hairs, apex glabrous, hyaline, bilobed, or entire and shortly toothed; awn of lower floret 3.6–4.6 mm, fine, ±straight, inserted above midpoint, awn of second floret 7–8.8 mm, stouter, geniculate, inserted near base, dark brown in lower ½, projecting beyond spikelet; palea 0 in both florets. ♂ floret: lemma 1.5–1.8 mm, 5-nerved, firmly membranous, orbicular, brown, shining, glabrous; palea = lemma; anthers *c.* 3 mm; caryopsis 1.1–1.3 × 0.6–0.7 mm.

Naturalised from southern Europe.

N.: Auckland City, Lake Taupo, Wellington City; S.: Canterbury (Paparua). Waste ground, uncommon.

2. **A. odoratum** L., *Sp. Pl.* 28 (1753). sweet vernal

Tufted perennials, 12–50–(90) cm, strongly coumarin-scented. Leaf-sheath firmly membranous, glabrous or sparsely hairy, especially towards ligule, ± bearded on margins near collar, sometimes with one minute lateral auricle. Ligule 0.6–5 mm, truncate to somewhat tapered, minutely ciliate. Leaf-blade (1)–3.5–18–(25) cm × 2–6.5–(10) mm, glabrous or hairy, ribs smooth or minutely scaberulous, tip acute to acuminate. Culm erect or spreading, simple, internodes glabrous. Panicle 1.5–10–(16) × (0.5)–1–2–(4) cm, cylindric to ± ovate, dense, branches very rarely obvious; rachis glabrous, branches and pedicels short-hairy or very rarely glabrous. Spikelets 7–9.5 mm, green or brownish green, sometimes proliferous. Glumes ± hairy, very rarely glabrous or minutely scabrid, acute to shortly mucronate; lower *c.* ½ length of upper, 1-nerved, ovate, upper > florets and enclosing them, 3-nerved, ovate-lanceolate. Ø florets: lemma 3–3.5 mm, oblong, firmly membranous, brown and hairy below, hyaline above, apex bifid, with short obtuse lobes; awn of lower floret 2.4–3.8 mm, straight, inserted above midpoint, awn of second floret 6.3–8.5 mm, curved near base, inserted near base of lemma and often slightly projecting beyond upper glumes; palea 0 in both florets. ♀ floret: lemma 1.4–2.2 mm, orbicular, smooth; palea = lemma, narrow-lanceolate; anthers (2.5)–3.5–4.7 mm; caryopsis 1.4–1.9 × 0.5–1 mm. Plate 4E.

Naturalised.

N.; S.: throughout; St.; K., Ch., A., C. Roadsides, waste land, pasture and tussock grassland, in scrub or forest clearings, sometimes in boggy, or sandy or rocky, stony ground; sea level to subalpine.

Indigenous to Europe, northern Asia and north-west Africa.

Some plants from Horseshoe Bay, Stewart Id, CHR 369379 *H. D. Wilson* 2.10.1979, with an exceedingly long panicle, 24 cm, grew amongst plants of *A. odoratum* in which the panicles were of average length.

HIEROCHLOE R.Br., 1810 *nom. cons.*

Perennials, coumarin-scented. Leaf-sheath rounded. Ligule membranous to chartaceous. Leaf-blade flat. Culm several-noded. Inflorescence a ± lax panicle. Spikelets ovate, ± laterally compressed, usually 3-flowered with the two lower florets ♂ and the upper ♀ (in N.Z.); rachilla not prolonged. Glumes subequal, ≈ florets, ovate, 1–3-nerved midnerve percurrent, keeled, chartaceous to membranous. Lemmas of ♂ florets 3–5-nerved, firmly membranous to coriaceous, acute to bilobed, margins ciliate, dorsally to subapically awned, or awn 0, callus long hairy; lemma of ♀ floret 5-nerved, cartilaginous, margins convolute and covering the palea, emarginate, glabrous to hairy above, subapically short-awned to muticous, callus glabrous. Palea of ♂ florets 2-nerved, of ♀ or ♀ floret 1–2-nerved, < lemma. Lodicules 2. Stamens 3 in ♂ florets, 2 in ♀ floret. Ovary glabrous; styles 2, free to connate at base; stigmas long, plumose. Caryopsis ± elliptic, ± laterally compressed; embryo ¼–½ caryopsis; hilum linear, basal. Fig. 11.

Type species: *H. odorata* (L.) P.Beauv.

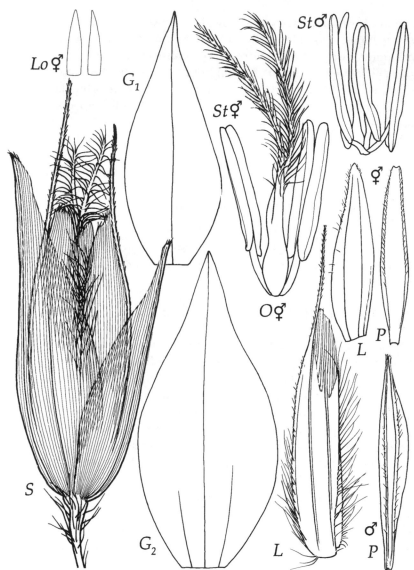

Fig. 11 *Hierochloe redolens*

S, spikelet, × 12;   **G₁**, **G₂**, glumes, × 11;   **L**, lemma of lower ♂ floret, × 11;   **L**, Lemma of upper ♀ floret, × 11;   **P**, palea of lower ♂ floret dorsal view, × 11;   **P**, palea of upper ♀ floret, × 11;   **Lo**, lodicules of ♀ flower, × 24;   **O♀**, ovary, styles and stigmas of ♀ flower, × 16; **St♂**, stamens of ♂ flower, × 16;   **St♀**, stamens of ♀ flower, × 16.

c. 30 spp. of temperate to arctic or subantarctic regions; in woods, marshy places, open grassland and tundra. Endemic spp. 6, indigenous sp. 1, extending to Australia, New Guinea and South America.

Although Schouten, Y. and Veldkamp, J. F. *Blumea 30*: 319–351 (1985) advocated the inclusion of *Hierochloe* in *Anthoxanthum* L. we are unable to concur with their conclusions for morphological and genetic reasons. Exceptions to their formal transfers of species of *Hierochloe* to *Anthoxanthum* were the species recognised for New Zealand by Zotov, V. D. *N.Z. J. Bot. 11*: 561–580 (1973); Schouten and Veldkamp did so because they felt "... that another revision of the material is needed".

Zotov (1973 op. cit.) recognised seven species. Here we conform to his taxa and complete his deficient floral descriptions. We acknowledge the validity of Schouten and Veldkamp's call for a complete revision.

1 ♀ floret clearly awned 2
   ♀ floret unawned or mucronate 5
2 Awns of ♂ florets inserted near lemma apex *3*
   Awns of ♂ florets inserted near lemma base 4
3 Lemma of ♀ floret conspicuously bearded ....**2. cuprea**
   Lemma of ♀ floret scarcely bearded .... **4. fusca**
4 Awn of upper ♂ floret straight.. **7. redolens**
   Awn of upper ♂ floret with twisting column, curved arista ...... **6. recurvata**
5 Glumes ≫ florets; adaxial leaf-blade prominently costate .... **1. brunonis**
   Glumes ≤ florets; adaxial leaf-blade shallowly costate 6
6 Awn straight; adaxial leaf-blade scabrid .... **5. novae-zelandiae**
   Awn column twisting, arista reflexed; adaxial leaf-blade hairy.... **3. equiseta**

1. H. brunonis Hook.f., *Fl. Antarct. 1*: 93, t. 52 (1845) ≡ *H. antarctica* var. *brunonis* (Hook.f.) Zotov, *T.R.S.N.Z. 73*: 234 (1943); Holotype: K! [*J. D. Hooker*] in tufts at the top of the Mts above Rendezvous harbour, Lord Auckland Islands, Nov. 1840.

Lax tufts. Leaf-sheath glabrous, keeled, ± striate. Ligule 3–5 mm, membranous, rounded, abaxially hairy. Collar thick, glabrous. Leaf-blade 7–30 cm × 4–8 mm, tapering, ± strict, subconvolute, subcoriaceous, abaxially and on margins glabrous, adaxially prickle-toothed on edges of and between prominent flat-topped ribs. Culm to 50 cm, internodes glabrous. Panicle 5–12 cm, rather dense, scarcely spreading, ±nodding; branches usually binate at nodes, glabrous, lower branches subtended by lanceolate bracts; pedicels to 5 mm, scabrid to pilose above. Glumes subequal, ovate-lanceolate, acute, membranous, almost entirely scarious, keel toothed above, 3-nerved; upper 10–14 mm. Florets included by glumes, light brown at maturity. ♂ florets: lemma 6–7 mm, ovate-oblong, lobes erose 2–4 mm and scarious-tipped, ±appressed long hairs on keel above shorter below, margins sparsely ciliate; awns *c.* 6 mm, slender, ± straight, insertion *c.* 3 mm above base; palea 5–6 mm, ciliate apex bifid, keels hairy above; lodicules 1–1.25 mm, ovate-lanceolate, irregularly 1–2-lobed; callus short, hairs 0.5 mm appressed; anthers 2–3.3 mm. ♀ floret: lemma 4–5 mm, broadly ovate, glabrous, appressed shortly hairy at apex with slender subapical mucro 0–0.5–1 mm; palea ≈ lemma, ovate-oblong, 1-keel, hairy above; lodicules 0.6–1 mm, glabrous; anthers 1.3–2 mm; gynoecium: ovary *c.* 1 mm, stigma-styles 4.5–6.5 mm; caryopsis *c.* 2 mm, embryo *c.* 0.5 mm, hilum *c.* 0.5 mm.

Endemic.

S.: Waipapa Point near Invercargill; A., C. Coastal.

Proliferation occurs in many Campbell Id plants.

2. H. cuprea Zotov, *N.Z. J. Bot. 11*: 571 (1973); Holotype: CHR 7509! *V. D. Zotov* West of Kime Hut, *c.* 4500 ft, Tararua Mts, 31.12.1933.

Laxly tufted, ±robust. Leaf-sheath ±striate, glabrous. Ligule 1–2 mm, long ciliate, abaxially hairy, chartaceous, scarious. Collar thick, glabrous. Leaf-blade 20–40 cm × 6–12 mm, tapered, abaxially glabrous, adaxially prickle-toothed on prominent rounded ribs; margins finely toothed. Culm 30–60 cm, internodes glabrous. Panicle 10–20 cm, lax, spreading; branches 1–2 at each node, naked below, glabrous, ± slender and often drooping, spikelets few towards tip; pedicels to 10 mm, scabrid to villous above. Glumes subequal, ovate, acute, membranous, scarious, glabrous, 3-nerved; upper glume 6–7.5 mm. Florets included by glumes, ± glistening coppery at maturity. ♂ florets: lemma 4–6.5 mm, lobes 0.5–1 mm, erose, tip scarious, scabrid, appressed long hairs on keel to sinus, margins densely long-ciliate; awns 2–6 mm, slender, ± straight, insertion *c*. 1 mm below apex; palea 3–4 mm, membranous, keels ciliate; lodicules *c*. 1 mm, ± ovate, acute, irregularly 1–2-lobed, glabrous; anthers (1)–1.5 mm. ♀ floret: lemma 3–3.5 mm, ovate, glabrous and shining, apex with conspicuous long hispid hairs; mucro slender, 0.6–1.5 mm, subapical, ≈ hairs; palea ≈ lemma, ovate-lanceolate, apex pointed ciliate, keel ciliate above; lodicules 0.6–1 mm, glabrous; anthers 0.9–1.6 mm; gynoecium: ovary *c*. 1 mm, stigma-styles *c*. 3.5 mm; caryopsis 1.5–1.8 × 0.7 –0.9 mm, shortly beaked, embryo 0.5 mm, hilum *c*. 0.5 mm.

Endemic.

N.: Mt Egmont, Ruahine and Tararua Ranges; S.: along main divide and to west, southwards from Arthurs Pass, Canterbury. Alpine.

Anthers are apiculate and in some anthers fine processes < 0.1 mm are extensions; it is difficult to find these processes at all times.

3. H. equiseta Zotov, *N.Z. J. Bot. 11*: 568 (1973); Holotype: CHR 9679! *V. D. Zotov* Bold Peak, Humboldt Mts, 5.1.1936.

Fairly robust, lax tufts. Leaf-sheath glabrous, ± striate, mostly deep purple. Ligule *c*. 0.5 mm, abaxially hairy. Collar rather thick, often hairy. Leaf-blade 15–30 cm × 4–8 mm, tapering, abaxially glabrous, adaxially abundantly hairy on prominent flat-topped ribs; margins toothed. Culm 30–60 cm, internodes glabrous, ridged. Panicles 8–15 cm, very lax and spreading; branches mostly binate at each node, glabrous, naked below, slender, drooping, spikelets few towards tips; pedicels to 10 mm, scabrid to villous above. Glumes subequal, ovate, subacute, membranous, margins wide scarious above, keel prickle-toothed above or glabrous; upper 6–8 mm, 3-nerved. Florets > lower glume, ♂ florets light brown to brown, ♀ floret paler, often golden and glistening at maturity. ♂ florets: lemma 5–6 mm, oblong, subacute, lobes 0.5–1 mm, villous below, scabrid above, margins densely long-ciliate; awns of both florets ± equal, rather stout, column 2–3.5–5 mm twisted, arista 3–4–(5) mm, insertion *c*. 2 mm above base of lemma; palea *c*. 4 mm, membranous, apex bifid, keels finely ciliate, interkeel hairy; lodicules 1.2–2.0 mm, ± ovate, acute, irregularly lobed, ciliate; callus hairs to 1 mm; anthers 2.3–3.4 mm.

♀ floret: lemma 3.5–4.5 mm, ovate, long hairs on keel below, finely scabrid and with long hispid golden brown hairs exceeding apex, muticous or subapically mucronate 0–0.5 mm; palea ≈ lemma, ovate-lanceolate, keel short hairy above; lodicules 1–1.8 mm, ciliate; anthers 1.1–2.5 mm; gynoecium: ovary 1 mm, stigma-styles 3.5–6 mm; caryopsis 2–2.5 × 0.75 mm, embryo 0.5 mm, hilum *c.* 0.75 mm.

Endemic.

S.: Nelson and Marlborough sparse, Main Divide southwards from Canterbury, and to east; St. Alpine.

Buchanan, J. *Indig. Grasses N.Z.* t. 35 (1879) illustrated *H. equiseta* (as *Danthonia buchananii*) from a plant from the Matukituki Valley. Kirk, T. *T.N.Z.I. 14*: 385 (1882) by referring to Buchanan's t. 35 allied our plant with Victorian *H. submutica* F.Muell. which he reduced in rank as *H. alpina* var. *submutica* (F.Muell.) Kirk. As the basionym for Kirk's combination is Australian it is not included in the synonymy of endemic *H. equiseta*.

4. H. fusca Zotov, *N.Z. J. Bot. 11*: 576 (1973); Holotype: CHR 119477! *V. D. Zotov* Beeman Hill, Campbell I, 3.1.1961.

Lax, robust tufts. Leaf-sheath glabrous, ± striate. Ligule 3–5 mm, ± chartaceous, hairy, erose to variously lobed. Leaf-blade 30–75 cm × 8–12 mm, tapering, abaxially glabrous, adaxially scabrid on prominent ribs; margins thick, toothed. Culm 60–120 cm, internodes glabrous, ridged. Panicle 15–30 cm, ± erect; branches 1–2 at each node, glabrous, naked below, spikelets densely crowded above; pedicels scabrid to villous. Glumes subequal, ± membranous with wide scarious margin and tip, ovate-lanceolate, acute, glabrous, keeled; upper 7–9 mm, 3-nerved. Florets included by glumes, brown to dark brown at maturity. ♂ florets: lemma 6–8 mm, oblong-ovate, lobes 1–1.25 mm erose and scarious-tipped, sparsely finely scabrid, long hairs at base, margins sparsely long-ciliate; awns 3–7 mm, slender, ±straight, insertion 1–2 mm below apex; palea 5–6 mm, membranous, keels ciliate; lodicules 0.5–1 mm, ± ovate, acute, irregularly 1–2-lobed, glabrous; callus hairs to 1 mm; anthers 2.5–3.5 mm. ♀ floret: lemma *c.* 5 mm, narrow-ovate, glabrous, apex minutely hairy, muticous or subapically mucronate with mucro 0–0.5 mm; palea ≈ lemma, keel toothed ± to base, ovate-lanceolate; lodicules *c.* 0.5 mm, glabrous; anthers 1.5–2 mm; gynoecium: ovary 1 mm, stigma-styles 5–6 mm; caryopsis 2.0–2.5 × 0.75 mm, embryo 0.5 mm, hilum 0.75–1 mm.

Endemic.

N.: Kapiti Id off western Wellington coast; S.: western and southern coasts, southwards from Westport, rare in Westland, more common in Fiordland and Southland; St.; Ch., A., C. Coastal.

Zotov (1973 op. cit.) suggests that the dispersal of *H. fusca* may be linked with the nesting sites of sea birds.

Proliferating spikelets occur in CHR 301473 *W. B. Brockie* Campbell I, Jan 1947; CHR 301469 *idem* has normal flowers.

5. H. novae-zelandiae Gand., *Bull. Soc. Bot. France 66*: 300 (1919); Holotype: LY! *G. M. T[homson]* Dunedin, 1896.

Slender, dense to lax tufts, to somewhat rhizomatous. Leaf-sheath glabrous, ±striate. Ligule 0.5–1 mm, membranous. Collar thick recurved occasionally with a few hairs. Leaf-blade 2–10 cm × 1.5–3 mm, tapering, scabrid or glabrous, shallowly ribbed; margins toothed. Culm 20–40 cm, internodes glabrous. Panicle 5–10 cm, very lax, spreading, branches glabrous filiform, 1–2 at each node, pedicels glabrous 1–5 mm, spikelets congested in 3s or 4s. Glumes obovate, apiculate, subacute to obtuse, membranous with scarious margins above, 3-nerved; upper 3.5–4 mm. Florets > glumes, light brown at maturity. ♂ florets: lemma 3.25–4 mm, oblong, ovate, glabrous below, finely scabrid to pubescent above, margins silky ciliate; awns 1–3 mm, slender, ±straight, insertion 2.5–3.5 mm above base; palea 2.5–3.5 mm, hairs on keels to base and elsewhere, bifid apex ciliate; lodicules 0.75–1.25 mm, ± ovate, acute, irregularly lobed, ciliate; callus hairs to 1 mm; anthers 1.5–2–2.5 mm. ♀ floret: lemma 2.5–3.5 mm, ovate, truncate, glabrous below hairy, muticous or mucronate apex; mucro subapical, 0.5–1 mm; palea 2.5–3 mm, oblong-ovate, 1–(2) keels hairy to base; lodicules 0.7–1.25 mm, ciliate; anthers (0.9)–1.2–1.9 mm; gynoecium: ovary 1 mm, stigma-styles 2.75–3.5 mm; caryopsis 2 × 0.75 mm, embryo *c.* 0.5 mm, hilum 0.5 mm.

Endemic.

N.: Mt Egmont; S.: Nelson and along main divide, also scattered to the east and west; St. Alpine.

Names earlier applied to this sp. were *H. borealis sensu* Hook.f., *H. alpina sensu* Hook.f. and *H. fraseri sensu* Cheeseman, none of them correctly.

Anthers are apiculate, and in some these continue as fine processes *c.* 0.1 mm long; these processes do not appear to be universal.

6. H. recurvata (Hack.) Zotov, *N.Z. J. Bot. 11*: 566 (1973) ≡ *H. fraseri* var. *recurvata* Hack., *in* Cheeseman, *Man. N.Z. Fl.* 856 (1906) ≡ *H. alpina* var. *recurvata* (Hack.) Zotov, *T.R.S.N.Z. 73*: 235 (1943); Holotype: AK 1330! *D. Petrie* Ruahine Range, 4000–5000 ft, Jan. 1889 (No 1039 to Hackel); duplicate WELT 76772.

Robust, lax tufts, ± rhizomatous. Leaf-sheath glabrous, ± striate, mostly deep purple. Ligule *c.* 1 mm. Collar thick, glabrous, recurved. Leaf-blade 7–15 cm × 3–6 mm, tapering, glabrous to ± scabrid on both surfaces, adaxially shallowly ribbed; margins prickle-toothed. Culm 20–40 cm, internodes glabrous, ridged. Panicle 8–16 cm, very lax, spreading; branches glabrous, 1–2 at each node, slender, naked below, spikelets few towards tips; pedicels 1–5 mm, ± scabrid to villous below spikelet. Glumes obovate, subacute to obtuse, glabrous, often purplish with wide hyaline margins, 3-nerved; upper glume 4–6 mm. Florets > lower glume, brown, uppermost somewhat lighter, ± shining. ♂ florets: lemma 4–5.5 mm, oblong, lobes *c.* 0.5 mm, long villous, rounded, finely scabrid, margins long-ciliate; awns unequal, rather stout, awn of lower floret 1.5–3 mm, scarcely twisted or bent, awn of second floret 4–6 mm, column 2–3.5 mm, twisted below geniculate arista 2–3 mm, insertion 2–3.5 mm above base; palea 4–4.5 mm, keels hairy to base finely pubescent above; lodicules *c.* 1 mm, ± ovate, acute, irregularly 1–2-lobed often ciliate; callus hairs sparse to 1 mm; anthers 2.0–2.5 mm. ♀ floret: lemma 3.5–4.5 mm, ovate, glabrous below, finely pubescent above, usually long hairy at apex; awn rather stout, 2–3 mm, subapical; palea *c.* 3.5 mm, ovate-lanceolate, keels 2, hairy throughout, minutely scabrid above; lodicules 1–1.2 mm; gynoecium: ovary 0.75–1.25 mm, stigma-styles 3.75–4.75 mm; anthers 1.4–1.8 mm; caryopsis 2 × 0.75 mm, embryo 0.3–0.4 mm, hilum 0.5–0.75 mm.

Endemic.

N.: mountains from East Cape southwards; S.: along main divide and to west, scattered to the east in Canterbury and Otago. Alpine.

7. H. redolens (Vahl) Roem. et Schult., *Syst. Veg.* 2: 514 (1817) ≡ *Holcus redolens* Vahl, *Symb. Bot.* 2: 101 (1791) ≡ *Avena redolens* (Vahl) Pers., *Syn. Pl. 1*: 100 (1805) ≡ *Anthoxanthum redolens* (Vahl) D.Royen, *Alpine Fl. N. Guinea 2*: 1185, t. 382 (1980) ≡ *Hierochloe antarctica* var. *redolens* (Vahl) Raspail, *Ann. Sci. Obs. 1*: 83 (1829) *comb. illeg.*; Holotype: C! *Forsteri* (dedit Fabricius). = *Holcus redolens* R.Br., *Prodr. 1*: 209 (1810) non Vahl 1791 nec Solander ex G.Forst. 1786, *nomen* ≡ *Torresia redolens* Roem. et Schult., *Syst. Veg.* 2: 516 (1817) ≡ *Hierochloe banksiana* Endl., *Ann. Wiener Mus. Naturgesch. 1*: 156 (1836); Lectotype: WELT 63940! *Banks & Solander* New Zealand (designated by Zotov 1973 op. cit. p. 574).   kāretu

Robust, lax tufts. Leaf-sheath glabrous, ± striate, lower ± purplish. Ligule 2–3 mm, chartaceous, ± irregularly rounded. Leaf-blade to 70 cm × 8–12 mm, abaxially ± glabrous, adaxially scabrid on prominent ribs; margins glabrous or prickle-toothed. Culm to 130 cm, internodes glabrous. Panicle 20–30 cm, erect, nodding above, and lower branches also nodding; branches binate at nodes, very slender, naked for ½ to ¾ length, spikelets crowded distally; pedicels 0.5–2 mm, villous. Glumes unequal, membranous, scarious, glabrous, ovate, acute, keeled, 3-nerved; lower 6–7 mm, mostly ≥ lower floret, upper 7–8 mm, > second floret. Florets pale straw-coloured. ♂ florets: lemma 5–6.5 mm, oblong-ovate, lobes *c.* 1 mm, chartaceous with scarious tips, long hairs on keel below, minutely scabrid above, margins ciliate with soft, silvery hairs; awns 3.5–6 mm, slender, ± straight, insertion 3–4 mm above base; palea 4–5.5 mm, membranous, irregularly finely scabrid on keels; lodicules 1–1.6 mm, ± ovate, lobed, acute, glabrous; callus hairs to 1.25 mm; anthers 2–3 mm. ♀ floret: lemma 4.5–6 mm, narrow-ovate, glabrous, apex minutely hairy, muticous to subapically mucronate 0.25–0.5 mm; palea 4–5.5 mm, ovate-lanceolate, keel 1–(2) finely irregularly ciliate; lodicules 0.75–1 mm, ovate-oblong, abruptly tapering, often lateral lobed, glabrous; anthers 1–1.5 mm; gynoecium: ovary *c.* 1 mm, stigma-styles 4–5 mm; caryopsis *c.* 2 mm, embryo 0.5 mm, hilum 0.75 mm. Fig. 11. Plate 5D.

Indigenous.

N.; S.; Ch.: throughout. In tussock grassland.

Also indigenous to Australia, New Guinea, and South America.

de Paula, M. E. *Darwiniana 19*: 422–457 (1975) admitted three varieties of *H. redolens*; we are unable to equate any with New Zealand material and therefore accord no infraspecific rank.

Nomenclatural concerns in, and the typification of, *H. redolens* as raised by Zotov (1973 op. cit.) were argued against by de Paula (1975 op. cit.); Schouten and Veldkamp (1985 op. cit.) found the argument over the *locus classicus* of Forsters' specimens immaterial because *H. redolens* is common to South America, Australasia and Malesia.

Disarrenum antarcticum Labill., *Nov. Holl. Pl. 2*: 83 t. 232 (1807), and its subsequent combinations, is based on an Australian provenance. *Melica magellanica* Desr. *in* Lamarck *Encycl. Méth. Bot. 4*: 72 (1797) and its subsequent combinations, is based on Commerson's specimen from the Straits of Magellan. Both names are treated by all authorities as synonymous with *H. redolens*. We have not seen the types for either, and having limited ourselves to material originating in New Zealand, have not included these names in our synonymy.

Hierochloe banksiana Endl. would be the appropriate name in *Hierochloe* should New Zealand plants no longer be referable to *H. redolens*.

FLORAL BIOLOGY   In all New Zealand plants the floral biology is identical viz two lower florets each with 2 lodicules and 3 stamens, and an upper ♀ floret with 2 lodicules, 2 stamens with polliniferous anthers shorter than those in ♂ florets, and a gynoecium where the two broad styles are connate below; this is an andromonoecious system. Protogyny occurs in ♀ flowers and the long stigmas are apically emergent; they are manifestly persistent even after caryopses have formed. No plants bore purely ♀ flowers; all plants have dimorphic anthers; all flowers are lodiculate. The debate, confusion, or misinterpretation of stamens in the apical floret (see de Paula 1975 op. cit.; Schouten and Veldkamp 1985 op. cit.) does not apply in New Zealand plants. In the few Chilean specimens that we have seen the apical floret bears two small (0.3–0.5 mm) sterile anthers, and the flower is thus female; the two lower flowers are ♂; this is a monoecious system.

Proliferation in spikelets in *H. brunonis* is described as predominant on Campbell Id by Zotov, V. D. *Rec. Dom. Mus. 5*: 101–146 (1965) at the time of his visit in 1961. Specimens are in CHR. In *H. fusca* on Campbell Id CHR 301473 *W. B. Brockie* Jan 1947 is the lone proliferating specimen in CHR.

In New Zealand species of *Hierochloe* often seem sympatric, as judged from specimens in CHR. The risks of natural hybridism seem high because of protogyny in ♀ apical florets and duodichogamy within anthers of the three florets.

PHALARIS L., 1753

Annuals or perennials. Leaf-sheath rounded. Ligule conspicuous, membranous, glabrous, oblong. Leaf-blade flat. Culm internodes glabrous. Panicle dense, stiff, spike-like, ovoid to cylindric or sometimes ±lobed with branches spreading at maturity. Spikelets shortly pedicelled, strongly compressed laterally, usually with 3 florets, the lower (1)–2 usually ∅ or ♂, the upper ♀, or spikelets sometimes entirely ♂ or ∅; disarticulation above glumes; rachilla not or obscurely prolonged. Glumes subequal and equalling spikelet, laterally compressed, flat, > lemmas and enclosing them, chartaceous, 3–5-nerved, often light green with darker green bands along nerves, keeled, keel often winged. Lemmas of ∅ florets small, linear to lanceolate, membranous, 1-nerved, hardened towards base, sometimes scale-like, epaleate; lemma of ♀ floret 5-nerved, keeled, ovate, acute, awnless, becoming crustaceous and shining. Palea of ♀ floret similar in texture to lemma, slightly smaller, nerves 2, close together, internerve minutely hairy. Lodicules 2, hyaline. Stamens 3. Ovary glabrous; styles connate at base, elongated; stigmas plumose. Caryopsis laterally compressed, ovate; embryo *c.* ¼ length of caryopsis; hilum oblong, short; endosperm solid. Chasmogamous.

Type species: *P. canariensis* L.

c. 20 spp. of warm temperate regions. Naturalised spp. 5; transient sp. 1.

1 Spikelets of 2 kinds, the ♀ spikelets surrounded by a cluster of ∅ or ♂ spikelets falling as a group at maturity; lemma of ♀ florets glabrous ...... **5. paradoxa**
  Spikelets all similar, sometimes reduced at base of panicle; lemma of ♀ florets pubescent 2

2 Lemmas of ∅ florets equal or subequal and well developed *3*
  Lemmas of ∅ florets very unequal, the lower much reduced or 0 .... *5*

3 Perennial with long rhizome; panicle cylindric above, lobed below; glumes with narrow wing evenly tapering to apex, or wingless ...... **2. arundinacea**
  Annual, tufted; panicle evenly cylindric to ovate, rarely lobed; glumes with wing widened in upper ⅓ *4*

4 Panicle ovoid-oblong, 1.5–2.5 cm wide; spikelets 7–8.5 mm .. **3. canariensis**
  Panicle narrow-oblong, to 1 cm wide; spikelets *c.* 4 mm ... **angusta**ζ (p. 355)

5 Perennial, rhizomatous; glumes with evenly curved wing-margin ... **1. aquatica**
Annual, tufted; glumes with uneven erose or denticulate wing-margin ....
.... **4. minor**

1. P. aquatica L., *Cent. Pl. 1*: 4 (1755).   phalaris

Strong perennial clumps, 60–200 cm, from short, knotted rhizome, shoots slightly swollen and ± tuberous at base. Leaf-sheath firm, glabrous, striate, light brown or brownish green. Ligule 4–9 mm, rounded. Leaf-blade 15–60 cm × 5–11 mm, very minutely scabrid, adaxially becoming smooth; margins slightly thickened, minutely scabrid, tapered to long, fine, acuminate tip. Culm 55–180 cm. Panicle (2.5)–3.5–12–(20) × 0.8–2.5–(3) cm, cylindric, oblong or slightly tapered at upper or both ends, occasionally lobed; rachis smooth, branches very sparsely scabrid, hidden. Spikelets 4.5–6.5 mm, light green. Glumes ± equal, 3-nerved, elliptic-lanceolate with deeper green ciliolate wing in upper ½, glabrous, evenly curved towards subacute tip. Ø florets: lemmas unequal, lower 0.5–1 mm, upper 1.2–1.7 mm, lanceolate, acute, sparsely hairy. ♂ floret: lemma 3–3.7 mm, ovate, subacute, with ± sparse, short, fine, silky hairs; palea narrower than lemma; anthers (2.2)–3–4 mm; caryopsis 2.2–2.6 × 1–1.3 mm.

Naturalised from Mediterranean.

N.: North and South Auckland, elsewhere scattered; S.: east of Main Divide. Waste ground, roadsides especially on drain margins, and coastal hillsides.

Known for many years as *P. tuberosa* L., a later synonym.

2. P. arundinacea L., *Sp. Pl.* 55 (1753).   reed canary grass

Robust perennials, 60–200 cm, with long-creeping rhizomes. Leaf-sheath chartaceous, glabrous, striate, light brown. Ligule 2.5–7.5–(10) mm, entire, but soon lacerate. Leaf-blade 20–40 cm × 8–20 mm, ribs numerous, fine, adaxially smooth but scabrid near tip, abaxially with strong midrib near base, ribs densely, minutely scabrid; margins minutely scabrid, long-narrowed to scabrid, acute tip. Culm 50–180 cm. Panicle 9–30 × 1.5–4 cm, lanceolate or oblong, lobed below; rachis smooth below, scabrid above, branches scabrid, spreading at anthesis.

Spikelets 4–5.5 mm, pale green or purplish. Glumes ± equal, 3-nerved, lanceolate, keeled but not winged, acute to acuminate, minutely scabrid, rarely lower glume with minute hairs near margin. ∅ florets: lemmas equal, 1.3–1.6 mm, narrow, short-hairy. ♀ floret: lemma 3–4 mm, broadly keeled, lanceolate, acute, firm and shining below, short-hairy above; palea much narrower than lemma; anthers (2)–2.5–3.2 mm; caryopsis *c.* 2 × 1 mm.

Naturalised.

N.: North Auckland, Waikato, Manawatu, southern Wairarapa; S.: Canterbury, Otago, local in Nelson and Westland. Waste land, road margins, and damp or swampy ground.

Indigenous to temperate Eurasia and North America.

The variegated form, *P. arundinacea* var. *picta* L. with cream-striped leaves, is cultivated as an ornamental and has escaped in many places.

3. P. canariensis L., *Sp. Pl.* 54 (1753).   canary grass

Slender to stout, tufted annuals, 35–90 cm. Leaf-sheath firm, minutely scabrid or smooth, striate, green to light brown, upper sheaths slightly inflated. Ligule 2–6 mm, tapered, erose. Leaf-blade 7.5–30 cm × 4–10 mm, finely ribbed, ribs minutely scabrid; margins finely scabrid, long-narrowed to acute tip. Culm 30–80 cm, often branched near base, internodes scabrid just below panicle. Panicle 2–3.5–(6) × 1.5–2.5 cm, ovoid to ovoid-oblong, subtended by a hard, scabrid, very short bract; rachis and branches hidden, finely scabrid-papillose. Spikelets 7–8.5 mm, pale greenish cream, rarely purplish. Glumes ± equal, 3–5-nerved, oblanceolate with abrupt acute tip, sparsely hairy, cream with green bands along nerves, keel minutely scabrid, broadly winged in upper ½. ∅ florets: lemmas equal, 2.5–4 mm, lanceolate, acute, very sparsely hairy. ♀ floret: lemma 4.5–5.6 mm, ovate, acute, shining, silky-haired; palea narrower than lemma; anthers 2–3.8 mm; caryopsis *c.* 4 × 1.4 mm.

Naturalised from western Mediterranean.

N.; S.: occasional throughout; Ch. Waste land.

4. P. minor Retz., *Observ. Bot.* 8 (1753). lesser canary grass

Tufted annuals 10–100 cm, very slender to robust. Leaf-sheath firm, minutely scabrid or smooth, striate, light green sometimes purplish; upper sheaths slightly inflated. Ligule 2–6.5 mm, obtuse, slightly tapered. Leaf-blade 3–30 cm × 2–10 mm, long-tapered, finely ribbed, ribs finely scabrid especially near finely scabrid margins, tip acuminate. Culm 8–95 cm, internodes ridged. Panicle 1.5–6.5 × 1–2 cm, ovoid-oblong to cylindric; rachis and branches hidden, papillose, scabrid, sometimes only sparsely so. Spikelets 4.5–5.5 mm, light green. Glumes ± equal, 3-nerved, smooth, oblanceolate with sharply acute tip, keel finely scabrid, winged in upper ½, wing usually toothed near abruptly narrowed apex. ∅ florets: lemmas very unequal, lower *c.* 0.3 mm, a glabrous shining minute scale, upper 0.9–1.3 mm, lanceolate, acute, hairy. ♀ floret: lemma 2.5–3 mm, ovate, acute, silky-haired; palea narrower than lemma, internerve long hairy or occasionally almost glabrous, apex ciliate; anthers 1–2 mm; caryopsis 2–2.3 × 0.8–1.2 mm.

Naturalised from Mediterranean.

N.: scattered; S.: scattered to east of Main Divide, usually near coast; K., Ch. Waste ground, roadsides, poor pasture, shingle and ballast.

Now naturalised and widespread throughout temperate regions of both Hemispheres.

5. P. paradoxa L., *Sp. Pl.* ed. 2, 1665 (1763). gnawed canary grass

Annuals, 20–80 cm. Leaf-sheath firm with narrow hyaline margin, ribs many, ± finely scabrid; uppermost sheaths conspicuously inflated, glabrous. Ligule 2–4–(6) mm, ± tapered. Leaf-blade 6–20 cm × 3–5.5 mm, tapered, ribs and margins finely scabrid, tip fine, soft, acute. Culm 14–56 cm, branched at base. Panicle 2.5–6 × 0.8–1.5 cm, cylindric, oblong, or widened above; rachis and very short branches scabrid, hidden. Spikelets in groups of 5–7–(9), with one central almost sessile ♀ spikelet 5–6.5 mm, surrounded by an involucre-like cluster of shorter, unequal, usually ∅, sometimes ♂ spikelets on short glabrous pedicels, two of the ∅ spikelets generally larger. Glumes of all spikelets hard, keel sparsely scabrid, wing margin minutely scabrid; glumes of ∅ spikelets sometimes sparsely hairy, margins scabrid in upper ⅓, apex rounded or truncate, or occasionally with acute flange; glumes of ♂ spikelets with 2–6 prominent lateral nerves and often shorter nerves as well, tapering above to acute tip, narrowly winged apart from an acute projecting middorsal flange. Ø florets: lemmas minute, scale-like, equal, glabrous. ♀ floret: lemma 2.9–3.3 mm, ovate, acute, almost glabrous; palea entirely glabrous or internerve sometimes sparsely hairy; anthers 1–1.6 mm; caryopsis 2–2.6 × 1.2–1.5 mm.

Naturalised.

N.: Upper Hutt; S.: North Canterbury (Omihi, near Rangiora), and near Christchurch. In waste and cultivated land.

Indigenous to the Mediterranean region and south-western Asia.

ζ**P. angusta** Trin.   Fibrous-rooted annual, 110 cm. Uppermost leaf-sheath ± inflated. Leaf-blade scabrid. Panicle narrow-oblong, to 10 × 1 cm. Spikelets 4 mm. Glumes oblong, ± apiculate, almost toothed at apex, hardly winged except near apex, keel scabrid. Ø florets: lemmas ± equal, 1.2–1.5 mm, linear, sparsely hairy. ♀ florets: lemma *c.* 3 mm, silky-hairy. A species of North America and southern South America, known from one collection, CHR 225478 *Smeaton* Tanekaha in Hikurangi Swamp, Whangarei County, in pasture on land subject to flooding, 1971.

10. BROMEAE

Annuals or perennials. Ligule membranous. Inflorescence a panicle. Spikelets of several–many ♀ florets, with Ø florets above, laterally compressed; rachilla prolonged. Glumes < lowest lemma, persistent. Lemma ± bidentate with straight or recurved apical awn, or awn 0. Palea 2-keeled. Stamens (2)–3. Ovary with a terminal, fleshy, lobed, hairy appendage; styles 2, subterminal. Caryopsis narrowly ellipsoid to linear, hollowed on hilar face, adherent to lemma and palea; hilum linear, as long as caryopsis.

BROMUS L., 1753

Annuals or perennials, tufted or occasionally rhizomatous. Leaf-sheath often pubescent; margins connate. Ligule membranous. Leaf-blade flat. Inflorescence an open or contracted panicle, or rarely a raceme. Spikelets 2–many-flowered, ± laterally compressed; florets chasmogamous or cleistogamous; disarticulation above glumes and between lemmas. Glumes unequal or subequal, acute or shortly awned; lower 1–7-nerved, upper 3–9-nerved. Lemma rounded or keeled, 5–13-nerved, bidentate, awned from near apex; awn straight, bent or recurved, or 0. Palea usually < lemma, 2-keeled, keels ciliate or scabrid. Lodicules 2, entire, connate below, glabrous. Stamens (2)–3. Ovary with a terminal, hairy appendage and styles inserted laterally. Fig. 12.

Type species: *B. secalinus* L.

c. 150 spp. of temperate regions especially in Northern Hemisphere. Naturalised spp. 15; transient spp. 5.

Bromus L. has been variously divided into six sections [Smith, P. *Notes Roy. Bot. Gard. Edinb. 30*: 361–375 (1970)], seven subgenera [Stebbins, G. L. *Bot. Jahrb. Syst. 102*: 359–379 (1981)], or five subgenera [Tsvelev, N. N. *Grasses of the Soviet Union 1*: 298–343 (1976)]. No consensus has been reached on the treatment of the complex but many authors agree with Smith, P. (op. cit.) that sectional rank is "most appropriate for the infrageneric elements within *Bromus* s.l."

In an expanded checklist of naturalised species of *Bromus* and their first records in New Zealand, Forde, M. B. and Edgar, E. *N.Z. J. Bot. 33*: 35–42 (1995) discussed the nomenclature of species in *Bromus* sect. *Ceratochloa*, in particular of South American *B. catharticus*, *B. valdivianus* and *B. willdenowii*; they concluded that *B. willdenowii* Kunth was the correct name for prairie grass, rather than *B. catharticus* or *B. unioloides*; they accepted *B. brevis* (Nees) Steud. as specifically distinct from *B. catharticus*. They also agreed with Muñoz-Schick, M. *Fl. Parque Nac. Puyehue* 488–489 (1980) that *B. valdivianus* was distinct from *B. stamineus*.

Stebbins, G. L. (op. cit.) discussed the levels of polyploidy within *Bromus*. Species in sect. *Bromus*, sect. *Bromopsis* and sect. *Genea* are mainly diploid ($2n = 14$) and tetraploid ($2n = 28$). Within sect. *Ceratochloa* the South American species are hexaploid ($2n = 42$), whereas North American species are octoploid ($2n = 56$) and 12-ploid ($2n = 84$).

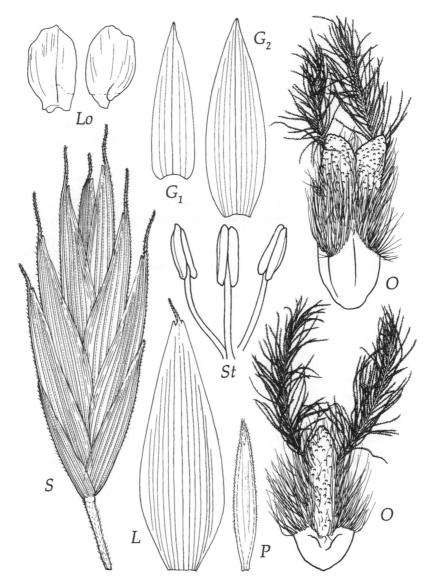

Fig. 12 *Bromus willdenowii*
S, spikelet, × 3.2; **G₁**, **G₂**, glumes, × 4; **L**, lemma dorsal view and flattened, × 4; **P**, palea dorsal view, × 4; **Lo**, lodicules, × 20; **O**, ovary, hairy appendage and stigmas, frontal and dorsal views, × 20; **St**, stamens, × 20.

SYNOPSIS

A. Annuals or rarely biennials

　　1. Sect. BROMUS. Spikelets ovate or lanceolate, terete or ± compressed, tapering to tip. Lower glume 3–5-nerved, upper 5–7-nerved. Lemma rounded; awn ≤, or > lemma: 1. *arenarius,* 3. *commutatus,* 5. *hordeaceus, japonicus* var. *vestitus,* 9. *racemosus, secalinus*

　　2. Sect. GENEA. Spikelets cuneiform, broader at tip. Lower glume 1-nerved, upper 3-nerved. Lemma rounded; awn > lemma: 4. *diandrus,* 8. *madritensis, rubens,* 12. *sterilis,* 13. *tectorum*

B. Perennials

　　3. Sect. BROMOPSIS. Spikelets parallel-sided, terete. Lower glume 1–(3)-nerved, upper 3-nerved. Lemma rounded or weakly keeled; awn < lemma or 0: *erectus,* 6. *inermis*

　　4. Sect. CERATOCHLOA. Spikelets strongly compressed, narrowed to tip. Lower glume 3–5-nerved, upper 5–9-nerved. Lemma strongly keeled; awn < lemma or 0: 2. *brevis, catharticus,* 7. *lithobius,* 10. *sitchensis,* 11. *stamineus,* 14. *valdivianus,* 15. *willdenowii*

1 Awns 0, or < 10 mm *2*
　 Awns 10–60 mm *14*
2 Palea *c.* ⅔ length of lemma or shorter *3*
　 Palea slightly <, or ≥ lemma *4*
3 Spikelets 2.5–4 cm; awns 0.5–4–(5) mm .... **15. willdenowii**
　 Spikelets 1–2 cm; awns 0, or 0.5–0.8 mm .... **2. brevis**
4 Lemmas rounded, not compressed or only slightly so *5*
　 Lemmas keeled, strongly compressed *10*
5 Lower glume ovate- or oblong-lanceolate, 3–5-nerved; upper glume 5–9-nerved *6*
　 Lower glume subulate or very narrowly lanceolate, usually 1–rarely 3-nerved; upper glume 3-nerved *9*
6 Lower leaf-sheaths glabrous; margins of lemmas becoming tightly incurved .... **secalinus**ζ (p. 372)
　 Lower leaf-sheaths villous; margins of lemmas flat, overlapping *7*
7 Lemma papery, prominently nerved, usually shortly pubescent throughout or in upper ½, rarely glabrous or scabrid throughout; anthers rarely > 1 mm, usually much less .... **5. hordeaceus**
　 Lemma firm, obscurely nerved, very minutely scabrid, especially in upper ½; anthers > 1 mm *8*

8 Panicle lax; spikelets 2–3 cm; palea flanks very finely scabrid ... **3. commutatus**
Panicle usually ± contracted; spikelets 1.5–2 cm; palea flanks smooth ....
.... **9. racemosus**

9 Plants rhizomatous; leaf-blades glabrous; lemma entire or with very short
awn to 2 mm .... **6. inermis**
Plants tufted; leaf-blades hairy; lemma awn *c.* 5 mm .... **erectus**ζ (p. 371)

10 Lemmas with numerous distinct though very short hairs throughout *11*
Lemmas very minutely scabrid or scabrid-pubescent in upper ½ to
glabrous throughout, sometimes with a few scattered minute hairs near
base *12*

11 Lower glume 6–7.5 mm; lemma evenly light green to usually purple-suffused
.... **7. lithobius**
Lower glume 4–4.5–(5) mm; lemma yellowish green below, brighter green
above with nerves, margins and awn often reddish-tinged ....
.... **14. valdivianus**

12 Leaf-sheaths glabrous apart from a few hairs at margins below ligule ....
.... **10. sitchensis**
Leaf-sheaths softly hairy throughout *13*

13 Spikelets 2.5–4 cm; awns 3.5–9 mm .... **11. stamineus**
Spikelets 2–2.5 cm; awns 1.5–2.5 mm .... **catharticus**ζ (p. 371)

14 Awns 30–60 mm; lemmas > 20 mm .... **4. diandrus**
Awns 10–25–(30) mm; lemmas usually < 20 mm *15*

15 Lower glume 1–(3)-nerved; upper glume 3-nerved *16*
Lower glume 3-nerved; upper glume 5–7-nerved *19*

16 Panicle branches short-hairy to densely pubescent; spikelets 2–3.5 cm *17*
Panicle branches scabrid; spikelets 3.5–6.5 cm *18*

17 Panicle lax or contracted, secund, nodding; lower branches of panicle bearing
4–8 spikelets .... **13. tectorum**
Panicle very dense, fan-shaped, erect; lower branches of panicle bearing 1–2
spikelets .... **rubens**ζ (p. 372)

18 Panicle branches drooping, widely spreading, ≥ spikelets .... **12. sterilis**
Panicle branches erect or slightly spreading, ≪ spikelets ...... **8. madritensis**

19 Spikelets loosely hairy; palea interkeels glabrous .... **1. arenarius**
Spikelets densely appressed-pubescent; palea interkeels short-hairy...........
.... **japonicus** var. **vestitus**ζ (p. 372)

1. B. arenarius Labill., *Nov. Holl. Pl. 1*: 23, t. 28 (1805).

sand brome

Annuals, 20–80 cm, loosely tufted or consisting of a solitary culm, basal leaves withering early. Leaf-sheath densely villous. Ligule 1–2.6 mm, lacerate. Leaf-blade 7–30 cm × 1.7–5 mm, densely villous, hairs shorter near acute tip. Culm 15–60 cm, erect or geniculate-ascending, internodes ± pubescent below panicle. Panicle 7–26 cm, lax, nodding; branches filiform, curving, short-hairy. Spikelets 3–4 cm, 5–8-flowered, loosely hairy, oblong-lanceolate to wedge-shaped. Glumes unequal, acute to acuminate, chartaceous, with long fine hairs; lower 7.2–9.5 mm, 3-nerved, narrow oblong-lanceolate, upper 9.5–13 mm, 5–7-nerved, narrow elliptic-lanceolate. Lemma 11–14 mm, 7–9-nerved, rounded, chartaceous, oblong- to narrowly elliptic-lanceolate, with long fine hairs; apex hyaline, sometimes entire or with 2 acute lobes *c.* 0.5 mm; awn 14–20 mm, arising *c.* 2 mm below apex of lemma or at base of sinus. Palea *c.* ¾ length of lemma, keels sparsely long-ciliate, interkeel glabrous. Callus with minute hairs. Rachilla 1.2 mm, shortly pubescent. Anthers 0.8–1.5 mm. Caryopsis 7.5–8 × 1.3–1.6 mm.

Naturalised from Australia.   = B. pectinatus Thunb.

N.: North Auckland, Auckland City, and offshore islands, Hamilton, Bay of Plenty, Wellington Harbour; Three Kings Is, Ch. Rocky places near coast.

Two early specimens from Canterbury were probably collected by J. F. and J. B. Armstrong in 1870s, Pegasus Bay [Canterbury] (WELT 76365), and near Christchurch, "rare" (WELT 70982).

2. B. brevis Nees, *in* Steudel *Syn. Pl. Glum. 1*: 326 (1854).

pampas brome

Short, light green, rather stiffly erect, narrow-leaved perennial tufts, 30–75 cm. Leaf-sheath keeled above, ± densely hairy with long, soft hairs, sometimes reddish purple at base. Ligule 1.5–2.5 mm, denticulate, abaxially with long hairs near margins in lower ½. Leaf-blade 9–20 cm × 1.5–4.5 mm, with scattered to dense, long, soft, fine hairs; margins finely scabrid, tip fine, acute. Culm erect, internodes glabrous. Panicle 8–15 cm; branches ± compressed, erect, minutely hairy on angles. Spikelets 1–2 cm, narrow-oblong, light yellowish green.

Glumes ± unequal, acute to acuminate, glabrous or sometimes with scattered short hairs, keel ciliate-scabrid; lower *c.* 4–4.5 mm, 3–5-nerved, upper 5–6 mm, 7-nerved. Lemma 8.5–10 mm, 7–9-nerved, keeled, with very short, silky, scattered, appressed hairs in upper ⅔, acute or shortly mucronate or with very short awn 0.5–0.8 mm. Palea ⅔ length of lemma and much narrower. Callus almost glabrous, with a few, very minute hairs. Rachilla with very short hairs. Anthers 0.2–0.5 mm in cleistogamous flowers, 1.5 mm in chasmogamous flowers. Caryopsis *c.* 6 × 2 mm.

Naturalised from South America.

S.: Canterbury (near Amberley, at Lincoln as a cultivation escape), Central Otago (near Ophir, on Raggedy Range, and Lake Roxburgh). Roadsides, and waste ground.

Planchuelo, A. M. *Kurtziana 21*: 243–257 (1991) and Petersen, P. M. and Planchuelo, A. M. *Novon 8*: 53–60 (1998) considered that the recognition of *B. brevis* at varietal rank within *B. catharticus sens. lat.* seemed warranted, but earlier Naranjo, C. A. *Darwiniana 31*: 173–183 (1992) supported species rank for *B. brevis*.

Bromus brevis behaved as an annual in trials at Christchurch (A. V. Stewart, pers. comm.).

3. B. commutatus Schrad., *Fl. Germ. 1*: 353 (1806).

Annuals to biennials, 60–120 cm, loosely tufted or consisting of a solitary culm. Leaf-sheath ± densely villous with soft spreading hairs; upper sheaths short-pubescent to glabrous. Ligule 0.7–2 mm, shortly denticulate to later lacerate, abaxially with some long hairs. Leaf-blade 8–16 cm × 2–6 mm, pubescent; margins scabrid near acute tip. Culm 40–100 cm, erect; nodes minutely hairy; internodes glabrous or rarely pubescent below panicle. Panicle 8–24 cm, erect, lax; branches and pedicels scabrid, mostly > spikelets. Spikelets 2–3 cm, 7–10-flowered, lanceolate, light green or purple-suffused. Glumes unequal, subacute to obtuse, chartaceous to subcoriaceous, smooth, but midnerve and hyaline margins usually minutely scabrid; lower 6–7 mm, 3–5-nerved, oblong-lanceolate, upper 7.5–8 mm, 5–9-nerved, elliptic. Lemma 8.5–9.5 mm, 7–9–(11)-nerved, rounded, subcoriaceous, nerves ±

obscure, obovate, very minutely scabrid especially in upper ½; margins flat, overlapping; apex obtuse to truncate, to very slightly emarginate; awn from just below apex, straight, 6–9 mm. Palea very slightly < lemma, keels sparsely long-ciliate and with minute hairs as well, flanks very finely scabrid. Callus glabrous. Rachilla *c.* 1.5 mm, finely scabrid. Anthers (1.2)–1.5–2 mm. Caryopsis *c.* 4.5 × 1 mm.

Naturalised.

N.: Auckland City (Remuera, early collection); S.: Canterbury (inland North and Mid Canterbury). On grassy roadsides and in sown pasture.

Indigenous to Europe, North Africa and western Asia.

4. B. diandrus Roth, *Bot. Abh. Beobacht.* 44 (1787).　　　ripgut brome

Annuals, 20–90 cm, loosely tufted or consisting of a solitary culm. Leaf-sheath with soft, short, spreading hairs; upper sheaths ± glabrous near base. Ligule 1.5–4 mm, denticulate to lacerate. Leaf-blade 4–30 cm × 2–7 mm, with short, ± scattered hairs, scabrid near acute tip. Culm 10–65 cm, erect or ascending, internodes with very short hairs below nodes and below panicle. Panicle 10–28 cm, very lax and nodding; branches spreading, slender, scabrid, tipped by 1–(2) spikelets. Spikelets 5.5–9.5 cm, 5–8-flowered, oblong-lanceolate to later wedge-shaped and gaping, greyish green to purplish. Glumes unequal, very narrow, acuminate, membranous, glabrous, but sparsely scabrid on keels and nerves in upper ½; lower 13–22 mm, 1–(3)-nerved, upper 20–30 mm, 3–(5)-nerved. Lemma 20–35 mm, 7-nerved, rounded, lanceolate, minutely scabrid, tapering to 2 acute to acuminate, hyaline, smooth to finely scabrid lobes 4–5 mm; awn 30–60 mm. Palea *c.* ⅔ length of lemma, keels sparsely ciliate. Callus with minute hairs. Rachilla 4–6 mm, finely, minutely scabrid. Stamens 2–3; anthers 1–3 mm. Caryopsis 11–13 × 1.5–2.8 mm. Plate 3A.

Naturalised from Mediterranean.

N.; S.: throughout; St.; K., Ch. Roadsides and waste ground, often in sandy ground.

Ripgut brome may injure livestock because the strong, rough awns penetrate the mouth, eyes or intestines.

5. B. hordeaceus L., *Sp. Pl.* 77 (1753).   soft brome

Annual, greyish green tufts, 10–100 cm. Leaf-sheath villous with long, soft hairs; upper sheaths sometimes glabrous. Ligule 0.5–1.5 mm, denticulate, abaxially short-hairy. Leaf-blade (1.5)–3–18 cm × (1)– 1.5–5 mm, flaccid, velvety with soft, short hairs; tip subacute. Culm erect or geniculate at base, internodes minutely pubescent. Panicle (1.5)–2.5–15 cm, erect, ± contracted; branches pubescent, bearing 1–few spikelets; sometimes panicle reduced to a single spikelet. Spikelets 1.5–3 cm, 4–12–(16)-flowered, ovate to oblong, greyish green or purplish. Glumes subequal, acute, membranous, ± closely pubescent; lower 4.5–7.5 mm, 3–5-nerved, ovate or oblong-lanceolate, upper 5.5–8.5 mm, 5–7-nerved, elliptic. Lemma 6.5–10.5 mm, 7-nerved, rounded, papery with prominent nerves, obovate to elliptic, shortly pubescent throughout or ± glabrous below, or rarely entirely glabrous or very rarely scabrid; margins hyaline, flat, overlapping, usually narrow; awn 3–8 mm. Palea slightly < lemma, keels with sparse short stiff hairs sometimes interspersed with minute hairs. Callus glabrous. Rachilla minutely pubescent. Anthers 0.5–1.5–(2.5) mm. Caryopsis 4.5–6.5 × 1.5–2 mm.

Naturalised.

N.; S.: throughout; K., Three Kings Is, Ch. Roadsides and waste land.

Indigenous to Europe and western Asia; now naturalised in temperate regions of both Hemispheres.

This sp. was long known worldwide as *B. mollis* L. Smith, P. *Watsonia* 6: 327–344 (1968), resolved the uncertainty involving the application of the earlier name *B. hordeaceus*. He also demonstrated that in Britain the hybrid *B.* ×*pseudothominii* P.M.Sm. (*B. hordeaceus*× *B. lepidus*) was comon in disturbed habitats and distinguished from *B. hordeaceus* by the usually glabrous, shorter lemmas, 6.5–8 mm, and by the caryopsis usually = palea. This hybrid is also found in Europe, western Asia, and North Africa. A high proportion of N.Z. material has short lemmas but only very rarely are the lemmas entirely glabrous; *B. hordeaceus* is treated here as an aggregate of *B. hordeaceus sens. strict.* and *B.* ×*pseudothominii*, and the description includes both.

6. B. inermis Leyss., *Fl. Halens.* 16 (1761).   smooth brome

Light green, perennial tufts to 100 cm, with creeping rhizomes. Leaf-sheath glabrous; basal sheaths sometimes reddish purple. Ligule 0.5–2 mm, truncate, ciliate. Leaf-blade 10–16 cm × 4–9 mm, glabrous; margins finely scabrid. Culm erect, internodes glabrous. Panicle 10–20 cm, open or contracted, erect; rachis slightly compressed, scabrid on angles, branches filiform, finely scabrid. Spikelets 15–25 mm, 3–5-flowered, parallel-sided, compressed, brownish green. Glumes unequal, subobtuse, glabrous, keel minutely scabrid near apex; lower 4–4.5 mm, 1-nerved, upper 6–6.5 mm, 3-nerved. Lemma 8–10 mm, 5–7-nerved, rounded, usually glabrous but sometimes finely pubescent at base, apex obtuse, hyaline; awn 0, or rarely 1–2 mm. Palea ≈ lemma. Callus glabrous. Rachilla finely scabrid or with minute hairs. Anthers 4.3–5.5 mm. Caryopsis *c.* 7.5 × 2 mm. Chasmogamous.

Naturalised from Eurasia.

N.: South Auckland (Ruapuke), Bay of Plenty (Whakatane); S.: Central Otago (near Luggate, near Alexandra). Roadsides.

AK 98319, from the H. B. Matthews Herbarium, without any details of locality, seems to match *B. inermis* but the awns are 2–4 mm.

Recorded as an impurity in imported seed by Cockayne, A. H. *J. Agric. N.Z. Dept Agric. 13*: 210 (1916).

7. B. lithobius Trin., *Linnaea 10*: 303 (1835).   Chilean brome

Soft-leaved, green to bluish green perennial tufts, 3–100–(165) cm, with spreading culms. Leaf-sheath rounded, with dense, fine, soft, horizontally projecting hairs. Ligule 1–1.5 mm, rounded or truncate, shortly denticulate, sometimes with scattered fine hairs at margins or abaxially; in uppermost culm leaves tapered and more lacerate. Leaf-blade 10–20 mm × 3–7 mm, villous; margins finely scabrid, tip subobtuse. Culm usually geniculate at base, usually ascending, to ± erect, internodes glabrous. Panicle 10–20–(35) cm, nodding; branches slender, finely scabrid, lower branches becoming strongly reflexed and drooping, upper branches ascending. Spikelets 2–3.5 cm, 5–7-flowered, linear-lanceolate, usually purple-suffused, sometimes light green. Glumes unequal, acute, with dense, short, soft, straight hairs throughout; lower 6–7.5 mm, 3–5-nerved, narrow-lanceolate, upper 8–9.5 mm, 5–7-nerved, ovate-lanceolate. Lemma 9–12 mm, 7–9-nerved, keeled, densely, shortly hairy throughout, evenly light green to usually purple-suffused; awn 4–6 mm. Palea = or slightly > lemma. Callus with minute hairs. Rachilla scabrid to minutely pubescent. Anthers 0.5–0.8 mm in cleistogamous flowers, *c.* 3 mm in chasmogamous flowers. Caryopsis 8.1–8.8 × 1.5–1.8 mm.

Naturalised from South America.

N.: southwards from Little Barrier Id and Auckland City; S.: Nelson, Marlborough, Canterbury; C. Waste ground.

Bromus lithobius was earlier equated in N.Z. with North American *B. breviaristatus* and *B. carinatus* and latterly with South American *B. fonkii,* but Forde, M. B. and Edgar, E. (1995 op. cit.) followed Matthei, O. *Gayana Bot. 43*: 47–110 (1986) who concluded that *B. fonkii* Phil. was a later synonym of *B. lithobius.*

Bromus lithobius occurs on low fertility lighter, summer dry soils, often on roadsides, but does not survive trampling or drought; two different forms have been collected in Canterbury, one prostrate and one upright with drooping panicles (A. V. Stewart, pers. comm.).

8. B. madritensis L., *Cent. Pl. 1*: 5 (1755).

Annuals, 30–55 cm, loosely tufted. Leaf-sheath softly short-pubescent; upper sheaths glabrous. Ligule 1–2 mm, denticulate. Leaf-blade 5–10 cm × 1.5–3 mm, short-hairy or ± glabrous; tip acute. Culm 5–45 cm, erect or ascending, internodes glabrous. Panicle 5–10 cm, erect, rather dense, fan-shaped, or nodding and slightly spreading; branches usually ≪ spikelets, filiform, minutely scabrid, tipped by 1–2 spikelets. Spikelets 3.5–6 cm, 5–8-flowered, oblong-lanceolate to later wedge-shaped and gaping, light green to purple-suffused. Glumes unequal, very narrow, acuminate, membranous, glabrous or with short to long hairs; lower 8–11 mm, 1-nerved, subulate, upper 10–14 mm, 3-nerved, linear-lanceolate. Lemma 12–20 mm, 7-nerved, rounded, narrowly oblong-lanceolate, scabrid or with short to long hairs, membranous with hyaline margins, narrowed above to 2 acute to acuminate lobes, 2–3 mm; awn *c.* 20 mm. Palea *c.* ¾ length of lemma, keels sparsely ciliate, interkeel glabrous or minutely hairy. Callus with minute hairs. Rachilla 2–2.5 mm, scabrid. Anthers 0.7–1 mm. Caryopsis *c.* 13 × 1.8 mm.

Naturalised from Mediterranean.

N.: Napier, Palmerston North and Wellington City. Waste ground and ballast.

9. B. racemosus L., *Sp. Pl.* ed. 2, 114 (1762).

Annuals or biennials, 12–50 cm, loosely tufted or consisting of a solitary culm. Leaf-sheath villous, with soft spreading hairs; upper sheaths with shorter, ±appressed hairs. Ligule 0.5–1 mm, denticulate. Leaf-blade 3–10 cm × 1.5–2.5 mm, pubescent; margins scabrid near acute tip. Culm 8–30 cm, erect, internodes pubescent. Panicle 4–6.5 cm, erect to later sometimes nodding, usually ±contracted; branches scabrid, some > spikelets. Spikelets 1.5–2 cm, 5–10-flowered, ovate- to elliptic-lanceolate, pale green. Glumes unequal, acute, firmly membranous to chartaceous, smooth, but scabrid on keel; lower 5–6.5 mm, 3–(5)-nerved, oblong-lanceolate, upper 6–8 mm, 5–7-nerved, ovate to elliptic. Lemma 7.5–9 mm, rather obscurely 7–9-nerved, rounded, firmly chartaceous, elliptic to ± obovate, very minutely scabrid especially in upper ½; margins flat, overlapping; apex very slightly emarginate; awn from just below apex, straight, 6–8 mm. Palea slightly < lemma, keels sparsely long-ciliate, flanks glabrous. Callus glabrous. Rachilla *c.* 1 mm, minutely scabrid. Anthers 1–1.5–(2) mm. Caryopsis not seen.

Naturalised from Europe.

N.: Auckland City, South Auckland (Pukeatua, near Rotorua), near Wellington; S.: Canterbury (Cheviot), southern Westland (Gillespies Beach). Sand dunes and river mouths and in waste land.

10. B. sitchensis Trin., *Mém. Acad. Imp. Sci. Saint-Pétersbourg VI*, Math. Phys. Nat. 2: 173 (1832).   Alaska brome

Robust, erect, perennial tufts, 90–150 cm, with wide, bright green leaves. Leaf-sheath glabrous apart from a few hairs at margins just below ligule, green, becoming brownish or purplish. Ligule 3–3.5 mm, denticulate, abaxially with long hairs near margins. Leaf-blade 18–25 cm × 5–12 mm, abaxially minutely scabrid on nerves, adaxially smooth, margins extremely finely scabrid, sometimes with a few scattered hairs. Culm erect, internodes smooth, but finely scabrid just below panicle. Panicle 20–30 cm, with erect, scabrid branches, lower branches later spreading, becoming reflexed. Spikelets 4–5 cm, 7–10-flowered, linear, green to purple-suffused, rachilla exposed at maturity. Glumes subequal, acute to acuminate, smooth, with scabrid keel; lower *c.* 12 mm, 3–5-nerved, upper *c.* 13 mm, 7–9-nerved. Lemma 14–16 mm, 7–9-nerved, keeled, very minutely scabrid to almost smooth, keel minutely scabrid, sometimes keel and flanks with a few scattered, minute hairs near base; awn 4.5–9 mm. Palea very slightly < lemma. Callus with minute hairs. Rachilla scabrid to minutely pubescent. Anthers 0.4–0.5 mm in cleistogamous flowers, 3–5 mm in chasmogamous flowers. Caryopsis *c.* 7 × 1.5 mm.

Naturalised from North America.

S.: Canterbury. Waste ground and gardens.

Bromus sitchensis frequently becomes heavily rust-infected whereas species in sect. *Ceratochloa* naturalised from South America are not susceptible (A. J. Healy *in litt.*).

11. B. stamineus Desv., *in* Gay, *Fl. Chil. 6*: 440 (1856).

spikey brome, grazing brome
Robust, light green, rather coarse-leaved perennial tufts, 60–200 cm, with spreading culms and heavy panicles. Leaf-sheath softly villous with long, silvery hairs. Ligule 1.5–2.5 mm, denticulate, abaxially glabrous. Leaf-blade 16–40 cm × 5–11 mm, with few, scattered, fine hairs; margins minutely scabrid, tip fine, acute. Culm ascending to ± erect, internodes glabrous. Panicle 8–30 cm, with stiffly erect to later spreading, finely scabrid branches. Spikelets 2.5–4 cm, 7–8-flowered, light green. Glumes unequal, acute; lower 7.5–10 mm, 5–7-nerved, very shortly appressed-pubescent, upper 11–12 mm, 7–9-nerved, usually more glabrous but finely scabrid on keel. Lemma 11.5–14.5 mm, 9–13-nerved, keeled, usually appressed scabrid-pubescent in upper ½, more glabrous and sometimes purple-suffused below; awn 3.5–9 mm. Palea very slightly < lemma. Callus with minute hairs. Rachilla scabrid. Anthers *c.* 0.5 mm in cleistogamous flowers. Caryopsis *c.* 9.5 × 2 mm.

Naturalised from South America.

S.: Marlborough (near Blenheim), Canterbury, North Otago; St. Roadsides and waste ground usually near the coast in areas of low rainfall.

The name *B. carinatus* was earlier applied in N.Z. to plants of *B. stamineus.*

Sterile plants with shorter awns and reddish purple panicles may occur (A. V. Stewart, pers. comm.).

12. B. sterilis L., *Sp. Pl.* 77 (1753).   barren brome

Annuals or biennials, 30–75 cm, loosely tufted or consisting of a solitary culm. Leaf-sheath densely short-pubescent; upper sheaths often ± glabrous. Ligule 1.5–2.5 mm, lacerate. Leaf-blade 6–25 cm × 2–6 mm, flaccid to firm, with short soft hairs; margins and acute tip finely scabrid. Culm (10)–25–55 cm, erect or geniculate-ascending, internodes glabrous or with minute fine hairs. Panicle 15–20 cm, nodding; branches usually >, or = spikelets, few, scabrid, widely spreading and drooping, tipped by 1–3 spikelets. Spikelets 3.5–6.5 cm, 4–14-flowered, oblong-lanceolate to later wedge-shaped and gaping, light green to purple-suffused. Glumes unequal, membranous, glabrous, but sparsely scabrid on keel and nerves above; lower 8–12.5 mm, 1–(3)-nerved, subulate, acuminate, upper 12.5–16 mm, 3-nerved, linear-lanceolate. Lemma 14–22 mm, 7-nerved, rounded, narrowly lanceolate, minutely scabrid, narrowed above to 2 acute to acuminate, hyaline, glabrous to finely scabrid lobes, 1–2 mm; awn 15–25–(30) mm. Palea *c.* ⅔ length of lemma, keels sparsely ciliate. Callus with minute hairs. Rachilla *c.* 2 mm, very sparsely minutely scabrid. Anthers 1–2 mm. Caryopsis 8.5–12 × 1–1.8 mm.

Naturalised.

N.: North Auckland (Mt Camel, near Houhora Harbour and Cuvier Id), southwards from Auckland City; S.: throughout except in wet western and southern areas; Ch. Waste ground.

Indigenous to Europe, south-western Asia and North Africa.

13. B. tectorum L., *Sp. Pl.* 77 (1753).   downy brome

Annuals, 9–55–(80) cm, loosely tufted or consisting of a solitary culm. Leaf-sheath densely, shortly pubescent, with some longer soft hairs; upper sheaths sometimes ± glabrous. Ligule 0.4–3.5 mm, lacerate. Leaf-blade 1–14 cm × 1.2–5.5 mm, shortly pubescent but with some longer hairs on margins towards base, tip subacute. Culm 6–35–(65) cm, slender, erect, internodes glabrous or minutely hairy. Panicle 3.5–25 cm, lax or contracted, nodding, secund; branches filiform, flexuous, spreading, short-hairy, lowermost bearing 4–8 spikelets. Spikelets 2.2–3.5 cm, 4–8-flowered, lanceolate to later gaping, light green to purple-suffused. Glumes very unequal, lanceolate, acute, membranous, usually pubescent with minute and longer hairs, or rarely glabrous; lower 6.5–9.5 mm, 1-nerved, upper 8.5–12 mm, 3-nerved. Lemma 10–16 mm, 7-nerved, rounded, lanceolate, later wedge-shaped, pubescent or sometimes only slightly scabrid, membranous, with shining hyaline margins tipped by 2 acute lobes, to *c.* 1 mm; awn 12.5–19 mm. Palea *c.* ¾ length of lemma, keels ± sparsely ciliate. Callus with extremely minute hairs. Rachilla 2–3 mm, with ± scattered, minute, appressed hairs. Anthers 0.6–0.9 mm. Caryopsis 6–8.5 × 0.9–1.5 mm.

Naturalised from Mediterranean.

N.: Auckland City (Onehunga); S.: Marlborough, Canterbury, Otago. Waste ground, modified tussock grasslands.

Specimens in which the lemmas are finely scabrid rather than pubescent have been collected in North Canterbury near Medbury and Waipara, and also at Bottle Lake, Christchurch.

Although Kirk, T. *T.N.Z.I. 3*: 148–161 (1871) listed *B. tectorum* for Auckland, no herbarium specimens were found by Esler, A. E. and Astridge, S. J. *N.Z. J. Bot. 25*: 523–537 (1987); it was recently collected in Onehunga, Auckland *P. J. de Lange* 6 Nov 1996 (AK 233145).

14. B. valdivianus Phil., *Linnaea 29*: 102 (1857).   stripey brome

Light green, rather slender, perennial tufts, 40–90 cm, with drooping panicles bearing noticeably bicoloured cream and green spikelets. Lower leaf-sheaths rounded, upper sheaths obviously keeled above, very densely villous. Ligule 1.5–2 mm, denticulate, abaxially often with long hairs near base and margins, or scabrid. Leaf-blade 12–22 cm × 3–7 mm, with ± scattered, fine, soft, straight hairs; margins very finely scabrid, tip subobtuse. Culm geniculate at base to erect, internodes glabrous, but slightly pubescent just below nodes. Panicle 12–30 cm, nodding, with slender, finely scabrid branches. Spikelets 1.5–3.5 cm, 5–9-flowered, elliptic-lanceolate to ± oblong, pale cream with noticeably green-tipped lemmas. Glumes very unequal, acute or subacute, with very short, appressed, silvery hairs, hyaline margins minutely scabrid; lower 4–4.5 mm, 3–(5)-nerved, elliptic-lanceolate, upper 6–6.5 mm, 7-nerved, more broadly ovate-lanceolate. Lemma 9.5–10 mm, 7–9-nerved, keeled, very shortly appressed-pubescent throughout, later papillose on basal cream to yellowish green portion, brighter green above, upper ⅓ of nerves reddish or bordered with green, margins and awn often reddish-tinged; awn 4.5–5 mm. Palea ≥ lemma. Callus with extremely minute hairs. Rachilla extremely minutely appressed-pubescent. Anthers 0.7 mm in cleistogamous flowers. Caryopsis 5.8–6.4 × 1.1–1.4 mm.

Naturalised from South America.

N.: Cuvier Id, The Noises, Auckland City (two early collections), scattered in South Auckland, Wanganui, Wellington (Ward Id); S.: Marlborough (near Blenheim), Canterbury (Christchurch); Ch. Roadsides, grassland, and garden weed.

This sp. was listed by Healy, A. J. *Standard Common Names for Weeds in New Zealand*, ed. 2, 134 (1984), as *B. coloratus* Steud.

Bromus valdivianus was collected by D. Petrie at Mt Eden, Auckland in 1910 and 1914 (WELT 71003, WELT 71007a,b), probably from Petrie's garden.

15. B. willdenowii Kunth, *Révis. Gram. 1*: 134 (1829).

prairie grass

Coarse green to yellowish green perennial tufts, 30–120 cm, with heavy panicles. Leaf-sheath keeled above, lower sheaths with fine, silky hairs, the hairs more scattered in upper sheaths. Ligule 3–5 mm, ± tapered, denticulate. Leaf-blade 15–40 cm × 8–15 mm, abaxially smooth, or minutely scabrid on ribs, adaxially with scattered fine hairs; margins finely scabrid, tip acute to acuminate. Culm stout, ascending to erect, internodes glabrous. Panicle 15–35 cm, erect or nodding above, pyramidal with scabrid branches in threes or pairs from the distant nodes; lower branches ascending, horizontal or sometimes drooping, upper branches ascending to erect. Spikelets 2.5–4 cm, 5–9-flowered, elliptic-oblong to ± cuneate, light green, sometimes purple-suffused. Glumes ± unequal, subacute to acute, keel minutely scabrid; lower 10–14 mm, 5–7-nerved, upper *c.* 15 mm, 9-nerved. Lemma 17–23 mm, 9–13-nerved, keeled, nerves scabrid, internerves smooth to slightly scabrid; awn 0.5–4–(5) mm. Palea *c.* ⅔ length of lemma or shorter. Callus with minute hairs. Rachilla scabrid or sparsely minutely pubescent. Anthers 0.5–1 mm in cleistogamous flowers, *c.* 4.5 mm in chasmogamous flowers. Caryopsis 7.1–7.2 × 1.5–2.2 mm. Fig. 12.

Naturalised from South America.

N.; S.: throughout; St.; K., Three Kings Is, Ch. Roadsides and waste ground.

Forde, M. B. and Edgar, E. (1995 op. cit.) supported Raven, P. H. *Brittonia 12*: 219–221 (1960) in treating prairie grass as *Bromus willdenowii* Kunth. They recognised a slender Andean segregate as a distinct species, *B. catharticus* Vahl; Raven also accorded this brome specific rank but referred it to *B. unioloides* Kunth. Other agrostologists favour a single taxon, *B. catharticus* Vahl: e.g., Matthei, O. *Gayana Bot. 43*: 47–110 (1986); Veldkamp, J. F., Eriks, M. and Smit, S. S. *Blumea 35*: 483–497 (1991); and Petersen, P. M. and Planchuelo, A. M. *Novon 8*: 53–60 (1998).

ζ**B. catharticus** Vahl   Slender, tufted, bluish green perennial, *c.* 60 cm, with narrowly branched panicles. Leaf-sheath with very fine, soft, white, straight, often dense hairs. Ligule 1–3 mm, rounded to tapered, denticulate. Leaf-blade 10–22 cm × 1.5–4.5 mm, softly hairy. Culm *c.* 40 cm, erect. Panicle 10–22 cm, with few, slender, scabrid branches. Spikelets 2–2.5 cm, 7–8-flowered, narrow-elliptic, light green or purple-suffused. Glumes ± unequal, finely scabrid on keel in upper ½, acute, sometimes mucronate; lower *c.* 10 mm, 5–7-nerved, upper 11–13 mm, 9-nerved. Lemma 12–13 mm, 9-nerved, keeled, very minutely scabrid; awn 1.5–2.5 mm. Palea *c.* ⅘ length of lemma. Callus with some sparse minute hairs. Rachilla minutely scabrid. Anthers in cleistogamous florets 0.6–0.8 mm. Caryopsis *c.* 7 × 1.5 cm. A South American species collected from waste land on roadsides in Alexandra, Central Otago *A. J. Healy 57/147* 25.1.1957 (CHR 98462) and *A. J. Healy 66/492* 18.10.1966 (CHR 174129).

ζ**B. erectus** Huds., upright brome   The single specimen consists of the upper part of a flowering culm. Panicle 12.5 cm, dense, erect; branches short, scabrid. Spikelets 3–3.5 cm, narrow-lanceolate. Glumes unequal; lower 1–3-nerved, subulate, upper 3-nerved. Lemma 7-nerved, faintly keeled, finely scabrid near apex; awn fine, 5–6 mm. Palea slightly < lemma. A Eurasian species recorded for Auckland by Kirk, T. *in* Hooker, J. D. *Handbk N.Z. Fl.* 763 (1867), and for Canterbury by Armstrong, J. F. *T.N.Z.I. 4*: 284 (1872). Only one specimen has been seen, *T. Kirk* Auckland, near St John's College, side of road, amongst gorse, 24.8.1888 (AK 98379).

ζ**B. japonicus** var. **vestitus** (Schrad.) Halácsy   Leaves villous. Culm to 25 cm, sparsely hairy below panicle. Panicle *c.* 15 cm, very lax, nodding. Spikelets 3–3.5 cm, 8–10-flowered, greyish- to silver-green. Glumes unequal, acute to acuminate, densely appressed-pubescent; lower 9–10.5 mm, 3-nerved, upper 11–12.5 mm, 5-nerved. Lemma 12–13 mm, 7–9-nerved, rounded, densely appressed-pubescent; awn 10–15 mm. Palea *c.* ¾ length of lemma, keel sparsely long-ciliate with shorter hairs interspersed, interkeel and flanks short-hairy. Anthers *c.* 1 mm. South African species once collected at Wellington, ballast heap, *T. Kirk* 1893 (CHR 3440, CHR 112334–5, WELT 70997, WELT 76367).

ζ**B. rubens** L.   Tufts 12–25 cm. Leaves shortly pubescent. Culm erect, internodes shortly pubescent below panicle. Panicle 3–8 cm, very dense, stiffly erect, fan-shaped; branches pubescent. Spikelets 2.5–3.5 cm. Glumes unequal, acute; lower 6–7 mm, 1-nerved, upper 10–11 mm, 3-nerved. Lemma *c.* 15 mm, rounded, minutely pubescent with some longer hairs, tipped by 2 finely acuminate lobes, *c.* 4 mm; awn 20–25 mm. Palea *c.* ¾ length of lemma. Anthers *c.* 0.6 mm. A Eurasian species once collected from Camp Russell, McKay's Crossing, 3 miles north of Paekakariki, a small patch on sandy ground near old vehicle park *A. J. Healy* 9.10.1944 (CHR 47231–4, CHR 47247).

ζ**B. secalinus** L., rye brome   Tufts slender, 60–70 cm. Lower leaf-sheaths glabrous. Panicle lax, 6–10 cm. Spikelets *c.* 2 cm, glabrous. Lemma rounded with tightly incurved margins. Palea ≈ lemma. A Eurasian and North African species once collected near Christchurch, "rare" [*c.* 1870] *J. B. Armstrong* (WELT 70987).

REPRODUCTIVE BIOLOGY   Flowers of *Bromus* may be chasmogamous or facultatively cleistogamous and both types may occur on the same plant; stamens and lodicules are shorter in cleistogamous flowers. Chasmogamous flowers are produced in spring in favourable conditions with ample soil moisture whereas cleistogamous flowers are induced by longer day-length and adverse conditions [Harlan, J. R. *Amer. J. Bot. 32*: 66–72 (1945); Langer, R. H. M. and Wilson, P. *New Phytol. 64*: 80–85 (1965)]. Some species are wholly chasmogamous or cleistogamous; McKone, M. J. *Amer. J. Bot. 72*: 1334–1338 (1985) found that *Bromus inermis* was self-sterile and chasmogamous and that *B. tectorum*, known to be highly self-fertile, produced mostly cleistogamous flowers.

11. BRACHYPODIEAE

Annuals or perennials. Leaf-sheath open or closed. Ligule membranous. Leaf-blade linear. Inflorescence a spike-like raceme. Spikelets very shortly pedicelled, terete to laterally compressed, several-flowered, florets ♀; disarticulation above glumes and between florets; rachilla prolonged. Glumes unequal, persistent. Lemma rounded, awned from apex. Palea 2-keeled, hyaline, pectinately ciliate. Lodicules 2, ciliate. Stamens 3. Ovary with terminal hairy appendage; styles 2. Caryopsis ellipsoid, hollowed on hilar face, adherent to lemma and palea; hilum linear, as long as caryopsis.

King, G. J. and Ingrouile, M. J. *New Phytol. 107*: 633–644 (1987), favour tribe Brachypodieae for *Brachypodium* as in Tutin, T. G. et al. (Eds) *Fl. Europaea 5*: 189–190 (1980) and Hilu, K. and Wright, K. W. *Taxon 31*: 9–36 (1982); but some authors place *Brachypodium* in tribe Hordeeae or in Poeae.

Molecular studies [Catalan, P. et al. *Bot. J. Linn. Soc. 117*: 263–280 (1995); see also Soreng, R. J. and Davis, J. I. *Bot. Rev. 64*: 1–88 (1998)] indicate that *Brachypodium* is best placed in its own tribe Brachypodieae.

BRACHYPODIUM P.Beauv., 1812

Perennials, rarely annuals, rhizomatous or tufted. Ligule membranous. Leaf-blade flat or involute. Raceme linear, loose, with rather few, shortly pedicelled spikelets borne singly and alternately on either side of non-articulating rachis. Spikelets 5–20-flowered, subterete to slightly laterally compressed; disarticulation above glumes and below each lemma. Glumes unequal, distinctly 3–9-nerved, < lemma, firmly membranous, lanceolate, rounded, obtuse to shortly awned. Lemma 7–9-nerved, herbaceous to firmly membranous, sometimes becoming coriaceous, rounded, entire, usually with a terminal awn. Palea ≥ lemma, keel ciliolate. Lodicules spathulate, ciliolate. Stamens 3. Ovary pubescent at apex; styles free. Caryopsis narrowly ellipsoid; embryo small; hilum ± linear, ≈ caryopsis. Fig. 13.

Type species: *B. pinnatum* (L.) P.Beauv.

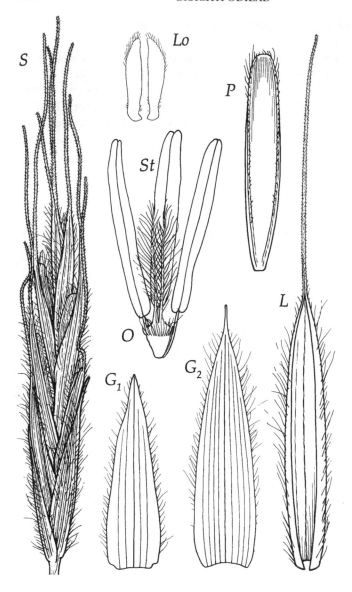

Fig. 13 *Brachypodium sylvaticum*
S, spikelet, × 5;   **G₁**, **G₂**, glumes, × 7;   **L**, lemma ventral view, × 7;   **P**, palea dorsal view, × 7;
Lo, lodicules, × 14;   **O**, ovary, styles and stigmas, × 14;   **St**, stamens, × 14.

16 spp., of temperate Eurasia and in mountains in the tropics from Mexico to Bolivia. Naturalised spp. 2; transient sp. 1.

1 Awn ≪ lemma; plant strongly rhizomatous .... **1. pinnatum**
   Awn ≥ lemma; plant tufted, sometimes very shortly rhizomatous *2*

2 Plants > 30 cm, perennial; anthers 3–4.5 mm .... **2. sylvaticum**
   Plants < 30 cm, annual; anthers < 1 mm .... **distachyon**ζ

1. B. pinnatum (L.) P.Beauv., *Ess. Agrost.* 101, 155 (1812).

Tufted perennials 30–65 cm, spreading by wiry, scale-covered rhizomes. Leaf-sheath rounded, glabrous or with long fine hairs, shortly pubescent at junction with leaf-blade. Ligule 1.4–1.6 mm, truncate, minutely ciliate. Leaf-blade 10–15 cm, *c.* 3.5 mm wide if flat, or inrolled and 1–1.5 mm diam., rather stiff, finely pointed, abaxially finely scabrid, adaxially with a few scattered hairs; margins finely scabrid. Culm erect, stiff, minutely hairy at, and near nodes, internodes otherwise glabrous. Raceme (3)–6–12 cm, spike-like, with spikelets partly overlapping and often projecting outwards; pedicels minutely pubescent, *c.* 1 mm. Spikelets 2–3 cm, 7–16-flowered. Glumes linear-lanceolate, acute, firm, glabrous; lower 4–4.6 mm, 3–5-nerved, upper *c.* 6 mm, 5–7-nerved. Lemma 8–8.5 mm, 5–7-nerved, oblong-lanceolate, subcoriaceous, abaxially smooth, or scabrid on nerves near apex, adaxially scabrid or with minute hairs just below awn base; awn 3.5–4.5 mm. Palea ≈ lemma, oblong, slightly emarginate to obtuse, keels and apex ciliate. Anthers 3.5–5 mm. Caryopsis not seen.

Naturalised.

N.: South Auckland (Aotea and Waimaori in Raglan County); S.: Otago (Dunedin, Tinwald Creek on Mt Pisa). Roadsides and waste ground.

Indigenous to Europe, south-western Asia and north-western Africa.

2. B. sylvaticum (Huds.) P.Beauv., *Ess. Agrost.* 101, 115 (1812).

Soft-leaved, densely tufted perennials 45–90 cm, occasionally very shortly rhizomatous. Leaf-sheath distinctly ribbed, becoming keeled above, usually with long soft hairs. Ligule 1.5–3.5 mm, truncate to rounded and ± erose, short- to long-ciliate, abaxially minutely pubescent and often with longer hairs. Leaf-blade 12–24 cm × 5.5–8.5 mm, flat, soft, loosely hairy, long-narrowed to fine tip. Culm erect or spreading, rather slender, hairy at, and near nodes, internodes otherwise glabrous. Racemes 8–18 cm, erect or nodding with spikelets slightly to closely overlapping on slender rachis; pedicels 0.5–1 mm, shortly pubescent. Spikelets 2–4.5 cm, 4–12-flowered. Glumes lanceolate, acute to acuminate, sometimes shortly awned, firm, pilose especially above near margin; lower 6–9 mm, 5–7-nerved, upper 7–11 mm, 7–(9)-nerved. Lemma 9–11.5 mm, 7-nerved, oblong-lanceolate, subcoriaceous, abaxially with short, stiff hairs, adaxially glabrous; awn 10–15 mm. Palea ≤ lemma, oblong, keels stiff-ciliate, apex truncate, shorter-ciliate. Anthers 2.8–3.8 mm. Caryopsis 6–7 × 0.6–1.2 mm. Fig. 13.

Naturalised.

N.: North and South Auckland (Whangarei, Auckland City, Tauranga), Hawkes Bay (Yeoman's Track, east foot of Ruahine Range, Tikokino); S.: Marlborough (near Kaikoura, and seaward Kaikoura Range). Lowland to montane in shade of trees or shrubs.

Indigenous to Europe, temperate Asia and north-western Africa.

ζ**B. distachyon** (L.) P.Beauv.   Plants annual, tufted, 20–28 cm. Leaf-blade with scattered long coarse hairs. Culm nodes densely, shortly pubescent. Racemes 4–7 cm, stiffly erect, of 2–4 ± laterally compressed, overlapping, shortly pedicelled spikelets. Spikelets 2.5–3.5 cm, 8–12-flowered. Glumes 6–7-nerved. Lemma *c.* 8.5 mm, 7-nerved, finely scabrid; awn to *c.* 15 mm. Palea ≈ lemma, keels with large, widely spaced prickle-teeth. Anthers 0.4 mm in cleistogamous flowers, 0.7–0.9 mm in chasmogamous flowers. A Mediterranean species once collected on Great Barrier Id, North Auckland *E. K. Cameron 5186* 2.1.1989 (AKU 21482).

12. HORDEEAE

Annuals or perennials. Leaf-sheath often bearing auricles at junction with leaf-blade. Ligule membranous. Leaf-blade flat, or involute-setaceous. Inflorescence spicate, a single bilateral raceme or quasi-raceme, with spikelets alternately arranged in two opposite rows and occurring singly or in groups of 2–3 (rarely more) at each node, broadside to rachis; rachis tough, or fragile and then disarticulating above each node leaving the spikelet attached to the internode below. Spikelets 1–many-flowered, laterally compressed, usually ♀ and sessile and all alike, or with the lateral spikelets of a triad pedicelled and ♂, or Ø and much reduced; disarticulation above glumes. Glumes persistent, < or narrower than lemma leaving much of it exposed, usually coriaceous, sometimes awn-like. Lemma coriaceous, 5–11-nerved, keeled or rounded on back, with straight or recurved awn from tip, or awnless. Palea 2-keeled. Lodicules 2, entire or ± toothed, ciliate. Stamens 3. Ovary with small membranous corona, and hispid hairs at apex; styles 2. Caryopsis ellipsoid, hollowed on hilar face; hilum linear, = caryopsis.

Löve, Á. *Feddes Repert. 95*: 425–521 (1984) is a conspectus of genera and species in the tribe, and Löve, Á. *Biol. Zentralbl. 101*: 199–212 (1982), is of their generic evolution.

Genomic symbols and/or designations follow Wang, R. R.-C. et al. *Proc. 2nd Int. Triticeae Symposium* 29–34 (1996). Although Seberg, O. *Pl. Syst. Evol. 166*: 159–171 (1989) and Seberg, O. and Petersen, G. *Bot. Rev. 64*: 372–417 (1998) disagree with almost all conclusions based on genome analysis we include designations, but for indigenous taxa only.

1 Spikelets solitary at each inflorescence node 2
  Spikelets 2 or more at each inflorescence node 8
2 Plants perennial *3*
  Plants annual 7
3 Plants rhizomatous *4*
  Plants caespitose, tufted 5
4 Rachis tough; glumes falling with spikelet .... **Elytrigia**
  Rachis fragile; glumes persistent .... **Thinopyrum**
5 Glumes absent or reduced and awn-like; spikelet edgewise to rachis ....
  .... **Stenostachys**
  Glumes always present, variously shaped; spikelet broadside to rachis 6
6 Spikelet sessile; inflorescence a spike .... **Elymus**
  Spikelet shortly pedicelled; inflorescence a spike-like raceme ....
  .... **Australopyrum**
7 Spikelet with 2 florets; lemma keel pectinately spinulose .... **Secale**
  Spikelet with 3–9 florets; lemma keel scabrid .... **Triticum**
8 Spikelets in groups of 2–4–(7), 4–5-flowered .... **Leymus**
  Spikelets in triads, 1-flowered *9*

9 Rachis fragile, readily fracturing beneath each triad .... **Critesion**
　Rachis tough, triads persistent .... **Hordeum**

AUSTRALOPYRUM (Tzvelev) Á.Löve, 1984

≡ *Agropyron* sect. *Australopyrum* Tzvelev, *Nov. Sist. Vyssh. Rast. 10*: 35 (1973) ≡ *Agropyron* subgen. *Kamptopyrum* Potztal, *Willdenowia 5*: 473 (1969).

Caespitose, extravaginally branching perennial. Leaf-sheath hairy. Auricles short clasping. Ligule short, lacerate. Leaf-blade hairy. Culm short stiff erect, internodes usually hairy. Inflorescence short, spike-like raceme of imbricate, pedicelled spikelets broadside to tough hairy rachis; rachis prolonged. Spikelets appressed or becoming reflexed, laterally compressed; disarticulation flat, and above glumes. Glumes persistent, rigid, shining or dull, indurate or foliaceous, patent or reflexed; awn-like; adaxially antrorsely hairy. Lemma 5–7-nerved, variously ornamented, terminated by sharp, shining or dull awn. Palea keels denticulate. Anthers 1–3 mm. Ovary with hairy apical corona 0.5–1 mm, apex with hispid hairs. Caryopsis linear, adhering to anthoecium. Chasmogamous or cleistogamous.

A diploid monogenomic genus in which spikelet and/or glume orientation alters as the inflorescence matures.

Type species: *A. pectinatum* (Labill.) Á.Löve

5 spp. of Australia, New Guinea and New Zealand. Endemic sp. 1; naturalised spp. 2.

Treatment follows Connor, H. E., Molloy, B. P. J. and Dawson, M. I. *N.Z. J. Bot. 31*: 1–10 (1993). The explicit exclusion of *Australopyrum* from the tribe by Kellogg, E. A. and Watson, L. *Bot. Rev. 59*: 273–343 (1993), and considered improbable by Connor, H. E. *N.Z. J. Bot. 32*: 125–154 (1994), is unsupported by any molecular study, [Kellogg, E. A., Appels, R. and Mason-Gamer, R. J. *Syst. Bot. 21*: 321–347 (1996)].

1 Abaxial leaf-blade glabrous except for prickle-teeth at apex; awn-like tip of lemma scabrid and dull .... **1. calcis**

    1a Leaf-blade minutely prickle-toothed; rachis glabrous .... subsp. **calcis**

    Leaf-blade pilose; rachis short hairy .... subsp. **optatum**

  Abaxial leaf-blade coarsely hirsute; awn-like tip of lemma smooth and shining *2*

2 Culm internodes hairy; glumes and florets reflexed from rachis .... **2. pectinatum**

  Culm internodes glabrous; glumes reflexed, florets appressed to rachis .... **3. retrofractum**

1. A. calcis Connor et Molloy, *N.Z. J. Bot. 31*: 2 (1993); Holotype: CHR 468517a! *B. P. J. Molloy* & *K. W. Ryan L2* Marlborough, Leatham River, near limestone quarry, 600 m, 29 Jan. 1991.   limestone wheatgrass

Slender, flat-leaved, extravaginally branching, stoloniferous, tuft-forming perennial with rough patent spikelets on a short inflorescence. Leaf-sheath 2.5–5 cm, glabrous or retrorsely short hairy, thin, becoming fibrous below, sometimes reddened, margin hyaline. Auricles 0.1–0.5 mm, glabrous, barely clasping. Ligule 0.6–1.25 mm, lacerate. Collar dark brown or purpled, glabrous. Leaf-blade 6–25 cm × 0.5–3.5 mm, soft, flat, thin, adaxially ribs hirsute or minutely scabrid; margins smooth below, prickle-teeth above; abaxially with conspicuous white midrib and 2 white lateral ribs, evident prickle-teeth on ribs above. Culm 17–60 cm, slender, geniculate and decumbent below; nodes conspicuous, swollen, dark brown or almost black; internodes glabrous or antrorsely lanate immediately below inflorescence. Inflorescence a spike-like raceme 3–13 cm, subtended by a small hair-fringed bract; spikelets 4–13, imbricate; rachis internodes short, sulcate, antrorsely toothed on margins, elsewhere antrorsely lanate or glabrous. Spikelets 10–14 mm, of 4–6 florets on short (0.3 mm) hairy pedicels, broadside to tough rachis, becoming patent. Glumes unequal, narrowly or broadly triangular-acute, < spikelet, nerves 3–5, evident white, keels eccentric, abaxially irregularly prickle-toothed, adaxially clothed in short white stiff antrorse hairs, margin very narrowly chartaceous, toothed; lower 2.5–4 mm, upper 3.7–5.5 mm. Lemma 9–11 mm, indefinite canaliculate awn 1.5–4 mm, 5-nerved, keel obscure below, abaxially with abundant antrorse, ± appressed prickle-teeth, margin very narrowly chartaceous, toothed. Palea 5.3–7 mm; keels shortly denticulate; apex shallowly bifid, ciliate; interkeel with abundant short, white, stiff, antrorse hairs; flanks with short hairs at apex and along margin. Callus 0.25–0.4 mm, surrounded adaxially by short stiff hairs; disarticulation oblique. Rachilla 1.0–1.25 mm, with stiff appressed white hairs. Lodicules 0.4–0.6 mm, ciliate. Anthers 2.4–2.75 mm. Gynoecium: ovary 0.8–1 mm; stigma-styles 1.75–2 mm. Caryopsis 3.5–4 mm, linear; embryo 0.75 mm; hilum linear, 3.5–3.75 mm. $2n = 14$.

Endemic.

The species is currently known from three areas: (1) inland Marlborough, Leatham River, (2) inland North Canterbury between Whitewater and Flock Hill streams, and (3) coastal North Canterbury at Mt Cass, and appears to be confined to limestone habitats. The ecology and conservation of *A. calcis* are described by Molloy, B. P. J. *N.Z. J. Bot. 32*: 37–51 (1994). It is a self-compatible and chasmogamous species.

subsp. **calcis**

Tall, stout. Leaf-blade adaxially long hairy; culm internodes glabrous except short hairy below inflorescence; rachis antrorsely short hairy; lemma abaxially with long stiff teeth.

S.: Marlborough (Leatham River). In light shade of hardwood scrub/ forest on limestone debris and derived soil; 600 m.

subsp. **optatum** Connor et Molloy, *N.Z. J. Bot. 31*: 4 (1993); Holotype: CHR 468526a! *B. P. J. Molloy FH 1a* Canterbury, Flock Hill, above cave, Cave Stream; 780 m, 23 Jan 1991.

Shorter and finer than typical subspecies. Leaf-blade shortly toothed on ribs; culm internodes glabrous; rachis glabrous but margins toothed; lemma abaxially shortly stiff toothed.

S.: North Canterbury: inland at headwaters of Waimakariri River, slopes of Mt Torlesse, Castle Hill, Flock Hill, Prebble Hill, and coastal at Mt Cass. Sparse grass-herb communities on skeletal to weakly developed soils at shaded bases of overhanging limestone tors; 500–900 m.

In cultivation the differences between subspecies are maintained.

2. A. pectinatum (Labill.) Á.Löve, *Feddes Repert.* 95: 443 (1984).

Compact coarsely hairy, extravaginally branched perennial with shining reflexed spikelets in a pectinate raceme. Leaf-sheath 3–4 cm, striate, minutely densely and retrorsely villous. Auricles 1.5 mm, glabrous, clasping. Ligule 0.5 mm, glabrous. Collar broad, with some long hairs. Leaf-blade 7 cm × 3 mm, flat, rough, abaxially, adaxially and on margin coarsely stiff hairy with scattered broad-based hairs to 0.5 mm, apex blunt. Culm to 35 cm, internodes antrorsely long stiff hairy below inflorescence, retrorsely soft hairy pubescent elsewhere, nodes evident, geniculate below; culm-sheath glabrous. Inflorescence a spike-like raceme 3–6 cm; spikelets 14–17; rachis internodes 2–5 mm, sulcate, antrorse long and short hairs mixed; pedicels *c.* 1 mm, shortly hairy, swollen and pulvinus-like diverting spikelets to reflexed habit. Spikelet 14 mm, of 5 florets on short sparsely hairy pedicel broadside to tough rachis, patent or becoming widely reflexed. Glumes ± equal, 5–7 mm, coriaceous, asymmetric, patent or retrorse, apex awn-like, sharp, glabrous, linear triangular acute, abaxially toothed on keel, adaxially very short hairy. Lemma 11 mm, 5–7-nerved, rounded, abaxially with scattered long stiff hairs, adaxially with short scattered hairs; awn stiff glabrous shining indurate ± 4 mm. Palea 5–6 mm, antrorse denticles (0.3 mm) spaced on keel, apex ciliate, interkeel and flanks adaxially with scattered short hairs. Callus 0.2 mm, glabrous, disarticulation oblique. Rachilla 1.5 mm, finely short stiff hairy. Lodicules 0.8 mm, ligulate, ciliate. Anthers 1.3–1.5 mm. Gynoecium: ovary 1 mm; stigma-styles 2.5–3.0 mm, stigmas sparsely branched. Caryopsis 2.8–3.25 mm, linear; embryo 0.8–1.2 mm; hilum linear = caryopsis.

Naturalised from Australia.

N.: Hawkes Bay (Wakarara); S.: Central Otago (Cromwell).

Other than WELT 76472 *D. Petrie* Cromwell, no plants referrable to *A. pectinatum* have been seen in South Island.

3. A. retrofractum (Vickery) Á.Löve, *Feddes Repert.* 95: 443 (1984).

Leaf-sheath 4–5 cm, coarsely pilose with long dense retrorse or mixed hairs, striate. Auricles 1.0 mm, clasping, glabrous. Ligule 1.5 mm, erose, glabrous. Collar white, margin sparsely hairy. Leaf-blade 3–6 cm ×

2 mm, rough, hairs coarse long (0.5 mm) and shorter, margin long-hairy, prow-tipped. Culm 10–25 cm, internodes glabrous, nodes evident some geniculate, inflorescence node with long soft hairs. Inflorescence a spikelike raceme 2–5 cm, spikelets *c*. 6, rachis retrorse hairy on margin, elsewhere hairs mixed, internodes *c*. 2 mm. Spikelets to 15 mm, florets 4–8, ± appressed broadside to rachis; pedicel 0.5 mm, hairy. Glumes ± unequal, 6–8 mm, nerves 1–3, indurate, shining, glabrous, becoming reflexed, swollen at base, awn-like above, linear triangular, asymmetrical, adaxially with short white hairs, margins chartaceous. Lemma 7.5–10 mm, narrowing into glabrous shining indurate awn ± 4 mm, keeled, ± glabrous or with scattered prickle-teeth, adaxially with short white hairs above, margin chartaceous. Palea 5–6 mm, denticles long and stout on upper keels, interkeel with few short white hairs. Callus 0.15 mm, broad, disarticulation flat. Rachilla 0.75 mm, finely short stiff hairy. Lodicules 0.7–0.8 mm, hairy. Anthers 0.7–1.5 mm. Gynoecium: ovary 1.0 mm; stigma-styles 2–2.5 mm. Caryopsis *c*. 4 mm; embryo 0.6 mm; hilum = caryopsis.

Naturalised from Australia.

N.: Hawkes Bay, *T. Kirk 1202*; S.: inland Marlborough; inland Waitaki and Central Otago basins.

See Jacobs, S. W. L. *Telopea 3*: 601–603 (1990) for discussion on the correctness of the binomial *A. retrofractum.*

Cockayne, A. H. *N.Z. J. Agric. 3*: 1–8 (1911) and Petrie, D. *N.Z. Dept Agric., Ind. Comm. Bull. No. 23* [i.e. 28] (1912) recorded this species from several localities in Central Otago, and the Waitaki Valley; in Marlborough Cockayne (1911 op. cit.) listed it as present on Upcott, Middlehurst and Molesworth stations in the Awatere Valley. In all these places it was said to be well established on badly depleted semi-arid areas. Modern records indicate that there are now no well established populations at any of those sites. Other than *T. Kirk 1202*, no plants referrable to *A. retrofractum* have been seen in North Island.

Plants referred to *A. pectinatum* by Hamilton, A. *List of the genera and species of New Zealand plants* (1899); Cheeseman, T. F. [*Man. N.Z. Fl.* (1906), *Man. N.Z. Fl.* (1925)]; Cockayne, A. H. (1911 op. cit.); Petrie (1912 op. cit.); Thomson, G. M. *The naturalization of animals and plants in N.Z.* (1922); and Allan, H. H. *N.Z. DSIR Bull. 49* (1936), *N.Z. DSIR Bull. 83* (1940), are *A. retrofractum.*

Hamilton, A. (1899 op. cit.) listed *Agropyron velutinum* (currently *Australopyrum*) as present in New Zealand, but we can find no supporting evidence.

GENOME ANALYSIS   *Australopyrum*, the only diploid triticoid in Australasia, to which Löve (1984 op. cit.) had allocated the genomic formula **W**, has contributed that genome to Australasian hexaploid species of *Elymus* [Torabinejad, J. and Mueller, R. J.   *Genome 36*: 147–151 (1993)]; that association is unrecorded in Kellogg et al. (1996 op. cit.).

CRITESION Raf., 1819

Tufted annuals or perennials. Leaf-sheath membranous to chartaceous, often with auricles. Ligule membranous. Leaf-blade flat. Inflorescence an oblong to linear, spike-like raceme, with spikelets in triads alternately at nodes of rachis, each triad consisting of a sessile to pedicelled ♀ central spikelet flanked by 2 pedicelled lateral spikelets, ♂ or ∅; rachis flattened; spikelets disarticulating as triads. Spikelets 1-flowered, dorsally compressed; rachilla prolonged. Glumes 2, anterior to spikelet, 1-nerved or faintly 3-nerved, linear, often awn-like, sometimes expanded at base, often with fine prickles in the 3 serrate lines. Lemma of central floret obscurely 5-nerved, rounded, awned or awnless; lemmas of lateral florets usually smaller to much reduced, sometimes larger. Palea ≈ lemma, 2-nerved, folded. Stamens 3. Lodicules 2. Ovary with hairy corona and long hispid hairs at apex. Caryopsis free, or adherent to anthoecium, elliptic, deeply furrowed adaxially; embryo *c.* ⅕–¼ caryopsis. Fig. 14.

Type species: *C. jubatum* (L.) Nevski

c. 35 spp. in temperate regions throughout the world. Naturalised spp. 6.

Formerly included in *Hordeum* as subgen. *Hordeastrum*; its segregation from *Hordeum* was completed by Löve, Á. (1984 op. cit.). Names in *Hordeum* are given for all taxa in New Zealand, as such treatment is still followed by some authorities e.g., Baden, C. and Bothmer, R. von   *Nord. J. Bot. 14*: 117–136 (1994); Jacobsen, N. and Bothmer, R. von   *Nord. J. Bot. 15*: 449–458 (1995); Bothmer, R. von et al.   *Nord. J. Bot. 9*: 1–10 (1989). The four common spp. in N.Z. are members of sect. *Trichostachys* (Dumort.) Á.Löve; all lie in the same part of the key below. Sect. *Critesion* is represented by *C. jubatum* localised in Central Otago, and sect. *Stenostachys* (Nevski) Á.Löve by *C. secalinum* known only from Hawkes Bay.

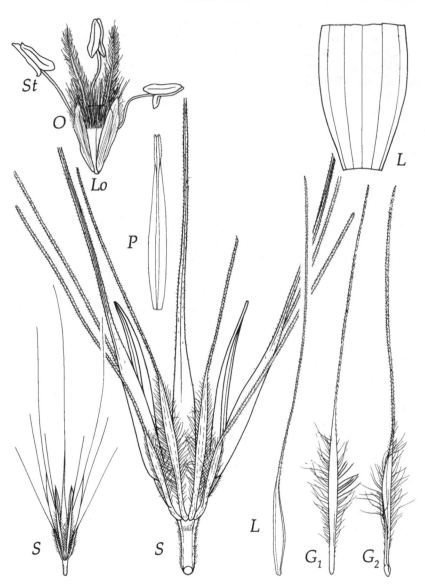

Fig. 14 *Critesion murinum* subsp. *murinum*
S, triad of spikelets, × 2, × 4; **G₁**, **G₂**, glumes, × 4; **L**, lemma, × 2, base × 8; **P**, palea dorsal view, × 4; **Lo**, lodicules, × 13; **O**, ovary, styles and stigmas, × 13; **St**, stamens, × 13.

Allen, F. C. and Popay, A. I., *in* Healy, A. J. *Identification of Weeds and Clovers* 280–291 (1982), provided an account of the vegetative and floral characters which distinguish the spp. of barley grasses in New Zealand.

1 Glumes of central spikelet in each triad long hairy on margins below, scabrid above; leaves with conspicuous clasping auricles *2*

Glumes hairless, scabrid throughout; leaves without auricles or rarely lowermost leaves with short auricles *3*

2 Rachilla prolongation of lateral florets (2.5)–3–5 mm, greenish; anthers of central floret 0.7–1.6 mm .... **5. murinum**

2a Lemmas of lateral spikelets ≈ lemma of central spikelet; awns of lateral lemmas < awn of central lemma .... subsp. **murinum**

Lemmas of lateral spikelets > lemma of central spikelet; awns of lateral lemmas overtopping awn of central lemma .... subsp. **leporinum**

Rachilla prolongation of lateral florets 1.5–2 mm, bright yellow-orange, dark brown when mature; anthers of central floret 0.3–0.6 mm .... **1. glaucum**

3 Plants annual; glumes of lateral spikelets slightly to strongly broadened at base *4*

Plants perennial; glumes of lateral spikelets awn-like throughout.............. *5*

4 One glume in each lateral spikelet usually slightly swollen at base but not winged; leaf-sheaths pilose with hairs 0.5–1 mm .... **2. hystrix**

One glume in each lateral spikelet obviously winged at base; leaf-sheaths glabrous, or puberulent with hairs < 0.3 mm .... **4. marinum**

5 Awn-like glumes 40–70 mm; sheath of uppermost leaf inflated.. **3. jubatum**

Awn-like glumes 10–13.5 mm; sheath of uppermost leaf appressed to culm .... **6. secalinum**

1. C. glaucum (Steud.) Á.Löve, *Taxon 29*: 350 (1980).

barley grass

Loosely tufted, bluish green annuals, 12–60 cm. Leaf-sheath straw-coloured, sometimes purplish, with auricles hyaline, glabrous, 1–3 mm, uppermost sheaths ± inflated. Ligule 0.4–0.6–(1.4) mm, truncate, denticulate and sparsely ciliate. Leaf-blade 2–12 cm × 1.5–5 mm, soft, flat, ± sparsely long hairy, abaxially scabrid on ribs near acute tip, margins finely scabrid. Culm 7.5–50 cm, erect or geniculate below, internodes glabrous. Racemes erect, very dense-flowered, green to purplish, sometimes dark brown when mature, 5–9.5 × 1.5–3 cm including the erect to somewhat spreading awns; rachis fragile, margins long hairy.

Central spikelet ♀, pedicelled, the pedicel glabrous, polished, *c.* 1 mm, lateral spikelets ♂ or more often ∅, pedicels shortly pubescent, *c.* 1 mm. Lateral spikelets: glumes unequal and dissimilar, outer glume 20–24 mm, very narrow, awn-like, scabrid, inner glume 14–18 mm, slightly widened near base and with long (0.75 mm) hairs on margins, narrow, scabrid and awn-like above; lemma 10–12 mm, elliptic-lanceolate, abaxially smooth but with sparse prickle-teeth on nerves above, adaxially densely villous, awn 15–25 mm; palea ≈ lemma, villous, keels scabrid near bifid scabrid tip; rachilla prolongation 1.5–2.5 mm, sturdy, sparsely hairy below, yellow to orange; lodicules 1 mm; anthers 1.5–2 mm, yellowish. Central spikelet: glumes equal, similar, 12–15 mm, widened near base, margins long (0.75 mm) hairy, awn-like and scabrid above; lemma 7–10 mm, elliptic-lanceolate, smooth except for sparse prickle-teeth abaxially on outer lateral nerves just below awn, whitish green, to purplish above, adaxially glabrous except for hairs at apex, awn 13–25 mm; palea ≈ lemma, folded interkeel abaxially sparsely pubescent, nerves scabrid near produced ciliate tip, adaxially smooth but ciliate-scabrid apically; rachilla prolongation 3–5 mm, slender, pale green to dark green, short stiff hairy. Lodicules 1.0 mm, glabrous. Anthers 0.3–0.6 mm. Gynoecium: ovary 0.6–1.0 mm; stigma-styles 0.5–1 mm. Caryopsis 4–6 × 1–2 mm; embryo 0.8 mm.

Naturalised from Mediterranean and south-western Asia.

S.: Canterbury (Amuri basin, Waipara, Yaldhurst, Christchurch), Otago (Waitaki, Taieri and Clutha basins). Pastures and stock camps. FL Sept–Feb.

The correct name in *Hordeum* is *H. glaucum* Steud., *Syn. Pl. Glum. 1*: 352 (1854).

2. C. hystrix (Roth) Á.Löve, *Feddes Repert. 95*: 440 (1984).

Mediterranean barley grass

Closely tufted, glaucous annuals, 7.5–40–(85) cm. Leaf-sheath green or straw-coloured, without auricles, lower covered with short, soft, horizontally spreading hairs, upper glabrous, the uppermost ± inflated. Ligule 0.2–0.6 mm, truncate, erose to ciliate. Leaf-blade 2–7–(14) cm × 1.5–2.5–(3.5) mm, flat, narrowed to acute tip, margins minutely scabrid, pilose or abaxially sometimes less densely hairy to

± glabrous. Culm 5–35–(80) cm, often decumbent, internodes glabrous. Raceme erect, glaucous, dense-flowered, ± oblong, 2–6 × 1–2 cm; rachis fragile but not disarticulating readily, margins finely ciliate. Central ♂ spikelet sessile; lateral spikelets Ø, on slender pedicels 0.5–1 mm. Glumes of all spikelets ± equal, 10–15 mm, scabrid, awn-like, glumes of lateral spikelets overtopping lemma awns, inner occasionally widened at base. Lemmas of lateral spikelets 2.5–5 mm, abaxially glabrous, with scabrid awn 2–7 mm, adaxially with scattered hairs; rachilla prolongation 1–3 mm. Lemma of central spikelet 5.5–7.5 mm, abaxially glabrous, scabrid awn 6–15 mm overtopping glumes, adaxially with scattered soft hairs above. Palea = lemma, folded, keels shortly excurrent and scabrid near tip, abaxially glabrous, adaxially with a few short hairs apically. Rachilla prolongation *c.* 2.5–4 mm. Lodicules 1 mm, glabrous. Anthers 0.75–1.3 mm. Gynoecium: ovary *c.* 0.5 mm; stigma-styles 1.2–1.5 mm. Caryopsis 3–5 × 1–1.5 mm; embryo 0.5 mm.

Naturalised from Mediterranean and south-western Asia.

N.: South Auckland, Hawkes Bay, Wellington (Raetihi); S.: Marlborough, North and Mid Canterbury, Central Otago, local in South Otago and Southland. In low-lying coastal saline flats, and inland at salty sites. FL (Sept)–Oct–Feb–(Mar).

Although *C. hystrix* was not recorded until 1968 it was first collected in 1941.

The correct name in *Hordeum* is *H. hystrix* Roth, *Catalecta Bot. 1*: 23 (1797), but Bothmer et al. (1989 op. cit.) interpret it as *H. marinum* subsp. *gussoneanum* (Parl.) Thell.

3. C. jubatum (L.) Nevski, *Fl. SSSR 2*: 721 (1934).

squirrel tail grass

Slender, loosely tufted, light green to purplish perennials, 30–75 cm, with plume-like racemes. Leaf-sheath without auricles, with scattered fine, soft, straight hairs, *c.* 0.5 mm, in basal leaves, but sheaths glabrous in upper cauline leaves, uppermost dilated, often enclosing base of raceme. Ligule 0.3–0.7 mm, truncate, ciliolate, abaxially smooth. Leaf-blade 5–20 cm × 1.5–3 mm, ribs fine and finely scabrid, sometimes sparsely hairy abaxially, margins finely scabrid, tip acicular. Culm

20–65 cm, erect or geniculate below, internodes glabrous. Raceme nodding, dense-flowered, 7–14 cm, including fine, light yellow-green to purplish awns, and 9–12 cm wide when awns are fully spread; rachis very fragile, margin conspicuously hairy. Central ♀ spikelet sessile; lateral spikelets ∅, pedicelled. Glumes of all spikelets about equal, very fine, awn-like, scabrid, spreading, 4–7 cm. Lemma of lateral spikelets much reduced, scabrid, 2–3.5 mm; awn ≪ glumes, or 0; palea vestigial, 0–0.5 mm. Lemma of central spikelet 4.8–6.2 mm, narrow-elliptic, abaxially finely scabrid above, adaxially with a few scattered hairs above, tapered to a very long, fine awn 3.5–7 cm. Palea ≈ lemma, ± folded, keels ciliate above near bifid apex, abaxially glabrous elsewhere, adaxially with scattered hairs above. Rachilla prolongation = lemma, sometimes recurved. Lodicules 0.7–0.6 mm, sparsely ciliate. Anthers (1)–1.2–1.8 mm. Gynoecium: ovary 0.5–0.7 mm; stigma-styles 1.25–2.0 mm. Caryopsis 2.8–3.5 × 1–1.3 mm; embryo 0.75 mm.

Naturalised from North America.

S.: Canterbury (Timaru), Central Otago. Scattered plants or in colonies on roadsides and in paddocks, often in association with *C. hystrix.* FL Dec–Mar.

The correct name in *Hordeum* is *H. jubatum* L., *Sp. Pl.* 85 (1753).

4. C. marinum (Huds.) Á.Löve, *Taxon 29*: 350 (1980).

salt barley grass

Loosely tufted, rather stiff, green to bluish green annuals, 10–30–(45) cm, sometimes reduced to a single shoot. Leaf-sheath without auricles, lower densely and minutely pubescent, upper glabrous, the uppermost inflated. Ligule 0.3–0.4 mm, ciliate. Leaf-blade 1.5–7 cm × 1–3.5 mm, flat, acute, extremely minutely pubescent. Culm (4)–8–25–(35) cm, erect, or geniculate below, internodes glabrous. Raceme erect, dense-flowered, green, purplish, or brownish, oblong or ovate, (0.8)–2–6–(8) × 1.5–2.5–(3) cm, including awns; rachis fragile, margins hairy. Central ♀ spikelet sessile; lateral spikelets ∅, on strong pedicels *c.* 1 mm. Lateral spikelets: glumes ± equal in length but dissimilar in shape, outer entirely awn-like, 13–22 mm, inner with a broad (0.5–1 mm) basal wing 3–5.5 mm, tipped by an awn 10–18 mm, both glumes overtopping lemma awns; lemma 3–6 mm, abaxially glabrous, adaxially with scattered hairs, tipped by scabrid awn 2.5–6 mm; palea = lemma, abaxially smooth but scabrid on keels near apex, adaxially minutely pubescent in upper ½; rachilla prolongation 1–1.5 mm. Central spikelet: glumes ± equal, similar, awn-like, scabrid, 13–20 mm, < awned lemma; lemma 6–7.5 mm, abaxially glabrous, adaxially minutely pubescent near base of scabrid awn 6–17 mm; palea = lemma, folded, scabrid on keels above, adaxially prickle-toothed near produced apex; rachilla prolongation 3.5–5 mm. Lodicules 1.0–1.25 mm, shortly ciliate, narrow. Anthers 1–1.5–(1.8) mm. Gynoecium: ovary *c.* 1.0 mm; stigma-styles 1.2–1.5 mm. Caryopsis 3–4–5 × 1–1.5 mm; embryo 0.5 mm.

Naturalised from Eurasia.

S.: Marlborough, Canterbury. In low-lying moist coastal flats, in association with *C. hystrix.* FL Oct–Feb.

The correct name in *Hordeum* is *H. marinum* Huds., *Fl. Angl.* ed. 2, *1*: 57 (1778).

6. C. murinum (L.) Á.Löve, *Taxon 29*: 350 (1980).   barley grass

Loosely tufted, light green annuals, 10–70 cm. Leaf-sheath green to light straw-coloured, with hyaline, clasping, glabrous auricles 1.5–3 mm, upper glabrous, uppermost ± inflated. Ligule 0.3–1 mm, rounded, minutely ciliate. Leaf-blade soft, flat, with scattered, long, fine hairs, ribs glabrous or very finely scabrid; margins very finely scabrid, tip acute. Culm 6–50 cm, erect or geniculate; internodes glabrous. Raceme very dense-flowered, oblong, 4–10 × 1–3.5 cm, including the erect or sometimes spreading awns; rachis fragile. Central ♀ spikelet sessile or pedicelled; lateral spikelets ♂ or ∅, pedicels 1–1.5 mm. Lateral spikelets: glumes unequal, ± dissimilar, outer very narrow, scabrid, ± awn-like throughout, inner slightly wider near base awn-like above; lemma adaxially copiously hairy, awned; palea = lemma, keels scabrid near minutely bifid ciliate apex; rachilla prolongation 2.5–3.5 mm; lodicules ciliate. Central spikelet: glumes equal, similar, 15–30 mm, widened near base, margins ciliate, awn-like and scabrid above; lemma 7–14 mm, lanceolate, long awned; palea = lemma, folded, keels scabrid above; rachilla prolongation 4–5 mm; lodicules

1.0–1.8 mm, ciliate. Gynoecium: ovary 0.6–1 mm; stigma-styles 1–2 mm. Caryopsis 5–9.5 × 1.5–2 mm; embryo 1.5 mm. Fig. 14. Plate 3C.

subsp. **murinum**

Lower leaf-sheaths glabrous, or with scattered, long, soft hairs. Leaf-blade 3–15 cm × 2–8 mm. Raceme erect or inclined, light green, sometimes with purplish or golden tinge; rachis margins ciliate. Central ♀ spikelet sessile or on pedicel *c.* 0.3 mm; lateral spikelets shortly pedicelled. Lateral spikelets: outer glumes 15–32 mm, inner 13–25 mm, sparsely hairy on the curved margin near base; lemma abaxially slightly ciliate on margins near base and scabrid above, awn 8.5–40 mm; palea adaxially copiously long hairy; lodicules 1.5–1.75 mm; anthers 0.4–1.3 mm. Central spikelet: lemma abaxially glabrous but scabrid on nerves near apex, adaxially short-ciliate towards apex, awn 15–45 mm, overtopping all glumes and awns; palea folded, keels scabrid near produced, minutely bifid, ciliate apex, interkeel glabrous. Anthers 0.7–1.6 mm.

Naturalised from Europe.

N.; S.: throughout; Ch. In dry districts in lowland and hill country pasture especially on sheep-camps, in crops, in waste land and on roadsides. FL (Sept)–Oct–May–(Jun).

The correct name in *Hordeum* is *H. murinum* L., *Sp. Pl.* 85 (1753) subsp. *murinum.*

subsp. **leporinum** (Link) Á.Löve, *Taxon 29*: 350 (1980).

barley grass

Lower leaf-sheaths with ± dense, long, soft hairs. Leaf-blade 2–22 cm × 1.5–8 mm. Racemes erect, green to purplish, sometimes dark brown at maturity; rachis margins long hairy. Central spikelet ♀, on glabrous pedicel 0.5–1 mm; lateral spikelets on shortly pubescent pedicels, 1–1.5 mm. Lateral spikelets: outer glumes 17–24 mm, inner 15–18 mm and with long hairy margins near base; lemma abaxially glabrous and straw-coloured below, green and more distinctly nerved and scabrid on nerves above, awn 11–27 mm; palea adaxially villous; lodicules 2.5 mm, hairy; anthers 0.8–1.6 mm. Central spikelet: lemma smooth but sparsely scabrid below awn, pale whitish green, greenish to purplish above, awn 18–30 mm, ± reaching to top of awns of lateral spikelets; palea keels with a few prickle-teeth near rounded, produced or bifid, ciliate apex, interkeel long hairy, adaxially short-ciliate. Anthers 0.8–1 mm.

Naturalised from Mediterranean and Asia.

N.: throughout except in Taranaki; S.: throughout except in Westland, Fiordland and Southland, and rare in Otago; St.; K. In depleted tussock grassland, pasture, waste land and roadsides, occasionally in cereal crops.

The correct name in *Hordeum* is *H. murinum* subsp. *leporinum* (Link) Arcang., *Comp. Fl. Ital.* 805 (1882).

7. C. secalinum (Schreb.) Á.Löve, *Taxon 29*: 350 (1980).

meadow barley grass

Loosely tufted light green or greyish green perennials, to 50 cm. Leaf-sheath pale greyish brown or straw-coloured, lower with soft, shaggy long hairs and narrow auricles *c.* 0.5 mm, upper glabrous, not inflated, often without auricles. Ligule 0.4–0.6 mm, truncate, denticulate to ciliate, abaxially with sparse prickle-teeth. Leaf-blade 4–10 cm × 1–3.5 mm, abaxially very minutely scabrid, adaxially scabrid on ribs, with scattered hairs, tip acute. Culm 25–45 cm, erect or geniculate, slender, internodes glabrous. Raceme ± erect, dense-flowered, narrow, 2–4 × *c.* 0.8 cm, including the scarcely spreading awns, pale greyish green; rachis fragile, margins ciliate. Central ♂ spikelet sessile; lateral spikelets ∅ and much reduced, pedicels slender, *c.* 1 mm. Glumes of all spikelets about equal, fine, awn-like, scabrid, 10–13.5 mm. Lemmas of lateral spikelets 2.5–5.25 mm, shortly stiff hairy below the fine scabrid awn 1.5–3 mm; palea vestigial *c.* 0.8 mm, or = lemma. Lemma of central spikelet 7–8 mm, oblong-lanceolate, abaxially glabrous, adaxially villous, tapered above to fine scabrid awn 7.5–11 mm, overtopping glumes. Palea ≈ lemma, folded, minutely bifid, keels glabrous, but interkeel short stiff hairy near ciliate apex; rachilla prolongation 3–3.5 mm. Lodicules 1.5 mm, hairy, narrow. Anthers *c.* 3 mm. Gynoecium: ovary 1.0 mm; stigma-styles 1.5 mm. Caryopsis 4–6 × 1–1.3 mm; embryo 1.0 mm.

Naturalised from western and southern Europe and north-western Africa.

N.: only known from Hawkes Bay (near Porangahau); S.: one early collection from near Christchurch. In wetter coastal paddocks. FL Nov–Feb.

The correct name in *Hordeum* is *H. secalinum* Schreb., *Spicil. Fl. Lips.* 148 (1771).

ELYMUS L., 1753

Tussocky or open tall, tufted, coarse or small grasses, with intra- or extravaginal branching. Spikelets sessile, laterally compressed, solitary (in New Zealand species), broadside to tough rachis, appressed to or diverging from it; rachis not prolonged; spikelet imbricate of many long-awned to mucronate florets; disarticulating above glumes. Glumes firm, long triangular, margins membranous, 1–3–5 parallel nerves, obtuse to shortly and distinctly awned. Lemma rounded, smooth to somewhat scabrid, coriaceous, 5-nerved, keeled above. Lodicules ciliate. Ovary with apical corona and hispid hairs. Caryopsis adherent to palea.

Type species: *E. sibiricus* L.

A cosmopolitan genus of 20 or 150 species depending on generic interpretation. Endemic spp. 6, indigenous sp. 1; naturalised sp. 1.

Treatment follows Connor, H. E. *N.Z. J. Bot. 32*: 125–154 (1994), derived in part from Connor, H. E. *N.Z. J. Sci. Tech. 35B*: 315–343 (1954) and Löve, Á. and Connor, H. E. *N.Z. J. Bot. 20*: 169–186 (1982). All species are self-compatible and chasmogamous but cleistogamy may occur in *E. falcis, E. multiflorus* and *E. tenuis*.

1 Long awned florets in spikelets appressed to rachis of ascending or nodding inflorescences *2*

  Long awned florets in spikelets divergent from rachis of erect spikes or appressed to rachis of inflorescences trailing on the ground, or florets short-awned *5*

2 Leaf-blades flat, margins involute; auricles long, clasping *3*

  Leaf-blades involute; auricles minute or absent, sheath margins becoming apical lobes *4*

3 Leaf-blades usually glaucous and glabrous abaxially; awn recurved; palea apex pointed, bifid .... **7. solandri**

    Leaf-blades coarsely hirsute; awn straight, rarely recurved; palea apex blunt, truncate to retuse .... **5. rectisetus**

4 Leaf-blades short, variously falcate, sinuous, or straight, abaxially glaucous usually with scattered erect long (1 mm) hairs; culms prostrate to ascending .... **3. falcis**

    Leaf-blades long thin and strict, abaxially ribbed, glabrous and glaucous, adaxially with dense weft of hairs at base; culms erect .... **6. sacandros**

5 Awn long, straight or recurved, 2–3 × lemma *6*

    Awn short, straight, up to 1 × lemma *7*

6 Culm trailing along ground, uppermost internode greatly elongating after anthesis; spikelets appressed to rachis; anthers 2–3 mm, yellow .... **8. tenuis**

    Culm erect, uppermost internode short; spikelets widely divergent from rachis; anthers 4–9 mm, purple to yellow .... **1. apricus**

7 Inflorescence of long widely spaced many-flowered spikelets; palea apex blunt, truncate to retuse; lemma apex bifid if mucronate ... **4. multiflorus**

    Inflorescence of small close-set, few-flowered shining spikelets; palea apex pointed, bifid; lemma apex bifid .... **2. enysii**

1. E. apricus Á.Löve et Connor, *N.Z. J. Bot. 20*: 182 (1982); Holotype: CHR 370822! *H. E. Connor* hillsides above Roxburgh town, Central Otago, 10 February 1947.

Erect glaucous tussock from a narrow base, with intravaginal branching though sometimes rhizomatous; spikelets widely divergent from rachis, and anthesis occurring after complete spike emergence reveals long purple anthers. Leaf-sheath 7–10 cm, keeled, striate, becoming fibrous, glabrous or occasionally sparsely hairy or pubescent mixed with hairs 1 mm; margins chartaceous. Ligule 0.3–0.5 mm, ciliate. Collar thickened short stiff hairy, occasionally glabrous, margin hairs 1–1.5 mm. Auricles 0.3–0.5 mm, clasping, or absent, long (1–2 mm) hairs occasional. Leaf-blade 20–30 cm × 2–4 mm, glaucous, flat, ribbed, sometimes involute, abaxially glabrous, occasionally with hairs to 1 mm, adaxially densely hairy at base; margin prickle-toothed or becoming so, occasionally with hairs to 1 mm below. Culm 50–100 cm, erect, nodes conspicuous black to red brown, internodes glabrous. Inflorescence 18–25 cm, stiff, of 3–7 widely divergent spikelets; margin sparsely antrorsely toothed or short hairy. Spikelets 30–50 mm, of 6–12 florets occasionally more. Glumes ± equal, 5–10 mm, 3-nerved, acute or becoming shortly awned, margin chartaceous and ciliate. Lemma 10–14 mm, glabrous except for prickle-teeth above and on margin below, keeled above, apex occasionally bifid, awn 22–45 mm, ± recurved or straight. Palea 6–13 mm, apex bifid. Rachilla 2–3 mm, short stiff hairy. Callus 0.75 mm, incompletely and shortly bearded. Lodicules 1–2.5 mm. Anthers 4–9 mm, usually purple often yellow. Gynoecium: ovary 2 mm; stigma-styles 2–3 mm. Caryopsis 7–10 mm; embryo 1.6–2 mm. $2n = 42$.

Endemic.

S: inland basins of Central Otago. Open dry short-tussock grassland; 150–500 m.

2. E. enysii (Kirk) Á.Löve et Connor, *N.Z. J. Bot. 20:* 183 (1982) ≡ *Agropyron enysii* Kirk, *T.N.Z.I. 27:* 359 (1895) ≡ *Asprella aristata* Petrie, *T.N.Z.I. 26:* 272 (1894) non Kuntze (1891) ≡ *Agropyron aristatum* Cheeseman, *Illus. Fl. N.Z. 2:* t. 234 (1914) *nom. superfl.*; Lectotype: WELT 68390! *D. Petrie* Mt Torlesse, Canterbury Alps, *c.* 3600 ft, Jan 1893 (designated by Connor 1994 op. cit. p. 131).

Small tufted or open, bronze flat-leaved grass rooting and shooting at nodes on stolons; culm slender and drooping, inflorescence often hidden in tussocks; branching extravaginal. Leaf-sheath 6–8 cm, long hairs retrorse or erect or pubescent; lobed at apex. Ligule 0.5–0.75 mm, erose, sometimes triangular. Collar slightly curved and thickened. Auricles small (0.75 mm), or minute. Leaf-blade 20 cm × 2–4 mm, flat, soft, abaxially densely retrorsely long hairy on and between ribs or glabrous, adaxially antrorsely prickle-toothed together with long hairs; margin prickle-toothed. Culm 25–80 cm, nodes few, ± geniculate below, internodes smooth or very slightly scabrid; upper internode elongating after flowering, sometimes reaching 60 cm. Inflorescence 5–13 cm, compact, of 10–18 spikelets. Spikelets 10–16 mm, shining, of 3–4 florets. Glumes ± equal, narrow, 6–9 mm, 1–3-nerved, scabrid, asymmetrically keeled, becoming awn-like, margin prickle-toothed. Lemma 7–10 mm, > palea, glaucous, glabrous except for prickle-teeth on nerves above; apex bifid; awn 1–3.5 mm; conspicuously indented at base. Palea 6–9 mm, keels toothed almost to base, apex shallowly bifid. Rachilla 1–2 mm, minutely stiff hairy. Callus 0.25 mm blunt, surrounded by minute hairs; disarticulation almost flat. Lodicules 1–1.25 mm, few apical hairs. Anthers 1.8–3 mm, yellow. Gynoecium: ovary 1.5 mm; stigma-styles 2–2.25 mm. Caryopsis 6 mm; embryo 1 mm. $2n = 28$.

Endemic.

S.: Nelson, eastern Marlborough and Canterbury east of Main Divide. Flushed places in tussock grasslands; 850–1675 m.

3. E. falcis Connor, *N.Z. J. Bot. 32*: 132 (1994); Holotype: CHR 19687! *V. D. Zotov* Mt Edwards, Canterbury, 5.1.1938.

Small loosely hummock-forming low grass with short glaucous falcate or sinuous leaf-blades, and prostrate to ascending culms; branching intra- and extravaginal sometimes forming short stolons. Leaf-sheath 2–5 cm, striate, retrorsely pubescent or glabrous, sometimes with a few long hairs; forming a short lobe at apex. Ligule 0.3–0.5 mm, erose or finely toothed. Collar thickened, curved, glabrous or short hairy, margin hairs to 1.5 mm. Auricles absent or minute, with 1–2 long hairs. Leaf-blade 5–15 cm × 0.5–0.7 mm diam., ± terete, involute, keeled but only faintly ribbed, falcate, sinuous or irregularly curled or ± straight, abaxially glaucous with scattered to abundant stiff erect or somewhat retrorse hairs 0.5–1 mm, becoming glabrous, or pubescent below becoming glabrous, or glabrous throughout; adaxially with abundant hairs or shorter hairs and/or prickle-teeth; margin prickle-toothed or with hairs to 0.5 mm. Culms 10–35 cm, prostrate to ascending; internodes smooth, glaucous. Inflorescence 2–10 cm, of 1–4 spikelets; rachis margin often glabrous. Spikelets 40–50 mm, of 4–6 shining florets. Glumes ± equal, 4–9 mm, keeled, 3-nerved, sometimes becoming awned, glabrous, margin chartaceous. Lemma smooth except for prickle-teeth on keel above and on margins below extending into reflexed awn 30–50 mm; pruinose. Palea 7–10 mm, apex pointed, bifid. Rachilla 1.5–2 mm, short stiff hairy. Callus 0.75 mm, very shortly bearded, hairs ≤ callus. Lodicules 1.2–1.4 mm. Anthers 2.4–2.5 mm, yellow. Gynoecium: ovary 1.5 mm; stigma-styles 2.5 mm. Caryopsis 6–6.5 mm; embryo 1 mm. $2n = 42$.

Endemic.

S.: inland Waimakariri, Ashburton, Waitaki and Taieri Basins. Dry, open, short tussock grassland, river beds, and rocky sites; 490–1250 m.

This species is based on Group Tekapo II of *Agropyron scabrum* as discussed in Connor (1954 op. cit.); it is the *Elymus* of dry open river beds and outwash plains.

4. E. multiflorus (Hook.f.) Á.Löve et Connor, *N.Z. J. Bot. 20*: 183 (1982) ≡ *Triticum multiflorum* Hook.f., *Fl. N.Z. 1*: 311 (1853) ≡ *Agropyron multiflorum* (Hook.f.) Cheeseman, *Man. N.Z. Fl.* 921 (1906) ≡ *A. multiflorum* (Hook.f.) Cheeseman var. *multiflorum* (autonym, Hackel 1906 op. cit. p. 921) ≡ *A. kirkii* Zotov, *T.R.S.N.Z. 73*: 233 (1943) *nom. superfl.*; Lectotype: BM! *J. Banks & D. Solander* Mercury Bay, 1769 (left and centre designated by Connor 1994 op. cit. p. 134). = *A. multiflorum* var. *longisetum* Hack., *in* Cheeseman *Man. N.Z. Fl.* 922 (1906) ≡ *A. kirkii* var. *longisetum* (Hack.) Zotov, *T.R.S.N.Z. 73*: 233 (1943) *nom. superfl.* ≡ *Elymus multiflorus* var. *longisetus* (Hack.) Á.Löve et Connor, *N.Z. J. Bot. 20*: 183 (1982); Holotype: W *n.v.*, *T. F. Cheeseman* Matamata, Thames Valley (No 1506 to Hackel, AK!).

   var. **multiflorus**

Tufted many-noded stoloniferous open grass with scabrid wide green leaf-blades often shorter than smooth culms; branching extravaginal. Leaf-sheath 6–10 mm, striate, margin chartaceous, glabrous or retrorsely short hairy. Ligule 0.2–0.5 mm, erose. Collar curved, thickened, infrequently bearing long hairs. Auricles 0.5–1.5 mm, clasping, glabrous or bearing long hairs. Leaf-blade 10–20 cm × 2–4 mm, flat usually bright green but sometimes glaucous, abaxially shortly antrorsely scabrid or occasionally glabrous, adaxially with antrorse short hairs or prickle-teeth on ribs or sometimes with longer hairs; margin shortly prickle-toothed. Culm 30–60 cm, often stout, internodes scabrid below inflorescence, otherwise smooth, nodes ± geniculate below; culm leaves often very broad. Inflorescence 10–25 cm, of 6–15 spikelets occasionally paired at base, rachis convex surface often sparsely prickle-toothed. Spikelets 14–25 mm, of 7–12 florets. Glumes ± equal, 5–9 mm, 3–5-nerved, eccentrically keeled, often broad, margins chartaceous, ciliate; keel and nerves prickle-toothed above elsewhere smooth, sometimes produced into short awn. Lemma usually with small scattered prickle-teeth above, glabrous below except on margin, apex often bifid when mucronate, awn absent or ≈ lemma, variable in one spikelet. Palea 9–12 mm, apex truncate or retuse, ciliate. Rachilla 1–2.5 mm, hairy. Callus 0.75–1 mm, incompletely shortly bearded. Lodicules 0.8–1.25 mm, simple, linear. Anthers 3–5 mm, purple or yellow. Gynoecium: ovary 1.5–2 mm; stigma-styles 1.75 mm. Caryopsis 7–10 mm; embryo 1 mm. $2n = 42$.

Indigenous.

N.: offshore islands and coastal throughout, inland in Hawkes Bay; S.: coastal from Nelson to Banks Peninsula; Three Kings Is. Cliffs and rocks of various substrates, frequently limestone, in coastal and inland areas; sea level to 600 m.

Also present in eastern Australia.

The Norfolk Island endemic *E. multiflorus* var. *kingianus* (Endl.) Connor can be distinguished from New Zealand plants by the multiplicity of small prickle-teeth on the lemmas and on the glumes. Awns, approaching 17 mm, are strict and about as long as in some *E. multiflorus* × *E. solandri* hybrids, although the ratio awn:lemma at 1.3–1.6 is less than in hybrids. The truncate palea apex is manifestly ciliate.

5. E. rectisetus (Nees) Á.Löve et Connor, *N.Z.J. Bot. 20*: 183 (1982).

Open coarsely hairy grass with long drooping inflorescences of many straight-awned spikelets; branching usually intravaginal but often extravaginal. Leaf-sheath to 10 cm, hirsute with erect or retrorse hairs 0.5–0.75 mm. Ligule 0.1 mm, truncate. Collar recurved, glabrous except for 1–2 very long hairs especially near auricles. Auricles 0.5–1 mm, clasping, glabrous. Leaf-blade 18 cm × 1–4 mm, everywhere coarsely hirsute, occasionally rough with prickle-teeth. Culm to 1.5 m, decumbent below, drooping above at maturity, internodes usually smooth but occasionally rough, irregularly hairy below lower nodes. Inflorescence 10–20 cm, up to 10 spikelets, convex surfaces fairly scabrid. Spikelets 40–55 mm, of 5–10 florets often purple flecked. Glumes unequal, 4–10 mm, 3–5-nerved, irregularly long triangular, centrally keeled, margins membranous, ciliate at apex, becoming awned and sometimes scabrid above. Lemma smooth below becoming prickle-toothed above, keeled above continuing into very scabrid straight awn 30–50 mm. Palea 9–11 mm, apex truncate to retuse. Rachilla 1.5–2.5 mm, hairs becoming long at apex. Callus 0.75 mm, surrounded by hairs > callus. Lodicules 1–1.25 mm. Anthers 1.8–2.3 mm, yellow or purple streaked. Gynoecium: ovary 1.25 mm; stigma-styles to 2.5 mm. Caryopsis 6–6.5 mm; embryo 1–1.2 mm.

Naturalised from Australia.

N.: throughout except unrecorded in the Waikato; S.: throughout except rare in Westland, absent from Fiordland and much of Southland. In poor pastures, waste places and roadsides; sea level to 1250 m; unacceptable to livestock.

Hair, J. B., *Heredity 10*: 129–160 (1956) described, as unique in the tribe, pseudogamous apospory in *E. rectisetus* (as *Agropyron scabrum*). Recent extensions of this work are Crane, C. F. and Carman, J. G. *Amer. J. Bot.* 74: 477–496 (1987); Carman, J. G., Crane, C. F. and Rieva-Lizarazu, O. *Crop Science 31*: 1527–1532 (1991); Peel, M. D. et al. *Crop Science 37*: 717–723 (1997).

The type specimen of *Festuca scabra* Labill. (1805) non Vahl ≡ *Triticum scabrum* R.Br. (1810) ≡ *Agropyron scabrum* (R.Br.) P.Beauv. (1812) ≡ *Anthosachne scabra* (R.Br.) Nevski (1934) ≡ *Elymus scaber* (R.Br.) Á.Löve (1984) ≡ *Roegneria scabra* (R.Br.) J.L.Yang et C.Yen (1991) is at Florence (FI!). The awned lemmas are shorter than in *E. rectisetus*, c. 25 mm, and the awn:lemma ratio is *c.* 2. The palea apex is described by Labillardière [*Nov. Holl. Pl. 1*: 26 (1805)] as emarginate. For an account of this and other Australian taxa see Murphy, M. A. and Jones, C .E. *Aust. Syst. Bot. 12*: 593–604 (1999).

6. E. sacandros Connor, *N.Z. J. Bot. 32*: 138 (1994); Holotype: CHR 279320! *A. P. Druce* Isolation C[ree]k; NW of Ben More, Marlborough, 800 ft., Dec. 1975.

Tufted or open branched with long, stout, many-noded shoots; branching intra- and extravaginal and shooting and rooting at nodes; prophyll keels hairy. Leaf-sheath 3–12 cm, abaxially striate, glabrous, adaxially with minute prickles, margins membranous often terminated by small lobe. Ligule 0.2–0.3 mm, united to leaf-sheath margins. Collar thickened, curved, usually sparsely villous. Auricles usually absent but often 1 small (0.25 mm) or 1 large (1 mm) and clasping, occasionally with 1–2 long hairs. Leaf-blade 10–80 cm × 0.5–0.7 mm diam., filiform, abaxially glaucous, ribbed, adaxially and on margin with dense weft of 1 mm hairs at base, often projecting, becoming shorter and less dense above. Culm

Plate 9. *Chionochloa* tiller base features. Leaf-blade persistent and sheath entire: **A** *C. rubra* subsp. *cuprea* (Southland); **B** *C. macra* (Dunstan Mts); **C** *C. australis* (north-west Nelson); **D** *C. oreophila* (Mt Cook).

Plate 10. *Chionochloa* tiller base features. Leaf-blade falling with upper sheath; sheath fracturing: **A** *C. flavescens* subsp. *brevis* (Mt Cook); **B** *C. rigida* subsp. *rigida* (Dunstan Mts). Leaf-blade disarticulating at ligule, sheath entire: **C** *C. conspicua* subsp. *conspicua* (Fiordland); **D** *C. pallens* subsp. *cadens* (Fiordland).

15–40 cm, erect, internodes glabrous. Inflorescence 5–20 cm, of up to 8 spikelets. Spikelets 40–60 mm, of 6–8 florets. Glumes unequal, usually produced into prickle-toothed awn up to 3 mm; lower 4.5–6.5 mm, 3–5-nerved, upper 7–11 mm, 5–7-nerved. Lemma with central nerve prominent and extending into recurved awn 25–60 mm, smooth except for prickle-teeth at base of awn and on margins. Palea 10–11.5 mm, apex pointed and bifid. Rachilla 2–3 mm, hairy. Callus 1–1.5 mm heavily bearded, hairs 1.5–2.5 mm $\geqslant$ callus. Lodicules 1.5–2 mm. Anthers 3.8–5.5 mm. Gynoecium: ovary 1.5 mm; stigma-styles 2.5–3.5 mm. Caryopsis to 6 mm; embryo 1.4–1.5 mm.

Endemic.

S.: Marlborough, mostly coastal. Limestone cliffs and river terraces; sea level to 900 m.

7. E. solandri (Steud.) Connor, *N.Z. J. Bot. 32*: 140 (1994) $\equiv$ *Triticum solandri* Steud., *Syn. Pl. Glum. 1*: 347 (1854) $\equiv$ *T. squarrosum* Hook.f., *Lond. J. Bot. 3*: 417 (1844) non Roth (1802); Lectotype: BM! *J. Banks & D. Solander* "in rupibus prope Totaranui" (specimens on left and right designated by Connor 1994 op. cit. p. 140). = *T. youngii* Hook.f., *Handbk N.Z. Fl.* 343 (1864) $\equiv$ *Agropyron youngii* (Hook.f.) P.Candargy, *Archiv. Biol. Végét. pure appliquée 1*: 39 (1901); Holotype: K! *Haast* grassy flats, sources of the Waitaki.

Open glaucous grass, prostrate to decumbent to ascending, erect, stoloniferous or tufted, intra- and extravaginally branching, rooting and shooting at nodes, leaf-blades flat to inrolled. Leaf-sheath to 5 cm, keeled, pubescent especially when young, hairy or glabrous, becoming fibrous. Ligule to 0.5 mm, truncate. Collar thickened, recurved, hairs (0.75 mm) dense or short and sparse; occasionally with long hairs on margin. Auricles 1–1.5 mm, clasping, hairs 1–1.5 mm. Leaf-blade 21 cm × 2–4 mm, glaucous, flat or folded or inrolling, abaxially ribbed, occasionally with some long retrorse hairs, adaxially short hairy or prickled on ribs, some long hairs forming a weft at base becoming short above; margin involute, glabrous or faintly toothed. Culm 40–100 cm, prostrate, erect or ascending, internodes glabrous. Inflorescence 8–20 cm, of 3–15 spikelets. Spikelets 25–80 mm, of 4–10 florets.

Glumes unequal, keeled, margins membranous, sometimes produced to a point or becoming awned, prickle-toothed above, lower 3–11 mm, 3-nerved, upper 5–12 mm, 3–5-nerved. Lemma with central nerve prominent and extending into recurved awn 35–75 mm, smooth except for a few prickle-teeth on nerves below awn and along margins, sometimes toothed at apex. Palea 9.5–12 mm, apex pointed and bifid. Rachilla 1.5–3 mm, shortly stiff hairy. Callus 0.75 mm, hairs just reaching lemma. Lodicules 1.25–1.5 mm, ligulate. Anthers 3–5 mm. Gynoecium: ovary 1.5 mm; stigma-styles 2.5–3 mm. Caryopsis to 6.5 mm; embryo 1.5 mm. $2n = 42$.

Endemic.

N.; S.: throughout. Rocky sites on cliffs on coast, tussock grasslands inland, and colonising riverbeds, screes, and moraines; sea level to 1500 m.

8. E. tenuis (Buchanan) Á.Löve et Connor, *N.Z. J. Bot. 20*: 183 (1982) ≡ *Agropyron scabrum* var. *tenue* Buchanan, *Indig. Grasses N.Z.* t. 57b et Add. & Corrig. 11 (1880) ≡ *A. tenue* (Buchanan) Connor, *N.Z. J. Sci. Tech. B35*: 318 (1954); Holotype: WELT 59620 (Buchanan's folio)!, no locality and no date.

Prostrate flat, bronze-leaved stoloniferous open grass, rooting and shooting from nodes on stolons; branching extravaginal; upper culm internode greatly elongating after anthesis and the long weak culm trails along the ground. Leaf-sheath 5–7 cm, glabrous or sometimes retrorsely pilose. Ligule 0.25–0.3 mm, very finely ciliate. Collar glabrous, curved and thickened. Auricles ± 1 mm or smaller, clasping, with 1–2 long hairs. Leaf-blade 10–15 cm × 2 mm, bronzed, flat sometimes ± involute, abaxially glabrous, smooth, ribs evident, adaxially with short antrorse prickle-teeth; margins prickle-toothed. Culm to 2 m, slender, trailing along ground, upper internode elongating after flowering sometimes reaching 1.5 m. Inflorescence 10–15 cm, of up to 15 spikelets. Spikelets 25–40 mm, of 6–8 florets. Glumes 9–15 mm, produced into barbed awn up to 3 mm, 3-nerved, keeled; margins shortly toothed. Lemma 6–10 mm, smooth except for prickles on central nerve, near bifid apex and on margins below; awn 15–35 mm.

Palea 7–10 mm, apex pointed and bifid. Rachilla 1.5–2.5 mm, short hairy. Callus 0.5 mm, surrounded by short hairs. Lodicules 1.5–1.7 mm, linear, irregularly hair tipped. Anthers 2–3 mm, yellow. Gynoecium: ovary 1.25–1.5 mm; stigma-styles 2–2.5 mm. Caryopsis 4.5–5 mm; embryo 1 mm. $2n = 56$.

Endemic.

N.: Volcanic Plateau and Ruahine Range; S.: north-west Nelson, Marlborough and east of Main Divide to Southland. Open fescue-tussock grassland; sea level to 900 m.

This species is characterised by awned glumes and greatly elongating flowering culms.

Elymus canadensis L. of North America was collected from Earnscleugh, Central Otago, CHR 121418 *A. J. Healy 58/213* 16 Feb 1958. It had persisted in an old experimental plot, but is unknown as an escape.

INCERTAE SEDIS   CHR 1613, no collector [possibly B. C. Aston], Omaka River [Marlborough], December 1915. "Has a blue tint when fresh, also plentiful on Waihopai where however *Danthonia pilosa* is the chief native grass". CHR 514886 *G. Jane* Flaxmere River, 4 Dec 1997.
Plants are short and tufted. Leaf-sheath softly hairy or glabrous. Leaf-blade flat or involute, adaxially shortly prickle-toothed; auricles long, clasping. Culm to 20 cm; glabrous, decumbent to erect. Inflorescence to 8 cm, of 5 overlapping spikelets 25 mm long with up to 6 florets. Glumes adaxially fine hairy. Lemma 9 mm, keeled almost to base, keel and nerves with small prickle-teeth above, apex toothed, awn 10 mm, straight. Palea < lemma, keels conspicuously toothed, apex truncate or retuse. Lodicules 1–1.3 mm, ciliate. Anthers 3 mm, purple. Gynoecium: ovary 1 mm; stigma-styles 2–2.5 mm.

These are unlike any other specimens of *Elymus* gathered in New Zealand, but the impression from Aston's label is of abundance; G. Jane's collection does not indicate frequency. They are certainly Australian in origin, and may be veritable *E. scaber* (R.Br.) Á.Löve.

CHR 214647 *R. Mason & A. E. Esler 11643* Kerikeri Inlet Road [Northland], roadside, 29 Nov. 1970.
Stout, hairy 1.25 m. Leaf-sheath antrorsely long coarse hairy; auricles long, clasping. Leaf-blade green, with long coarse hairs or prickle-toothed on both surfaces. Culm 1 m, internodes slightly rough. Inflorescence 25 cm, of up to 20 overlapping spikelets; rachis margins and both surfaces prickle-toothed. Spikelets to 25 mm, of 6–7 florets. Glumes 6–7 mm, 5-nerved, prickle-toothed above.

Lemma 7–9 mm, prickle-toothed throughout, apex toothed, awn straight, 15–25 mm. Palea 8–9 mm, apex ± cleft. Callus bearded only on upper margins. Anthers 2.7 mm; pollen fertile. Seed is set.

The habit is of *E. rectisetus,* but does not conform in palea apex, awn length and bearding of callus. It is probably introduced here.

CHR 327990 *B. Matthews 33/78* Te Paki Trig, North Auckland, grassland 1000 ft, 29 Dec. 1978.
Leafy shoots from many-noded stems. Leaf-sheath glabrous; auricles 1.5 mm, clasping. Leaf-blade long-hairy or prickle-toothed. Culm to 50 cm, internodes becoming smooth. Inflorescence 11 cm, of 6–8 spikelets, rachis margin very strongly prickle-toothed and upper (convex) surface equally so. Spikelets to 35 mm, up to 5 florets. Glumes 4–6 mm, broad, sometimes emarginate, 3–5 evident prickle-toothed nerves, prickle-teeth above. Lemma 9.5 mm, abundantly prickle-toothed, awn to 22 mm. Palea 9.5 mm, apex truncate. Lowest floret in spikelet awnless, often < palea. Seeds setting.

This specimen possesses many *E. multiflorus* characters but the awn is very long and awn:lemma = 2.32. It may possibly be a segregate from a hybrid *E. multiflorus* × *E. solandri* or a backcross generation.

HYBRIDISM

Natural Hybrids: The frequency of natural hybrids is not high and does not reflect the ease with which experimental hybrids can be made, or the frequency with which spontaneous hybrids can occur in experimental plots with close planting [Connor, H. E. *Evolution 10*: 415–420 (1956)].

Elymus × *Stenostachys*

E. solandri (*n* = 21) × *S. gracilis* (*n* = 14), × *S. laevis* (*n* = 14), and × *S. deceptorix*. All are vegetatively similar, and are alike in culm, inflorescence and spikelet morphology.

Slender, soft, stoloniferous grass with extravaginal branching. Leaf-sheath hairy. Auricles 0.75 mm, clasping, usually with a few long hairs. Ligule 0.3 mm. Leaf-blade thin, abaxially and adaxially with long hairs on ribs. Culm to *c.* 50 cm, smooth. Inflorescence to 10 cm, of 8–10 spikelets broadside to rachis. Spikelets to 20 mm, of 2–4 florets. Glumes ± equal 4–8 mm, narrow. Lemma 8–10 mm; awn 10–20 mm. Palea 8 mm. Rachilla 1.5–2.5 mm, hairy; rachilla prolongation to 3 mm. Anthers 2–2.5 mm; pollen sterile. Gynoecium: ovary to 1 mm; stigma-styles to 2.25 mm.

E. solandri × *S. deceptorix* CHR 358145 *A. P. Druce* Ada Valley, Spencer Mts, 3000 ft., Jan. 1979. *E. solandri* × *S. gracilis* CHR 323564 *A. P. Druce* Nina V(alley), S. of Lewis Pass, 2000 ft., Jan. 1978. *E. solandri* × *S. laevis* CANU 10831 *C. J.*

Burrows Head of the Grebe, Jan. 1967, comprised *E. solandri, S. laevis,* and two inflorescences the spikelets and florets of which are longer than those in the combinations above. All are mounted separately in CANU.

E. enysii × *E. solandri*

E. ×*wallii* (Connor) Á.Löve et Connor, *N.Z. J. Bot. 20*: 183 (1982) ≡ *Agropyron* ×*wallii* Connor, *T.R.S.N.Z. 84*: 757 (1957); Holotype: CHR 91921! *H. E. Connor* Porters Pass, Canterbury, 3000 ft, 5/1/56.

Perennial, stoloniferous. Branching intra- and extravaginal. Leaves up to 2 mm wide, blue-green, glaucous, villous; sheaths of basal leaves villous. Ligule short, truncate. Auricles clasping, villous. Culm 1 m, lax, ascending to drooping. Inflorescence to 12 cm, of 3–7 spikelets. Spikelets 25–40 mm, of 2–7 florets, > internodes. Glumes subequal, awn-tipped, 3–5-nerved. Lemma rounded on back, sharply indented at base, keeled and prickle-toothed at apex, 7.5–11.5 mm; awn 12–20 mm; apex of lemma and the awn frequently purple. Palea ≈ lemma, apex pointed, bifid. Callus bearded, obtuse, 0.5 mm. Rachilla 1.5–2.5 mm, finely setulose. Anthers 1.5–3 mm, yellow to purple. Pollen sterile. $2n = 35$.

S.: Canterbury, Porters Pass.

E. multiflorus × *E. solandri* Leaves are as in *E. multiflorus.* The spikelets are long with several florets. Lemmas bear scattered prickle-teeth, and awns are straight, 15–26 mm, and usually 1.5–2.5 × lemma length, except on the lowest floret in a spikelet. Palea apices are truncate to retuse. Mostly North Id.

Experimental Hybrids: Intergeneric hybrids are reported in Löve and Connor (1982 op. cit.). Interspecific hybrids: see Connor, H. E. 1956 op. cit., *N.Z. J. Sci. 5*: 116–119 (1962); Löve and Connor (1982 op. cit.); Lu, B.-R. and Bothmer, R. von *Pl. Syst. Evol. 185*: 33–53 (1993); Lu, B.-R. *Pl. Syst. Evol. 186*: 193–212 (1993). A list of intraspecific hybrids is given in Connor, H. E. *N.Z. J. Sci. 5*: 95–115 (1962) and Löve and Connor (1982 op. cit.).

GENOME ANALYSIS   The interpretation of the genome constitution of Australasian taxa by Torabinejad, J., Carman, J. G. and C. F. Crane *Genome 29*: 150–155 (1987), *Agron. Abstr. 1989*: 102 (1989); Torabinejad, J. and Mueller, R. J. *Genome 36*: 147–151 (1993), *Theor. Appl. Gen. 86*: 288–294 (1993); Lu and Bothmer (1993 op. cit.) all supersede that in Löve and Connor (1982 op. cit.). They show that the hexaploids contain the **S^{t}Y** *Elymus* genome, even though the homology is often low, and the **W** genome of *Australopyrum* as the third component, and that the unique genomic formula is **S^{t}YW** where **Y** is of unknown diploid origin. For tetraploid *E. enysii* the **H** genome is proposed as one component but the second is unknown; it is not **S** [Svitashev, S. et al. *Genome 39*: 1093–1101 (1996)]. The octoploid *E. tenuis* has not been included in modern studies.

ELYTRIGIA Desv., 1810

Rhizomatous perennials. Leaf-sheath with or without auricles. Ligule membranous, truncate. Leaf-blade linear, ± narrow, flat or rolled. Inflorescence a single spike with solitary ± sessile spikelets, alternate on tough persistent rachis. Spikelets 3–7–(10)-flowered, ± laterally compressed, falling entire with glumes; florets ♀♂; rachilla prolonged. Glumes very unequal to subequal, similar, < adjacent lemmas, 3–11-nerved, lanceolate, rounded or slightly keeled towards tip, acute to shortly awned. Lemmas similar to glumes, 5-nerved, pilose, glabrous, or scabrid, awnless, mucronate or awned; awn, if present, apical or from a sinus, straight. Palea ≈ lemma, 2-keeled, ciliate or scabrid on keels. Lodicules 2, membranous, ciliate, usually entire, not or scarcely nerved. Stamens 3. Ovary with apical corona, and hispid hairs; styles 2, free. Caryopsis oblong, adhering to lemma and/or palea, shallowly grooved; embryo small.

Type species: *E. repens* (L.) Nevski

c. 8 spp. of Eurasia. Naturalised spp. 2.

1 Leaf-blades often tightly inrolled, adaxially with prominent, close-set scabrid ribs .... **1. pycnantha**
Leaf-blades usually flat, adaxially with slender, inconspicuous ribs and long scattered hairs, or glabrous .... **2. repens**

1. E. pycnantha (Godr.) Á.Löve, *Taxon 29*: 351 (1980).   sea couch

Bluish grey, tufted, rigid perennial, 35–125 cm, sometimes forming large patches from strongly spreading wiry rhizomes. Leaf-sheath glabrous, subcoriaceous, greyish to light brown, auricles 0.5–1–(1.5) mm, submembranous sometimes long hairy. Ligule *c.* 0.5 mm, scarious. Collar finely hairy. Leaf-blade 5–20 cm, usually tightly inrolled, sometimes flat and up to 7 mm wide, stiff, abaxially smooth, adaxially with prominent close-set scabrid ribs, tip acute, hard, margins scabrid. Culm 25–75 cm, erect, or geniculate below, internodes glabrous. Spike 5–20 × 0.5–1–(2) cm, erect, with spikelets very close-set, but 1–2 lowermost more distant; rachis with a few prickle-teeth on margins. Spikelets 10–20 mm, pale green, oblong or elliptic-oblong, 3–10-flowered, > internodes. Glumes ± equal, disarticulating, 7–12 mm, 5–

7-nerved, lanceolate-oblong, obtuse or acute, coriaceous and very tough, scabrid on mid-nerve almost throughout or near tip. Lemma 7.5–11 mm, 5-nerved, coriaceous, lanceolate-oblong, keeled above, smooth apart from a few prickle-teeth on keel near tip, obtuse to subacute, often mucronate, rarely with fine straight awn to 7 mm. Palea ≤ lemma, keels finely, closely ciliate-scabrid, glabrous elsewhere except adaxially hairy apex. Callus *c.* 0.3 mm, glabrous. Rachilla to 1.5 mm, shortly stiff hairy. Lodicules 1.5 mm, triangular, shortly ciliate. Anthers 5–6 mm. Gynoecium: ovary 0.7–1.0 mm; stigma-styles 2.5–3 mm. Caryopsis 4–6 mm; embryo 0.5 mm.

Naturalised from western and southern Europe.

N.: North Auckland, Bay of Plenty, Gisborne, Hawke's Bay, Wellington (Waikanae, Wairarapa); S.: Canterbury (Avon R. Estuary). Coastal on foreshore waste land, consolidated sand near dunes, mudflats and roadsides.

2. E. repens (L.) Nevski, *Tr. Bot. Inst. AN SSSR*, Ser. 1, *1*: 14 (1933).

twitch, couch

Perennial, 30–150–(200) cm, forming dull green, sometimes bluish green tufts from long-creeping rhizomes. Leaf-sheath glabrous or with ± scattered fine hairs, firmly membranous, light brown to light green. Auricles to 2 mm very fine, membranous. Ligule rim-like, 0.25–0.5 mm, scarious at tip. Leaf-blade 15–40 cm × (2)–3–10 mm, flat, soft to rather stiff, abaxially smooth or sparsely scabrid above, adaxially with slender inconspicuous ribs, usually sparsely long hairy, or glabrous, linear, ± abruptly narrowed to acute tip, margins white, prickle-toothed. Culm 20–150 cm, erect or geniculate below, internodes glabrous. Spike (5)–10–30 × 0.5–1.5 cm, erect, spikelets very close-set or sometimes more distant below; rachis scabrid on margins, otherwise smooth or sometimes abundantly softly hairy. Spikelets 10–20 mm, green or sometimes bluish green, oblong-elliptic or wedge-shaped, 3–8-flowered. Glumes ± equal, 7–12 mm, 5–7-nerved, lanceolate, obtuse or acute, coriaceous, scabrid on midnerve near tip, adaxially with many small white hairs. Lemma 8–13 mm, 5-nerved, coriaceous, rounded to ± keeled above, abaxially glabrous, adaxially clothed in small white hairs, awnless, mucronate or awned to 8 mm. Palea ≤ lemma, keels scabrid, apex entire, ciliate, flank margins wide, membranous ± contiguous, interkeel glabrous, adaxially clothed in short white hairs. Callus 0.2–0.3 mm, shortly bearded. Rachilla to 1.25 mm, shortly hairy. Lodicules to 1.6 mm, shortly ciliate, > ovary. Anthers 3.5–6 mm. Gynoecium: ovary 0.5–0.7 mm; stigma-styles 2.5–3.0 mm. Caryopsis 3.5–5.5 mm; embryo 1–1.25 mm; hilum = caryopsis.

Naturalised from Eurasia.

N.; S.: common; St. and islands offshore; Ch.; C. Waste land, roadsides, cultivated ground and pasture.

HORDEUM L., 1753

Tufted erect annuals. Leaf-sheath glabrous. Leaf-blade flat, auriculate. Inflorescence a 5–10 cm erect or nodding spike-like raceme with a tough rachis; spikelets in 2–6 rows of persistent triads all ♀, or lateral pedicelled spikelets ∅. Spikelets 1-flowered; rachilla prolonged. Glumes linear, equal, shortly awned. Lemma glabrous, awn long, tough. Palea ≈ lemma, glabrous, folded below. Lodicules ciliate. Anthers 2–4 mm. Ovary apex coronate and hispid hairy. Caryopsis broad, stout.

Type species: *H. vulgare* L.

Monotypic diploid genus. The one species naturalised.

1. H. vulgare L., *Sp. Pl.* 84 (1753). barley

Erect or somewhat spreading annual tufts. Leaf-sheath green to light brown to straw-coloured, glabrous, with colourless to sometimes purplish auricles 2.5–5 mm. Ligule 0.5–3 mm, truncate, sparsely ciliate. Leaf-blade 5–35 cm × 3–15 mm, ribs finely scabrid or only so adaxially, margins glabrous, tip finely acuminate. Culm internodes glabrous. Raceme erect or nodding; rachis tough, margins ciliate. Spikelets all ♀, all sessile, or the florets of lateral spikelets ∅ and shortly pedicelled. Glumes usually equal and similar, 4–10 mm, linear-lanceolate, glabrous or short-pubescent to villous, produced to a fine scabrid awn 1.5–8 mm. Lemma of ♀ spikelets broadly ovate-lanceolate, smooth with a few prickle-teeth abaxially on outer lateral nerves near awn-base; awn very tough, scabrid. Palea ≈ lemma, keels very finely scabrid near shallowly bifid or truncate apex. Rachilla prolongation 2–4 mm, ciliate. Lodicules 1 mm, hairy, ciliate. Anthers 2.5–4 mm. Caryopsis elliptic or elliptic-obovate, white, yellow, bluish grey, reddish violet, brown or black.

1 Spikes with 4–6 rows of fertile spikelets .... subsp. **vulgare**
 Spikes with 2 rows of fertile spikelets .... subsp. **distichon**

  subsp. **vulgare**   four- and six-rowed barley

Tufts 50–120 cm. Culm 45–110 cm. Raceme 5–9 × 1–1.5 cm, lax- or dense-flowered, in section square (4-rowed barley), or hexagonal (6-rowed barley). Spikelets all sessile, ♂. Lemma 9–12 mm; awn 9–17 cm. Caryopsis 8–10 × 2–3.5 mm.

Cosmopolitan crop of temperate regions.

N.; S.: occasional. An escape from cultivation; on roadsides or on waste land.

Cultivars with short pyramidal spikes 4–6 × *c.* 2.5 cm and spreading awns may also be found growing wild, e.g., CHR 92133 *A. J. Healy 56/190* near Rolleston, waste land, railway reserve, 29.10.1956; such cultivars belong to subgroup Zeocriton.

  subsp. **distichon** (L.) Körn., *Zeitschr. Ges. Brauw. 5*: 125 (1882).

                                        two-rowed barley

Tufts 35–100 cm. Culm 30–90 cm. Racemes 4.5–12 × 0.6–1.5–(2.5) cm, dense-flowered, distichous, with 2 rows of sessile ♂ spikelets, each spikelet flanked by 2 ∅ lateral spikelets in which the florets are shortly pedicelled. Lemma of lateral spikelets 5.5–8 mm, elliptic or linear-oblong, obtuse, abaxially smooth, adaxially with a few long, soft hairs along the central nerve, awnless. Rachilla prolongation 2 mm, glabrous. Lemma of central spikelets 8–12.5 mm; awn 10–17 cm. Rachilla prolongation 4 mm, long hairy. Caryopsis 8–11.5 × 2–3 mm.

Cultigen.

N.: near Auckland City, Waikato, Bay of Plenty, Mt Egmont and near Wanganui; S.: occasional in Nelson, Canterbury, Otago and Southland. An escape from cultivation onto roadsides and waste land.

LEYMUS Hochst., 1848

Rhizomatous perennials. Leaf-blade stiff, usually glaucous, with ± pungent tip. Inflorescence a linear spike with spikelets in groups of (1)– 2–4–(7), at nodes of the tough rachis. Spikelets (1)–2–5–(7)-flowered; disarticulation between florets. Glumes 1–5-nerved, usually > ½ length of lemma, linear, subulate, or narrow-lanceolate, rounded or keeled, acute to shortly awned. Lemma 3–7-nerved, clothed in hairs, rounded or keeled near tip, acute to shortly awned. Palea folded, flank margins contiguous, hairy. Lodicules ciliate, simple. Ovary with conspicuous hairy corona. Caryopsis oblanceolate to oblong, ± grooved; embryo ⅕–¼ caryopsis.

Type species: *L. arenarius* (L.) Hochst.

c. 40 spp. of temperate regions in Northern Hemisphere, and 1 sp. in Argentina. Naturalised spp. 2.

> *1* Spikelets paired at rachis nodes; glumes with scattered fine hairs................
> .... **1. arenarius**
> Spikelets in 3s to 5s at rachis nodes, occasionally paired; glumes glabrous
> .... **2. racemosus**

1. L. arenarius (L.) Hochst., *Flora (Regensb.) 31*: 118 (1848).

lyme grass

Green- or blue-grey perennials, in thick tufts 60–140 cm, with long, strong rhizomes. Leaf-sheath glabrous, firmly chartaceous, shredding into fibres at maturity, cream to light brown, with chartaceous auricles to 3 mm. Ligule 0.6–1 mm, rim-like, obviously ciliate. Leaf-blade 12–75 cm × 6–12 mm, flat or involute, rigid, abaxially smooth, adaxially finely scabrid on ribs, long-narrowed to pungent tip. Culm 45–110 cm, erect or spreading, internodes glabrous. Spike 14–30 × 1.5–2 cm, stiff, erect, with paired spikelets appressed to rachis; rachis angled, margin finely hairy or glabrous. Spikelets 16–25 mm, 3–6-flowered, yellowish green. Glumes equal, ≈ spikelet, 3–5-nerved, narrow-lanceolate, rigid, abaxially with scattered fine hairs, especially on keel and towards margins near acuminate or mucronate tip, adaxially finely hairy. Lemma 13–20 mm in lower florets, upper lemmas shorter, 7-nerved, lanceolate, acuminate to mucronate, firm, abundantly silky-pubescent throughout. Palea ≈

lemma, folded, keels finely toothed, interkeel minutely silky-pubescent near apex, membranous margins of flanks contiguous, conspicuously short hairy; adaxially hairy, shortly ciliate at notched apex. Callus 0.5 mm, bearded with long hairs. Rachilla 3.5–4 mm, abundantly silky-pubescent. Lodicules 2.5–2.75 mm, ciliate. Anthers 6–7.2 mm. Gynoecium: ovary 1.5–2 mm, orange-brown, hair covered; stigma-styles 2.8–3.5 mm. Caryopsis 7–9 × 2–2.5 mm.

Naturalised from northern and western Europe.

S.: Canterbury (Hawarden, Banks Peninsula), Otago (Lowburn, Clyde, Dunedin); Ch. On coastal sand dunes, or inland on gravels.

2. L. racemosus (Lam.) Tzvelev, *Bot. Mat. (Leningrad) 20*: 429 (1960).
Siberian lyme grass

Very robust, yellow-green or glaucous tufts, 75–150–(200) cm, with stout rhizomes, often forming large patches. Leaf-sheath glabrous or very sparsely very fine hairy, firmly chartaceous, yellowish, adaxially shining, shredding into fibres at maturity, with chartaceous glabrous auricles 1.5–3 mm. Ligule 1–2–(3.5) mm, rim-like, minutely ciliate. Leaf-blade 10–65 cm × 4.5–16 mm, flat, to involute above, abaxially sometimes with very short fine hairs above collar, otherwise glabrous, adaxially strongly ribbed and finely pubescent-scabrid there, narrowed to smooth, hard, pungent tip. Culm 50–110–(160) cm, stout, erect, internodes glabrous except below nodes. Spike 25–40–(50) × (1.5)–2–3.5 cm, stiff, erect, tapered above, spikelets in clusters of 3s to 5s at nodes, or occasionally some paired, well-spaced near base and imbricate above; rachis angled with fine hairs on angles. Spikelets 15–25 mm, 4–6-flowered, straw-coloured, included by glumes. Glumes equal, very narrow-lanceolate, 1–3-nerved, smooth, coriaceous, the long-acuminate tip produced into fine shining awn, becoming ± patent, asymmetric at base, margins sometimes sparsely scabrid, adaxially finely hairy especially centrally. Lemma 10–15.5 mm, 5–7-nerved, elliptic, abaxially silky-pubescent below, glabrous near acute apex, adaxially with appressed very short silky hairs. Palea ≈ lemma, interkeel glabrous, shortly ciliate at notched apex. Callus 0.5 mm, bearded with long hairs. Rachilla 3–4 mm, hairy.

Lodicules 2 mm, very long ciliate. Anthers 4.5–6.5 mm, sometimes pollen-sterile. Gynoecium golden: ovary 2 mm, hair covered; stigma-styles 2.5 mm. Caryopsis *c.* 6.5 × 2 mm.

Naturalised from Eurasia.

S.: Canterbury (Gore Bay, Hawarden, near Christchurch, Banks Peninsula, north of Timaru), Central Otago (Cromwell), Southland (Bluff). Coastal dunes, foreshore, sandy places and waste land.

SECALE L., 1753

Annuals, biennials, or perennials. Leaf-sheath auriculate. Ligule membranous, truncate, often lacerate. Leaf-blade flat or inrolled. Raceme dense, laterally compressed, oblong to linear, with single spikelets distichously arranged; rachis fragile, or tough in cultivated spp. Spikelets 2–(3)-flowered, ♀, rachilla prolongation short. Glumes ± equal, 1-nerved, linear, sharply keeled, acuminate or awned. Lemma 5-nerved, coriaceous, sharply keeled, keel and outer margin pectinate-ciliate; awn long, straight, scabrid. Palea membranous, 2-keeled, keels ciliate-scabrid above. Lodicules 2, large, elliptic-obovate or cuneate, ciliate. Stamens 3. Ovary apex hairy; styles 2, free. Caryopsis obovoid or ellipsoid-cylindric, dorsiventrally compressed, longitudinally sulcate; embryo small, ¼–⅓ caryopsis; hilum linear, ≈ caryopsis.

Type species: *S. cereale* L.

c. 5 spp. of Mediterranean, eastern Europe to central Asia and South Africa. Naturalised sp. 1.

1. S. cereale L., *Sp. Pl.* 84 (1753)   rye, ryecorn

subsp. **cereale**

Annual or biennial tufts, (25)–50–150 cm. Leaf-sheath chartaceous, obviously striate, glabrous or very shortly pilose, with short auricles. Ligule 0.5–1.5 mm. Leaf-blade (5)–10–25 × 0.2–2 cm, minutely pubescent-scabrid throughout or abaxially glabrous, tapered above, margins scabrid, tip acute. Culm (20)–40–120 cm, erect, usually villous just below spike or glabrous throughout. Spike (4)–10–30×0.7–2 cm; rachis tough, non-disarticulating, flattened, margins densely hairy, longer tufts above overlapping proximate glume bases. Spikelets 2-flowered. Glumes 7–12.5 mm, narrow linear-lanceolate, sparsely scabrid, keel minutely ciliate-scabrid; awns 0.5–5.5 mm. Lemma 10–20 mm, firm, oblong-lanceolate, smooth or rarely minutely sparsely scabrid, keel and outer margin above pectinate-scabrid, inner margin sparsely scabrid above, adaxially finely hairy; awn 1.5–6.5 cm. Palea ≈ lemma, folded, keels with sparse prickle-teeth or smooth. Lodicules 3 mm, entire, long ciliate. Anthers 5–8 mm. Gynoecium: ovary 2 mm; stigma-styles to 4 mm. Caryopsis 6.5–7.5 × *c*. 2.5 mm; embryo *c*. 1.25 mm.

Naturalised from Asia Minor.

N.; S.: casual escape from cultivation onto roadsides and waste land.

STENOSTACHYS Turcz., 1862

= *Cockaynea* Zotov, *T.R.S.N.Z. 73*: 233 (1943).

Slender hairy stoloniferous grass with extravaginal branching, and with long graceful glabrous culms with drooping or nodding inflorescences. Spikelets solitary, sessile at nodes, edgewise to sinuous, antrorsely toothed rachis, appressed, imbricate, in narrow slender spikes of 10–30 spikelets with 1–3 awned to mucronate florets disarticulating above glumes. Glumes parallel to rachis, scabrid, aristate, > rachis internodes, or absent or reduced to small stumps; rachis prolonged. Lodicules ciliate. Ovary with apical corona and hispid hairs. Caryopsis adherent to palea; hilum = caryopsis.

Type species: *S. narduroides* Turcz.

Endemic genus of three species.

Treatment follows Connor, H. E. *N.Z. J. Bot. 32*: 125–154 (1994) excluding it from *Elymus* as treated by Löve, Á. and Connor, H. E. *N.Z. J. Bot. 20*: 169–186 (1982).

1 Glumes absent or minute or very occasionally awn-like and mixed in
   inflorescence; lemma conspicuously prickle-toothed .... **2. gracilis**
   Glumes awn-like, ≥ rachis internodes; lemma smooth except below awn,
   occasionally sparsely prickle-toothed elsewhere *2*

2 Lemma mucronate or shortly awned between lateral teeth at apex; palea
   apex bifid .... **3. laevis**
   Lemma long-awned, rarely laterally toothed at apex; palea apex produced
   or retuse .... **1. deceptorix**

1. S. deceptorix Connor, *N.Z. J. Bot. 32*: 144 (1994); Holotype: CHR
249929! *A. P. Druce* Stone C[reek], S.W. Mt Luna, 3800 ft; tussockland,
Jan. 1971.

Tall robust perennial stoloniferous grass with loosely open to compact
shoots of flat leaves, and a long nodding narrow inflorescence. Leaf-
sheath 5–10 cm, keeled, frequently densely pubescent with short and
long hairs, sometimes glabrous; becoming fibrous. Auricles 0.5–0.7 mm,
scarcely clasping, very occasionally 1–2 long hairs. Ligule 0.25–0.5 mm,
very faintly erose. Leaf-blade 10–30 cm × 1–2.5 mm, flat, thin, with
abundant small prickle-teeth on ribs, very occasionally with hairs 0.5 mm
between ribs abaxially and sometimes near ligule; margin prickle-
toothed. Culm stout 40–190 cm, internodes glabrous, shining; nodes
sometimes geniculate. Inflorescence slender, nodding, 10–20 cm, of
20–30 spikelets > internodes; rachis prolongation 2–6 mm. Spikelets
to 15 mm, of 1–3 florets, shining; rachilla prolongation 3–3.5 mm.
Glumes 2, 5–10 mm, equal, narrow canaliculate below soon becoming
awn-like, prickle-toothed, closely appressed to floret above, < spikelet.
Lemma 8–10 mm, smooth except for prickle-teeth below awn and near
callus, sometimes pruinose, infrequently bifid at apex; scabrid awn
5–6.5 mm. Palea 8–10 mm, ≥ lemma; apex produced or retuse,
ciliate; keels toothed. Callus short, surrounded by short stiff hairs;
disarticulation flat. Rachilla 1.8–2 mm, with abundant stiff hairs.
Lodicules 0.7–0.8 mm. Anthers 3–3.5 mm. Gynoecium: ovary 0.75–1 mm;
stigma-styles to 2.5 mm. Caryopsis 5–5.5 mm; embryo 1–1.5 mm.
Chasmogamous or cleistogamous.

Endemic.

S.: Nelson, north-western mountains. In tussock grassland and on
river terraces; 800–1525 m.

2. S. gracilis (Hook.f.) Connor, *N.Z. J. Bot. 32*: 146 (1994)
≡ *Gymnostichum gracile* Hook.f., *Fl. N.Z. 1*: 312, t. 70 (1853) ≡ *Hystrix gracilis* (Hook.f.) Kuntze, *Rev. Gen. Pl. 2*: 778 (1891) ≡ *Asprella gracilis* (Hook.f.) Kirk, *T.N.Z.I. 27*: 353 *in obs.* (1895) ≡ *Cockaynea gracilis* (Hook.f.) Zotov, *T.R.S.N.Z. 73*: 234 (1943) *comb. illeg.*; Lectotype: K! *Colenso 1611* Patea Village (designated by Connor 1994 op. cit. p. 146). = *Stenostachys narduroides* Turcz., *Bull. Soc. Nat. Moscou 35*: 331 (1862) ≡ *Elymus narduroides* (Turcz.) Á.Löve et Connor, *N.Z. J. Bot. 20:* 184 (1982); Holotype: LE *n.v.*, *J. E. Home* New Zealand. = *Agropyron subeglume* P.Candargy, *Archiv. Biol. Végét. pure appliquée 1*: 64 (1901); Holotype: P *n.v.*, *Raoul 241* Akaroa, 1843.

Perennial stoloniferous grass forming open, wide and flat-leaved patches with narrow nodding inflorescences; often quite stout in forests. Leaf-sheath 5–15 cm, with long (0.5–1 mm) hairs irregularly retrorse or erect, occasionally few or glabrous. Auricles to 0.5 mm or minute, scarcely clasping. Ligule 0.3–1 mm, erose. Leaf-blade 10–20 cm × 1.5–2 mm, flat, thin, usually with hairs 0.5–1 mm or with sparse short prickle-teeth adaxially and glabrous abaxially; margins glabrous. Culm 70–100 cm, slender, nodes evident sometimes ± geniculate, internodes glabrous. Inflorescence slender, narrow, 10–20 cm, of 15–30 spikelets > internodes; internodes 2–5 mm but longer at base; rachis prolongation 2–6 mm. Spikelets to 10 mm, of 1–2–3 florets, on 1–1.5 mm stipes in the absence of glumes; rachilla prolongation 1.5–3 mm, conspicuously short stiff hairy. Glumes usually 0, sometimes 1 or 2 and awn-like, 0–2–3 mm, very occasionally 5–6 mm above, 1-nerved, prickle-toothed, ≪ spikelets. Lemma 7–10 mm, prickle-teeth abundant, weakly keeled, infrequently bifid at apex, canaliculate above and tapering to awn 1.5–6 mm. Palea 5–7 mm, < lemma, apex usually produced but sometimes retuse; keels toothed and usually inrolled. Callus short, 0.5 mm, surrounded by abundant short stiff hairs; disarticulation ± oblique. Rachilla 1.5–2.5 mm, shortly prickle-toothed. Lodicules 0.75–1 mm. Anthers 1.5–2 mm, often retained on apex of caryopsis. Gynoecium: ovary 1.25–1.4 mm; stigma-styles 1.5–2 mm. Caryopsis 4–4.25 mm; embryo 1–1.2 mm. Chasmogamous and commonly cleistogamous. $2n = 28$.

Endemic.

N.: throughout except North Cape; infrequent or absent in central southern areas; S.: Nelson, Marlborough and Canterbury; sparsely distributed from inland South Canterbury to Fiordland; St. Forests and shrublands, occasionally in grasslands; sea level to 1300 m.

3. S. laevis (Petrie) Connor, *N.Z. J. Bot. 32*: 146 (1994) ≡ *Asprella laevis* Petrie, *T.N.Z.I. 27*: 406 (1895) ≡ *Cockaynea laevis* (Petrie) Zotov, *T.R.S.N.Z. 73*: 234 (1943) *comb. illeg.* ≡ *Hystrix laevis* (Petrie) Allan, *N.Z. DSIR Bull. 49*: 88 (1936) ≡ *Elymus laevis* (Petrie) Á.Löve et Connor, *N.Z. J. Bot. 20*: 184 (1982) non Hoover (1966); Lectotype: WELT 68353! *D. Petrie* Matukituki Valley, 3/1893 (designated by Connor 1994 op. cit. p. 147).

Perennial grass with long stolons, rooting and shooting at nodes, inflorescences narrow and nodding or drooping on long slender culms. A species of grasslands. Leaf-sheath 5–10 cm, with retrorse hairs 0.5 mm, or glabrous; keeled. Auricles absent or minute, 0.15–0.25 mm, scarcely clasping. Ligule 0.5–1 mm, lacerate; prominent. Leaf-blade 15–25 cm × 1.5–2 mm, flat, thin, abaxially with distinct midrib, glabrous or occasionally with hairs 1 mm beside ribs or shortly retrorse prickle-toothed, adaxially with prickle-teeth on ribs occasionally with hairs 0.5 mm; margin glabrous or shortly toothed. Culm 50–75 cm, nodes inconspicuous, internodes glabrous, slender. Inflorescence very narrow, slender, 8–15 cm, of 10–25 spikelets > internodes; internodes 2.5 mm to 6 mm below; rachis prolongation 3–6 mm. Spikelets 8–12 mm, of 1–2 florets, appearing shortly stipitate above glumes; rachilla prolongation 3–4 mm, shortly prickle-toothed. Glumes 2, awnlike, 2–6 mm, ± equal, > internodes, 1-nerved, prickle-toothed on keel and margins, < spikelet. Lemma 6.5–8 mm, smooth except for prickle-teeth towards bifid apex and evident on lateral nerves in upper half; keel ± prominent; scabrid awn 0.5–1 mm. Palea 6.5–7 mm, ≤ lemma; apex bifid; keels shortly toothed and inrolled. Callus 0.5–0.75 mm, flat; disarticulation almost flat. Rachilla to 1.3 mm if 2-flowered, shortly prickle-toothed. Lodicules 0.75–1.75 mm, lobed or entire. Anthers 1.6–2.6 mm, often retained on apex of caryopsis. Gynoecium: ovary 1–2 mm; stigma-styles 1.75–2.25 mm. Caryopsis 4 mm; embryo 1–1.3 mm. Chasmogamous and cleistogamous. $2n = 28$.

Endemic.

N.: north-west Ruahine Range (Reparoa Bog); S.: throughout except Nelson, Buller and Westland; St. Grasslands, often in flushed sites; sea level to 1300 m.

HYBRIDISM   For natural intergeneric hybrids with *Elymus* see *Elymus*. Experimental interspecific and intergeneric hybrids are treated in Löve and Connor (1982 op. cit.). Although genomic formulae are allocated there they lack the precision of modern genomic interpretation.

Connor, H. E. *N.Z. J. Sci. Tech. 38A*: 742–751 (1957) showed that *S. gracilis* is self-compatible.

THINOPYRUM Á.Löve, 1980

Rigid, erect, glaucous, rhizomatous perennials. Auricles absent. Ligule membranous, truncate. Leaf-blade linear, narrow, usually involute, aromatic when crushed. Inflorescence a simple spike, with solitary ± sessile spikelets, alternate on stout very fragile rachis. Spikelets 2–10-flowered, laterally compressed, appressed to rachis; disarticulation above glumes; florets ♀, rachilla prolonged. Glumes ± equal, similar, < adjacent lemmas, 4–12-nerved, coriaceous, oblong, obtuse, acute or truncate, awnless. Lemmas similar in texture to glumes, 5-nerved, glabrous, rounded below keeled near tip, obtuse, awnless. Palea ≈ lemma, 2-keeled. Lodicules 2, membranous, ciliate, usually asymmetrically lobed. Stamens 3. Ovary hairy, hispid hairs at apex; styles 2, free. Caryopsis free from lemma and palea; embryo small; hilum long, linear.

Type species: *T. junceum* (L.) Á.Löve

c. 6 spp. of Eurasia. Naturalised sp. 1; transient sp. 1.

> *1* Leaf-blades adaxially densely minutely hairy; glumes 9–20 mm, glabrous ...
> .... **1. junceiforme**
> Leaf-blades adaxially glabrous or with sparse long hairs; glumes 5–8 mm,
> scabrid on keel near apex .... **intermedium**ζ

1. T. junceiforme (Á.Löve et D.Löve) Á.Löve, *Taxon 29*: 351 (1980).

sand couch

Bluish grey tufts 25–60 cm, often forming large colonies from long-creeping wiry rhizomes. Leaf-sheath abaxially glabrous, adaxially abundantly finely hairy, subcoriaceous, straw-coloured or often purplish, auricles 0. Ligule to 1 mm, scarious, ciliate, hairy. Leaf-blade (7)–10–30 cm, often involute, or flat and up to 4 mm wide, stiff or ± flexuous, abaxially smooth, adaxially prominently ribbed, ribs minutely densely hairy, long-tapered to fine, acute tip, margins minutely ciliate. Culm 10–35 cm, erect or drooping, internodes glabrous. Spike 4–20 × 1–2 cm, erect or curved, with close-set or distant spikelets; rachis very fragile, glabrous. Spikelets 15–25 mm, oblong, elliptic or wedge-shaped, 3–7-flowered; rachilla prolongation short. Glumes ± equal, 9–20 mm, 7–11-nerved, oblong, obtuse, coriaceous and very tough, asymmetrically keeled, or rounded on back, abaxially glabrous, adaxially clothed in fine hairs, mucronate or blunt, margins membranous. Lemma 10–15 mm, coriaceous, 5–7-nerved, abaxially glabrous and shining, adaxially clothed in fine hairs, apex notched, mucro between 2 short lobes. Palea to 14 mm, ≤ lemma, keels finely ciliate, interkeel with fine hairs towards margins, flanks finely hairy; apex produced, ciliate. Callus short, glabrous, articulation oblique. Rachilla 2 mm, finely hairy. Lodicules to 2 mm, lobed, ciliate, hairy. Anthers 6–7 mm, orange, filaments brown. Gynoecium: ovary 2 mm, turbinate, brown; stigma-styles to 3 mm.

Naturalised from northern and western Europe.

S.: North Canterbury and from Ashley River to Christchurch. Coastal in sand dunes and sandy ground.

Formerly known as *Agropyron junceum.*

ζ**T. intermedium** (Host) Barkworth et D.R.Dewey   The specimen consists of a single culm broken off above the base of the plant. Leaf-blade inrolled, adaxially glabrous. Culm internodes smooth. Spike 24 × 1 cm, spikelets imbricate except near base. Spikelets *c.* 15 mm, *c.* 5-flowered, elliptic-oblong. Glumes 5–8 mm, scabrid on keel near obtuse tip. Lemma 9–10 mm, obtuse. Palea slightly < lemma. This Eurasian species was collected once from near Christchurch *J. B. Armstrong* (CHR 5401) probably in the 1870s.

Triticum L., 1753

Annuals. Leaf-sheath auriculate. Ligule membranous. Leaf-blade flat. Spike dense, linear, with single sessile spikelets distichously arranged; rachis fragile, or tough in cultivated spp. Spikelets 2–6–(9)-flowered,

♂; rachilla prolonged. Glumes ± equal, usually < lemmas, 5–11-nerved, coriaceous, chartaceous or rarely membranous, asymmetrical, ± keeled, obtuse, truncate, or bidentate, mucronate or awned. Lemma coriaceous, 7–11-nerved, rounded, or keeled above, apex similar to that of glumes. Palea hyaline to membranous, 2-keeled, entire or splitting at maturity. Lodicules 2, ovate, ciliate. Stamens 3. Ovary apex with hispid hairs; styles 2, plumose. Caryopsis free, or adherent to anthoecium, ellipsoid, deeply grooved adaxially; embryo ⅕–⅓ length of caryopsis; hilum linear, ≈ caryopsis.

Type species: *T. aestivum* L.

10–20 spp. from eastern Mediterranean to Iran. Naturalised sp. 1; transient sp. 1.

> *1* Spikes linear-oblong; spikelets obliquely vertical on rachis .... **1. aestivum**
> 　Spikes ovate; spikelets strongly divergent from rachis .... **compactum**ζ

1. T. aestivum L., *Sp. Pl.* 85 (1753).　　　　　　　　wheat

Erect, robust, green or ± glaucous annual tufts, 30–100 cm. Leaf-sheath firm, subcoriaceous, shortly pubescent or glabrous, green or straw-coloured. Auricles 2–3.5 mm, pale green to pinkish, usually ciliate-margined. Ligule 0.6–4 mm, hyaline, truncate, shortly denticulate and minutely ciliate. Leaf-blade 4–30 cm × 2–12 mm, abaxially smooth, adaxially minutely scabrid, margins finely scabrid, tip acuminate and slightly hooded. Culm 25–80 cm, internodes smooth or minutely scabrid. Spike distichous, linear-oblong, lax- to dense-flowered, 4–9 × 1–2.5 cm; rachis tough, margins ± ciliate. Spikelets 3–6-flowered. Glumes 7–10 mm, coriaceous, or pubescent, especially on margins below, truncate, keel with sparse prickle-teeth produced above to a short tooth or awn. Lemma 9.5–12 mm, coriaceous and rigid, 7-nerved, acute and awnless or awned between 2 small lobes; awn 1–50–(70) mm, scabrid. Palea ≈ lemma, not folded, submembranous, keels hairy, interkeel with some small hairs especially below, margins membranous, meeting, glabrous. Rachilla 1–1.5 mm, hairy. Lodicules 1.5–1.75 mm, long ciliate. Anthers 2.5–4 mm. Gynoecium: ovary 2 mm; stigma-styles short 1–1.5 mm. Caryopsis 7–7.5 × 3.4–3.8 mm, ellipsoid, ovate-ellipsoid or ovoid, variously coloured; embryo 2.5 mm. Cleistogamous, but briefly chasmogamous.

Cultigen.

N.; S.; St.: scattered throughout; C. On roadsides and waste ground.

ζ**T. compactum** Host   Plant 60 cm. Leaf-sheath and leaf-blade shortly pubescent. Spikes 4 × 2 cm, oval, light brown, with spikelets very close-set on the rachis and strongly divergent from it. A Eurasian species collected once, Wellington railway yards *V. D. Zotov* Jan. 1942 (CHR 25460).

IV SUBFAMILY ARUNDINOIDEAE

Spikelets usually 2–many-flowered, sometimes 1-flowered, laterally compressed, with fragile rachilla; glumes 2, well-developed; lemma (1)–3–9-nerved; palea 2-nerved; lodicules 2, cuneate, fleshy, truncate; stamens 3, or 2; stigmas 2. Inflorescence usually a panicle. Ligule a line of hairs.

Discussion on the subfamily Arundinoideae, its tribal structure, and generic complement is extensive. We have adopted here a very simple suprageneric classification for the eight genera occuring in New Zealand, among which only *Chionochloa* and *Rytidosperma* are polytypic; it is inconsistent with many, but not all, recent treatments. The most important recent assessments affecting New Zealand taxa are: Barker, N. P., Linder, H. P. and Harley, E. H. *Syst. Bot. 20*: 423–435 (1995), *Syst. Bot. 23*: 327–350 (1998); Clayton and Renvoize (1986 op. cit.); Conert, H. J. *in* Soderstrom, T. R. et al. (Eds) *Grass Systematics and Evolution* 239–250 (1987); Hilu, K. W. and Esen, A. *Pl. Syst. Evol. 173*: 57–70 (1990); Hsiao, C. et al. *Aust. Syst. Bot. 11*: 41–52 (1998); Renvoize, S. A. *Kew Bull. 41*: 323–342 (1986).

Tribes: 13. Arundineae; 14. Danthonieae

13. ARUNDINEAE

Plants reed-like with cauline leaves. Inflorescence a large plumose panicle. Spikelets 2–4-flowered, florets all ♀ or the lower ♂, or all florets unisexual; rachilla prolonged beyond floret or not. Glumes persistent, ≤ lemma, ± hyaline. Lemma with long silky hairs near base, ± hyaline, awned or awnless, entire or with lateral, ± awned lobes. Palea < lemma. Lodicules ciliate or glabrous. Stamens 3, or 2. Caryopsis ellipsoid; embryo *c.* ⅓ caryopsis; hilum oblong or short.

1 Lemma hairy; callus short, ± glabrous; all flowers perfect .... **Arundo**
Lemma glabrous; callus long, conspicuously villous; lower flower ♂, others
perfect .... **Phragmites**

ARUNDO L., 1753

Tall, robust, perennial reeds, with stout creeping rhizomes, stout, hollow culms, wide flat cauline leaves, and large panicles. Ligule membranous, ciliate. Leaf-blade linear, long-tapering. Culm erect, sometimes almost woody. Inflorescence a terminal, dense, plume-like panicle, with fascicled lower branches. Spikelets 2–4-flowered, laterally compressed, all florets ♀; disarticulation above glumes and between florets; rachilla prolonged. Glumes 3–5-nerved, persistent, ± equal. Lemma 3–7-nerved, covered with long, soft hairs from below, usually bidentate, with central awn from the sinus. Palea < lemma, keels densely ciliate. Lodicules 2, glabrous, irregularly toothed. Stamens 3. Fig. 15.

Type species: *A. donax* L.

3 spp., of tropical and warm temperate regions, on river banks and in wet places. Naturalised sp. 1.

1. A. donax L., *Sp. Pl.* 81 (1753).   giant reed

Large robust perennial clumps to 5 m, with very wide leaves, and stout, woody, thick and knotty rhizomes up to 3 cm diam. Leaf-sheath rounded, coriaceous, smooth and glabrous, long hairy at apex. Ligule 1–2 mm, truncate, short-ciliate, collar brownish with some long soft tangled hairs. Leaf-blade 30–60 × 2.5–5 cm, cordate at base, scabrid on margins and abaxially on the much-narrowed, long-tapered tip, otherwise smooth. Culm erect, internodes smooth. Panicle 30–50 cm, plume-like, very densely branched; branches obliquely ascending, scabrid from fine teeth. Spikelets 10–12 mm, 2–4-flowered, purplish later becoming brownish. Glumes ± equal, the lower including lemmas, 3-nerved, glabrous, lanceolate, acute to acuminate. Lemma 7–9 mm, 5–7-nerved, densely, soft long (*c.* 5 mm) hairy from below, lanceolate, tapered, lateral lobes 2, very short, awn-tipped, sinus with awn 2–3 mm, straight. Palea < lemma, apex ciliate, keels shortly

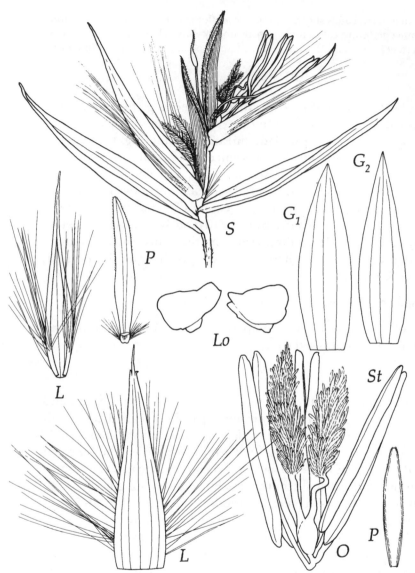

Fig. 15 *Arundo donax*
S, spikelet, × 7; **G₁**, **G₂**, glumes, × 7; **L**, lemma ventral view and dorsal view flattened to show nervation, × 7; **P**, palea ventral view with callus and hairs, dorsal view, × 7; **Lo**, lodicules, × 24; **O**, ovary, styles and stigmas, × 14; **St**, anthers, × 14.

densely ciliate. Callus sharp, small, glabrous, sparse hair tufts *c.* 1 mm. Rachilla and prolongation *c.* 2 mm, glabrous. Lodicules 2, *c.* 0.5 mm, cuneate, 3-nerved, glabrous, irregularly toothed. Anthers 2.5–3.5 mm. Gynoecium: ovary 0.5 mm, stigma-styles to 2 mm. Caryopsis not seen. Fig. 15.

Naturalised from Eurasia.

N.; S.: occasional garden escape. Roadsides and sandy waste land.

A. donax is grown as an ornamental; a variegated form is also grown and may escape into the wild.

In N.Z. many megagametophytes fail to reach maturity [Philipson, M. N. and Connor, H. E. *Bot. Gaz. 145*: 78–82 (1984)]; seeds will rarely be found.

PHRAGMITES Adans., 1763

Tall, robust perennial reeds with flat cauline leaves, hollow culms and creeping rhizomes. Ligule a membranous truncate rim, long-ciliate. Leaf-blade linear, long-tapering, deciduous. Inflorescence a large plumose panicle. Spikelets 2–6-flowered, lowest floret ♂ or ∅, others ♀; disarticulating above glumes and between lemmas; rachilla long-hairy, hardly prolonged. Glumes unequal, < lowest lemma, 3–5-nerved, membranous. Lemma 1–3-nerved, narrow-lanceolate, rounded, hyaline, glabrous, lemma of lowest floret similar to glumes, others caudate-acuminate. Palea ≪ lemma. Lodicules 2, obovate. Callus plumose. Stamens 3.

Type species: *P. australis* (Cav.) Trin. ex Steud.

3–4 spp., cosmopolitan, in marshes and along rivers. Naturalised sp. 1.

1. P. australis (Cav.) Trin. ex Steud., *Nomencl. Bot.* ed. 2, *2*: 324 (1841).
reed, phragmites

Robust perennials 1.5–3 m, with wide flat leaves borne along stout reed-like stems and with stout creeping rhizomes to 2 cm diam. Leaf-sheath rounded, chartaceous, glabrous, with numerous, minute cross-veinlets. Ligule *c.* 1 mm, truncate, short membranous, long ciliate.

Collar brownish on both surfaces with dense, appressed, minute hairs, and ephemeral hairs to 6 mm on margin. Leaf-blade 40–60 × 1–3 cm, coriaceous, linear, somewhat narrowed at base, much-tapered above, smooth, but abaxially scabrid towards long, filiform, scabrid tip, margins scabrid above; disarticulating at ligule. Culm erect, rigid, usually simple, internodes glabrous. Panicle 20–30 cm, erect or finally drooping, purplish to brownish; branches numerous, glabrous except tufts of long (4 mm), silky hairs at nodes. Spikelets 10–14 mm, 3–6-flowered, finally widely gaping, lowest floret ♂ or ∅, others ♀. Glumes unequal, persistent, 3–5-nerved, chartaceous, glabrous, ovate-lanceolate, acute; lower *c.* 4 mm, upper *c.* 6 mm. Lemma of basal floret 9–10 mm, 3-nerved, narrow-lanceolate, membranous, acute, persistent; lemmas of perfect florets shorter, narrower, tip long-acuminate to caudate. Palea 3–4 mm, ≪ lemma, keels ciliate. Callus 0.5–1 mm, with dense long (6–8 mm) silky hairs. Rachilla *c.* 0.2 mm, glabrous, prolongation 0.5 mm, glabrous. Lodicules 0.3 mm, obovate, nerved, glabrous. Anthers 1.5–2 mm. Gynoecium: ovary 0.5 mm; stigma-styles 2 mm. Caryopsis not seen.

Naturalised from temperate zones in both Hemispheres.

N.: several locations near Napier, in Hawkes Bay; S.: once in Westland. Stream margins, roadsides, waste places.

Generally known as *P. communis* Trin. until about 1975.

In N.Z. mature megagametophytes are rare (Philipson and Connor 1984 op. cit.); seeds will rarely be found. Connor, H. E. et al. *N.Z. J. Bot. 36*: 465–469 (1998) reported $2n = 4x = 48$ and $2n = 8x = 96$ in some populations together with mixoploid numbers $2n = 53, 54$, and *c.* 264.

A variegated form is also grown as an ornamental grass and may escape into the wild.

14. DANTHONIEAE

Plants tufted or tussock-forming. Inflorescence varying from a single spikelet to a small panicle, or large plumose panicle. Spikelets 2–several-flowered, florets ♀ or unisexual; disarticulation above glumes and between florets. Glumes 2, persistent, well-developed.

Lemma entire to sinus, entire or lobed above, with a straight or geniculate awn from tip or sinus, or awnless. Lodicules 2, hair-tipped. Fruit a caryopsis or rarely an achene.

DANTHONIINAE

Plants tufted with basal leaves. Inflorescence a panicle, rarely reduced to a single spikelet. Spikelets several-flowered, florets ♀, rarely male-sterile; rachilla prolonged. Glumes persistent, 1–13-nerved, membranous to chartaceous, ≈ or < spikelet. Lemma usually 7–9-nerved, usually hairy, membranous to coriaceous, bilobed or minutely toothed, lateral lobes awned or awnless, awned from sinus, awn straight or geniculate, with ± twisted column. Stamens 3. Fruit a caryopsis or rarely an achene.

1 Lemma hairs mostly in tufts in 2 ± complete upper and lower transverse series, rarely lemma glabrous .... **Rytidosperma**
 Lemma hairs ± uniform in length, not gathered in tufts *2*

2 Lemma lobes conspicuous; awn long, with flattened and ± twisted column; leaf-blades persistent or falling .... **Chionochloa**
 Lemma lobes reduced to teeth *c.* 0.5 mm, or lemma minutely 3-toothed; awn minute; leaf-blades persistent *3*

3 Fruit an achene; plant long-rhizomatous .... **Pyrrhanthera**
 Fruit a caryopsis; plant densely tufted .... **Sieglingia**

CHIONOCHLOA Zotov, 1963

Perennial tussocks, solitary or occasionally sward forming; branching intravaginal or sometimes extravaginal. Leaf-sheath long persistent and clothing shoots, entire or fracturing into irregular segments, glabrous or with internerve hairs, tuft of hairs at apex. Ligule a ring of hairs usually *c.* 1 mm. Leaf-blade 5 cm to 1.5 m long, glaucous, flat to V-shaped, or junceous, persistent or disarticulating at ligule or falling with part of sheath, abaxially usually glabrous, occasionally hairy, adaxially prickle-toothed or papillate, rarely unornamented, margins usually long hairy below. Culm to 2 m, erect, exceeding leaves. Inflorescence an open or compact panicle of few to many spikelets, glabrous to conspicuously long hairy, usually with long hairs at branch axils. Spikelets of few to several ♀ florets, disarticulating above glumes and between florets. Glumes unequal, linear, acute or rarely awned, usually glabrous; lower 1–3-nerved, upper 3–5–7-nerved. Lemma (7)–9-nerved, nerves anastomosing below sinus, lateral lobes awned or triangular-acute, densely hairy on margins and in columns in all, few, or no internerves; central awn straight or reflexed from twisting, flat or indistinct column from sinus. Palea > lemma sinus, keels ciliate, interkeel infrequently prickle-toothed, flanks long hairy below. Callus short, blunt, hairs long covering lemma; disarticulation oblique. Rachilla usually glabrous. Lodicules 2, irregularly rhomboidal and lobed, nerved, margins long hairy especially apically. Stamens 3; anthers caudate, shorter in male-sterile flowers. Gynoecium bistylar, ovary glabrous, < lodicules. Caryopsis free, obovate, smooth or rugose; embryo ⅓–½ caryopsis length; hilum linear, ½–⅔ caryopsis length. Fig. 16.

Type species: *C. rigida* (Raoul) Zotov

24 spp. of Australasia. Endemic spp. 22.

N.Z. spp. were revised by Connor, H. E. *N.Z. J. Bot. 29*: 219–283 (1991), and synoptically treated by Zotov, V. D. *N.Z. J. Bot. 1*: 78–136 (1963). They are collectively known as snow-tussock or snow-grass except for a few.

SYNOPSIS

A. Leaf-blade persistent on sheath
    1. Lemma scabrid with prickle-teeth
        (a) Leaf-blade margin smooth
            (i) Mat-forming; alpine: 3. *australis*
            (ii) Caespitose; coastal cliffs: 4. *beddiei*, 5. *bromoides*
        (b) Leaf-blade margin scabrid: 6. *cheesemanii*, 11. *flavicans*
    2. Lemma lacking prickle-teeth
        (a) Leaf-blade gramineous
            (i) Sheath long, margin flat: 14. *macra*, 17. *pallens*
            (ii) Sheath short, margin undulating: 15. *oreophila*
        (b) Leaf-blade junceous: 12. *juncea*, 19. *rubra* subsp. *cuprea*

B. Leaf-blade disarticulating at ligule

    1. Leaf-blade gramineous

        (a) Leaf-blade margin glabrous

           (i) Leaf-blade tip blunt: 2. *antarctica*, 13. *lanea*

           (ii) Leaf-blade tip pungent: 16. *ovata*

        (b) Leaf-blade margin hairy below

           (i) Inflorescence hairy: 7. *conspicua*, 8. *crassiuscula*

           (ii) Inflorescence glabrous: 17. *pallens* subsp. *cadens*, 22. *vireta*

    2. Leaf-blade junceous: 1. *acicularis*, 21. *teretifolia*

C. Leaf-blade falling with part of sheath below ligule

    1. Leaf-blade gramineous: 10. *flavescens*, 18. *rigida*

    2. Leaf-blade junceous

        (a) Leaf-sheath strict: 9. *defracta*, 19. *rubra*

        (b) Leaf-sheath coiling up: 20. *spiralis*

1 Leaf-blade junceous *2*
Leaf-blade gramineous *9*

2 Leaf-blade disarticulating at ligule or falling with part of sheath *3*
Leaf-blade persistent on sheath *7*

3 Leaf-blade falling with part of sheath *4*
Leaf-blade disarticulating at ligule *6*

4 Abaxial leaf-blade evidently hairy; inflorescence hairy .... **9. defracta**
Abaxial leaf-blade glabrous; inflorescence glabrous *5*

5 Leaf-sheath erect or strict; adaxial leaf-blade with prickle-teeth or papillae,
margin scabrid or smooth .... **19. rubra**
    5a Leaf-sheath entire .... subsp. **cuprea**
    Leaf-sheath fracturing into short segments... *5b*
    5b Adaxial leaf-blade with short hairs above ligule or 0 .... subsp. **rubra**
        5c Leaf-blade margin prickle-toothed .... var. **rubra**
        Leaf-blade margin smooth .... var. **inermis**
    Adaxial leaf-blade with weft of long hairs above ligule ..... subsp. **occulta**
Leaf-sheath coiling up; adaxial leaf-blade with prickle-teeth, margin smooth
.... **20. spiralis**

6 Abaxial leaf-blade with two columns of antrorse hairs .... **21. teretifolia**
Abaxial leaf-blade glabrous .... **1. acicularis**

7 Leaf-sheath glabrous, margin undulating, long entangling hairs at apex; leaf-
blade to 10 cm .... **3. australis**
Leaf-sheath hairy, margin often coiling up; leaf-blade to 1 m *8*

8 Adaxial leaf-blade with long hairs at midpoint; margin with long hairs below and prickle-teeth above .... **19. rubra** subsp. **cuprea**
Adaxial leaf-blade with prickle-teeth at midpoint; margin with long hairs below, becoming glabrous .... **12. juncea**

9 Leaf-blade persistent on sheath *10*
Leaf-blade disarticulating at ligule or falling with part of sheath *16*

10 Adaxial leaf-blade scabrid at midpoint *11*
Adaxial leaf-blade glabrous at midpoint *13*

11 Leaf-sheath to 5 cm, thin, margins wavy; mat-forming .... **15. oreophila**
Leaf-sheath 15–20 cm, firm, margins flat; caespitose *12*

12 Leaf-sheath keeled; leaf-blade V-shaped .... **17. pallens**
  12a Leaf-blade disarticulating at ligule .... subsp. **cadens**
  Leaf-blade persistent on sheath *12b*
  12b Adaxial leaf-blade with prickle-teeth or papillae .... subsp. **pallens**
  Adaxial leaf-blade with long hairs and prickle-teeth or papillae ....
  .... subsp. **pilosa**
Leaf-sheath rounded; leaf-blade flat to shallowly U-shaped .... **14. macra**

13 Leaf margin scabrid *14*
Leaf margin smooth *15*

14 Branching intravaginal; sheath densely hairy .... **6. cheesemanii**
Branching extravaginal; sheath glabrous .... **10. flavicans**

15 Adaxial leaf-blade with weft of short hairs above ligule .... **4. beddiei**
Adaxial leaf-blade with weft of long hairs above ligule .... **5. bromoides**

16 Adaxial leaf-blade with weft of hairs above ligule or with interlocking long hairs *17*
Adaxial leaf-blade lacking weft of hairs or interlocking hairs *20*

17 Leaf-sheaths persistent and entire *18*
Leaf-sheaths breaking into short segments *19*

18 Abaxial leaf-blade very scabrid above; sheath compressed and flat; innovations extravaginal (North, South and Stewart Is) .... **7. conspicua**
  18a Leaf-sheath conspicuously long hairy; adaxial leaf-blade with weft of long hairs at base .... subsp. **conspicua**
  Leaf-sheath and adaxial leaf-blade glabrous .... subsp. **cunninghamii**
Abaxial leaf-blade smooth above; sheath terete; innovations intravaginal (Auckland and Campbell Is) .... **2. antarctica**

19 Leaf-blade to 10 mm wide, tapering, often reddened above; sheath to 20 cm, coloured or pale; shoots to 10 mm diam., stout .... **11. flavescens**

   19a   Leaf-blade with weft of long hairs above ligule *19b*

       Leaf-blade with weft of short hairs or lacking hairs above ligule *19c*

   19b   Leaf-sheath glabrous except on margin .... subsp. **lupeola**

       Leaf-sheath hairy .... subsp. **hirta**

   19c   Leaf-blade lacking hairs above ligule .... subsp. **flavescens**

       Leaf-blade with weft of short hairs above ligule .... subsp. **brevis**

   Leaf-blade to 7 mm wide, narrowing, often golden above; bronze sheath to 30 cm; shoots to 5 mm diam., slender .... **18. rigida**

   19d   Inflorescence glabrous except for long hairs at branch axils; leaf-blade margin hairy below becoming scabrid .... subsp. **rigida**

       Inflorescence hairy or scabrid; leaf-blade margin becoming smooth below .... subsp. **amara**

20 Leaf-blade margin smooth above, long hairs below *21*

   Leaf-blade margin prickle-toothed above, long hairs below *23*

21 Adaxial leaf-blade papillate; sheath stout and stiff .... **8. crassiuscula**

   21a   Leaf-blade curved, coriaceous, tapering abruptly to a very pungent point (Stewart Id) .... subsp. **crassiuscula**

       Leaf-blade curved or twisting above, ± flexible, tapering gradually to a long, thin, pungent apex (South Id) *21b*

   21b   Shoots from short stout stems; leaf-blade stout, to 6 mm wide, curled or twisting above .... subsp. **torta**

       Shoots from very long prostrate stems; leaf-blade slender, to 3 mm wide, straight or somewhat twisting above .... subsp. **directa**

   Adaxial leaf-blade prickle-toothed *22*

22 Ligule to 3.5 mm; sheath becoming fibrous, short hairy .... **16. ovata**

   Ligule to 1 mm; sheath intact, very long (4 mm) hairy .... **13. lanea**

23 Plants slender; sheath margins thin and undulating; leaf-blade ± flat, to 2.5 mm wide .... **22. vireta**

   Plants stout; sheath margins firm and flat; leaf-blade V-shaped, to 10 mm wide .... **17. pallens** subsp. **cadens**

1. C. acicularis Zotov, *N.Z. J. Bot. 1*: 101 (1963); Holotype: CHR 70454! *V. D. Zotov* Leslie Clearing, Expectation Stream, Caswell Sound, 30.3.49.

Tall, slender, erect, pedicelled tussock, with sharp-pointed, glabrous, deciduous leaves. Leaf-sheath to 12 cm, dark above, pale shining below, persistent, becoming chartaceous, keeled, clothed with long (3 mm) abundant deciduous hairs, apical tuft of hairs to 7 mm. Ligule to 0.5 mm.

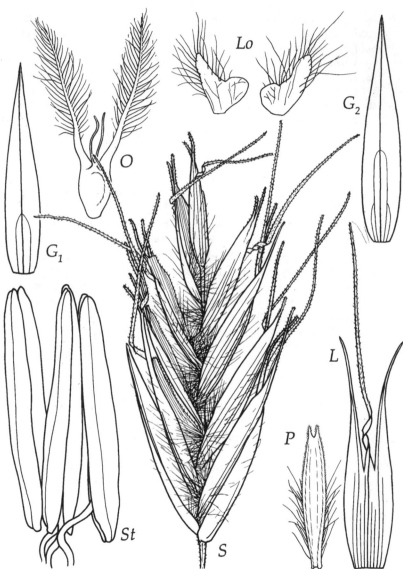

Fig. 16 *Chionochloa rigida*

S, spikelet, × 5;   **G₁, G₂**, glumes, × 5;   **L**, lemma dorsal view, × 5;   **P**, palea dorsal view, × 5;   **Lo**, lodicules, × 16;   **O**, ovary, styles and stigmas, remnant third style, × 16;   **St**, anthers, × 16.

Leaf-blade to 50 cm × 1 mm diam., acicular junceous, thickened at midrib, disarticulating at ligule, abaxially glabrous, adaxially with dense very short stiff hairs below, papillae and occasionally some prickle-teeth above; margin smooth. Culm to 60 cm, long and short hairs below inflorescence otherwise glabrous. Inflorescence to 10 cm, spikelets few; rachis, branches and pedicels abundantly long hairy especially at branch axils. Spikelets of up to 7 golden florets. Glumes becoming acute or mucronate, glabrous, > adjacent lemma lobes; lower to 11 mm, 3-nerved, upper to 13 mm, 5-nerved. Lemma to 5 mm; dense hairs at margin and erect hairs aside central nerve or rarely between all nerves, ± reaching sinus; lateral lobes to 4 mm, triangular-acute; central awn to 14 mm from divergent flat column to 2 mm. Palea to 7 mm. Callus to 1 mm, hairs to 2.5 mm. Rachilla to 0.5 mm. Lodicules to 1 mm. Anthers to 4 mm. Gynoecium: ovary to 0.8 mm; stigma-styles to 2 mm. Caryopsis to 3 mm. $2n = 42$. Plate 8F.

Endemic.

S.: South Westland and Fiordland. On peaty soils in grasslands, and scattered through scrub; 750–1225 m.

In a specimen from cultivation some scattered, weak, long (3 mm) hairs, like those of the sheath, were present on the abaxial leaf-blade though this was never seen in plants from the wild.

2. C. antarctica (Hook.f.) Zotov, *N.Z. J. Bot. 1*: 99 (1963) ≡ *Bromus antarcticus* Hook.f., *Fl. Antarct. 1*: 97, t. 54 (1844) ≡ *Danthonia antarctica* (Hook.f.) Hook.f., *Fl. N.Z. 1*: 302 (1853) non Spreng. (1825) ≡ *D. flavescens* var. *hookeri* Zotov, *T.R.S.N.Z. 73*: 234 (1943); Holotype: K! *J. D. Hooker* Lord Auckland Islands, November 1840.

Tall, slender, often peat-stained, pale tussock with pungent leaves often spirally twisting above and deciduous. Leaf-sheath to 20 cm, dark above, pale shining below, persistent, glabrous though often with many very short hairs between ribs. Ligule to 2 mm. Leaf-blades to 30 cm × 6 mm, flat or U-shaped, disarticulating at ligule, abaxially many glabrous ribs, adaxially below with interlocking hairs from near margins, abundant papillae above; margin glabrous. Culm to 1 m, internodes glabrous. Inflorescence to 15 cm, congested, ± triangular, glabrous except for long hairs at branch axils and below spikelets. Spikelets of up to 7 lightly purpled florets. Glumes sometimes purpled, acute or shortly awned, < adjacent lemma lobes; lower to 12 mm, 1-nerved or shortly 3-nerved, upper to 15 mm, 3–5-nerved, margin consistently long hairy below otherwise glabrous. Lemma to 8 mm; hairs dense at margin and aside central nerve, sometimes in other internerves but then fewer, ± reaching sinus; lateral lobes to 7 mm including awn to 3 mm or long triangular-acute; central awn to 20 mm reflexed from flat column up to 3 mm. Palea to 10 mm. Callus to 1 mm, hairs to 3 mm. Rachilla to 1 mm. Lodicules to 0.75 mm. Anthers to 3.5 mm. Gynoecium: ovary to 0.75 mm; stigma-styles to 4 mm. Caryopsis to 2.5 mm. $2n = 42$. Plate 1A.

Endemic.

C., A. Tussock grassland and cushion bogs; sea level to 450 m.

3. C. australis (Buchanan) Zotov, *N.Z. J. Bot. 1*: 103 (1963) ≡ *Danthonia raoulii* subsp. *australis* Buchanan, *T.N.Z.I. 4*: 224 (1872) ≡ *D. australis* (Buchanan) Buchanan, *Indig. Grasses N.Z.* t. 31 (1879); Lectotype: WELT 59576! *J. Buchanan* Kaikoura Mountains (designated by Zotov 1963 op. cit. p. 103).   carpet grass

Mat-forming grass with persistent, short leaves densely crowded in tight shoots among many old sheaths, and much shorter than flowering culms; hairy prophylls evident. Leaf-sheath to 7 cm, imbricate, persistent, entire, becoming fibrous and separating into two distinct layers when old, shiny, glabrous, long (7 mm) entangled hairs at apex usually spreading across proximate sheath, adaxially with antrorse, soft, internerve hairs to 0.25 mm; margins undulating and usually glabrous. Ligule to 1.5 mm, often obscured by hairs of sheath and leaf-blade. Leaf-blade to 10 cm × 0.8 mm diam., acicular junceous, persistent, navicular, ± terete, distichous rarely mono-stichous, abaxially shining, glabrous, adaxially with a dense weft of long hairs at base, hairs gradually becoming fewer and shorter above, usually projecting beyond leaf margin, abundant papillae above; margin smooth. Culm to 40 cm, internodes glabrous except for short hairs below nodes and long hairs below inflorescence. Inflorescence

Plate 11. *Rytidosperma* florets: dorsal view of **A** *R. clavatum*, **B** *R. merum*, **C** *R. biannulare*, **D** *R. corinum*, **E** *R. setifolium*, **F** *R. viride*, and **G** *R. petrosum* (all × 6).

Plate 12. *Rytidosperma* florets: dorsal view of **A** *R. maculatum,* **B** *R. buchananii,* **C** *R. tenue,* **D** *R. pulchrum,* **E** *R. nudum,* **F** *R. unarede,* **G** *R. thomsonii,* **H** *R. nigricans,* and **I** *R. gracile* (all × 6).

to 5 cm; 3–5 solitary spikelets on short, flexuous, long hairy, pulvinate branches. Spikelets of up to 8 often purpled florets. Glumes to 14 mm, glabrous, acute or mucronate, frequently purpled, > adjacent lemma lobes; lower 3-nerved, upper 5-nerved. Lemma to 5 mm; hairs dense at margin and erect aside main nerve, usually absent or fewer elsewhere, ≈ sinus; lateral lobes to 7 mm including awn up to 4 mm, sometimes shortly lobed again, glabrous except for prickle-teeth above; central awn to 15 mm, reflexed from flattened to strongly twisting column to 4 mm. Palea to 7 mm. Callus to 1 mm, hairs to 2.5 mm. Rachilla to 1 mm. Lodicules to 1 mm. Anthers to 4 mm. Gynoecium: ovary to 0.8 mm; stigma-styles to 3 mm. Caryopsis to 3 mm. $2n = 42$. Plates 8H, 9C.

Endemic.

S.: Nelson to Canterbury. Forming dense mats in herbfields and tussock grasslands, and occasionally on cushion bogs; 700–1675 m.

Distinctive in the presence of adaxial sheath hairs, projecting leaf-blade hairs, and hanks of hairs at sheath apex.

4. C. beddiei Zotov, *N.Z. J. Bot. 1*: 90 (1963); Holotype: CHR 7830! *A. D. Beddie* Mouth of Mukumuku River, Palliser Bay, 1935.

Short, shiny tussock with stiff and widely spreading shoots and persistent leaves. Leaf-sheath to 15 cm, slightly keeled, pale yellow, persistent, becoming fibrous, margins above with long hairs, apical tuft of hairs to 2 mm. Ligule to 0.5 mm. Leaf-blade to 60 cm × 4 mm, flat to U-shaped, persistent, deflexed at collar, abaxially glabrous, adaxially with a weft of short hairs at base, abundant prickle-teeth above; margin with long hairs below, becoming shaggy, hairs mostly antrorse and appressed. Culm to 75 cm, internodes glabrous except for short, dense hairs below inflorescence. Inflorescence to 15 cm, congested with short branches, bristling with awns; rachis, branches and pedicels short soft hairy. Spikelets of up to 5 florets. Glumes to 13 mm, shortly awned, < adjacent lemma lobes, prickle-teeth above, otherwise glabrous; lower 1–3-nerved, upper 1–3–5-nerved. Lemma to 6 mm; hairs dense at margin and in all internerves, or sometimes only aside central nerve, > sinus; lateral lobes to 11 mm including strict awn to 8 mm, prickle-teeth adaxially and abaxially and on lemma margins; central awn to 22 mm much deflexed from twisting column to 4 mm. Palea to 8.5 mm, produced into two conspicuous narrow processes; prickle-teeth abaxially above. Callus to 1 mm, hairs to 5 mm. Rachilla to 1 mm. Lodicules to 1.75 mm. Anthers to 4.5 mm. Gynoecium: ovary to 1 mm; stigma-styles to 4 mm. Caryopsis to 3 mm. $2n = 42$.

Endemic.

N.: Wellington and Wairarapa. On coastal bluffs and cliffs and inland for about 10 km; sea level to 850 m.

The tight, awn-bristled panicle is characteristic, as are prickle-teeth on glumes and lemma lobes.

Connor (1991 op. cit. p. 231) referred plants from Editor Hill and Lookout Peak, Marlborough Sounds, to *C. beddiei*; these plants do have *C. beddiei* characters but are perhaps better treated as *C. beddiei* × *C. flavescens* although *C. beddiei* is unknown in this area.

5. C. bromoides (Hook.f.) Zotov, *N.Z. J. Bot. 1*: 90 (1963) ≡ *Danthonia bromoides* Hook.f., *Fl. N.Z. 1*: 303 (1853); Holotype: K! *W. Stephenson 102* New Zealand, 1843–4.

Stout, pendent often sprawling, bright green tussock with persistent leaves and sheaths. Leaf-sheath to 15 cm, shining yellow, keeled, persistent and entire, becoming fibrous, margin abundantly long hairy below, apical tuft of hairs to 4 mm; adaxially with many minute interrib hairs. Ligule to 1.5 mm. Leaf-blade to 50 cm × 10 mm, flat or shallowly U-shaped, smooth, persistent, adaxially glabrous except for long hairs on margin below and some short or long hairs, sometimes dense, at base. Culm to 70 cm, internodes glabrous. Inflorescence to 20 cm, very congested; rachis and main branches glabrous but with some long hairs at axils; pedicels short and densely hairy. Spikelets of up to 6 florets. Glumes acute or slightly awned, < adjacent lemma lobes, many prickle-teeth abaxially and a few adaxially; lower to 12 mm, 1–3-nerved, upper to 16 mm, 5-nerved. Lemma to 9 mm; hairs dense at margin and in all internerves though sometimes absent from all or some, ≤ sinus, prickle-teeth abundant abaxially and adaxially on lobes and margins; lateral lobes to 5 mm including awn to 3 mm or acute, rarely dividing from awn at sinus; central awn to 22 mm from indistinct straight column. Palea to 10 mm, prickle-teeth abaxially and on flanks. Callus to 1.5 mm, hairs to 5 mm. Rachilla to 0.5 mm. Lodicules to 1.75 mm. Anthers to 5.5 mm in male-fertile flowers, up to 3 mm in male-sterile flowers. Gynoecium of male-fertile flowers with stigma-styles to 3.5 mm, ovary to 1.5 mm, and of male-sterile flowers to 5 mm, ovary 1.5 mm. Caryopsis to 3.5 mm. Gynodioecious. $2n = 42$. Plate 8G.

Endemic.

N.: south to Leigh (east coast) and Mt Manganui (on west) and offshore islands. On coastal cliffs and bluffs.

Spikelets with 3 glumes seem frequent; the uppermost glume may then be 7-nerved and the margins long hairy. Minute hairs on the adaxial sheath surface are characteristic.

Natural populations comprise 50% ♂ and 50% ♀ plants on offshore islands; on the mainland females were absent or at a low frequency except at Whangaruru North Head.

6. C. cheesemanii (Hack.) Zotov, *N.Z. J. Bot. 1*: 95 (1963) ≡ *Danthonia raoulii* var. *cheesemanii* Hack., *in* Cheeseman *Man. N.Z. Fl.* 887 (1906) ≡ *D. flavescens* var. *cheesemanii* (Hack.) Zotov, *T.R.S.N.Z. 73*: 234 (1943); Holotype: AK 1633! *T. F. C[heeseman]* forest at the source of the Takaka River, Nelson, alt. 3000 ft, Jan 1881 (No 1250 to Hackel).

Tall tussock with dark green, persistent, scabrid leaves on brown, hairy sheaths. Leaf-sheath to 20 cm, entire, persistent, becoming fibrous, abundant deciduous hairs to 4 mm between nerves, on margins and at apex. Ligule to 1 mm. Leaf-blade to 80 cm × 4 mm, flat or folded, many ribs, persistent, abaxially with abundant prickle-teeth above, very occasionally with long hairs below, adaxially glabrous; margin with long hairs below soon becoming very scabrid. Culm to 2 m, internodes glabrous or with dense short hairs below nodes. Inflorescence to 35 cm; rachis and pedicels short stiff hairy, but rachis sometimes glabrous in North Id plants. Spikelets of up to 6 distant florets. Glumes hyaline, scabrid, often bifid at apex, < adjacent lemma lobes; lower to 7 mm, 1-nerved, upper to 10 mm, 3–5-nerved. Lemma to 6 mm, scabrid especially above; hairs long at margin and usually aside central nerve but absent or fewer elsewhere, < sinus; lateral lobes up to 3 mm, scabrid, triangular or shortly awned; central awn to 15 mm reflexed from 1 mm flat or infrequently twisting column. Palea to 6.5 mm, ≈ lemma, interkeels with prickle-teeth above and often short hairs below. Callus to 1.5 mm, hairs to 4 mm. Rachilla to 1 mm, occasionally short stiff hairy. Lodicules to 1.5 mm. Anthers to 4.5 mm. Gynoecium: ovary to 0.75 mm; stigma-styles to 2 mm. Caryopsis to 3 mm. $2n = 42$.

Endemic.

N.: Lake Waikaremoana, Ruahine Range, and mountain ranges in southern Wellington; S.: Nelson and northern Marlborough, Kaikoura Mountains. Beech forests, and grasslands; 900 to 1225 m but to 1500 m in Marlborough.

7. C. conspicua (G.Forst.) Zotov, *N.Z. J. Bot. 1*: 92 (1963) ≡ *Arundo conspicua* G.Forst., *Prodr.* 9 (1786) ≡ *Calamagrostis conspicua* (G.Forst.) J.F.Gmel., *Syst. Nat. 2*: 172 (1791) ≡ *Achnatherum conspicuum* (G.Forst.) P.Beauv., *Ess. Agrost.* 20, 142, 152 (1812) ≡ *Agrostis conspicua* (G.Forst.) Roem. et Schult., *Syst. Veg. 2*: 364 (1817) ≡ *Cortaderia conspicua* (G.Forst.) Stapf, *Handlist Herbac. Pl. Roy. Bot. Gard. Kew ed. 2*: 137, 333 (1902) ≡ *Deyeuxia conspicua* (G.Forst.) Zotov, *T.R.S.N.Z. 73*: 234 (1943); Holotype: GOET! *[G. Forster] 29, Arundo conspicua Prodr. 48.*

Elegant, tall, scabrid tussock of many flabellate sectors with drooping leaves tardily deciduous and much shorter than flowering culms; branching extravaginal with numerous cataphylls. Leaf-sheath to 20 cm, persistent, entire, becoming fibrous, strongly keeled and compressed, very dark brown below, straw above, apical tuft of hairs to 2 mm. Ligule to 1.5 mm. Leaf-blade to 75 cm × 7 mm, folded and strongly keeled, tardily disarticulating at ligule, many distinct ribs; abaxially becoming very scabrid above with many rows of prickle-teeth, adaxially glabrous or with weft of long hairs at base or with interlocking long hairs near margin becoming sparser and glabrous above; margin with long hairs below, becoming strongly scabrid. Culm to 2 m; nodes dark, conspicuous and severely constricted; internodes glabrous, compressed. Inflorescence to 35 cm, open, rachis smooth below becoming shortly stiff hairy on branches and pedicels, some long hairs at branch axils. Spikelets of up to 7 florets. Glumes to 9 mm, apex ciliate acute or shortly (1 mm) awned, sometimes bilobed, < adjacent lemma lobes, margin of upper or both often with long hairs below. Lemma to 6.5 mm, scabrid; hairs long at margin absent elsewhere or very occasional, < sinus; central awn to 12 mm reflexed from flat column to 1 mm. Palea to 7.5 mm. Callus to 1.5 mm, hairs to 4 mm. Rachilla to 0.75 mm, shortly stiff hairy or glabrous. Lodicules to 2 mm. Gynoecium: ovary to 1 mm; stigma-styles to 2.5 mm. Caryopsis to 3 mm. $2n = 42$.

N.: localised on mountain ranges; S.: throughout except for lower rainfall areas in east; St.: throughout.

subsp. **conspicua**

Leaf-sheath long hairy. Leaf-blade with 3 conspicuous ribs, abaxially with some long hairs below near keel, adaxially with weft of long hairs at base. Glumes with prickle-teeth above; lower 1–3-nerved, upper 3–5-nerved. Lemma with prickle-teeth on lobes and margins; lateral lobes to 6.5 mm including awn to 2.5 mm. Palea internerve with few prickles or hairs at apex. Anthers to 3.5 mm. Plates 8C, 10C.

Endemic.

S.: in west from Nelson to Fiordland, and east of Main Divide in higher rainfall areas, Banks and Otago peninsulas; St. Forest, scrub, and some cleared areas; sea level to 1225 m.

On some plants in Nelson the leaf-sheath may be glabrous or hairs may be restricted to the keel, and the leaf-blade may lack the adaxial weft of long hairs e.g., CHR 187676 *I. M. Ritchie*; CHR 190997 *A. P. Druce*; CHR 262908 *A. W. Purdie*. In floret characters these conform to subsp. *conspicua*.

subsp. **cunninghamii** (Hook.f.) Zotov, *N.Z. J. Bot. 1*: 94 (1963) ≡ *Danthonia cunninghamii* Hook.f., *Handbk N.Z. Fl.* 332 (1864) ≡ *D. antarctica* var. β *laxiflora* Hook.f., *Fl. N.Z. 1*: 302 (1853); Lectotype: K! *Colenso 607* N. Zealand [Woods near Summit, Ruahine Range] (designated by Zotov 1963 op. cit. p. 94). = *D. antarctica* var. γ *parviflora* Hook.f., *Fl. N.Z. 1*: 302 (1853); Lectotype: K! *Colenso 2053* New

Zealand [Woods near summit, Ruahine Range] (designated by Zotov 1963 op. cit. p. 94). = *D. pentaflora* Colenso, *T.N.Z.I. 16*: 343 (1884); Holotype: WELT 21871! *A. Hamilton* Ruahine Range.

Leaf-sheath glabrous. Leaf-blade with up to 5 conspicuous ribs, glabrous but occasionally with some long hairs adaxially. Glumes with abundant prickle-teeth abaxially and adaxially; lower 3–5-nerved, upper 5–7-nerved. Lemma with prickle-teeth abaxially and adaxially; lateral lobes to 2.5 mm including awn to 1 mm. Palea internerve with conspicuous prickles or hairs. Anthers to 4.5 mm.

Endemic.

N.: Northland, in east on Coromandel Peninsula, mountain ranges of East Cape, Ruahine Range south to Cook Strait, in west on Herangi Range and Mt Pirongia. Forest, scrub, cliff faces and rocky clearings; sea level to 1500 m.

Some small, soft plants bear hairs near the keel on the sheath e.g., CHR 275127 *A. P. Druce* Palliser Bay; CHR 325823 *A. P. Druce* Pirongia; CHR 393991 *A. P. Druce* Herangi Range.

Two gatherings, CHR 190868 *A. P. Druce* Tararua Range, 1969, and CHR 190570 *A. P. Druce* Rimutaka Range, 1968, are typical of *C. conspicua* subsp. *cunninghamii* in: cataphylls, flabellate shoots, compressed entire sheath, conspicuous darkened collar, leaf-blade with several strong ribs, glabrous adaxially but sometimes with a few long hairs, and leaf-blade margins with long hairs below. But they lack the extensive prickle-teeth which confer the marked scabridity of the abaxial leaf-blade and leaf-blade margin. Prickle-teeth are present, but sparingly, on the leaf margin of CHR 190570; the apex of the leaf-blade is sparsely prickle-toothed in both. Both are slender, and shorter than is typical of the subspecies. A. P. Druce labelled both as *C. conspicua* var. [subsp.] *cunninghamii* × *C. flavescens sens. strict.* remarking that such "...hybrids were not uncommon in the [Tararua Range] area".

8. C. crassiuscula (Kirk) Zotov, *N.Z.J. Bot. 1*: 103 (1963) ≡ *Danthonia crassiuscula* Kirk, *T.N.Z.I. 17*: 224 (1885); Holotype: WELT 36481! *T. Kirk* Mount Anglem, Stewart Isd.   curly snow tussock

Short, shining tussock with smooth, pungent leaves, twisting or straight but deciduous and leaving many imbricate old sheaths. Leaf-sheath to 12 cm, shining above, darkened below, sometimes purpled or reddened, persistent, entire, compressed, sometimes apical tuft of hairs to 1 mm. Leaf-blade to 30 cm × 6 mm, keeled, pungent, conduplicate, disarticulating at ligule, abaxially glabrous except for prickle-teeth towards apex, adaxially papillate but with some prickle-teeth especially near ligule; margin thickened, long hairs below or sometimes with prickle-teeth. Culm to 65 cm, often purpled, often hairy below inflorescence. Inflorescence to 15 cm, open, pulvinate; spikelets often paired on flexuous branches; rachis, branches and pedicels with mixed long and short hairs or rachis hairy on margins only. Spikelets of up to 7 golden florets. Glumes thin, purpled, acute or mucronate from between teeth or aristate to erose, ≥ adjacent lemma lobes; lower to 12 mm, shortly 1–3-nerved, glabrous, upper to 16 mm, 5–7-nerved, margin often long hairy below otherwise glabrous. Lemma to 6 mm, shining; hairs dense at margin and aside central nerve, < sinus; lateral lobes to 6.5 mm including awn to 3 mm, or triangular-acute or long acute; central awn to 12 mm slightly twisting and markedly reflexed from flat column to 2 mm. Palea to 7 mm. Callus to 1 mm, hairs to 2.5 mm. Rachilla to 1.5 mm. Lodicules to 2 mm. Anthers to 4 mm. Gynoecium: ovary to 0.75 mm; stigma-styles to 3.5 mm. Caryopsis to 3 mm.

Endemic.

S.: Mountains of Main Divide from Canterbury to Fiordland and Southland; St.

   subsp. **crassiuscula** = *Danthonia pungens* Cheeseman, *Man. N.Z. Fl.* 887 (1906) ≡ *Chionochloa pungens* (Cheeseman) Zotov, *N.Z. J. Bot. 1*: 103 (1963); Holotype: AK 1640! *T. Kirk* Smith's Lookout, Stewart Island, Jan. 1887.

Short, stout, robust tussocks. Leaf-sheath to 8 cm, glabrous, pale or dark brown. Ligule to 1 mm. Leaf-blade to 30 cm × 6 mm, very coriaceous, curved, shortly tapering to very pungent apex, adaxially with abundant prickle-teeth at base. Lemma to 4.5 mm; hairs dense on margin and aside central nerve, few or absent elsewhere.

St. Herbfields, boggy meadows, and scrub; 700–1000 m.

The holotype of *C. crassiuscula* is tall, strict, very pungent; CHR 21860 *L. B. Moore & L. M. Cranwell*; CHR 262860a *A. W. Purdie*; CHR 321817 *C. J. Webb & T. H. Webb* all resemble it.

Tussocks may be small and prostrate at open, wet, windswept sites with leaf-blades up to 10 cm, but taller (30 cm) in favourable or sheltered conditions. Kirk's specimen from Smith's Lookout is like the former. There is a suggestion from cultivated material that short stature may be genetically determined, or that growth and change is very slow; CHR 182390 after 2 years in cultivation had leaf-blades to 7 cm, sheath to 1 cm, and an emerging inflorescence on a culm to 6 cm. CHR 212510 after 2 months in cultivation had leaf-blades to 5 cm on sheaths to 2 cm, and a panicle 5 cm on 12 cm of subtending culm. CHR 212523 after 9 months in a pot bore leaf-blades to 13 cm on sheaths to 4 cm, and a panicle 7 cm on a subtending culm to 20 cm.

subsp. **directa** Connor, *N.Z. J. Bot. 29*: 236 (1991); Holotype: CHR 321064! *A. W. Purdie* Bald Hill, Longwood Range, Southland, wet grassland, 7.2.1974.

Short, slender tussock with long prostrate stems often forming large patches. Leaf-sheath to 10 cm, glabrous, frequently reddened. Ligule 0.25 mm. Leaf-blade to 25 cm × 3 mm, drawn out into long tapering, thin, straight or often somewhat twisting, pointed apex. Culm to 50 cm, internodes glabrous. Lemma to 5.5 mm with dense hairs only on margin and aside central nerve. Plate 8E.

Endemic.

S.: Southland on Longwood Range and Takitimu and Hunter Mts. Poorly drained or boggy grasslands; 700–1400 m.

subsp. **torta** Connor, *N.Z. J. Bot. 29*: 237 (1991); Holotype: CHR 9613! *V. D. Zotov* L[ake] Harris, 4 Jan. 1936.

Short, stout, robust tussock. Leaf-sheath to 12 cm, sometimes purpled, usually with many short interrib hairs. Ligule to 0.5 mm. Leaf-blade 30 cm × 6 mm, drawn out into long tapering, twisting or curling, pointed apex; adaxially with few prickle-teeth near ligule. Culm to 65 cm, internodes often hairy. Lemma to 6 mm; hairs dense on margin and in all internerves though sometimes absent from some. Plate 1B.

S.: east and west of Main Divide from lat. 42°S to Fiordland. Alpine grasslands with impeded drainage, and bogs; 600–1850 m.

Glabrous leaf-sheaths are more frequent in Fiordland; leaf-sheaths with conspicuous or ± conspicuous hairs occur north of there.

9. C. defracta Connor, *N.Z. J. Bot. 25*: 164 (1987); Holotype: CHR 322084a! *A. W. Purdie* Dun Mt, Nelson, 2600 ft, open slopes between Windy Point and Coppermine Saddle, 14.1.74.

Tussocks of variable stature, some robust with long leaves, others slender with shorter, narrow leaves; culm internodes hairy; leaves stiff, scabrid, deciduous. Leaf-sheath to 25 cm, fracturing into short segments, shining light brown at base, long (4 mm) and short deciduous interrib hairs above, margin long hairy above, apical tuft of hairs to 4 mm. Ligule to 0.8 mm. Leaf-blade to 75 cm × 1.3 mm diam., acicular junceous, midrib evident, falling with part of sheath, abaxially with long hairs aside midrib below becoming very scabrid above, adaxially with weft of long hairs at base, abundant prickle-teeth and papillae above; margin thick, below with long (up to 4 mm) hairs interlocking across leaf-blade, scabrid above. Culm to 65 cm; lower internodes densely hairy, less so above, hairy below inflorescence. Inflorescence to 20 cm, narrow and strict, very scabrid and with a few long hairs at branch axils and below spikelets. Spikelets of up to 6 lightly purpled florets. Glumes acute or shortly awned, scabrid above, ≤ adjacent lemma lobes; lower to 11 mm, 3-nerved, upper to 13 mm, 5–7-nerved. Lemma to 5.5 mm; hairs dense at margin and aside central nerve, often in all internerves or sometimes few or absent, ≥ sinus; lateral lobes to 4.5 mm including awn to 2 mm, or acute, scabrid above; central awn to 11 mm from 2.5 mm twisting column. Palea to 8 mm. Callus to 1 mm, hairs to 4 mm. Rachilla to 1 mm, very occasionally short hairy. Lodicules to 1 mm. Anthers to 3.5 mm. Gynoecium: ovary to 1 mm; stigma-styles to 3.5 mm. Caryopsis to 3.5 mm.

Endemic.

S.: Nelson and Marlborough. In grassland, scrub, and forest on ultramafic rocks and soils; to 1000 m.

Sometimes the hairs fall from the leaves and cannot be detected. Plants from Red Hills, Marlborough, lack hairy culm internodes.

10. C. flavescens Zotov, *N.Z. J. Bot. 1*: 97 (1963); Holotype: CHR 62345! *V. D. Zotov* Mount Dennan, *c.* 4000 ft, Tararua Range, January 1948.   broad-leaved snow-tussock

Tall, stout tussock with fracturing sheaths and long, broad, drooping, deciduous leaves which form a dense litter. Leaf-sheath glabrous or internerves or margins hairy, keeled, pale or purpled, fracturing into short segments, apical tuft of hairs to 1 mm. Ligule to 1 mm. Leaf-blade flat or shallowly U-shaped, glaucous, falling with part of sheath, abaxially glabrous but sometimes with some long hairs, adaxially with prickle-teeth or papillae; margin with long hairs below becoming glabrous or infrequently scabrid. Culm to 1.5 m, internodes glabrous or very infrequently long hairy. Inflorescence to 30 cm, spikelets dense on stiff, pulvinate branches. Spikelets of up to 8 purpled florets. Glumes glabrous, acute but sometimes shortly awned, > adjacent lemma lobes; lower to 14 mm, 1–3-nerved, upper to 18 mm, 3–5–7-nerved. Lemma to 6 mm; hairs dense on margin and in internerves, < sinus; lateral lobes to 6 mm including awn to 3 mm often long triangular-acute; central awn to 16 mm from twisting column to 4 mm. Palea to 10 mm. Callus to 1 mm, hairs to 4 mm. Rachilla to 1 mm. Lodicules to 1.5 mm. Gynoecium: ovary to 1 mm; stigma-styles to 3 mm. Caryopsis to 3.5 mm. $2n = 42$.

Endemic.

N.: Tararua Range; S.: mountains in Nelson, Marlborough, Westland and Canterbury.

CHR 358478 *A. P. Druce* 1980, Mt Garibaldi, bears leaves with abundant, dense, long hairs abaxially and although they become less dense up the blade, they are always present. The same specimen bears on the adaxial sheath surface short hairs in rows similar to those in *C. australis*. Adaxially at the base of the leaf-blade and above there is an enormous thick weft of long (2 mm) hairs. No other specimen seen resembles it for hairiness. CHR 179675 *A. P. Druce* 1968, Tararua Range, is of juveniles; these too bear long hairs on the abaxial leaf-blade surface.

subsp. **flavescens**

Leaf-sheath to 20 cm, pale, incurved, internerves hairy below, glabrous above or appearing so. Leaf-blade to 80 cm × 10 mm, adaxially with abundant prickle-teeth or papillae. Culm to 70 cm, glabrous.

Inflorescence to 15 cm, glabrous except for long hairs at branch axils and hairs below spikelet, rarely hairy elsewhere. Lemma to 5.5 mm; hairs dense on margins, fewer in internerves, < sinus. Anthers to 5 mm.

N.: Tararua and Rimutaka Ranges; S.: Mt Stokes. Grasslands from 750 to 1550 m.

subsp. **brevis** Connor, *N.Z. J. Bot.* *29*: 240 (1991); Holotype: CHR 94668! *H. E. Connor* Starvation Gully, Porters Pass, 8.1.1954.

Leaf-sheath to 20 cm, mostly purpled, internerves and margins hairy. Leaf-blade to 130 cm × 10 mm, adaxially with weft of short hairs at base and abundant prickle-teeth above. Inflorescence to 30 cm, glabrous except for long hairs at branch axils and some stiff hairs below spikelets. Lemma to 6 mm; hairs dense on margins and in internerves, > sinus; lateral lobes usually awned. Anthers 6 mm. Plate 10A.

S.: Southern Marlborough to headwaters of Waitaki River. On open rubbly sites in grassland and scrub; to 1500 m.

subsp. **hirta** Connor, *N.Z. J. Bot.* *29*: 241 (1991); Holotype: CHR 174585! *L. B. Moore* Mt Watson, Paparoa Range, Westland, scrubby slope just below top *c.* 3600 ft, 14.1.67.

Leaf-sheath to 20 cm, purpled, internerves and margins hairy. Leaf-blade to 70 cm × 10 mm, abaxially glabrous but sometimes with long hairs and forming contra-ligule, adaxially with dense weft of long hairs at base and abundant prickle-teeth above. Inflorescence to 25 cm, very long hairy. Glumes often awned. Lemma to 6 mm; hairs dense on margins, fewer in internerves, ≈ sinus. Palea to 7 mm, usually shorter than in other subspp. Anthers to 4.5 mm.

S.: Westland south to Clarke River, and higher passes of Canterbury. In grasslands and scrub; to 1400 m.

subsp. **lupeola** Connor, *N.Z. J. Bot.* *29*: 242 (1991); Holotype: CHR 277734! *A. P. Druce* Mt Arthur, NW Nelson, 4200 ft, Marble rocks, "Unnamed sp." S13 09-38-, Jan. 1975.

Leaf-sheath to 25 cm, pale or purpled, glabrous except on margins. Leaf-blade to 85 cm × 10 mm, adaxially with a dense weft of long hairs at base and abundant prickle-teeth above. Inflorescence to 25 cm, densely hairy. Lemma to 6 mm; hairs dense on margin, elsewhere sparse if present, < sinus. Anthers to 4.5 mm.

S.: Nelson and parts of northern Westland. Almost exclusive to limestone; to 1550 m.

Specimens of plants on "Altimarlock", Black Birch Range, Marlborough, (e.g. CHR 165488 *A. P. Druce* 1967; CHR 247228 *A. W. Purdie* 1973; CHR 262923 *A. W. Purdie* 1974) share with subsp. *lupeola* leaf-sheaths hairy on the margin only, and hairy inflorescences; in the absence of a weft of hairs of any sort on the adaxial surface of the leaf-blade they align with subsp. *flavescens*.

11. C. flavicans Zotov, *N.Z. J. Bot. 1*: 91 (1963); Holotype: CHR 21901! *V. D. Zotov* East Cape, 22.2.1940. = *Danthonia antarctica* var. α *elata* Hook.f., *Fl. N.Z. 1*: 302 (1853); Holotype: K! *Colenso 1149* [Arapawanui River, Hawkes Bay], N. Zealand.

Tall, stout, often sprawling, flabellate tussock with numerous coloured cataphylls and extravaginal shoots, and persistent leaves and sheaths. Leaf-sheath to 15 cm, pinkish or purplish, chartaceous, entire, becoming fibrous, keeled, glabrous or with a few long hairs, apical tuft of hairs to 1 mm. Ligule to 0.7 mm. Leaf-blade to 75 cm × 8 mm, keeled, persistent, glabrous except for some short hairs above ligule and prickle-teeth on margins and abaxially at apex. Culm to 1.5 m, internodes glabrous. Inflorescence to 30 cm, clavate, dense and compact, not naked below; rachis smooth below, branches and pedicels densely scabrid and with some long hairs at branch axils. Spikelets of up to 4 distant florets. Glumes to 7 mm, broad, shallowly bifid, sometimes purpled, margins ciliate, prickle-teeth adaxially above, < nearest lemma lobes; lower 3-nerved, upper 5-nerved. Lemma to 6 mm; hairs dense on margin, usually fewer or none aside central nerve, rarely reaching sinus, prickle-teeth above adaxially and abaxially on nerves; lateral lobes up to 2.5 mm, conspicuously awned adjacent to a small lobe; central awn to 16 mm, reflexed, column absent. Palea to 6 mm, interkeel with prickle-teeth above. Callus to 1.5 mm, hairs to 4 mm. Rachilla to 0.25 mm. Lodicules to 1 mm. Anthers to 4 mm. Gynoecium: ovary to 1 mm; stigma-styles to 3 mm. Caryopsis to 3 mm. $2n = 42$.

Long calluses conspicuously separate florets in a spikelet.

forma **flavicans**

As for species; sizes for floret characters are at the higher end of the range.

Endemic.

N.: Coromandel Peninsula to Hawkes Bay. Cliff and rock faces; sea level to 975 m.

forma **temata** Connor, *N.Z. J. Bot. 29*: 245 (1991); Holotype: CHR 16542! *A. P. Druce* Te Mata Pk, Havelock North, *c.* 1000 ft, limestone cliff, Jan. 1967.

Spikelets of up to 4 florets. Glumes to 4 mm. Lemma to 4 mm, shorter and broader than typical form, hairs dense on margin, few aside central nerve; lateral lobes to 0.2 mm; central awn to 6 mm, reflexed, column absent. Palea to 4.5 mm, interkeel with a few short hairs.

Endemic.

N.: localised on Te Mata, Havelock North. Limestone cliffs.

12. C. juncea Zotov, *N.Z. J. Bot. 1*: 101 (1963); Holotype: CHR 9667! *V. D. Zotov* Denniston, 20.1.1936. = *Danthonia raoulii* var. *teretifolia* Petrie, *T.N.Z.I. 54*: 571 (1923) ≡ *D. rigida* var. *teretifolia* (Petrie) Zotov, *T.R.S.N.Z. 73*: 234 (1943); Lectotype: WELT 40099! *D. Petrie* Cedar Creek near Denniston, S.W. Nelson, 11.2.1913 (designated by Zotov 1963 op. cit. p. 101).

Tall, junceous, red tussock with swollen bases from rootstock clothed in old leaf-sheaths, leaves persistent, ultimately falling below sheath with one fracture. Leaf-sheath to 15 cm, very dark brown, entire, persistent, internerves hairy but often appearing glabrous, margin long hairy above, apical tuft of hairs to 4 mm. Ligule to 0.5 mm. Leaf-blade to 70 cm × 1 mm diam., acicular junceous, finally falling with top part of sheath, abaxially glabrous sometimes with long hairs below aside prominent, shining, hollow keel, becoming glabrous, adaxially with weft of long hairs at base, prickle-teeth above; margin with long hairs below glabrous above. Culm to 90 cm, internodes glabrous. Inflorescence to 20 cm, open, spikelets on long, pulvinate, flexuous branches; rachis and branches sparsely short hairy below becoming glabrous except for few hairs at axils and below spikelets. Spikelets of up to 6 purpled florets. Glumes glabrous, < adjacent lemma lobes; lower to 11 mm, 3-nerved, upper to 12 mm, 5-nerved. Lemma to 5.5 mm; hairs dense on margin, very few aside central nerve, glabrous elsewhere, < sinus; lateral lobes up to 4 mm, including awn up to 2.5 mm; central awn up to 10 mm, reflexed from 1 mm flat column. Palea to 6 mm. Callus to 1 mm, hairs to 3 mm. Rachilla to 0.8 mm. Lodicules to 0.5 mm. Anthers to 3.5 mm. Gynoecium: ovary to 0.5 mm; stigma-styles to 2 mm. Caryopsis to 2 mm.

Endemic.

S.: Westland, especially on the Denniston Plateau. Scrub, tussock and swampy ground; 450 to 950 m.

13. C. lanea Connor, *N.Z. J. Bot. 25*: 165 (1987); Holotype: CHR 166731! *P. Wardle* Trig D, Tin Range, Stewart I., 1600 ft, grassland, 30.3.66.

Tall, slender tussock with abundantly long hairy, dull sheaths, and shiny, slender, deciduous leaves. Leaf-sheath to 15 cm, compressed, entire, persistent, covered with long (4 mm) interlacing internerve hairs, apical tuft of hairs to 4 mm. Ligule to 1 mm. Leaf-blade to 45 cm × 2 mm, V-shaped, or convolute, disarticulating at ligule, abaxially glabrous, adaxially with very short hairs at base and prickle-teeth and papillae above; margin smooth. Culm to 60 cm, internodes glabrous, slightly compressed, sheath with few or no long hairs. Inflorescence to 10 cm, pulvinate, few spikelets, abundantly long hairy. Spikelets of up to 6 florets. Glumes glabrous, acute, > adjacent lemma lobes; lower to 10 mm, 3-nerved, upper to 14 mm, 5-nerved. Lemma to 5 mm; hairs dense at margin and few or none aside central nerve, glabrous elsewhere, < sinus; lateral lobes up to 6 mm, linear-triangular; central awn up to 15 mm from twisting column to 2 mm. Palea to 6 mm. Callus to 0.8 mm, hairs to 3 mm. Rachilla to 1 mm. Lodicules to 1 mm. Anthers to 3.5 mm. Gynoecium: ovary to 1 mm; stigma-styles to 4 mm. Caryopsis to 1.5 mm.

Endemic.

St. Grassland and scrub in wet, boggy, or peaty sites; almost at sea level near some cliffs; to 1000 m.

14. C. macra Zotov, *N.Z. J. Bot. 8*: 91 (1970); Holotype: CHR 950215b! *H. E. Connor* Starvation Gully, Porters Pass, 2900 ft, 8.1.1954.

slim snow-tussock

Tussock of modest stature with persistent sheaths and soft leaf-blades weathering in situ. Leaf-sheath to 15 cm, dark brown below lighter above, persistent, interrib hairs minute or short, apical tuft of hairs to 2 mm. Ligule to 0.6 mm. Leaf-blade to 60 cm × 5 mm, flat or shallowly U-shaped, persistent, abaxially glabrous except for prickle-teeth towards apex and occasionally some long hairs towards base, adaxially with rows of short hairs at base, abundant prickle-teeth; margin with long hairs below, prickle-teeth above. Culm to 60 cm, glabrous, or sheath and internode long hairy, sometimes empurpled. Inflorescence to 25 cm, open, pulvinate, glabrous except for few long hairs at branch axils and some hairs occasionally below spikelet. Spikelet of up to 7 florets. Glumes subequal, to 13 mm, glabrous, acute or occasionally bifid or mucronate, > adjacent lemma lobes; lower 3-nerved, upper 5-nerved. Lemma to 5 mm; hairs dense at margin and aside central nerve, shorter in other internerves or rarely absent, > sinus; lateral lobes to 7.5 mm including 3 mm awn, infrequently long triangular-acute; central awn to 12 mm from twisting column to 3.5 mm. Palea to 7.5 mm. Callus to 1 mm, hairs to 3 mm. Rachilla to 1 mm. Lodicules to 1 mm. Anthers to 4 mm. Gynoecium: ovary to 1 mm; stigma-styles to 3 mm. Caryopsis to 3.25 mm. $2n = 42$. Plate 9B.

Endemic.

S.: east of Main Divide from southern Marlborough to Southland. Grasslands 500–1700 m; at lower altitudes present on shaded aspects of rolling terrain.

Apart from short stature, and frequent evidence of grazing, the long awn column is distinctive. Culm-sheaths and internodes are rarely long hairy south of Rakaia River.

Some specimens from high altitudes matching *C. macra* in vegetative characters but with elongate, fine, leaf-blades with twisting tips and long abaxial hairs are *C. oreophila* × *C. macra*.

15. C. oreophila (Petrie) Zotov, *N.Z. J. Bot. 1*: 104 (1963) ≡ *Danthonia oreophila* Petrie, *T.N.Z.I. 27*: 406 (1895) ≡ *D. pallida* Petrie, *T.N.Z.I. 26*: 271 (1894) non R.Br. (1810); Holotype: WELT 36441! *D. P[etrie]* Kelly's Hill, *c.* 4000 ft, Otira River, Westland, Jan. 1893.   snow-patch grass

Small, pale tussock with persistent sheaths, and leaves very much shorter than slender flowering culms. Leaf-sheath to 5 cm, pale, shiny, glabrous, chartaceous, persistent, margins undulating, apical tuft of hairs to 3 mm. Ligule 0.5–1 mm. Leaf-blade to 15 cm × 2 mm, U- or V-shaped, persistent, spiralling, pointed, narrower than sheath, keeled, veins few, abaxially with prickle-teeth in upper ⅓, long interrib hairs frequent; adaxially with scattered prickle-teeth; margins with prickle-teeth below becoming glabrous above. Culm to 30 cm, internodes glabrous. Inflorescence to 6 cm, up to 8 spikelets on flexuous, pulvinate branches, glabrous except for few long hairs at branch axils and below spikelets. Spikelets small, solitary, of up to 6 purpled or golden florets. Glumes ± equal, to 10 mm, apex sometimes shortly awned, > adjacent lemma lobes; lower 3–5-nerved, upper 5–7-nerved, margin often long hairy below. Lemma to 4.5 mm; hairs long on margins, few aside central nerves and often few in all internerves, < sinus; lateral lobes to 4.5 mm including awn to 2.5 mm, or shortly acute-triangular; central awn to 5.5 mm reflexed from flat 1.5 mm column. Palea to 5.5 mm. Callus to 1.5 mm, hairs to 2.5 mm. Rachilla to 0.75 mm. Lodicules to 0.6 mm. Anthers to 2.5 mm. Gynoecium: ovary to 0.7 mm; stigma-styles to 3 mm. Caryopsis to 2 mm. $2n = 42$. Plate 9D.

Endemic.

S.: east and west of Main Divide from Nelson to Fiordland. In tussock grasslands, snow hollows and banks; 1000–1850 m.

Eventually some leaf-blades disarticulate at the ligule. Plants in Fiordland sometimes have long hairs on the abaxial surface of emerging leaf-blades. CANU 31438 *H. D. Wilson & C. D. Meurk* Sealy Lakes, Mt Cook, has glabrous adaxial leaf-blade surfaces.

16. C. ovata (Buchanan) Zotov, *N.Z. J. Bot. 1*: 104 (1963) ≡ *Danthonia ovata* Buchanan, *Indig. Grasses N.Z.* t. 29 (2) (1879); Holotype: WELT 59578! *J. Morton* Mt Eglinton, South Island. = *D. planifolia* Petrie, *T.N.Z.I. 33*: 328 (1901); Holotype: WELT 36455! *D. Petrie* Clinton Saddle near Lake Te Anau, Jan. 1892.

Sparse tussock with shoots clothed in many old, fibrous, entire sheaths; flowering shoot taller than the leaves which eventually disarticulate at ligule; sheaths, culms, spikelets often purpled or golden. Leaf-sheath to 10 cm, persistent, entire, prominently ribbed, interrib hairs minute, apical tuft of hairs to 2 mm. Ligule to 3.5 mm. Leaf-blade to 25 cm × 5 mm, flat, disarticulating at ligule, spiralling above, pungent, abaxially glabrous, adaxially with scattered prickle-teeth; margin smooth or sometimes with prickle-teeth. Culm to 45 cm, internodes glabrous except for hairs below inflorescence. Inflorescence to 10 cm, shortly branched; rachis, branches and pedicels long hairy. Spikelets of up to 8 strongly purpled florets. Glumes > adjacent lemma lobes; lower to 11 mm, 1–5-nerved, acute, sometimes bifid, upper to 16 mm, 3–5–7-nerved, acute. Lemma to 8 mm; hairs dense on margin and aside central nerve, usually glabrous elsewhere, < sinus; lateral lobes to 7 mm acute, or produced into awn up to 5 mm, or long triangular-acute; central awn to 20 mm divergent from 1.5 mm flat column. Palea to 9 mm. Callus to 1 mm, hairs to 3 mm. Rachilla to 0.75 mm. Lodicules to 1 mm. Anthers to 3 mm. Gynoecium: ovary to 1 mm; stigma-styles to 2.5 mm. Caryopsis to 3.25 mm. $2n = 42$.

Endemic.

S.: Fiordland. Wet places in grasslands, and rock crevices; to 1400 m.

17. C. pallens Zotov, *N.Z. J. Bot. 1*: 99 (1963); Holotype: CHR 7317! *V. D. Zotov* Pukematawai, *c.* 5000 ft, Tararua Mountains, 22.1.1933.

midribbed snow-tussock

Tall, pallid tussock with inflorescences pink-tinged; leaf-blade persisting on leaf-sheath or caducous at ligule. Leaf-sheath glabrous or internerves with long or short hairs, entire, becoming fibrous, keeled, usually with apical tuft of hairs. Ligule to 1 mm. Leaf-blade keeled, flat or U-shaped or involute, abaxially glabrous but sometimes long hairy on emergent leaf-blades, scabrid apically, adaxially usually with prickle-teeth or less frequently papillae; margin usually with long hairs below, becoming prickle-toothed. Culm to 1 m, internodes glabrous. Inflorescence glabrous, or glabrous below becoming scabrid or stiff hairy above, some long hairs at branch axils and some short hairs below spikelet, sometimes branches and pedicels with mixed short and long hairs. Spikelets of up to 9 lightly purpled or straw coloured florets. Glumes chartaceous, glabrous, acute, < nearest lemma lobes; lower to 11 mm, 1–3–5-nerved, upper to 14 mm, 5–7–9-nerved. Lemma to 6 mm; hairs dense on margin and aside central nerve, fewer or absent elsewhere, < sinus; lateral lobes to 6 mm including awn to 3 mm, or usually long triangular-acute; central awn to 16 mm from 2 mm twisting or flat column. Palea to 8 mm. Callus to 1.5 mm, hairs to 3 mm. Rachilla to 1 mm. Lodicules to 1 mm. Anthers to 5 mm. Gynoecium: ovary to 1 mm; stigma-styles to 3 mm. Caryopsis to 3.5 mm. $2n = 42$.

Endemic.

N.: mountains south from Raukumara Range; S.: mountains.

subsp. **pallens**

Stout tussock with persistent leaf-blades. Leaf-sheath to 20 mm, pale straw coloured, internerves and margins glabrous, apical tuft of hairs small or absent. Leaf-blade to 1 m × 6 mm, persistent, adaxially with prickle-teeth or papillae. Culm to 1 m, sheath margins glabrous.

Endemic.

N.: mountain ranges from Raukumara Range south; S.: mountain ranges in northern Marlborough and Nelson. Grasslands to 1550 m.

Included here are the many plants of short stature collected from the Tararua Range (CHR 262891, 82216, 86290) and from other North Id mountain ranges (CHR 25155, 74363, 63188, 389809) which in this respect seem to be different from the holotype.

CHR 366206a *A. P. Druce* Tararua Range, has flowers with ligulate lodicules up to 2 mm, as also in CHR 62367 (Tararua Range). In other plants lodicules were of the normal rhomboid shape and *c.* 1 mm e.g., CHR 6551 *V. D. Zotov*, CHR 262891a *A. P. Druce*.

In specimens from the Ruahine Range occasionally the culm-sheath margins are hairy; this is very occasionally so in South Id.

subsp. **cadens** Connor, *N.Z. J. Bot. 29*: 251 (1991); Holotype: CHR 320475! *W. G. Lee* Plateau Creek, Murchison Mts, Fiordland S140 594403, young, fertile sites beneath bluff system, Feb. 1979.

Stout tussock with many persistent, old, bladeless, straw-coloured or purple-suffused, keeled, entire, hairy sheaths. Leaf-sheath to 15 cm, apical tuft of hairs to 2 mm. Leaf-blade to 60 cm × 10 mm, twisting above, adaxially occasionally with some long hairs and abundant prickle-teeth, disarticulating at ligule. Culm to 1 m, sheath margins hairy. Inflorescence to 20 cm. Plate 10D.

S.: Mountains of Canterbury, Westland, Otago and Fiordland south of 43° 30'. Grasslands to 1700 m, occasionally to 2000 m.

subsp. **pilosa** Connor, *N.Z. J. Bot. 29*: 252 (1991); Holotype: CHR 176087! *I. M. Ritchie* Ridge to Manakau, Kowhai R. *c.* 3500 ft. Seaward Kaikoura Ra., Marlborough, 13.1.1967.

Leaf-sheath to 20 cm, often purple; internerves and margins hairy, apical tuft of hairs to 2 mm. Leaf-blade to 70 cm × 8 mm, persistent, adaxially with scattered long hairs (4 mm) and prickle-teeth. Culm to 1 m, sheath margin hairy.

S.: Mountain ranges in southern Marlborough and Nelson, North Canterbury, and Westland. Grasslands; to 1650 m, but at higher altitudes in Nelson.

At and near Nelson Lakes the adaxial leaf-blade surface lacks long hairs, but the culm-sheath margins and the leaf-sheath internerves are hairy. Populations at "Altima.lock", Black Birch Range, Marlborough, are variable in stature, just as in subsp. *pallens* on the Tararua Range.

18. C. rigida (Raoul) Zotov, *N.Z. J. Bot. 1*: 96 (1963) ≡ *Danthonia rigida* Raoul, *Ann. Sci. Nat.* Ser. 3, Bot. *2*: 116 (1844) ≡ *D. raoulii* Steud., *Syn. Pl. Glum. 1*: 246 (1854); Holotype: P! *E. F. L. Raoul* Nouvelle Zélande, Prom. de Banks, 1843. = *D. raoulii* var. *flavescens* (Hook.f.) Hack., *in* Cheeseman *Man. N.Z. Fl.* 886 (1906) ≡ *D. flavescens* Hook.f., *Handbk N.Z. Fl.* 332 (1864); Lectotype: K (Photo!) *Hector & Buchanan No. 8* Otago Lakes District (designated by Connor 1991 op. cit. p. 253).

narrow-leaved snow-tussock

Tall, stout, golden tussock with hairy fracturing sheaths and narrow, sweeping, long, deciduous leaves which form a dense litter. Leaf-sheath to 30 cm, dark brown below, keeled, interribs with abundant long and/ or short hairs occasionally forming a contra-ligule, fragmenting into short segments, apical tuft of hairs to 2 mm. Ligule to 2 mm. Leaf-blade tough, flat or shallowly U-shaped, adaxially usually with dense weft of short or long hairs at base, abundant prickle-teeth above; margin thick, long hairy below; falling with part of sheath. Culm to 1.5 m, internodes infrequently long hairy; sheath hairy. Inflorescence open, pulvinate. Spikelets of up to 8 florets. Glumes glabrous, acute or very shortly awned, ≥ nearest lemma lobes; lower to 14 mm, 1–3-nerved, upper to 16 mm, 5–7-nerved, very occasionally long hairy on margin below. Lemma densely hairy on margin and in all internerves, hairs rarely absent, > sinus. Palea to 8.5 mm. Callus to 1.5 mm, hairs to 3.5 mm. Rachilla to 1.5 mm. Lodicules to 0.75 mm. Gynoecium: ovary to 1 mm; stigma-styles to 3 mm. Caryopsis to 3.5 mm. $2n = 42$. Fig. 16.

Endemic.

S.: east and west of Main Divide south of latitude 43°S.

subsp. **rigida**

Leaf-sheath to 30 cm. Leaf-blade to 80 cm × 7 mm, abaxially glaucous, very occasionally with a few hairs aside midrib at base; margin thick, long hairy below sometimes hairs interlocking across leaf-blade, becoming scabrid. Inflorescence to 30 cm, glabrous except for some long hairs at branch axils. Lemma to 6 mm; lateral lobes to 7.5 mm including awn to 3.5 mm, or long triangular-acute; central awn to 12 mm from twisting column to 4 mm. Anthers to 5 mm. Plate 10B.

S.: east of Main Divide from Banks Peninsula and Rakaia River to Southland. Grasslands to 1700 m.

In Otago and at Mt Cook the adaxial weft of hairs extends further up the leaf-blade than at other sites. Inflorescences are hairy on CHR 9963, 9899, 9333 Ben Lomond; CHR 262867a, Waimate; CHR 29290 Waikouaiti; CHR 20043 Lindis Pass; CHR 75113b, CHR 75116 Skipper Saddle.

Plants of *C. rigida* subsp. *rigida* with narrow leaves and of short stature may be confused with *C. macra*, but the sheath fracturing habit readily distinguishes them. Such plants may be found on sites of lowered fertility or inhospitable exposure as on the Ben Ohau Range (CHR 465773–465777 *H. E. Connor* 1968). Plants found in North Canterbury, Marlborough, and Buller resembling *C. rigida* are usually *C. flavescens* × *C. rubra*.

   subsp. **amara** Connor, *N.Z. J. Bot. 29*: 254 (1991); Holotype: CHR 309513! *P. A. Williams & P. Wardle* Woolsack, Okuru district Sth Westland, abundant on range of slopes, 4800 ft, below limit of *C. pallens* and *C. crassiuscula*, 1.4.79.

Leaf-sheath to 20 cm. Leaf-blade to 85 cm × 7 mm, abaxially with long hairs aside midrib at base; margin thick, long hairy below, becoming glabrous above. Inflorescence to 25 cm; rachis, branches and pedicels hairy or scabrid, sometimes glabrous below and ornamented above, long hairs below spikelets. Lemma to 6.5 mm; lateral lobes to 6 mm, triangular-acute or with awn to 2.5 mm, sometimes shortly lobed again; central awn to 16 mm from twisting column to 2 mm. Anthers to 4.5 mm.

S.: west of the Main Divide from Lake Kaniere to Fiordland, and south-east to Longwood Range; St.: Mt Anglem. Low scrub, and tussock grassland on poorly drained sites and bogs; 500–1450 m.

19. C. rubra Zotov, *N.Z. J. Bot. 1*: 96 (1963); Holotype: CHR 41591!
V. D. Zotov Tongariro National Park, 8.1.1944. = *Danthonia antarctica*
var. δ *minor* Hook.f., *Fl. N.Z. 1*: 302 (1853); Lectotype: K (Photo!)
Colenso 4420 ["Puke Taramea, halfway up Ruahines"] (designated
by Connor 1991 op. cit. p. 254).   red tussock

Tall, slender, red tussock with crowded, erect, stiff, junceous leaves.
Leaf-sheath to 30 cm, dark brown, keeled, incurving, remaining entire
or fracturing into short segments, margin separating and coiling,
apical tuft of hairs to 3 mm. Ligule to 1 mm. Leaf-blade to 1 m ×
1.2 mm diam., acicular junceous, splitting longitudinally, keel hollow,
abaxially glabrous but infrequently with long hairs near base, prickle-
teeth towards apex, adaxially with rows of short hairs or wefts of long
hairs at base, prickle-teeth and sometimes papillae above; margin
with long hairs below, prickle-teeth above, rarely smooth. Culm to
1.5 m, internodes glabrous but sometimes long hairy, sheath glabrous.
Inflorescence to 45 cm, open on pulvinate branches, glabrous except
for long hairs at branch axils and short stiff hairs below spikelets,
rarely becoming scabrid above. Spikelets of up to 9 florets. Glumes
glabrous, acute, infrequently awned, ≤ adjacent lemma lobes, lower
to 12 mm, 1–3–5-nerved, upper to 14 mm, 3–5–7-nerved. Lemma to
6 mm; hairs dense on margin, usually absent or sparse elsewhere, or
present in all internerves (subsp. *cuprea*), < sinus; lateral lobes to
6 mm including awn to 3 mm, rarely unawned; central awn to 13 mm
from twisting column to 3 mm. Palea to 8 mm. Callus to 1.5 mm,
hairs to 4 mm. Rachilla to 0.75 mm. Lodicules to 1 mm. Anthers to
5 mm. Gynoecium: ovary to 1 mm; stigma-styles to 4 mm. Caryopsis
to 3.5 mm. $2n = 42$.

Endemic.

N.: Volcanic Plateau to Tararua Range; S.: throughout except in
lowland Westland; St.

subsp. **rubra**

Leaf-sheath to 30 cm, fracturing into short segments, dark brown,
internerves glabrous or long hairy. Leaf-blade to 1 m, falling with
part of sheath, adaxially with rows of short hairs at base. Culm to

1.5 m, internodes and sheath glabrous. Lemma with long dense hairs on margin, very sparse or absent elsewhere, < sinus; lateral lobes to 6 mm including awn to 2 mm.

var. **rubra**

Leaf-blade red to red-brown, margin scabrid, adaxially with prickle-teeth. Lemma to 5 mm. Anthers to 3.5 mm.

N.: Volcanic Plateau southwards; S.: Marlborough and North Canterbury. Tussock grassland and bogs; 550–1400 m.

var. **inermis** Connor, *N.Z. J. Bot. 29*: 255 (1991); Holotype: CHR 158671! *A. P. Druce* Mt Egmont: Holly Flat, 3200 ft, Scrub-tussockland, Jan. 1963.

Leaf-blade greenish, adaxially papillate, margin smooth. Anthers to 5 mm.

N.: Mt Egmont. Tussock grassland and bogs; 900–1500 m.

subsp. **cuprea** Connor, *N.Z. J. Bot. 29*: 256 (1991); Holotype: CHR 132481! *H. E. Connor* Pigroot, 16 miles from Ranfurly, 22.1.1955.

Leaf-sheath to 30 cm, entire, dark brown, interribs with minute hairs sometimes glabrous. Leaf-blade to 1 m, persisting on sheath, adaxially with dense weft of long hairs at base extending up leaf-blade often with short hairs as well, papillae or prickle-teeth above. Culm to 1.5 m, internodes glabrous, sheath glabrous. Lemma to 6 mm; hairs dense on margins less so in internerves, < sinus; lateral lobes to 7 mm, awn to 3.5 mm, infrequently long triangular-acute. Plates 8D, 9A.

S.: North Canterbury and south and west to Fiordland; St. Bogs and tussock grassland; sea level to 1500 m.

subsp. **occulta** Connor, *N.Z. J. Bot. 29*: 257 (1991); Holotype: CHR 247260! *P. Wardle* Mark's Flat, Clarke River, Westland, Swamp 3100 ft., 23.2.72.

Leaf-sheath to 30 cm, fracturing into short segments, dark brown, interribs with minute and long hairs. Leaf-blade to 1 m, falling with part of sheath; adaxially with dense weft of long hairs extending up leaf-blade or short hairs in rows, papillate or prickle-toothed above.

Culm to 1.5 m, internodes sometimes long hairy, sheath glabrous. Lemma to 6 mm; hairs dense on margin, absent or very sparse elsewhere, < sinus; lateral lobes to 6 mm including awn to 2.5 mm.

S.: Nelson and western coast to Cascade Plateau. Scrubland, or tussock grassland; 600–1700 m.

Plants in Nelson Lakes National Park are often of larger stature than at many other sites.

Plants of the west coast lowland populations at Skiffington Swamp (300 m) and Lake Gault (300 m) do not conform to subsp. *occulta* in that the leaf-sheaths remain entire and leaf-blades are persistent on them. Abundant long hairs are present on the abaxial leaf-blade. For adaxial leaf-blade hairs, glabrous inflorescences, and lemma hairs there is good agreement. CHR 508626a,b *P. Wardle 94/129 & R. P. Buxton* near Smythe–Wanganui confluence, Westland, bears long hairs on the abaxial leaf-blade surface but has fracturing leaf-sheaths.

20. C. spiralis Zotov, *N.Z. J. Bot. 1*: 100 (1963); Holotype: CHR 110792! *M. J. A. Simpson 1915* Head of Lake Monk Valley, 3200 ft, Fiordland, 19.1.1960.

Slender tussock with narrow leaves falling and leaving inwardly spiralling sheaths. Leaf-sheath to 20 cm, glabrous, pale, chartaceous, spiralling and breaking into short segments, margin hairy above, apical tuft of hairs to 3 mm. Ligule to 1 mm. Leaf-blade to 50 cm × 1 mm diam., acicular junceous, falling with part of sheath, abaxially with occasional long (2 mm) hairs below, adaxially with dense weft of long (3 mm) hairs at base projecting over smooth margin, scattered prickle-teeth above. Culm to 65 cm, internodes glabrous. Inflorescence to 12 cm, narrow, glabrous except for occasional long hairs at axils. Spikelets of up to 7 florets. Glumes glabrous, ≥ adjacent lemma lobes, acute or shortly awned; lower to 12 mm, 3-nerved, upper to 13 mm, 3–5-nerved. Lemma to 5 mm; hairs dense on margin fewer aside central nerve, glabrous or sparsely hairy elsewhere, < sinus; lateral lobes up to 4.5 mm, long triangular-acute; central awn up to 13 mm divergent from 2.5 mm flat column. Palea to 6 mm. Callus 0.5 mm, hairs to 3 mm. Rachilla to 0.8 mm. Lodicules to 0.75 mm. Anthers to 4 mm. Gynoecium: ovary to 0.75 mm; stigma-styles to 4.5 mm. $2n = 42$.

Endemic.

S.: Known only from Takahe Valley (Murchison Mts); Lake Monk (Cameron Mts); Mt Luxmore (Kepler Mts). Limestone bluffs; to 1000 m.

21. C. teretifolia (Petrie) Zotov, *N.Z. J. Bot. 1*: 100 (1963) ≡ *Danthonia teretifolia* Petrie, *T.N.Z.I. 46*: 36 (1914); Lectotype: WELT 40323! *D. Petrie* Longwood Range 2200 ft, Southland, 1.1.1913 (designated by Zotov 1963 op. cit. p. 100).

Small tussock of few pale shoots from darkened persistent sheaths, with hairy leaves twisting above but deciduous. Leaf-sheath up to 15 cm, pale green, sometimes reddened or purpled, persistent, interribs hairy, apical tuft of hairs to 2.5 mm. Ligule to 1 mm. Leaf-blade to 40 cm × 1 mm diam., acicular junceous, disarticulating at ligule, abaxially with long (1.5 mm) white, antrorse, hairs aside evident main nerve or between all nerves below, adaxially clothed in dense antrorse short (0.5 mm) hairs becoming sparser; margin with long hairs below becoming smooth above. Culm to 50 cm, internodes glabrous. Inflorescence to 15 cm, open, pulvinate, long hairy, branches flexuous. Spikelets few, solitary, of up to 6 purpled florets. Glumes < adjacent lemma lobes; lower to 14 mm, 3-nerved, upper to 16 mm, 5-nerved. Lemma to 6 mm, purpled; hairs dense at margins and sparse aside central nerve rarely elsewhere, ≥ sinus; lateral lobes to 6 mm including awn to 3 mm; central awn to 11 mm from twisting column to 4 mm. Palea to 8 mm. Callus to 1 mm, hairs to 2.5 mm. Rachilla to 1 mm. Lodicules to 2 mm. Anthers to 4.5 mm. Gynoecium: ovary to 1 mm; stigma-styles to 2.5 mm. Caryopsis to 2.5 mm. $2n = 42$. Plates 1B, 8E.

Endemic.

S.: Fiordland and Southland. Grasslands and herbfield on leached slopes or peaty soils; 300 to 1300 m.

22. C. vireta Connor, *N.Z. J. Bot. 29*: 261 (1991); Holotype: CHR 218503! *P. Wardle* Landsborough up valley from Spencer Glacier, Sth Westland, 4100 ft. on sunny moraine terrace, with *Chion[ochloa] oreophila*, 14.3.1971.

Slender tussock with stramineous, hairy sheaths. Leaf-sheath to 10 cm, hairy, entire, margins often undulating, apical tuft of hairs to 3.5 mm. Ligule to 0.5 mm. Leaf-blade to 30 cm × 2.5 mm, flat or ± U–V-shaped, keeled, twisting above, disarticulating at ligule, abaxially prickle-toothed above, adaxially with prickle-teeth; margin with long hairs below, becoming prickle-toothed or rarely becoming smooth. Culm to 60 cm, slender, internodes glabrous. Inflorescence to 13 cm, glabrous except for long hairs at branch axils. Spikelets golden and purple, of up to 8 florets. Glumes usually > nearest lemma lobes; lower to 10 mm, 1–3–5-nerved, upper to 11 mm, 3–5–7-nerved. Lemma to 5.5 mm; hairs dense on margin and aside central nerve, sparse or absent elsewhere; lateral lobes to 4 mm, shortly awned or triangular-acute; central awn to 7 mm from a flat, rarely twisting, column to 2 mm. Palea to 7 mm. Callus to 1 mm, hairs to 2.5 mm. Rachilla to 1 mm. Lodicules to 1 mm. Anthers to 3.5 mm. Gynoecium: ovary to 1 mm; stigma-styles to 2 mm. Caryopsis to 2 mm.

Endemic.

S.: Close to Main Divide from south Canterbury to Fiordland. Grasslands; 1200–1600 m.

INCERTAE SEDIS   The exact identity of specimens gathered by A. F. Mark and J. A. Wells from Mt Burns, Hunter Mts, Fiordland, at 5400 ft, in May 1967, and described as "Dominating extensive areas above about 4,800 ft, except on exposed ridge crests" (OTA 18879, OTA 18880; duplicate CHR 132494) is somewhat uncertain. They have keeled, flat, prickle-toothed, persistent leaf-blades up to 5 mm wide, on entire, minutely hairy sheaths; the inflorescences of late season origin are glabrous. They are hybridogenous. Similar plants were gathered on Mt Tiriroa by K. Lloyd in 1997.

HYBRIDS

Chionochloa×elata (Petrie) Connor, *N.Z. J. Bot. 29*: 272 (1991) ≡ *Danthonia oreophila* var. *elata* Petrie, *T.N.Z.I. 45*: 274 (1913) ≡ *Chionochloa oreophila* var. *elata* (Petrie) Zotov, *N.Z. J. Bot. 1*: 104 (1963) ≡ *Chionochloa elata* (Petrie) Connor, *N.Z. J. Bot. 19*: 166 (1981); Lectotype: WELT 36436! *D. P[etrie]* Sealy Range, Mt Cook, 5000 ft, Feb. 1911 (designated by Zotov 1963 op. cit. p. 104).

Because this hybrid has a formal name it is described in full:

Slender yellow tussock of modest stature with shining leaves disarticulating at ligule. Leaf-sheath to 15 cm, entire, apical tuft to 3.5 mm. Ligule to 0.5 mm. Leaf-blade to 30 cm × 3 mm, ± flat or U-shaped, keeled, drawn out to long thin spiralling point, abaxially smooth and glabrous except for prickle-teeth above and sometimes long internerve hairs, adaxially with abundant prickle-teeth; margin with long hairs below, prickle-teeth above. Culm to 60 cm, internodes glabrous. Inflorescence to 15 cm. Spikelets of up to 7 golden and lightly purpled florets. Glumes > nearest lemma lobes; lower to 9 mm, 3-nerved, upper to 11 mm, 1–3–5-nerved, glabrous. Lemma to 5.5 mm; hairs dense at margin and aside main nerve, and in most internerves; lateral lobes to 4.5 mm including awn to 1 mm or long triangular-acute; central awn to 10 mm reflexed from flat column up to 2 mm. Palea to 7 mm. Rachilla to 1.5 mm. Anthers to 3.75 mm. Gynoecium: ovary to 0.75 mm; stigma-styles to 2.75 mm. Caryopsis to 2.5 mm.

S.: Sealy Range, Mt Cook. Alpine and subalpine grasslands; 1200–1700 m.

Connor (1991 op. cit.) restricted the name solely to plants from the Sealy Range, Mt Cook National Park; they are local *C. oreophila* × *C. pallens* subsp. *cadens*.

To thirty five interspecific hybrid combinations described in detail by Connor (1991 op. cit.), *C. beddiei* × *C. flavescens* and *C. conspicua* × *C. flavescens* are now added. Localised distribution of some species limits the extent of interspecific hybridisation: *C. acicularis* × *C. oreophila* (Fiordland), *C. acicularis* × *C. rubra* (Fiordland), *C. australis* × *C. conspicua* (Nelson), *C. australis* × *C. crassiuscula* (Canterbury), *C. australis* × *C. macra* (Marlborough), *C. australis* × *C. oreophila* (Nelson and Canterbury), *C. australis* × *C. pallens* (Nelson and Marlborough), *C. australis* × *C. rubra* (Nelson), *C. beddiei* × *C. cheesemanii* (Wellington), *C. beddiei* × *C. flavescens* (Marlborough Sounds), *C. cheesemanii* × *C. conspicua* (Urewera N.P., Manawatu, Nelson), *C. cheesemanii* × *C. rubra* (north-west Nelson), *C. conspicua* × *C. flavescens* (Tararua Range), *C. conspicua* × *C. rubra* (Egmont, central North Id, Nelson, Canterbury, Mt Cook, Stewart Id), *C. conspicua* × *C. pallens* (Gisborne), *C. crassiuscula* × *C. lanea* (Stewart Id), *C. crassiuscula* × *C. oreophila* (Canterbury), *C. crassiuscula* × *C. pallens* (Canterbury), *C. crassiuscula* × *C. rigida* (south Westland, Stewart Id), *C. crassiuscula* × *C. rubra* (Stewart Id), *C. crassiuscula* × *C. teretifolia* (Southland), *C. flavescens* × *C. macra* (Canterbury), *C. flavescens* × *C. oreophila* (Mt Cook), *C. flavescens* × *C. pallens* (Tararua Range, Marlborough, Westland), *C. flavescens* × *C. rubra* (Tararua Range, Nelson, Buller, Canterbury), *C. macra* × *C. oreophila* (Southern Alps), *C. macra* × *C. pallens* (Southland), *C. macra* × *C. rigida* (Canterbury), *C. macra* × *C. teretifolia* (Southland), *C. oreophila* × *C. pallens* (Nelson, Westland), *C. oreophila* × *C. teretifolia* (Fiordland), *C. pallens* × *C. rigida* (Fiordland), *C. pallens* × *C. rubra* (central North Island, Nelson, Marlborough), *C. rigida* × *C. rubra* (Canterbury, Otago), *C. rigida* × *C. teretifolia* (Fiordland, Southland), *C. rubra* × *C. spiralis* (Fiordland), *C. spiralis* × *C. teretifolia* (Fiordland).

Experimental interspecific hybrids were fertile, as were those which occurred spontaneously in the experimental garden [Connor, H. E. *N.Z. J. Bot.* 5: 3–16 (1967)]. Seeds, often present in natural interspecific hybrids, are consistent with experimental results.

REPRODUCTIVE BIOLOGY   Reproduction is by seeds. All species are chasmo-gamous and anthers and stigmas are simultaneously exserted at anthesis. Most of the pollen produced by *C. pallens* subsp. *pilosa* comes from the largest plants [McKone, M. J. *New Phytol. 116*: 555–562 (1990)]. All species tested are self-compatible [Connor, H. E. *N.Z. J. Agric. Res. 3*: 728–733 (1960); Connor, H. E. *N.Z. J. Bot. 5*: 3–16 (1967)]; McKone, M. J., Thom, A. L. and Kelly, D. *N.Z. J. Bot. 35*: 259–262 (1997) experimentally demonstrated self-compatibility in *C. macra* and *C. pallens*. Self-pollination is not exclusive as the widespread occurrence of natural hybrids indicates (*vide supra*). Connor, H. E. *N.Z. J. Bot. 5*: 3–16 (1967) suggested that *C. flavicans* might be apomictic but there is no confirming evidence. Gynodioecism occurs in *C. bromoides*; female frequency is very variable — 58.2 % on Poor Knights Islands and zero in several populations on the mainland [Connor, H. E. *N.Z. J. Bot. 28*: 59–65 (1990)].

Flowering is irregular and there are mast years [Connor, H. E. *N.Z. J. Bot. 4*: 392–397 (1966); Mark, A. F. *N.Z. J. Bot. 3*: 180–193 (1965), *N.Z. J. Bot. 3*: 300–319 (1965), *Vegetatio 18*: 289–306 (1969); Payton, I. J. and Mark, A. F. *N.Z. J. Bot. 17*: 43–54 (1979); Kelly, D. *Trends in ecology and evolution 9*: 465–470 (1994); Kelly, D. et al. *N.Z. J. Bot. 30*: 125–133 (1992); McKone, M. J., Kelly, D. and Lee, W. G. *Global Change Biology 4*: 591–596 (1998)].

CYTOLOGY $2n = 6x = 42$ in all 17 species examined: Beuzenberg, E. J. and Hair, J. B. *N.Z. J. Bot. 21*: 13–20 (1983); Calder, J. W. *J. Linn. Soc. Bot. 51*: 1–9 (1937); Dawson, M. I. *N.Z. J. Bot. 27*: 163–165 (1989); just what plants were growing at the Royal Botanic Gardens, Kew, in which Singh, D. N. and Edmonds, M. B. E. *Heredity 18*: 538–540 (1963) reported $2n = 12$ cannot be determined, but it is unlikely that they were *Danthonia flavescens* Hook. as was suggested. Somatic chromosome counts of $2n = 36$ by Calder, J. W. *J. Linn. Soc. Bot. 51*: 1–9 (1937) are unsupported by current counts.

CHEMOTAXONOMY   (1) Epicuticular leaf wax composition: (a) lipids [Cowlishaw, M. G., Bickerstaffe, R. and Young, H. *Phytochem. 22*: 119–124 (1983); Cowlishaw, M. G., Bickerstaffe, R. and Connor, H. E. *Biochem. Syst. Evol. 11*: 247–259 (1983); Savill, M. G., Bickerstaffe, R. and Connor, H. E. *Phytochem. 27*: 3499–3507 (1988)]; 20 spp.; (b) triterpene methyl ethers: [Russell, G. B., Connor, H. E. and Purdie, A. W. *Phytochem. 15*: 1933–1935 (1976); Connor, H. E. and Purdie, A. W. *N.Z. J. Bot. 14*: 315–326 (1976), *N.Z. J. Bot. 19*: 161–170 (1981); Pauptit, R. A. et al. *Aust. J. Chem. 37*: 1341–1347 (1984); Rowan, D. D. and Russell, G. B. *Phytochem. 31*: 702–703 (1992)]; data are summarised in Connor (1991 op. cit.); triterpene methyl ethers are not universally found in the 22 spp. analysed. (2) Structural polysaccharide composition [Connor, H. E. et al. *N.Z. J. Agric. Res. 13*: 534–554 (1970); Bailey, R. W. and Connor, H. E. *N.Z. J. Bot. 10*: 533–544 (1972); Connor, H. E. and Bailey, R. W. *N.Z. J. Bot. 10*: 515–532 (1972)]; carbohydrates [Payton,

I. J. and Brasch, D. J. *N.Z. J. Bot. 16*: 435–460 (1978)]. (3) Concentration of inorganic elements [Connor, H. E., Bailey, R. W. and O'Connor, K. F. *N.Z. J. Agric. Res. 13*: 534–554 (1970); Molloy, B. P. J. and Connor, H. E. *N.Z. J. Bot. 8*: 132–152 (1970); O'Connor, K. F., Connor, H. E. and Molloy, B. P. J. *N.Z. J. Bot. 10*: 205–224 (1972); Williams, P. A. *N.Z. J. Bot. 15*: 399–442 (1977); Williams, P. A., Mugambi, S. and O'Connor, K. F. *N.Z. J. Bot. 15*: 761–765 (1977); Williams, P. A. et al. *N.Z. J. Bot. 16*: 255–260 (1978); Williams, P. A. et al. *N.Z. J. Bot. 16*: 479–498 (1978); Lee, W. G. and Fenner, M. *J. Ecol. 77*: 704–716 (1989); Payton, I. J. et al. *N.Z. J. Bot. 24*: 529–537 (1986)].

PYRRHANTHERA Zotov, 1963

Monotypic and endemic to N.Z. Characters as for the sp.

1. P. exigua (Kirk) Zotov, *N.Z. J. Bot. 1*: 126 (1963) ≡ *Triodia exigua* Kirk, *T.N.Z.I. 14*: 378 (1882) ≡ *Sieglingia exigua* (Kirk) Kuntze, *Rev. Gen. Pl. 2*: 789 (1891) ≡ *Danthonia exigua* (Kirk) Zotov, *T.R.S.N.Z. 73*: 234 (1943) ≡ *Rytidosperma exiguum* (Kirk) H.P.Linder, *Telopea 6*: 614 (1996); Lectotype: WELT 40227! *T. Kirk* by the Thomas River, Waimakariri, Feb. 1881 (designated by Zotov 1963 op. cit. p. 126).

Low-growing tufts of slender shoots usually spaced singly along a comparatively stout, far-creeping, woody rhizome; leaves setaceous, curved outwards, bright green becoming yellow, apparently glabrous except at sheath apex, much < mature culms; branching intravaginal. Leaf-sheath tightly appressed to shoot, mostly below ground-level, light brown to whitish, pinkish-tinged at first, entirely glabrous, shining, soft, ± transparent, ribs more distinct in older sheaths, apical tuft and collar hairs to 1 mm. Ligule *c.* 0.1 mm. Leaf-blade to 6 cm, folded, with smooth incurved margins, glabrous, tip obtuse, with a few minute prickle-teeth. Culm 2–7–(8.5) cm, erect, filiform, internode glabrous. Inflorescence a reduced panicle, to 1.5 cm, of 1–2–(4) brownish green, ovoid, 2–3–(6)-flowered spikelets *c.* 6 mm; pedicels minutely hairy; florets ⚥; disarticulating above glumes and between florets. Glumes ± equal, (3.5)–5–6.5–(9.5) mm, ≈ florets, stiffly concave, ovate, green at centre with purple lateral band and broad light brown margins, lower 7-nerved, upper 5-nerved, glabrous apart from minute prickle-teeth on midrib near apex and finely ciliate margins. Lemma (2.5)–3.5–4–(5) mm, (7)–9-nerved, light green, purple-tinged above, hard and shining; hairs short and scattered, longer on margins and in 2 bands ± centrally, scarious towards tip with rather dense minute prickle-teeth; apex tridentate with central scabrid mucro 0.2–0.5–(0.8) mm between shorter lobes. Palea < lemma, (2.5)–3–3.5–(4.5) mm, keel and apex conspicuously densely ciliate, several longer hairs on margin. Callus 0.3–0.5 mm, glabrous. Rachilla glabrous. Lodicules 2, ± truncate, apically long hairy. Stamens 3, anthers (0.8)–1.3–1.8 mm, scarlet or cream. Gynoecium: ovary 0.5 mm; stigma-styles 2, *c.* 2 mm. Fruit an achene, *c.* 1.5 × 0.8–1 mm; embryo ⅓ to ½ length achene; hilum basal, ⅓ length achene.

Endemic.

S.: Canterbury and North Otago. Lowland to montane in short tussock grassland and dry or sandy ground.

Inflorescences may be reduced to 1 spikelet on hairy pedicel above aborted spikelet(s). On the ovary apex the remnant of aborted third style is evident.

RYTIDOSPERMA Steud., 1854

= *Notodanthonia* Zotov, *N.Z. J. Bot. 1*: 104 (1963) ≡ *Thonandia* H.P.Linder, *Telopea 6*: 612 (1996) *nom. illeg.*

Tufted perennials, short or moderately tall, with intra- or extravaginal branching. Leaf-sheath with apical tuft of hairs sometimes extending dorsally, abaxially ± shining, glabrous or hairy. Ligule a rim of hairs. Leaf-blade persistent or disarticulating at ligule, flat or inrolled, ± flaccid, or stiff, abaxially glabrous or hairy, adaxially minutely prickle-toothed, occasionally with long hairs, margins smooth or scabrid. Inflorescence a raceme, racemose panicle or panicle. Rachis and pedicels slender, often hairy at branch-axils and usually scabrid to hairy below erect few- to several-flowered spikelets; florets ♀; disarticulating above glumes and between florets. Glumes ± equal, firmly membranous, keeled above, nerves 3–13, anastomosing, occasionally internerves long-hairy, adaxially with minute prickle-teeth. Lemma with two scabrid lateral lobes usually tipped by a minute to long, straight awn; central awn from sinus longer and stronger, usually geniculate, column ± twisted; nerves 5–9, anastomosing;

abaxially pilose in two complete or incomplete transverse series of tufts of rigid hairs, sometimes reduced to marginal tufts. Palea membranous, keels and apex finely ciliate, interkeel glabrous or hairy, margins glabrous or with a few long hairs below. Callus abaxially ± flat or rounded, disarticulation oblique, margins variously long hairy. Rachilla and prolongation glabrous. Lodicules 2, cuneate, apically long-hairy, or glabrous. Stamens 3. Gynoecium: ovary glabrous; styles 2, free, swollen below. Caryopsis free, obovate to elliptic, planoconvex; embryo ⅓–½ caryopsis; hilum ± ⅓–⅕ caryopsis, round to somewhat elliptic. Chasmogamous and cleistogamous. Plates 6A, 11, 12.

Type species: *R. lechleri* Steud.

c. 45 spp. from South America, New Guinea and Australasia. Endemic spp. 15, indigenous spp. 3, shared with Australia; naturalised spp. 9 (all from Australia).

Connor, H. E. and Edgar, E. *N.Z. J. Bot. 17*: 311–337 (1979), presented a full treatment of *Rytidosperma* Steud. in place of the synoptic arrangement by Zotov, V. D. *N.Z. J. Bot. 1*: 78–136 (1963) as *Notodanthonia*. *Rytidosperma*, as indicated by Nicora, E. *Darwiniana 18*: 80–106 (1973), is the earliest correct name for taxa that Zotov (loc. cit.) included in *Notodanthonia* Zotov, which earlier were known here as *Danthonia* DC. [Cheeseman, T. F. *Man. N.Z. Fl.* 171–179 (1925)]. See also Vickery, J. W. *Contrib. N.S.W. Natl Herb.* 2: 249–325 (1956); Baeza P., C. M. *Sendtnera 3*: 11–93 (1996); and Astegiano, M. E., Anton, A. M. and Connor, H. E. *Flora Fanerogámica Argentina 22*: 15–19 (1996). Connor, H. E. and Edgar, E. *N.Z. J. Bot. 25*: 115–170 (1987), included spp. of *Erythranthera* Zotov within *Rytidosperma* as was indicated by Clayton and Renvoize (1986 op. cit. p. 175).

Linder, H. P. *Telopea 7*: 269 (1997) proposed *Austrodanthonia* as a new genus to accommodate those species of Zotov's *Notodanthonia* sect. *Semiannularia* and sect. *Notodanthonia* subsect. *Clavatae*, retaining *Notodanthonia sens. strict.* for Zotov's section *Notodanthonia* ≡ *Thonandia* H.P.Linder *nom. illeg.* The original proposals of Linder, H. P. and Verboom, G. A. *Telopea 6*: 591–626 (1996) are thereby made nomenclaturally correct. As they affect New Zealand that arrangement seems confusing compared with our earlier synoptic subdivisions of *Rytidosperma* which ally themselves logically with Zotov's infrageneric treatment; none of his sections and subsections has been formally transferred from *Notodanthonia.*

SYNOPSIS

A. Tufts often low-growing, branching mainly intravaginal. Leaf-blades ± inrolled, disarticulating at ligule; lemma lobes, including fine, short or minute awn, ≤, sometimes > lemma, adaxially glabrous, rarely slightly scabrid; palea narrow-spathulate to elliptic; 10 spp. endemic, 2 indigenous shared with Australia: 2. *australe*, 4. *buchananii*, 7. *corinum*, 12. *maculatum*, 15. *nudum*, 17. *petrosum*, 19. *pulchrum*, 20. *pumilum*, 22. *setifolium*, 23. *tenue*, 25. *thomsonii*, 27. *viride*

B. Tufts stiff, often hoary, branching mainly intravaginal. Leaf-blades ± flat, persistent; lemma lobes, including conspicuous awn, 2× lemma, adaxially scabrid; palea broad-ovate or elliptic; mostly naturalised Australian spp.: 1. *auriculatum*, 3. *biannulare*, 5. *caespitosum*, 8. *erianthum*, 9. *geniculatum*, 11. *laeve*, 18. *pilosum*, 24. *tenuius*

C. Tufts ± lax, leafy, often large, branching mainly extravaginal. Leaf-blades ± flat, persistent; lemma lobes, including conspicuous awn, 2× lemma, adaxially scabrid or glabrous; palea narrow-elliptic; native and naturalised spp.: 6. *clavatum*, 10. *gracile*, 13. *merum*, 14. *nigricans*, 16. *penicillatum*, 21. *racemosum*, 26. *unarede*

These informal groups, A, B, and C, correspond with sections formally described by Zotov (1963 op. cit.) as *Notodanthonia* sect. *Buchanania*, sect. *Semiannularia*, and sect. *Notodanthonia* respectively, though *R. pilosum* may be more appropriately placed in group C. The spp. referred by Zotov (1963 op. cit.) to *Erythranthera* are placed in group A.

The common name, danthonia, is applied to all spp. of *Rytidosperma* in N.Z.

1 Lateral lemma lobes exceeding upper lemma hairs and at least twice as long as lemma *2*

  Lateral lemma lobes concealed by upper lemma hairs, or visible above them and at most only slightly longer than lemma *15*

2 Upper lemma hairs in a few isolated tufts, short, at most hardly exceeding lemma *3*

  Upper lemma hairs in a continuous dense or sparse row, long, much exceeding lemma *7*

3 Callus hairs short, rarely reaching lower row of lemma hairs; or lower row in marginal tufts only *4*

Callus hairs overlapping lower row of lemma hairs *5*

4 Lemma hairs in continuous lower row and in an interrupted, tufted upper row; callus long, narrow, up to 1.5 mm .... **21. racemosum**

Lemma hairs in two pairs of marginal tufts only, occasionally a few single hairs in lower row; callus short, stout, up to 1 mm .... **13. merum**

5 Branching intravaginal; palea usually just reaching awn sinus ...... **18. pilosum**

Branching extravaginal; palea usually exceeding awn sinus *6*

6 Culm scabrid just below contracted panicle or quite glabrous; awn column usually flat at base of sinus and concealing top of palea .... **6. clavatum**

Culm minutely hairy to scabrid for 10 mm or more below elongated panicle; awn column twisted and divergent at base of sinus and revealing top of palea .... **16. penicillatum**

7 Upper lemma hairs exceeding or equalling awn column *8*

Upper lemma hairs shorter than awn column *12*

8 Lower row of lemma hairs short, or just reaching upper row *9*

Lower row of lemma hairs distinctly overlapping upper row, or lower row in marginal tufts only *10*

9 Lemma with only a few scattered hairs between two rows of hairs; awn column equalling or just exceeding palea-tip; branching extravaginal .... **26. unarede**

Lemma with numerous scattered hairs between two rows of hairs; awn column much exceeding palea-tip; branching intravaginal .... **3. biannulare**

10 Lower row of lemma hairs in marginal tufts only; palea interkeel glabrous .... **11. laeve**

Lower row of lemma hairs continuous; palea interkeel long-hairy........... *11*

11 Lateral lemma lobes entire at base; callus (0.8)–1–1.3 mm .... **24. tenuius**

Lateral lemma lobes shallowly lobed again at base; callus 0.4–0.7 mm .... **1. auriculatum**

12 Upper row of lemma hairs sparse to almost continuous, equalling or not reaching tip of palea *13*

Upper row of lemma hairs continuous, exceeding tip of palea *14*

13 Rachis and pedicels closely short-scabrid, with hairs at branch axils only slightly longer; upper lemma hairs ± continuous .... **7. corinum**

Rachis and pedicels almost glabrous, and usually with small tufts of long hairs at axils; upper lemma hairs very sparse .... **17. petrosum**

14 Palea narrow-lanceolate, much exceeding awn sinus; lower lemma hairs almost reaching upper .... **5. caespitosum**
Palea broadly obovate, ± equalling awn sinus; lower lemma hairs short, and lemma body shining between two rows .... **8. erianthum**

15 Upper lemma hairs very much exceeding tip of palea *16*
Upper lemma hairs ± equalling or shorter than tip of palea *19*

16 Callus hairs short, sparse, not reaching to lower row of lemma hairs or scarcely overlapping them; branching extravaginal *17*
Callus hairs long, abundant, greatly overlapping lower row of lemma hairs; branching intravaginal *18*

17 Lemma short-hairy between two rows of hairs; awn column overtopped by or equalling tip of palea .... **10. gracile**
Lemma ± glabrous between two rows of hairs; awn column exceeding or equalling tip of palea .... **14. nigricans**

18 Upper and lower lemma hair rows equally dense and continuous; glumes linear-lanceolate, gradually narrowed to tip, overtopped by awns .... **27. viride**
Upper row of lemma hairs continuous, lower less dense, sometimes ill-defined; glumes broad at base, shortly narrowed to tip, equalling or overtopping awns .... **9. geniculatum**

19 Panicle branches glabrous, or with few scattered teeth and occasional long hairs at branch axils; anthers (1.2)–1.5–3 mm .... **22. setifolium**
Panicle branches ± closely short-scabrid, usually with some slightly longer hairs below spikelets and occasionally at branch axils; anthers 0.2–1–(1.4) mm *20*

20 Lemma hairs sparse in both rows, or in marginal tufts only, sometimes with a few scattered hairs at centre, or glabrous *21*
Lemma hairs in dense continuous rows, or dense in upper row only *25*

21 Lateral lemma lobes well developed, > 1 mm; awn usually divergent; palea ± reaching top of awn column .... **4. buchananii**
Lateral lemma lobes minute, up to 0.5 mm; awn erect, or reduced to a mucro; palea at most reaching base of awn *22*

22 Lemma glabrous apart from upper and lower pairs of marginal tufts .... **25. thomsonii**
Lemma with sparse hairs between marginal tufts or with scattered hairs in upper half, or with a single pair of marginal tufts, or lemma entirely glabrous *23*

23 Lemma with awn 1–2 mm .... **15. nudum**
Lemma mucronate *24*

24 Lemma glabrous, or rarely with a single pair of marginal tufts; leaf-blades flat to folded .... **2. australe**
Lemma ± hairy; leaf-blades setaceous .... **20. pumilum**

25 Upper row of lemma hairs noticeably shorter than lemma lobes; pedicels with dense long stiff hairs below spikelet .... **12. maculatum**

Upper row of lemma hairs ± equalling or exceeding lemma lobes; pedicels uniformly short-scabrid, occasionally long hairs below spikelet 26

26 Lemma glabrous between rows of hairs; callus hairs overlapping lower marginal hair tufts ....**19. pulchrum**

Lemma with scattered hairs between rows of hairs; callus hairs not, or a few just reaching lower marginal hair tufts ....**23. tenue**

1. R. auriculatum (J.M.Black) Connor et Edgar, *N.Z. J. Bot. 17*: 322 (1979).

Small, grey-green, hoary, stiff tufts, shortly rhizomatous; leaves < culms; branching intravaginal, shoots crowded, ± swollen at base. Leaf-sheath pale stramineous, light purple above, scattered long hairs increasing towards collar, culm sheaths especially hairy; apical tuft of hairs 1.5–3 mm, extending dorsally below blade. Ligule 0.4–0.5 mm. Leaf-blade to 5 cm, densely and conspicuously covered in long stiff hairs, inrolled with thickened sparsely scabrid margins, tip obtuse. Culm to 15 cm, internodes glabrous but long-hairy below inflorescence. Panicle small, compact, erect, to 2 cm of few large spikelets on short pedicels; rachis and pedicels conspicuously clothed in long stiff hairs. Spikelets (4)–6–8-flowered, awns exserted from glumes. Glumes irregularly purple-tinged, lanceolate, subobtuse, ± equal, (7.5)–9.5–13 mm; lower 11–13-nerved, upper 9–11-nerved. Lemma (2.2)–2.8–3.0 mm, 9-nerved, upper and lower row of hairs very dense, continuous, upper row < lemma lobes, lower row overlapping upper, glabrous elsewhere; lobes (6)–8–10 mm, tapering to strong awns ≤ central awn, lobes with a short (0.3–0.4 mm) acute membranous subsidiary lobe, adaxially tufts of hairs near base of awn sinus; central awn (6.5)–9–12 mm, column *c.* 3 mm ≤ upper lemma hairs. Palea (2.2)–3–3.4 mm, broad, apex rounded, < upper lemma hairs, interkeel hairs short, scattered, margins with longer fine hairs. Callus 0.4–0.7 mm, marginal hair tufts just reaching lower lemma hairs. Rachilla 0.2–0.3 mm. Anthers 0.3–0.4 mm. Caryopsis *c.* 1.5 × 1 mm; embryo 0.7 mm; hilum 0.4 mm.

Naturalised from Australia.

S.: known only from dry sites at Wither Hills, Marlborough.

2. R. australe (Petrie) Connor et Edgar, *N.Z. J. Bot. 25*: 166 (1987)
≡ *Triodia australis* Petrie, *T.N.Z.I. 22*: 442 (1890) ≡ *Erythranthera australis* (Petrie) Zotov, *N.Z. J. Bot. 1*: 125 (1963) ≡ *Danthonia petriei* Zotov, *T.R.S.N.Z. 73*: 234 (1943) [a combination in *Danthonia* preoccupied by *D. australis* (Buchanan) Buchanan (1872)]; Holotype: WELT 40330! *D. Petrie* Mt Ida Range, Maniototo Co., *c.* 3500 ft. = *Triodia australis* var. *mucronulata* Hack., *in* Cheeseman *Man. N.Z. Fl.* 897 (1906) ≡ *Danthonia petriei* var. *mucronulata* (Hack.) Zotov, *T.R.S.N.Z. 73*: 234 (1943); Holotype: AK 1731! *T. F. Cheeseman* swamps in Tasman Valley, Jan. 1898 (No 1328 to Hackel).

Low-growing and wiry, or taller, softer and more slender tufts, shortly- to sometimes long-rhizomatous; leaves ≤ culms, disarticulating at ligule; branching intravaginal, shoots rather stout below. Leaf-sheath pale to dark grey, cream to purplish above, much broader than leaf-blade, entirely glabrous, or more often several scattered long fine hairs above; apical tuft of hairs to 1.5 mm, occasionally extending dorsally below blade. Ligule to 0.5 mm. Leaf-blade to 8–(18) cm, flat to folded; margins inrolled, glabrous, or with a few scattered hairs below. Culm (1.5)–3–15–(35) cm, erect or spreading, internodes glabrous. Panicle 0.5–2–(3.5) cm; rachis and pedicels scabrid, sometimes with fine hairs at nodes and on pedicels. Spikelets 3–4-flowered, upper florets often exceeding glumes. Glumes green to purple-tinged, ovate, rounded, slightly notched or subacute, rarely a few scattered hairs, ± equal, (2)–2.5–3–(4) mm; lower 5-nerved, upper 3–5-nerved. Lemma 1.5–1.8 mm, 7–(9)-nerved, ± orbicular, glabrous, or rarely a single pair of marginal hair tufts at level of ciliate rachilla apex, notched, with a minute mucro *c.* 0.1 mm in the notch. Palea 1.5–1.8–(2) mm, = lemma, ± elliptic, interkeel glabrous. Callus 0.1–0.3 mm, glabrous. Rachilla *c.* 0.5 mm. Anthers 0.4–0.7 mm, red. Caryopsis *c.* 0.8 × 0.5 mm; embryo to 0.5 mm; hilum *c.* 0.2 mm.

Indigenous.

N.: East Cape and Ruahine and Tararua Ranges; S.: north-west Nelson and to east of Main Divide. Montane to alpine in tussock grassland, herbfield, usually in damp ground.

Also indigenous to Australia.

3. R. biannulare (Zotov) Connor et Edgar, *N.Z. J. Bot. 17*: 324 (1979) ≡ *Notodanthonia biannularis* Zotov, *N.Z. J. Bot. 1*: 116 (1963) ≡ *Austrodanthonia biannularis* (Zotov) H.P.Linder, *Telopea* 7: 270 (1997); Holotype: CHR 85021! *V. D. Zotov* Waitangi Forest, Northland, roadside, abundant, 28.11.1953.

Narrow-leaved green tussocks, leaves much < culms; branching intravaginal. Leaf-sheath pale stramineous, often purplish above, glabrous; apical tuft of hairs to 1.7 mm, sparse or lacking. Ligule 0.3–0.5 mm. Leaf-blade to 30–(40) cm, usually rolled, ± glabrous at maturity, young leaves with scattered long, fine, silky hairs, margins and tip scabrid. Culm to 85 cm, internodes glabrous. Panicle often rather dense, strict, 10–(20) cm, branches short, of many spikelets on slender pedicels, or panicle smaller and fewer spikelets; rachis and pedicels densely short-scabrid, occasionally slightly longer hairs at branch axils. Spikelets 6–7-flowered, awns usually exserted from glumes. Glumes purple-margined, lanceolate, subacute, 7.6–11–(13.2) mm, ± equal; lower 5–7-nerved, upper 5-nerved. Lemma 1.8–2.2–(2.8) mm, 7–9-nerved, upper and lower rows of hairs dense, continuous, upper row ≤ lemma lobes, lower row usually not reaching base of upper, sometimes not well-defined, elsewhere numerous short scattered hairs, glabrous at times just below upper row, margins hair-fringed; lobes (3.5)–4–5–(8.5) mm, narrowed to fine awn, adaxially often a few longer hairs at base; central awn (6)–7–9.5–(12.5) mm, column 2.5–3 mm = upper lemma hairs ≫ palea. Palea (2.5)–2.7–3.3–(4.6) mm, ≪ upper lemma hairs, interkeel hairs few, short at base, margins usually long-hairy. Callus 0.5–0.7 mm, marginal hair tufts reaching to base of lower lemma hairs. Rachilla 0.3–0.4 mm. Anthers 0.8–1.5 mm. Caryopsis 1.2–1.9 × 0.6–0.8 mm; embryo 0.5–0.8 mm; hilum 0.3–0.6 mm. Plate 11C.

Endemic.

N.: Surville Cliffs to Auckland, also recorded from Taumaranui and Wellington; S.: north-west Nelson, D'Urville Id and Wairau Valley. Lowland modified sites in both islands.

4. R. buchananii (Hook.f.) Connor et Edgar, *N.Z. J. Bot. 17*: 320 (1979) ≡ *Danthonia buchananii* Hook.f., *Handbk N.Z. Fl.* 333 (1864) ≡ *Notodanthonia buchananii* (Hook.f.) Zotov, *N.Z. J. Bot. 1*: 110 (1963);

Holotype: K! *Hector and Buchanan* Otago. = *Danthonia semiannularis* var. *breviseta* Hook.f., *Fl. N.Z. 1*: 304 (1853); Lectotype: K! *Colenso 1609* scarce grass near Patea (designated by Zotov 1963 op. cit. p. 110).

Loose, dull green tussocks; leaves semi-pungent, usually ≪ mature culms, disarticulating at ligule; branching intravaginal. Leaf-sheath pale stramineous, glabrous, rarely long scattered hairs; apical tuft of hairs (0.5)–1–2–(3) mm. Ligule 0.2–0.3–(0.6) mm. Leaf-blade to 30 cm, folded or tightly rolled, sometimes almost filiform, glabrous or rarely some long scattered hairs, margins scabrid near base and again towards tip. Culm to 50–(70) cm, internodes smooth but minutely scabrid below inflorescence. Panicle erect, to 10 cm, usually ± compact, of several spikelets; rachis and filiform pedicels closely short-scabrid with slightly longer hairs below spikelets and sometimes at branch axils. Spikelets (3)–4–5–(6)-flowered, awns slightly exserted from glumes. Glumes purplish centrally, lanceolate, subobtuse, (6)–7–10–(11) mm, ± equal; lower 5–7-nerved, upper 3–5-nerved. Lemma 2–3–(3.5) mm, 7–9-nerved, upper and lower rows of hairs usually sparse or lacking between dense marginal tufts, upper row < lemma lobes, lower row overlapping upper, glabrous and shining elsewhere, rarely one or two hairs centrally; lobes 1–2.5–(3) mm, tapering to a minute to short fine awn; central awn 2.5–5.5–(6.5) mm, column 1–2 mm, occasionally not developed, ≈ palea and upper lemma hairs. Palea 3–4 mm, ≈ upper lemma hairs, interkeel glabrous, margins rarely with a few long hairs. Callus *c.* 0.5 mm, marginal hair tufts much overlapping lower lemma hairs. Rachilla 0.5–1 mm. Anthers 0.5–1.3 mm. Caryopsis 1.2–1.8 × 0.5–0.7 mm; embryo 0.5–0.7 mm; hilum 0.2–0.3 mm. $2n = 72$. Plate 12B.

Endemic.

N.: central and southern mountains, also from near Kawhia (The Lady); S.: mountainous areas throughout except in Buller and north Westland. Montane and subalpine grasslands, cliffs and rocky sites.

5. R. caespitosum (Gaudich.) Connor et Edgar, *N.Z. J. Bot. 17*: 325 (1979).

Strict, crowded tufts with erect leaves and culms, occasionally shortly rhizomatous; leaves semi-pungent, ≤ culms; branching intravaginal. Leaf-sheath whitish cream, glabrous, or few to many hairs usually near collar;

apical tuft of hairs 2–3.5 mm. Ligule 0.3–0.5 mm. Leaf-blade to 17 cm, dull green, narrowly inrolled, or flatter, scattered to dense, short, stiff hairs, sometimes glabrous, margins closely scabrid. Culm to 45–(65) cm, internodes glabrous but with a few short hairs just below inflorescence. Panicles erect, to 5–(10) cm, compact, few, large spikelets on short pedicels; rachis and pedicels closely short-scabrid, often with stiff longer hairs at branch axils and below spikelets. Spikelets 4–9-flowered, awns much exserted from glumes. Glumes green, purple-tinged, lanceolate, subacute, ± equal, 11–15–(19) mm, both 5–7-nerved, occasionally a few long hairs or densely hairy. Lemma 2.7–4 mm, 9-nerved, with two, rarely sparse to usually continuous, often dense rows of hairs, upper row < lemma lobes, lower row usually reaching base of upper, occasionally almost overlapping, usually glabrous elsewhere, occasionally a few short scattered hairs centrally and on margins, adaxially usually few longer hairs close to base of awn sinus; lobes (6.5)–7–10–(11) mm, tapering to long awns; central awn 10.5–20 mm, column (3)–4–5 mm usually ≫ upper lemma hairs. Palea (3.2)–5–6 mm, narrow-lanceolate, < upper lemma hairs, occasional interkeel hairs, margins usually with long fine hairs. Callus (0.6)–1–1.4 mm, very dense marginal hair tufts often reaching to base of lower lemma hairs. Rachilla 0.1–0.2–(0.4) mm. Anthers 0.3–0.6 and 0.7–2 mm. Caryopsis *c.* 2 × 1 mm; embryo *c.* 1 mm; hilum *c.* 0.7 mm.

Naturalised from Australia.

N.: Kawakawa to Maunganui Bluff and south to Kaipara and Raglan Harbours, to the east on Mokohinau Is, Cuvier Id and Coromandel Peninsula, Wellington coast (Mana Id); S.: eastern Marlborough, North and Mid Canterbury, North Otago. Dry sites in lowland grassland, and waste places.

Some specimens from Harewood (Canterbury) and one from Waipara (CHR 45955 *A. J. Healy* 1946) fit *R. caespitosum* except that the awn column is about equal to the lemma hairs rather than noticeably exceeding them.

6. R. clavatum (Zotov) Connor et Edgar, *N.Z. J. Bot. 17*: 326 (1979)
≡ *Notodanthonia clavata* Zotov, *N.Z. J. Bot. 1*: 119 (1963)
≡ *Austrodanthonia clavata* (Zotov) H.P.Linder, *Telopea 7*: 271 (1997);
Holotype: CHR 79778! *V. D. Zotov* Upper Hutt, railway cutting,

12.12.1952. = *Danthonia pilosa* var. *stricta* Buchanan, *Indig. Grasses N.Z.* t. 33 (2) A (1879) ≡ *Notodanthonia stricta* (Buchanan) Zotov, *N.Z. J. Bot. 1*: 121 (1963); Holotype: WELT 59581 (Buchanan's folio)! *Buchanan.* = *Danthonia nervosa* Colenso, *T.N.Z.I. 28*: 612 (1896) non (R.Br.) Hook.f. (1858); Holotype: K! *Thomas Hallett* Dry Hills, 1000–3000 ft, Hawkston, Hawke's Bay, 1894–95.

Variable in size and habit, from short and slender to larger and stout, dull green, densely soft-hairy or glabrous, shortly rhizomatous, spreading tufts; leaves ≪ culms; branching extravaginal. Leaf-sheath light grey-brown, densely strict-hairy or glabrous; apical tuft of hairs to 5 mm, often extending dorsally below blade. Ligule 0.1–0.7 mm. Leaf-blade flat or sometimes folded, to 30 cm × 3–5 mm, abundant hairs, or glabrous, margins minutely scabrid. Culm to 90 cm, internodes smooth, or finely scabrid below inflorescence. Raceme or racemose panicle contracted, to 5.5–(9.5) cm, spikelets few to several on very short pedicels; rachis and pedicels shortly and densely scabrid with longer hairs below spikelets and at branch axils. Spikelets 5–10-flowered, awns exserted from glumes. Glumes occasionally purple-margined, lanceolate, subobtuse, ± equal, 8–13.5–(17.5) mm; lower (5)–7–9–(11)-nerved, upper 5–7–(9)-nerved, sometimes scattered long hairs. Lemma 2.5–3.5–(4.5) mm, 9-nerved, upper row of hairs ≥ lemma, in marginal tufts only, or two additional central tufts, lower row with short weak central tufts, or continuous, with longer marginal tufts barely, or not, reaching upper marginal tufts, elsewhere glabrous, shining; lobes 6–13 mm, gradually narrowed to long awns; central awn 9–14–(16) mm, column (2)–2.5–3.5 mm. Palea 3.3–5.5–(6) mm, > upper lemma hairs, interkeel glabrous, margins with several long hairs or glabrous. Callus 0.6–1.2 mm, strong marginal hair tufts overlapping lower lemma hairs. Rachilla 0.2–0.4 mm. Anthers 0.4–1.0 and 1.5–2.2 mm. Caryopsis 1.9–2.3–(3) × 0.8–1.1–(1.6) mm; embryo 0.7–1–(1.5) mm; hilum 0.4–0.5–(0.7) mm. Plate 11A.

Endemic.

N.: southwards from Auckland and Coromandel Peninsula; S.: throughout east of Main Divide, and Westland (Cook River Flats); Ch. Grasslands, and modified sites in lowland and montane zones especially of drier areas.

7. R. corinum Connor et Edgar, *N.Z. J. Bot. 17*: 317 (1979); Holotype: CHR 318008! *H. E. Connor* Ahuriri River, North Otago, rock outcrop near disused bridge, *c.* 650 m, 13.1.1978.

Small, compact, light green or blue-green tussocks; leaves pungent, acicular, ≈ or usually slightly < culms, disarticulating at ligule; branching intravaginal. Leaf-sheath pale stramineous; apical tuft of hairs to 1.5–(2.5) mm. Ligule (0.3)–0.5 mm. Leaf-blade to 15 cm, glabrous, stiff, inrolled, margins scabrid. Culm to 25 cm, internodes smooth but scabrid below inflorescence. Panicle racemose, small, compact, erect, to 4–(6) cm, of very few spikelets on short slender pedicels; rachis and pedicels short-scabrid with few longer hairs below spikelets and at branch axils. Spikelets 4–5–(6)-flowered, awns equalling or slightly exserted from glumes. Glumes sometimes purplish, lanceolate, acute or subacute, 9–11–(15) mm, ± equal; lower (3)–5–(7)-nerved, upper 3-nerved. Lemma 2–2.5–(2.8) mm, 9-nerved, upper row of hairs continuous but sparse, with one or two stronger marginal tufts and tufts of few hairs between, ≪ lemma lobes, lower row of strong marginal tufts and scattered tufts between, just reaching to upper row, glabrous elsewhere but occasional hairs centrally and on margins; lobes 3–5–(6) mm, tapering to fine awn; central awn 7–8.5 mm, column (1.5)–2–2.5 mm, usually > upper lemma hairs. Palea 3.5–4 mm, usually > upper lemma hairs, interkeel glabrous, margins very rarely with 1–2 long hairs. Callus (0.3)–0.4–(0.5) mm, sparse marginal hair tufts reaching lower lemma hairs. Rachilla 0.6–0.9 mm. Anthers 0.6–1–(1.6) mm, red or pinkish red. Caryopsis 1.5–1.8 × 0.6–0.8 mm; embryo 0.6–0.8 mm; hilum 0.4–0.5 mm. Plate 11D.

Endemic.

S.: Banks Peninsula, central and South Canterbury, and Otago. Rocky sites in grasslands of montane and subalpine zones.

8. R. erianthum (Lindl.) Connor et Edgar, *N.Z. J. Bot. 17*: 323 (1979).

Small hoary-leaved tufts; leaves ≪ culms; branching intravaginal. Leaf-sheath light purple, of basal leaves glabrous, of culm leaves abundantly long-hairy; apical tuft of hairs to 3.5 mm. Ligule 0.7 mm. Leaf-blade to 9 cm, ± flat, to 2 mm wide, densely and conspicuously covered in long stiff hairs, margins thickened, scabrid. Culm to 45 cm, internodes glabrous but long-hairy below inflorescence. Panicle short, broad, to 4.5 cm, of few large spikelets on short pedicels; rachis and pedicels covered in long fine hairs. Spikelets 5–8-flowered, awns exserted from glumes. Glumes purplish, lanceolate, obtuse, 10–14 mm, ± equal; lower 9–11-nerved, upper 9-nerved, occasionally long scattered hairs. Lemma 3.0–3.7 mm, 9-nerved, upper and lower rows of hairs very dense, continuous, upper hairs ≪ lemma lobes, marginal tufts of lower row overlapping upper but central tufts shorter, glabrous, shining elsewhere; lobes 8–9.8 mm, tapering to long conspicuous awns, adaxially scattered short hairs and two thick longer tufts near awn sinus; central awn 12.8–15 mm, column 4 mm ≫ upper lemma hairs. Palea 3.3–4 mm, obovate-oblong, broad, apex rounded, < upper lemma hairs, interkeel hairs few and very short at base, margins with 1–2 long fine hairs. Callus 0.8–1 mm, thick marginal hair tufts overlapping lower lemma hairs. Rachilla 0.1 mm. Anthers 0.4, and 0.7–0.9 mm. Caryopsis *c.* 1.5 mm; embryo 0.6 mm; hilum 0.5 mm.

Naturalised from Australia.

S.: known only from dry sites at Wither Hills, Taylor Valley, and Blind River, Marlborough, and Taieri Ferry, Otago.

Although not recognised for N.Z. until 1979, specimens of *R. erianthum* were collected almost ninety years earlier (WELT 40259 *D. Petrie* Taieri Ferry, 1892).

9. R. geniculatum (J.M.Black) Connor et Edgar, *N.Z. J. Bot. 17*: 323 (1979).

Small, compact, stiff tussocks; leaves semi-pungent, usually < culms; branching intravaginal, occasionally ± swollen at base. Leaf-sheath pale stramineous, usually a few scattered long stiff hairs; apical tuft of hairs to 2 mm. Ligule to 0.5 mm. Leaf-blade to 12 cm, tightly inrolled, rather scattered long stiff hairs, margins scabrid. Culm to 25 cm, internodes smooth, or minutely scabrid below inflorescence. Panicle short and broad, to 2.5 cm, of few broad spikelets on short pedicels; rachis and pedicels with rather dense short hairs. Spikelets 3-flowered, awns included by glumes. Glumes purplish, ovate-lanceolate, broad at base, abruptly narrowed towards subobtuse tip, 6.5–9 mm, ± equal; lower 9-nerved, upper 7–9-nerved. Lemma 2–3 mm, 7–9-nerved, upper row of hairs continuous, very dense, usually = lemma lobes, lower row usually indistinct, elsewhere abundant short scattered hairs, adaxially a triangular band of scattered hairs below sinus; lobes 3–3.5 mm, tapering to very short awn, occasionally lobes with a shallow subsidiary lobe; central awn 3.5–5 mm, column 1–1.5 mm < upper lemma hairs. Palea 3–3.5 mm, ≪ upper lemma hairs, interkeel hairs few short scattered below, margins with 1–2 long hairs. Callus 0.5–1.0 mm, marginal hair tufts long, abundant, much overlapping lower hairs. Rachilla 0.1–0.2 mm. Anthers 0.7–1 mm. Caryopsis 1.5–1.7 × 0.7–0.9 mm; embryo 0.6–0.9 mm; hilum to 0.5 mm.

Naturalised from Australia.

S.: Marlborough (Wither Hills), Canterbury (Port Hills). On dry sites.

In N.Z. specimens the lateral lemma lobes barely, if at all, exceed the upper lemma hairs, unlike typical Australian material.

10. R. gracile (Hook.f.) Connor et Edgar, *N.Z. J. Bot. 17*: 330 (1979) ≡ *Danthonia gracilis* Hook.f., *Fl. N.Z. 1*: 303, t. 69B (1853) ≡ *Notodanthonia gracilis* (Hook.f.) Zotov, *N.Z. J. Bot. 1*: 123 (1963) ≡ *Thonandia gracilis* (Hook.f.) H.P.Linder, *Telopea* 6: 612 (1996) *comb. illeg.* ≡ *Danthonia semiannularis* var. *gracilis* (Hook.f.) Hook.f., *Handbk N.Z. Fl.* 333 (1864); Lectotype: K! *D. Monro 130/2* Rotoiti Lake, Aglionby Plains, New Zealand, 1851 (designated by Zotov 1963 op. cit. p. 123).

Tufts leafy, bright green, sometimes stoloniferous with prostrate culms rooting and sending up new tufts; leaves about ½ length of culms; branching extravaginal. Leaf-sheath light brown often purplish, scattered soft fine hairs towards collar or sometimes throughout, or quite glabrous; apical tuft of hairs (2)–3–4–(6) mm. Ligule (0.2)–0.4–0.7–(1) mm. Leaf-blade up to 25–(50) cm, flat with ± inrolled margins, 2–(4) mm wide, or completely inrolled, glabrous or often a few scattered long hairs near collar, sometimes almost all leaves with some long hairs throughout, margins finely scabrid. Culm to 50–(75) cm, internodes smooth but minutely scabrid below inflorescence. Panicle erect to ± loose and drooping from above, to 10–(14) cm, spikelets numerous on short pedicels; rachis and pedicels densely short-scabrid, occasionally with

1–2 longer hairs at branch axils. Spikelets 3–4–(6)-flowered, awns exserted from glumes. Glumes sometimes variously purple-tinged, lanceolate, acute to acuminate, 7–10–(12) mm; lower 5–7-nerved, upper (3)–5-nerved. Lemma 1.6–2.8 mm, 9-nerved, upper row of hairs continuous, ≈ lemma lobes, lower row not always well-defined and not reaching upper, elsewhere covered in numerous short hairs; lobes 2.5–4.5 mm, finely short-awned at tip; central awn 4.5–8.5–(9) mm, column 0.5–1.2 mm ≤ palea, ≪ upper lemma hairs. Palea 2.5–4 mm, ≪ upper lemma hairs, interkeel hairs few, margins long-hairy. Callus 0.3–0.5 mm, marginal hair tufts not reaching lower lemma hairs or scarcely overlapping them. Rachilla 0.4–1 mm, often with minute hairs at apex. Anthers 0.5–1.3 mm. Caryopsis 0.9–1.4 × 0.5–0.8 mm; embryo 0.4–0.6 mm; hilum 0.2–0.3 mm. $2n = 24$. Plate 12I.

Indigenous.

N.: throughout; S.: throughout, only occasional in dry inland basins, and the plains, of Canterbury and Otago; St.; Ch. Grassland, scrub, forests, and modified sites in lowland to alpine zones.

Also indigenous to Tasmania.

11. R. laeve (Vickery) Connor et Edgar, *N.Z. J. Bot. 17*: 325 (1979).

Slender, wiry, green tufts, shortly rhizomatous; leaves semi-pungent, ≪ culms; branching both intra- and extravaginal, swollen at base. Leaf-sheath glabrous, pale, becoming grey-brown, some with scattered, long hairs above; apical tuft of hairs to 2.5 mm. Ligule 0.05–0.1 mm. Leaf-blade up to 22 cm, thick, flat to inrolled, usually scattered to fairly dense, long, fine hairs, margins minutely scabrid. Culm to 80 cm, internodes smooth, or finely scabrid below inflorescence. Panicle racemose, erect, to 7 cm, of few large spikelets on slender pedicels; rachis and pedicels densely clothed with short stiff hairs, and occasional longer hairs below spikelets and at branch axils. Spikelets 5–6-flowered, awns much exceeding glumes. Glumes usually purplish near margin and sometimes elsewhere, lanceolate, narrowing to long acute tip, ± equal, 13–20 mm; lower 11–13-nerved, upper 9-nerved. Lemma 3–4 mm, 9-nerved, upper row of hairs very dense, continuous, < lemma lobes, lower row lacking except for short marginal tufts, elsewhere shining, glabrous, margins shortly hair-fringed above; lobes 8–13 mm, soon tapering to long awns; central awn 12–17 mm, column to 2–4.5 mm ≤ upper lemma hairs. Palea 5–6.2 mm, abruptly narrowed above, < upper lemma hairs, interkeel and margins glabrous. Callus 1.0–1.4 mm, marginal hair tufts reaching lower marginal lemma tufts. Rachilla 0.1–0.2 mm. Anthers 0.5–0.7 and 1.9–3 mm. Caryopsis 2–2.7 × 1–1.4 mm; embryo 1–1.5 mm; hilum *c.* 0.6 mm.

Naturalised from Australia.

N.: North Cape, Auckland City, and the Wairarapa; S.: Nelson (Maitai R.). Pastures and waste places in lowland zone.

Although not recorded for N.Z. until 1963 (as *Notodanthonia laevis*), it was collected as early as 1919 (WELT 40296 *D. Petrie* Auckland City).

12. R. maculatum (Zotov) Connor et Edgar, *N.Z. J. Bot. 17*: 320 (1979) ≡ *Notodanthonia maculata* Zotov, *N.Z. J. Bot. 1*: 108 (1963); Holotype: CHR 3660! Galloway, Central Otago, 1929.

Low-growing, flattish, or erect, often silky-hairy, green to yellow-green, tight tufts, shortly rhizomatous; leaves usually ≪ culms, stiffly long-hairy but not invariably so, disarticulating at ligule; branching intravaginal, shoots crowded, ± swollen at base. Leaf-sheath mostly hairy; apical tuft of hairs 1.5–2.5 mm, often extending dorsally below blade. Ligule *c.* 0.4 mm. Leaf-blade to 12 cm, acicular, inrolled, often with long hairs, margins thickened, scabrid. Culm to 30 cm, internodes often slightly short-hairy below inflorescence otherwise glabrous. Racemose panicle or raceme small, to 3.5 cm, erect, of few spikelets on short stiff pedicels; rachis and pedicels closely short-scabrid with longer stiff hairs below spikelets and at branch axils. Spikelets 4–7-flowered, awns exserted from glumes. Glumes purplish centrally, often with regular horizontal bars, ovate-lanceolate, obtuse or subobtuse, (6)–7–10 mm, ± equal; lower 7-nerved, upper 5-nerved, occasionally long-hairy between nerves. Lemma 2–2.5 mm, 9-nerved, upper and lower rows of hairs dense, continuous, upper row < lemma lobes, lower row ± overlapping upper, some few hairs elsewhere; lobes 1.5–3 mm, tapering, usually tipped by fine awn; central awn 3–5 mm, column 1–1.5 mm, > upper lemma hairs. Palea 3–5 mm, ≈ upper lemma hairs, interkeel glabrous or occasional long hairs, margins long-hairy. Callus 0.4–0.6 mm, marginal hair tufts overlapping lower lemma hairs. Rachilla 0.5–0.6 mm. Anthers 0.3–0.5 mm. Caryopsis 1.3–1.5 × 0.7–0.8 mm; embryo *c.* 0.7 mm; hilum *c.* 0.3 mm. Plate 12A.

Endemic.

S.: dry inland basins of Canterbury and Otago, occasionally on plains and downs in Marlborough and Canterbury. Grassland of lowland and montane zones.

13. R. merum Connor et Edgar, *N.Z. J. Bot. 17*: 328 (1979) ≡ *Austrodanthonia mera* (Connor et Edgar) H.P.Linder, *Telopea 7*: 272 (1997); Holotype: CHR 309413! *P. A. Williams* Elliot Stream, Marlborough, hollows amongst steep to rolling country, N. facing slopes, Feb. 1977.

Light green, fine-leaved dense tufts, shortly rhizomatous, later with straw-coloured, long, drooping or trailing culms; leaves ≪ culms; branching extra- or intravaginal. Leaf-sheath pale stramineous, glabrous; apical tuft of hairs few, to 1.5 mm. Ligule 0.1–0.2 mm. Leaf-blade to 25–(40) cm, ± flat, glabrous or occasionally a few scattered short hairs, margins scabrid. Culm very slender, tawny, internodes smooth but minutely scabrid below inflorescence, elongating to *c.* 140 cm at maturity. Raceme, or rarely racemose panicle, with 1–2 short branches at base, slender, to 8 cm, of few, narrow, almost sessile spikelets; rachis and pedicels finely scabrid with longer fine hairs on margin especially below spikelets. Spikelets 4–6-flowered, awns exserted from glumes. Glumes light green, occasionally purplish, lanceolate, subobtuse, (7)–9.5–13–(15) mm, ± equal, both 5–(7)-nerved. Lemma (2.5)–3–4 mm, 7–9-nerved, upper row of hairs in small marginal tufts only, ≈ lemma, or occasionally upper tufts absent, lower row in small marginal tufts only, or a few small tufts or single hairs between, elsewhere glabrous and shining, occasional hairs on margin; lobes (3)–6–7.5–(8.5) mm, soon tapering to fine awn; central awn (6.5)–8–11.5–(14) mm, column (2.5)–3–3.5 mm. Palea (2.5)–3.5–4.5–(5) mm, > upper lemma hairs, interkeel and margins glabrous. Callus 0.6–1 mm, marginal hair tufts rarely reaching base of lower lateral tufts. Rachilla 0.5–0.6 mm. Anthers 0.3–0.9 and 1.2–1.5 mm. Caryopsis 1.7–2.5 × 0.9–1.2 mm; embryo 0.7–1 mm; hilum 0.5–0.7 mm. Plate 11B.

Endemic.

N.: sporadically in central mountains, eastern Wairarapa, and Aorangi Range, Wellington; S.: sporadically in Marlborough, Canterbury, and Sutton Salt Lakes near Middlemarch. Grassland in lowland and montane zones.

14. R. nigricans (Petrie) Connor et Edgar, *N.Z. J. Bot. 17*: 331 (1979) = *Danthonia semiannularis* var. *nigricans* Petrie, *T.N.Z.I. 46*: 37 (1914) ≡ *D. gracilis* var. *nigricans* (Petrie) Zotov, *T.R.S.N.Z. 73*: 234 (1943) ≡ *D. nigricans* (Petrie) Calder, *J. Linn. Soc. Bot. 51*: 8 (1937) ≡ *Notodanthonia nigricans* (Petrie) Zotov, *N.Z. J. Bot. 1*: 123 (1963) ≡ *Thonandia nigricans* (Petrie) H.P.Linder, *Telopea 6*: 612 (1996) *comb. illeg.*; Holotype: WELT 40273A! *D. P[etrie]* Mt Hector, Tararuas, 3000–4000 ft, 29.1.1907.

Open tufts, rather stiffly erect; leaves about ½ length of culms to only slightly shorter; branching extravaginal. Leaf-sheath brownish or often purplish above, glabrous or long-hairy; apical tuft of hairs 1.5–2.5 mm. Ligule ± 0.2 mm. Leaf-blade to 15–(20) cm, usually folded, sometimes flat and margins inrolled, to 2 mm wide, glabrous or often a few scattered long hairs near collar, sometimes almost all leaves with some long hairs, margins occasionally finely scabrid. Culm to 30–(50) cm, internodes smooth but minutely scabrid below inflorescence. Panicle usually erect, dark purple, to 3.5–(8.5) cm; rachis and pedicels densely but shortly scabrid, often with very slightly longer hairs above branch axils. Spikelets 3–(4)-flowered, awns exserted from glumes. Glumes usually dark purplish centrally, lanceolate, acute to acuminate, 6.5–9 mm, ± equal, 3–(5)-nerved. Lemma 1.5–2.5 mm, 9-nerved, upper row of hairs continuous, > or ≈ lemma lobes, lower row of marginal tufts and a few weak central tufts, or ± continuous, elsewhere predominantly glabrous, occasionally some few isolated hairs between rows, margins hair-fringed; lobes 2.7–

3.5 mm, tapering, awnless or very finely awn-tipped; central awn 6–
8.5 mm, column 1–1.5 mm < upper lemma hairs. Palea 2.5–3 mm,
≪ upper lemma hairs, interkeel and margins glabrous or occasionally
1–2 long hairs. Callus 0.2–0.3–(0.5) mm, a few short scattered
marginal hairs not, or barely, reaching lower lemma hairs. Rachilla
0.8–1.2 mm. Anthers 0.5–0.8–(1.1) mm. Caryopsis 1.1–1.3 × 0.6 mm;
embryo 0.5–0.7 mm; hilum 0.2–0.3 mm. $2n = 24$. Plate 12H.

Endemic.

N.: Tararua and Rimutaka Ranges; S.: along and west of Main Divide
from Nelson to Southland, occasionally east of Divide; St. Wet or
boggy sites in montane and alpine zones.

15. R. nudum (Hook.f.) Connor et Edgar, *N.Z. J. Bot. 17*: 322 (1979)
≡ *Danthonia nuda* Hook.f., *Fl. N.Z. 2*: 337 (1855) ≡ *Notodanthonia nuda*
(Hook.f.) Zotov, *N.Z. J. Bot. 1*: 112 (1963); Holotype: K! *Colenso 4140*
from summit of Ruahine.

Open small tufts, leafy at ends of elongating shoots, rooting at
nodes; leaves somewhat acute, ≈ culms, disarticulating at ligule;
branching intravaginal. Leaf-sheath pale stramineous to light grey,
glabrous; apical tuft of hairs few, up to 1–2 mm. Ligule to 0.4 mm.
Leaf-blade to 8 cm, folded, narrow, rather soft, glabrous. Culm to
10 cm, internodes glabrous. Panicle small, erect, to 3 cm, of few
spikelets on short pedicels; rachis and pedicels slender, closely
short-scabrid and some longer hairs at branch axils. Spikelets
usually 3-flowered, awns barely exserted from glumes. Glumes
purplish centrally, ovate, obtuse, ± equal, 3.5–5–(5.7) mm; lower
3–(5)-nerved, upper 3-nerved. Lemma 2.3–3.5 mm, 5–(9)-nerved,
upper and lower rows of hairs very sparse, upper row usually of
small marginal tufts with a few scattered hairs between, < lemma
lobes, lower row of small marginal tufts and few scattered hairs,
elsewhere glabrous, margins with scattered hairs; lobes 0.2–0.5 mm,
acute to acuminate, not awn-tipped; central awn 1–2 mm, erect,
column scarcely developed. Palea 2.4–2.8 mm, reaching base of
awn sinus, = upper lemma hairs, interkeel glabrous, margins
occasionally with a few long hairs. Callus *c.* 0.3 mm, a few marginal

hairs not reaching lower lemma hairs, or marginal tufts denser, slightly overlapping lower lemma tufts. Rachilla 0.5–0.8 mm. Anthers 0.7–0.8 mm. Caryopsis not seen. Plate 12E.

Endemic.

N.: Ruahine and Tararua Ranges. Bogs of alpine zone.

16. R. penicillatum (Labill.) Connor et Edgar, *N.Z. J. Bot. 17*: 327 (1979).

Tall, slender, light green, in dense, crowded or loose tufts, shortly rhizomatous; leaves to about ½ height of culms; branching extravaginal. Leaf-sheath pale stramineous to grey-brown, rather dense or scattered short hairs; apical tuft of hairs to 2.5–(3) mm. Ligule *c.* 0.1 mm. Leaf-blade to 25–(35) cm, flat, to 2 mm wide, or folded and inrolled, with scattered long hairs, margins scabrid. Culm to 1 m, internodes glabrous but densely, minutely hairy to scabrid for usually *c.* 10 mm below inflorescence. Panicle slender, elongate, to 12 cm of few to several spikelets on short pedicels; rachis and pedicels densely short-scabrid with some longer hairs, especially below spikelets and above branch axils. Spikelets (3)–4–6-flowered, awns exserted from glumes. Glumes usually purplish centrally or near margin, lanceolate, subobtuse, 9–12–(13.5) mm, ± equal; lower 7–(9)-nerved, upper 5–(7)-nerved. Lemma 2.2–3.5 mm, 9-nerved, upper row of hairs ≥ lemma, in marginal and occasional adjacent tufts with usually two, occasionally 0 central tufts, lower row of marginal tufts not reaching to upper tufts, and shorter ± continuous to extremely sparse hairs between tufts, elsewhere glabrous or 1–2 hairs centrally; lobes 6–8.5 mm, narrowed to fine awns; central awn (9)–10–12 mm, column (2)–2.5–3.5–(4) mm. Palea 3.5–4.5–(5) mm, > upper lemma hairs, interkeel and margins glabrous. Callus 0.5–0.8–(1) mm, thick marginal hair tufts usually overlapping lower lemma hairs. Rachilla 0.5–0.6 mm. Anthers 0.5–0.7 and 1.2–1.8 mm. Caryopsis *c.* 2 × 0.8–1.1 mm; embryo 0.7–0.9 mm; hilum 0.3–0.5 mm.

Naturalised from Australia.

N.: throughout but rare on Volcanic Plateau; S.: Nelson, Marlborough, Canterbury (Kaitorete Spit), coastal Otago and Bluff. Grassland and modified sites in lowland and montane zones.

17. R. petrosum Connor et Edgar, *N.Z. J. Bot. 17*: 317 (1979); Holotype: CHR 273105! *A. P. Druce* Cape Palliser, Wairarapa, 100 ft, cliff, Dec. 1973.

Stiff, wiry, small tussocks, tufts formed at ends of elongating shoots, rooting at nodes; leaves ≤ culms, semi-pungent, disarticulating at ligule; branching intravaginal. Leaf-sheath glabrous, light grey to dark brown; apical tuft of hairs 1–1.5 mm. Ligule *c.* 0.5 mm. Leaf-blade to 15 cm, glabrous, stiff, inrolled, margins glabrous. Culm to 40 cm, internodes glabrous. Raceme or racemose panicle small, erect, to 4 cm, of very few large spikelets on short pedicels; rachis and pedicels almost glabrous with only a few scattered teeth and usually small tufts of long hairs at branch axils. Spikelets 4–6-flowered, awns and sometimes lemma lobes exserted from glumes. Glumes green, lanceolate, acute, 8.5–15 mm, ± equal; lower 5–(7)-nerved, upper 3-nerved. Lemma 2.5–3.5 mm, 9-nerved, upper row of hairs interrupted, dense marginal tufts and a few additional hairs adjacent, hairs ≪ lemma lobes, lower row almost continuous, with dense strong marginal tufts and less dense tufts between, overlapping upper, glabrous elsewhere; lobes 4.5–8 mm, narrowing to strong awn and often shortly lobed at awn base; central awn 7.5–14 mm, column 2.5–4 mm, > upper lemma hairs. Palea 3.6–6 mm, = or slightly > upper lemma hairs, interkeel glabrous, margins with 1–2 long hairs. Callus 0.7–1 mm, strong marginal hair tufts overlapping lower lemma hairs. Rachilla 0.5–0.6 mm. Anthers 0.7–1.1 and 1.5 mm, yellow. Caryopsis *c.* 2.5 × 1.0 mm; embryo *c.* 1.0 mm; hilum 0.5 mm. Plate 11G.

Endemic.

N.: Wellington (Kapiti Id, South Wellington coast); S.: Nelson (Stephens Id, D'Urville Id). Coastal rocks.

18. R. pilosum (R.Br.) Connor et Edgar, *N.Z. J. Bot. 17*: 326 (1979).

Rather coarse, stiff, dense tufts, shortly rhizomatous; leaves ≪ culms; branching intravaginal. Leaf-sheath pale, to light grey, glabrous or with scattered to dense hairs; apical tuft of hairs stiff, thick, to 2.5–(3.5) mm. Ligule 0.1–0.2 mm. Leaf-blade to 17 cm, rather harsh, grey-green, flat to inrolled, long stiff scattered hairs, margins closely scabrid. Culm stout, to 85 cm, internodes glabrous but minutely hairy below inflorescence. Panicle ± compact and broad to somewhat laxer, to 6 cm, of few large spikelets on short pedicels; rachis and pedicels rather sparsely, stiffly scabrid, with longer hair below spikelets and near branch axils. Spikelets 5–8-flowered, awns much exserted from glumes. Glumes pale, sometimes faintly purplish, lanceolate, subacute, ± equal, 8.5–12 mm, both 5–7-nerved, occasionally a few scattered hairs. Lemma (2)–3–4 mm, 9-nerved, upper row of hairs ≈ lemma, in a few isolated tufts, usually (0)–2–4 central tufts and 1–(2) denser marginal tufts, lower row dense and continuous, lower marginal tufts reaching or overlapping upper marginal tufts, elsewhere glabrous; lobes 7–10 mm, tapering to fine awns; central awn 12.5–15 mm, column 3–5 mm. Palea 2.5–3.7 mm, = upper marginal tufts, ≥ awn sinus, interkeel glabrous, margins glabrous or with a few long fine hairs. Callus 0.8–1 mm, thick marginal hair tufts overlapping lower lemma hairs. Rachilla 0.2–0.3 mm. Anthers *c.* 0.4 and 0.8–1.7 mm. Caryopsis 1.2–2 × 0.7–1 mm; embryo 0.5–0.9 mm; hilum 0.4–0.5 mm.

Naturalised from Australia.

N.: southwards to Raglan in west and to near Opotiki in east, also in Wairarapa; S.: coastal areas of Nelson, Marlborough, and North and Mid Canterbury. Grasslands and modified sites in lowland zone.

19. R. pulchrum (Zotov) Connor et Edgar, *N.Z. J. Bot. 17*: 321 (1979) ≡ *Notodanthonia pulchra* Zotov, *N.Z. J. Bot. 1*: 111 (1963); Holotype: CHR 42376! *V. D. Zotov* Waikamaka River, *c.* 3500 ft, Ruahine Mts, river terrace in light *Nothofagus cliffortioides* forest, 4.1.1944.

Small dense tufts, or often open, with tufts of leaves at ends of elongating shoots, rooting at nodes; leaves somewhat acute, ≪ culms, disarticulating at ligule; branching intravaginal. Leaf-sheath pale stramineous to light brown, occasionally purplish, glabrous; apical tuft of few hairs, sometimes 0, up to *c.* 1.5 mm. Ligule to *c.* 0.5 mm. Leaf-blade to 10 cm, folded, narrow, rather soft, glabrous, rarely with a few hairs. Culm to 30 cm, internodes glabrous. Panicle small, to 5 cm, of few spikelets on very slender pedicels; rachis and pedicels closely short-scabrid, a few long hairs at branch axils and occasionally 1–2 below spikelets. Spikelets 3–(5)-flowered, awns exserted from glumes. Glumes blackish centrally, ovate to occasionally ovate-lanceolate, obtuse, (5)–6–7.5–(8) mm, ± equal; lower 5-nerved, upper 3-nerved. Lemma (1.8)–2–2.6 mm, 5–(7)-nerved, upper and lower rows of hairs continuous or occasionally upper row hairs fewer, upper hairs slightly < lemma lobes, lower row overlapping upper, elsewhere glabrous, some few hairs along margins; lobes (0.5)–1.8–2.8 mm, usually awn-tipped; central awn 3–4.5 mm, column 0.5–1 mm, hardly twisting, ≈ palea, < upper lemma hairs. Palea 3–3.8 mm, ≤ upper lemma hairs, interkeel glabrous, margins sometimes with a few long hairs. Callus 0.2–0.4 mm, a few marginal hairs overlapping lower lemma hairs. Rachilla 0.5–0.8 mm. Anthers 0.5–1.1–(1.4) mm. Caryopsis 1.2–1.4 × 0.5–0.7 mm; embryo 0.6–0.7 mm; hilum 0.3 mm. Plate 12D.

Endemic.

N.: Mt Egmont, central mountains, and Tararua Range; S.: north-west Nelson (Haupiri Bog and Adelaine Tarn, Douglas Range). Boggy places in subalpine and alpine zones.

20. R. pumilum (Kirk) Connor et Edgar, *N.Z. J. Bot. 25*: 166 (1987) ≡ *Atropis pumila* Kirk, *T.N.Z.I. 14*: 379 (1882) ≡ *Triodia pumila* (Kirk) Hack., *in* Cheeseman *Man. N.Z. Fl.* 896 (1906) ≡ *Erythranthera pumila* (Kirk) Zotov, *N.Z. J. Bot. 1*: 124 (1963) ≡ *Danthonia kirkii* Zotov, *T.R.S.N.Z. 73*: 234 (1943) [a combination in *Danthonia* preoccupied by *D. pumila* Nees (1841)]; Lectotype: WELT 39891! *D. Petrie 920* McRaes [Macraes], Otago, 1800 ft (designated by Zotov 1963 op. cit. p. 125).

Stiff, slender, bright green, usually low-growing tufts; leaves setaceous, ≪ to sometimes = culms, disarticulating at ligule; branching intravaginal. Leaf-sheath green to purple-tinged, later whitish, not much wider than leaf-blade, glabrous, or with scattered long fine hairs; apical tuft of hairs to 1.5 mm, sometimes very minute. Ligule to 0.4 mm. Leaf-blade to 8 cm, setaceous, usually glabrous, rarely a few long fine hairs near base. Culm 1.5–25 cm, internodes smooth but finely scabrid just below inflorescence. Panicle to 3–(4) cm, shortly branched below, racemose above, of few, comparatively large, narrow spikelets; rachis and pedicels scabrid, with some short hairs on pedicels. Spikelets (3)–4–6–(8)-flowered, glumes enclosing florets. Glumes usually purplish at centre, sometimes with darker horizontal blotches, usually narrow-lanceolate, ± acute, but often shorter, more ovate and subobtuse, ± equal, (2.5)–3.5–6.5–(8.5) mm; lower 3–5-nerved, upper 3-nerved. Lemma 1.5–2.2 mm, 5–7-nerved, ovate, scattered hairs on upper ½, sometimes an obvious marginal hair tuft on either side about the level of rachilla apex, occasionally almost glabrous or at times more densely hairy with two pairs of marginal tufts and ± distinct upper and lower row of hairs, with hairs scattered between rows and a row of hairs along either margin, tip notched, mucro 0.1–0.2–(0.9) mm in notch, sometimes a shorter mucro on either lateral lobe beside notch. Palea 1.4–2 mm, < lemma, interkeel glabrous. Callus 0.1–0.3 mm, glabrous, or, in lemma with well-developed hairs, thick marginal hair tufts to 0.5 mm overlapping lemma hairs. Rachilla 0.4–0.7 mm. Anthers 0.2–0.6 mm. Caryopsis 0.7–1 × 0.3–0.5 mm; embryo 0.4–0.6 mm; hilum 0.1–0.3 mm.

Indigenous.

N.: Volcanic Plateau; S.: along Main Divide and to the east. Montane to alpine in tussock grassland and herbfield.

Also indigenous to Australia.

Specimens from south Ashburton R. and from near L. Tekapo have lemmas hairier than usual.

21. R. racemosum (R.Br.) Connor et Edgar, *N.Z. J. Bot. 17*: 327 (1979).

Variable in size and habit from short, loose, slender, to large, stout, wiry tufts, bright green, ± shortly rhizomatous; leaves ≤ culms; branching extravaginal. Leaf-sheath light brown, glabrous or with ± stiff, scattered hairs; apical tuft of hairs to 3–(4) mm. Ligule 0.2–0.5 mm. Leaf-blade to 25 cm, ± flat, to 2 mm wide, or narrow and rolled, scattered or abundant, ± stiff hairs, or glabrous, margins closely scabrid. Culm to 90 cm, internodes smooth or minutely scabrid below inflorescence. Panicle to 15 cm, erect, racemose with few short branches below. Spikelets few to many on short pedicels; rachis uniformly short-scabrid, pedicels often bare below spikelets or with slightly longer hairs. Spikelets 5–10-flowered, awns, and sometimes upper florets, exserted from glumes. Glumes light green, margins purple above, 8–12 mm, lanceolate, subacute, ± equal; lower (5)–7-nerved, upper 5–(7)-nerved, sometimes a few scattered hairs. Lemma 2.5–3.5 mm, 9-nerved, upper row of hairs ≤ lemma, in large marginal and adjacent tufts with usually two, sometimes more or occasionally 0 central tufts, lower row of hairs continuous, reaching upper hairs, elsewhere glabrous or a few scattered hairs, obviously narrowed and prominently nerved above callus; lobes 5–10 mm, abruptly narrowed to awns; central awn 11–14 mm, column (1.5)–2–3–(4) mm. Palea 3.5–5 mm, = upper marginal tufts, interkeel and margins glabrous. Callus very narrow, (0.6)–0.9–1.5 mm, marginal hair tufts short, usually < lower lemma hairs. Rachilla minute, 0.1–0.2 mm. Anthers 0.3–1.2 and 1.4–2.0 mm. Caryopsis 1.7–2.1 × 0.8–1.1 mm; embryo 0.8–0.9 mm; hilum 0.4–0.5 mm.

Naturalised from Australia.

N.: throughout except in Taranaki; S.: Nelson, Marlborough, coastal Canterbury to Rakaia River, basins of inland Otago, and near Dunedin; K.: Raoul and Macauley Is; Three Kings Is. Grassland and modified sites in lowland zone.

Buchanan, *Indig. Grasses N.Z.* t. 33 (2) B (1879) formally named this Australian taxon *Danthonia pilosa* var. *racemosa*; Holotype: WELT 59580 (Buchanan's folio)! *Buchanan.*

22. R. setifolium (Hook.f.) Connor et Edgar, *N.Z. J. Bot. 17*: 316 (1979) ≡ *Danthonia semiannularis* var. *setifolia* Hook.f., *Fl. N.Z. 1*: 304 (1853) ≡ *D. setifolia* (Hook.f.) Cockayne, *N.Z. J. Agric. 23*: 146 (1921) ≡ *Notodanthonia setifolia* (Hook.f.) Zotov, *N.Z. J. Bot. 1*: 108 (1963); Lectotype: K! *Colenso 2366* sides of R. Whangaehu (designated by Zotov 1963 op. cit. p. 108). = *Danthonia semiannularis* var. *alpina* Buchanan, *Indig. Grasses N.Z.* t. 34 (2) A (1879); Holotype: WELT 59583 (Buchanan's folio)!

Stiff, yellow-green or bright green tussocks, often with tufts of leaves at ends of elongating shoots, rooting at nodes; leaves semi-pungent, ≤ culms (occasionally very much shorter), disarticulating at ligule; branching intravaginal. Leaf-sheath pale stramineous, glabrous; apical tuft of hairs (0.2)–1.0–3.0 mm. Ligule 0.2–0.8 mm. Leaf-blade to 25–(35) cm, stiff, inrolled, rounded, glabrous, margins glabrous. Culm to 50 cm, internodes glabrous. Panicle ± erect, to 6 cm, of few large spikelets on very slender pedicels; rachis and pedicels almost glabrous or with a few scattered prickle-teeth usually along edges and below spikelets, a few long hairs at branch axils. Spikelets (3)–4–6-flowered, awns exserted from glumes. Glumes becoming light red-purple to purple-black centrally, lanceolate, subobtuse to acute, (5.5)–7–11–(20) mm, ± equal; lower 3–5-nerved, upper (3)–5–(7)-nerved. Lemma (2)–2.5–3.5–(4) mm, 7–9-nerved, upper and lower rows of hairs continuous, upper row slightly < lemma lobes, lower row usually reaching to base of upper row, glabrous elsewhere but occasional hairs centrally, margins often hair-fringed; lobes (2.3)–3–4–(5) mm, tapering to short fine awn; central awn 8–12 mm, widely divergent, column 1–2 mm usually < upper lemma hairs. Palea 3–5 mm, ≥ upper lemma hairs, interkeel glabrous or occasional scattered short hairs, margins with long hairs below or glabrous. Callus *c.* 0.5 mm, short marginal hair tufts overlapping lower lemma hairs. Rachilla 0.6–1 mm, occasionally crowned with minute hairs. Anthers (1.2)–1.5–3 mm. Caryopsis 1.1–1.5×0.6–0.7 mm; embryo 0.5–0.8 mm; hilum 0.2–0.3 mm. $2n = 24$. Plates 6A, 11E.

Endemic.

N.: Moehau Range, Coromandel Peninsula and on mountains from Raukumara Ranges southwards, and at Mt Pirongia; S.: throughout on mountain ranges; St. Grassland and rocky places in montane to alpine zones, sea level in far south.

Some plants from Fiordland (West Cape, and Preservation and Chalky Inlets) have glumes much longer than usual, 12–20 mm, and lemma lobes to twice as long as lemma. Leaf-blades are often curled towards the tip, and apical tuft of hairs very short or lacking.

Specimens collected from the Manawatu Gorge also have acute-tipped glumes *c.* 12 mm, and fine, almost filiform leaves.

23. R. tenue (Petrie) Connor et Edgar, *N.Z. J. Bot. 17*: 321 (1979)
≡ *Danthonia buchananii* var. *tenuis* Petrie, *T.N.Z.I. 46*: 37 (1914)
≡ *Notodanthonia tenuis* (Petrie) Zotov, *N.Z. J. Bot. 1*: 111 (1963);

Lectotype: WELT 39920! *D. P[etrie]* Upper Waipori, Otago, Dec. 1893 (No. 1281 to Hackel) (designated by Zotov 1963 op. cit. p. 111).

Slender tufts, often with tufts of leaves at ends of elongating shoots, rooting at nodes; leaves < culms; branching extra- and intravaginal. Leaf-sheath light brown, usually with scattered long hairs, sometimes smooth; apical tuft of hairs to 2 mm. Ligule 0.2–0.3 mm. Leaf-blade to 15 cm, inrolled, mostly glabrous but margins slightly scabrid near base. Culm to 25 cm, internodes smooth but finely scabrid below inflorescence. Panicle erect, contracted, narrow, 1–7 cm, of few spikelets on short pedicels; rachis and pedicels uniformly closely short-scabrid. Spikelets 4–5–(6)-flowered, awns ± included by glumes. Glumes purplish centrally, lanceolate, subacute, 6–7.4 mm, ± equal, 5-nerved. Lemma *c.* 2 mm, 9-nerved, upper row of hairs dense, continuous, ≥ lemma lobes, lower row ± indistinct except for marginal tufts, elsewhere short scattered hairs; lobes 0.7–1.2 mm, tapering to minute awn or membranous; central awn 1.5–2.5 mm, column *c.* 0.5 mm, sometimes absent, < palea and upper lemma hairs. Palea *c.* 2.5 mm, ≤ upper lemma hairs, interkeel glabrous, margins with a few long hairs. Callus 0.2–0.3 mm, very short marginal hair tufts not, or a few just reaching lower row of lemma hairs. Rachilla *c.* 0.7–1 mm. Anthers 0.6–0.9 mm. Caryopsis not seen. Plate 12C.

Endemic.

S.: known only from Upper Waipori, Old Man Range, West Lake Mahinerangi, and Flagstaff Hill, Otago.

Only four specimens match Petrie's type from Upper Waipori: WELT 39938 *D. Petrie* Flagstaff Hill, 1908; OTA 219224 (duplicate CHR 155850) *G. T. Daly* West L. Mahinerangi, 1400 ft, 1966; CHR 518303 *B. P. J. Molloy* Rock and Pillar Range, 900 m, 1999; and CHR 149750 *B. P. J. Molloy* Old Man Range, *c.* 3000 ft, 1964. Two other sheets, one of *R. gracile,* CHR 3665, Rees Valley, collector unknown, the other *R. buchananii,* WELT 39945 *D. Petrie 3094* Mt Ida, have pieces of *R. tenue* mounted on them, but because the sheets are mixed there is no certainty that the *R. tenue* was in fact collected from these localities.

Connor, H. E. *N.Z. J. Bot. 26*: 163–167 (1988), reported that *R. tenue* seemed sterile and concluded that "the reproductive behaviour of the taxon does not inspire confidence for its future".

24. R. tenuius (Steud.) A.Hansen et Sunding, *Fl. Macaronesia Checklist Vasc. Pl.* pt. 1, ed. 2, 93 (1979).

Grey-green or greyish purple, coarse erect tussocks, occasionally shortly rhizomatous; leaves ≫ culms; branching intra- and extravaginal. Leaf-sheath white or sometimes greyish, sometimes with few scattered fine hairs above; apical tuft of hairs 1–2–(3) mm. Ligule 0.1–0.4–(0.6) mm. Leaf-blade to 35 cm, flattish, becoming inrolled, to 3 mm wide, sometimes with scattered hairs, margins sparsely scabrid. Culm stout, to 1 m, internodes smooth, minutely scabrid below inflorescence. Panicle large, strict, to 15 cm, of many spikelets on ± rigid pedicels; rachis and pedicels closely short-scabrid, often longer hairs intermixed. Spikelets 5–6–(7)-flowered, awns included by or exserted from glumes. Glumes usually broad, purple-margined, lanceolate, subacute, (12)–15–17–(19) mm, ± equal, both 7-nerved. Lemma 2.5–3 mm, 9-nerved, upper and lower rows of hairs very dense, upper row < lemma lobes and much overlapped by lower, glabrous elsewhere, occasional hairs centrally and on margins; lobes (6.3)–7–9 mm, narrowed to fine awn; central awn (8.5)–10–14 mm, column 1–2.5 mm < upper lemma hairs. Palea 3.5–4.5 mm, < upper lemma hairs, interkeel hairs long, numerous, in lower ⅔, and on margins. Callus (0.8)–1–1.3 mm, marginal hair tufts overlapping lower lemma hairs. Rachilla *c.* 0.2 mm. Anthers 0.8–1.2 and 1.7–2.1 mm. Caryopsis 1.6–2 × 0.8 mm; embryo 0.7 mm; hilum 0.5 mm.

Naturalised from Australia.

N.: North Cape to Auckland City and Thames. Grassland and modified sites in lowland zone.

Although not recorded for N.Z. until 1963, as *Notodanthonia purpurascens*, specimens of *R. tenuius* were collected as early as 1901 (WELT 39688, OTA 16911).

25. R. thomsonii (Buchanan) Connor et Edgar, *N.Z. J. Bot. 17*: 322 (1979) ≡ *Danthonia thomsonii* Buchanan, *Indig. Grasses N.Z.* t. 36 (2) (1879) ≡ *Triodia thomsonii* (Buchanan) Petrie, *T.N.Z.I. 44*: 188 (1912) ≡ *Notodanthonia thomsonii* (Buchanan) Zotov, *N.Z. J. Bot. 1*: 112 (1963); Holotype: WELT 59624 (Buchanan's folio)! *D. Petrie* Mount St Bathans, Otago, 1000–2000 ft.

Variable in size and habit, from low-growing tussocks to moderately tall tufts, rather dull green, ± shortly rhizomatous; leaves semi-pungent, usually ≫ culms; branching intravaginal, shoots crowded, ± swollen at base. Leaf-sheath pale to greyish, glabrous, sometimes with scattered long hairs; apical tuft of hairs 1–3 mm. Ligule *c.* 0.3 mm. Leaf-blade to 15–(22) cm, stiff, inrolled with scabrid margins, glabrous, or with scattered long hairs. Culm to 45 cm, internodes smooth but minutely scabrid below inflorescence. Panicle strict, to 6.5 cm, shortly branched below, of few to many spikelets on short pedicels; rachis and pedicels closely short-scabrid, with numerous minute hairs interspersed, longer hairs below spikelets and at branch axils. Spikelets 5–8-flowered, sometimes upper florets exceeding glumes. Glumes purplish centrally, ovate-lanceolate, subacute, (3.5)–5–7–(9) mm, ± equal; lower 5–(7)-nerved, upper 3–(5)-nerved. Lemma 2.5–3.5–(4) mm, finely 7–(9)-nerved, both rows of hairs reduced to short dense marginal tufts, elsewhere glabrous, upper hairs < lemma lobes; lobes reduced to minute bristles, 0.1–0.4–(0.6) mm, rarely almost wanting; central awn (0.2)–0.5–1–(2.5) mm, straight, erect, fine, or reduced to a mucro from a bifid lemma-tip, column absent. Palea 2.3–3 mm, ≤ base of sinus, ≫ upper lemma hairs, interkeel and margins glabrous. Callus 0.2–0.3 mm, thick marginal hair tufts reaching base of lower marginal lemma tufts. Rachilla *c.* 0.5 mm. Anthers 0.4–0.7 mm. Caryopsis 1–1.3 × 0.4–0.7 mm; embryo to 0.5 mm; hilum *c.* 0.2 mm. Plate 12G.

Endemic.

S.: east of the Main Divide especially in inland basins of Canterbury and Otago. Grassland in dry lowland and montane zones.

26. R. unarede (Raoul) Connor et Edgar, *N.Z. J. Bot. 17*: 328 (1979) ≡ *Danthonia unarede* Raoul, *Ann. Sci. Nat.* Ser. 3, Bot. 2: 116 (1844) ≡ *D. semiannularis* var. *unarede* (Raoul) Hook.f., *Fl. N.Z. 1*: 304 (1853) ≡ *Notodanthonia unarede* (Raoul) Zotov, *N.Z. J. Bot. 1*: 122 (1963) ≡ *Thonandia unarede* (Raoul) H.P.Linder, *Telopea 6*: 612 (1996) *comb. illeg.*; Holotype: P! *Raoul* Nouvelle Zélande — Presqu'île de Banks, 1843. = *Danthonia cingula* Steud., *Syn. Pl. Glum. 1*: 246 (1853); Holotype: P! *Kampmann 99* Nova Zeelandia.

Strong tufts spreading outwards from ± open centre, very shortly rhizomatous, mostly glabrous apart from apical tuft of hairs but not invariably so; leaves slightly < mature culms; branching extravaginal. Leaf-sheath pale stramineous, usually glabrous but some with long, scattered to very occasionally abundant, fine hairs, rarely all sheaths hairy; apical tuft of hairs to 4 mm. Ligule 0.2–0.8 mm. Leaf-blade flat, or margins inrolled, to 40 cm × 3.5 mm, glabrous, occasionally a few scattered long hairs, margins finely scabrid. Culm to 75–(85) cm, internodes smooth but finely scabrid below inflorescence. Panicle narrow, long-branched, to 15–(18) cm, of large spikelets on short slender pedicels; rachis and pedicels densely short-scabrid, occasionally with a few longer hairs below spikelets and at branch axils. Spikelets (3)–4–6–(8)-flowered, awns exserted from glumes. Glumes purplish centrally, lanceolate, acute or subacute, ± equal, (7.5)–10–12 mm, both 5–(7)-nerved. Lemma 2–3 mm, 9-nerved, both upper and lower rows of hairs dense, continuous, upper row < lemma lobes, lower row only reaching base of upper, some few scattered hairs elsewhere and on margins; lobes (4)–5–7 mm, gradually tapering to fine awns; central awn (6)–9–12 mm, column (1)–2–3 mm, ≥ palea, usually ≈ upper lemma hairs. Palea 4–5 mm, slightly < upper lemma hairs, interkeel glabrous or rarely a few hairs, margins a few long hairs or glabrous. Callus 0.5–0.8 mm, with marginal hair tufts overlapping lower lemma hairs. Rachilla 0.4–0.6–(0.7) mm. Anthers 0.3–0.8 and 1–1.6 mm. Caryopsis 1.6–1.8 × 0.8–0.9 mm; embryo 0.8 mm; hilum 0.5 mm. Plate 12F.

Endemic.

N.: throughout; S.: throughout, but rare to west of Main Divide; St.; K., Three Kings Is, Ch. Grassland and modified sites in lowland and montane zones.

27. R. viride (Zotov) Connor et Edgar, *N.Z. J. Bot. 17*: 316 (1979) ≡ *Notodanthonia viridis* Zotov, *N.Z. J. Bot. 1*: 108 (1963); Holotype: CHR 2484! *V. D. Zotov* Onetapu Desert, Volcanic Plateau, 5.4.1931.

Stiff, rather dark purple-green tussocks; leaves semi-pungent, somewhat < culms, disarticulating at ligule; branching intravaginal. Leaf-sheath pale stramineous below, purplish above, usually glabrous or with some scattered long hairs; apical tuft of hairs 3–4–(6) mm. Ligule (0.2)–0.8–1.5 mm. Leaf-blade to 35 cm, stiff, inrolled, glabrous or occasionally scattered long hairs, margins often scabrid above ligule and near tip. Culm to 45–(55) cm, internodes smooth or sometimes fine-scabrid below inflorescence. Panicle ± erect, to 10 cm, of few large spikelets on very slender pedicels; rachis and pedicels closely short-scabrid with some long hairs occasionally at branch axils. Spikelets 3–4–(5)-flowered, awns exserted from glumes. Glumes blackish purple centrally, linear-lanceolate, subobtuse, (6)–8–12 mm, ± equal; lower 5–(7)-nerved, upper 3–(5)-nerved. Lemma 2.5–3 mm, 9-nerved, upper and lower rows of hairs very dense and continuous, upper row ⩾ lemma lobes, lower row overlapping upper, shorter hairs scattered elsewhere; lobes (3.7)–4–4.7 mm, tapering to fine awn; central awn 8–11 mm, column 1–1.5 mm, ≈ palea, < upper lemma hairs. Palea 3.5–4.5 mm, reaching about ½ way up upper lemma hairs, long hairs on interkeel below and on margins. Callus *c.* 0.5 mm, marginal hair tufts long, abundant, much overlapping lower lemma hairs. Rachilla 0.7–1 mm. Anthers (0.8)–1–1.5–(2) mm. Caryopsis 1.4–1.7 × 0.6–0.7 mm; embryo 0.5–0.8 mm; hilum 0.3 mm. Plate 11F.

Endemic.

N.: Mt Egmont, central mountains, and Raukumara and Coromandel Ranges; S.: north-west Nelson (Cobb Valley, Spey River, Lake Aorere). Rocky places in montane and subalpine zones.

INCERTAE SEDIS   Two specimens from Canterbury, CHR 89588   *A. J. Healy* Harewood, on the dry shingly land of an old river terrace, Oct 1955, and CHR 92848 *R. D. Dick* Darfield - Hororata foothill country, Dec 1955, are not referrable to any known taxon; both appear to be naturalised plants.

SIEGLINGIA Bernh., 1800

Monotypic. Characters as for sp. Naturalised in N.Z.

1. S. decumbens (L.) Bernh., *Syst. Verz. 20*: 44 (1800).   heath grass

Densely tufted, dull green, rather stiff perennials, 12–50 cm; innovations extravaginal; cataphylls shining, cream. Leaf-sheath rounded to keeled above, usually with scattered long fine hairs or glabrous, strongly ribbed, light green or creamy brown, darker and fibrous at maturity; apical tufts of soft hairs to 2 mm. Ligule a short ring of hairs 0.3–0.7 mm, and often ± forming a contra-ligule. Leaf-blade 4–15–(30) cm × 1.5–3 mm, flat or involute, firm, with long soft hairs or glabrous, scabrid on nerves and on margins near curved boat-shaped tip. Culm 10–45 cm, erect or spreading, internodes smooth except a few prickle-teeth on angles below panicle. Panicle 2–6 cm, compact, with 3–9 spikelets on erect or somewhat pulvinately spreading finely scabrid branches, inflorescence sometimes a raceme. Spikelets 6–14 mm, 4–6-flowered, elliptic or oblong, plump, purplish or green, flowers ♂. Glumes persistent, equal, ≤ spikelet, 3–5-nerved, ovate-lanceolate, firmly membranous with wide transparent margins, rounded below, keeled above and finely scabrid on midnerve, acute or obtuse. Lemma 5–8 mm, 7–9-nerved, elliptic, subcoriaceous, rounded and glabrous, but with a line of hairs at each margin from base to about midway, shortly 3-lobed above and finely scabrid there on margins and apex, awnless. Palea coriaceous, < lemma; keels and obtuse apex short-ciliate, keels thickened near base. Callus c. 0.2 mm, lateral thick tuft of hairs to 1.25 mm. Rachilla prolongation c. 0.5 mm. Lodicules 2, 1 mm, nerved, truncate, hair tipped, 0 in cleistogamous florets. Stamens 3, anthers 1.25 mm in chasmogamous flowers, 0.3–0.5 mm in cleistogamous flowers. Gynoecium: ovary 0.5–0.75 mm, glabrous; stigmastyles 1.5 mm, subterminal, sparsely branched. Caryopsis 2–2.5 × 0.8–1.4 mm; embryo c. 0.8 mm; hilum linear, almost 1 mm.

Naturalised from Europe.

N.: scattered throughout except Gisborne and Hawkes Bay; S.: Westland, from Westport to Hokitika; Canterbury, one collection from Banks Peninsula. Waste ground, roadsides, tracks, poor pasture.

Cleistogenes are sometimes developed in basal leaf-sheaths.

Sieglingia is sometimes included within *Danthonia sens. strict.*, see Clayton and Renvoize (1986 op. cit.).

CORTADERIINAE

Large tussocky perennials with basal leaves. Inflorescence a large plumose panicle. Spikelets 3–4-flowered; florets ♂ or ♀ on separate plants (gynodioecious). Glumes persistent, transparent, 1-nerved, ≥

spikelet. Lemma 3-nerved, long hairy; awn straight between lemma lobes. Palea < lemma. Stamens 3; anthers shorter in male-sterile flowers. Fruit a caryopsis.

CORTADERIA Stapf, 1897 *nom. cons.*

Tall perennial tussocks with robust intravaginal shoots (extravaginal in *C. splendens*). Leaf-sheaths heavily coated in epicuticular wax or not, breaking into short segments or remaining entire. Ligule a rim of short hairs. Leaf-blade to 2 m long, tapering, 1–many evident ribs, scabrid and sharply cutting from many antrorse prickle-teeth or scarcely so, homo- or heterochromous. Panicle large, plumose, variously coloured subtended by long, hair-fringed bract, branches stiff to flexible. Glumes 1-nerved, transparent, minutely prickle-toothed. Spikelets with 3–7 florets. Flowers dimorphic on separate plants, ♀ and ♀, gynodioecious. Lemma 3-nerved, hairs abundant and radiating (except on *C. selloana* ♀) awned from between bifid lobes or entire and mucronate. Lodicules 2, cuneate and irregularly lobed, hair-tipped. Androecium of male-fertile flowers with 3 long anthers, of male-sterile flowers reduced to small staminodes or shorter sterile anthers. Gynoecium with 2 plumose stigmas (and the remnant of a third), ♀ > ♀. Caryopsis free; hilum linear *c.* ½ caryopsis. Fig. 17.

Type species: *C. selloana* (Schult. et Schult.f.) Asch. et Graebn.

25 spp., most in South America from Venezuela to Chile and Argentina. Endemic spp. 5; naturalised spp. 2, from South America.

For names and types in *Cortaderia* see Connor, H. E. and Edgar, E. *Taxon 23*: 595–605 (1974) and Connor, H. E. *Taxon 32*: 633–634 (1983). DNA sequences reported by Barker, N. P., Linder, H. P. and Harley, E. H. *Syst. Bot. 23*: 327–350 (1998) indicated that New Zealand and South American spp. are not monophyletic.

The taxonomy here follows that of Zotov, V. D. *N.Z. J. Bot. 1*: 78–136 (1963) with the addition of taxa recognised since then.

Native species are collectively known as toetoe, and the naturalised species as pampas grass.

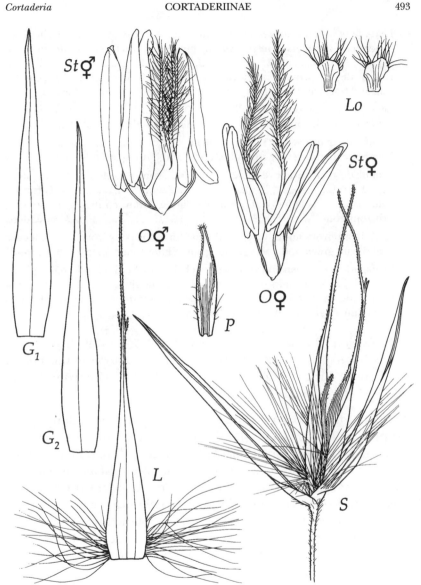

Fig. 17 *Cortaderia fulvida*
S, spikelet, × 10; **G₁**, **G₂**, glumes, × 8; **L**, lemma dorsal view, × 8; **P**, palea dorsal view, × 8;
Lo, lodicules, × 20; **O♂**, ovary, styles and stigmas ♀ flower, × 15; **O♀**, ovary, styles and
stigmas of ♀ flower, × 15; **St♂**, anthers of ♂ flower, × 15; **St♀**, anthers of ♀ flower, × 15.

1 Leaf-blade with prominent midrib and several conspicuous lateral ribs; leaf-sheath evidently glaucous, remaining entire and strict (sect. *Bifida;* endemic spp.) *2*

  Leaf-blade with prominent midrib only; leaf-sheath not glaucous, later curling up and fracturing into short segments (sect. *Cortaderia;* naturalised spp.) ... *6*

2 Leaf-sheath with long hairs; all flowers ♂ .... **7. turbaria**

  Leaf-sheath glabrous; flowers on separate plants all ♂, or all ♀ *3*

3 Leaf-blade margins very slightly scabrid *4*

  Leaf-blade margins very strongly scabrid and cutting at mid-point *5*

4 Leaf-blade glabrous above ligule; ligule 1 mm; contra-ligule absent; caespitose .... **1. fulvida**

  Leaf-blade densely hairy above ligule; ligule 3 mm; contra-ligule present; rhizomatous .... **5. splendens**

5 Leaf-sheath ivory under waxy coating; culm internodes ivory .... **6. toetoe**

  Leaf-sheath green under waxy coating; culm internodes green ...... **3. richardii**

6 Leaf-blade blue-green above, dark green below; rachis finely silky hairy; plants ♀ and ♂; lemma hairs of ♀ floret arising throughout .... **4. selloana**

  Leaf-blade dark green on both surfaces; rachis minutely scabrid; all plants ♀; lemma hairs mostly arising above palea height .... **2. jubata**

1. C. fulvida (Buchanan) Zotov, *N.Z. J. Bot. 1*: 84 (1963) ≡ *Arundo fulvida* Buchanan, *T.N.Z.I. 6*: 242 (1874) ≡ *A. conspicua* var. *fulvida* (Buchanan) Kirk, *T.N.Z.I. 10*: app. xliii (1879); Holotype: WELT 59573! *J. Buchanan* Wellington, 1873; ♀.

Very tall, stout tussock with sharp, flexible leaves, and inflorescences with pendent branches. Leaf-sheath green under white epicuticular wax, glabrous. Ligule to 1 mm. Collar light brown, glabrous. Leaf-blade to 2.5 m × 2 cm, soft, flexible, tapering to thin point; abaxially glabrous below becoming very scabrid from rows of abundant prickle-teeth, adaxially glabrous except for caducous long hairs near margins; margins long hairy below becoming lightly prickle-toothed. Culm to 3.5 m, internodes glabrous. Inflorescence to 1 m, branches pendent; rachis scabrid, branches and pedicels short stiff hairy, long hairs at axils and below spikelet. Spikelets to 20 mm, of 2–3 florets. Glumes ± equal, to 15 mm, ≤ florets, 1-nerved. Lemma to 1 mm, 3-nerved, scabrid above; lateral lobes to 2 mm; hairs to 7 mm radiating from below and < lemma; central awn to 13 mm. Palea to 4.5 mm, attenuated, apex hair-tipped, keels ciliate, interkeel and flanks scabrid above. Callus hairs to 1.5 mm.

Rachilla to 0.5 mm, glabrous. Lodicules to 0.75 mm. Anthers of ♂ flowers to 3.7 mm, yellow; of ♀ flowers to 2.2 mm, white. Gynoecium: of ♂ flowers ovary to 0.6 mm, stigma-styles to 1.75 mm, of ♀ flowers ovary to 0.75 mm, stigma-styles to 2.5 mm. Caryopsis 1.5–2 mm; embryo to 0.75 mm; hilum to 1 mm. $2n = 90$. Fig. 17. Plate 8A.

Endemic.

N.: Stream, lake, and forest margins, and in wet places and on hillsides, to 1300 m; S.: very occasional in Nelson-Marlborough. FL late Nov–Dec.

Natural populations comprise 66% ♂ and 34% ♀ plants.

Usually a species of open places with large inflorescences of drooping branches. Specimens gathered in forests at Kerikeri and nearby Puketi Forest have narrow panicles; CHR 333099 *H. Carse* Kaiaka, Nov 1906, is typical of the former: CHR 35158 *H. H. Allan* Kerikeri River, Dec 1941; CHR 469825a,b *A. P. Druce 900* Puketi Forest, Dec 1989; CHR 477325 *P. J. de Lange 1282* Kerikeri, Nov 1991 are typical of the latter. Among these specimens spikelet length, awns exceeding glumes, anthoecium sizes, anther length, caryopsis size correspond to *C. fulvida*.

The usual geographic range for *C. fulvida* is North Id, but it is popularly planted in gardens in South Id from which it can escape, e.g., at Oaro, Kaikoura coast, and elsewhere nearby e.g. Okarahia Stream, Hundalees. In northern South Id on D'Urville Id and Maud Id a few plants occur; the latter may have been introduced.

2. C. jubata (Lemoine) Stapf, *Bot. Mag.* t. 7607 (1898).

Very tall, stout, tussock with dark green, sharp, drooping leaves, and inflorescence initially violet but drying dirty brown. Leaf-sheath thin, rolling up and eventually fracturing into short segments; abaxially with scattered long hairs, adaxially with short soft hairs below ligule. Ligule to 4 mm. Collar white, sparsely prickle-toothed. Leaf-blade to 2 m × 2 cm, surfaces homochromous, arching and pendent from sheath, abaxially sparsely long hairy, adaxially with weft of short hairs on nerve at base, minute prickle-teeth throughout; margins long (2 mm) hairy below becoming scabrid from rows of prickles. Culm to 3.5 m; internodes finely scabrid. Inflorescence to 75 cm, plumose, dense, erect to drooping with flexible branches falling to one side, violet drying dirty dull brown; rachis finely scabrid, branches and pedicels violet and abundantly finely short stiff hairy, pedicels lacking long hairs below spikelets. Spikelets 20 mm, of up to 7 florets. Glumes unequal, to 10–13 mm, > florets, scabrid, produced into awn-like process, nerve becoming violet. Lemma to 12 mm, scabrid, drawn out into long awn-like process, nerves 3, violet; hairs to 10 mm, radiating from upper lemma; central awn a mucro to 0.3 mm, from between bifid lemma apex. Palea to 4 mm, attenuated, apex hair-tipped, violet, keels ciliate, interkeel and flanks scabrid. Callus to 1 mm, curved, hairs to 2 mm. Rachilla to 1.5 mm. Lodicules to 0.5 mm, lobed or simple, ciliate, nerved. Androecium of 3 staminodes to 0.15 mm on filaments to 0.8 mm. Gynoecium: ovary to 1.5 mm; stigma-styles to 2 mm. Caryopsis to 2.5 mm; embryo to 1 mm; hilum linear to 1 mm. $2n = 108$.

Naturalised from South America.

N.: south from Lake Taupo; waste places, lowlands, plantation forests and scrub to 800 m; S.: Nelson, Marlborough, and Canterbury. Cultivated in many gardens in both Is. FL Jan–mid Mar.

All plants are ♀ as in Bolivia, Ecuador and Peru — the area of origin — and reproduction is by autonomous apomixis.

3. C. richardii (Endl.) Zotov, *N.Z. J. Bot. 1*: 84 (1963) ≡ *Arundo richardii* Endl., *Ann. Weiner Mus. Naturgesch. 1*: 158 (1836) ≡ *A. kakao* Steud., *Syn. Pl. Glum. 1*: 194 (1854) ≡ *A. australis* A.Rich., *Ess. Fl. N.Z.* 121 (1832) non Cav. (1799); Holotype: P! Herb. Richard no. 29 N'lle Zelande. Havre de l'Astrolabe; ♀; (fragment CHR 236584). = *Gynerium zeelandicum* Steud., *Syn. Pl. Glum. 1*: 198 (1854); Holotype: P (Herb. Steudel)!, N. Zeelandia; ♂; (fragment CHR 236583).

Very tall, stout tussock with narrow, erect, sharp leaves, and narrow inflorescences. Leaf-sheath glabrous beneath white epicuticular wax. Ligule to 3.5 mm. Collar brown, abaxially glabrous, adaxially with ribs short stiff hairy. Leaf-blade to 2 m × 2.5 cm, stiff, abaxially very scabrid in upper ⅓, adaxially with thick weft of hairs at base becoming fewer up midrib, abundant minute prickle-teeth throughout; margins long (5 mm) hairy below becoming very scabrid. Culm to 2.5 m, internodes glabrous. Inflorescence to 1 m, plumose, stiff, erect to pennant-shaped and drooping; rachis glabrous, branches and pedicels scabrid, long hairs at axils and below spikelet. Spikelets to 25 mm, of 3 florets. Glumes equal, ≥ florets, transparent, acuminate, 1-nerved occasionally shortly 3-nerved; keel scabrid; sometimes long hairy on lower margin. Lemma to 10 mm, scabrid, long (7 mm) hairs radiating from base; lateral lobes to 3 mm; central awn to 15 mm from between lobes. Palea to 6 mm, attenuated, apex hair-tipped, keels ciliate, interkeel and flanks scabrid above. Rachilla to 1 mm, glabrous. Callus hairs to 2 mm. Lodicules to 0.7 mm, hair-fringed. Anthers of ♂ flowers to 4.5 mm, of ♀ flowers to 3 mm. Gynoecium: of ♂ flowers ovary to 1 mm, stigma-styles to 2.5 mm; of ♀ flowers ovary to 1.3 mm, stigma-styles to 4 mm. Caryopsis 3–4 mm; embryo to 1 mm; hilum to 2.0 mm. $2n = 90$. Plate 8B.

Endemic.

S.: throughout on river beds, lake and stream margins and wet places; sea level to 900 m.

Natural populations comprise 62% ♂ and 38% ♀ plants; populations exclusively of ♂ plants were found in Mackenzie Country.

Naturalised in Tasmania [Curtis, W. M. and Morris, D. I. *Student's Fl. Tasmania 4B* p. 319 (1994)].

4. C. selloana (Schult. et Schult.f.) Asch. et Graebn., *Syn. Mitteleuropa Fl. 2*: 325 (1900).

Very tall, stout, erect tussock with sharp leaves of different colours on the two surfaces, and large plumose inflorescences. Leaf-sheath to 50 cm, thin, cross-veins evident, rolling up and eventually fracturing into short segments, abaxially with long internerve hairs denser below collar and abundant short interrib hairs, adaxially clothed above in minute hairs becoming fewer below; apical tufts 3 mm. Ligule to 3 mm. Collar white, adaxially with minute hairs as on sheath. Leaf-blade to 2 m × 2 cm, erect, arching when older, surfaces heterochromous, abaxially with caducous long (2 mm) hairs especially below and near midrib and abundant small interrib hairs, keel smooth below soon becoming very scabrid, adaxially with small weft of short hairs at base and abundant small internerve hairs; margins with some long hairs below becoming very scabrid from close-set rows of prickles. Culm to 6 m, internodes clothed in short, soft, silky hairs. Inflorescence to 1 m, plumose, variously coloured, branches erect in ♂, rachis densely clothed in minute, soft, silky hairs longer and denser at nodes, branches and pedicels short finely stiff hairy, long hairs at axils. Spikelets of up to 6 florets. Glumes ± equal, to 12 mm, < florets, 1-nerved, violet suffused, acute or sometimes bifid, scabrid or toothed. Lemma to 15 mm, 3-nerved, produced and awn-like above but awn and mucro to 0.3 mm from between 2 small teeth, hairs to 10 mm, abundant and radiating from whole lemma in ♀ flowers but fewer, 7 mm long and radiating from base of lemma in ♀̄ flowers. Palea to 4 mm, produced, keels and apex ciliate, interkeel and flanks sparsely scabrid. Callus to 1 mm, hairs to 2 mm. Rachilla to 0.25 mm. Lodicules to 0.7 mm. Anthers of ♀̄ flowers to 4.5 mm, of ♀ flowers staminodes to 0.1 mm. Gynoecium: of ♀̄ flowers ovary to 0.6 mm and stigma-styles to 1 mm; of ♀ flowers ovary to 1 mm, stigma-styles to 2 mm. Caryopsis to 2.5 mm; embryo 0.6 mm; hilum 1 mm. $2n = 72$.

Naturalised from central South America.

N.; S.: widespread on roadsides, waste places, scrubland, and of more recent times in plantation pine forests; sea level to 800 m. FL mid Mar–late Apr.

Natural populations comprise 51% ♀̄ and 49% ♀ plants.

Although seeds are formed on ♀̄ plants they are never abundant and are unlikely to give rise to many new plants; the reproductive system, thus, is chiefly dioecious.

As a weedy species in plantation forests see Gadgil, R. L., Knowles, R. L. and Zabkiewicz, J. A. *Proc. 37th N.Z. Weed and Pest Control Conf.*: 187–190 (1984); Gadgil, R. L. et al. *N.Z. J. Forest. Sci. 20*: 176–183 (1990).

Once widely planted as shelter and to a lesser extent as animal fodder.

5. C. splendens Connor, *N.Z. J. Bot. 9*: 519 (1971); Holotype: CHR 184354! *R. Bell* Ruapuke Beach, 20 December 1967; ♀.

Exceedingly tall, stout tussock with long rhizomes in sand hills and extravaginal branching. Leaf-sheath clothed in long hairs or occasionally only at margins, pale green under heavy white epicuticular wax; contra-ligule present. Ligule to 3 mm. Leaf-blade to 3 m × 3–5 cm, flexible, abaxially glabrous except for prickle-teeth at apex, adaxially with dense weft of long hairs at base becoming fewer on midrib, minute hairs throughout; margins long hairy below becoming slightly scabrid with blunt teeth. Culm to 6 m, white shining, internodes scabrid. Inflorescence to 1 m, erect or nodding, dense, shining, plumose; rachis scabrid below becoming shortly, stiff hairy above, branches and pedicels short stiff hairy, long hairs at axils and below spikelets. Spikelets to 4 cm, of 2–3 florets. Glumes equal, to 40 mm, produced to awn-like apex, > florets; short stiff hairy on keel, prickle-toothed elsewhere. Lemma to 11 mm, 3-nerved, scabrid; lateral lobes to 7.5 mm including awn to 3 mm; hairs to 12 mm, radiating from lower ⅓; central awn to 30 mm. Palea to 9 mm, attenuate, apex scarcely produced, keels ciliate, interkeel and flanks scabrid above. Callus hairs to 4 mm. Rachilla to 1 mm. Lodicules to 0.75 mm, hair-fringed. Anthers of $\male$ flowers to 6 mm, of $\female$ flowers to 4 mm. Gynoecium: of $\male$ flowers ovary to 0.7 mm, stigma-styles to 2 mm; of $\female$ flowers ovary to 1 mm, stigma-styles to 4 mm. Caryopsis 4–5 mm; embryo to 1 mm; hilum to 2 mm. $2n = 90$.

Endemic.

N.: North Cape to Marakopa in the west and Opotiki on the east; Three Kings Is, and on eastern offshore islands. On sand hills, consolidated sands, rocks and cliffs; sea level to 150 m. FL Dec.

Natural populations comprise 61% $\male$ and 39% $\female$ plants.

This is the only endemic species with prominent, long, stout rhizomes; they bind sand in dunes.

OFFSHORE IS　Plants of small stature on some offshore islands in eastern and northern waters were discussed by Connor (1971 op. cit.). CHR 132754, CHR 391835 *I. A. E. Atkinson* Keyhole Rock near Big Chicken, Hen and Chicken Islands, Mar 1971, are two specimens of small stature. Inflorescences are 12 and 13 cm respectively, and their appropriate subtending culms are 40 cm and 37 cm. On CHR 313383a,b,c *A. P. Druce* from the same locality and grown on in an experimental garden, the inflorescence is 27 cm, and the culm 1.63 m. CHR 409059 *A.E. Wright 4497*, also from Keyhole Rock is taller than the Atkinson specimens; here the inflorescence is 23 cm and the subtending culm 65 cm.

CHR 132752 *I. A. E. Atkinson* cliff on Little Ohena Id, off the eastern coast of the Coromandel Peninsula, is of small stature; CHR 268502a,b from an open ridge on the same island is from a large, robust, wide-leaved specimen.

CHR 354493 *A. E. Esler 5907, W. M. Hamilton, R. E. Beever & M. L. Scott* from cliffs on Little Barrier Id is correctly described as depauperate; the inflorescence is 19 cm and its culm 39 cm. On CHR 312323a *R. L. Bieleski* from cliffs, the inflorescence is 18 cm and its culm 48 cm — a little taller than the former. From the foot of a cliff, CHR 312313 *R. E. Beever 77058* is much taller; here the inflorescence was *c.* 50 cm and its culm > 1.70 m; CHR 312322 *R. E. Beever 77057,* same locality, is smaller — 1.3 m.

CHR 321653a,b *A. E. Esler 5701* was originally collected by W. W. E. Sanders in April 1978 from Simmonds Id [*Tane 21*: 101–102 (1975)]. Esler's specimen from cultivation has culms 2 m; the inflorescence is 45 cm. CHR 260936, 260977 *W. W. Sanders* Motu Puruhi, the identical locality, are of the same stature.

Among 12 specimens in AK from offshore islands and growing on cliffs or rocks none differs from those described from CHR. On AK 159705 *A. E. Wright 4512* Hen & Chickens, the culm is 1.80 m; AK 201226 *A. E. Wright 10958* Hen & Chickens, is of short stature but has a short stout rhizome, a condition rare in these habitats compared with the sand of western areas.

There is no necessity to recognise plants of small stature in a formal way when they are simply responses to the difficult environment of cliff faces. Plants from less demanding sites are taller; those transferred to experimental gardens soon correspond to mainland plants.

6. C. toetoe Zotov, *N.Z. J. Bot. 1*: 85 (1963); Holotype: CHR 95457! *V. D. Zotov* Wainui-o-mata Valley, near sea, Feb 1955; ♀.

Very tall, stout tussock with ivory sheaths and sharp leaves below a shining white culm bearing a large plumose panicle. Leaf-sheath glabrous, ivory beneath epicuticular wax, midrib green. Ligule to 4 mm. Collar dark brown, adaxially clothed in short hairs. Leaf-blade to 2 m × 3 cm; stiff, abaxially long hairy towards margins below, sharply scabrid from many prickle-teeth in upper ⅓, adaxially with thick weft of hairs at base becoming fewer above, minute stiff hairs throughout; margins long (5 mm) hairy below, becoming very scabrid with rows of prickle-teeth. Culm to 4 m, shining, waxy, internodes glabrous. Inflorescence to 1 m, stiff, erect, densely plumose, rachis smooth, branches and pedicels short stiff hairy, long hairs at axils and below spikelets. Spikelets to 25 mm, of 2–3 florets. Glumes equal, to 25 mm, > florets, 1-nerved, acuminate and awn-like above, scabrid. Lemma to 10 mm, 3-nerved, finely scabrid; hairs to 8 mm radiating from base; lateral lobes to 5 mm; central awn to 15 mm. Palea to 6.5 mm, attenuated, keels and apex ciliate, interkeel and flanks hairy. Callus hairs to 1.5 mm. Rachilla to 0.5 mm, glabrous. Lodicules to 0.5 mm. Anthers of ♂ flowers to 4.75 mm, of ♀ flowers to 2.75 mm. Gynoecium: of ♂ flowers ovary to 1 mm, stigma-styles to 1.8 mm; of ♀ flowers ovary to 1.3 mm, stigma-styles to 3.5 mm. Caryopsis to 2.5–3 mm; embryo to 0.75 mm; hilum to 1 mm. $2n = 90$.

Endemic.

N.: Wellington north to Rotorua and Tauranga. In swamps and wet places; sea level to 800 m. FL late Jan–Feb.

Natural populations comprise 57% ♂ and 43% ♀ plants.

Banks and Solander specimens from Tolaga Bay 1769–1770 (AK 110395, AK 110396) are referable to *C. toetoe* as Zotov (1963 op. cit.) said. Connor (1971 op. cit.) indicated that a similar specimen at WELT was *C. splendens*; this is incorrect.

7. C. turbaria Connor, *N.Z. J. Bot. 25*: 167 (1987); Holotype: CHR 417471! *D. R. Given 13899* Rakeinui, east end of lake, Chatham Island, 24 Feb 1985; ♂.

Very tall, stout, glaucous tussock growing on peat or in bogs. Leaf-sheath internerves and margins long hairy. Ligule to 2 mm. Collar glabrous abaxially, some few short hairs adaxially. Leaf-blade to 1.5 m × 1.5 cm, tapering to long, thin point; abaxially with long interrib hairs below becoming fewer and prickle-teeth becoming dense above, adaxially with dense weft of long interrib hairs at base becoming fewer and generally glabrous except near margins; margins scabrid below becoming very scabrid above. Culm to 2 m, internodes glabrous. Inflorescence to 70 cm, dense, plumose, branches and pedicels densely long hairy, rachis less so, hairs longer below spikelet. Spikelets with 2 ♂ florets. Glumes ± equal, to 25 mm, including florets, 1-nerved, thin; upper with hairs to 10 mm from below, lower with fewer or none, elsewhere minute scattered teeth. Lemma to 9 mm, 3-nerved, scattered prickle-teeth; lateral lobes to 2 mm; hairs to 10 mm radiating from below and reaching tip of lemma, central awn to 9 mm. Palea to 7 mm, attenuated, long-hairy, apex hair-tipped, keels ciliate, interkeel and flanks glabrous above. Callus hairs to 3 mm. Rachilla to 1 mm, glabrous. Lodicules irregularly rhomboidal to 0.5 mm, hairs to 0.75 mm. Anthers 1.7–2.6 mm. Gynoecium: ovary to 0.8 mm; stigma-styles to 2.5 mm. Caryopsis ovate, rugose, shortly stipitate, to 3 mm; embryo to 1 mm; hilum to 1.5 mm.

Endemic.

Ch.: Chatham and Pitt Is. On peat and in sphagnum bogs.

All plants bear ♀ flowers.

For comments on endangered status see Given, D. R. and Williams, P. A. *Botany Division DSIR Report* 1985; for commentary on reproductive biology see Connor, H. E. *N.Z. J. Bot. 26*: 163–167 (1988).

REPRODUCTIVE BIOLOGY   All hermaphrodites of indigenous species are self-compatible, but those of *C. selloana* are self-incompatible.

(1)   Floral dimorphisms between H and F in gynodioecism: Connor, H. E. *N.Z. J. Bot. 1*: 258–264 (1963); Connor, H. E. *Evolution 27*: 663–678 (1974). (2)   Sex form frequencies: (a) natural populations: Connor (1963 op. cit.); Connor, H. E. *N.Z. J. Bot. 3*: 17–23 (1965); Connor, H. E. and Penny, E. D. *N.Z. J. Agric. Res. 3*: 725–727 (1960); Connor, H. E. *N.Z. J. Bot. 9*: 519–525 (1971); (b) in experimental intraspecific hybrids: Connor, H. E. *N.Z. J. Bot. 3*: 233–242 (1965); Connor, H. E. and Charlesworth, D. *Heredity 63*: 373–382 (1989); these papers also contain analyses of the genetics of male-sterility; (c) in experimental interspecific hybrids: Connor, H. E. *Heredity 51:* 395–403 (1983). (3)   Embryology: Philipson, M. N. *N.Z. J. Bot. 16*: 45–59 (1978); Philipson, M. N. and Connor, H. E. *Bot. Gaz. 145*: 78–82 (1984).

Although hybrids are easily made in controlled conditions, and occur spontaneously in experimental gardens, and are quite fertile, none has been found in the wild.

CHEMOTAXONOMY   Most investigations are on triterpene methyl ethers, their kinds, interspecies distribution, and their inheritance in epicuticular leaf wax, but also on alkanes and fatty acids: Bryce, T. A. et al. *Phytochem. 6*: 727–722 (1967); Connor, H. E. and Purdie, A. W. *Phytochem. 15*: 1937–1939 (1976); Connor, H. E. and Purdie, A. W. *N.Z. J. Bot. 19*: 171–172 (1981); Eglinton, G. et al. *Tetrahedron Letters 34*: 2323–2327 (1964); Martin-Smith, M., Subramanian, G. and Connor, H. E. *Phytochem. 6*: 559–572 (1967); Martin-Smith, M., Ahmed, S. and Connor, H. E. *Phytochem. 10*: 2167–2173 (1971); Purdie, A. W. and Connor, H. E. *Phytochem. 12*: 1196 (1973). Cyanogenesis: Tjon Sie Fat, L. *Proc. Kon. Nederl. Akad. Wetensch. C82*: 165–170 (1979).

V   SUBFAMILY ARISTIDOIDEAE

Spikelet 1-flowered, rachilla prolongation absent; glumes 1–many-nerved; lemma 1–3-nerved, concealing palea, 3-awned apically with or without a column; inflorescence paniculate.

Tribe: 15. Aristideae

15. ARISTIDEAE

Annual or usually perennial tufted herbs. Ligule ciliate. Leaf-blade narrow, flat or convolute. Inflorescence an open to densely contracted panicle with pedicelled 1-flowered spikelets; rachilla not prolonged. Florets ♀. Glumes usually persistent, usually 1-nerved, equal or unequal, membranous to scarious, usually > lemma. Lemma faintly 1–3-nerved, terete, chartaceous to ± coriaceous, convolute or involute and concealing small palea, terminal awn 3-fid. Lodicules 2–3, rarely 0, glabrous, strongly nerved. Stamens 3, rarely 1. Ovary glabrous; styles 2; stigmas plumose. Caryopsis usually fusiform; embryo ¼–⅔ length of caryopsis; hilum linear.

ARISTIDA L., 1753

Annual or perennial tufts, usually with wiry culms and leaves either basal or mostly cauline. Ligule shortly ciliate. Leaf-blade narrow, linear, flat or rolled. Panicle contracted or open, sometimes spike-like. Spikelets 1-flowered, ♀, narrow, pedicelled; disarticulating above glumes. Glumes equal or unequal, usually 1-nerved, keeled, ± mucronate to shortly awned. Lemma 3-nerved, ± coriaceous, narrow, convolute or involute, terminating in a 3-fid awn, with or without a column. Palea ≪ lemma, 2-nerved. Callus sharp, shortly bearded. Stamens 3, or 1. Lodicules 2 or 3. Caryopsis terete, or grooved; embryo small; hilum linear, almost = caryopsis. Fig. 18.

Type species: *A. adscensionis* L.

c. 250 spp., of tropics and subtropics in dry places or in regions of low fertility. Naturalised sp. 1; transient sp. 1.

Esen, A. and Hilu, K. W. *Taxon 40*: 5–17 (1991), suggested an isolated position for *Aristida* outside the Arundinoideae; as also do Soreng, R. J. and Davis, J. I. *Bot. Rev. 64*: 1–85 (1998); currently subfamilial rank is favoured.

1 Culms much branched from upper nodes; lower glume 6–7 mm; upper glume 7–9 mm .... **1. ramosa**
Culms with few branches from lower nodes; lower glume 4–5 mm; upper glume 6–7 mm .... **vagans**ζ

1. A. ramosa R.Br., *Prodr.* 173 (1810).   purple wire-grass

Tufts about 50 cm with stiff, wiry, divergent culms each branched from upper glabrous, swollen nodes; branching extravaginal, cataphylls cream, shining. Leaf-sheath striate, finely scabrid, purplish, terminating in fine twisting hairs, *c.* 2 mm. Ligule *c.* 0.4 mm, membranous, shortly ciliate. Leaf-blade 6–16 cm × *c.* 1 mm, involute, abaxially glabrous, adaxially densely toothed with some longer hairs. Culm rigid, cane-like, branches above subtended by loose sheaths, internodes finely scabrid. Panicle 8–12 cm, narrow, spike-like or ±loose; branches ±erect or sometimes slightly spreading, branches and pedicels scabrid. Spikelets 9–15 mm. Glumes unequal, < lemma, keeled, 1-nerved, acuminate to shortly (2 mm) awned; lower glume 5–7 mm extensively prickle-toothed, upper 7–9 mm ± glabrous. Lemma 4.5–8.5 mm, narrow-linear, convolute, purple-mottled, smooth but scabrid on midnerve above, tipped by 3 spreading straight awns 6–15 mm, lacking columns. Callus *c.* 1 mm, densely short-hairy. Palea 2 mm, < lemma. Lodicules 2, *c.* 2 mm, nerved, glabrous, dentate. Stamens 3, anthers 1.3–1.5 mm, shortly tailed. Gynoecium: ovary 1 mm; stigma-styles 2 mm. Caryopsis not seen. Fig. 18.

Naturalised from Australia; occurs in Northern Territory, Queensland and New South Wales.

N.: Mt Maunganui near Tauranga. Northern slopes at base of rock outcrop.

A. ramosa was first collected on Mt Maunganui in March, 1938 by *M. Hodgkins* (CHR 19373). Plants have persisted in this locality, and again it was collected in July, 1984, by *M. D. Wilcox* (NZFRI 14643, duplicate CHR 416548A,B).

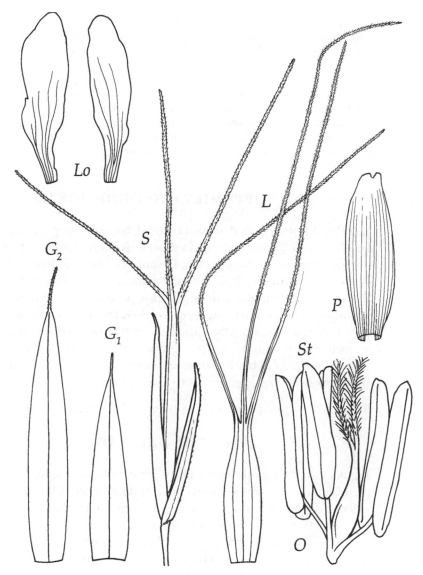

Fig. 18 *Aristida ramosa*
S, spikelet, × 7; **G₁**, **G₂**, glumes, × 12; **L**, lemma dorsal view, × 9; **P**, palea dorsal view, × 12;
Lo, lodicules, × 30; **O**, ovary, styles and stigmas, × 30; **St**, stamens, × 30.

ζ**A. vagans** Cav. Tufts 30–50 cm, with wiry culms and with few branches from lower nodes; ligule ciliate, 0.2 mm; collar with a few longer hairs to 1 mm; leaf-blade involute, abaxially glabrous, adaxially shortly pubescent with a few longer hairs; culm 18–40 cm, internodes glabrous; panicle 6–9 cm, with few, short branches; spikelets 7–8 mm; glumes purple, the lower 4–5 mm, upper 6–7 mm; lemma convolute, > glumes, 3-awned, with awns 9–12 mm. Although this Australian species was in 1949 attributed a New Zealand origin from exported wool [Probst, R. *Wolladventivflora Mitteleuropas* 11 (1949)] only one specimen has been seen, CHR 71840 *J. P. Beggs* Fairhall, near Blenheim, 11.6.1951; a note on the label states that it was locally believed to have been present for about 10 years.

VI SUBFAMILY CHLORIDOIDEAE

Spikelets 1-flowered, or 2–many-flowered, laterally compressed; rachilla fragile; florets ♀, or ♂ and ♀ on dioecious plants. Glumes 2, or rarely the lower minute or suppressed. Lemma 1–3-nerved. Palea 2-nerved. Lodicules 2, cuneate, fleshy, usually truncate. Stamens (1)–3. Stigmas 2. Caryopsis ovoid, fusiform or globose; pericarp adhering to seed or becoming loose; embryo large, often ⅓–¾ length of caryopsis; hilum punctiform or occasionally elliptic. Annual or perennial herbs. Ligule very variable, membranous, or membranous with a ciliate fringe, or a line of hairs. Leaf-blade linear or convolute; Kranz leaf anatomy. Inflorescence a panicle, or of 1 to several racemes.

Tribes: 16. Chlorideae; 17. Leptureae; 18. Eragrostideae

16. CHLORIDEAE

Ligules usually membranous and ciliate, sometimes predominantly membranous or predominantly ciliate. Inflorescence of tough unilateral racemes digitate or scattered along an axis and often deciduous. Spikelets with 1 ♀ floret, with or without ♂ or ∅ florets, laterally or dorsally compressed, disarticulating above glumes but not between florets, or falling entire. Glumes usually persistent, or caducous, herbaceous to membranous, 1–3–(5)-nerved, < floret or enclosing floret; lower sometimes 0. Lemma membranous to coriaceous, (0)–3-nerved, entire or 2–3–(5)-lobed with or without 1–3–(5) terminal or subapical, usually straight awns. Achene sometimes with free pericarp.

1 Glumes, 1 or both, < lowest floret, thinner than or of same texture as lemma *2*

Glumes both > lowest floret, coriaceous and much thicker than ± membranous lemma *4*

2 Spikelets with 1 ♀ floret and several, usually ∅, upper florets .... **Chloris**

Spikelets 1-flowered (rarely with a vestigial floret) *3*

3 Inflorescence of digitate racemes .... **Cynodon**

Inflorescence of subdigitate racemes disposed along an axis .... **Spartina**

4 Leaf tip straight; florets ♀ in 1–several deciduous spikelets on a cylindrical raceme .... **Zoysia**

Leaf tip curled; florets unisexual; ♀ spikelets crowded together and burr-like, ♂ spikelets on a pectinate raceme .... **Buchloe**ζ (p. 518)

CHLORIS Sw., 1788

Tufted or stoloniferous perennials or annuals. Leaf-sheath glabrous, striate. Ligule ciliate. Collar often with hairs. Leaf-blade flat or folded, somewhat scabrid. Culm usually branched, often 1 m, or more, internodes glabrous. Inflorescence of 2–several digitate or subdigitate, erect or spreading, spike-like persistent racemes. Spikelets sessile or shortly pedicelled in 2 secund rows on continuous rachis, disarticulating above glumes; lowest floret ♀, upper 1–6 florets usually ∅, often truncate, and if more than one, ± enclosed by first ∅ floret. Glumes unequal with the lower shorter, narrow, keeled, acute to acuminate. Lemma of ♀ floret keeled, usually broad, 1–5-nerved, outer nerves submarginal, keel or marginal nerves villous or long-ciliate, apex entire, or bilobed with midnerve extended to slender awn, sometimes reduced to a mucro; lemma of upper florets reduced, awnless or awned. Palea ≈ lemma, 2-keeled, narrow. Callus inconspicuous. Lodicules 2 in ♀ florets, fleshy, glabrous. Stamens 3. Ovary glabrous, styles free. Caryopsis ± compressed, ellipsoid and trigonous to lanceolate and subterete; hilum punctiform or shortly elliptic.

Type species: *C. cruciata* (L.) Sw.

c. 40 spp. from tropical and warm temperate regions. Naturalised spp. 2.

1 Culms branched above; racemes erect, bunched together at maturity; base
   of leaf-blade adaxially sometimes with a row of long hairs; collar hairs 3–
   5 mm .... **1. gayana**
   Culms usually unbranched; racemes widely spreading at maturity; base of
   leaf-blade adaxially glabrous, rarely with a few long hairs; collar glabrous,
   sometimes with minute (0.1–0.5 mm) hairs, or rarely with a few long (0.8–
   5 mm) hairs .... **2. truncata**

1. C. gayana Kunth, *Révis. Gram. 1*: 89 (1829).   Rhodes grass

Tufted perennials, brownish green, tough, branched above, to 1 m;
stolons strong, wiry. Leaf-sheath keeled, chartaceous, often faintly
purplish. Ligule 0.4–0.6–(0.8) mm, minutely ciliate. Collar hairs 3–
5 mm. Leaf-blade 10–30 cm × (1.5)–3–6 mm, folded to flat, stiff,
narrow linear-lanceolate, strongly keeled, minutely scabrid, adaxially
sometimes with a row of long hairs at base; margins thickened, much-
narrowed to fine acicular tip. Culm 60–80 cm, ± flattened and
geniculate at base, internodes slightly scabrid below inflorescence.
Racemes 7–18, spike-like, ± digitate, tawny brown, 4.5–7.5–(10) cm
× 3–3.5 mm, erect, bunched together, bearing close-set shortly
pedicelled green to golden-brown or brownish spikelets; rachis
slender, short-scabrid, pubescent near base. Glumes unequal,
membranous, 1-nerved, keel sparsely prickle-toothed; lower 1.2–
1.7 mm, narrow-lanceolate, glabrous, acute, upper 2–3 mm, elliptic-
lanceolate, sparsely scabrid, ± truncate above, midnerve excurrent
forming awn 0.2–0.5 mm. Spikelets 3-flowered. Basal ♀ floret: lemma
2.5–3.5 mm, 3-nerved, lateral nerves and callus sparsely hairy with
longer hairs near apex, midnerve scabrid above, awn fine, scabrid,
1.4–3 mm; palea 2.5–3 mm, keels purple, minutely scabrid in upper
½, interkeel ± smooth, minutely scabrid near notched apex; anthers
1.4–1.8 mm; gynoecium with ovary *c.* 0.4 mm, stigma-styles *c.* 1.5 mm;
caryopsis not seen. Median ∅ floret: lemma 2–3 mm, 3-nerved,
oblong, scabrid near almost truncate apex, awn 0.4–1.5 mm, scabrid;
palea 2–3 mm, similar to palea of basal floret but ± entire; anthers
when present, pollen-sterile, 0.7–1.3 mm. Apical ∅ floret: lemma
empty, 0.3–1.2 mm, inrolled, truncate, minutely scabrid above,
sometimes mucronate, borne at tip of glabrous rachilla 1.1–1.5 mm.

Naturalised from Africa.

N.: Auckland Province, scattered; K.: Raoul Id. Coastal, behind dunes in grassy flats, also in pasture.

2. C. truncata R.Br., *Prodr.* 186 (1810). windmill grass

Light green, wiry, stoloniferous perennials to 50 cm; widely spreading racemes of inflorescence conspicuous at maturity. Leaf-sheath keeled, chartaceous, light green to light brown. Ligule 0.2–0.4 mm, minutely ciliate. Collar glabrous, sometimes with minute (0.1–0.5 mm) hairs, rarely with a few long (1–5 mm) hairs. Leaf-blade 6.5–12 cm × 2–2.5 mm, folded, linear, ± scabrid near base, adaxially rarely with sparse long (0.8–5 mm) hairs; margins finely scabrid, tip curved. Culm 20–40 cm, unbranched. Racemes 4–11, spike-like, digitate, 10–17 cm × 2.5–3.5 mm, bunched together and ± erect at first, later spreading, lowest racemes almost horizontal; rachis slender, pubescent near base, minutely scabrid above, bearing close-set, shortly pedicelled, light greenish spikelets maturing dark brown, almost black. Glumes unequal, hyaline, 1-nerved, narrow-lanceolate, acute, smooth, midnerve sparsely scabrid; lower 1.2–1.9 mm, upper 3–3.8 mm, deeper purple, margins scabrid near apex. Spikelets 2–3-flowered. Basal ♀ floret: lemma 2.7–3.6 mm, 3-nerved, sparsely, minutely scabrid, lateral nerves with short stiff hairs, callus hairs long, stiff, apex notched, bilobed, awn from notch 6.5–14 mm, slender, straight, scabrid, purple; palea 2.2–2.8 mm, narrow, entire, keels ciliate; anthers 0.5–0.8 mm, purple; gynoecium with ovary 0.4–0.5 mm, stigma-styles 1.2–1.5 mm; caryopsis 1.5–1.9 × 0.3–0.5 mm. Median floret: lemma 1.5–2.5 mm, 3–(5)-nerved, glabrous, truncate, awn *c.* 12 mm; palea shorter and much narrower; anthers when present, 0.5 mm, polliniferous. Apical ∅ floret when present: lemma *c.* 1 mm, 3-nerved, truncate, always barren, awn 2.5–3.5 mm; rachilla glabrous, 0.5–0.8 mm.

Naturalised from Australia.

N.: North Auckland (Waitangi), Hawkes Bay (Napier, Hastings); S.: Nelson (Rabbit Id, Richmond), Canterbury (Sumner, near Christchurch); Three Kings Is. Coastal, waste land, stony or sandy ground.

Two Australian species, *C. acicularis* Lindl., curly windmill grass, and *C. pectinata* Benth., comb chloris, have been recorded as seed impurities in imported commercial seed but are unknown in the wild in N.Z. [Healy, A. J. *Standard Common Names for Weeds in N.Z.* ed. 2, 141 (1984)].

CYNODON Rich., 1805 *nom. cons.*

Perennials, rhizomatous and stoloniferous, rooting at nodes and sending up tufts of erect flowering shoots. Leaf-sheath glabrous, short, ± rounded. Ligule membranous, often ciliate. Leaf-blade flat, glabrous to loosely short-hairy; margins scabrid. Culm erect or geniculate at base, internodes glabrous. Inflorescence of 2–6 narrow, digitate, slender, persistent, spike-like racemes; rachis with 2 secund rows of imbricate spikelets. Spikelets ± sessile, laterally compressed, 1-flowered, ♀, rarely with a second vestigial floret; disarticulation above or between glumes. Glumes ± equal, acute, keeled; lower 1-nerved, subpersistent, upper 1–3-nerved, usually deciduous with lemma. Lemma > glumes, naviculate, firmly membranous, 3-nerved, lateral nerves close to margins, awnless, keel ciliate, wingless or narrowly winged. Palea ≈ lemma, 2-keeled. Lodicules 2, minute, obovate-cuneate, glabrous. Stamens 3. Ovary glabrous; styles free to base; stigmas plumose. Caryopsis oblong, subterete; embryo *c.* ⅓ length of caryopsis; hilum linear, *c.* ⅔ length of caryopsis.

Type species: *C. dactylon* (L.) Pers.

c. 8 spp. of tropical and warm temperate regions. Naturalised sp. 1.

1. C. dactylon (L.) Pers., *Syn. Pl. 1*: 85 (1805).   Indian doab

Mat-forming, green or glaucous, 10–55 cm, with long, scaly rhizomes and long-creeping, much-branched, strong, wiry stolons rooting and tufted at nodes. Leaf-sheath chartaceous, light green to straw-coloured, ± keeled. Ligule 0.1–0.2 mm, truncate, ciliate. Collar hairs 1–2 mm. Leaf-blade 1.6–9 cm × 1–3 mm, sometimes slightly wider, abaxially glabrous or with scattered hairs, adaxially finely ribbed, papillose to scabrid on ribs, with row of hairs at base to 2 mm, midrib scabrid near blunt tip. Culm 10–50 cm, erect from geniculate base, internodes glabrous. Racemes (3)–4–6–(7), 1.5–6 cm × 1–3 mm, erect to spreading; rachis 3-angled, short-scabrid on angles, pubescent at base, bearing scarcely pedicelled imbricate spikelets in 2 secund rows. Spikelets 2–2.5 mm, green or purplish. Glumes 1–2 mm, reflexed from rachis at maturity, membranous, narrow-lanceolate, keel scabrid. Lemma = spikelet, curved to the ± hooded, shortly mucronate apex, glabrous, keel and sometimes margins with short hairs. Palea keels minutely scabrid, interkeel glabrous. Anthers 1–1.5 mm. Gynoecium: ovary 0.3–0.5 mm; stigma-styles 1–1.2 mm. Caryopsis *c.* 1 × 0.5 mm.

Naturalised.

N.: throughout but rare in Wellington and Wairarapa; S.: Nelson, Marlborough, Westland, Canterbury, Central Otago (Alexandra); K. Usually coastal in sandy or gravelly places, in pasture, scrub, or on tracksides, often between dunes or in reclamation areas; also a weed of lawns, and in waste land, on roadsides or gutter crevices.

Widespread in tropical and subtropical regions of both Hemispheres.

SPARTINA Schreb., 1789

Often stout, strongly rhizomatous perennials. Leaf-sheath firm, open. Ligule a dense fringe of fine straight hairs. Leaf-blade coriaceous, flat or involute. Inflorescence of 2–many congested persistent spike-like racemes arranged racemosely along an axis; spikes bearing spikelets in 2 rows on one side of 3-angled rachis, rachis usually prolonged beyond spikelets. Spikelets 1-flowered, very rarely 2-flowered, strongly laterally compressed, falling entire with glumes; rachilla not prolonged. Glumes unequal, firmly membranous, strongly keeled, acute to shortly awned; lower shorter, 1–3-nerved, upper usually > floret, 1-nerved or obscurely 3–(6)-nerved. Lemma keeled, firmly membranous with wide margins, 1–3-nerved, sometimes with additional indistinct nerves, tapering to narrow, subacute, awnless apex. Palea ≈ lemma, 2-nerved. Lodicules 0, or sometimes 2. Stamens 3. Caryopsis fusiform; embryo large; hilum elliptic. Fig. 19.

Type species: *S. schreberi* J.F.Gmel.

c. 15 spp., of coasts of North and South America, Europe and Africa, in temperate to subtropical regions. Naturalised spp. 3.

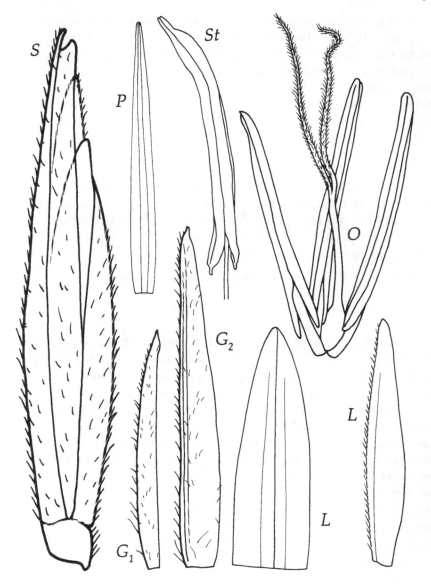

Fig. 19 *Spartina anglica*
S, spikelet, × 9; **G₁**, **G₂**, glumes, × 6; **L**, lemma dorsal and lateral view, × 6; **P**, palea ventral
view, × 6; **O**, ovary, styles, stigmas, and stamens × 12; **St**, single anther, × 9.

Some spp., especially *S. anglica*, have been used to stabilise mudflats. *Spartina* was introduced into N.Z. for this purpose, but Lee, W. G. and Partridge, T. R. *N.Z. J. Bot. 21*: 231–236 (1983), note that the spread of *Spartina* rapidly destroys habitats formerly occupied by a range of faunal spp.

Partridge, T. R. *N.Z. J. Bot. 25*: 567–575 (1987), outlined the introduction, history, planting and spread of *Spartina* in N.Z., presenting descriptions of, and keys to the 3 spp. in N.Z. This account is largely based on his treatment.

1 Plants forming open clumps; shoots 8–30 mm diam. at base; spikelets almost glabrous on rarely flowering plants .... **1. alterniflora**
  Plants forming close clumps; shoots 3–8–(10) mm diam. at base; spikelets hairy 2

2 Ligule 2–3 mm; anthers 8–12 mm, dehiscent................................ **2. anglica**
  Ligule 1–2 mm; anthers 5–8 mm, indehiscent and shrivelled ... **3. ×townsendii**

1. S. alterniflora Loisel., *Fl. Gall.* 719 (1807).   American spartina

Robust perennials, (20)–50–100–(200) cm, with thick fleshy far-creeping rhizomes, forming rather open clumps. Leaf-sheath ± glabrous; cross-veinlets scattered, minute; margins often minutely ciliate below ligule. Ligule 1–2 mm, densely ciliate. Leaf-blade 8–45 cm × 6–15 mm, persistent, coriaceous, flat, glabrous, adaxially ribbed, much narrowed to fine hard tip. Culm erect, 8–30 mm diam. near base including closely ensheathing leaves. Panicles in N.Z. usually not developed, this description is from the two floriferous specimens seen: panicle *c.* 25 cm, erect, contracted; spikes 7–14, 5–7 cm; rachis of each spike glabrous, tipped by bristle 10–12 mm. Spikelets 14–17 mm, lanceolate, 1-flowered. Glumes very unequal, lanceolate, glabrous, subacute; lower slightly > ½ length of upper and much narrower, 1–3-nerved, upper = spikelet, 5–9-nerved. Lemma ≤ upper glume, 3–5-nerved, lanceolate, glabrous. Palea ≥ lemma, hyaline, glabrous. Anthers 3.5–6 mm. Caryopsis not seen.

Naturalised from North America.

N.: North Auckland, South Auckland, Poverty Bay, Wellington (Hutt R.). Estuaries, mostly around mid-tide.

Only 2 specimens have been seen in flower, CHR 252099 and CHR 252100, both from Kaipara Harbour *C. R. Veitch* 2.7.1975.

2. S. anglica C.E.Hubb., *Bot. J. Linn. Soc. 76*: 364 (1978).   spartina

Stiff perennials 50–100 cm, with fleshy rhizomes, forming large dense clumps or swards. Leaf-sheath glabrous. Ligule 2–3 mm, densely ciliate. Leaf-blade 5–40 cm × 5–12 mm, persistent or caducous, coriaceous, flat or inrolled, glabrous, adaxially ribbed, narrowed to fine, hard tip. Culm erect, 5–8–(10) mm diam. at base, including closely ensheathing leaves. Panicle 10–30 cm, erect, contracted; spikes 2–8, 5–15 cm; rachis of each spike glabrous, tipped by bristle 1–4 cm. Spikelets 15–20 mm, 1-flowered, very rarely 2-flowered, oblong, pubescent, falling entire at maturity. Glumes unequal, lanceolate-oblong, acute, with short scattered hairs, keels ciliate; lower ⅔–⅘ length of upper and narrower, 1-nerved, upper = spikelet, 3–6-nerved, asymmetrical. Lemma ≤ upper glume, 1–3-nerved, lanceolate-oblong, obtuse, with short hairs above and on keel. Palea ≥ lemma, hyaline, glabrous. Anthers 8–12 mm, dehiscent. Caryopsis 10–12 × *c.* 1.5 mm. Fig. 19. Plate 2C.

Naturalised from England.

N.: North Auckland, Auckland City, South Auckland (near Thames, Tauranga, near Raglan, near Kawhia), Gisborne, west and south Wellington coast; S.: Nelson, Marlborough (Havelock), Canterbury (Christchurch), south-east coast from Waikouaiti to Invercargill; St. Estuaries, especially around mid-tide and in salt marsh.

The natural sterile hybrid *S. ×townsendii* gave rise to the fertile amphidiploid *S. anglica* by chromosome doubling. In N.Z. early reports of fertile plants or of seed, recorded as *S. ×townsendii* must refer to *S. anglica*, e.g., Bryce, J. *Kew Bull. 1936*: 32 (1936), recorded that seeds of *S. ×townsendii* were sent to N.Z. in 1928 and tested for germination. Partridge (1987 op. cit. p. 571) reported that *S. anglica* is now the most widespread and abundant spartina in N.Z. and has almost completely replaced *S. ×townsendii*.

3. S. ×townsendii H.Groves et J.Groves, *Rep. Bot. Exch. Club Brit. Is 1880*: 37 (1881).

S. alterniflora × *S. maritima*   spartina

Stiff perennials 10–50 cm, with fleshy rhizomes forming dense clumps or swards. Leaf-sheath glabrous. Ligule 1–2 mm, densely ciliate. Leaf-blade 5–35 cm × 4–10 mm, persistent or caducous, coriaceous, flat or inrolled, glabrous, adaxially ribbed, narrowed to fine hard tip. Culm erect, 3–8 mm diam. near base including closely ensheathing leaves. Panicle 8–30 cm, erect, contracted; spikes 2–5–(14), 4–15–(18) cm; rachis of each spike glabrous, tipped by bristle 1.5–3.5 cm. Spikelets 10–20 mm, narrowly oblong, 1-flowered, minutely hairy. Glumes unequal, oblong-lanceolate, acute, finely hairy; lower ⅔–¾ length of upper and much narrower, 1-nerved, upper = spikelet, 1–3-nerved. Lemma < upper glume, 1–3-nerved, oblong-lanceolate, minutely hairy. Palea ≥ lemma, hyaline, glabrous. Anthers 5–8 mm, very narrow, shrivelled, indehiscent, with sterile pollen. Caryopsis not developed.

Naturalised from England.

N.: North Auckland (Helensville, Waitemata Harbour), South Auckland (Tauranga), Wellington (Foxton); S.: Invercargill. In estuaries, especially around mid-tide.

The natural sterile hybrid *S.* ×*townsendii* (*S. alterniflora* × *S. maritima*) arose at Hythe, near Southampton, England, about 1870. In N.Z. Partridge (1987 op. cit.) found that *S.* ×*townsendii* is now very rare with the only recently verified stand being at Te Atatu, Waitemata Harbour, Auckland.

ZOYSIA Willd., 1801 *nom. cons.*

Perennials with extensive monopodial rhizomes. Culms few noded. Leaves distichous. Ligule ciliate; collar distinct, ± long-ciliate with cushion-based hairs. Leaf-blade linear to linear-lanceolate, tapering. Inflorescence a cylindrical raceme or a solitary terminal spikelet. Spikelets 1-flowered, disarticulating above pedicels and falling entire at maturity. Lower glume usually 0, occasionally present as small bract, very rarely complete; upper laterally compressed, enfolding floret, indurated, ± awned. Lemma entire, firmly membranous, 1-nerved. Palea nerveless. Lodicules 0. Stamens 3. Ovary glabrous; styles 2; stigmas long, emerging apically. Caryopsis elliptic, laterally compressed; embryo *c.* ½ length of caryopsis, pericarp free.

Type species: *Z. pungens* Willd.

c. 10 spp. of tropical and subtropical Asia and Australasia. Endemic spp. 2.

Zotov, V. D. *N.Z. J. Bot. 9*: 639–644 (1971), recognised 3 spp. for N.Z. His treatment is amended here and *Z. pauciflora* Mez includes *Z. planifolia* Zotov.

> *1* Spikelets solitary, rarely 2–3 on some peduncles; collar glabrous, very rarely with some long hairs .... **1. minima**
> Spikelets (2)–3–7–9 in a loose or congested raceme, rarely a few solitary; collar with several to many hairs 2–4 mm, very rarely almost glabrous ......
> .... **2. pauciflora**

1. Z. minima (Colenso) Zotov, *T.R.S.N.Z. 73*: 237 (1943) ≡ *Gaimardia minima* Colenso, *T.N.Z.I. 22*: 491 (1890); Lectotype: K! *K. Hill* Mount Tongariro, dry open grounds, 1889, (designated by Zotov 1971 op. cit. p. 640).

Rhizome stout, long, to 1.5 mm diam., whitish, smooth; cataphylls smooth, grey or whitish; shoots to 5–(10) cm, branching several times intravaginally at or above ground. Leaf-sheath shining white below, purplish above, striate, margins membranous, wide, glabrous. Ligule 0.1–0.2 mm, ciliate. Collar glabrous, swollen, recurved, very rarely with 1–2 long hairs. Leaf-blade 5–30 mm, distinctly divergent, folded, abaxially smooth, adaxially scabrid on ribs; margins incurved, thickened and scabrid. Culm to 2 cm, terete to usually laterally compressed, ridged, usually scabrid on ridges below spikelet otherwise glabrous, bearing 1 leaf or leaf reduced to sheath, and often a minute bract above. Spikelet 1, occasionally 2–3, peduncles narrowly ovate, lanceolate, *c.* 1 mm wide. Glumes indurated, lower usually absent, if present much reduced, occasionally complete and awned, to *c.* 4.5 mm, upper glume (3)–3.5–7 mm, minutely scabrid above and on margins, ± golden to golden-brown, nerves 7–9, very weak; awn 0.2–2.5 mm, scabrid. Lemma 3–4.5 mm, ovate-lanceolate, 1-nerved, firm-hyaline, shallowly bifid, glabrous. Palea 3–4 mm. Anthers *c.* 2 mm. Gynoecium: ovary 0.5–0.75 mm; stigma-styles 7–7.5 mm. Caryopsis 1.5–1.75 mm; embryo 0.75 mm; pericarp detaching with 0.25–0.35 mm beak. Protogynous.

Endemic.

N.: South from Auckland and Coromandel; S.: scattered in localities in Nelson, Marlborough, Westland, Canterbury and Central Otago. Coastal and inland sand dunes, sands and gravel; sea level to 600 m.

Plants with predominantly solitary spikelets may bear some racemes of 2–3 spikelets, occasionally some plants have a few collar hairs. Single spikelet inflorescences persist in cultivation (CHR 156943, 223438 Amberley Beach).

CHR 332396 *H. Carse* Tauroa [Peninsula] Northland, Jan 1912, is a sheet of four separate pieces; mostly inflorescences of single spikelets but not all; collar hairs are not always evident though they are present. Veritable *Z. pauciflora* is mounted on the sheet, and is found in the locality, see: CHR 331401a,b *H. Carse* Jan 1917; CHR 469630 *A. P. Druce* Jan 1990. CHR 331400a,b *H. Carse* Jan 1913 is luxuriant.

2. Z. pauciflora Mez, *Feddes Repert. 17*: 145 (1921); Lectotype: WELT 68500! *D. Petrie* Opotiki, Dec. 1903, Herb. L. Cockayne 2315. Duplicate of original at B destroyed in World War II [designated by Chase, A. and Niles, C. D. *Index to Grass Species* (1962)]. = *Z. planifolia* Zotov, *N.Z. J. Bot. 9*: 641 (1972); Holotype: CHR 68413! *T. J. M. Wordley* Northern Wairoa, Northland, hillside, 14.11.1949.

Rhizome stout, long, 1–1.5 mm diam., shining, light brown; cataphylls smooth, grey or whitish; shoots 10–15 cm, branching several times intravaginally at or above ground. Leaf-sheath shining, white below, purplish above, striate, margins membranous, wide, glabrous. Ligule 0.2–0.3 mm, ciliate. Collar hairs few to many, 2–4 mm, rarely glabrous. Leaf-blade to 10 cm, divergent to erect, folded to flat and 1–2–4 mm wide, abaxially smooth, adaxially finely scabrid on ribs or glabrous; margins thickened, sparsely scabrid, tip blunt. Culm 3–10 cm, terete to laterally compressed, ridged, glabrous, bearing 1 leaf or leaf reduced to sheath, and a minute bract above. Raceme 0.5–2 cm, of (1)–2–5–9 lanceolate to ovate-lanceolate, imbricate, spikelets *c.* 1–1.5 mm wide, frequently congested, rachis and pedicels flattened, scabrid on edges. Glumes indurated, lower glume if present a reduced bract, upper glume 4–6–(7) mm, minutely hairy above and on margins, ± golden ochreous often tinged pinkish at first, nerves 7 very weak; awn 0.5–2–(5) mm, scabrid. Lemma 3–3.75 mm, ovate-lanceolate, 1-nerved, firm-hyaline, glabrous. Palea 2.75–3.5 mm, transparent. Anthers *c.* 2–2.6 mm, apiculate, filaments ligulate. Gynoecium: ovary 0.75–1 mm; stigma-styles 3.5–5 mm. Caryopsis 1.5–2.0 × 0.75 mm; embryo 0.6–0.75 mm; pericarp thin, detaching with beak. Protogynous.

Endemic.

N.: North Cape south to Raglan Harbour and Bay of Plenty; Three Kings Is. In sandy and rocky places, often in scrub or under trees, mainly coastal.

Typification of *Z. pauciflora* Mez in Zotov (1971 op. cit. p. 641) contains the transcription error "L. Cockayne 2513". WELT 16007 is a duplicate of WELT 68500 *D. Petrie* Opotiki Dec. 1903.

Extremely variable in habit sometimes forming a turf to 3 cm tall with leafy shoots with very short internodes, and at other times with a loose, open habit from long shoot internodes and reaching to 40 cm. The specific criteria in Zotov (1971 op. cit.) are size related. Leaf length, inflorescence length, and number of spikelets/inflorescence vary as the overall habit. CHR 83554 *L. B. Moore* Pukenui, Houhora Harbour [Northland] Jan 1954 on a sunny roadside bank, and CHR 83553 *L. B. Moore ibid.* in shade of *Leptospermum* scrub, illustrate the three points. The former would be referable to *Z. pauciflora* and the latter to *Z. planifolia.* CHR 29265 *T. Kirk* Pumice Hills by Cambridge May 1870 is tall, slender, with shoots to 15 cm, and CHR 29265a *ibid.* has shoots to 3 cm; CHR 29265b *ibid.* is intermediate for stature and leaf width; the former was determined as *Z. planifolia,* and the two latter as *Z. pauciflora.* There are similar parallels in AK specimens. *Zoysia planifolia* Zotov is here reduced to synonymy in *Z. pauciflora* Mez; the two are united by the presence of long cushion-based hairs on leaf-blade collars. The distribution given by Zotov was coincident for both taxa, and remains unchanged.

TERATOLOGY　　Growing shoots may become enlarged and thickened and up to 10–15 mm wide; they comprise distichous, very close, leaf-sheaths. Whole shoots may be affected. Examples in *Z. minima* are Farewell Spit CHR 133928b *V. D. Zotov,* and AK 143035 *A. E. Wright 2495,* and in *Z. pauciflora* are CHR 42710 *K. W. Allison* Bay of Plenty, Mar 1941, CHR 178251 *G. C. Kelly* Scott Pt, Jan 1967, CHR 331398 *H. Carse* Ranganui Harbour, Jan 1900.

ζ**Buchloe dactyloides** (Nutt.) Engelm.　　The single specimen consists of rather dense, reddish grey, curly leaved tufts, 10–14 cm, with one tuft bearing a long stolon giving rise to 3 smaller tufts. Basal bracts short, straw-coloured, some reddish at base. Leaf-sheath glabrous with scattered long hairs above. Ligule *c.* 0.6 mm, ciliate. Leaf-blade 3–10 cm × 1–1.5 mm, flat, with scattered long hairs; margins finely scabrid, tip long-filiform, curled. A North American sp., known there as buffalo grass, collected once from Earnscleugh, Central Otago, on grassland on dry flats *A. J. Healy 59/185* 16.2.1958, escaped from old experimental plots (CHR 121871); the specimen is vegetative only.

Bouteloua gracilis (Kunth) Steud., blue grama, of North America, was once collected from Northburn Run, Dunstan Mts, Otago, in Cockayne's Plot 3a *J. A. Douglas* 7.6.1967 (CHR 107184) where it had persisted since the plot was sown in 1948 [Douglas, J. A. *Proc. N.Z. Ecol. Soc. 17*: 18–24 (1970)].

17. LEPTUREAE

Ligules membranous, ciliate. Inflorescence a single bilateral raceme with alternate spikelets borne edgeways on and embedded in hollows in the fragile rachis. Spikelets 1-flowered, dorsally compressed, falling entire; florets ♂. Lower glume 0 or suppressed, well-developed in terminal spikelet; upper glume appressed to rachis, concealing the sunken floret. Lemma membranous, 3-nerved. Caryopsis ellipsoid with free pericarp.

LEPTURUS R.Br., 1810

Perennials. Ligule membranous, ciliate. Inflorescence a cylindrical bilateral raceme with spikelets arranged alternately on opposite sides of a thickened fragile rachis, edgeways on and embedded in hollows of the rachis. Spikelets 1-flowered, dorsally compressed, falling at maturity with portion of rachis; rachilla prolonged. Lower glume minute or 0, except in uppermost spikelet; upper glume exceeding the rachis cavity and concealing the floret, appressed to rachis, coriaceous, 5–12-nerved, acute to caudately awned. Lemma 3-nerved, membranous, rounded, obtuse to acute. Palea ≤ lemma, 2-keeled. Caryopsis ellipsoid. Fig. 20.

Type species: *L. repens* (G.Forst.) R.Br.

c. 8 spp., from shores of Indian and Pacific oceans. Indigenous sp. 1.

1. L. repens var. **cinereus** (Burcham) Fosberg, *B. P. Bishop Mus. Occ. Papers 21*: 292 (1955) ≡ *L. cinereus* Burcham, *U.S. Natl Herb. Contrib. 30*: 424 (1948); Holotype: US *n.v.*, *Burcham 169* Russell Island, Pavuru (*fide* Burcham 1948 op. cit.).

Semiprostrate perennial, forming mats with long trailing stolons rooting at nodes; branching intravaginal. Leaf-sheath coriaceous, very finely scabrid between ribs, sometimes with a few scattered long fine hairs. Ligule *c.* 0.5 mm, membranous, minutely ciliate. Collar finely scabrid. Leaf-blade 2–6 cm × 2–6 mm, lanceolate, coriaceous, glaucous, abaxially smooth, but scabrid near acute tip, adaxially scabrid; margins scabrid. Culm erect, rigid, internodes glabrous. Racemes borne singly

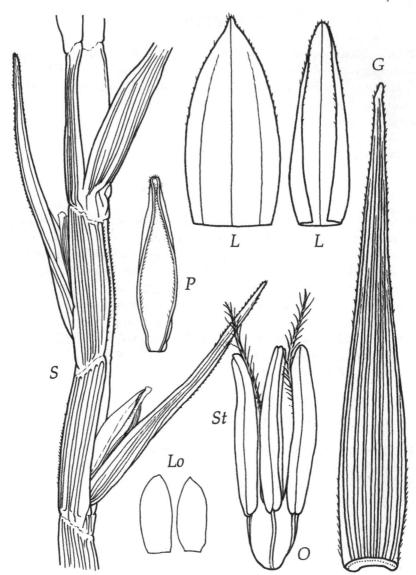

Fig. 20 *Lepturus repens* var. *cinereus*
S, spikelet on portion of raceme-rachis, × 12; **G**, glume, × 25; **L**, lemma dorsal and ventral view, × 16; **P**, palea dorsal view, × 16; **Lo**, lodicules, × 40; **O**, ovary, styles appearing between (**St**) stamens, × 40.

at tips of lateral branches, 3.5–7 cm × *c.* 3 mm, cylindric; rachis fragile, with numerous, fine, finely scabrid nerves. Spikelets 5–7 mm, the single floret embedded in the rachis and concealed by the upper glume. Glumes coriaceous, lanceolate, scabrid above, tapered to acute tip; lower 0 except on uppermost spikelet, upper > spikelet and exceeding the rachis cavity, 7-nerved. Lemma *c.* 3 mm, 3-nerved, firmly membranous, oblong-lanceolate, glabrous below, usually hairy in upper ½. Palea ≈ lemma, keels minutely ciliate, interkeel minutely pubescent. Lodicules oblong, denticulate, distally fleshy. Rachilla prolongation *c.* ⅓ length of lemma, minutely ciliate. Anthers *c.* 1.2 mm. Caryopsis not seen. Fig. 20.

Indigenous.

K.: North Chanter Islet (Herald Islets). Confined to a small area of lava rock a few metres above high tide level in open ground in petrel colonies on coralline soil in small pockets among coral outcrops.

Also indigenous to islands in the Pacific Ocean.

Sykes, W. R. *N.Z. DSIR Bull. 219*: 171 (1977) regarded *L. repens* as indigenous to Kermadec Is "because it was collected on the remote North Chanter Islet, rarely visited by man before 1966". Sykes, W. R. and West, C. J. *N.Z. J. Bot. 34*: 447–461 referred the Kermadec Is plant to *L. repens* var. *cinereus* because of its semiprostrate habit.

18. ERAGROSTIDEAE

Ligule a line of hairs or sometimes membranous. Inflorescence a panicle or composed of tough unilateral racemes digitate or scattered along an axis. Spikelets 1–several–many-flowered, lower florets ♀ and uppermost ± reduced, usually laterally compressed; disarticulation between florets. Glumes usually persistent, membranous, 0–1-nerved, < lowest lemma. Lemma membranous to coriaceous, 1–3-nerved, entire or 2–3-lobed with or without 1–(3) straight or flexuous terminal awns. Caryopsis sometimes with free pericarp.

1 Spikelets 1-flowered .... **Sporobolus**
   Spikelets 2–many-flowered 2
2 Inflorescence of racemes *3*
   Inflorescence a panicle, or reduced to a few spikelets *4*

ELEUSINE Gaertn., 1788

Annual or perennial tufts. Leaf-sheath strongly keeled. Ligule
membranous, usually ciliate. Leaf-blade usually folded. Inflorescence
of several digitate or subdigitate racemes; rachis < longest raceme,
narrowly winged, bearing spikelets in 2 rows on one side, and
terminating in a single spikelet. Spikelets several-flowered, imbricate;
disarticulation above glumes and usually between florets; rachilla
prolonged. Florets ♀. Glumes persistent, often keeled, awnless; lower
1–3-nerved; upper 1–3–(7)-nerved. Lemma strongly keeled, membranous, glabrous, obtuse or acute, sometimes shortly mucronate; keels
sometimes thickened and containing 1–3 closely spaced subsidiary
nerves. Palea < lemma, 2-keeled, keels often narrowly winged. Lodicules
± cuneate. Stamens 3. Caryopsis ellipsoid to subglobose, ± trigonous,
flat or concave on hilar face, rugose; pericarp free.

Type species: *E. coracana* (L.) Gaertn.

Spp. 9, mostly of eastern and north-eastern tropical Africa, but 1 sp.
widespread, *E. indica*, and 1 sp. of South America, *E. tristachya*.
Naturalised spp. 2.

1 Racemes (1)–2–8–(10), 2.5–8–(12) × 0.4–0.6 cm .... **1. indica**
   Racemes 2–(3), 1–2 × (0.5)–0.8–1.2 cm .... **2. tristachya**

1. E. indica (L.) Gaertn., *Fruct. Sem. Pl. 1*: 8 (1788).

crowfoot grass

Compact annual tufts 10–60–(90) cm, shoots at first spreading out very
close to the ground, culms later ascending; branching often intravaginal
near base. Leaf-sheath coriaceous; margins with a few long, soft hairs
near ligule. Ligule 0.3–1 mm, truncate, ciliate. Leaf-blade 6–30 cm ×

2–7 mm, usually folded, sometimes flat, coriaceous, often with a few, scattered, long, fine hairs; margins, midrib and subobtuse tip minutely scabrid. Culm 2–50 cm, usually geniculate at base, internodes glabrous. Inflorescence of (1)–2–8–(10) subdigitate spikes, the lowermost ± distant; spikes (1.5)–5–8–(12) cm × 4–6–(8) mm; rachis scabrid on margins, with hairs 0.5–2 mm at base. Spikelets 4.5–6 mm, 3–6-flowered, light green to brownish olive. Glumes < spikelet, membranous, subacute to subobtuse, keel scabrid, thickened, narrowly winged; lower 1.5–2.5 mm, 1-nerved, upper 2–4.5 mm, (3)–5-nerved. Lemma 2.5–5 mm, (1)–3-nerved, subacute to subobtuse, keel thickened, scabrid, with a subsidiary nerve on each side above. Palea 2–3 mm, keels winged. Lodicules 0.4–0.6 mm, membranous, glabrous, nerveless, ± truncate to acutely asymmetrically bilobed. Anthers 0.3–0.9 mm. Caryopsis 1.2–1.5 × 0.7–1 mm, ellipsoid to ellipticoblong and truncate, red-brown to almost black, obliquely rugose, concave on hilar face; hilum small, basal, round.

Naturalised.

N.: North and South Auckland, Bay of Plenty, Taranaki, Manawatu; S.: Canterbury (Christchurch, Lincoln); K. Waste ground, roadsides, farm tracks, poor damp pasture, weed of gardens and lawns.

A pantropical sp. now almost cosmopolitan and weedy.

2. E. tristachya (Lam.) Lam., *Tabl. Encycl. 1*: 203 (1792).

Tufted perennials, 14–20 cm, sometimes shortly rhizomatous; branching intravaginal. Leaf-sheath light green to pale creamy brown, glabrous, rounded at base and keeled above, coriaceous; margins wide, hyaline. Ligule 0.3–0.5 mm, truncate, ciliate. Leaf-blade 4–8 cm × 2–3 mm, coriaceous, folded, glabrous; margins scabrid and sometimes with a few, scattered long, very fine hairs, midrib scabrid near obtuse tip. Culm 9–18 cm, compressed, internodes glabrous. Inflorescence of 2 oblong, digitate spikes, occasionally a third slightly distant spike; spikes 1–2 × (0.5)–0.8–1.2 cm; rachis long-hairy at base, glabrous above but with 1 or 2 long hairs at base of some spikelets. Spikelets 5–6.5 mm, 4–10-flowered, usually purplish. Glumes unequal, < spikelet, membranous, subacute, keel stiffly scabrid, narrowly winged; lower 1.5–2 mm, 1–3-nerved, upper

2–3 mm, 5–(7)-nerved. Lemma 3.5–4 mm, membranous, subacute, 3-nerved, keel thickened, scabrid, with 2–3 subsidiary nerves near margin above, each marginal nerve with a short, close, subsidiary vein. Palea 2.5–3 mm, keels scabrid, hardly winged, apex ciliate, acute or bilobed. Lodicules 0.6–0.9 mm, nerved, asymmetrically 3-lobed. Anthers 0.6–0.8 mm. Caryopsis *c.* 1 × 0.8 mm, ± oblong, truncate, red-brown to almost black, with parallel ± transverse ridges, concave on hilar face; hilum small, basal, round.

Naturalised from South America.

N.: Gisborne, Havelock North, near Napier, also one early collection in ballast at Wellington. Waste land, lawns and golf courses.

ERAGROSTIS Wolf, 1781

Annuals or perennials, varying in habit but usually tufted. Leaf-sheath ± coriaceous, usually open, except near base. Ligule a rim of minute hairs, very rarely membranous. Leaf-blade narrow, flat, involute or convolute. Inflorescence an open or contracted, spiciform or glomerate panicle. Spikelets solitary, pedicelled, rarely sessile, usually laterally compressed, or subterete, 3–many-flowered, florets usually closely imbricating, disarticulation above glumes and between lemmas, or glumes and lemmas deciduous but paleas ± persistent. Glumes ± unequal, < lemmas, usually membranous, keeled, acute or acuminate; lower shorter, 1-nerved, upper 1–(3)-nerved. Lemma membranous to chartaceous, keeled or rounded, 3-nerved with lateral nerves ± obscure, obtuse to acuminate, entire or rarely mucronate. Palea ≈ lemma, 2-keeled, keels sometimes ciliate. Lodicules 2, small, ± fleshy. Stamens (2)–3. Ovary glabrous; styles free; stigmas plumose. Caryopsis ovate-oblong or cylindric, sometimes angled; embryo to ½ length of caryopsis or longer; hilum basal, punctiform. Chasmogamous or cleistogamous. Fig. 21.

Type species: *E. eragrostis* (L.) Wolf *nom. illeg.*

c. 350 spp., widespread in tropics and subtropics. Naturalised spp. 4; transient spp. 5.

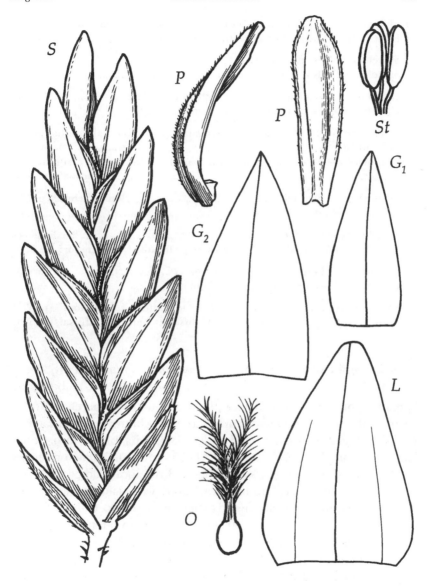

Fig. 21 *Eragrostis brownii*
S, spikelet, × 22; **G₁**, **G₂**, glumes, × 32; **L**, lemma dorsal view, × 32; **P**, palea lateral view showing rachilla, and ventral view, × 32; **O**, ovary, styles and stigmas, × 56; **St**, stamens, × 56.

1 All branch axils of panicle glabrous, or sometimes with hairs to 0.5 mm .. *2*
Lower branch axils of panicle with hairs 2–4 mm *7*

2 Spikelet pedicels eglandular *3*
Spikelet pedicels glandular *5*

3 Upper glume ≪ adjacent lemma; anthers 1.2–1.8 mm .... **4. plana**
Upper glume ≈ adjacent lemma; anthers 0.2–0.5 mm *4*

4 Leaf-blade to 2.5 mm wide; lemma ± opaque, submembranous; plants
perennial .... **1. brownii**
Leaf-blade 3–6 mm wide; lemma transparent, membranous; plants annual
.... **mexicana**ζ (p. 530)

5 Leaf margins eglandular .... **leptostachya**ζ (p.529)
Leaf margins with raised glands *6*

6 Spikelets 2–4 mm wide; caryopsis subglobular .... **2. cilianensis**
Spikelets *c.* 1.5 mm wide; caryopsis oblong-ellipsoid .... **minor**ζ (p. 530)

7 Lemma 2.4–2.8 mm; anthers 0.9–1.6 mm .... **3. curvula**
Lemma *c.* 1.5–2 mm; anthers 0.2–0.3 mm *8*

8 Spikelets *c.* 1 mm wide; panicle axil hairs to 4 mm .... **pilosa**ζ (p. 530)
Spikelets *c.* 1.5 mm wide; panicle axil hairs to *c.* 2 mm ...... **diffusa**ζ (p. 529)

1. E. brownii (Kunth) Wight, *Catalogue Indian Plants* 105 (1834).

bay grass

Spreading or occasionally narrow, light yellow-green, perennial tufts,
15–60–(70) cm, sometimes with prostrate culms; branching intravaginal.
Leaf-sheath glabrous, rounded. Ligule ciliate, hairs 0.1–0.2 mm. Collar
hairs to 3.5 mm. Leaf-blade 4–20 cm × 1–3.5 mm, usually flat, sometimes
involute, abaxially smooth, sometimes scabrid above, adaxially finely
ribbed and finely scabrid on ribs, sometimes with scattered long hairs
near base; margins scabrid, tapered to filiform, acute tip. Culm 5–60 cm,
internodes smooth. Panicle 3–30 cm, contracted to often very lax with
delicate widespread branches, bearing clustered to well-spaced pedicelled
spikelets; rachis smooth below, becoming scabrid above, branches and
pedicels scabrid, branch-axils glabrous. Spikelets (3)–6–10 × 2–2.5 mm,
(5)–7–16-flowered, glabrous, compressed, linear-lanceolate or narrow-
oblong, grey-green, leaden green, or purplish. Glumes subequal or
unequal, 1-nerved, submembranous, acute, lanceolate to ovate-

lanceolate, keeled, scabrid on keel; lower 1.0–1.9 mm, subulate, upper 1.5–2.4 mm, lanceolate. Lemma 1.7–2.4 mm, 3-nerved with lateral nerves distinct, submembranous, ± opaque, ovate-lanceolate, smooth, but keel finely scabrid near subacute to obtuse apex. Palea < lemma, persistent, keels and truncate apex closely ciliate. Rachilla 0.3–0.4 mm, glabrous. Stamens 3; anthers 0.3–0.5 mm. Caryopsis 0.6–0.7 × 0.4–0.5 mm. Fig. 21.

Naturalised from Australia.

N.: throughout; S.: Nelson, Marlborough. Waste ground, roadsides, modified grassland, sometimes in swampy ground or near hot springs.

Allan, H. H. *N.Z. DSIR Bull. 49*: 67 (1936) noted that plants found in N.Z. agreed with *E. brownii* var. *patens* Benth. This lax-panicled variety was regarded for a time as specifically distinct, as *E. benthamii* Mattei [Jacobs, S. W. L. and McClay, K. L. *in* Harden, G. J. (Ed.) *Fl. N. S. W. 4*: 534–544 (1993); Edgar, E., Connor, H. E. and Shand, J. E. *N.Z. J. Bot. 29*: 117–129 (1991)]. However, Lazarides, M. *Aust. Syst. Bot. 10*: 77–187 (1997) treated *E. benthamii* as a synonym of *E. brownii* considering it to be "a variant without discontinuous features".

2. E. cilianensis (All.) Janch., *Mitt. Naturw. Ver. Wien 5(9)*: 110 (1907).

stink grass

Tufted stinking annuals, 12–60 cm, with stiff panicle of large spikelets much overtopping leaves; branching intravaginal. Leaf-sheath submembranous, strongly ribbed, keeled above, light brown, glabrous, with scattered, sunken glands especially on keel. Ligule ciliate, hairs 0.3–0.7 mm. Collar hairs 1–2.5 mm. Leaf-blade 1.5–15 cm × 2–5.5 mm, flat, somewhat narrowed to subobtuse tip, abaxially glabrous, adaxially very minutely prickle-toothed; margins with prominent warty glands. Culm 10–45 cm, geniculate at base, then erect, internodes glabrous, with ring of glandular tissue below nodes. Panicle 3.5–10.5 × 2–5 cm; branches few, stiff, short, minutely scabrid, bearing large close-set, short-pedicelled spikelets, branches and pedicels often with wart-like glands, branch-axils sometimes with a few short (to 0.5 mm) hairs. Spikelets 3–10 × 2.4–4 mm, (3)–10–20-flowered, minutely scabrid, compressed, ovoid-oblong, shining, grey, sometimes purplish. Glumes *c.* equal, 1.5–2.2 mm, ovoid, membranous, sometimes with warty glands on midnerve near tip, deciduous at maturity; lower 1-nerved, upper

3-nerved. Lemma 1.6–2.5 mm, distinctly 3-nerved, membranous, ovoid, obtuse, very minutely scabrid near apex and between lateral nerves and margins. Palea < lemma, keels ciliate, apex obtuse, ciliate. Rachilla glabrous, 0.3–0.4 mm. Stamens 3; anthers 0.3–0.5 mm. Caryopsis 0.5–0.7 × 0.4–0.6 mm.

Naturalised from Mediterranean.

N.: scattered and local. Waste land, roadsides, mainly in sandy ground.

The authorities for the name were confirmed by Perry, G. and McNeill, J. *Taxon* *35*: 696–701 (1986).

3. E. curvula (Schrad.) Nees, *Fl. Afr. Austral. Ill.* 397 (1841).

African love grass
Stiff, densely tufted perennials, to 70 cm; branching intravaginal, shoots ± thickened at base. Leaf-sheath coriaceous, strongly ribbed, light creamy brown at base, purplish above, with short, scattered, stiff hairs, lower sheaths tomentose near base. Ligule ciliate, hairs 0.8–1.4 mm. Collar hairs 2–5.5 mm. Leaf-blade 10–55 cm × 0.5–1.3 mm diam., narrowly involute or convolute, abaxially glabrous, adaxially ribbed, ribs minutely scabrid; margins minutely scabrid, long-narrowed to filiform, acute, scabrid tip. Culm 20–80 cm, rarely branched above, erect, internodes glabrous. Panicle 12–22 cm, lax, at first narrow-sagittate, later more open; branches ascending to later spreading, solitary or binate, ± scabrid, branch-axils at lower nodes with hairs (to 4 mm). Spikelets 4.5–6–(8) mm, 4–6–(8)-flowered, ± smooth, not very compressed, linear-lanceolate, olive-grey. Glumes unequal, hyaline, 1-nerved, oblong-lanceolate, apex subacute, minutely scabrid; lower 1.6–2 mm, upper 2.2–2.8 mm. Lemma 2.5–2.8 mm, 3–(5)-nerved, membranous, elliptic-oblong, obtuse, hardly keeled, minutely scabrid. Palea ≈ lemma, keels sparsely scabrid, apex truncate, ciliate. Rachilla glabrous, 0.6–0.8 mm. Stamens 3; anthers 0.9–1.6 mm. Caryopsis 1.4–1.6 × 0.6–0.8 mm.

Naturalised from Africa.

N.: scattered and local; S.: Canterbury (Christchurch), North Otago (Kurow), Central Otago (Ophir, Wanaka-Cromwell Rd, Earnscleugh); Three Kings Is. Roadsides, in sandy soil, or on banks, depleted tussock grassland, stony flats and waste land. Some records are of escapes from experimental plots.

4. E. plana Nees, *Fl. Afr. Austral. Ill.* 390 (1841).

Stiff, narrow, perennial tufts, to 100 cm. Leaf-sheath coriaceous, strongly keeled, striate, light creamy brown, glabrous. Ligule ciliate, hairs 0.2–0.4 mm. Collar hairs few, 1–1.8 mm. Leaf-blade 20–30 cm × 1–1.5 mm diam., coriaceous, folded, abaxially strongly densely ribbed, with few, very fine, scattered hairs 1–1.8 mm, adaxially finely ribbed, ribs minutely scabrid; margins minutely scabrid, long-narrowed to filiform tip. Culm 50–80 cm, erect, internodes glabrous. Panicle 15–27 cm, narrow; branches short, erect, stiff, almost filiform, sparsely scabrid to smooth, tipped by very narrow spikelets, branch-axils glabrous. Spikelets 9–14 × 1.5–2 mm, 9–12-flowered, glabrous, linear to narrow elliptic-oblong, greyish green. Glumes hyaline, ≪ lemmas, nerve 1, smooth or scabrid; lower 0.4–0.6 mm, ovate-lanceolate, obtuse to acute, upper 0.8–1.4 mm, ± oblong, obtuse to truncate. Lemma *c.* 2.5 mm, 3-nerved, firmly membranous, elliptic, apex obtuse, sometimes very minutely scabrid. Palea ≈ lemma, glabrous. Rachilla 0.8–1 mm, sparsely scabrid. Stamens 3; anthers 1.2–1.8 mm. Caryopsis not seen.

Naturalised from Africa.

N.: South Auckland (Firth of Thames, Coromandel). Hill pasture, on sunny faces.

ζ**E. diffusa** Buckley　　The single specimen consists of 6 separate culms. Collar hairs numerous, to 3 mm. Leaf-blade 4–6 cm × 2–3 mm; margins finely scabrid. Panicle 9–15 mm; branches widely spreading, hairs in lower branch-axils few (*c.* 2 mm). Spikelets 4–6 × *c.* 1.5 mm, 5–9-flowered, linear, dull olive-green, purple-tinged. Lemma 1.5–2 mm. Stamens 3; anthers 0.2 mm. Caryopsis *c.* 0.8 mm, elliptic-oblong. A North American species once collected from Matapihi, Bay of Plenty, single clump along railway line　*M. Hodgkins* Apr 1961 (CHR 327210).

ζ**E. leptostachya** (R.Br.) Steud.　　The single specimen consists of a slender tuft, *c.* 75 cm. Leaf-sheath keel minutely glandular. Leaf-blade to *c.* 25 cm × 6 mm; margins eglandular. Culm with glandular ring below nodes. Panicle 18–20 × 7 cm; branches ± horizontally spreading, bearing few spikelets on short glandular pedicels. Spikelets to 7 × 1.5 mm, 4–10-flowered, leaden green, linear-lanceolate. Lemma *c.* 1.7 mm. Stamens 3; anthers 0.5 mm. Caryopsis *c.* 0.8 mm, rectangular, truncate. An Australian species once collected on Mahuki Id (near Great Barrier Id)　*A. E. Wright 6878* 2 Jan 1985 (AK 171282).

ζ**E. mexicana** (Hornem.) Link   Tufted, to 30 cm. Leaf-blade to 8 cm × 4 mm; margins eglandular. Panicle 6–12 cm, narrow, bearing numerous spikelets on eglandular pedicels. Spikelets 2–4.5 × 1–1.5 mm, 2–6-flowered, light brownish green. Lemma *c.* 1.5 mm, membranous, broadly ovate. Stamens 3; anthers 0.2 mm. Caryopsis immature, *c.* 0.5 mm. A species of North and South America: one 19th Century collection from ballast at Wellington *T. Kirk* (AK 98483, CHR 4100).

ζ**E. minor** Host   Plants slender, tufted, 10–25 cm. Leaf-sheath glandular, sometimes with scattered hairs. Collar hairs to 3 mm. Leaf-blade to 11 cm × 4 mm; margins glandular. Culm with ring of glandular tissue below nodes. Panicle 4–10 cm; branches and pedicels minutely scabrid, sometimes with a few glands, branch-axils glabrous. Spikelets 3–4.5 mm, < 1.5 mm wide, 3–7-flowered, greyish green. Lemma 1.4–1.7 mm, ovate. Stamens 3; anthers 0.2–0.3 mm. Caryopsis 0.6–0.8 × 0.5–0.6 mm, ellipsoid-oblong, truncate. A species of Eurasia and tropical Africa twice collected in North Id: one 19th Century collection from ballast at Wellington *T. Kirk* (WELT 70583, WELT 77636), and again at Rotorua *R. Murray* 1965 (CHR 156758).

ζ**E. pilosa** (L.) P.Beauv.   The single specimen consists of a tuft *c.* 60 cm. Leaf-blade to 10 cm × 2 mm; margins eglandular. Panicle to *c.* 18 cm, narrowly branched; lower branch-axils with a few fine hairs to 4 mm. Spikelets *c.* 6 × 1 mm, numerous, linear, 6–10-flowered, greyish brown. Lemma *c.* 1.5 mm, green, tip purplish. Stamens 3; anthers 0.2 mm. Caryopsis *c.* 0.8 mm, oblong-ellipsoid. A Eurasian species once collected at Rotorua, FRI nursery *A. Vanner* 8.2.1990 (NZFRI 18731). *Eragrostis pilosa* was recorded for N.Z. in *Iowa Geol. Survey Bull. no. 4, Weed Fl. Iowa* 732 (1913) but no specimens were seen to support this record.

SPOROBOLUS R.Br., 1810

Perennials or annuals, tufted or creeping. Ligule a short-ciliate rim. Leaf-blade narrow, flat or convolute, sometimes setaceous. Culm usually erect. Inflorescence an open or contracted, sometimes spiciform panicle. Spikelets often small, fusiform, mostly glabrous, 1-flowered, usually olivaceous or olive-grey; disarticulation above glumes; rachilla rarely prolonged. Glumes equal or unequal, deciduous or persistent, 1-nerved or nerveless, thinly membranous, seldom keeled. Lemma ≥ glumes, rounded, 1–3-nerved, thinly membranous, soft, awnless. Palea ≤ lemma, of similar texture and colour to lemma, 2-keeled, apex obtuse, truncate, emarginate, or minutely bilobed. Lodicules hyaline, truncate. Callus 0. Stamens 2–3. Ovary often globose; styles free, short; stigmas plumose. Caryopsis with loose pericarp, globose to ellipsoid, rounded or truncate; embryo *c.* ½ length of caryopsis or less; hilum basal, punctiform.

Type species: *S. indicus* (L.) R.Br.

c. 160 spp. of tropical and subtropical, sometimes warm temperate regions. Naturalised spp. 2; transient sp. 1.

> *1* Panicle densely spicate and branches ± appressed, often some ± distant near base; stamens 3 .... **1. africanus**
> Panicle interrupted throughout and branches spreading in lower part or very distant near base; stamens 2 2
>
> *2* Panicle branches ascending; spikelets in dense clusters .... **2. elongatus**
> Panicle branches ± spreading; spikelets loosely spaced .... **diandrus**ζ

1. S. africanus (Poir.) Robyns et Tournay, *Bull. Jard. Bot. Brux. 25*: 242 (1955).   ratstail

Stiff, perennial tufts, 20–80 cm, with convolute or flat leaves overtopped by narrow, olive-green spiciform panicles; branching intravaginal. Leaf-sheath very light green to creamy brown, subcoriaceous or somewhat chartaceous, glabrous, margins often short-ciliate below ligule. Ligule 0.2–0.5 mm. Collar hairs to 1 mm. Leaf-blade 10–25 cm × 0.8–2 mm diam., inrolled, or flat and up to 5 mm wide, abaxially glabrous, adaxially scabrid on ribs; margins finely scabrid, long-tapered to fine tip. Culm 15–50 cm, internodes glabrous. Panicle 8.5–22 cm × 4–8 mm, ± continuous but with some ± distant, appressed branches, up to 2 cm, near base; rachis ± smooth, branches sparsely scabrid, bearing densely crowded spikelets. Spikelets 2.2–3 mm, leaden green. Glumes unequal, < lemma; lower 0.7–1 mm, nerveless, elliptic, obtuse, upper 1.4–1.7 mm, 1-nerved, lanceolate, subacute. Lemma = spikelet, faintly 3-nerved, ovate-elliptic, acute. Palea ≤ lemma, apex narrowly truncate. Stamens 3; anthers 0.7–1.2 mm, often purplish. Caryopsis 1–1.3 × 0.6–0.7 mm, ellipsoid, truncate.

Naturalised from Africa.

N.: common in North Auckland, Auckland City, and Rotorua, scattered further south; S.: Nelson, Marlborough, Canterbury (Banks Peninsula); K. Coastal in pasture, and in grassland on sunny hill slopes, and on cliffs, also in scrub, waste land and on roadsides.

Sometimes treated as *S. indicus* var. *capensis* (P.Beauv.) Engl. but Simon, B. K. and Jacobs. S. W. L. *Aust. Syst. Bot. 12*: 375–448 (1999) note that *S. africanus* is distinct from *S. indicus* in its narrow denser inflorescence and larger spikelets, and also by the method in which the mature caryopsis is held in the floret.

2. S. elongatus R.Br., *Prodr.* 170 (1810).

Perennials, in dense strict clumps > 1 m; branching intravaginal. Leaf-sheath very light green to creamy brown, subcoriaceous or somewhat chartaceous, glabrous; margins sometimes short-ciliate below ligule. Ligule *c.* 0.3 mm. Collar hairs to 1 mm. Leaf-blade 15–45 cm × *c.* 1 mm diam., inrolled, or flat and 2–4 mm wide, abaxially glabrous, adaxially finely scabrid on ribs; margins finely scabrid, long-tapered to fine tip. Culm internodes glabrous. Panicle 14–30 cm, ± lax with obvious, ascending branches, lowest branches often very distant and up to 2 cm; rachis ± smooth, branches sparsely scabrid bearing spikelets in dense spike-like clusters. Spikelets 1.5–2.5 mm, leaden green. Glumes unequal, < lemma; lower 0.5–0.8 mm, nerveless, oblong-elliptic, obtuse, upper 0.8–1.5 mm, 1-nerved, ovate-elliptic, subacute. Lemma = spikelet, faintly 3-nerved, elliptic-lanceolate, subacute. Palea ≤ lemma, narrowly truncate. Stamens 2; anthers 0.4–0.7 mm, yellow. Caryopsis 0.7–1.2 × 0.5–0.7 mm, oblong-ellipsoid, truncate.

Naturalised from Australia.

N.: Coromandel Peninsula, Bay of Plenty (near Te Kaha). Rough pasture and on roadside.

Found also at Hamilton as a weed in a pot originally planted as *Cyperus, J. R. Murray* 25.3.1969 (CHR 284194).

A specimen of *S. elongatus* from Kuaotunu, Coromandel Peninsula *G. Batten* 18.3.1964 (CHR 146347) was misidentified as *S. natalensis*, in a checklist of naturalised chloridoid grasses [Edgar, E., Connor, H. E. and Shand, J. E. *N.Z. J. Bot. 29*: 117–129 (1991)].

ζS. diandrus (Retz.) P.Beauv.   The only specimen seen consisted of the upper part of a culm topped by a lax panicle. Panicle 22 cm; rachis smooth, branches ascending to spreading, very sparsely prickle-toothed, bearing ± evenly spaced spikelets almost to base. Spikelets ≤ 1.5 mm, leaden green. Glumes unequal, < spikelet; lower 0.3 mm, upper 0.7–0.8 mm. Lemma = spikelet. Palea slightly < lemma. Stamens 2; anthers 0.5–0.6 mm, white. Caryopsis 0.8 × 0.5 mm, truncate. A species of India, Malesia to Polynesia, once collected at Hatepe, Lake Taupo, on footpath by river *D. J. Nicolle* 17 Jan 1994 (K, CHR 518485). *Sporobolus diandrus* was treated as *S. indicus* var. *flaccidus* Veldkamp, in a revision of *Sporobolus* in Malesia [Baaijens, G. J. and Veldkamp, J. F. *Blumea 35*: 393–458 (1991)].

Sporobolus cryptandrus (Torr.) A.Gray, of North America, was once collected from Northburn Run, Dunstan Mts, Otago, in Cockayne's Plot 3a, *J. A. Douglas*, 7.6.1967 (CHR 133514) where it had persisted since the plot was sown in 1948 [Douglas, J. A. *Proc. N.Z. Ecol. Soc. 17*: 18–24 (1970)].

Sporobolus virginicus (L.) Kunth was recorded as occurring in N.Z. by Curtis, W. M. and Morris, D. I. *Student's Fl. Tasmania 4B*: 328 (1994). No specimens have been seen.

ζDiplachne fusca (L.) P.Beauv.   The single specimen is damaged, with culms broken from the basal tuft; overall height is estimated as *c.* 25 cm. Leaf-sheath scabrid on ribs near base. Ligule *c.* 3 mm, membranous, lacerate. Leaf-blade to *c.* 20 cm, convolute, greyish green, scabrid. Panicle 9–14 cm; rachis scabrid, racemes numerous, ascending, slender, bearing closely appressed spikelets. Spikelets 6–8 mm, *c.* 7-flowered, dark greenish grey, almost sessile to shortly pedicelled. Glumes unequal, 1-nerved, membranous, acute, keel scabrid; lower *c.* 1.5 mm, very narrow-lanceolate, upper *c.* 2 mm, ovate-oblong. Lemma *c.* 3 mm, 3-nerved, chartaceous, lateral nerves with long fine hairs in lower ½, apex minutely ciliate, notched, shortly mucronate from notch. Palea ≤ lemma and of same texture; keels finely scabrid. Stamens 3; anthers *c.* 0.4 mm. A palaeotropical species once collected from Thames *C. W. Walker* 12.2.1941 (CHR 25084).

ζDistichlis spicata (L.) Greene   Two specimens composed of widely spaced tufts to 15 cm, from a strong rhizome bearing stout, grey, bract-like sheaths. Ligule ciliate. Collar with a few hairs. Leaf-blade 1.5–2.5 cm, involute, coriaceous, glabrous, tip finely pungent. Inflorescence 1.5–2 × 0.5–0.8 cm, of 2–3 closely overlapping spikelets. Spikelets 12–14 × 4–7 mm, 8–10-flowered, greenish brown, glabrous; florets ♂ (the sp. is dioecious). Glumes unequal, acute; lower 2.5 mm, 3-nerved, upper 4.5 mm, 5-nerved. Lemma 5.5 mm, 11-nerved, rounded, chartaceous, acute. Palea ≈ lemma, membranous with narrow entire wings; margins curved near base. Stamens 3. A North American species once collected by J. B. and J. F. Armstrong near Christchurch (CHR 8719, and in Armstrong Herbarium CHBG unnumbered; the specimens are undated, but may have been collected in the 1870s).

VII SUBFAMILY PANICOIDEAE

Spikelets solitary, or paired, or in triplets, usually dorsally compressed, all alike, or differing in sex, size, shape and structure. Fertile spikelets usually 2-flowered, the lower floret ♂, ∅, or ♀, the upper floret ♀ or ♀; rachilla not prolonged; rarely plants dioecious. Glumes or upper lemma indurated. Lodicules 2, cuneate, fleshy, truncate. Stamens usually 3, rarely fewer. Stigmas usually 2. Caryopsis with large embryo. Ligule usually a membranous rim with dense ciliate fringe, but sometimes entirely membranous (*Axonopus, Digitaria, Paspalum*) and occasionally suppressed (*Echinochloa*).

The mainly tropical to subtropical subfamily Panicoideae is represented by only 4 indigenous spp. in N.Z. — 3 spp. in tribe Paniceae, 1 sp. in tribe Isachneae; and 1 endemic sp. in tribe Andropogoneae. However, 42 taxa of Paniceae, and 9 taxa of Andropogoneae are fully naturalised in N.Z. In addition, 9 spp. in Paniceae and 4 spp. in Andropogoneae have been recorded once or twice as introduced to N.Z.

Tribes: 19. Paniceae; 20. Isachneae; 21. Andropogoneae

19. PANICEAE

Inflorescence a panicle, or of variously arranged often fragile racemes or spikes. Spikelets 2-flowered, falling entire at maturity, solitary or paired, or sometimes in groups of 3 or more; lower floret ♂ or ∅, upper floret ♀; occasionally dioecious. Glumes usually herbaceous or membranous; lower usually the smaller, sometimes much reduced or 0, upper usually = spikelet, sometimes shorter, rarely 0. Lower floret: lemma similar to upper glume, at least in texture, rarely indurated; palea present or 0. Upper floret: lemma and palea similar in texture, firmer than glumes, usually indurated, usually awnless; caryopsis firmly enclosed by hardened anthoecium; embryo usually $\frac{1}{3}$–$\frac{1}{2}$ length of caryopsis; hilum usually basal, punctiform.

1 Inflorescence globular or hemispheric, of numerous spatheate racemes; flowers unisexual .... **Spinifex**
   Inflorescence an open or spike-like panicle, or of 1–2, or numerous, espatheate racemes along a linear rachis; flowers bisexual *2*

2 Ligule absent, or as scattered short hairs .... **Echinochloa**
   Ligule usually well developed, entirely membranous, or a membranous ciliate rim *3*

3 Inflorescence an open panicle, or contracted, cylindric and spike-like *4*
   Inflorescence of variously arranged racemes *10*

4 Spikelets subtended by bristle-like branchlets *5*
   Spikelets ± sessile or on simple pedicels, bristle-like branchlets 0 .... *7*

5 Bristles subtending spikelets fused, forming a barbed involucre .... **Cenchrus**
   Bristles subtending spikelets free except at base, not involucrate *6*

6 Spikelets deciduous, falling together with bristles .... **Pennisetum**
   Spikelets deciduous but bristles persistent .... **Setaria**

7 Panicle contracted or spike-like .... **Sacciolepis**
   Panicle open *8*

8 Panicle branches ending in a bristle .... **Setaria**
   Panicle branches ending in a spikelet.. *9*

9 Lemma margins in upper floret inrolled .... **Panicum**
   Lemma margins in upper floret flat .... **Melinis**ζ (p. 595)

10 Ligule entirely membranous, glabrous, or rarely minutely ciliate, hairs < 0.5 mm *11*
   Ligule a truncate membranous rim, densely ciliate, hairs 3–5 mm *13*

11 Lemma margins in upper floret hyaline, flat .... **Digitaria**
   Lemma margins in upper floret rigidly inrolled *12*

12 Back of lemma of upper floret facing outwards from rachis .... **Axonopus**
   Back of lemma of upper floret facing towards rachis .... **Paspalum**

13 Glumes awned or mucronate .... **Oplismenus**
   Glumes muticous *14*

14 Spikelets on racemes sunken in a disarticulating rachis .... **Stenotaphrum**
   Spikelets on digitate racemes *15*

15 Lemma of upper floret pubescent .... **Entolasia**
   Lemma of upper floret glabrous, transversely rugose .... **Urochloa**

AXONOPUS P.Beauv., 1812

Annuals, or rarely biennials, or perennials, herbaceous, long-stoloniferous, sometimes mat-forming or tufted. Ligule membranous, glabrous or very rarely minutely ciliate. Inflorescence of spike-like racemes bearing secund, ± sessile, adaxial spikelets ± embedded in the flattened, hollowed or winged rachis and mostly solitary at nodes, falling entire at maturity. Spikelets 2-flowered, dorsally compressed; lower floret ∅, upper floret ♀. Lower glume 0, upper = spikelet, membranous, 4–5-nerved. Lower floret: lemma ≤ glume, membranous. Upper floret: lemma shorter, chartaceous to crustaceous, ± mucronate, glabrous, 4-nerved; palea ≤ lemma. Lodicules 2, fleshy, glabrous. Stamens 3. Ovary apex glabrous; styles subterminal, free. Caryopsis dorsiventrally compressed, glabrous; embryo large, at least ⅓ length of caryopsis; hilum punctiform or shortly elliptic, < ½ length of caryopsis.

Type species: *A. compressus* (Sw.) P.Beauv.

c. 35 spp. of tropical South America. Naturalised sp. 1.

1. A. fissifolius (Raddi) Kuhlm., *Relat. Commiss. Linhas. Telegr. Estratég. Matto Grosso Amazonas 11*: 87 (1922).

narrow-leaved carpet grass
Narrow, glabrous, shortly rhizomatous, ± fan-shaped tufts, from rather woody, creeping, sometimes arching stolons, to 2 mm diam.; often forming dense mats to 1 m diam., with rather short, wide leaves and wiry, ± filiform, very narrow, ± drooping culms; branching intravaginal. Leaf-sheath whitish green to light brown, sometimes purplish, coriaceous, glabrous, compressed and strongly keeled; margins wide, membranous. Ligule 0.2–0.3 mm, minutely ciliate, hairs < 0.5 mm, sometimes extending as a contra-ligule. Leaf-blade (2)–5–12 cm × 2–7 mm, linear, scarcely narrowed above, flat or folded, often keeled towards tip, stiffly herbaceous, glabrous, ribs few, well-spaced; margins scarcely thickened, with a few scattered long hairs near base and scabrid near obtuse tip. Culm (6)–20–55 cm, nodes glabrous, or rarely villous, internodes glabrous. Inflorescence of 2–3, rarely 4, slender, ± erect racemes. Racemes 3.5–7.5 cm; rachis triangular, minutely scabrid on narrow wings, bearing close-set, subsessile, alternate spikelets in 2 rows. Spikelets *c.* 2 mm, green

or purplish, oblong-elliptic, obtuse or subacute. Lower glume 0. Upper glume = spikelet, firmly membranous, 4-nerved, the 2 pairs of nerves near each margin, outer internerves often with a few soft hairs. Lower floret: lemma similar to upper glume. Upper floret: lemma 1.6–1.8 mm, elliptic, obtuse, thinly crustaceous, glabrous, shining, cream, very minutely punctulate-striolate; palea of same texture, slightly shorter and narrower than lemma, keels very rounded; anthers 0.7–1 mm; caryopsis *c.* 1.2 × 0.7 mm.

Naturalised from America.

N.: northern half; S.: Nelson (Farewell Spit, north-western coast to Big R., Pupu Springs near Takaka); K.: Raoul Id. Disturbed ground, waste land, pastures, lawns, roadsides, tracks in open forest and margins of scrub.

Found also on open heated soils on trackside in thermal area at Wairakei *D. R. Given, L. Coham & C. Ecroyd* 13 July 1976 (CHR 276963).

Field, T. R. O. and Forde, M. B. *Proc. N.Z. Grasslands Assoc. 51*: 47–50 (1990) reported that narrow-leaved carpet grass had recently become an alarmingly invasive lawn weed in northern North Id.

Until recently *A. fissifolius* was known in New Zealand and elsewhere as *A. affinis* Chase.

CENCHRUS L., 1753

Annuals or perennials. Leaf-sheath compressed. Ligule reduced to ciliate rim. Leaf-blade flat or involute. Culm terete, solid. Inflorescence a solitary spike-like panicle of prickly glomerules, composed of coalescing spines; spikelets sessile, solitary, or in clusters of 2–5 surrounded by, and deciduous with the sessile or minutely peduncled involucre of ± connate, often rigid, antrorsely or retrorsely scabrid spines or bristles. Spikelets 2-flowered; lower floret ♂, or ∅ and reduced to a lemma, upper floret ♀. Glumes usually dissimilar, rarely subequal, hyaline or thinly membranous; lower often 0, or if present, 1–5-nerved, upper ≤ spikelet, 1–5–(7)-nerved. Lower floret: lemma usually similar in texture to upper glume and ≈ spikelet, 3–7-nerved; palea ≈ lemma, 2-keeled. Upper floret: lemma = spikelet, firmly membranous, chartaceous or coriaceous, 5–7-nerved; palea = lemma; lodicules 0; stamens 3; ovary apex glabrous, styles terminal, free, or connate below; caryopsis dorsiventrally compressed, glabrous, embryo *c.* ⅔ length of caryopsis, hilum basal, punctiform.

Type species: *C. echinatus* L.

c. 25–30 spp., of tropical and warm temperate dry regions of America, Africa, India, south-western Asia, Polynesia, and Australia. Indigenous sp. 1.

Some spp. of *Cenchrus* are common weeds and may become serious pests on account of the scabrid spines of their involucres.

1. C. caliculatus Cav., *Icon.* 5: 39, t. 463 (1799); Holotype: MA *n.v.*, *L. Née* Society Islands.   large burr grass

Lax clumps from a very short woody rhizome, with trailing culms rooting at lower nodes; branching extravaginal, and intravaginal above. Leaf-sheath longer than internodes, firm, papery, keeled, smooth, minutely scabrid on narrow membranous margin and on ribs below collar. Ligule 1–1.5 mm, a truncate densely ciliate rim. Leaf-blade 7–20 cm × 4–9 mm, linear-lanceolate, rounded-truncate at base, flat, firm, abaxially smooth, adaxially minutely scabrid on ribs; margins somewhat thickened, minutely scabrid, tapered to long, acuminate tip. Culm to 60 cm × *c.* 2–3.5 mm diam., terete to ± angled, internodes minutely pubescent-scabrid on ridges below panicle. Panicle (10)–15–25 × *c.* 1.5 cm, spike-like; rachis triangular, slightly winged, densely minutely pubescent-scabrid, bearing ovoid clusters of 1–3 spikelets, hidden among bristles and densely, spirally arranged along rachis; clusters 7–10 × 4–7 mm, including involucre of stiff bristles, at first appressed, but finally borne at right angles to rachis; bristles retrorsely barbed, very variable in length, 0.5–11 mm, outermost smaller, terete, inner more planoconvex with margins densely softly long-ciliate, one bristle in each cluster usually exceeding the others. Spikelets 5–6 mm, 2-flowered, < inner bristles, sessile, glabrous, light green, almost colourless. Glumes hyaline; lower 2–3–(4.5) mm, 1-nerved, upper 4–5 mm, 5-nerved. Lower floret: lemma 4.8–6 mm, 5-nerved, scabrid; palea keels ciliate, interkeel minutely hairy, margins scabrid; anthers 1.5–2 mm, brown with thick orange-yellow filaments. Upper floret: lemma 5–5.5 mm, 5-nerved, finely scabrid-papillose; palea very finely scabrid-papillose; anthers as in lower floret; caryopsis 2.2–2.7 mm, ± ellipsoid.

Indigenous.

K. Open places near coast on grassy headlands.

Also indigenous to Norfolk Id, New Caledonia, New Guinea and much of Polynesia.

DIGITARIA Haller, 1768 *nom. cons.*

Annual or perennial, often stoloniferous tufts, of moderate height or low-growing. Leaf-sheath submembranous or stiffly striate, rounded, midrib obvious above. Ligule membranous. Leaf-blade flat to involute, flaccid. Culm erect, or decumbent and rooting at nodes. Inflorescence a panicle of digitate or subdigitate racemes on a short rachis. Racemes slender, secund; rachis 3-angled, or flattened and winged, usually serrate, persistent, bearing along the two abaxial sides, alternate, appressed groups of 2, 3, or rarely solitary spikelets, usually one spikelet of each group ± sessile, and 1–2 shortly pedicelled; spikelets disarticulating below glumes and falling entire at maturity. Spikelets dorsally compressed, lanceolate or elliptic, ± planoconvex, variously pubescent, rarely glabrous, awnless, 2-flowered; lower floret ∅, upper floret ♀. Lower glume minute or 0, upper ≤ lemma of lower floret. Lower floret: lemma usually = spikelet; palea usually a minute scale, or 0. Upper floret: lemma chartaceous to cartilaginous, nerveless, or faintly 3-nerved, glabrous, hyaline margins enfolding palea; palea of similar texture to lemma, nerveless, or faintly 2-nerved; lodicules cuneate, 3-nerved; callus 0; stamens 3; ovary apex glabrous, styles 2, apical, shortly connate below; caryopsis planoconvex, embryo large, to ⅓ length of caryopsis, hilum punctiform.

Type species: *D. sanguinalis* (L.) Scop.

c. 220 spp., cosmopolitan, but mainly tropical and subtropical. Naturalised spp. 6.

The genus was revised for Malesia by Veldkamp, J. F. *Blumea 21*: 1–80 (1973), and Australian spp. were revised by Webster, R. D. *Brunonia 6*: 131–213 (1984).

1 Lemma of lower floret with minutely scabrid nerves .... **4. sanguinalis**
   Lemma of lower floret with smooth nerves *2*

2 Spikelets in pairs; lemma of upper floret yellowish, light brown or somewhat
   purplish *3*
   Spikelets usually in triplets, rarely some in pairs; lemma of upper floret dark
   chestnut-brown *5*

3 Lower glume distinct, (0.2)–0.3–0.5 mm, upper *c.* ⅔ spikelet .... **2. ciliaris**
   Lower glume 0 or *c.* 0.1 mm, upper to ⅓ or ≈ spikelet *4*

4 Upper glume to ⅓ length of spikelet, 0–3-nerved; lemma of upper floret
   yellowish to light brownish; anthers 1–1.3 mm .... **5. setigera**
   Upper glume ≈ spikelet, 5–7-nerved; lemma of upper floret greyish; anthers
   0.4–0.5 mm .... **1. aequiglumis**

5 Spikelets *c.* 2 mm, hair tips swollen .... **3. ischaemum**
   Spikelets 1.5–1.8 mm, hair tips pointed .... **6. violascens**

1. D. aequiglumis (Hack. et Arechav.) Parodi, *Revista Fac. Agron. Veterin. 4*: 47 (1922).

Stoloniferous annual, rooting at lower nodes, forming erect tufts to *c.* 50 cm. Leaf-sheath glabrous, or with a few scattered long hairs. Ligule to 2 mm, membranous, glabrous, rounded. Leaf-blade 3.5–6 cm × 1.5–2.5 mm, linear, glabrous; margins minutely prickle-toothed, occasionally with a few short soft hairs, narrowed to fine acute tip. Culm nodes blackish brown, internodes glabrous. Racemes 2–5, 4–9 cm, slender, digitate; rachis 3-angled, minutely prickle-toothed on angles. Spikelets *c.* 3 mm, in pairs, light green, scarcely overlapping, lanceolate, acuminate. Lower glume 0; upper ≈ spikelet, 5–7-nerved, membranous, acuminate, with soft hairs on outer internerves and margins. Lower floret: lemma similar to upper glume, 7-nerved, nerves glabrous. Upper floret: lemma slightly < spikelet, glabrous, shining, greyish, acuminate; palea ≈ lemma, similar in texture and shape; anthers 0.4–0.5 mm, deep purple; stigmas deep purple; caryopsis *c.* 1.5 × 0.6 mm, oblong.

Naturalised from South America.

N.: North Auckland (Te Aupouri Peninsula), South Auckland (Rotorua). Coastal and thermal areas.

2. D. ciliaris (Retz.) Koeler, *Descr. Gram.* 27 (1802). summer grass

Sprawling annual tufts, rooting at nodes to form loose mats. Leaf-sheath folded, striate, stiff, light green with numerous, spreading to retrorse, fine tubercle-based hairs. Ligule 1–2 mm, membranous, glabrous, truncate to rounded, erose. Leaf-blade 3–6 cm × (3)–3.5–5–(6) mm, soft, linear, usually with scattered to dense fine hairs, ribs usually scabrid especially above, sometimes smooth; margins scabrid, narrowed rather abruptly to acute tip. Culm (15)–20–40 cm, erect to decumbent, internodes glabrous. Racemes 2–9, (3)–6–8.5–(11.5) cm, slender, closely digitate to approximate, spreading; rachis 3-angled, winged, 0.4–0.8 mm wide, scabrid on angles and often near base; pedicels scabrid on angles with longer prickle-teeth. Spikelets 2.5–3 mm, in pairs, lanceolate, acute, light green or purplish, laterally hairy, hairs fine, acute-tipped. Lower glume (0.2)–0.3–0.5 mm, triangular, acute to obtuse, nerveless, glabrous, upper 1.5–2.2 mm, *c.* ⅔ length of spikelet, 3-nerved, margins long-hairy. Lower floret: lemma = spikelet, 3–5-nerved, nerves glabrous, outer internerves long-hairy. Upper floret: lemma ≈ spikelet, acute, yellowish or purplish; palea ≈ lemma, acute; anthers 0.8–1.3 mm, purple; stigmas purple; caryopsis *c.* 1.5–2 mm, oblong.

Naturalised from tropical and subtropical Asia.

N.: scattered localities; S.: one early record; K.: Raoul Id. Common to very common in waste and open places.

Pantropical; formerly known as *D. adscendens* Kunth, a later synonym.

One specimen from South Id CHR 5144, was collected by J. B. Armstrong from "Near ChCh — common" late in the 19th Century.

Sykes, W. R. *N.Z. DSIR Bull. 219:* 166 (1977), discussed early records of *D. sanguinalis* (as *Panicum sanguinale*) from Kermadec Is and concluded that most of these were based on *D. setigera*, and that *D. ciliaris* was later introduced to Kermadec Is.

3. D. ischaemum (Schreb.) Muhl., *Descr. Gram.* 131 (1817).
                                                smooth summer grass
Summer annuals, forming loose or compact dark purple-green tufts. Leaf-sheath submembranous, striate, glabrous or rarely with a few hairs on margins, light green, folded. Ligule 1–1.5–(2) mm, membranous,

glabrous, ± truncate, erose. Leaf-blade 3–6 cm × 3–5 mm, lanceolate, ± rounded at base, glabrous; margins minutely scabrid throughout with a few long hairs near ligule. Culm 7–20–(30) cm, slender, geniculate-ascending, spreading or prostrate, internodes glabrous. Racemes 2–4–(5), 2.5–10 cm, subdigitate, very slender, spreading at maturity; rachis flat, 0.5–1 mm wide, narrowly winged, sparsely long-hairy at base, margins finely scabrid; pedicels ± 3-angled, rarely flattened, sparsely, finely prickle-toothed. Spikelets *c.* 2 mm, close-set in pairs or triplets, shortly, unequally pedicelled, elliptic or oblong-elliptic, hairs minute, numerous, swollen-tipped. Lower glume 0, or a minute scarious rim, upper *c.* 2 mm, ≈ spikelet, submembranous, 3–(5)-nerved, internerves with bands of dense short fine hairs. Lower floret: lemma = spikelet, submembranous, 5-nerved, nerves glabrous, outer internerves sparsely to densely fine-hairy; palea minute. Upper floret: lemma = spikelet, dark chestnut-brown, glabrous, acute, margins wide, hyaline, enfolding palea; palea of similar texture to lemma, margins widened below to enfold caryopsis; anthers 0.4–0.7 mm; stigmas purple; caryopsis *c.* 1.5 × 0.8 mm, oblong.

Naturalised from Eurasia.

N.: Auckland (early records, but not seen since 1970), scattered localities southwards from Waikato and Bay of Plenty; S.: Marlborough, Otago (Dunedin, Alexandra, Clyde). Sandy or dry ground, pasture, roadsides, metal drives.

Webster (1984 op. cit. p. 187) noted that *D. ischaemum* is often confused with *D. violascens* especially when both invade habitats which are frequently trampled or cut.

4. D. sanguinalis (L.) Scop., *Fl. Carniol.* ed. 2, *1:* 52 (1772).

summer grass

Summer annuals, forming loose, dull green to purplish tufts, ± creeping and rooting at lower nodes. Leaf-sheath submembranous, striate, light green, with few to numerous tubercle-based hairs. Ligule 0.5–1 mm, membranous, glabrous, truncate, erose. Leaf-blade (1.5)–3–6.5 cm × 3–9 mm, soft, narrow-lanceolate, rounded at base, flat, glabrous, or with long fine hairs; margins minutely scabrid, rather abruptly narrowed to acute tip. Culm (15)–20–30–(115) cm, ascending to erect from prostrate base, nodes loosely hairy, internodes glabrous. Racemes

(2)–3–6–(8), (3.5)–5–15–(20) cm, slender, digitate or subdigitate, finally spreading; rachis 3-angled, winged, 0.7–1 mm wide, scabrid on angles; pedicels 3-angled, angles scarcely scabrid. Spikelets 2.5–3 mm, in pairs, ovate to oblong-elliptic, subacute, green to purplish, close-set, unequally short-pedicelled. Lower glume 0.2–0.5 mm, a minute ± triangular rim, rarely with a few hairs, upper 1–1.5 mm, to ½ length of spikelet, lanceolate, acute, 3-nerved, membranous, with fine hairs near margins. Lower floret: lemma = spikelet, 7-nerved, nerves minutely scabrid, a band of fine hairs just inside margin; palea minute. Upper floret: lemma = spikelet, grey-brown, firm, glabrous, acute, margins hyaline, enfolding palea; palea similar to lemma in texture but slightly smaller; anthers 0.7–1 mm; stigmas purple or brown; caryopsis 1.5–2 mm, oblong.

Naturalised from Eurasia.

N.: common throughout; S.: scattered in Nelson, Marlborough, and Westland, more common in Canterbury, in Central Otago and at Oamaru and Dunedin; K. Cultivated ground (gardens, lawns, street berms), waste land.

Almost cosmopolitan and weedy.

Plants of *D. sanguinalis* in which the nerves on the lemma of the lower floret are very minutely scabrid or almost smooth, are difficult to distinguish from *D. ciliaris*. In *D. sanguinalis* the upper glume is usually < 1.5 mm, but in *D. ciliaris* it is 1.5–2.2 mm.

Field, T. R. O. and Forde, M. B. *Proc. N.Z. Grasslands Assoc. 51*: 47–50 (1990) reported that summer grass had shown a phenomenal recent increase in many places in North Id. Very aggressive and difficult to eradicate because of early development of many, very strong roots (A. E. Esler pers. comm.).

5. D. setigera Roem. et Schult., *Syst. Veg. 2*: 474 (1817).

Annuals, with stems rooting at lower nodes, sprawling to form loose mats. Leaf-sheath folded, strongly ribbed, light green, with several long fine spreading tubercle-based hairs. Ligule (1)–1.5–2–(2.5) mm, membranous, glabrous, truncate, erose. Leaf-blade (3.5)–5.5–10 cm × 2–7 mm, soft, linear, usually with scattered fine hairs, sometimes glabrous, ribs and margins scabrid, tapering to long, finely pointed tip. Culm

(10)–20–45 cm, ascending-erect or straggling, internodes glabrous. Racemes 3–5, (3.5)–5–9.5 cm, slender, digitate (on Kermadec Is plants normally remaining bunched together) rarely spreading; rachis 3-angled, winged, 0.3–0.8 mm wide, scabrid on wing margins and sometimes on central ridge; pedicels 3-angled, scabrid on angles, short-hairy at base. Spikelets 2.5–3.5 mm, in pairs, lanceolate, long-acute, laterally with fine acute-tipped hairs. Lower glume 0, or a minute rim or scale *c.* 0.1 mm, upper 0.6–1–(1.5) mm, *c.* ¼–(⅓) length of spikelet, 0–3-nerved, triangular, obtuse to subacute, glabrous, or with long hairs on margins overtopping glume. Lower floret: lemma = spikelet, (5)–7-nerved, nerves inaequidistant, glabrous, outer internerves with long hairs. Upper floret: lemma ≈ spikelet, glabrous, acuminate, yellowish or light brownish at maturity; palea ≈ lemma, similar in texture and shape; anthers 1–1.3 mm; stigmas brown; caryopsis *c.* 1.5 mm, oblong.

Naturalised.

K. Waste places, bird colony areas, soil pockets among boulders at cliff bases.

A variable species indigenous to and widespread in tropical Asia, Malesia, Pacific to French Polynesia and northern Australia; formerly widely referred to in Pacific treatments as *D. pruriens* (Trin.) Büse. Plants with very long racemes (11–22 cm) were also referred to *D. microbachne* (C.Presl) Henrard.

A raceme from a specimen at US, collected at Bay of Islands during the U.S. Exploring Expedition 1838–42 and determined by Dr Agnes Chase as *Digitaria microbachne* was sent by Dr Chase to Mr V. D. Zotov (CHR 97607). It agrees well with Kermadec Is specimens of *D. setigera.* The envelope containing the raceme is labelled "Z New Zealand" in Dr Chase's hand. No other specimens are known from the main islands of N.Z.

6. D. violascens Link, *Hort. Berol. 1*: 229 (1827).

Annual leafy tufts, sometimes rooting at nodes and forming large mats. Leaf-sheath submembranous, striate, light green or suffused purple, glabrous, or occasionally with very fine hairs mostly near margin. Ligule 1–2 mm, truncate, erose, glabrous. Leaf-blade 3.5–7.5 cm × 3–6 mm, soft, almost entirely glabrous except for occasional long hairs on margins near ligule, midrib obvious, lateral ribs numerous, very fine; margins minutely prickle-toothed, abruptly narrowed to acute minutely scabrid tip. Culm 15–25–(40) cm, internodes glabrous. Racemes (2)–3–4–(6), 3.5–6 cm, slender, closely digitate to ± approximate, spreading or curved at maturity; rachis flat, 0.7–1 mm wide, winged, margins closely prickle-toothed; pedicels subterete to angular, smooth to slightly scabrid. Spikelets 1.5–1.8 mm, in triplets, close-set, acute, elliptic, minute hairs acute-tipped. Lower glume 0, or rarely a minute scarious rim, upper 1–1.7 mm, slightly < spikelet, 3–5-nerved, acute, internerves with bands of dense hairs. Lower floret: lemma = spikelet, (3)–5–7-nerved, acute, nerves glabrous, outer nerves weak, outer internerves hairy. Upper floret: lemma ≈ spikelet, acute, dark chestnut-brown at maturity; palea ≈ lemma, dark brown; anthers 0.5–0.7 mm; stigmas deep purple; caryopsis 1–1.3 mm, oblong.

Naturalised.

N.: North Auckland (Pukenui, Whangaroa, Kawakawa) Auckland City and environs, Tauranga. Grassy roadsides. FL: late Jan–Apr.

Indigenous to tropical and subtropical America and Asia; naturalised in Australia.

ECHINOCHLOA P.Beauv., 1812 *nom. cons.*

Annual, or rarely perennial, sometimes very tall tufts, or straggling or decumbent, rooting at nodes, sometimes floating. Ligule 0, or a fringe of hairs, the ligule area always well defined. Leaf-blade broad, flat. Panicle of racemes densely or loosely arranged along a central rachis. Racemes bearing densely packed, secund spikelets almost to base; rachis hispid, 3-angled; spikelets ± sessile or pedicelled, clustered in pairs or in short secondary racemelets, falling entire at maturity. Spikelets often hispid, often cuspidate or awned, 2-flowered; lower floret ♂ or ∅, upper floret ♀. Glumes membranous to membranous-herbaceous, unequal; lower much shorter, ± broadly ovate from clasping base, 3–5-nerved, often mucronate, upper ≈ spikelet, very convex, 3–7-nerved, ± acute, mucronate, cuspidate, or shortly awned. Lower floret: lemma ≈ upper glume and similar in texture, rarely smooth and shining; palea = lemma, hyaline, 2-keeled, or reduced or 0 in ∅ florets. Upper floret: lemma subcoriaceous or crustaceous, faintly 5-nerved, awn straight or curved, or 0; palea ≈ lemma and similar in texture, keels rounded, margins broadened towards base; lodicules 2, cuneate, fleshy, glabrous; stamens 3; ovary apex glabrous, styles subterminal, free, stigmas pink, red, purple or black; caryopsis compressed, glabrous, embryo > ½ length of caryopsis, hilum punctiform, subbasal.

Type species: *E. crusgalli* (L.) P.Beauv.

c. 30 spp. of tropical and temperate regions in both Hemispheres. Naturalised spp. 4; transient spp. 2.

1 Panicle very compact; racemes close-set, erect; upper florets usually rotundate; caryopsis yellowish or brownish, very turgid .... **3. esculenta**
Panicle ± open, often stiff; racemes ± spreading and/or ± distant; upper florets elliptic to ovate; caryopsis greenish or whitish, not turgid *2*

2 Panicle erect, with branches erect and/or stiffly spreading; ligular area usually glabrous, rarely faintly pubescent *3*
Panicle drooping, or horizontal with racemes drooping to one side; ligular area with short hairs or diffusely minutely pubescent especially at centre, or glabrous *5*

3 Lemmas in lower ∅ florets awned or some awnless; upper glumes cuspidate and at least some distinctly awned .... **4. telmatophila**
Lemmas of lower ∅ florets awnless or a few with distinct awns; upper glumes acute to cuspidate, rarely slightly awned *4*

4 Upper floret narrowing abruptly to scabrid, deltoid, submembranous tip; palea apex soft, subobtuse .... **1. crus-galli**
Upper floret narrowing gradually to smooth, acute, stiff tip; palea apex hard, acute .... **microstachya**ζ (p. 550)

5 Panicle drooping; racemes flexuose .... **2. crus-pavonis**
Panicle ± horizontal; racemes secund .... **oryzoides**ζ(p. 550)

1. E. crus-galli (L.) P.Beauv., *Ess. Agrost.* 53, 161, t. 11, fig. 2 (1812).
                                                                        barnyard grass

Strong, tufted, bright green, rarely glaucous, summer annuals, with stout culms, usually ± prostrate at first, particularly in open situations, becoming erect at flowering. Leaf-sheath light green or purplish, chartaceous, keeled, glabrous, or occasionally with soft tubercle-based hairs just inside margin and scattered elsewhere. Ligule 0; ligular area glabrous, rarely faintly pubescent. Leaf-blade 15–50 cm × 5–15 mm, soft, flat, linear, glabrous, or sometimes with scattered soft, tubercle-based hairs, midrib distinct; margins and sometimes midrib scabrid, tapering to acute tip. Culm (20)–30–130 cm, slender to stout, often branching and compressed below, internodes glabrous. Panicle (6)– 10–20 cm, erect, ± open, very variable in shape and in presence or absence of awned florets, with few to many spike-like racemes borne singly, or in pairs or clusters at nodes; rachis 3-angled, scabrid. Racemes (1)–2.5–6 cm, erect or spreading, distant or approximate, but not extremely close-set, bearing spikelets crowded in clusters of 2 or 3; rachis 3-angled, scabrid, hairs few to many, bristle-like tubercle-based hairs clustered at base; pedicels scabrid, 0.5–1 mm. Spikelets 3–5 mm, greenish or often purplish, ovate-elliptic, acute to acuminate. Glumes quite unequal, submembranous, acute to cuspidate, scabrid; lower 1.2–1.7 mm, 3–5-nerved, broadly ovate, enwrapping base of spikelet, upper 3–4 mm, 5-nerved, very convex, with bristle-like, tubercle-based hairs on nerves, very rarely slightly awned. Lower floret: lemma similar to upper glume, 5–7-nerved, flat or dorsally depressed with central nerves often obscure, nerves, especially lateral nerves, with tubercle-based spinules, apex acute, cuspidate or distinctly awned (2–25 mm) in some florets; palea < lemma, hyaline, broad, keels finely ciliate. Upper floret: lemma 2.7–3.5 mm, ovate-elliptic, very convex, whitish or yellowish, glabrous, shining, thinly crustaceous, cuspidate, apex greenish, scabrid, submembranous, deltoid; palea ≈ lemma, apex subobtuse, soft; anthers 0.5–0.7 mm, brownish yellow or greyish; caryopsis 1.2–2 × 1.4–1.8 mm, orbicular, greenish or whitish, not turgid.

Naturalised from Europe.

N.: throughout; S.: Nelson, Marlborough, Canterbury, rare in Westland and Otago; K., Ch. Waste land, cultivated land, often coastal and sometimes in swampy ground.

Troublesome, mainly because of its large bulk produced rapidly in late spring. Now a widespread weed in many temperate regions in both Hemispheres.

2. E. crus-pavonis (Kunth) Schult., *Mant. 2*: 269 (1824).

Gulf barnyard grass

Stout, summer annuals, to *c.* 100 cm, in bright green tufts with drooping panicles, stems either erect, or decumbent and rooting from lower nodes; branching extravaginal. Leaf-sheath bright purple, chartaceous, keeled above, glabrous. Ligule 0; ligular area often with scattered to more dense short hairs especially towards centre. Leaf-blade 10–35 cm × 6–11 mm, flat, linear, glabrous apart from a few occasional tubercle-based hairs on margins near ligular area, abaxially with distinct whitish midrib; margins whitish, ± thickened, obviously prickle-toothed, long-tapering to fine-acuminate tip. Culm (15)–50–80 cm, stout, internodes glabrous. Panicle (8)–12–20 cm, soft and drooping, narrow-pyramidal, with numerous flexuose racemes densely covered by spikelets for much of their length, florets usually awned; rachis 3-angled, finely scabrid. Racemes (2)–4–6.5 cm, spreading, clustered closely, especially above, often more distant below, bearing spikelets in clusters along rachis; pedicels scabrid, < 1 mm. Spikelets light green, sometimes purple-suffused, ovate-elliptic. Glumes quite unequal, submembranous, scabrid-pubescent, with stiff spinules on nerves, acute to acuminate, cuspidate or shortly awned; lower 1.3–1.8 mm, 3–5-nerved, broadly ovate, enwrapping base of spikelet, upper 3–4 mm, ovate-elliptic, 5-nerved. Lower floret: lemma similar to upper glume in size, texture and scabridity, 7-nerved, usually awned, awn 2.5–15 mm; palea ≈ lemma, elliptic, acute. Upper floret: lemma 2.5–3.5 mm, faintly 5-nerved, glabrous, shining, ovate-elliptic, thinly crustaceous, tapering to an almost membranous, faintly scabrid cusp; palea ≈ lemma; anthers 0.5–1 mm; caryopsis *c.* 1.5 × 1 mm, orbicular-oblong.

Naturalised from South America.

N.: Auckland City and environs, scattered elsewhere in North and South Auckland, and in Hawke's Bay (Hastings); S.: Canterbury (Christchurch and environs, Ashburton). Uncommon, always on disturbed ground; roadsides, waste land.

Habitat as for *E. crus-galli* and usually growing with it, but less plentiful.

3. E. esculenta (A.Braun) H.Scholz, *Taxon 41*: 523 (1992).

Japanese millet

Robust, erect, light green annuals. Leaf-sheath light green to pale creamy brown, chartaceous, glabrous, rather loosely enfolding culm, slightly keeled above, finely striate. Ligule 0; ligular area glabrous. Leaf-blade (13)–17–30 cm × 10–16 mm, chartaceous, flat, linear, smooth, or adaxially rarely slightly scabrid on primary lateral ribs, midrib whitish, very distinct; margins whitish, slightly thickened, finely scabrid, tapering to subacute tip. Culm (35)–40–60 cm, stout, internodes glabrous. Panicle 6.5–12 cm, erect, very dense, usually with numerous close-set, sessile, often subverticillate racemes; rachis angular, ridged, scabrid, with a dense ring of bristle-like hairs at base and at nodes. Racemes 1.5–3–(4) cm, of many small dense clusters of spikelets; pedicels finely scabrid, < 1 mm. Spikelets 3–4 mm, purplish, sometimes whitish green, broadly ovate to subglobose, subacute. Glumes quite unequal, firmly membranous, 5-nerved, nerves scabrid, internerves usually finely scabrid-pubescent; lower 1–1.5 mm, enwrapping base of spikelet, upper 2.5–3.5 mm, ≈ spikelet, reflexed at maturity exposing upper part of ripening ♀ floret. Lower floret: lemma similar to upper glume, *c.* 3 mm, 7-nerved, rotund-ovate, shortly acuminate or shortly cuspidate, nerves scabrid to hispid, internerves minutely scabrid-pubescent; palea < lemma, hyaline, keels minutely scabrid near apex. Upper floret: lemma *c.* 3 mm, broadly elliptic to rotundate, very convex, crustaceous, obscurely 5-nerved, glabrous, shining, with a minute herbaceous cusp; palea *c.* 2.5 mm; anthers 0.8–1 mm, yellowish brown to blackish; caryopsis *c.* 1.5 × 1.5 mm, orbicular, very turgid, yellowish or brownish.

Naturalised.

N.: scattered throughout; S.: Nelson City, Blenheim, Christchurch; K. Stony waste land, coastal sands, roadsides (mostly from seed spillages), crops.

Japanese millet is considered to have originated in eastern Asia and Japan; it has often been confused with *E. frumentacea* of India.

Ohwi, J. and Yabuno, T. *in* Ohwi, J. *Acta Phytotax. Geobot. 20*: 50 (1962) distinguished Japanese millet from *E. frumentacea* calling it *E. utilis*, but Scholz, H. *Taxon 41*: 522–523 (1992), observing that Ohwi and Yabuno had overlooked an earlier synonym *Panicum esculentum* A.Braun (1861), made the combination *E. esculentum.*

4. E. telmatophila P.W.Michael et Vickery, *Telopea 1*: 44 (1975).

Erect annual, to 60 cm. Leaf-sheath green to purplish, chartaceous, keeled above, glabrous. Ligule 0; ligular area glabrous. Leaf-blade to 20 cm × 7 mm, flat, linear, glabrous, midrib distinct; margins scabrid, tapering to acute tip. Culm to 50 cm, sometimes branching below, internodes glabrous. Panicle to 14 cm, rather dense, erect, with spike-like racemes borne singly at lower nodes, shorter and more clustered above, florets strongly awned; rachis 3-angled, scabrid, long tubercle-based hairs at base of each raceme. Racemes 0.5–3 cm, erect, the lowest 1–2 distant, the rest approximate, bearing crowded spikelets; rachis 3-angled, very finely scabrid, with a few, long, tubercle-based hairs; pedicels very short, finely scabrid. Spikelets greenish to purple-suffused. Glumes quite unequal, submembranous, scabrid; lower *c.* 1 mm, 3–5-nerved, ovate, enwrapping base of spikelet, upper 3–4 mm, convex, 5-nerved with some bristle-like hairs on nerves, acuminate, cuspidate, or sometimes awned, awn to 3 mm. Lower floret: lemma similar to upper glume, 5–7-nerved, flat, nerves with tubercle-based spinules, almost always awned, awn 10–40 mm, purplish; palea < lemma, hyaline, keels ciliate. Upper floret: lemma to 3 mm, faintly 5-nerved, elliptic, convex, yellowish white, glabrous, shining, thinly crustaceous, shortly cuspidate; palea ≈ lemma and similar in texture, apex subobtuse; anthers *c.* 0.6 mm, brownish; caryopsis *c.* 2 × 1.3 mm, oblong, cream.

Naturalised from Australia.

N.: North and South Auckland (Port Albert, Mauku and Waikato R. near Wellington Beach). Coastal on damp sand flats on river bank; or in roadside gutters.

ζ**E. microstachya** (Wiegand) Rydb.   Tufts *c.* 30 cm, erect. Ligular area glabrous. Panicles stiff, not exserted from uppermost leaves. Spikelets *c.* 3 mm. Glumes cuspidate. Lower floret: lemma very shortly awned. Upper floret: lemma ovate-elliptic, gradually tapering to glabrous, acute, stiff apex; palea apex hard, acute. A North American species collected once at Nelson in waste land, *A. J. Healy 71/59* 25.4.1971, "... In damp hollows: heads strictly erect and fruit falling before panicles open" (CHR 225004).

ζ**E. oryzoides** (Ard.) Fritsch   One specimen (CHR 21099) consists of a slender plant *c.* 30 cm. Ligular area glabrous. Panicles 2.5–6 cm, slender; racemes secund, short, *c.* 1 cm. Spikelets *c.* 3.5 mm, pale cream. Upper glumes cuspidate. Lower floret: lemma awned; awns 2–4 mm. Upper floret: lemma ovate-elliptic; caryopses immature. Another specimen (CHR 22597), consisting of the upper parts of culms, is similar, but has larger panicles to 10 cm, bearing close-set racemes *c.* 2 cm; spikelets 4 mm; awns 10–20 mm; caryopses *c.* 2.5 × 2 mm. This Eurasian species has been twice collected at Tauranga *M. Hodgkins* 5.1938 (CHR 21099) and *M. Hodgkins* April 1939 (CHR 22597) but there are no later records.

ENTOLASIA Stapf, 1920

Long-rhizomatous to tufted perennials. Ligule ± conspicuously hair-fringed. Leaf-blade usually disarticulating at maturity. Inflorescence spike-like or paniculate with sessile racemes; florets awnless. Racemes spike-like; rachis flattened, hollowed or winged, with close-set, secund, pedicelled spikelets, borne singly on the rachis or in appressed secondary racemelets, falling entire at maturity. Spikelets usually laterally compressed, 2-flowered; lower floret ∅, upper floret ♀. Glumes very dissimilar; lower minute, hyaline, 0–3-nerved, upper membranous, = spikelet, 3–7-nerved. Lower floret: lemma ≈ or ≫ lemma of upper floret, both lemmas either similar in texture to glumes, or firmer; palea 0. Upper floret: lemma conspicuously hairy, 3–5-nerved; palea ≤ lemma, nerves 2, well-separated; lodicules 2, fleshy, glabrous, cuneate; stamens 3; ovary apex glabrous, styles subterminal, free; caryopsis dorsiventrally compressed, glabrous, embryo large, at least ⅓ length of caryopsis, hilum punctiform or shortly elliptic, < ½ length of caryopsis.

Type species: *E. olivacea* Stapf

c. 5 spp. of tropical Africa and eastern Australia. Naturalised sp. 1.

1. E. marginata (R.Br.) Hughes, *Kew Bull. for 1923*: 331 (1923).

bordered panic grass
Scrambling, shortly rhizomatous, variable in size, with ± trailing culms often rooting at lower nodes, branched near base; branching intravaginal. Leaf-sheath light green, becoming pale dull brown, closely enwrapping lower half of internode, later loose, submembranous, striate, with fine tubercle-based hairs, or glabrous with a fringe of tubercle-based hairs just inside margin. Ligule composed of hairs 0.3–1 mm, hairs extending as a contra-ligule. Collar hairs few, short. Leaf-blade 2.5–8–(15) cm × 3–8 mm, firm to rather rigid, linear-lanceolate, acuminate, truncate to ± rounded at base, flat, glabrous or with a few scattered tubercle-based hairs, adaxially minutely scabrid on ribs, shortly pubescent above ligule; margins somewhat thickened, minutely scabrid. Culm 20–85 cm, slender and wiry, nodes usually pubescent, internodes smooth, to finely scabrid above, with scattered, fine, tubercle-based hairs below panicle. Panicle (2)–3–10–(16) cm, ± erect, linear or lanceolate, with short,

appressed to slightly spreading branches covered with spikelets to base; rachis ± planoconvex, occasionally with a few tubercle-based hairs near base; branches bearing a few short racemes or reduced to a single raceme, axils pubescent. Racemes 1–4 cm; rachis flattened, angled, finely scabrid. Spikelets *c.* 2.5–3 mm, pale green, elliptic-lanceolate, acute to subacuminate. Lower glume 0.7–1.2 mm, upper glume = spikelet. Lower floret: lemma = spikelet, 5-nerved, ≫ lemma of upper floret and curved in above towards acute tip. Upper floret: lemma 1.7–2.3 mm, coriaceous, pubescent with fine white hairs; palea similarly pubescent as lemma; anthers 0.5–1 mm; caryopsis *c.* 1 × 0.8 mm, elliptic-oblong.

Naturalised from Australia.

N.: North and South Auckland, and Wellington (Tawa). Open ground in pine plantations, forest margins, pasture, logging tracks.

Once collected in South Id from Banks Peninsula *J. B. Armstrong* 1877 (CHBG 5452).

Frequently growing with the support of taller grasses and rushes; aggressive and troublesome in pastures at Warkworth (and possibly elsewhere); usually overlooked (A. E. Esler pers. comm.).

OPLISMENUS P.Beauv., 1810 *nom. cons.*

Perennials or annuals, ± weak, trailing, often carpet-forming, decumbent and rooting at lower nodes with slender ascending leafy culms. Leaf-sheath < internodes, keeled above, often with minute cross-veinlets, open; margins often hair-fringed, overlapping. Ligule ciliate. Leaf-blade thin, flat, ovate to lanceolate, usually variously pubescent; margins finely scabrid. Inflorescence a racemose panicle with spikelets solitary, or in pairs or racemelets, falling entire at maturity; rachis, branches and pedicels slender, angular. Spikelets 2-flowered; lower floret ♂ or ∅, upper floret ♀. Glumes ± equal, < florets, membranous, 3–7-nerved, ± keeled above, both, or at least the lower, apically awned. Lower floret: lemma similar to glumes in texture, 5–9-nerved, mucronate to shortly awned; palea = lemma, or reduced, or 0. Upper floret: lemma chartaceous to subcoriaceous, oblong, glabrous, margins enfolding palea, awnless; palea = lemma; lodicules 2, broadly cuneate;

stamens 3; ovary apex glabrous, styles terminal to subterminal, free to base; caryopsis tightly enclosed by anthoecium, embryo large, hilum linear, variable in length. Fig. 22.

Type species: *O. africanus* P.Beauv.

c. 7 tropical to subtropical spp. extending into warm temperate regions in both Hemispheres; usually found on shaded forest floors. The 1 sp. indigenous to N.Z. is also widespread in the tropics, South Pacific and Australia.

Oplismenus was reviewed by Davey, J. C. and Clayton, W. D. *Kew Bull. 33*: 147–157 (1978); and monographed by Scholz, U. *Phan. Monogr. 13*: 1–213 (1981).

Cheeseman (1925 op. cit.) referred N.Z. plants to *O. undulatifolius* P.Beauv., but this is a conspicuously long-awned sp. with spikelets clustered in fascicles, occurring in warm temperate regions of the Northern Hemisphere and extending southwards only into Africa and India. Under Davey and Clayton's treatment (1978 op. cit.) N.Z. mainland plants would fall into *O. hirtellus* (L.) P.Beauv.; their "hirtellus group" comprised *O. hirtellus, O. aemulus,* and *O. imbecillis.* Scholz (1981, op. cit.) referred N.Z. material to *O. aemulus* var. *flaccidus* (R.Br.) Domin and treated Kermadec Is plants as *O. hirtellus* subsp. *imbecillis* (R.Br.) U.Scholz forma *imbecillis* whereas Fosberg, F. R. and Sachet, M. H. *Micronesica 18*: 45–102 (1982) preferred varietal rank for the latter as *O. hirtellus* var. *imbecillis* (R.Br.) Fosberg et Sachet.

Sykes, W. R. and West, C. J. *N.Z. J. Bot. 34*: 447–462 (1996) recognised two taxa on the Kermadec Is, *O. hirtellus* subsp. *hirtellus* and *O. hirtellus* subsp. *imbecillis*; both subspecies may also be present in northern North Id, but only subsp. *imbecillis* occurs further south. Their treatment is followed here.

1. O. hirtellus (L.) P.Beauv., *Ess. Agrost.* 54, 168, 170 (1812) ≡ *Panicum hirtellum* L., *Syst. Nat.* ed. 10, *2*: 870 (1759) ≡ *Oplismenus hirtellus* (L.) P.Beauv. var. *hirtellus* (autonym, Fosberg and Sachet 1982 op. cit. p. 78); Holotype: LINN *n.v., Browne 133* Jamaica (*fide* Davey and Clayton 1978 op. cit. p. 156).

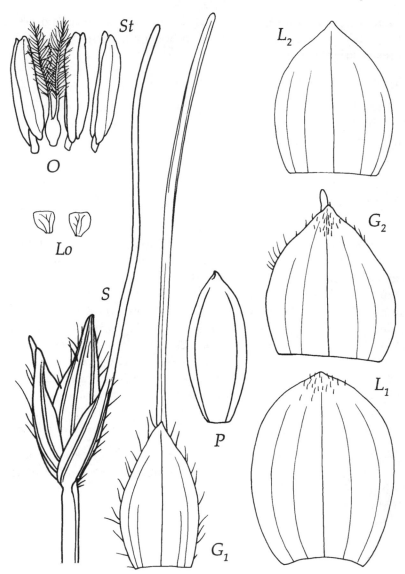

Fig. 22 *Oplismenus hirtellus* subsp. *imbecillis*

S, spikelet, × 12; **G₁**, **G₂**, glumes, × 24; **L₁**, lemma of lower Ø floret, × 24; **L₂**, lemma of upper ♂ floret, × 24; **P**, palea dorsal view, × 24; **Lo**, lodicules, × 24; **O**, ovary, styles and stigmas, × 24; **St**, stamens, × 24.

Perennials with decumbent stems, forming a loose mat; branching intravaginal. Leaf-sheath light green to purplish, membranous, striate, often with scattered short hairs or long tubercle-based hairs, margins often darker purplish brown, densely short-ciliate or hairs scattered, short. Ligule 0.6–1.5 mm, truncate, erose, ciliate, or hairs few, short centrally and longer near margins. Leaf-blade with short, fine, evenly spaced hairs throughout or only on one surface, and often with scattered, long, tubercle-based hairs, rarely glabrous, gradually narrowed to acute tip; midrib obscure in upper ½, variously pubescent or glabrous. Culm nodes pubescent, internodes finely ridged, minutely scabrid on ridges below panicle. Panicle raceme-like; rachis bearing ± distant to approximate small clusters of spikelets and sometimes racemelets below and paired to solitary spikelets above; racemelets composed of a few contiguous spikelet-pairs usually < 3 mm apart; rachis and branches finely scabrid on angles, long tubercle-based hairs at spikelet axils or on pedicels. Spikelets light green to purple above, awned; awns short, straight, glabrous. Glumes 1.5–2.5 mm, membranous, often with scattered tubercle-based hairs, often ciliate above and at base; lower 3–5-nerved, always awned (awn (1.5)–3–9 mm), upper (3)–5–(7)-nerved, acute, or mucronate to awned. Lower floret ∅: lemma 2–3 mm, 5–7–(9)-nerved, = spikelet, shortly awned, shortly mucronate or acute; palea ½ length of lemma or 0. Upper floret ♀: lemma and palea chartaceous, ≈ spikelet; anthers *c.* 1–2 mm; gynoecium: ovary 0.5 mm, stigma-styles 2–3.5 mm; caryopsis *c.* 1.5 mm.

1 Leaf-blade (1.5)–4–10 cm × (2)–4–12 mm; upper glumes awned, awn 1–1.5–(2) mm .... subsp. **hirtellus**
 Leaf-blade (0.8)–1.5–4 cm × 2–6 mm; upper glume acute or ± mucronate, sometimes with awn < 1 mm .... subsp. **imbecillis**

subsp. **hirtellus**

Leaf-blade (1.5)–4–10 cm × (2)–4–12 mm, ovate-lanceolate. Panicle (6)–8–17 cm; rachis of racemelets and of spikelet-clusters 5–15–(30) mm. Spikelets 4–12 in lower racemelets and clusters, paired or solitary near apex of rachis. Awn of upper glume 1–1.5–(2) mm.

Indigenous.

N.: scattered localities from North Cape to Auckland; K.: Raoul Id, Meyer Islets. Shaded slopes and coastal forest.

Also indigenous to tropical America and Asia, South Pacific and Australia.

subsp. **imbecillis** (R.Br.) U.Scholz, *Phan. Monogr. 13*: 127 (1981) ≡ *Orthopogon imbecillis* R.Br., *Prodr.* 194 (1810) ≡ *Oplismenus imbecillis* (R.Br.) Roem. et Schult., *Syst. Veg. 2*: 487 (1817) ≡ *Panicum imbecille* (R.Br.) Trin., *Gram. Icon. 2*: t. 141 (1829) ≡ *Oplismenus hirtellus* var. *imbecillis* (R.Br.) Fosberg et Sachet, *Micronesica 18*: 78 (1982) ≡ *O. hirtellus* subsp. *imbecillis* (R.Br.) U.Scholz forma *imbecillis, Phan. Monogr. 13*: 127 (1981); Holotype: BM! *Brown 6133* Australia, Port Jackson [*fide* Green, P. S. *Fl. Australia 49*: 480 (1994)].

Leaf-blade (0.8)–1.5–4 cm × 2–6 mm, narrow elliptic to ovate-lanceolate. Panicle (1)–2–9 cm; rachis of racemelets and longer spikelet clusters usually 1–4 mm, rarely longer. Spikelets up to 4 in small clusters and racemelets below, paired to solitary above or throughout rachis. Upper glume acute, mucronate, or with very short awn < 1 mm. $2n = 54$. Fig. 22.

Indigenous.

N.: throughout; S.: Nelson (near Farewell Spit, D'Urville Id), Canterbury (one 19th Century collection); K., Three Kings Is. Open forest and forest clearings, second growth bush and coastal scrub.

Also indigenous to South Pacific and Australia.

Once collected in Canterbury in late 19th Century (Banks Peninsula *J. B. Armstrong* CHR 5786 and CHBG 5451).

On Kermadec Is *Oplismenus hirtellus* subsp. *hirtellus* flowers mainly in winter and spring, whereas subsp. *imbecillis* flowers mainly in summer (Sykes and West 1996 op. cit. p. 457).

Oplismenus compositus (L.) P.Beauv. was recorded as naturalised on Kermadec Is [Cheeseman, T. F. *T.N.Z.I. 20*: 151–181 (1888); and also Sykes, W. R. *N.Z. DSIR Bull. 219*: 172 (1977), who at that time regarded both *O. compositus* and *O. imbecillis* as naturalised there]. Edgar, E. and Shand, J. E. *N.Z. J. Bot. 25*: 343–353 (1987) cited *O. compositus* in their checklist as native to Kermadec Is. All these records of *O. compositus* are now regarded as misidentifications for *O. hirtellus* (Sykes and West 1996 op. cit.).

Hekaterosachne elatior Steud., *Syn. Pl. Glum. 1*: 118 (1854) from "N. Zeeland", was referred to *Oplismenus* by Buchanan, J. *Indig. Grasses N.Z.* t. 11 (1878) and by Cheeseman (1906 op. cit. p. 849; 1925 op. cit. p. 141). Scholz (1981 op. cit. p. 185) considered that the original description of *Hekaterosachne elatior* was not that of a species of *Oplismenus* but Clayton and Renvoize (1986 op. cit. p. 127) equate it with *O. hirtellus*. No specimens have been seen.

PANICUM L., 1753

Perennials, or annuals, of various habit. Leaf-sheath open, or closed below. Ligule usually a membranous ciliate rim. Leaf-blade usually flat, rolled in bud. Culm simple or branched. Inflorescence a panicle of variable form, often spreading and much-branched, with fine branches tipped by single pedicelled spikelets falling entire at maturity. Spikelets lanceolate to oblong, elliptic or orbicular, usually ± dorsally compressed, 2-flowered, florets awnless; lower floret ∅, or ♂ and then with distinct palea; upper floret ♀. Glumes ± herbaceous to membranous, usually very unequal; lower usually shorter, 1–7-nerved, upper usually = spikelet, 3–9-nerved. Lower floret: lemma similar to upper glume in size and texture, (3)–5–9–(11)-nerved; palea subhyaline, ≤ lemma, or 0. Upper floret: lemma subcoriaceous, usually glabrous, obtuse to subacute, rarely minutely apiculate, only faintly nerved, margins inrolled over palea; palea < lemma and similar in texture, lower margins incurved over flower; lodicules 2, broadly cuneate; stamens 3; ovary apex glabrous, styles subterminal, free; caryopsis tightly enclosed by anthoecium, dorsally compressed, biconvex to almost planoconvex; embryo large, at least ⅓ length of caryopsis; hilum punctiform or shortly elliptic, < ½ length of caryopsis.

Type species: *P. miliaceum* L.

Possibly over 500 spp. of tropical, subtropical and warm temperate regions. Naturalised spp. 7; transient sp. 1.

1 Leaf-sheaths ± densely villous *2*
   Leaf-sheaths glabrous, sometimes ciliate on margin *5*
2 Spikelets 2.3–3 mm *3*
   Spikelets < 2 mm or > 4 mm *4*

3 Spikelets glabrous except at tip; lower floret Ø; annual .... **1. capillare**
　　Spikelets pubescent throughout; lower floret ♂; perennial ....
　　.. **4. maximum** var. **trichoglume**

4 Panicle (10)–20–45 cm; spikelets > 4 mm, pointed, glabrous ...... **5. miliaceum**
　　Panicle 2.5–8 cm; spikelets < 2 mm, obtuse, pubescent .... **3. huachucae**

5 Spikelets glabrous, 2.2–3.3 mm 6
　　Spikelets pubescent or puberulent, < 2.0 mm 7

6 Spikelets acute; lower floret Ø .... **2. dichotomiflorum**
　　Spikelets obtuse; lower floret ♂ .... **6. schinzii**

7 Ligule < 1 mm; culm nodes appressed-pubescent .... **7. sphaerocarpon**
　　Ligule 3–5 mm; culm nodes glabrous .... **lindheimeri**ζ (p. 564)

1. P. capillare L., *Sp. Pl.* 58 (1753).　　　　　　　　　　　witchgrass

Tufted summer annuals. Leaf-sheath light green to purplish, rounded, densely villous, hairs fine, soft, tubercle-based. Ligule 1–2 mm, a truncate membranous ciliate rim. Leaf-blade 7.5–15–(30) cm × 3–9–(12) mm, rounded at base, gradually narrowed above, light green, rather thin, usually villous with scattered, fine, tubercle-based hairs, midrib obvious; margins finely scabrid, often undulate, tip acuminate. Culm (7.5)–15–30–(45) cm, erect, slightly compressed, nodes densely pubescent, internodes villous with long tubercle-based hairs, or glabrous. Panicle 15–40 × 4–15 cm, often ½ length of plant, very diffuse, with fine branches at first ascending, later spreading; rachis and branches with scattered hairs, finer branchlets and pedicels scabrid; whole panicle detaching at maturity and blown by the wind. Spikelets 2.3–3 mm, glabrous except at tip, light green or purplish, elliptic, narrowed to acute or acuminate tip. Lower glume < ½ length of spikelet, ovate-triangular, 5-nerved, acute, tipped by minute hairs, upper = spikelet, 7–9-nerved, elliptic, tapered above to acute, very minutely ciliate apex. Lower floret Ø: lemma very slightly < upper glume, 5-nerved, elliptic, dorsally flattened, laterally incurved, tapered above to acute, very minutely ciliate apex; palea 0. Upper floret ♀: lemma 1.5–2 mm, elliptic, subacute, faintly striolate, glabrous, shining, light cream; palea narrow; anthers 0.8–1.3 mm; caryopsis *c.* 1.5 mm.

Naturalised from North America.

N.: scattered throughout, mainly in Waikato and Bay of Plenty; S.: scattered in Nelson, Marlborough, Canterbury, and Otago (Earnscleugh). Weed of pastures, orchards, gardens, roadsides and crops.

2. P. dichotomiflorum Michx., *Fl. Bor.-Amer. 1*: 48 (1803).

smooth witchgrass

Almost glabrous, tufted summer annuals. Leaf-sheath light green to commonly purplish, submembranous, flattened, rounded, glabrous. Ligule 1–2 mm, a membranous densely ciliate rim. Leaf-blade 7–30 cm × 3–5.5 mm, ± rounded at base, narrowed above, dull green to purplish, glabrous, abaxially slightly keeled, adaxially channelled, rarely with scattered, long, fine hairs; margins finely scabrid, tip acuminate. Culm 20–60 cm, often geniculate at base, almost prostrate and rooting at lower nodes, or ascending or ± erect, internodes glabrous or slightly hairy near nodes. Panicle 8.5–35.5 × 1.5–7–(15) cm, usually not fully emergent from leaf-sheath; branches fine, smooth, ascending to later spreading, branchlets and pedicels shorter, finer, scabrid, tipped by ± close-set spikelets. Spikelets 2.8–3.3 mm, glabrous, light green to purplish, narrow elliptic-oblong, much narrowed to acute tip. Glumes very unequal; lower *c.* 1 mm, ovate, sometimes slightly notched at broad acute tip, 3-nerved, upper = spikelet, (7)–9-nerved, elliptic, tapering above to acute tip. Lower floret ∅: lemma slightly < upper glume, (7)– 9-nerved; palea shorter and much narrower than lemma, 2-keeled, keels glabrous. Upper floret ♀: lemma *c.* 2 mm, elliptic, subacute, faintly striolate, glabrous, shining, light cream; palea narrower; anthers 1–1.5 mm; caryopsis *c.* 1.5 × 0.8 mm.

Naturalised from North America.

N.: North and South Auckland (common in Waikato), occasional further south. Weed of crops (especially maize, onions), and disturbed soil in orchards, gardens, roadsides, waste places and pasture.

3. P. huachucae Ashe, *J. Elisha Mitchell Sci. Soc. 15*: 51 (1898).

Small perennial tufts, basal leaves ≪ culm leaves, in winter only a basal tuft, in spring sending out simple leafy culms each topped by a delicate panicle, later stems branching freely above, with small panicles at first hidden among leaves. Leaf-sheath greenish or often purplish, stiffly membranous, rounded; margins with many long soft hairs. Ligule ciliate, hairs soft, 3–4.5 mm, extending as a contra-ligule. Leaf-blade of basal leaves 1–2.5 cm × 4–6 mm, ovate-lanceolate, acute, of cauline leaves 4–8.5 cm × 4–8 mm, linear-lanceolate, subacuminate, all leaves firm, erect or ascending, flat, strongly ribbed without obvious midrib, abaxially short-pubescent, adaxially sparsely finely pilose to almost glabrous, but usually hairy near base; margins minutely scabrid. Culm 30–45 cm, very slender near panicle, ± erect, nodes pubescent, internodes striate with scattered long fine hairs. Panicle lax, pyramidal to ovoid, 2.5–8 × 2–5 cm; rachis with scattered fine hairs, branches and branchlets few, slender, glabrous, bearing few spikelets on rather long pedicels. Spikelets 1.5–1.8 mm, pubescent, obtuse, green to purplish. Glumes very unequal; lower 0.5–1 mm, ovate, obtuse or almost truncate, nerves 0, upper = spikelet, 7-nerved, ovate, obtuse, with short scattered hairs. Lower floret ∅: lemma similar to upper glume, 7-nerved; palea 0. Upper floret ♀: lemma 1.2–1.4 mm, elliptic, subacute, faintly striolate, glabrous, shining, light cream; palea narrower; anthers 0.2–0.9 mm; caryopsis *c.* 1 × 1 mm.

Naturalised from North America.

N.: North Auckland (Kaitaia, Great Barrier Id), Auckland City, South Auckland. Open waste land, rough mown grassland, roadsides and tracks, margins of hot springs.

Some N.Z. specimens of *P. huachucae* were determined as the closely related *P. lindheimeri*, which differs in having glabrous leaf-sheaths and has only been collected once, at Kaitaia.

4. P. maximum var. **trichoglume** Robyns, *Mém. Inst. Roy. Col. Belge 116*: 31 (1932). green panic

Strong, light green perennial tufts to 120 cm, from very short rhizomes, with open fans of ascending to erect culms. Leaf-sheath firmly chartaceous, folded, keeled and ribbed, softly villous above, with longer tubercle-based hairs near margin. Collar hairs long, tubercle-based. Ligule *c.* 1 mm. Leaf-blade *c.* 25 cm × 5–8 mm, flat, linear, finely, minutely pubescent, abaxially strongly keeled, adaxially with central furrow; margins finely, closely scabrid, tapered in upper ⅓ to long fine tip. Culm to 100 cm, flattened below, with wide shallow furrow on one side. Panicle 15–25 cm, ± erect, very open, with spikelets borne towards tips of long, filiform to capillary branches. Spikelets *c.* 3 mm, elliptic-oblong, somewhat turgid, minutely pubescent, purplish. Glumes faintly nerved, thin; lower *c.* 1 mm, ⅓ length of spikelet, rounded, obtuse, hyaline, 1-nerved, upper = spikelet, elliptic, obtuse, 5-nerved. Lower floret ♂: lemma ≤ 3 mm, 5-nerved; palea slightly shorter; anthers 1.5–2 mm. Upper floret ♀: lemma ≥ 2 mm, elliptic-oblong, shortly acute, finely transversely rugulose; anthers 1–1.4 mm; caryopsis not seen.

Naturalised from Africa.

N.: Auckland City (Ellerslie). On motorway berm.

P. maximum var. *trichoglume* was first collected from a single clump in 1979 by A. E. *Esler 6007* (CHR 357511). In spite of annual herbicide treatments, plants had spread along the edge of the motorway for more than 100 m by 1986.

Webster, R. D. *Australian Paniceae* 241 (1987) transferred *P. maximum* to *Urochloa* because of the rugulose lemma of the upper floret.

5. P. miliaceum L., *Sp. Pl.* 58 (1753).   broomcorn millet

Tufts leafy, annual, stout, ± erect, or decumbent at base and rooting at lower nodes, to 80 cm. Leaf-sheath light green or straw-coloured, submembranous, rounded, with numerous, long, fine, conspicuously tubercle-based hairs. Ligule 2–4 mm, membranous, truncate, ciliate. Leaf-blade 10–45 cm × 6–20 mm, thin, ± curved at base, almost glabrous to loosely hairy with long fine tubercle-based hairs, midrib prominent; margins very finely scabrid, often somewhat undulating, narrowed to filiform, scabrid tip. Culm 10–40 cm, nodes usually closely pubescent, internodes glabrous or sometimes softly hairy below lower nodes. Panicle (10)–20–45 cm, contracted and rather dense, or open, nodding, often not fully emergent from upper sheath; rachis and primary branches smooth, branchlets and filiform pedicels finely scabrid on angles. Spikelets 4–5 mm, numerous, glabrous, green to greenish brown, ovate-lanceolate, turgid, apiculate-acuminate. Glumes prominently nerved; lower 3–3.4 mm, ½ to ¾ length of spikelet, ovate, acute, 5-nerved, upper = spikelet, 11-nerved, ovate-lanceolate, tip finely scabrid, mucronate. Lower floret ∅: lemma ≈ spikelet, (11)–13-nerved;

palea ≪ lemma, ovate, truncate or emarginate, hyaline, with 2 faint, winged keels. Upper floret ♀, falling readily at maturity: lemma 2.8–3.4 mm, elliptic, subacute to obtuse, crustaceous, obscurely nerved, very turgid and convex, variously coloured white, yellow, red, brown or black, glabrous, shining; palea narrower; anthers 1.6–2.2 mm; caryopsis 2–2.5 × 1.5–2 mm.

Naturalised from Asia.

N.: scattered; S.: scattered throughout except in Fiordland and Southland; Ch. Waste places, gardens.

Often occurring as a result of seed spillages and the disposal of cage-bird seed.

6. P. schinzii Hack., *Verh. Bot. Ver. Brand. 30*: 142 (1888).

swamp panicum
Strong, erect, glabrous, light or dull green to purplish annuals, to 120 cm, in diffuse tufts with stems much branched near base. Leaf-sheath submembranous, rounded, somewhat keeled above, glabrous. Ligule a membranous rim *c.* 0.5 mm, tipped by a ciliate fringe usually *c.* 1 mm. Leaf-blade 8–20 cm × 4–8 mm, long-tapering, glabrous, firm, midrib prominent; margins sometimes undulate below, tip acuminate. Culm (20)–35–80–(100) cm, terete or compressed, nodes brownish, internodes glabrous. Panicle 15–35 cm, very lax, branches later widely spreading; rachis and branches smooth, branchlets and pedicels finer, minutely scabrid on angles. Spikelets 2.2–2.7 mm, glabrous, ovate-elliptic, obtuse, rather dull. Glumes very unequal; lower 0.6–1 mm, clasping spikelet base, truncate, sometimes with a central subacute tip, upper = spikelet, (7)–9–11-nerved, elliptic, subacute. Lower floret ♂: lemma 9-nerved; palea narrower, narrowly winged above, 2-keeled, keels glabrous; anthers 1–1.8 mm. Upper floret ♀: lemma 1.8–2 mm, elliptic, subacute, faintly striolate, glabrous, shining, light cream; palea narrower; anthers 1.2–1.6 mm; caryopsis *c.* 1.5–2 × 1 mm.

Naturalised from south-west Africa.

N.: North Auckland (Kerikeri), Auckland City, South Auckland (Hamilton); S.: Nelson (Motueka, Moutere, Nelson City), Canterbury (Darfield). Waste land, usually damp places, crop weed.

Often forming large spreading patches and becoming troublesome in the Motueka district; spreading especially in irrigated orchards.

Though first collected in 1943 at Auckland (CHR 7765), this sp. was confused for many years with *P. dichotomiflorum* from which it can be distinguished by the staminate lower florets, obtuse rather than acute-tipped spikelets, the more robust habit and laxer panicle. In Australasia the later synonym *P. laevifolium* Hack. was applied for a time to swamp panicum, e.g., Vickery, J. W. *Fl. N.S.W. No. 19, Gramineae 2*: 184 (1975); Edgar, E. and Shand, J. E. *N.Z. J. Bot. 25*: 343–353 (1987) in a checklist of panicoid grasses.

7. P. sphaerocarpon Elliott, *Bot. S. Carolina and Georgia 1*: 125 (1816).

Stiff, light green tufts, 12–20 cm; basal leaves differing from culm leaves and forming a winter rosette. Leaf-sheath light green to purplish, glabrous but margins ciliate. Ligule 0.2–0.5 mm, ciliate, hairs sparse. Collar hairs long, tubercle-based. Leaf-blade 3–10 cm × 7–13 mm, firm, rather thick, broadly linear to elliptic-oblong, narrowed above, ribs fine, inconspicuous; abaxially smooth, adaxially finely scabrid on ribs; margins thickened, scabrid, ciliate near cordate base, tip acute to acuminate. Culm erect to spreading, nodes appressed-pubescent, internodes glabrous. Panicle 5–8 × 2–4 cm; branches very slender, glabrous; pedicels almost filiform, tipped by solitary spikelets. Spikelets < 2 mm, green to purplish, ovoid, subacute. Glumes very unequal; lower *c.* 0.5 mm, membranous, glabrous, obtuse, upper = spikelet, 7-nerved, elliptic-ovate, subacute, puberulent. Lower floret ∅: lemma similar to upper glume, 7-nerved; palea 0. Upper floret ♂: lemma 1.6–1.8 mm, elliptic, subacute, glabrous, faintly striolate, white, shining; palea narrower; anthers 0.3–0.4 mm; gynoecium: ovary *c.* 0.3 mm, stigma-styles *c.* 1.8 mm, stigmas deep purple; caryopsis not seen.

Naturalised from America.

N.: South Auckland, Thames County (Whenuakite), Lake Whangape. Damp ground.

In all specimens seen culms bore only terminal panicles, spikelets were ovoid and no seed was set; in American plants spikelets containing mature seed become spherical.

ζ**P. lindheimeri** Nash   Perennial; basal leaves differing from culm leaves and forming a winter rosette; culms produced in spring simple, terminating in a dense panicle, culms produced in autumn much branched above with smaller axillary panicles often almost hidden by leaf-sheaths. Leaves ± glabrous. Ligules 3–5 mm. Culm nodes glabrous. Panicle 4–9 cm. Spikelets 1.5 mm, orbicular, pubescent. A North American sp. once grown on at Kaitaia by *H. B. Matthews* from seed from a plant growing at Kerikeri (specimens were made 11.12.1913, AK 99300, 93301, 93302, with duplicate CHR 3448 labelled *H. B. Wilson* as collector presumably in error for *H. B. Matthews*).

MISAPPLIED NAME   *Panicum glumare* Trin., *Gram. Pan.* 143 (1826) was stated by Trinius to come from New Zealand. Veldkamp, J. F. *Blumea 41*: 413–437 (1996) considered that the holotype at LE was mislabelled as to locality and treated *P. glumare* as the basionym of *Urochloa glumaris* (Trin.) Veldkamp, a species of Indomalesia, Polynesia and New Caledonia.

PASPALUM L., 1759

Perennials, of various habit. Leaf-sheath open or closed below. Ligule membranous, glabrous. Collar often bearded. Leaf-blade flat or convolute, rolled in bud. Culm simple or branched. Inflorescence paniculate, of spike-like racemes, these solitary, digitate or scattered along a simple common rachis. Racemes bearing subsessile to short-pedicelled, ± appressed, single or paired spikelets in two rows along one side of narrow or wing-like rachis. Spikelets lanceolate to oblong or orbicular, awnless, dorsally compressed, usually planoconvex, falling entire at maturity, 2-flowered, florets awnless; lower floret ∅, reduced to a lemma; upper floret ♀. Glumes membranous; lower usually 0, rarely a small scale, upper ≈ spikelet. Lower floret: lemma similar in size and texture to upper glume but less convex, or flat; palea 0 or vestigial. Upper floret: lemma ± crustaceous, glabrous, usually obtuse with firm inrolled margins; palea ≈ lemma and similar in texture; lodicules 2, minute, broadly cuneate, rather fleshy; stamens 3; ovary apex glabrous, styles subterminal, free; caryopsis ± biconvex, embryo < ½ length of caryopsis, hilum subbasal, punctiform.

Type species: *P. dimidiatum* L.

c. 250 spp. of tropical and warm temperate regions. Naturalised spp. 6; transient spp. 2.

1 Inflorescence of 2 racemes, ± digitate at culm apex, rarely a third raceme below, or rarely a solitary raceme *2*
Inflorescence of 2–many racemes, ± distant along a common rachis *4*

2 Spikelets *c.* 1.5 mm, broad-elliptic to almost orbicular .... **1. conjugatum**
Spikelets 2.5–4.5 mm, elliptic *3*

3 Upper glume minutely appressed-puberulent; in seepages ..... **3. distichum**
Upper glume glabrous; in dry or coastal saline sites .... **6. vaginatum**

4 Spikelets glabrous .... **4. orbiculare**
Spikelets pubescent or hair-fringed *5*

5 Upper glume conspicuously fringed with long, silky hairs *6*
Upper glume pubescent but without a conspicuous fringe of hairs *7*

6 Culms geniculate-ascending; panicle drooping; racemes (2)–3–6; spikelets 3–3.5 mm .... **2. dilatatum**
Culms and panicles stiffly erect; racemes 10–20–(26); spikelets 2–2.7 mm .... **5. urvillei**

7 Racemes *c.* 30; spikelets *c.* 1.5 mm .... **paniculatum**ζ (p. 571)
Racemes 2–4; spikelets *c.* 3 mm .... **pubiflorum**ζ (p. 571)

1. P. conjugatum P.J.Bergius, *Acta Helv.-Phys. Math.* 7: 129, t. 8 (1762).

Strongly stoloniferous perennials with erect or sprawling shoots; stolons often with purple internodes. Leaf-sheath rather stiff, keeled, glabrous. Ligule 0.5–1.5 mm, slightly erose, truncate, narrowed centrally to a point. Collar hairs few, long. Leaf-blade 10–20 cm × 6–9 mm, linear-lanceolate, flat, soft, glabrous apart from long hairs near base, narrowed to acute or acuminate tip. Culm 20–45 cm, erect or ascending, slender, wiry, internodes compressed, glabrous. Panicle of 2–(3) widely divaricating slender racemes ± digitate at culm apex. Racemes 6–10 cm, rachis very narrowly-winged, 3-angled, sparsely scabrid, 0.6–0.8 mm wide, densely pubescent at base, bearing 2 rows of single shortly pedicelled spikelets; pedicels smooth. Spikelets *c.* 1.5 mm, imbricate, broad-elliptic to almost orbicular, acute, yellowish green. Lower glume 0, upper = spikelet, thin and closely appressed to upper floret, 2- or 3-nerved with median nerve often suppressed, glabrous apart from long fine hairs on lateral nerves. Lower floret: lemma = spikelet, similar in texture to upper glume; palea 0. Upper floret: lemma 1.2–1.4 mm, slightly indurated, 3-nerved; palea slightly narrower than lemma, membranous margins only slightly widened at base; anthers *c.* 0.5 mm, yellow; caryopsis not seen.

Naturalised.

N.: Ohinemutu (one record, AKU 18022, CHR 274232); K.: Raoul Id. Old grassy tracks in semi-shade on Raoul Id.

Widespread in tropical America and throughout the Old World tropics and subtropics.

2. P. dilatatum Poir., *in* Lamarck *Encycl. Méth. Bot.* 5: 35 (1804).

paspalum

Tufted perennials, sometimes forming dense stands, from short, strong rhizomes to *c.* 1 cm diam. Leaf-sheath rather stiff, lowermost densely long-hairy, others glabrous. Ligule (2)–3–5 mm, entire, rounded. Leaf-blade (6.5)–12.5–18.5–(25) cm × 5.5–8 mm, rather stiff, flat, glabrous apart from a few long hairs on margins and adaxially near ligule, midrib conspicuous; margins scabrid, narrowed above to acuminate tip. Culm 25–55–(180) cm, ± flattened, to 2.5 mm diam., erect or geniculate-ascending and drooping above, nodes glabrous or the lowermost sparsely pubescent, internodes glabrous. Panicle usually nodding, 8–15 cm, with (2)–3–6 often spreading or drooping, rather distant racemes; rachis slender, flattened. Racemes 4.5–6.5–(11) cm; rachis *c.* 1 mm wide, narrowly winged, scabrid-margined with numerous white hairs at base, bearing 2 rows of paired, shortly pedicelled spikelets; pedicels scabrid. Spikelets *c.* 3–3.5 mm, closely imbricate, broadly elliptic-ovate, ciliate, acute, green or purplish. Lower glume 0, upper = spikelet, 5–9-nerved, sparsely hairy, margins ciliate, hairs long, silky. Lower floret: lemma ≈ spikelet, 5–9-nerved, glabrous, or margins sometimes sparsely hairy; palea 0. Upper floret: lemma 2.2–2.5 mm, very broadly elliptic to orbicular, crustaceous, faintly 5-nerved, minutely papillose and longitudinally striate, light cream; palea ≈ lemma, incurved, membranous margins widened at base, forming wings enclosing flower; anthers 0.9–1.3 mm, dark purple; stigmas deep reddish purple; caryopsis *c.* 1.5 mm.

Naturalised from South America; Brazil to Argentina.

N.: very common; S.: scattered or locally dense as far south as Taieri R.; K., Ch. Shingly roadsides, waste land, ditch banks, pasture, consolidated sand dunes, also in heated soil at Karapiti, Wairakei.

Field, T. R. O. and Forde, M. B. *Proc. N.Z. Grasslands Assoc. 51*: 47–50 (1990) summarised the spread of paspalum in N.Z. It was abundantly naturalised in North Id by 1935 and reached northern South Id by 1940. By the 1970s it was dominant in pastures in Northland, Coromandel, and Bay of Plenty, and at that time increased spectacularly in southern North Id, and even more markedly since 1985. Substantial populations were also reported in South Canterbury and the Waitaki Valley.

Paspalum dilatatum has been widely sown as a pasture grass. Dormant in winter, it spreads vigorously in late spring and summer. A troublesome weed in lawns and common on shingly roadsides where it can form extensive patches. The flower heads are sometimes sticky through infection with ergot. Paspalum is sometimes sown in swamps and periodically flooded land in the north of North Id. It will live for 5–6 months under water and grows vigorously when the water subsides.

3. P. distichum L., *Syst. Nat.* ed. 10, 2: 855 (1759).   Mercer grass

Perennials, with numerous long-creeping branching stolons to 3 mm diam., rooting at nodes. Leaf-sheath submembranous, glabrous. Ligule 1–1.6 mm, erose, truncate. Collar hairs few, long, tubercle-based. Leaf-blade 2.5–9 cm × 2–8 mm, rather soft, flat, linear-lanceolate, curved at base, abaxially glabrous with obvious midrib, adaxially with scattered, long, fine tubercle-based hairs; margins scabrid, gradually narrowed to acuminate scabrid tip. Culm 7–35–(45) cm, erect or geniculate-ascending, compressed, nodes long-hairy, internodes glabrous. Panicle of (1)–2–(3) erect to spreading racemes, ± digitate at culm apex. Racemes 2–5 cm, sometimes one or both shortly pedunculate; axils with a few long white hairs; rachis 1–2 mm wide, concave, narrowly winged with scabrid margins, bearing 2 rows of single or paired, subsessile or shortly pedicelled spikelets; pedicels scabrid; paired spikelets usually near centre of rachis. Spikelets *c.* 2.5–3.5 mm, imbricate, elliptic, acute, light green. Lower glume a minute scale rarely to 1 mm, or 0, upper = spikelet, 5-nerved, minutely appressed-puberulent. Lower floret: lemma ≈ upper glume and slightly less membranous, 3–5-nerved, glabrous; palea rarely present, narrow, short, hyaline, 2-keeled. Upper floret: lemma 2–3 mm, faintly 5-nerved, cartilaginous-indurate, finely punctulate-striolate, broad-elliptic, acute, light cream, apex often with tuft of short hairs; palea ≈ lemma, hyaline margins only slightly widened near centre; anthers 1–1.7 mm, dark purple; stigmas dark purple; caryopsis 1.5–2 mm.

Naturalised from Europe.

N.: throughout; S.: Nelson (Nelson City, Takaka, near Lake Otuhie, Karamea), Canterbury (Christchurch, Lincoln); K. Damp soil at lake edges, sand flats, swampy roadsides, waste ground, also a weed of lawns and cultivated ground — often in fresh, not brackish water.

A long controversy concerning the correct application of the name *P. distichum* L. was ended by the decision of the Committee for Spermatophyta [*Taxon 32*: 281 (1983)] to retain the name for the sp. growing in freshwater. Because the type specimen was a mixed sheet the name *P. distichum* was formerly sometimes applied to saltwater paspalum (*P. vaginatum*), and the name, *P. paspalodes* (Michx.) Scribn., was applied to the sp. from freshwater. In N.Z. also, Mercer grass was known as *P. paspalodes* (as in the original spelling), and *P. paspaloides*.

Field, T. R. O. and Forde, M. B. *Proc. N.Z. Grasslands Assoc. 51*: 47–50 (1990) noted that Mercer grass appeared to be most abundant in Waikato and the Bay of Plenty where it was reported as clogging waterways.

4. P. orbiculare G.Forst., *Prodr.* 7 (1786).

Tufted perennials, with rather stiff, loosely sheathing, erect leaves. Leaf-sheath subcoriaceous, striate, keeled, brown to purplish brown, mainly glabrous, but lowermost sheaths with fine silky hairs, especially near base; sheath apex extended upwards at each margin and fused with ligule. Ligule 1–2 mm, truncate, entire. Leaf-blade 10–20–(30) cm × 3.5–5 mm, often narrower than summit of sheath, flat, rather rigid, midrib obvious, abaxially glabrous, adaxially thickly short-pilose near ligule, with a few very long hairs near margin below; margins finely scabrid, tapered to fine but firm tip. Culm (20)–35–70 cm, usually erect, sometimes semi-prostrate, slightly compressed, internodes glabrous, striate. Panicle erect, 6–12 cm, of 3–8 erect to somewhat spreading ± distant racemes; rachis slender, compressed to angular, finely scabrid on angles above. Racemes (2)–3–4 cm; rachis narrowly winged, with very scabrid margins, 1.2–1.7 mm wide, with short white hairs at base, bearing 2 rows of single or paired, shortly pedicelled or almost sessile spikelets; pedicels sparsely scabrid; paired spikelets usually near centre of raceme. Spikelets 2–2.5 mm, imbricate, ovoid-elliptic to ovoid-orbicular, glabrous, obtuse, light brown. Lower glume 0, upper = spikelet, closely appressed to fertile floret, 3–(5)-nerved, glabrous. Lower floret: lemma = spikelet, 3–(5)-nerved, glabrous; palea 0. Upper floret: lemma *c.* 2 mm, elliptic-orbicular, cartilaginous-indurate, shining, brown, finely punctulate-striolate; palea ≤ and narrower than lemma, membranous margins widened at base, forming wings enclosing flower; anthers *c.* 1 mm, yellow to brownish; stigmas dark purple; caryopsis slightly > 1 mm.

Naturalised from Pacific.

N.: North and South Auckland and offshore islands; K.: Raoul Id. In North Id on clay soil, open *Leptospermum* scrubland, dry banks and waste ground; also on damp ground, particularly seepages. On Raoul Id on thermally heated ground near active fumaroles in Denham Bay, partly shaded tracks, dry forest clearings and open ridges.

Paspalum orbiculare was equated with *P. scrobiculatum* L., scrobic, an annual Indian sp., by Cheeseman (1925 op. cit.) and treated as indigenous to N.Z. *Paspalum orbiculare* is widespread throughout the Pacific; Cameron, E. K. *Auck. Bot. Soc. J. 53*: 40–42 (1998) postulated that it was naturalised in N.Z. because it was initially collected at coastal sites frequently visited by ships and absent from sites rarely visited by humans. *Paspalum orbiculare* is also recorded from Australia but Vickery, J. W. *Contrib. N.S.W. Natl Herb. Flora Series 19(1)*: 118–119 (1961) regarded Australian plants as differing somewhat in spikelet size and shape from Pacific Is material. N.Z. specimens, however, match Pacific plants in spikelet characters.

5. P. urvillei Steud., *Syn. Pl. Glum. 1*: 24 (1854). Vasey grass

Perennials to 2 m or sometimes taller, forming dense clumps from very stout, short woody rhizome 1–1.5 cm diam., with persistent sheath bases. Leaf-sheath stiff, light purplish brown, lowermost often densely villous with long, fine, tubercle-based hairs, others glabrous, keeled above; apex extended upwards at each margin and fused with ligule. Ligule 3.5–9.5 mm, sparsely, shallowly dentate. Collar with some long fine hairs. Leaf-blade 8.5–30–(50) cm × 4–9 mm, flat, stiff, linear-lanceolate, long-tapered, midrib obvious, adaxially with row of long fine hairs near ligule; margins finely scabrid, often crenulate-undulate, tip filiform. Culm 50–175–(250) cm, to 4 mm diam., erect, internodes glabrous. Panicle erect, 15–30 cm, with 10–20–(26) erect to somewhat spreading ± distant racemes; rachis slender, ± angled, glabrous, a few long hairs in raceme axils. Racemes 6–13 cm in lower part of panicle, becoming shorter above; rachis *c.* 1 mm wide, narrowly winged with scabrid margins, bearing 2 rows of paired, shortly pedicelled spikelets; pedicels sparsely scabrid. Spikelets 2–2.7 mm, closely imbricate, ovate-elliptic, ciliate, somewhat tapered to acute tip, green or purplish. Lower glume 0, upper = lemma of lower floret, 3–(5)-nerved, fringed with long silky hairs, and also bearing sparse appressed silky hairs. Lower floret: lemma 3–(5)-nerved, ± glabrous, fringed by long silky hairs; palea 0. Upper floret: lemma 1.6–1.8 mm, ovate-elliptic, cartilaginous-indurate, creamy, finely punctulate-striolate; palea slightly < and narrower than lemma, margins membranous, widened at base, forming wings enclosing floret; anthers 0.8–1.1 mm, yellow to brownish; stigmas dark purple; caryopsis ≤ 1.5 mm.

Naturalised from South America; Brazil to Argentina.

N.: North and South Auckland, Gisborne, Manawatu (Sanson); K. Usually roadsides, sometimes in large patches; also waste land, manuka scrub.

P. J. de Lange (*in litt.*) reports Vasey grass as spreading in parts of Waikato.

6. P. vaginatum Sw., *Nov. Gen. Sp. Pl.* 21 (1788).

saltwater paspalum

Decumbent perennials, with numerous, long-creeping stolons, bearing very wide, loose, papery sheaths at nodes; erect shoots often in close tufts, much-branched near base. Leaf-sheath submembranous, glabrous; apex extended upwards at each margin and fused with ligule. Ligule 0.5–1 mm, scarcely tapered, shortly bluntly pointed, glabrous. Collar with small tufts of hairs. Leaf-blade 3–8 cm × 1–2 mm, rather stiff, much narrower than sheath, ± inrolled, narrow-linear, glabrous, tapering to acuminate tip. Culm (3)–7.5–18–(30) cm, erect or geniculate-ascending, slender, compressed, internodes glabrous. Panicle of 2–(3) erect to spreading racemes ± digitate at culm apex. Racemes 2–3 cm, distinctly pedunculate; rachis < 1 mm wide, scarcely winged, ± glabrous, bearing 2 rows of single, subsessile spikelets. Spikelets 3–4 mm, imbricate, elliptic, tapered above to acute or acuminate tip, light green to light brown, glabrous. Lower glume 0, upper = lemma of lower floret, 3-nerved. Lower floret: lemma 3-nerved; palea 0. Upper floret: lemma *c.* 3 mm, indurated, faintly 3-nerved, glabrous, shining, light creamy brown; palea = lemma, margins flattened, very slightly incurved; anthers 1.7–1.9 mm, yellow to brownish; caryopsis *c.* 1.5 mm.

Naturalised from Europe.

N.: North and South Auckland, Gisborne. Coastal, often brackish, areas; often forming swards near edge of mud flats, or on sandy and shingly shores and occasionally spreading into pasture nearby.

ζ**P. paniculatum** L.   Plant *c.* 75 cm. Leaf-sheath with long, tubercle-based hairs, especially near margin. Collar hairs long, tubercle-based. Leaf-blade *c.* 20 × 20 mm, with scattered hairs. Culm nodes densely villous. Panicle pyramidal, 12 cm; racemes *c.* 30, very slender, bearing numerous, short-pedicelled, orbicular, pubescent spikelets *c.* 1.5 mm. A Central American sp. once collected from Auckland City, Morningside *A. E. Esler* 8.6.1986 (AK 174172, duplicate CHR 421175).

ζ**P. pubiflorum** E.Fourn.   Very long-stoloniferous. Leaf-sheath glabrous below, pubescent above. Culm nodes pubescent. Panicle of 2–3 racemes, to 8 cm, with 2 rows of paired spikelets. Spikelets *c.* 3 mm, elliptic-acute, greenish to purplish. Upper glume 3-nerved, appressed-pubescent. Lower floret: lemma similar to upper glume. Upper floret: lemma chartaceous, glabrous, shining, creamy brown; anthers and stigmas dark reddish purple. A North American sp. once collected from Wanganui *A. R. Dingwall* 5.3.1938 (CHR 19374, 19375).

Paspalum pumilum Nees from South America, Brazil, was sown in trials at Kaikohe, Northland, and has persisted there for over 20 years and spread vegetatively (P. Woods, pers. comm.).

INCERTAE SEDIS   *Paspalum tenax* Trin., *Gram. Pan.* 122 (1826), was based on a specimen collected by Lindley from N.Z. which has not been traced.

PENNISETUM Rich., 1805

Tufted or rhizomatous or straggling perennials; or annuals. Leaf-sheath loose, usually open. Ligule usually a ciliate rim, rarely membranous. Leaf-blade linear to lanceolate, flat, folded or convolute. Culm simple or much-branched. Inflorescence usually a dense, spike-like panicle with very numerous, very short branches falling with spikelets at maturity, and usually leaving scars on the rachis; rarely reduced to a few spikelets.

Spikelets 2-flowered, solitary or in groups of 2 or 3 surrounded by an involucre of unequal glabrous or plumose bristles united at base; spikelets and bristles shed freely at maturity; lower floret ♂ or ∅, with or without palea, rarely ♀; upper floret ♀, or sometimes ♂ in lateral spikelets of a cluster. Glumes membranous, subequal, or the lower smaller, sometimes minute or 0; lower (0)–1–3-nerved, upper ≤ spikelet, (0)–1–9-nerved. Lower floret: lemma ≤ spikelet, membranous, (0)–3–9–(15)-nerved, enclosing narrow palea, or palea 0. Upper floret: lemma ≈ spikelet, chartaceous, (3)–5–7–(15)-nerved, glabrous or sometimes pubescent near margins; palea ≈ lemma and of similar texture; lodicules minute or 0; stamens 3, anthers occasionally penicillate; ovary apex glabrous, styles ± terminal, free or ± connate; caryopsis dorsiventrally compressed, embryo large, hilum punctiform.

Type species: *P. typhoideum* Rich. *nom. illeg.*

c. 130 spp. of tropical and warm temperate regions. Naturalised spp. 7; transient sp. 1.

1 Forming large patches or semiscandent; inflorescence of (1)–2–4 spikelets included (except at anthesis) within uppermost leaf-sheath on short branches .... **2. clandestinum**
Forming erect tufts, clumps, or tussocks; inflorescence of many spikelets in a conspicuous spike-like or compound panicle 2

2 Involucral bristles short-scabrid *3*
Involucral bristles plumose throughout, below, or on inner only *5*

3 Leaf-sheaths usually with conspicuous cross-veinlets; inflorescence a compound panicle with clusters of long-pedunculate spikes from axils of upper culm leaves .... **3. latifolium**
Leaf-sheaths without conspicuous cross-veinlets; inflorescence a simple spike-like panicle *4*

4 Panicle (16)–20–40 cm; involucral bristles ≥ spikelets and 6–10 mm, with a single stouter bristle to 15–(20) mm .... **4. macrourum**
Panicle 7–12 cm; involucral bristles ≫ spikelets, 15–25 mm ... **1. alopecuroides**

5 Panicle usually subglobose, usually pale green or silvery; spikelets 8–11 mm .... **7. villosum**
Panicle cylindric, usually tinged reddish purple; spikelets 5–7.5 mm 6

6 Involucral bristles < spikelets; plants annual .... **glaucum**ζ (p. 580)
Involucral bristles > spikelets; plants perennial 7

7 Leaf-blade 0.5–1.5 mm wide, folded or ± involute; involucral bristles 15–25 mm, a single longer bristle to 35 mm .... **6. setaceum**

Leaf-blade 20–30 mm wide, flat; most involucral bristles *c.* 10 mm, a single stouter bristle to 20 mm.. **5. purpureum**

1. P. alopecuroides (L.) Spreng., *Syst. Veg. 1*: 303 (1824).

Chinese pennisetum

Coarse grey-green tussocks of medium size, with narrow wiry leaves; branching intravaginal. Leaf-sheath hard, strongly keeled, glabrous, shining, brownish or often pinkish, margins membranous. Ligule 0.3–0.6 mm, a densely, shortly ciliate rim. Leaf-blade 15–30 cm × 1–1.5 mm diam., folded or involute, sometimes flat, linear, long-tapering, abaxially glabrous, adaxially often shallowly ribbed, ribs minutely scabrid, occasionally with a few soft hairs near ligule; margins minutely scabrid, tip acuminate. Culm 20–40 cm, erect, rounded or ± compressed, internodes loosely covered by fine, long, soft hairs near panicle. Panicle 7–12 cm, spike-like, cylindric, purplish; rachis and ± appressed branches and pedicels soft-hairy. Spikelets 5.5–6.5 mm, shortly pedicelled, densely clustered on rachis, narrow-lanceolate, acute, light creamy brown, minutely scabrid, greenish below, deep purple-brown towards apex, usually solitary within a much longer involucre of numerous, fine, erect, short-scabrid unequal bristles (15–25 mm), united at base; rarely 2 spikelets within involucre. Glumes ≪ spikelet, unequal, submembranous; lower 0.5–1 mm, varying from nerveless, ± cuneate or rimlike, to longer-lanceolate, 1-nerved, upper 2–3 mm, *c.* ⅓ length of spikelet, faintly 5–(7)-nerved. Lower floret ∅: lemma 5–6 mm, ≈ spikelet, submembranous, 7-nerved, almost enclosing lemma of upper floret; palea 0. Upper floret ♀: lemma 5.5–6.5 mm, somewhat chartaceous, 5–7-nerved, ± enclosing palea; palea similar to lemma but slightly shorter, slightly rounded, nerves 2, glabrous; lodicules 0.2–0.5 mm; anthers 2.7–3.5 mm; styles connate below feathery stigmas, or sometimes partly or wholly free; caryopsis *c.* 2–2.5 × 1 mm, dark brownish.

Naturalised from eastern Asia.

N.: Takatu near Warkworth, and lower Wanganui R.; S.: Nelson, Marlborough. Tracks, creek beds, coastal sand, wet low lying ground near swamps.

2. P. clandestinum Chiov., *Annuario Reale Ist. Bot. Roma 8*: 41, t. 5, fig. 2 (1903).   kikuyu grass

Decumbent to semiscandent perennials, with creeping, cataphyll-covered rhizomes and long stolons 2.5–5.5 mm diam., rooting at nodes; when grazed or mown forming a close, compact, light green turf; plants with ♀ florets, or monoecious, or dioecious. Leaf-sheath light brown or almost white, ± membranous, usually with scattered, fine, tubercle-based hairs, rarely glabrous. Ligule 1–2 mm, a densely ciliate rim. Leaf-blade (1.5)–3–8–(25) cm × 1.5–2.5–(9) mm, at first tightly folded, later flattened, linear, gradually tapering, soft, glabrous, or with a few, fine, tubercle-based hairs, sparingly scabrid on margins and midrib near narrow subobtuse tip. Culm internodes shining, glabrous, extending as long, rooting, flattened stolons, with short to long ascending branches often closely clustered into tufts. Inflorescences borne on the short branches, usually consisting of clusters of (1)–2–4 spikelets partly or almost entirely hidden in the leaf-sheaths, very rarely a short, reduced, spike-like panicle exserted beyond uppermost leaf-sheath. Spikelets 1.5–2 mm, subsessile, or usually the terminal and sometimes the lateral spikelets pedicelled and subtended by (3)–5–15 ciliate, whitish, unequal (1.5–12 mm), finely scabrid bristles. Glumes unequal; lower 0, or sometimes a minute, nerveless scale, upper often 0, occasionally a short (3 mm), transparent, 3-nerved scale. Lower floret ∅: lemma = spikelet, 9–13-nerved, papery; palea 0. Upper floret ♀, ♂ or ♀: lemma ≈ spikelet, 9–11-nerved, papery; palea of same texture as lemma, rounded, with 2–4 smooth nerves; lodicules 0; anthers of ♀ and ♂ florets: 4–5 mm, with filaments much extended at anthesis, to *c.* 3 cm, and exserted from leaf-sheath giving patches of the plant a silvery appearance, of ♀ florets: 3–4 mm, on short filaments retained within spikelets; style and stigmas long-exserted from floret, 1.5–3.5 mm, stigmas connate, finely plumose, often bifid at tip; caryopsis *c.* 2 × 15 mm, oblong, slightly laterally compressed, brown. Plate 4B.

Naturalised from North Africa.

N.: North Auckland, and elsewhere scattered near coast except in eastern Wellington Province; S.: Marlborough Sounds, Nelson, Westland (as far south as Greymouth), Canterbury (Lincoln); Three

Kings Is: Great Id. Pastures, coastal (sandy or rocky foreshore), reclamation areas, roadsides, lawns, stream banks.

A troublesome weed of high rainfall areas in many warm countries.

Field, T. R. O. and Forde, M. B. *Proc. N.Z. Grasslands Assoc. 51*: 47–50 (1990) reported that kikuyu grass was most common in pasture north of Auckland and also at Whakatane and was widespread as a road verge sp. in coastal Taranaki from Kotare to Opunake. It was reported as being much more dominant in the Bay of Plenty than usual in the summer of 1988/89.

This strongly creeping grass can form swards over several hectares in area and is difficult to eradicate. Because shading does not stop its growth, it can grow up through hedges and bushes, eventually shading them to extinction. Where kikuyu grass grows in open scrubland or on forest margins the seedlings of native trees and shrubs have very little chance of establishment. Kikuyu grass creates a considerable fire hazard as the old stolons die and accumulate because of the new growth scrambling over the top.

Although kikuyu grass is frost tender, a plant which appeared in a coke-breeze pathway at DSIR, Lincoln, persisted over two winters and flowered in its second season (CHR 143957). It originated from soil about *Cyperus* plants from North Auckland.

3. P. latifolium Spreng., *Syst. Veg. 1*: 302 (1824). Uruguay pennisetum

Tall, erect, tufted, bright green, shortly rhizomatous perennials, the stout (*c.* 2 cm diam.), short internodes of the rhizome giving a corm-like appearance to base of culm. Leaf-sheath loosely enveloping culm, rounded, subcoriaceous, with numerous usually conspicuous cross-veinlets, lowermost sheaths with a few scattered long hairs, others glabrous. Ligule 2–5.5 mm, a membranous densely long-ciliate rim. Leaf-blade 30–50 cm × 1–3.5 mm, flat, lanceolate, tapering to long fine point, narrowed, almost pseudopetiolate at base, ribs and margins minutely scabrid. Culm *c.* 2–2.5 m, nodes purplish, appressed-pubescent, internodes glabrous. Panicle 17–30 cm, compound, with clusters of 3–4 long-peduncled spikes arising in axils of 3–4 short, reduced leaves. Spikes 3–6.5 cm; rachis minutely pubescent; peduncles (5)–12–25 cm, triangular-compressed, with strongly pubescent-scabrid ridges. Spikelets 6–7 mm, elliptic-lanceolate, acuminate, densely clustered and ± sessile on rachis, surrounded by involucre of numerous finely scabrid bristles, usually one stouter, ≫ spikelet (10–17 mm), others usually < spikelet

(5–7 mm). Glumes very short, hyaline, with long appressed hairs, margins long-ciliate near acute tip; lower 0.5–1 mm, upper 1–2 mm, 1-nerved. Lower floret ∅: lemma = spikelet, 5-nerved, submembranous, minutely pubescent-scabrid; palea 0. Upper floret ♀: lemma ≈ spikelet, 5–6 mm, 5-nerved, very slightly hardened, glabrous below, finely minutely pubescent-scabrid near tip; palea 4.5–5 mm, of same texture as lemma, rounded, nerves 2, glabrous; lodicules 0.4–0.7 mm; anthers 1.4–2 mm; gynoecium: ovary *c.* 1 mm, styles free, glabrous, *c.* 1.5 mm, stigmas purple, plumose, *c.* 2.5 mm; caryopsis 1.7–2.3 mm.

Naturalised from South America.

N.: North Auckland (south of Kerikeri), Auckland City, scattered in South Auckland, one record from Wanganui. Garden escape; waste land, roadsides, pasture, damp places.

4. P. macrourum Trin., *Gram. Pan.* 64 (1826). African feather grass

Large, coarse, erect, perennial clumps with densely crowded leaves from stout, short, or sometimes far-creeping rhizomes; branching intravaginal. Leaf-sheath long, chartaceous, rounded, strongly ribbed, light creamy brown or sometimes pinkish, usually with scattered to dense, stiff, ± appressed hairs, sometimes glabrous. Ligule 1–2 mm, to 4.5 mm on upper leaves, a membranous densely long-ciliate rim. Leaf-blade (25)–40–60–(100) cm × 1–3 mm diam., inrolled or folded, or flat and up to 7 mm wide, linear, very long tapering, abaxially with some scattered hairs, scabrid above, adaxially deeply furrowed, ribs finely scabrid or smooth; margins scabrid above, tip long, fine, flexuous. Culm 75–120 cm, erect, terete, internodes very minutely scabrid below inflorescence. Panicle (16)–20–40 cm, spike-like, erect or ± nodding, straw-coloured, light brown to sometimes purple suffused; rachis densely finely scabrid. Spikelets 4.5–6 mm, densely crowded, almost sessile, lanceolate, acute or acuminate, solitary or rarely paired within an involucre of numerous, fine, minutely scabrid bristles, one bristle much stouter and longer 9–15–(20) mm, remaining bristles usually slightly > spikelet (6–10 mm). Glumes ≪ spikelet, hyaline, glabrous, nerveless or 1-nerved; lower 0.5–0.8 mm or 0, upper 0.8–1.7 mm. Lower floret ∅: lemma 4–6 mm, 5-nerved, hyaline, very minutely scabrid; palea

0. Upper floret ♀: lemma 4.5–5.5 mm, 5-nerved, submembranous, smooth below, often minutely scabrid towards acute apex; palea 4–5 mm, of similar texture to lemma, 2-nerved, faintly scabrid towards apex; lodicules 0.3–0.6 mm; anthers 2.5–4 mm; caryopsis *c.* 2 mm.

Naturalised from tropical and southern Africa.

N.: scattered throughout; S.: near Westport and to the east from Marlborough to North and Central Otago. Roadsides, grassy areas, lowland and hill country pasture, creek edges, swampy soil, sandy soil.

Pennisetum macrourum was used in ornamental plantings 30–50 years ago and has often escaped and spread vigorously.

5. P. purpureum Schumach., *Beskr. Guin. Pl.* 44 (1827).

elephant grass

Robust tufted perennial to 3 m or more, from stout rhizome to 6 mm diam. Leaf-sheath with fine ± appressed hairs. Ligule densely ciliate, hairs to 3 mm. Leaf-blade 30–75 × 2–3 cm, flat, with numerous fine appressed hairs. Culm internodes with fine hairs for some distance below inflorescence and long soft retrorse hairs above nodes. Panicle *c.* 14×2 cm, cylindric, spike-like with densely crowded spikelets; rachis villous. Spikelets 6.5–7.5 mm, solitary or in clusters of up to 5, straw-coloured, tinged reddish purple; involucre of numerous fine minutely scabrid bristles (*c.* 10 mm), some bristles plumose below and one much stouter and longer (to 20 mm). Lower glume 0, or minute, upper 0.5–1 mm, nerveless or 1-nerved, acute. Lower floret ∅: lemma < spikelet, 1–3-nerved, lanceolate, acute to acuminate; palea 0. Upper floret ♀: lemma = spikelet, 5-nerved, glabrous, acuminate; palea ≈ lemma and of similar texture; anthers *c.* 3 mm, penicillate; gynoecium: ovary 1–1.5 mm, stigma-styles to 9.5 mm; caryopsis not seen.

Naturalised from tropical Africa.

N.: North Auckland (Lake Tangonge, Ahipara, Waiharara). Escape from cultivation; on roadsides.

The lower floret in *P. purpureum* is usually described as ♂ or ∅, ≤ spikelet, and with or without a palea, e.g., Vickery, J. W. *Fl. N.S.W. No. 19, Gramineae* 2: 246–247 (1975); but in the very few specimens examined from the wild the lower floret was ∅, ≪ spikelet, and without palea.

6. P. setaceum (Forssk.) Chiov., *Boll. Soc. Bot. Ital. 1923*: 113 (1923).

fountain grass

Thickly tufted perennials, to 100 cm, light green at base with narrow, inrolled leaves; branching intravaginal. Leaf-sheath rounded, or slightly keeled above, subcoriaceous, margins long-ciliate. Ligule 0.3–0.7 mm, a membranous densely ciliate rim. Collar hairs to 2.5 mm. Leaf-blade 15–30 cm × 0.5–1.5 mm, rather stiff, folded or ±involute, long-tapering, adaxially very scabrid; margins very scabrid, with a few scattered long hairs, tip filiform, acute. Culm 30–45 cm, erect, internodes with many, fine, finely scabrid ridges. Panicle 8.5–20 cm, narrow-cylindric, ± dense, spike-like, often tinged reddish purple; rachis hairs short, soft. Spikelets 5–6–(6.5) mm, lanceolate, acute, pale green to purple, solitary or in clusters of 2–3 on ciliate pedicels to 3 mm; involucre of numerous, plumose, fine, unequal bristles (15–25 mm), one longer, though not stouter bristle (to 35 mm). Lower glume a hyaline scale to 1 mm, or 0, upper 2–2.5 mm, 1-nerved, hyaline, minutely scabrid. Lower floret ∅ or ♂: lemma 4.5–5 mm, 3–(5)-nerved, membranous, nerves minutely scaberulous near tip; palea ≈ lemma, hyaline, rounded, keeled near tip, minutely scaberulous, or palea 0; anthers 2–3 mm, or 0. Upper floret ♀: lemma 5.5–6 mm, 5-nerved, membranous, nerves scaberulous near mucronate tip; palea 4.5–5 mm, hyaline, keels scarcely scaberulous near tip; lodicules 0.3–0.4 mm; anthers 2–3 mm; styles connate, stigmas free; caryopsis *c.* 3 × 1.2 mm.

Naturalised from north-eastern Africa.

N.: North Auckland (Kaitaia, Warkworth), Auckland City, South Auckland (Pokeno, Thames, Hamilton), Wanganui; S.: Nelson (Richmond). Waste land, occasional near ornamental plantings.

Formerly grown in gardens under the name *Pennisetum ruppellii* Steud., a synonym of *P. setaceum*.

7. P. villosum Fresen., *Mus. Senckenberg. 2*: 134 (1837).   feathertop

Tufted perennials to *c.* 35 cm; branching intravaginal from a short branching rhizome to 4 mm diam.; plants sometimes stoloniferous. Leaf-sheath flattened, keeled, firmly chartaceous, mostly glabrous,

occasionally with a few scattered long tubercle-based hairs. Ligule 0.7–
1.5 mm, a membranous densely long-ciliate rim. Collar hairs few, long,
tubercle-based. Leaf-blade 8–30 cm × 2–4 mm, rather stiff, light green
or bluish green, linear, flat, or folded and 0.6–1.5 mm diam., tapering,
abaxially glabrous, strongly ribbed, ribs scabrid near tip, adaxially
shallowly ribbed, minutely scabrid on ribs, sometimes with a few
scattered, long, tubercle-based hairs; margins scabrid, tip acute. Culm
10–30–(45) cm, erect or geniculate-ascending, simple, or branching
at lower nodes, internodes softly long-hairy below inflorescence. Panicle
4–11 cm, conspicuously plumose, broadly cylindric to subglobose,
± dense, light silvery green or light brown, rarely purplish; rachis
softly long-hairy. Spikelets 8–11 mm, lanceolate, acuminate, cream,
sometimes purple-tipped, solitary, or in clusters of 2–4, on very short,
densely soft-hairy pedicels, surrounded by a plumose involucre of
many, cream, sometimes purple, unequal bristles (25–40 mm),
finely scabrid throughout with long soft hairs in lower ½. Glumes
membranous, unequal in size and shape; lower 0.5–1.2 mm, broad,
truncate-erose, nerveless, upper 2.5–5.5 mm, narrow, tapered,
1-nerved. Lower floret ♂ or ∅: lemma 7–9.3 mm, membranous, 7–
9-nerved, very minutely scabrid with longer prickle-teeth on
midnerve near acute tip; palea 6–8 mm, hyaline, keels with scattered
short hairs, interkeel with minute hairs near bifid apex, occasionally
palea 0; anthers 3.5–4.5 mm, occasionally 0. Upper floret ♀: lemma
8.5–11 mm, membranous, 7–9-nerved, glabrous, nerves minutely
prickle-toothed; palea ± rounded, keels becoming scabrid and more
prominent near bifid apex, interkeel minutely scabrid; lodicules
minute, or 0; anthers 3.5–4.5 mm; styles connate; caryopsis
c. 2.5 mm.

Naturalised from Ethiopia.

N.: scattered throughout, except Wellington Province; S.: Marlborough
(Blenheim, Tuamarina, lower Clarence Valley), Canterbury
(Christchurch), Fiordland (Te Anau). Roadsides, pasture, coastal waste
land, shingle banks, gravel bars, grassy street berms, lawns.

Garden escape often known in horticulture as *P. longistylum*, not persisting
vigorously.

ζ**P. glaucum** (L.) R.Br., pearl millet   Annual, with large, very dense spike-like panicles. The single specimen consists of one shoot, 65 cm. Ligule a densely ciliate, membranous rim, hairs 2–3 mm, extending as a contra-ligule round base of leaf-blade almost to midrib. Culm nodes fringed below by dense hairs, *c.* 2 mm, internodes very densely pubescent below panicle. Panicle 18 × 2 cm, cylindric. Spikelets 5 mm, tinged purple above, in clusters of 2–4 within an involucre of slightly shorter bristles; inner bristles plumose, outer shorter, scabrid; one stout bristle > spikelet. A species of America, Africa and India once collected as an escape from cultivation at Te Awamutu, on roadside (LEV 111.411.23.9, 28 Apr. 1986).

SACCIOLEPIS Nash, 1901

Annuals or perennials, herbaceous, often aquatic; of variable habit. Inflorescence a contracted spike-like or rarely open panicle with rather small, often densely clustered spikelets. Spikelets solitary, pedicelled, falling entire at maturity, 2-flowered; lower floret ♂ or ∅, upper floret ♀. Glumes very unequal in length, of similar texture, softly to rigidly membranous, or the lower reduced to a scale and the upper = spikelet, often basally gibbous, (5)–7–9–(13)-nerved. Lower floret: lemma ≈ upper glume but straighter; palea ≈ lemma, finely 2-keeled, or palea almost 0. Upper floret: lemma very convex, chartaceous to subcrustaceous, obscurely 5-nerved; palea = lemma, of similar texture, 2-nerved, scarcely keeled; lodicules 2, small, broadly cuneate, fleshy, glabrous; stamens 3; ovary apex glabrous, stigmas free, terminal or subterminal; caryopsis compressed, embryo *c.* half length of caryopsis, hilum punctiform.

Type species: *S. gibba* (Elliott) Nash

c. 40 spp. mostly in the tropics of both Hemispheres. Naturalised sp. 1.

1. S. indica (L.) Chase, *Proc. Biol. Soc. Wash. 21*: 8 (1908).

Slender almost glabrous annuals, in erect tufts, or decumbent and rooting at lower nodes. Leaf-sheath firmly membranous, rounded. Ligule 0.1–0.2 mm, a membranous, sparsely, short-ciliate rim. Collar hairs short. Leaf-blade 5–10 cm × 2.5–3–(4) mm, narrow-linear, rounded-truncate at base, tapered above, herbaceous, flat, midrib prominent, glabrous; margins very minutely scabrid, tip long, acuminate. Culm 40–60–(70) cm, slender, internodes glabrous. Panicle 3.5–8 cm × *c.* 5 mm, spike-like, cylindric, erect, with densely crowded spikelets; rachis angular, glabrous;

branches numerous, very short, closely appressed to adnate, with spikelets on short, glabrous pedicels widened at apex to minute, concave cup. Spikelets 2.5–3 mm, greyish green, elliptic, acute, ± gibbous. Glumes membranous, strongly nerved, very unequal, glabrous; lower 1.5–1.8 mm, (3)–5-nerved, upper 2.5–3 mm, 9-nerved, gibbous towards base. Lower floret ∅: lemma similar in texture to and ≈ upper glume, 7–9-nerved, oblong, obtuse, hardly inflated at base; palea < 1 mm, very narrow, stiff. Upper floret ♀: lemma *c.* 1.5 mm, elliptic, acute, very shining, cream, nerves not visible, thinly crustaceous, margins closely enfolding palea; palea of same texture as lemma but slightly shorter; anthers 0.6–0.8 mm, deep reddish purple; stigmas deep reddish purple; caryopsis *c.* 1.5 mm.

Naturalised.

N.: North Auckland (North Cape to Kaitaia); K.: Raoul Id, rare. Rough pasture (Raoul Id), damp ground near roadsides (North Auckland).

Indigenous to tropical Asia, Polynesia, western and northern Australia.

SETARIA P.Beauv., 1812 *nom. cons.*

Usually tufted annuals or perennials, of medium height. Leaf-sheath open, or closed near base, often keeled. Ligule a ciliate rim. Leaf-blade flat to folded, rolled in bud. Culm erect, or ascending from geniculate base. Inflorescence usually an erect, dense, cylindric, spike-like panicle, or more branched and open, each spikelet subtended by one to several persistent scabrid bristles (∅ branchlets) almost to entirely suppressed in plants with plicate leaves (*S. palmifolia*). Spikelets 2-flowered, usually falling entire at maturity; lower floret ♂ or ∅, upper floret ♀. Glumes membranous, unequal, glabrous, lower generally the shorter, upper ≤ spikelet. Lower floret: lemma of same texture as glumes, = spikelet, dorsally flattened to slightly depressed; palea sometimes 0. Upper floret: lemma ≤ spikelet, crustaceous, often rugose, naviculate; palea ≈ lemma and similar in texture; lodicules 2, broadly cuneate, rather fleshy; stamens 3; styles free; caryopsis tightly enclosed by hardened anthoecium, embryo *c.* ½ length of caryopsis, hilum basal, punctiform or orbicular.

Type species: *S. viridis* (L.) P.Beauv.

c. 140 spp., of tropical and warm temperate regions, often weedy. Naturalised spp. 7; transient sp. 1.

1 Leaf-blades strongly plicate; panicle spreading, with longer branches pendulous; bristles suppressed below most spikelets, except terminal one on each branchlet, or bristles 0 .... **3. palmifolia**
Leaf-blades flat; panicle contracted, spike-like, cylindric; bristles 1 or more subtending most or all spikelets *2*

2 Bristles 1–3 beneath each spikelet, or up to 5 in some spikelets; margins of leaf-sheath ciliate *3*
Bristles consistently 4 or more beneath each spikelet; margins of leaf-sheath glabrous *5*

3 Bristles retrorsely scabrid .... **6. verticillata**
Bristles antrorsely scabrid *4*

4 Spikelets falling entire at maturity .... **7. viridis**
Spikelets persistent, except for upper floret .... **2. italica**

5 Plant with knotted rhizome; bristles 1.5–6 mm; anthers 0.6–0.9 mm ....
.... **1. gracilis**
Plant tufted; bristles (2)–5–8.5 mm; anthers (0.7)–0.9–1.6 mm *6*

6 Plants perennial; panicle 15–35 cm .... **sphacelata**ζ (p. 589)
Plants annual; panicle (1)–2.5–10 cm *7*

7 Spikelets 2.5–3.5 mm; bristles usually greenish to golden brown ...... **5. pumila**
Spikelets < 2.5 mm; bristles usually reddish brown to purplish ..... **4. parviflora**

1. S. gracilis Kunth, *in* Humboldt, Bonpland and Kunth *Nov. Gen. Sp. 1*: 109 (1816).   knot-root bristle grass

Many-tufted perennials, with mostly inclined culms from an extremely short, slender, knotted, wiry rhizome. Leaf-sheath light green, or distinctly red, or purplish, keeled, submembranous, smooth, sometimes minutely scabrid above; margins glabrous. Ligule ciliate, hairs 0.5–1.2 mm. Collar hairs few, long, tubercle-based. Leaf-blade 5–15 cm × 1.5–6 mm, flat, linear, twisted in upper ½, tapering to scabrid, narrowly acuminate tip, often glaucous, abaxially with obvious midrib, smooth to somewhat scabrid, adaxially scabrid. Culm (2)–20–60 cm, slender, somewhat compressed, internodes very short-pubescent to scabrid just below panicle. Panicle (1.5)–2.5–8.5 cm × 3–8 mm, narrow-cylindric, dense-flowered; rachis densely, softly scabrid-pubescent, with numerous very short branches each bearing a single spikelet subtended by 4–7 unequal, whitish or yellowish, antrorsely scabrid bristles (1.5)–2.5–6 mm. Spikelets 2.2–3 mm, pale green, elliptic, acute. Glumes broadly ovate, obtuse to subacute or apiculate; lower 1–2 mm, 3-nerved, usually < ½ but occasionally to ⅔ length of spikelet, upper 1.5–2.5 mm, 5-nerved, usually ¾ length to occasionally ≈ spikelet. Lower floret: lemma 5–(7)-nerved, ovate; palea = lemma, keels glabrous, rounded. Upper floret: lemma ≈ spikelet, sometimes apiculate, indurated, transversely rugose with close, narrow ridges, convex, greenish at first, later purple-tipped; palea ≈ lemma, somewhat indurated, very finely rugose, keels thickened, smooth except for a few minute prickle-teeth near apex, margins hyaline; anthers 0.6–0.9 mm; caryopsis 1–1.5 mm.

Naturalised from North America.

N.: throughout; S.: Nelson, Marlborough (Wairau Valley), Westland (Greymouth). Pasture, roadsides, waste land, gardens, crops (maize, asparagus), often in sandy soil, often in lawns.

Formerly known as *S. geniculata* (Poir.) Kunth.

N.Z. plants have bristles 1–2–(3) times the length of the spikelets, in this respect resembling some Australian and South American plants [Vickery, J. W. *Fl. N.S.W. No. 19, Gramineae 2*: 236 (1975)], but in general bristles in *S. gracilis* are 3–6 times the length of the spikelet. In N.Z. just as in New South Wales and Victoria (Vickery 1975 op. cit.) some plants are found in which the lower glume is "...up to two thirds and the upper glume *c.* seven-eighths the length of the spikelet".

Field, T. R. O. and Forde, M. B. *Proc. N.Z. Grasslands Assoc. 51*: 47–50 (1990) reported that knot-root bristle grass appeared to be particularly abundant in the Bay of Plenty and Manawatu, and had increased in abundance over the last 2–3 years.

Very variable in height, knot-root bristle grass may grow to 60 cm, or in lawns be dwarfed to 2 cm and still produce seeds (A. E. Esler pers. comm.).

2. S. italica (L.) P.Beauv., *Ess. Agrost.* 51, 170, 178 (1812).

foxtail millet

Rather stout, tufted, wide-leaved annuals. Leaf-sheath light green, submembranous, rounded below to slightly keeled above, glabrous; margins ciliate; some lower sheaths with scattered fine hairs. Ligule ciliate, hairs 0.5–1.5 mm. Collar margins villous, hairs forming a contra-ligule. Leaf-blade (4)–6–15–(20) cm × 4–9 mm, flat, linear-lanceolate, long-tapering, abaxially smooth, adaxially minutely scabrid; margins minutely scabrid, tip filiform. Culm 15–35–(75) cm, internodes ridged, minutely scabrid near panicle. Panicle 5–15 cm × 8–20–(40) mm, dense, cylindric, usually lobed towards base, light green at first; rachis densely short-pubescent, with short-pubescent compound branches, the ultimate minute branchlets bearing 1–4 antrorsely scabrid bristles (4–15 mm) and 2–4 spikelets. Spikelets 2.5–3 mm, persistent except for upper floret. Lower glume 1.5–2–(2.5) mm, 1–3-nerved, ovate, acute or subacute, upper (2)–2.5–3 mm, 5–7-nerved, broad-elliptic, obtuse. Lower floret: lemma 5-nerved, elliptic, obtuse; palea a hyaline rim, or 0. Upper floret shining, becoming yellow to reddish brown or black at maturity and disarticulating from spikelet: lemma = spikelet, crustaceous, ± glabrous, convex, broadly elliptic or oblong; palea of same texture as lemma but somewhat shorter; anthers 0.8–1 mm; caryopsis *c.* 1.5–2 mm.

Naturalised.

N.; S.: scattered. Stony waste land, gardens — from discarded cage-bird seed and seed spillages.

Origin unknown, but probably derived from European *S. viridis*.

3. S. palmifolia (J.König) Stapf, *J. Linn. Soc. Bot. 42*: 186 (1914).

palm grass

Densely clumped perennials from short stout rhizome to *c.* 1 cm diam. Leaf-sheath coriaceous, rounded or slightly keeled, with numerous, spreading, stiff-acicular, tubercle-based hairs to 5 mm; margins densely ciliate or glabrous. Ligule ciliate, hairs to 15 mm, extending as a contra-ligule. Leaf-blade 40–75 × 4–8 cm, conspicuously plicate and palm-like, linear-elliptic, tapering to base, almost petiolate in lower leaves, sharply scabrid especially near and at margins, sometimes with soft hairs as well, tapering to long, acuminate tip. Culm 80–200 cm, terete, striate, with long hairs at nodes and often for some distance below nodes, internodes scabridulous below panicle and nodes. Panicle 25–60 × *c.* 10–20 cm, very loosely branched, the longer branches pendulous; rachis almost smooth below, increasingly scabrid above, occasionally with scattered fine tubercle-based hairs; primary branches (3)–10–16 cm, scabrid, with shorter, scabrid branchlets bearing short-pedicelled, light green to purplish spikelets. Spikelets 3–4 mm, at least some, and usually the terminal one on each branchlet, subtended by a single antrorsely scabrid bristle (5–9 mm). Lower glume 1.5–2.5 mm, 5-nerved, upper 2–3.5 mm, *c.* ⅔–¾ length of spikelet, 5–7-nerved. Lower floret ♂ or ∅: lemma 5-nerved, acuminate; palea *c.* ½ length of lemma, hyaline, nerveless. Upper floret ♀: lemma chartaceous to crustaceous, apiculate, finely rugulose, convex; palea of same texture as lemma but with hyaline margins, keels glabrous, rounded; anthers 1.3–1.6 mm; caryopsis not seen.

Naturalised from India.

N.: Great Barrier Id, Auckland City, South Auckland, Hawkes Bay (Hastings), Wellington (Wanganui). Escape from gardens; waste ground.

Easily distinguishable by the plicate leaves, but the branched panicle, and spikelets subtended by very few bristles give an appearance similar to *Panicum.*

4. S. parviflora (Poir.) Kerguélen, *Lejeunia 120*: 161 (1987).

Laxly tufted often straggling annuals, 18–85 cm, yellow-green, often purplish near base. Leaf-sheath compressed, keeled, submembranous to firm, striate, glabrous. Ligule ciliate, hairs to 1 mm. Leaf-blade (3)–8–15–(20) cm × 2–5 mm, flat, soft, abaxially glabrous, adaxially minutely prickle-toothed on ribs, with a few long (to 8 mm) hairs near base; margins and midrib minutely prickle-toothed above, tip long-tapered, acuminate. Culm 10–30–(75) cm, ±erect, nodes compressed, often dark brown, internodes ridged, glabrous, minutely prickle-toothed to pubescent on ridges just below panicle. Panicle 2–5–(8.5) cm × 6–18 mm, cylindric, dense-flowered; rachis hispidulous, with many, short, close-set, hispidulous branches bearing clusters of 2–3, almost sessile spikelets, each spikelet subtended by 6–11 tawny or reddish brown to purplish, antrorsely scabrid bristles (2–8 mm). Spikelets 2.2–2.4 mm, light green, falling entire at maturity. Lower glume 0.7–1.2 mm, ovate, acute or apiculate, 3-nerved, upper 1–2 mm, ovate-orbicular, obtuse, 3–(5)-nerved. Lower floret ∅: lemma 5-nerved, acute; palea ≈

lemma, hyaline. Upper floret ♀: lemma = spikelet, very convex, crustaceous, finely rugulose, pale cream, sometimes apiculate; palea of same texture but ≤ lemma, keels rounded, glabrous; anthers (0.7)–0.9–1.2 mm; caryopsis *c.* 1.5 × 1.0 mm.

Naturalised.

N.: northern half (scattered — Warkworth, Taipuha, Henderson, Tolaga Bay); K.: Raoul Id. Tracksides, road cuttings, waste land; also recorded as a weed in vineyards.

A polymorphic pan-tropical species, widespread in Malesia; formerly treated as *S. pallide-fusca* (Schumach.) Stapf et C.E.Hubb. or as *S. pumila* subsp. *pallide-fusca* (Schumach.) B.K.Simon based on *Panicum pallide-fuscum* Schumach. (1827). Veldkamp, J. F. *Blumea 39*: 373–384 (1994) equated *Panicum pallide-fuscum* with *Setaria parviflora*, which was based on an earlier name *Cenchrus parviflorus* Poir. (1804).

5. S. pumila (Poir.) Roem. et Schult., *Syst. Veg.* 2: 891 (1817).

yellow bristle grass

Loosely to densely tufted, erect annuals, 25–35–(85) cm, green or yellow-green, often purplish near base. Leaf-sheath strongly compressed, firm, but ± membranous towards margins, glabrous, flattened and keeled. Ligule ciliate, hairs 0.5–1.5 mm. Leaf-blade (3)–5–15–(23) cm × (2.5)–3–5–(6) mm, flat, soft, linear or linear-lanceolate, distinctly twisted, tapering above, glabrous, or adaxially slightly scabrid and with long, fine, soft, tubercle-based hairs towards base; margins very minutely scabrid, tip slender, acuminate. Culm 15–65–(75) cm, compressed, erect or geniculate-ascending, rooting at lower nodes, internodes scabrid, to ± densely minutely pubescent below panicle. Panicle (1)–2.5–10 cm × 5–15 mm, cylindric, dense-flowered; rachis densely pubescent, with numerous, very short, close-set, pubescent branches, each usually bearing a single spikelet, sometimes with 1–2 abortive spikelets, subtended by 5–10 greenish to golden brown, antrorsely scabrid, unequal bristles (5–8.5 mm). Spikelets 2.5–3.5 mm, pale green at first, later purple, falling entire at maturity. Lower glume 0.8–1.7 mm, ovate, acute or apiculate, 3-nerved, upper 1.7–2.3 mm, subacute, 5-nerved, ovate-orbicular, convex. Lower floret usually ∅, rarely ♂: lemma 5-nerved, acute; palea ≈ lemma, hyaline; anthers 0.9–1.4 mm, or 0. Upper floret ♀: lemma = spikelet, crustaceous, prominently transversely rugose, becoming yellowish or brownish, broadly naviculate, very convex, sometimes apiculate; palea of same texture but ≤ lemma, keels rounded and thickened, minutely hairy towards apex, otherwise glabrous; anthers 0.9–1.6 mm; caryopsis *c.* 1.5–2 mm.

Naturalised from southern Europe.

N.: North and South Auckland, Gisborne (Airport), Hawkes Bay (Wairoa), Taranaki (New Plymouth, Patea), Wairarapa (Carterton); S.: Nelson, Marlborough, Canterbury (near Christchurch, Ashburton); K. Open pasture, waste places, roadsides, footpaths, old gardens, cultivated ground.

Setaria pumila has also been known in N.Z. as *S. glauca* and *S. lutescens*. The long discussion over the correct name of yellow bristle grass seems, "to be resolved in favour of *S. pumila* (Poir.) Roem. & Schult. — the familiar *S. glauca* (L.) P. Beauv. being regarded as a species of *Pennisetum*" [Clayton, W. D. *Kew Bull. 33*: 501–508 (1978)].

Setaria pumila resembles *S. gracilis* in the number of bristles subtending the spikelet, but *S. pumila* is an annual whereas *S. gracilis* is a perennial with a short, wiry, twisted rhizome. Spikelets of *S. pumila* are often larger than those of *S. gracilis*.

A. E. Esler (*in litt.*) noted a red form of *S. pumila* growing at Pukekohe. Plants had mainly erect tillers compared to the mainly oblique tillers of the usual green form; ligule *c.* 2 mm; leaf-blade with a conspicuous red patch and smooth adaxially; awns < those in green plants.

6. S. verticillata (L.) P.Beauv., *Ess. Agrost.* 51, 178 (1812).

rough bristle grass

Narrowly tufted annuals, bright green, usually with few branches at base, usually 20–40–(100) cm. Leaf-sheath light green or purplish, submembranous, rounded, to slightly keeled above, smooth apart from minutely scabrid ribs below collar; margins ciliate, hairs tubercle-based. Ligule ciliate, hairs 0.6–1.5 mm. Leaf-blade 3–15 cm × 2–12 mm, flat, soft, linear-lanceolate, tapered to fine tip, adaxially often with scattered long hairs, ribs and margins minutely scabrid. Culm 15–30–(100) cm, internodes ridged, scabrid above on ridges. Panicle 3.5–8.5–(11) cm × 5–15 mm, often drooping, ± cylindric, with light green to purplish, rather loosely, often irregularly clustered spikelets; rachis hispidulous-scabrid on ridges with very short, minutely scabrid branches bearing shortly pedicelled spikelets, each spikelet subtended by 1–2–(3) retrorsely scabrid bristles (2.5–5.5 mm). Spikelets 2–2.4 mm, falling entire at maturity. Lower glume 1–3-nerved, *c.* 1 mm, upper 5–7-nerved, ≈ spikelet, almost covering upper floret at maturity. Lower floret: lemma 5–7-nerved; palea hyaline, nerveless, to ½ length of lemma. Upper floret: lemma = spikelet, elliptic-oblong, subacute, pale cream, convex, thinly crustaceous, finely rugose; palea of same texture as lemma, keels not thickened, interkeel flat; anthers 0.6–0.9 mm; caryopsis *c.* 1.5 mm.

Naturalised from Europe.

N.; S.: scattered throughout. Commonly a garden and crop weed (especially of maize in North Id), waste land, roadsides, footpaths.

Plants which may be the hybrid *S. verticillata* × *S. viridis* were collected in Auckland City (CHR 391110) and in Marlborough at Blenheim (CHR 234839) and Renwicktown (CHR 224995). All three specimens have leaf-sheaths with ciliate margins, spike-like panicles with somewhat distant spikelets and 1–2 bristles below each spikelet; all have mature caryopses. The rachis is hispidulous, a character which distinguishes *S. verticillata* from *S. viridis* in which the rachis is pilose. However, the putative hybrids resemble *S. viridis* in their antrorsely scabrid bristles. The hybrid occurs naturally in southern Europe.

7. S. viridis (L.) P.Beauv., *Ess. Agrost.* 51, 178 (1812).

green bristle grass

Loosely tufted annuals, (4)–30–70 cm. Leaf-sheath light green or yellowish, submembranous, rounded, to slightly keeled above, glabrous, but margins ciliate. Ligule ciliate, hairs 0.7–1.5 mm. Collar hairs few, long. Leaf-blade 6–10.5–(13.5) cm × 3–5 mm, flat, long-tapering, very soft, minutely scabrid; margins minutely scabrid, tip filiform. Culm (10)– 25–45–(60) cm, internodes ridged, scabrid above on ridges. Panicle 3–10 cm × 5–20 mm, very dense, cylindric, tapering above, with very light green or purplish bristles; rachis densely pilose, with short minutely pubescent-scabrid branches, bearing clusters of spikelets on very short discoid-tipped pedicels, each spikelet subtended by 1–3 antrorsely scabrid bristles (4.5–9.5 mm). Spikelets (2)–2.3–3.2 mm, light green, falling entire at maturity. Lower glume 1–3-nerved, *c.* 1.2–1.7 mm, upper 5-nerved, = spikelet, covering the upper ♀ floret. Lower floret: lemma = spikelet, 5–7-nerved; palea hyaline, nerveless, to ½ length of lemma.

Upper floret: lemma = spikelet, elliptic-oblong, subacute or obtuse, pale creamy, convex, thinly crustaceous, very finely rugose; palea of same texture as lemma, nerves faint, interkeel flat; anthers 0.5–0.9 mm; caryopsis *c.* 1.5 mm.

Naturalised from Eurasia.

N.; S.: scattered throughout; Ch. Crops (potatoes, carrots), waste ground, gardens.

ζ**S. sphacelata** (Schumach.) Stapf et C.E.Hubb.   Large perennial tufts. Leaf-sheath strongly keeled, glabrous. Panicle spike-like, to 35 cm; each spikelet subtended by 6–10 yellow-brown antrorsely scabrid bristles (*c.* 5–7 mm). Spikelets *c.* 3 mm, purplish. Anthers *c.* 1.5 mm. A South African species cultivated in experimental plots; one specimen was collected at Te Paki in 1972 from sandy soil around other *Setaria* plants *B. J. Hunt* 20 Jan 1972 (CHR 224715). It also persists at the Kaikohe Research Station in areas beyond where it was sown (W. Harris *in litt.*) and probably also at the Whatawhata Research Station (P. Woods pers. comm.).

SPINIFEX L., 1771

Stout, dioecious perennials, with much-branched, far-creeping stolons; or tussock-forming. Leaf-sheath loose, open. Ligule a ciliate rim. Leaf-blade long, ± convolute, often silky-hairy. ♂ and ♀ inflorescences very dimorphic, the ♂ hemispherical, the ♀ globular, both consisting of clusters of racemes subtended by large bract-like spatheoles, individual racemes subtended by smaller 2-keeled bracteoles; racemes in ♂ plants pedunculate, bearing several spikelets and terminating in a short bristle; racemes in ♀ plants bearing only one spikelet near base and rachis much prolonged to a scabrid bristle. Spikelets 2-flowered, sometimes 1-flowered in ♀ plants. ♂ spikelet: glumes ½ length to ≈ spikelet, unequal, 3–9-nerved, ± chartaceous, ovate-oblong, entire, glabrous or hairy; lemma of lower and upper ♂ florets similar, 3–7-nerved, lanceolate, coriaceous, glabrous or hairy; palea ± similar to lemma in size and texture, strongly 2-keeled; lodicules small, broadly obovate, ± dentate, often ± united; stamens 3 in each floret; ovary rudimentary. ♀ spikelet: glumes ≈ spikelet, 3–9-nerved, chartaceous, ovate to lanceolate, entire, glabrous or hairy; lemmas of lower and upper floret equal or ± unequal, = spikelet, nerves

3 or 0, ± chartaceous, glabrous; lower floret ∅ or 0, sometimes without palea, lodicules small, stamens 3, small, rudimentary on flattened filaments, rarely polliniferous; upper floret ♀, palea ± similar to lemma, lodicules larger, stamens 3, longer, on flattened filaments, pollenless or rarely polliniferous, ovary apex glabrous and long plumose stigmas, caryopsis enclosed in, but free from, hardened anthoecium; embryo large; hilum punctiform.

Type species: *S. squarrosus* L.

4 spp., 3 indigenous to Australia, 1 to Indomalesia. The 1 indigenous N.Z. sp. is found also in Australia.

1. S. sericeus R.Br., *Prodr.* 198 (1810); Holotype: BM! *R. Brown* Broad Sound, 10 Sept. 1802.

Stoloniferous, often forming colonies stretching to 80–(160) m along sand dunes, with much-branched, knotted, rope-like, hard, creeping culms. Leaf-sheath coriaceous, strongly-nerved, silky-hairy. Ligule a minute ciliate rim, hairs very dense to 6 mm. Leaf-blade *c.* 30 cm, inrolled and *c.* 1.5 mm diam., coriaceous, strongly nerved, silky-villous. Culm 2.5–6 mm diam., internodes glabrous, silky-villous below inflorescence. Dioecious: ♂ inflorescence with numerous pedunculate racemes, 5–12 cm, bearing up to 15 silky-villous spikelets, and terminated by a short bristle *c.* 1 cm; raceme clusters subtended by spathaceous bracts ≤ raceme. ♂ spikelets 10 cm; glumes ≤ spikelet, 7–9-nerved; lemmas similar to glumes but less villous, 5-nerved; each floret with 2 emarginate lodicules 0.6 × 0.3 mm, and 3 pollen-filled anthers to 6 mm, at anthesis flexible on long filaments from open florets. ♀ inflorescence very conspicuous, globular, appearing spiny with strict bracts to 15 cm, disarticulating from culm at maturity and wheeling along sand; spikelets solitary, hidden at base of bract, 15–18 mm; glumes = spikelet, 5–7-nerved, silky-villous; lemmas shorter, less villous, rather chartaceous, 3–5-nerved; lower floret ∅, consisting of 2 lodicules, and 3 pollenless white, rudimentary anthers, 0.7–1 mm, on stout filaments, gynoecium 0; upper floret ♀, larger, with 2 lodicules *c.* 1 × 1 mm, and 3 stamens with stout filaments bearing white, pollenless anthers up to 1.5 mm; ovary 1.5–2 mm, stigma-styles 17–20 mm; caryopsis free, *c.* 4.5–5 × 2.5 mm; embryo 3 mm; hilum 0.75–1.0 mm, in a broad depression. $2n = 18$. Plate 5A.

Indigenous.

N.: coastal throughout; S.: Nelson, Marlborough, early records from North Canterbury, near Christchurch and near Dunedin. Sand dunes, almost the sole component of foredunes at many sites in North Id; diminishing in South Id.

The most important indigenous sand binding grass in N.Z., *S. sericeus* was formerly referred to *S. hirsutus* Labill., but Craig, G. F. *Nuytsia 5*: 67–74 (1984), chose a West Australian specimen as type of *S. hirsutus* and reinstated *S. sericeus* for N.Z. and Australian plants.

Spinifex sericeus was recorded from Canterbury, South Id, by Armstrong, J. B. *T.N.Z.I. 12*: 355 (1880); there are specimens in the Armstrong Herbarium, CHBG 5408, 5409, but *S. sericeus* is no longer found in Canterbury. There is one specimen from Dunedin *Buchanan* undated (WELT 76473).

Dioecism in *Spinifex* was discussed by Connor, H. E. *N.Z. J. Bot. 22*: 569–574 (1984); and Connor, H. E. *Blumea 41*: 445–454 (1996). Connor (1984 op. cit.) estimated sex forms in *S. sericeus*, finding that the frequency of ♂ and ♀ colonies is about equal; there were no significant differences in colony size between ♂ and ♀ plants except at Wanganui where males formed significantly larger colonies than females.

STENOTAPHRUM Trin., 1822

Long-rhizomatous, long-stoloniferous or tufted perennials, or annuals. Leaf-sheath ± strongly compressed. Ligule a fringe of hairs. Leaf-blade usually disarticulating from sheath at maturity. Panicles spike-like, narrow, terminal, and often also lateral from the uppermost leaves, pedunculate, composed of numerous very short spike-like racemes embedded in hollows along a dorsally-flattened rachis, sometimes with long racemes closely appressed to rachis; panicle breaking up at nodes at maturity. Racemes bearing 1–5 sessile or pedicelled spikelets along a 3-angled rachis. Spikelets dorsiventrally compressed, falling entire at maturity, 2-flowered, florets awnless; lower floret ♂ or ∅, upper floret ♀. Glumes very dissimilar; lower minute, scale-like, upper ≈ spikelet, membranous,

5–7-nerved; or glumes similar, very minute. Lower floret: lemma = spikelet, chartaceous to coriaceous, 3–7-nerved, nerves sometimes very obscure; palea ≈ lemma, 2-keeled, or palea 0. Upper floret: lemma chartaceous to subcoriaceous, faintly 3–4-nerved; palea ≈ lemma and similar in texture, faintly 2-nerved; lodicules 2, fleshy, cuneate; stamens 3; ovary apex glabrous, styles free or fused near base, stigmas pink, red, purple or black; caryopsis dorsiventrally compressed, embryo large, to ½ length of caryopsis, hilum punctiform, basal.

Type species: *S. glabrum* Trin. *nom. illeg.*

c. 7 tropical and subtropical, coastal spp. Naturalised sp. 1.

1. S. secundatum (Walter) Kuntze, *Rev. Gen. Pl. 2*: 794 (1891).

buffalo grass

Perennial, coarse, glaucous, almost glabrous tufts arising from a short woody rhizome and from long, trailing, tough, arching, compressed, angular stolons; stolons deeply furrowed along one edge and rooting at lower nodes; branching extravaginal from rhizome, culms branching intravaginally above. Leaf-sheath light green to later creamy brown, firm, papery, strongly folded, keeled, narrow membranous margin edged with long soft hairs just below collar. Ligule a densely ciliate rim, hairs *c.* 0.5 mm. Collar villous. Leaf-blade (3)–10–20 cm × *c.* 2–4 mm diam., strongly folded, later flattened, firm, papery, linear, abaxially glabrous, adaxially with occasional long hairs; keel pronounced, narrow; margins often inrolled, somewhat thickened, smooth, but scabrid towards obtuse, almost truncate tip. Culm to 25 cm, nodes glabrous, compressed, internodes glabrous. Panicle 3.5–7 cm, spike-like, terminal, with 1–2 lateral, pedunculate panicles from axils of upper leaves, included within or shortly exserted from uppermost sheath; rachis erect or curved, stout, dorsally flattened and (1.5)–2–3.5 mm wide, bearing numerous, short, alternate, contiguous racemes on one side, often with depressions marking off internodes, disarticulating below each raceme at maturity; racemes bearing (1)–2–3–(5) spikelets, in hollows along 3-angled rachis; raceme-rachis scabrid-angled above and prolonged to a sharp point beside upper spikelet. Spikelets 4–5 mm, elliptic-lanceolate, acute or subacuminate, pale greenish, brownish or purplish; pedicels very reduced, stout. Glumes membranous, very dissimilar; lower 0.5–1 mm, nerveless, obtuse to truncate, glabrous, or apex ciliate, upper 3.5–5 mm, 7–9-nerved, acute, glabrous. Lower floret ♂: lemma = spikelet, 5–9-nerved, subcoriaceous, scabrid near apex; palea ≈ lemma, chartaceous, glabrous, keels rounded and thickened, margins hyaline; anthers 1.6–2.8 mm. Upper floret ♀: lemma 5-nerved, more lanceolate and more acute than lemma of lower floret; both lemma and palea scabrid in upper ½; anthers 2–2.5 mm; stigmas conspicuous, purplish; caryopsis not seen.

Naturalised.

N.: North and South Auckland and offshore islands, scattered further south; S.: Nelson and Canterbury, scattered; K. Usually coastal, often on banks; margins of hot springs.

Indigenous to tropical America and Africa but now naturalised in warm temperate and tropical regions in both Hemispheres.

Buffalo grass is often used as a lawn grass in coastal areas. It forms a coarse lawn and makes a thick, spongy turf if not mown frequently.

No seed has been seen; seed was not recorded from Australian material [Vickery, J. W. *Fl. N.S.W. No. 19, Gramineae* 2: 266 (1975)].

The names *Buchloe dactyloides* and *Bulbilis dactyloides* have been applied in error to *Stenotaphrum secundatum* in N.Z.; both names refer to a chloridoid North American grass which is known there as buffalograss. In North America *Stenotaphrum secundatum* has the common name St Augustine grass.

UROCHLOA P.Beauv., 1812

Annuals or perennials. Ligule a membranous shortly ciliate rim. Culm erect or decumbent, simple or sparingly branched. Inflorescence a panicle of racemes along a common rachis; branches appressed or spreading, terminating in a spikelet. Spikelets secund, ± appressed to rachis adaxially or abaxially, terete or dorsiventrally compressed, solitary, paired or clustered, 2-flowered; disarticulation below glumes; lower floret ∅, ♂, sometimes ♀, upper floret ♀. Lower glume nerveless, upper ≈ spikelet, 5-nerved. Lower floret: lemma 5–7-nerved, membranous to chartaceous, muticous; palea present or 0. Upper floret: lemma ≈ lemma of lower floret, chartaceous to cartilaginous, dull, rugose, margins incurved and clasping palea, apex muticous to mucronate; palea 2-nerved, similar to lemma in texture; lodicules 2, cuneate, truncate; stamens 3; ovary apex glabrous, styles free to base, stigmas plumose; caryopsis ovoid to ellipsoid, dorsiventrally compressed, embryo ½–¾ length of caryopsis, hilum subbasal, punctiform, elliptic or linear.

Type species: *U. panicoides* P.Beauv.

c. 120 pantropical spp. Naturalised sp. 1; transient sp. 1.

In recent years most species of the related genus *Brachiaria* (Trin.) Griseb. have been transferred to *Urochloa*; see especially Morrone, O. and Zuloaga, F. O. *Darwiniana 31*: 43–109 (1992) who made an intensive study of generic characteristics and limits and the relationship of *Urochloa* with other genera of Paniceae.

> *1* Panicle of up to 12 racemes; racemes to 7 cm; spikelets 3–3.5 mm .... **1. mutica**
> Panicle of 2–4 racemes; racemes 2–3 cm; spikelets 4–5 mm ..... **panicoides**ζ

1. U. mutica (Forssk.) T.-Q.Nguyen, *Novit. Syst. Pl. Vasc. Acad. Sci. URSS 13*: 13 (1966).   Para grass

Velvety-leaved clumps to 180 cm, from short, stout rhizome, stoloniferous, sometimes rooting below; branching extravaginal. Leaf-sheath light yellow-green to light brown, papery, rounded, to somewhat keeled above, densely pilose, hairs soft, tubercle-based. Ligule *c.* 1 mm, densely ciliate. Leaf-blade *c.* 20 cm × 5–15 mm, linear-lanceolate, rounded at base, papery, flat, with numerous tubercle-based hairs, ribs numerous, fine, midrib thicker; margins somewhat thickened with close-set, long, prickle-teeth, tapered above to fine acuminate tip. Culm 2–5 mm diam., simple, or narrowly branched, nodes densely villous, internodes glabrous. Panicle of up to 12 racemes on a rachis up to 16 cm. Racemes to 7 cm, of numerous paired spikelets secund on flattened rachis; rachis with scabrid margins and a few, long, bulbous-based hairs; pedicels with a few long hairs. Spikelets 3–3.5 mm, glabrous. Glumes dissimilar; lower *c.* 1.5 mm, 1-nerved, ± triangular, upper ≈ spikelet, 5-nerved, elliptic.

Lower floret ♂: lemma ≈ upper glume and similar in texture, 5-nerved; palea ≈ lemma; anthers *c.* 1.2 mm. Upper floret ♀: lemma *c.* 2.5 mm, 5-nerved, subcoriaceous, very finely papillose; palea *c.* 2 mm, of similar texture to lemma; anthers *c.* 1.5 mm; caryopsis *c.* 1.8 × 1.2 mm.

Naturalised from North Africa.

K.: Raoul Id. Long-abandoned orchard on terraces above The Farm.

Widely introduced in tropical areas as a pasture grass and now naturalised throughout America; formerly treated as *Brachiaria mutica* (Forssk.) Stapf.

Bosappa, G. P. et al. *Can. J. Bot.* 65: 2297–2309 (1987), noted that *U. (Brachiaria) mutica* is "most commonly and efficiently propagated vegetatively. Flowering is rare for this taxon". Only one flowering collection has been made on the Kermadec Is *Sykes 1695* (CHR 472385).

ζ**U. panicoides** P.Beauv.   Low-growing, 15–30 cm, spreading from rosette-like base, rooting at lower nodes. Leaf-sheath ciliate on margins. Ligule a densely ciliate rim. Leaf-blade soft, light green, with scattered short hairs; margins crimped, with a few longer hairs. Panicle of 2–4 racemes, 2–3 cm. Spikelets shortly pedicelled, 2-flowered, in 2 secund rows on flattened rachis. Lower glume < 1 mm, hyaline; upper *c.* 4–5 mm, herbaceous, = spikelet. Lower floret: lemma similar to upper glume but flattened, nerves anastomosing below apex. Upper floret: lemma finely, transversely rugose. This tropical annual sp. from South Africa, India and Mauritius has been collected from one locality, Auckland City, One Tree Hill, 5 Korokino Road, growing in waste land beside house *A. E. Wright 2658* 21.4.1978 (AK 144229; duplicate CHR 339914); it was still there at the same site eleven years later *A. E. Wright 8853* 12 Mar 1989 (AK 183947; duplicate CHR 450085).

ζ**Melinis repens** (Willd.) Zizka, red Natal grass   Tufts up to 40 cm. Culms erect, *c.* 2 cm diam., geniculate at base. Ligule rim-like, densely ciliate. Panicle *c.* 15 cm, open; branches and pedicels very slender, villous. Spikelets numerous, 2-flowered, with very distinct, copious, long, silky, white to pinkish or purplish hairs covering and > shortly awned upper glume and lower lemma. This species of tropical and South Africa was collected from the wild in North Auckland *H. B. Matthews* May 1909 (CHR 20568), and, *c.* 30 years later, it was collected from waste land in a garden at Takapau, Hawke's Bay *A. C. Malin* 1.6.1980 (MPN). Recorded as *Rhynchelytrum repens* (Willd.) C.E.Hubb. by Edgar, E. and Shand, J. E. *N.Z. J. Bot.* 25: 346 (1987) in a checklist of naturalised panicoid grasses.

20. ISACHNEAE

Infloresence terminal, a panicle, or composed of racemes. Spikelets 2-flowered, disarticulating at maturity above glumes, solitary, all similar. Glumes membranous, ≤ spikelets. Lower floret usually ♀ or sometimes ♂, upper floret ♀ or ♀. Lemmas membranous to coriaceous.

ISACHNE R.Br., 1810

Annuals or perennials, often aquatic, with culms often geniculate at base. Ligule reduced to a ciliate rim. Leaf-blade usually closely and prominently ribbed adaxially. Inflorescence a loose or open panicle, or contracted, with small or minute pedicelled spikelets borne singly along the branches. Spikelets dorsiventrally compressed, 2-flowered, florets awnless separated by well-developed rachilla; disarticulation above glumes and usually between florets; lower floret ♀ or ♂, upper floret ♀ or ♀. Glumes ± equal, from *c.* ⅔ to ≈ spikelet, membranous, glabrous or hairy; lower 3–9-nerved, upper 5–9-nerved. Lower floret: lemma = spikelet or < glumes, membranous, chartaceous or finely coriaceous, glabrous or minutely hairy, obscurely 5–7-nerved; palea = lemma, 2-nerved. Upper floret: lemma of similar size and texture to lemma of lower floret, or somewhat smaller and firmer, often hairy; palea similar to lemma with 2 well-separated nerves; lodicules 2, minute, cuneate, fleshy, glabrous; stamens 3; ovary apex glabrous, styles free, terminal; caryopsis dorsiventrally compressed, embryo large, ⅓ length of caryopsis, hilum basal, > ½ length of caryopsis. Fig. 23.

Type species: *I. australis* R.Br.

c. 80 spp. in tropical to warm temperate regions of both Hemispheres but mainly in tropical Africa. Indigenous sp. 1.

1. I. globosa (Thunb.) Kuntze, *Rev. Gen. Pl. 2*: 778 (1891) ≡ *Milium globosum* Thunb., *Fl. Jap.* 49 (1784); Holotype: UPS! *Thunberg 2041* Japan. = *Isachne australis* R.Br., *Prodr.* 196 (1810); Holotype: BM!

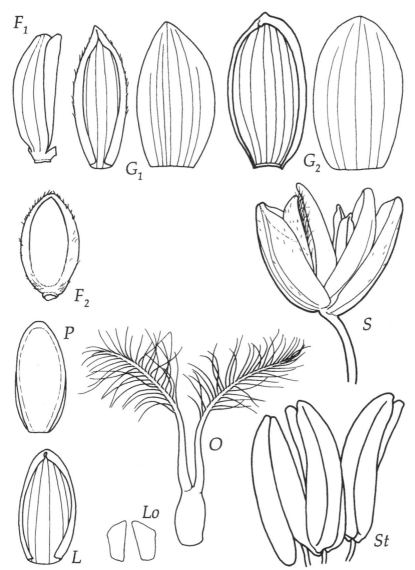

Fig. 23 *Isachne globosa*
S, spikelet, × 16; **G₁**, **G₂**, glumes ventral and dorsal views, × 16; **F₁**, lower ♂ floret lateral view, × 16; **F₂**, upper ♀ floret ventral view, × 16; **L**, lemma ventral view, × 16; **P**, palea dorsal view, × 16; **Lo**, lodicules, × 27; **O**, ovary, styles and stigmas of ♀ floret, × 27; **St**, anthers of ♂ floret, × 27.

R. Brown 6129 Nepean [New South Wales]. = *Panicum gonatodes* Steud., *Syn. Pl. Glum. 1*: 95 (1854); Holotype: P, *n.v.*, *Urville* N. Zeeland.

Semi-aquatic, ± glabrous perennials, with decumbent stems branching near base and becoming erect at flowering; branching intravaginal. Leaf-sheath light green to brownish, submembranous, folded closely around culm, glabrous but ciliate along one margin and along upper part of other. Ligule 0.4–1.8 mm, ciliate. Leaf-blade 3.5–9–(15) cm × 2.5–6.5 mm, flat, linear-lanceolate, gradually narrowed above, rather stiff, finely scabrid, tip acute. Culm (15)–30–60–(110) cm, slender, subterete, internodes glabrous. Panicle 5–15 cm, contracted at first, pedunculate and spreading when fully exserted from upper sheath, pyramidal; rachis and branches very slender, smooth or occasionally minutely, sparsely scabrid, often flexuous, with long-pedicelled spikelets borne singly towards tips of filiform branchlets. Spikelets 2–3 mm, oblong, to obovate to obovate-globose, greenish or slightly purplish especially towards very obtuse tip; disarticulation above glumes and between florets; rachilla 0.2–0.4 mm, remaining attached to base of upper floret. Glumes ± equalling spikelet, 7–9-nerved, chartaceous, smooth, or sparsely scabrid on outer nerves, especially near incurved tip. Lower floret ♂, ≈ spikelet: lemma 2–3 mm, texture similar to upper glume, ovate-elliptic, obtuse, glabrous; palea similar in size and texture to lemma; anthers 1–1.7 mm. Upper floret ♀: lemma 1.3–1.8 mm, oblong to suborbicular, obtuse, slightly indurated; palea = lemma, of similar texture; gynoecium: ovary 0.5–0.7 mm, stigma-styles 1.8–2 mm; caryopsis *c.* 1 × 0.8 mm. Fig. 23.

Indigenous.

N.: throughout, especially in northern half; S.: Westland (Mahers Swamp). Edges of swamps and lake margins, often partly submerged, sometimes in deep gullies, coastal.

Also indigenous to South-East Asia, Japan, and Australia.

21. ANDROPOGONEAE

Inflorescence of solitary or digitate or panicled spiciform racemes. Spikelets 2-flowered, usually in pairs, sometimes in threes or solitary, one spikelet in each pair or triplet sessile, the other(s) pedicelled on segments of raceme-rachis; falling entire at maturity; lower floret ♂ or Ø, upper floret ♀. Glumes firmer than lemmas, the lower ≥ spikelet. Lemmas membranous or hyaline; lemma of upper floret entire or 2-lobed or notched, usually with geniculate awn from sinus or tip. Paleas < lemmas, usually hyaline, often absent in lower or in both florets. Caryopsis loosely enclosed by glumes or lemma; hilum usually basal and punctiform.

1 Spikelets unisexual; inflorescence of 2 racemes, ♀ raceme within a bony globose utricle, ♂ raceme projecting from utricle .... **Coix**ζ (p. 613)
Spikelets, or at least one of each pair, bisexual; inflorescence lacking a bony utricle 2

2 Spikelets all pedicelled, one of each pair shortly pedicelled, the other long-pedicelled; pedicels free *3*
One spikelet of each pair sessile, the other pedicelled; pedicels free, or fused to internodes *4*

3 Leaf-blades linear; lemma of ♀ floret awned; plants tufted .... **Miscanthus**
Leaf-blades narrowed towards base; lemma of ♀ floret awnless; plants with long-creeping rhizome .... **Imperata**

4 Inflorescence branches bearing only spikelets or racemes *5*
Inflorescence branches bearing spathes which subtend groups of spikelets or racemes 6

5 Segments of raceme-rachis and pedicels with a translucent median line ....
.... **Bothriochloa**
Segments of raceme-rachis and pedicels without a translucent median line
.... **Sorghum**

6 Pedicels fused to internodes .... **Hemarthria**ζ (p. 613)
Pedicels not fused to internodes 7

7 Spathes subtending a group of spikelets, composed of 2 ♀ spikelets within an involucre of Ø spikelets .... **Themeda**
Spathes subtending racemes of paired spikelets .... **Andropogon**

ANDROPOGON L., 1753

Usually perennials, rather coarse, of variable habit; long-rhizomatous, or caespitose, or straggling, or decumbent and rooting at nodes. Leaf-sheath open, or margins fused near base. Ligule membranous, sometimes ciliate. Leaf-blade variable in shape. Inflorescence with numerous racemes aggregated on an exserted peduncle, or racemes single or paired or few, the common peduncle usually enclosed by a spathe-like sheath; several spatheate raceme-clusters often forming a narrow or sometimes dense compound panicle. Racemes with paired spikelets, one sessile to shortly pedicelled, one longer-pedicelled, each pair borne at a node of the rachis which disarticulates at maturity; rachis and spikelet pedicels often densely villous. Spikelets 2-flowered, lower florets ∅, upper floret ♀ in ± sessile spikelets, or ♂, ∅ or 0 in pedicelled spikelets. Pedicelled spikelet: ≈ sessile spikelet, florets awnless, sometimes reduced to 1 or 2 glumes or 0, with only pedicel present. Sessile spikelet: glumes very dissimilar, ± equal, subcoriaceous to membranous, glabrous or hairy; lower flat, concave or canaliculate, 2-keeled, 1–11-nerved, margins folded, upper naviculate, keeled above, 1–3-nerved, sometimes awned; lemmas < glumes, hyaline; lemma of upper ♀ floret awned, sometimes substipitate below awn; palea hyaline, nerveless, reduced or 0; lodicules fleshy, hairy or glabrous; stamens 3; ovary apex glabrous; caryopsis free from anthoecium, embryo large, *c.* ⅓ caryopsis, hilum punctiform or shortly elliptic, < ½ caryopsis.

Type species: *A. distachyos* L.

c. 100 spp. of tropical and subtropical regions. Naturalised sp. 1.

1. A. virginicus L., *Sp. Pl.* 1046 (1753).   broomsedge

Erect, coarse, perennial tufts, 50–70 cm, light greenish brown, tinged reddish; branching intravaginal. Leaf-sheath keeled, strongly folded, firmly chartaceous, light green to straw-coloured, glabrous but margins long-villous. Ligule *c.* 0.5 mm, a ciliolate rim. Leaf-blade 5–20 cm × *c.* 2–5 mm, linear, coriaceous, flat or folded, somewhat keeled with prominent midrib, abaxially glabrous, adaxially finely scabrid on ribs and long-villous towards base; margins finely scabrid. Culm *c.* 25 cm, compressed, internodes glabrous. Inflorescence 25–40 cm, narrow-linear, interrupted; branches slender, erect, subtended by spathe-like leaves and bearing a few spatheate clusters of racemes at apex; upper spathes 3–6 cm, acuminate, tawny. Racemes fragile, *c.* 2 cm, 2–4 on a very short common peduncle, shorter than and mostly included within spathe; rachis very slender, long-villous. Pedicelled spikelet: usually reduced to villous pedicel, rarely with 1–2 glumes. Sessile spikelet ♀: *c.* 3 mm, concealed by hairs, lanceolate, pale greenish; glumes very narrow, keels minutely scabrid; lemma of upper floret with very delicate straight awn, 10–15 mm; anthers and caryopsis not seen.

Naturalised from America.

N.: North Auckland (Karikari Peninsula, Warkworth, Albany, Glenbrook), Auckland City, South Auckland (Waiotapu). Roadsides and along railway line.

Andropogon is cited in Healy, A. J. *Standard Common Names for Weeds in N.Z.* ed. 2, 128 (1984), as being a genus whose seeds are listed as impurities in imported commercial seed.

BOTHRIOCHLOA Kuntze, 1891

Tufted, usually erect perennials with slender, simple or branched culms. Ligule membranous, ± ciliate. Leaf-blade linear, flat or revolute. Inflorescence of subdigitate or paniculate racemes on short peduncles. Racemes bearing paired spikelets, naked below; rachis fragile, rachis segments and spikelet pedicels filiform, ± hairy, longitudinally grooved, groove ± hyaline. Spikelets ± hairy, one sessile, one pedicelled, each spikelet pair deciduous with adjacent segment of rachis attached. Pedicelled spikelet ∅, lemma, if present awnless, hyaline and nerveless. Sessile spikelet dorsally compressed, with small, obtuse, bearded callus, 2-flowered; lower floret ∅, upper floret ♀, awned. Glumes very dissimilar, ± membranous; lower 2-keeled, 5–9-nerved, sometimes with circular depression (pit) about the middle, upper narrower, 1-keeled, 1–4-nerved. Lower floret: lemma hyaline, nerveless. Upper floret: lemma reduced to a hyaline base topped by a twisted awn; palea small, or 0; lodicules distally fleshy, glabrous; stamens 3, or 1; ovary apex glabrous, styles ± terminal, free; caryopsis turgid, embryo ½ length of caryopsis, hilum punctiform or shortly elliptic.

Type species: *B. anamitica* Kuntze

c. 25–30 spp. in tropical to warm temperate countries. Naturalised spp. 2.

1 Racemes paniculately arranged, with rachis naked below for 1–1.5 cm; sessile
　spikelets *c.* 3 mm .... **1. bladhii**
　Racemes subdigitate, with rachis naked near base for *c.* 0.5 cm; sessile spikelets
　4.5–5.5 mm .... **2. macra**

1. B. bladhii (Retz.) S.T.Blake, *Proc. Roy. Soc. Queensland 80*: 62 (1969).

Tufted, winter dormant, to 1 m, with greenish or glaucous leaves; branching extravaginal. Leaf-sheath straw-coloured, subcoriaceous, scarcely keeled above, glabrous. Ligule 0.5–1 mm, truncate, short-ciliate. Leaf-blade 10–20 cm × 2.5–3.5 mm, smooth or scabrid, keel and ribs prominent; margins minutely scabrid, sometimes a few, long, fine, tubercle-based hairs near ligule, long-tapered to fine, acicular tip. Culm to *c.* 90 cm, often branched above, nodes hairy, internodes glabrous. Racemes 5–8, very slender, dark reddish purple, paniculately arranged, 2–4 cm, naked below for 1–1.5 cm, short hairs in axils; spikelet pedicels *c.* 2 mm, margins densely ciliate. Pedicelled spikelet ♂: *c.* 3 mm, glumes, lower lemma and 3 stamens as in sessile spikelet; or Ø and composed of lower glume only and very reduced upper glume. Sessile spikelet: *c.* 3 mm surrounded by callus hairs *c.* 0.5 mm; glumes submembranous, lower = spikelet, 5–7-nerved, elliptic-oblong, obtuse, with hairs on keels and scattered in lower ½, elsewhere glabrous, slightly depressed at centre, upper slightly shorter, lanceolate, acute, finely scabrid on keel near tip and minutely sparsely hairy near margins, elsewhere glabrous; lemma of lower floret ≤ glumes, ± oblong, tip erose; lemma of upper floret narrow at base and topped by a stout brown geniculate 12–15 mm awn; palea 0; stamens 3, anthers *c.* 1 mm; caryopsis *c.* 1.3 × 0.5 mm, brownish; embryo *c.* 0.6 mm; hilum basal, 0.2 mm.

Naturalised.

K.: Raoul Id (Low Flat). Stenotaphrum meadow.

Indigenous to eastern Asia and Australia.

2. B. macra (Steud.) S.T.Blake, *Proc. Roy. Soc. Queensland 80*: 64 (1969). red-leg grass

Loosely or densely tufted, sometimes rather straggling, geniculate at base with erect culms; branching intravaginal. Leaf-sheath pale creamy green to light brownish, subcoriaceous, keeled, glabrous, rarely with one or two long, fine, tubercle-based hairs towards base near margin, or sparsely hairy throughout. Ligule 0.5–1 mm, finely ciliate. Leaf-blade 4.5–8 cm × 2–3 mm, lanceolate, scabrid-papillose and with long, fine, sparse, tubercle-based hairs, keel prominent; margins minutely scabrid, often with a few, long, fine, tubercle-based hairs near ligule; long-tapered to acuminate tip; culm-leaves subpseudopetiolate. Culm (15)–20–60–(80) cm, nodes usually pubescent, internodes occasionally with scattered hairs below inflorescence. Racemes (2)–3–(4), subdigitate, 2–6 cm, naked near base for *c.* 0.5 cm, erect or curving but not spreading, shortly pedunculate, peduncles 2–5.5 mm, occasionally sparsely hairy, usually with a few hairs in axils between racemes; spikelet pedicels 3–4 mm, margins densely ciliate. Pedicelled spikelet ∅: 2.5–4.5 mm, consisting of glumes only; lower glume 7–9-nerved, linear-lanceolate, subobtuse, margins and nerves near tip ciliate-scabrid, upper usually much shorter, hyaline; rarely lower pedicelled spikelets ♂ with small hyaline lemma and 3 stamens, anthers *c.* 2 mm. Sessile spikelet: 4.5–5.5 mm, greenish purple, surrounded at base by callus hairs to 1 mm, glumes thinly coriaceous, lower = spikelet, 5–7-nerved, linear-lanceolate, with scattered hairs especially in lower ½, tip obtuse with short stiff hairs, glume not pitted but sometimes with a depression at centre, upper slightly shorter, finely scabrid on keel near tip; lemma of lower floret *c.* ¾ length of glumes, ovate-lanceolate, tip erose; lemma of upper floret, narrow at base and topped by a stout, brown, geniculate 20–30 mm awn; palea 0; stamens 3, anthers 1.2–1.8 mm; gynoecium: ovary 1–1.2 mm, stigma-styles 3.5–4 mm; caryopsis *c.* 2.5 mm, purplish, embryo *c.* 1 mm, hilum basal *c.* 0.3 mm.

Naturalised from Australia.

N.: North Auckland to Auckland City, Cuvier Id, Gisborne; S.: Nelson, Marlborough. Dry grassy hillsides near coast.

Formerly known as *Dichanthium annulatum*.

Seeds of Australian *B. decipiens* (Hack.) C.E.Hubb., pitted bluegrass, were listed as impurities in imported commercial seed [Healy, A. J. *Standard Common Names for Weeds in N.Z.* ed. 2, 133 (1984)] but plants are not known to occur in the wild in N.Z. This 1-anthered sp. in which the lower glume of the sessile spikelet is usually pitted, is closely allied to *B. macra* and may be confused with it.

IMPERATA Cirillo, 1792

Perennials, tufted from long, many-noded rhizomes, with leaves crowded at base. Leaf-sheath firm. Ligule membranous, ± ciliate. Leaf-blade long, flat, stiff, sometimes pseudopetiolate. Culm erect, unbranched, few-noded. Inflorescence a spike-like or narrowly branched silvery panicle; branches bearing numerous, usually paired, unequally pedicelled spikelets. Spikelets 2-flowered; lower floret usually reduced to a lemma, rarely ♂, upper floret ♀. Glumes subequal, (0)– 3–7-nerved, membranous, firmer below, enveloped by long silky hairs from very obtuse basal callus and lower portions of glumes. Lemmas 0–(1)-nerved, often < glumes, hyaline, awnless; lemma of lower floret ± broadly oblong, often denticulate; lemma of upper floret shorter, lanceolate to oblong, rarely 0. Palea broad, hyaline, nerveless, denticulate. Lodicules 0. Callus hairy. Rachilla not prolonged. Stamens 1–2 (in N.Z.)–3. Ovary glabrous; styles fused near base. Caryopsis oblong, loosely enclosed by lemma and palea; embryo ½–⅔ length of caryopsis; hilum elliptic, basal.

Type species: *I. arundinacea* Cirillo *nom. illeg.*

10 spp. in tropical and subtropical regions of both Hemispheres. Endemic sp. 1; naturalised sp. 1.

> *1* Panicle narrow-lanceolate with lower branches ± spreading; stamen 1; culm nodes ± glabrous .... **1. cheesemanii**
> Panicle densely spiciform with appressed branches; stamens 2; culm nodes pubescent .... **2. cylindrica** var. **major**

1. I. cheesemanii Hack., *T.N.Z.I. 35*: 378 (1903); Holotype: W! *T. F. Cheeseman* Kermadec Islands, August, 1887 (No. 1001 to Hackel).

Forming dense, large mats or scattered clumps, 50–80 cm; rhizomes 2–3 mm diam., internodes covered by bracts; branching extravaginal. Leaf-sheath light cream to brown, glabrous. Ligule 1 mm, truncate, ciliolate. Leaf-blade 12–50 × 0.5–1.5 cm, linear from pseudopetiolate base, glabrous apart from hairs on margins near ligule; margins sparsely scabrid above, tip acute, pungent. Culm 50–80 cm, nodes and internodes glabrous. Panicle 10–20 × 2–4 cm, narrow-lanceolate, gradually narrowed upwards to an acute point, greyish white with soft, dull hairs, *c.* 6 mm, concealing spikelets; branches erect or the lower ± spreading, simple or branched again near base. Spikelets *c.* 3 mm, enveloped by hairs to 6 mm. Glumes = spikelet, purplish, with some scattered long hairs; lower 5-nerved, upper 3-nerved. Lemmas and palea < glumes, hyaline, nerveless, denticulate. Stamen 1; anther *c.* 2 mm. Stigmas purple. Caryopsis not seen.

Endemic.

K.: Raoul Id. Common to locally abundant in many open areas.

2. I. cylindrica var. **major** (Nees) C.E.Hubb. *in* Hubbard and Vaughan *Grasses of Mauritius and Rodriquez* 96 (1940).

lalanga, bladey grass

Forming loose or compact tufts or colonies from long-creeping, tough, scaly rhizomes *c.* 3 mm diam. Leaf-sheath light brown, sometimes purplish, glabrous or with scattered fine hairs above, shredding into stiff fibres at maturity. Ligule *c.* 1.5 mm, truncate, ciliate. Leaf-blade to 1 m × 4–10 mm, often > culms, linear from pseudopetiolate base, glabrous; margins scabrid, tip acuminate. Culm 20–50 cm, upper nodes villous, hairs fine, to 6 mm, internodes glabrous. Panicle 9–15 × *c.* 2 cm, dense, spiciform, silky white, shining; branches very numerous, very slender, erect, ±appressed to rachis; pedicels with some long fine hairs below spikelets. Spikelets *c.* 4 mm, enveloped by hairs 9–14 mm. Glumes = spikelets, brownish, 3–7-nerved. Lemmas and palea < glumes, hyaline, often denticulate, tips ciliate. Stamens 2; anthers 2–3 mm, orange or brownish. Stigmas purple or brown. Caryopsis not seen.

Naturalised from Palaeotropics.

N.: Northland (near Kaitaia, Kerikeri, Whangarei). Open gullies, sunny places on hills, flats on volcanic soil near coast.

MISCANTHUS Andersson, 1855

Robust, tufted, tall bamboo-like perennials. Leaf-sheath loose. Ligule a truncate densely ciliate rim. Leaf-blade long, rather narrow. Culm erect, solid, leafy. Inflorescence a large, fan-shaped or corymbiform, loose, open panicle; branches consisting of long racemes with very tough glabrous rachis bearing many paired spikelets on unequal glabrous pedicels. Spikelets 2-flowered, surrounded by an involucre of long fine hairs from the very short obtuse callus; lower floret ∅; upper floret ♀. Glumes ± chartaceous, muticous. Lower floret: lemma hyaline, awnless; palea 0. Upper floret: lemma shorter, hyaline, bidentate at apex with delicate awn arising in the notch; palea small, hyaline, nerveless; lodicules 2, cuneate, truncate, glabrous, or 0; stamens 2, or 3; ovary apex glabrous, styles free; caryopsis oblong, embryo ½ length of caryopsis, hilum punctiform or elliptic, < ½ length of caryopsis. Fig. 24.

Type species: *M. capensis* (Nees) Andersson

20 spp. in tropical regions, also in South Africa, Japan and the Philippines. Naturalised spp. 2.

> *1* Panicle golden-brown; florets ¼–⅓ involucral hairs .... **1. nepalensis**
> Panicle creamy- or pinkish-brown; florets ≈ involucral hairs .... **2. sinensis**

1. M. nepalensis (Trin.) Hack. *in* de Candolle, A. *Monogr. Phan. 6*: 104 (1889).   Himalayan fairy grass

Tall robust clumps. Leaf-sheath stiff, ± keeled, often conspicuously purple, with scattered long silky hairs, more densely hairy below collar. Ligule 2–3.5 mm, membranous, obtuse, short-ciliate. Leaf-blade 20–60 cm × 4–10 mm, stiff, flat or folded, gradually tapering, midrib pale green, conspicuous below, abaxially with scattered fine hairs, adaxially glabrous; margins closely, finely scabrid, tip very finely acute. Culm 40–80 cm, stiff, purple-green to yellow-green, nodes purplish, internodes with long fine silky hairs below panicle. Panicle 10–20 cm, golden brown, ± fan-

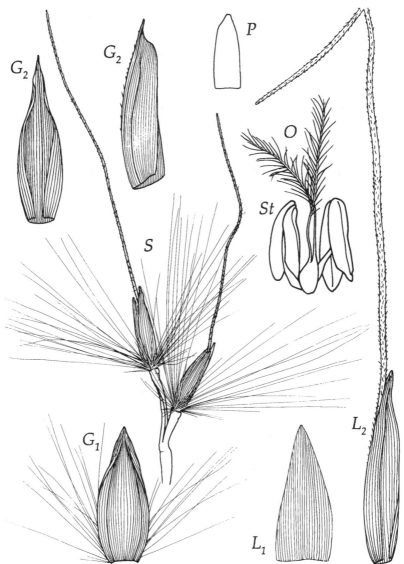

Fig. 24 *Miscanthus nepalensis*
S, spikelet pair, × 9; **G₁**, lower glume ventral view, × 24; **G₂**, upper glume ventral and lateral view, × 24; **L₁**, lemma of lower ∅ floret, × 24; **L₂**, lemma of upper ♀ floret lateral view, × 24; **P**, palea of upper floret, × 24; **O**, ovary, styles and stigmas, × 24; **St**, stamens, × 24.

shaped; rachis long silky hairy. Racemes 3.5–10.5 cm, numerous, silky, ≫ panicle rachis. Spikelets numerous; florets ¼–⅓ length of soft, brownish cream, spreading involucral hairs; pedicels purplish, short pedicels = spikelets, long pedicels ≫ spikelets, ± 3-angled. Glumes 1.6–2.8 mm, ± equal, oblong-lanceolate, acute, brownish, finely mucronate or entire, minutely scabrid above on nerves and often purple towards hyaline tip; lower 3–(5)-nerved, margins ciliate, upper 1–(3)-nerved. Lower floret ∅: lemma 1.6–2.6 mm, lanceolate, acute, sparsely, finely prickle-toothed at centre. Upper floret ♂: lemma 1.6–2.2 mm; awn 8.5–11 mm; palea 1–1.5 mm, apex sometimes irregularly bifid; lodicules 0; anthers 1–1.4 mm; gynoecium: ovary 0.5–0.7 mm, stigma-styles 2.6–3 mm; caryopsis 1–1.5 mm. Fig. 24.

Naturalised from India.

N.: northern half. Waste areas, roadsides.

A garden escape which is spreading.

2. M. sinensis Andersson, *Öfvers. Förh. Kongl. Svenska Vetensl.-Akad. 12*: 166 (1855).   Chinese fairy grass

Erect, robust clumps to 2 m, from stout, shortly branched rhizomes. Leaf-sheath stiff, papery, with scattered fine hairs, closely appressed to culm, slightly keeled above; margins with a few hairs near ligule. Ligule 3.5–6.5 mm, long-ciliate. Leaf-blade 30–65–(120) cm × 8–14 mm, tapering, abaxially midrib very wide, pale greenish cream, adaxially glaucous, with scattered appressed fine straight hairs; margins sharply, closely, finely scabrid, tip long, very fine. Culm 120–180 cm, tough, internodes glabrous. Panicle 15–25 cm, creamy- or pinkish-brown, ± fan-shaped, rachis glabrous or slightly soft-hairy. Racemes 8–15 cm, numerous, silky, ≫ panicle rachis. Spikelets numerous; florets almost equalling soft, white involucral hairs, short pedicels < spikelets, long pedicels > spikelets, ± 3-angled, finely prickle-toothed on angles. Glumes 4–5 mm, ± equal, straw-coloured, acuminate, rounded; lower 3-nerved, scabrid above, upper 1-nerved, margins membranous, incurved, finely ciliate. Lower floret ∅: lemma 3.5–4 mm, acute, margins ciliate. Upper floret ♂: lemma 3–4 mm, awn 5–7.5 mm; palea 1–1.5 mm; lodicules 2; anthers 2–2.3 mm; gynoecium: ovary *c.* 1 mm, stigma-styles *c.* 3 mm; caryopsis *c.* 2.5 × 0.7 mm.

Naturalised from eastern Asia.

N.: scattered throughout; S.: Nelson (Ngatimoti), Buller (Westport), Canterbury (Ashburton). Waste land, roadsides.

A garden escape; the variegated form *M. sinensis* var. *zebrinus* Beal with white bands at intervals across the leaves, is often grown in gardens and has also been collected from the wild.

SORGHUM Moench, 1794 *nom. cons.*

Usually robust, tall annuals or perennials. Leaf-sheath open. Ligule membranous, ciliate. Leaf-blade usually flat, often large. Culm simple. Panicle large, open, erect or ± nodding, or densely contracted in cultivated grain-bearing spp., with strong rachis and rather slender verticillate or scattered branches, each bearing several racemes with paired dissimilar spikelets, the terminal group of spikelets a triplet; each pair consisting of one sessile and one pedicelled spikelet, the sessile spikelet falling with the contiguous rachis segment and accompanying pedicelled spikelet, or at least with its pedicel, or persistent. Spikelets 2-flowered, lower floret in all spikelets ∅ and reduced to an empty lemma, upper floret ♀ in sessile spikelet, ♂, ∅ or 0 in pedicelled spikelets. Pedicelled spikelets: narrower than sessile spikelets; glumes lanceolate, herbaceous; lemma awnless; palea 0; sometimes reduced to both glumes or 1 glume. Sessile spikelets: glumes equal, coriaceous and shining at maturity, muticous; lower 5–11-nerved, flattened or rounded, margins involute, upper naviculate, 3–11-nerved, keeled above, margins narrow, hyaline, usually ciliate; lemmas and palea hyaline, enclosed by glumes; lemma of lower floret 2-nerved or nerveless; lemma of upper floret 1–3-nerved, apex ± lobed, awned or mucronate from sinus, rarely muticous; palea often minute or 0; lodicules 2, ciliate or glabrous; stamens 3; ovary apex glabrous, styles ± terminal, free, stigmas laterally exserted; caryopsis free in anthoecium, embryo *c.* ½ length of caryopsis, hilum punctiform or shortly elliptic, < ½ length of caryopsis.

Type species: *S. bicolor* (L.) Moench

c. 60 spp. of tropical and subtropical regions, with many cultivated forms and varieties. Naturalised spp. 2.

1 Plants annual, tufted .... **1. bicolor**
   Plants perennial, strongly rhizomatous .... **2. halepense**

1. S. bicolor (L.) Moench, *Methodus* 207 (1794).

forage sorghum, grain sorghum
Annual robust tufted plants to 1.5 m. Leaf-sheath chartaceous, somewhat keeled near collar, glabrous, or occasionally with sparse, long, fine hairs, straw-coloured, often suffused or blotched reddish purple. Ligule 2–3 mm, stiffly membranous, ciliate. Leaf-blade 30–90 × (1)–2.5–7 cm, with strong pale creamy midrib, linear-lanceolate, long-tapered, glabrous or rarely with a few sparse hairs, adaxially densely pubescent above ligule; margins and filiform tip finely scabrid. Culm internodes often blotched purplish red, pubescent below panicle. Panicle contracted or open, 18–25 × 4–7–(13) cm, branches stiffly ascending or spreading and sometimes pendulous, usually naked below; rachis stout. Racemes several along each branch. Pedicelled spikelet: glumes linear-lanceolate, subcoriaceous; florets ∅, or upper floret ♂; stamens 3, anthers 2.5 mm. Sessile spikelet: glumes 4–6 mm, elliptic-lanceolate to obovate, glabrous to densely pubescent; lemmas hyaline, pubescent, ≪ glumes, lemma of upper floret often awned; stamens 3, anthers 3 mm; caryopsis *c.* 2.5 × 1.5 mm.

Naturalised from Africa.

N.: North and South Auckland, Gisborne, Taranaki (Hawera); S.: Marlborough (Seddon). Cultivated ground, crops (often asparagus or maize), roadsides, railway lines.

Sorghum bicolor may form hybrids with *S. halepense.* Hybrid plants are ± rhizomatous and may be difficult to distinguish from Johnson grass.

2. S. halepense (L.) Pers., *Syn. Pl. 1*: 101 (1805).   Johnson grass

Robust perennials to 2 m, strongly rhizomatous with far-creeping, woody rhizome *c.* 0.5–10 mm diam., rooting at nodes, internodes covered by scale-like bracts. Leaf-sheath chartaceous, ± keeled above with strong midrib, glabrous, light green or straw-coloured, often streaked reddish purple, sometimes completely suffused red-purple near base. Ligule 3.5–5 mm. Leaf-blade to 90 × 1–2–(5) cm, linear, ± glabrous, adaxially densely pubescent near ligule, abaxially pubescent from margin, midrib prominent, pale cream; margins minutely scabrid. Culm simple, erect, internodes glabrous, often with red-purple streaks. Panicle large, rather open, (15)–20–30 cm, lower branches very long, narrowly ascending, sometimes ± pendulous, naked below. Racemes several towards ends of slender branches; rachis and pedicels finely short-hairy. Pedicelled spikelets: 5–7 mm, greenish or dark reddish, narrow elliptic-lanceolate, acute; anthers *c.* 3 mm. Sessile spikelets: glumes 4–5 mm, elliptic-lanceolate, almost glabrous to densely hairy; lemmas < glumes, hyaline, sparsely hairy, lemma of upper floret usually awned, awn geniculate, twisted, 10–15 mm; stamens 3, anthers 2–3 mm; caryopsis *c.* 3 × 1.5 mm.

Naturalised from Mediterranan.

N.: recorded from Whangarei, Auckland City, South Auckland, Gisborne (Muriwai), Hawke's Bay (Napier, Hastings), Taranaki (Hawera), Wellington (Wanganui, Bulls, Feilding); S.: recorded from Nelson (Motueka, Nelson City), Marlborough (Oaro, south of Kaikoura), Canterbury (Leithfield, Kaiapoi). Crops (especially maize), cultivated ground, waste ground, footpaths, railway lines.

Aggressive and extremely difficult to eradicate because of the extensive, very strong rhizomes.

THEMEDA Forssk., 1775

Rather tall and coarse, tufted perennials or annuals. Leaf-sheath variable. Ligule membranous, ± ciliate. Leaf-blade flat or folded. Culm leafy. Inflorescence a loose panicle, the axils of uppermost culm-leaves bearing lateral branches topped by a cluster of spikelets; each cluster subtended by a spathe and composed of 1–3 spikelet-fascicles; each fascicle further subtended by a smaller spathe-like bract and consisting of an involucre of 2 paired, ± sessile, ♂ or ∅ spikelets surrounding a pair (or terminal triplet) of spikelets; one spikelet sessile, ♀, the other 1 or 2 spikelets pedicelled, ♂ or ∅. Rachis of spikelet-cluster terete, glabrous, disarticulating above the involucre of spikelets and below the paired spikelets. All spikelets 2-flowered; lower floret often reduced to empty lemma, upper floret ♀ in sessile member of spikelet-pair, ♂ or ∅ in involucral spikelets and in pedicelled spikelets. Involucral spikelets and pedicelled spikelets with 2-keeled, ± herbaceous lower glume, and membranous, rarely 0, upper glume; both florets present, ♂ or ∅, or one or both suppressed, lemmas hyaline, and upper floret with or without palea. Sessile spikelet of spikelet-pair terete, often awned, with acute to pungent densely bearded callus; lower floret ∅, upper floret ♀; glumes equal, coriaceous, lower tightly involute, very obscurely nerved, upper 3-nerved, with deep longitudinal groove on each side; lemmas unequal, lower hyaline, nerveless, upper consisting of a short base and a stiff geniculate awn, rarely awn reduced; palea hyaline, nerveless, small, or 0; lodicules 2, rather large, glabrous; stamens 3; ovary apex glabrous, stigmas subterminal or median; caryopsis subterete, embryo *c.* ½ length of caryopsis, hilum punctiform or shortly elliptic.

Type species: *T. triandra* Forssk.

c. 16 spp., from tropical to warm temperate regions especially in South-East Asia and Australia. Naturalised sp. 1.

1. T. triandra Forssk., *Fl. Aegypt.-Arab.* 178 (1775).   kangaroo grass

Stiff, reddish glaucous perennial tufts often forming large patches; branching intravaginal. Leaf-sheath light greenish brown to reddish tinged, coriaceous, keeled, lower sheaths villous, hairs fine tubercle-based, upper sheaths more sparsely hairy to glabrous. Ligule 0.7–1 mm, membranous, truncate, erose or shortly ciliate, finally shredding into fine cilia. Leaf-blade 15–35 cm × *c.* 2 mm, folded to flat with revolute margins, strongly keeled, of similar texture to sheath, scabrid, often a few tubercle-based hairs near margins above ligule; margins and midrib pubescent-scabrid near long, fine, acicular tip. Culm 25–90 cm, internodes glabrous. Panicle 15–35 cm, spikelets in 2–4 clusters at tips of slender branches, each cluster slightly overtopped by leaf-like spathe and consisting of 2–4 spikelet-fascicles; each fascicle subtended by a spathe-like bract > spikelets but < awns; rachis 2–3.5 mm, callus-like and covered with dense fulvous hairs; segments falling with sessile and pedicelled spikelets. Involucral spikelets ♂ or ∅: 8.5–15 mm, dorsally compressed, persistent; glumes ± equal, acuminate; lower firmly membranous, closely 11-nerved, inflexed and keeled near narrow hyaline margins, upper subhyaline, 3-nerved, 2-keeled with ciliate margins; lemma of lower floret shorter, 1-nerved; lemma of upper floret minute, or 0; palea 0; anthers, when present, 2.5–5 mm. Pedicelled spikelets: 9.5–13 mm, similar to involucral spikelets but narrower; pedicels glabrous, *c.* 2 mm. Sessile spikelets ♀: 3.5–8 mm, narrow-elliptic; lower glume obscurely 7–9-nerved, fulvous, with stiff reddish hairs just below apex, upper shortly pubescent near apex; lemmas < glumes, glabrous, spike-like in upper floret, awn 3.5–5.5 cm, geniculate, column brown, pubescent, bristle scabrid; anthers *c.* 2–2.5 mm; caryopsis 4.5 mm.

Naturalised from Africa.

N.: Auckland City (early record only), Rangitikei; S.: Nelson (between Nelson City and Brightwater), Marlborough (Waihopai Valley, Wairau Valley, Redwood Pass, near Blenheim), Canterbury (once collected at Bowens Valley on Port Hills). Roadsides, hillslopes in modified tussock grassland.

ζ**Coix lachryma-jobi** L., Job's tears   Robust annual, up to 2 m. Culm branched, internodes glabrous. Leaf-blade wide, lanceolate, cordate at base. Inflorescence enclosed by a hard globular bead-like utricle and consisting of 1 sessile ♀ spikelet and 2 pedicelled ∅ spikelets, with a terminal raceme of ♂ spikelets protruding, together with stigmas of the fertile floret, through the opening at top of utricle; "beads" varying in colour from white to blue-grey to reddish to black. This Indian sp. has been collected in the wild at Tauranga Harbour, on bank descending to foreshore *C. Bonley* 18.2.1989 (CHR 401399) and also in Warkworth, in street-pavement, P. J. de Lange (*in litt.* 10 Jan. 1996).

ζ**Hemarthria altissima** (Poir.) Stapf et C.E.Hubb.   Perennial with creeping rhizome. Culms to 100 cm, branched, reddish. Leaf-blade narrow. Inflorescence a rigid curved raceme terminating a culm branch, but branches often in clusters, with several from each node subtended by a spathe-like bract; racemes *c.* 10 cm, with paired spikelets, 6–8 mm, all similar and appressed to rachis; pedicel fused with rachis internode and pedicelled spikelet appearing sessile. An African species collected once from Kaitaia *R. H. Matthews* Apr. 1935 (CHR 18503). *H. altissima* persists in the abandoned nursery at the Kaikohe Research Station (W. Harris *in litt.*).

Saccharum officinarum L., sugar cane, has been collected in a ± wild area in Bay of Islands, between Paihia and Opua near a mangrove walkway *E. K. Cameron 5915* 4 Dec 1989 (AKU 22421) and from a similar site at Rawene *P. J. de Lange* Sept. 1995 (AK 224270; CHR 529117).

Zea mays L., maize, has been twice collected from the wild; from a roadside near the Piako River, north of Matamata *E. K. Cameron 3435* 19.5.1985 (AKU 18409); and from Levin, beside railway line *F. C. Duguid* 20.5.1976 (CHR 510874). However, *Z. mays* is an entirely domesticated plant originating in Central America and cultivated since prehistoric times and is unlikely to survive long out of cultivation.

GENUS OF UNCERTAIN APPLICATION

Kampmannia Steud. *Syn. Pl. Glum. 1*: 34 (1854) based on *K. zeylandica* Steud. cannot be referred to any known taxon; the type of New Zealand origin has been lost.

ABBREVIATIONS

Abbreviations for names of authors follow Brummitt, R. K. and Powell, C. E. (Eds) *Authors of Plant Names* (1992). Abbreviations of titles of journals and books are not listed here; with a few exceptions (e.g., *T.N.Z.I.* and *T.R.S.N.Z.* for Transactions of the New Zealand Institute and of the Royal Society of New Zealand) they follow the *Kew Record of Taxonomic Literature* and *Flora Europaea.* Abbreviations for herbaria follow *Index Herbariorum.*

A.: Auckland Is
Add.: *addenda* (plural), items to be added
Ant.: Antipodes Is
app.: appendix
Art.: article
auct.: *auctorum,* of authors

C.: Campbell Id
c.: *circa,* about, approximately
Ch.: Chatham Is
comb. illeg.: *combinatio illegitimum,* illegitimate combination
Corrig.: *corrigenda* (plural), items to be corrected
cv.: cultivar; plural cvs

diam.: diameter

E.: east or eastern
Ed.: editor; plural Eds
ed.: edition
e.g.: *exempli gratia,* for example
emend.: *emendavit,* emended
et al.: *et alii,* and others

f.: *filius,* son (following author name)
Fig.: Figure (this Flora)
fig.: figure; plural figs
FL: flowering period (months by first three letters)
FT: fruiting period (months by first three letters)

Herb.: herbarium
hic. comm.: *hic communicatus,* communicated here

ibid.: *ibidem,* in the same place
ICBN: International Code of Botanical Nomenclature
Id: Island; plural Is
idem: the same
i.e.: *id est,* that is
in litt.: *in litteris,* in correspondence

K.: Kermadec Is

legit.: legitimate
loc. cit.: *loco citato*, in the place cited

M.: Macquarie Id
MS: manuscript
Mt: Mount

N.: North Id, north or northern
n: gametic chromosome count
2*n*: somatic chromosome count
no.: number
nom.: *nomen*, name
nom. altern.: *nomen alternativum*, alternative name
nom. cons.: *nomen conservandum*, conserved name
nom. illeg.: *nomen illegitimum*, illegitimate name
nom. invalid.: *nomen invalidum*, name not validly published
nom. nud.: *nomen nudum*, name published with no or insufficient description
nom. specific. rejic.: *nomen specificum rejiciendum*, rejected specific name
nom. superfl.: *nomen superfluum*, superfluous name
non: not of
non... nec: not of... nor of...
n.s.: new series
n.v.: *non visus*, not seen
N.Z.: New Zealand

op. cit.: *opere citato*, in the work cited

p.: page; plural pp.
pers. comm.: personal communication

q.v.: *quod vide*, which see

S.: South Id, south or southern
sect.: *sectio*, section
sens. lat.: *sensu lato*, in a broad sense
sens. strict.: *sensu stricto*, in a narrow sense
sensu: in the sense of
ser.: series
Sn.: Snares Is
sp.: species; plural spp.
St.: Stewart Id
subsp.: subspecies; plural subspp.
subvar.: subvariety

t.: *tabula*, plate
TS: transverse section

var.: *varietas*, variety; plural vars

vide supra: see above

W.: west or western

x: basic chromosome number

SIGNS AND MEASURES

♀	female
♂	male
☿	hermaphrodite
∅	neuter or sterile
0	absent
×	indicates hybrid plant or joins names of parents
=	equal to
≈	approximately equal to
>	greater than, more than
≫	much greater than, much more than
≥	greater than or equal to
<	less than, fewer than
≪	much less than, much fewer than
≤	less than or equal to
±	more or less, approximately
=	heterotypic synonym
≡	homotypic synonym
!	type specimen seen by the authors
ζ	discussed below

Metric measures are used, with the usual abbreviations. Numbers connected by the multiplication sign refer to length by breadth. Numbers enclosed in brackets indicate dimension beyond the usual range, e.g., 10–12–(18) × 5–7–(11) mm.

CONVENTIONS

In discussion in the text where reference is made to books and papers in journals, the genus is treated as the independent unit of reference.

Frequently cited books are:

Clayton, W. D. and S. A. Renvoize. 1986: *Genera Graminum; Grasses of the World.* London, H.M.S.O.
Cheeseman, T. F. 1906 (ed. 1), 1925 (ed. 2): *Manual of the New Zealand Flora.* Wellington, Government Printer.

GLOSSARY

abaxial: facing away from an axis, usually for lower leaf-blade surface, but for all organs; hence **abaxially**.

abbreviated: shortened.

achene: a small dry indehiscent one-seeded fruit with a thin free pericarp, as in *Pyrrhanthera*.

acicular: slender, or needle-shaped.

acuminate: tapering to a fine point.

acute: sharply pointed.

adaxial: facing towards an axis, usually for leaf-blade, but for all organs; hence **adaxially**.

adherent: separate parts or organs touching or in union, but not fused.

adjacent: lying near, next to.

adnate: attached to or united to an organ of a different kind.

adventitious: arising from an unusual position, as of roots.

alpine: an altitudinal zone ± above 1400 m.

alternate: placed singly along an axis, not opposite or whorled.

anastomosing: joining to form a network, as of nerves or veins.

androecium: male organs of a flower, i.e. stamens with their filaments, and anthers.

anemophilous: wind-pollinated.

annual: completing life-cycle in one year.

anterior: on the side away from the axis and thus appearing at the front.

anther: pollen-bearing organ of a stamen, usually filamented.

anthesis: opening of floret to allow emergence of anthers and stigmas.

anthoecium: floral envelope of lemma and palea; plural anthoecia.

antrorse: directed forward, usually of teeth or hairs.

apex: tip of an organ or part; hence **apical, apically**.

apiculate: ending in a short slender ± flexible point.

apomixis: process of producing seeds without fertilisation, as of *Elymus;* hence **apomictic**.

appressed: closely and flatly pressed against a surface.

approximate: close to, nearby.

aquatic: living in water.

arista: awn, as of lemma, sometimes borne on column q.v.; hence **aristate**.

ascending: directed upwards usually sharply.

aspergilliform: of stigma branches arranged in all directions, as of stigmata in *Ehrharta*.

asymmetric: lateral halves of dissimilar shape.

attenuate: gradually tapering, drawn out.

auricle: an ear-like appendage of a leaf-blade; hence **auriculate**.

auricular lobe: a short ± rounded extension of the leaf-sheath margin, as in *Festuca*.

autonym: a name established automatically at the subdivision of a taxon, whether or not published at the same time.

awn: a stiff or bristle-like projection from tip or back of an organ, as of lemma, glume; diminutive **awnlet**.

awn column: the base of an awn, of different form from the arista or bristle above.

awn sinus: notch in lemma whence the awn originates.

axillary: occurring in an axil between two parts, as of leaves or bracts.

bambusiform: having or resembling the habit of a bamboo.

beard: short stiff hairs usually surrounding the callus.

biaristulate: with two short awns.

bicuspid: having two sharp rigid points.

bidentate: having two teeth; hence diminutive **bidenticulate**.

bifid: divided into two parts; twice-cleft.

bilateral: occurring on both sides of an axis, as of bilateral racemes.

bilobed: with two lobes.

binate: occurring in pairs.

bisexual: both sexes present in a flower, perfect, hermaphrodite.

bract: a modified, often much reduced leaf, often scale-like.

bristle: stiff stout hair, or a very fine straight awn or arista.

caducous: falling off at an early stage.

caespitose: growing in tufts.

callus: hardened basal projection at base of floret, of differing length and ornamentation; point of disarticulation in a spikelet.

canaliculate: longitudinally channelled.

capillary: hair-like, fine.

cartilaginous: firm and tough but flexible.

caryopsis: dry indehiscent naked seed of grasses with a thin adherent pericarp.

cataphyll: a modified leaf reduced to a short sheath or bract surrounding the base of extravaginal innovations or rhizomes.

caudate: bearing a tail-like appendage.

cauline: belonging to the stem, as of leaves.

chartaceous: of thin papery texture.

chasmogamous: of flowers opening for pollination.

cilia: short hairs forming a fringe usually on margin; hence **ciliate, ciliolate**.

clavate: club-shaped, thicker towards apex.

cleistogamy: fertilisation without the opening of florets, always therefore self-fertilising; hence **cleistogamous**.

cleistogene: axillary hidden seed-forming spikelets or reduced inflorescences usually at the base of culms and differing from aerial spikelets; as of *Microlaena* or *Nassella*.

coalescence: the incomplete fusion of like parts, as of spines.

collar: junction of leaf-blade and leaf-sheath, anterior to ligule, often thickened.

column: base of arista or bristle of awn, sometimes twisted.

coma: ring of hairs at apex of lemma as of *Austrostipa*; hence **comate**.

compatibility: genetic reaction between pollen and stigma; hence **self-compatible**, **self-incompatible**, q.v.

compound: composed of several ± similar parts.

compressed: flattened; often as dorsally compressed and laterally compressed as of spikelets.

concave: having the surface curved inwards.

concolorous: of ± the same colour throughout.

conduplicate: folded together lengthwise along midrib, with the upper surface within, as of leaf-blades.

confluent: blending or running together as of nerves etc.

congested: crowded.

connate: joined together, especially of two similar parts as of lodicules, styles, etc.

contiguous: touching but not fused.

continuous: not breaking up or uninterrupted as of rachis in *Chloris*.

contra-ligule: line of hairs on the abaxial surface at junction of leaf-blade and leaf-sheath.

contracted: of panicles with erect branches close to rachis or almost so.

convex: bulging outward.

convolute: rolled together longitudinally, as of leaf-blades.

cordate: heart-shaped with the notch at the base.

coriaceous: of somewhat leathery texture, tough.

corona: membranous cup-shaped growth as of florets in *Nassella*; hence **coronate**.

corymb: flat-topped raceme; hence **corymbiform**.

costa: a rib, especially of the leaf-blade; plural costae; hence **costal**, **costate**.

coumarin: the scent of mown or dried holy grasses.

crenate: with shallow rounded teeth, the sinus acute; hence **crenulate**.

cross-veined: divided into lattices by short septa, especially of leaves.

crustaceous: of brittle texture.

culm: flowering stem, usually comprising nodes, leaves and internodes; in vegetative phase of Bambuseae multinoded stems bearing culm-sheaths and culm-leaves.

culm-sheath: sheath of a cauline leaf, as of Bambuseae.

cultivar: a selected and named form reproduced in cultivation; abbreviation **cv.**

cuneate, cuneiform: wedge-shaped, gradually and evenly narrowing to the base.

cupule: a small cup.

cusp: a sharp rigid point.

cuspidate: with the apex narrowed abruptly to a sharp rigid point.

cymbiform: boat-shaped.

deciduous: falling after the completion of normal function.

decumbent: lying along the ground with tip ascending.

decurrent: running or extending downwards.

deflexed: bent sharply downwards.

dehisce: opening to shed contents as of anthers; hence **dehiscence**.

deltoid: more or less triangular in shape.

dentate: toothed.

denticle: a very small tooth; hence **denticulate**, with very small sharp teeth perpendicular to the margin, the sinus ± open.

dichogamous: early maturity and exsertion of one sexual element; hence **duodichogamy**, 2 dichogamous events.

diffuse: of open or straggling growth.

digitate: spreading from the centre like fingers of a hand, as of inflorescences.

dilated: enlarged or expanded.

dimorphic: occurring in two forms, as of shoots.

dioecism: population comprising two sexual morphs one ♀ and the other ♂; hence **dioecious**.

disarticulation: separating at a joint as of florets in a spikelet, leaf-blades at ligules.

discolorous: of two different colours especially of upper and lower surfaces of leaves; see also **heterochromous**.

disjunct: used of plant distribution with large gaps between occurrences.

distal: towards the free end of an organ as opposed to the attached or proximal end.

distichous: arranged in two opposite rows so as to be in one plane.

divaricate: spreading at a very wide angle.

divergent: spreading from an axis at a rather wide angle, as of panicle branches, spikelets.

dorsal: relating to the back or outer surface of an organ.

dorsiconvex: convex on the back.

dorsiventral: of an organ with both dorsal and ventral surfaces, as of a leaf-blade.

e-, or **ex-:** prefix meaning lacking.

eccentric: offset from the centre; hence **eccentrically**.

ecotype: the genetical response of a plant to its environment; often simplistically treated as a morphological variant.

eglandular: lacking glands.

ellipsoid: of a solid object elliptic in section or outline.

elliptic: in the shape of an ellipse, rounded at both ends, widest in the middle.

elongate: lengthened, stretched out.

emarginate: with a shallow notch at the apex.

embryo: the rudimentary plant within the caryopsis.

endemic: native only to a particular country or region; in this Flora exclusive to the N.Z. Botanical Region.

endosperm: the starchy tissue of a caryopsis, mostly dry, sometimes doughy or liquid.
ensheathed: retained within leaf-sheath.
entire: continuous, not toothed or otherwise cut, whole.
epaleate: lacking a palea.
epicuticular: seated, resting or deposited on the cuticle of any organ, especially of wax on culms and leaves.
equilateral: having all sides equal.
erose: with an irregular margin as if gnawed.
espatheate: lacking a spathe.
excurrent: running out beyond the apex or margin.
exserted: projecting, as of awns beyond glumes, or stigmas outside an anthoecium.
extravaginal: of an innovation shoot bursting out through the leaf-sheath.

falcate: sickle-shaped; strongly curved.
fascicle: a close cluster or bundle; hence **fascicled, fasciculate.**
fastigiate: with branches ± erect and close to the axis.
female: plants or flowers bearing megagametophytes or ovules.
fertile floret: in the sense of caryopsis bearing, often as female-fertile floret, by extension to fertile spikelet.
-fid: suffix meaning cleft.
filiform: thread-like.
fimbria: hairs or processes forming a fringe; hence **fimbriate,** fringed.
fistula: hollow centre of a cylindric culm; hence **fistulate.**
flabellate: fan-shaped.
flaccid: limp, not rigid.
flexible: capable of being bent and restoring to original form.
flexuous, flexuose: having a wavy or zigzag form.
floret: lemma and palea with the enclosed flower.
floriferous: bearing flowers or flowering stems.
foliaceous: leaf-like.
forma: a minor variant of a species.
fulvous: tawny; dull yellow.
fuscous: of a brownish or greyish brown colour.
fusiform: spindle-shaped; of a solid object, ± swollen in the middle and narrowed to both ends.

geniculate: with a knee-like bend; hence **1-geniculate, bigeniculate, trigeniculate,** as of culms, awns, etc.
gibbous: somewhat swollen on one side, as of *Nassella trichotoma* caryopsis.
glabrous: without hairs of any sort; hence **glabrescent,** becoming glabrous or nearly so.
gland: a secreting organ or part, as of hairs; hence **glandular.**
glaucous: a distinct bluish-green colour not necessarily caused by a waxy bloom; hence **glaucescent.**

globose: nearly spherical.

glomerule: a very dense cluster, especially of coalescing spines or bristles in *Cenchrus;* hence **glomerate**.

glume: empty bract at the base of a spikelet, usually two; hence **glumaceous**.

gynodioecism: population comprising two sexual morphs, one ♀ and the other ♂, both seed producing; hence **gynodioecious**.

gynoecium: the female part of the flower, comprising ovary, style(s), and stigma(s).

gynomonoecism: having ♀ and ♀ flowers on the same plant in the same or different spikelets; hence **gynomonoecious**.

hermaphrodite: having both androecium and gynoecium in the same flower, perfect.

heterochromous: of different colours, as of leaf-blades.

hexagonal: having the form of a hexagon, with six angles and six sides.

hilum: the scar on a caryopsis, often punctiform or linear; hence **hilar**.

hirsute: with rather rough or coarse hairs.

hispid: bearing stiff bristle-like hairs or teeth; diminutive **hispidulous**.

holotype: the one specimen, or illustration, used by the author, or designated by the author, as the nomenclatural type.

homochromous: of the same colour, as of surfaces of leaf-blades.

hooded: with margins united for a very short distance below tip, as of leaf-blade, lemma.

hyaline: translucent or transparent.

hybrid: a plant resulting from a cross between parents which differ sufficiently to be accorded taxonomic recognition; hence **hybridism**.

hybrid, nomenclature: the name of a hybrid is preceded by × or if unnamed the parents (putative or known) are connected by ×.

imbricate: overlapping like roof-tiles, as of spikelets in a spike.

imperfect: sterile, as of florets.

inaequidistant: unequally separated, as of nerves.

included: not projecting beyond the enveloping structure, as of glumes and awns or spikelets.

incompatible: genetically determined failure of pollen grains to germinate on stigma; see also **self-compatibility** and **self-incompatibility**.

indigenous: native to a particular area or region, or shared between countries.

indurate: hardened and toughened.

inflated: swollen, as of the upper leaf-sheath in *Critesion* or *Lagurus*.

inflexed: bent inwards, incurved.

inflorescence: flower head terminating the stem; see also **panicle**, **raceme**, **spike**.

innovations: new vegetative shoots at the base.

insertion: attached to or growing upon; often the place or position of origin of an organ as of awn on lemma.

interkeel: area between two keels, as of palea.

internerve: area between nerves, as of palea in Stipeae.

internode: portion between two successive nodes, as of culm.

intravaginal: of an innovation shoot growing within an enveloping leaf-sheath.

involucre: one or more whorls of spines, bristles, bracts, sterile spikelets subtending or surrounding spikelets, and often falling with them.

junceous: rush-like.

keel: a ± sharp central fold or ridge, as of leaf-sheath, lemma, etc.; hence **keeled**.

lacerate: irregularly torn or cleft.

lamina: see **leaf-blade**.

lanate: woolly.

lanceolate: lance-shaped, much longer than wide, tapering gradually to apex and more rapidly to base.

lateral: on or near the side, distant from centre, as of lobes of lemma, ribs of leaves, nerves etc.

leaf-blade: part of leaf above the leaf-sheath; of varying shape and form, also known as lamina.

leaf-sheath: the lower part of the leaf surrounding culm or innovation.

lectotype: a specimen or illustration designated as the nomenclatural type when no holotype was indicated at the time of publication, when the holotype is found to belong to more than one taxon, or as long as it is missing.

lemma: lower of two bracts enclosing the flower, sometimes called flowering glume, of diverse shapes, division, and ornamentation.

lenticular: the shape of a ± biconvex lens.

ligulate: strap-shaped as of stamen filaments.

ligule: outgrowth of the inner junction of leaf-sheath and leaf-blade, of various forms, or a ligular area as in *Echinochloa*; hence **ligular**.

linear: very narrow with parallel margins.

lobe: a recognisable but not separated division of an inflorescence, leaf-sheath, lemma, etc.; hence **lobed**, **lobate**; diminutive **lobule**, **lobulate**.

lodicule: minute scale subtending the stamens and gynoecium; hence **lodiculate**.

longitudinal: lengthwise.

lowland: an altitudinal zone ± equating to sea level to 450 m.

membranous: thin and ± pliable.

middorsal: at the middle of a dorsal surface, as of awns; hence **middorsally**.

midrib: the main central rib of a leaf-blade, usually thickened.

monoecious: with unisexual flowers, ♂ and ♀, on the same plant.

monopodial: a stem with a single and continuous axis.

monostichous: in a single vertical row.

monotypic: having only one representative; as of a genus with a single species.

montane: an altitudinal zone ± equating to 450–900 m.

mottled: with spots or blotches on a surface.

mucro: a short sharp tip or excurrent nerve; hence **mucronate**.
multinoded: many nodes especially of stems and culms.
muticous: awnless, blunt.

naked below: referring particularly to lower part of panicle branches lacking
    spikelets.
naturalised: thoroughly established but originally coming from another country.
navicular: boat-shaped; hence **naviculate**.
neotype: a specimen or illustration selected to serve as a nomenclatural type as
    long as all the material on which the name of the taxon was based is missing.
nerve: applied to slender veins of glumes, lemmas, and paleas.
neuter: of a flower lacking functional sexual organs, ∅.
node: a point on stem or axis at which a leaf arises.

ob-: prefix signifying inversion, e.g. lanceolate and oblanceolate.
oblanceolate: the reverse of lanceolate q.v.
oblique: slanting.
oblong: longer than broad with the sides nearly or quite parallel for most of
    their length.
obsolete: rudimentary, vestigial.
obtuse: blunt.
olivaceous: olive-green.
opaque: not translucent, dull.
oral: at the mouth, as in oral bristles in Bambuseae.
orbicular: circular or nearly so.
ornamentation: the presence of hairs, teeth, prickles, cilia etc., on any organ.
ovary: that part of the gynoecium enclosing the ovule and surmounted by style(s)
    and stigma(s).
ovate: egg-shaped, attached by the broad end.
ovoid: of a solid body with an ovate outline.

pachymorph: sympodial growth with short thick rhizomes of Bambuseae.
palea: the upper of the two bracts enclosing the flower.
panicle: an indeterminate branched inflorescence with pedicelled spikelets on
    rachis branches; hence **paniculate**.
papilla: a minute pimple-like process; hence **papillate**, **papillose**.
patent: spreading, usually ± horizontally.
pectinate: divided in a comb-like manner with narrow, close-set teeth; hence
    pectinately.
pedicel: the stalk of a spikelet; hence **pedicelled**.
peduncle: a common axis bearing several spikelets; hence **pedunculate**.
pendent, pendulous: drooping.
penicillate: provided with a brush-like tuft of hairs (often of anthers).
percurrent: extending throughout the entire length.
perennial: with a life-span of two or more years.

perfect: of flowers with both male and female organs functional, ♀.

pericarp: the skin of a caryopsis, often free.

persistent: remaining attached to an axis after maturity, as of glumes, leaf-blades.

petiole: the stalk of a leaf, in grasses a false petiole; hence **petiolate**.

pilose: bearing soft shaggy hairs.

plano-: having one surface flat, as in planocompressed, planoconvex.

plicate: folded into pleats usually lengthwise, or as in a fan.

plumose: feather-like.

pollen-sterile: pollen grains lacking cytoplasm and deemed incapable of germination, often of hybrids.

posterior: on the side nearest the axis and thus appearing at the back.

prickle: robust, pointed structure with swollen base arising directly from the epidermis; hence **prickle-teeth**.

procumbent: trailing or lying over or along the ground.

produced: extended beyond.

proliferous: bearing vegetative buds or plantlets in inflorescences; hence **proliferation**.

prolongation: extending beyond the ultimate spikelet of an inflorescence (rachis prolongation), or floret of a spikelet (rachilla prolongation).

prolonged: see **prolongation**.

prophyll: the first leaf of an innovation usually reduced to a short bikeeled sheath.

prostrate: lying flat on the ground.

protogynous: with stigma emerged and receptive before the emergence of anthers.

proximal: towards the attached end of an organ as opposed to its free or distal end.

proximate: towards the attached end of an organ.

pruinose: bearing a waxy bloom on the surface.

pseudopetiolate: falsely petiolate.

puberulent: covered with exceedingly fine, short, dense hairs.

pubescent: clad in short soft hairs; hence **pubescence**.

pulvinus: a swelling or cushion at an axis often causing reflexed or deflexed branches, pedicels, or spikelets; hence **pulvinate**.

punctate: with dot-like markings often because of glands or wax; hence **punctiform**; diminutive **punctulate**.

pungent: ending in a stiff sharp point.

pyramidal: resembling a pyramid.

pyriform: pear-shaped.

quadrangular: four-angled.

quasi-raceme: resembling a raceme.

raceme: an unbranched, ± elongate, indeterminate inflorescence with stalked spikelets; hence **racemelets, racemose**.

rachilla: main axis of a spikelet bearing the florets; associated with disarticulation between florets, sometimes prolonged beyond ultimate floret.

rachis: main axis of inflorescence; plural rachides; sometimes elsewhere as rhachis.

recurved: curved backwards or downwards.

retrorse: bent or facing backwards or downwards.

retuse: notched slightly at a usually obtuse apex.

rhizome: underground stem bearing scale-like leaves usually spreading ± horizontally and often extensively; hence **rhizomatous**.

rhombic: diamond-shaped; hence **rhomboidal**.

rib: main or prominent nerves on leaf-blade; hence **ribbed, midrib**.

rotund: rounded, orbicular, nearly circular; hence **rotundate**.

ruderal: growing in waste places.

rudimentary: arrested at an early stage of development and thus imperfectly developed.

rugose: wrinkled; hence diminutive **rugulose**.

rupestral: growing on rocks.

sagittate: in the form of an arrowhead with the basal lobes pointing down.

scabrid: rough to the touch because of minute harsh projections; hence diminutive **scaberulous, scabridulous**.

scale: small ± leaf-like organ, often of rhizomes and stolons.

scandent: climbing.

scarious: very thin, dry, almost translucent.

sclerenchyma: subepidermal abaxial thickened fibrous tissue of leaf-blade, often in separate strands or continuous or nearly so, sometimes in ribs (costal sclerenchyma), especially of *Festuca* transverse section.

secund: one sided with leaves, inflorescences, or spikelets appearing to be arranged along one side of axis.

self-compatible: setting seeds after self-pollination; hence **self-compatibility**.

self-incompatible: unable to set seeds after self-pollination; hence **self-incompatibility**.

semi-: half, in part.

septum: a partition or crosswall; plural septa; hence **septate**.

sessile: lacking a stalk, sitting.

setaceous: bearing fine bristle-like structures.

sheath: see **leaf-sheath, culm-sheath**.

sheath-blade: leaf-blade of a culm leaf in Bambuseae; see also **culm-sheath**.

silicified: impregnated with silica.

sinuous: with shallow broad waves.

sinus: the space or recess between two lobes or segments, as of lemmas.

smooth: used to indicate the absence of asperities, teeth, roughness.

solitary: borne singly.

spathe: a large bract enclosing an inflorescence; hence **spatheate**; diminutive **spatheole**.

spathulate: spoon-shaped.

spike: an unbranched ±indeterminate inflorescence with sessile spikelets; hence **spicate, spiciform, spike-like**.

spikelet: unit of grass flower head (inflorescence), generally comprising two glumes, one or more florets (anthoecia), each with lemma and palea between which the flower is borne.

spine: a stout process with a sharp point; hence diminutive **spinule, spinulose**.

stamen: pollen bearing organ of a flower, comprising an anther with pollen sacs, or loculi, and the staminal filament; hence **staminate**.

staminode: a non-functional stamen often considerably reduced in size.

sterile: not producing seeds, pollen or ovaries; hence **pollen-sterile** in particular.

stigma-styles: united organs of gynoecium which receive pollen and transmit pollen tubes towards the egg apparatus; hence **stigmatic**.

stipitate: having a stipe or special stalk.

stolon: a stem, ± horizontal or arched or running along the ground rooting at nodes, these giving rise to vegetative shoots and culms; hence **stoloniferous**.

stramineous: straw-like or straw-coloured.

striate: with fine longitudinal lines or ridges; hence diminutive **striolate**, with finer lines or ridges.

strigose: covered with appressed, rigid, bristle-like, straight hairs.

style: see **stigma-style**.

sub-: prefix meaning somewhat less than, slightly, not quite; e.g. subapical, subcoriaceous.

subalpine: an altitudinal zone ± equating to 900–1400 m, usually above timber line.

subtending: standing below, but usually close to another organ.

subulate: awl-shaped, tapering from wide base to sharp apex, ± circular in cross-section.

sulcus: a small furrow; hence **sulcate**, with longitudinal grooves.

sympodial: of a rhizome where growth is continued by the activity of an axillary bud, as in Bambuseae.

synchronous: occurring at the same time.

synonym: a name which applies to the same taxon as another name.

tailed: caudate appendages of anthers.

tardily: of slow or late occurrence.

taxon: a taxonomic group of any rank, e.g. tribe, genus, species.

teratology: the study of malformations.

terete: circular in cross-section.

terminal: borne at the end of a stem.

tessellate: forming a lattice of cross-veins, especially in leaves of bamboos.

tiller: a new lateral shoot.

tortuous: twisting.

translucent: allowing the passage of light but diffusing it.

transverse: at right angles to an axis.

triad: in 3s, especially of spikelets in Hordeeae.

trifid: split into 3; 3-fid.

trigonous: a solid body triangular in section.

triplets: in 3s, especially of spikelets in Andropogoneae.

truncate: appearing as though cut off squarely across.

tubercle: a small wart-like swelling; hence **tubercled, tubercular, tuberculate**.

turbinate: top-shaped.

turgid: swollen or fully inflated.

tussock: a grass of dense tufted habit; hence **tussocky**.

type, nomenclatural: see **holotype, lectotype, neotype**.

ultramafic: parent material of rocks or soils in N.Z. with high iron and magnesium content.

unawned: lacking an awn or bristle.

uncinate: hooked obtusely at the tip.

undulate: waved in a plane at right angles to the surface.

unilateral: one-sided.

unisexual: of one sex only, ♂ or ♀.

utricle: a thin but hard covering enveloping inflorescences in *Coix*.

variegated: striped or blotched with various colours; usually of leaves.

vascular bundle: conducting tissue in leaf-blades, as of *Festuca*.

vein: a strand of conducting tissue.

ventral: of the inner or lower surface of an organ; hence **ventrally**.

verticel: a whorl of flowers, branches; hence **verticillate**.

vestigial: of a part now degenerate and non-functional.

vesture: covering, often of hairs.

villous: clad in long soft hairs not matted together.

wart: a swelling, or tubercle; hence **warty**.

weft: an area of closely interwoven or matted together hairs.

whorl: an arrangement of three or more parts, or organs, at the same level around an axis; hence **whorled**.

wing: a thin membranous expansion of an organ or part; hence **winged**.

INDEX

Accepted names are in roman; synonyms, doubtful names, and names in discussion are in *italic*; names of subfamilies and tribes are in CAPITALS.

Principal page references are in **bold**; § indicates a taxon between genus and species in rank, * denotes a text figure, and colour illustrations are indexed by Pl. number.

The index covers pp. 1–614; synopses and keys are not indexed.

Monerma P.Beauv. 211
  cylindrica 212
mountain oat grass 250

NARDEAE **61**
Nardus L. **62**
  stricta L. **62**, *63
narrow-leaved carpet grass 536
narrow-leaved snow-tussock 450
Nassella Desv. **79**
  neesiana (Trin. et Rupr.) Barkworth
    80
    var. neesiana **80**
  tenuissima (Trin.) Barkworth **81**
  trichotoma (Nees) Arechav. **82**, Pl. 3D
nassella tussock 82
New Zealand wind grass 272
nit grass 264
Notodanthonia Zotov 460, 461
  §Buchanania 462
  §Clavatae 461
  §Notodanthonia 461, 462
  §Semiannularia 461, 462
  biannularis Zotov 467
  buchananii (Hook.f.) Zotov 467
  clavata Zotov 469
  gracilis (Hook.f.) Zotov 473
  laevis 475
  maculata Zotov 475
  nigricans (Petrie) Zotov 477
  nuda (Hook.f.) Zotov 478
  pulchra Zotov 481
  purpurascens 487
  setifolia (Hook.f.) Zotov 484
  stricta (Buchanan) Zotov 470
  tenuis (Petrie) Zotov 485
  thomsonii (Buchanan) Zotov 487
  unarede (Raoul) Zotov 488
  viridis Zotov 489

oat 304
  Algerian 306
  sand 306
  slender 302
  wild 303
  winter wild 305

oat grass
  mountain 250
  tall 300
Oldham's bamboo 22
onion twitch 301
Oplismenus P.Beauv. 6, **552**, 557
  aemulus 553
    var. *flaccidus* (R.Br.) Domin 553
  compositus (L.) P.Beauv. 556
  hirtellus (L.) P.Beauv. **553**, 556, 557
    subsp. hirtellus 553, **555**, 556
    subsp. imbecillis (R.Br.) U.Scholz
      6, 553, *554, **556**
      forma imbecillis 553, 556
    var. hirtellus 553
    var. *imbecillis* (R.Br.) Fosberg et Sachet
      553, 556
  imbecillis (R.Br.) Roem. et Schult.
    553, 556
  undulatifolius P.Beauv. 553
Orthopogon imbecillis R.Br. 556
ORYZEAE 40, **57**
Oryzopsis lessoniana (Steud.) Veldkamp 67
  rigida (Steud.) Zotov 67

palm grass 584
pampas brome 360
pampas grass 492
PANICEAE **534**, 594
PANICOIDEAE **534**
Panicum L. **557**
  capillare L. **558**
  dichotomiflorum Michx. **559**, 563
  esculentum A.Braun 549
  glumare Trin. 564
  gonatodes Steud. 598
  hirtellum L. 553
  huachucae Ashe **559**
  imbecille (R.Br.) Trin. 556
  laevifolium Hack. 563
  lindheimeri Nash 560, 564
  maximum var. trichoglume Robyns **560**
  miliaceum L. **561**
  pallide-fuscum Schumach. 586
  sanguinale 541
  schinzii Hack. **562**
  sphaerocarpon Elliott **563**

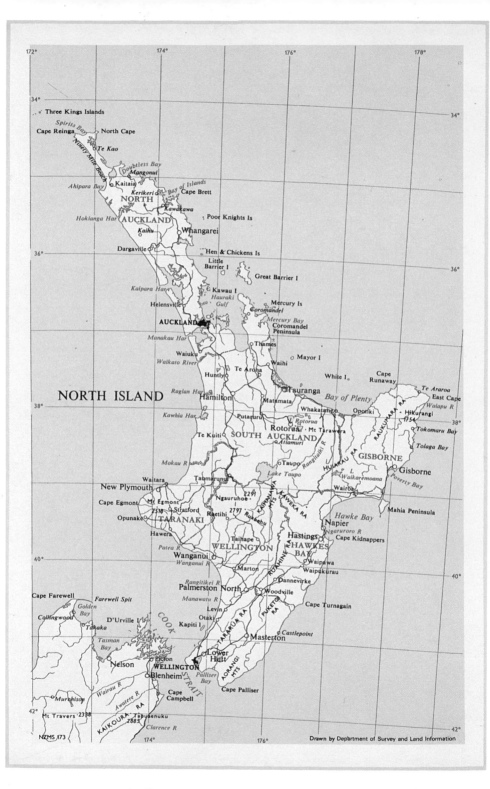

NORTH ISLAND